Streamlining Free Radical Green Chemistry

Streamlining Free Radical Green Chemistry

V. Tamara Perchyonok
VTPCHEM PTY LTD, Melbourne, Victoria, Australia

Ioannis Lykakis
Department of Chemistry, University of Crete, Voutes-Heraklion, Greece

Al Postigo
Faculty of Science, University of Belgrano, Buenos Aires, Argentina

RSCPublishing

ISBN: 978-1-84973-332-8

A catalogue record for this book is available from the British Library

Published by The Royal Society of Chemistry,
Thomas Graham House, Science Park, Milton Road,
Cambridge CB4 0WF, UK

Registered Charity Number 207890

For further information see our web site at www.rsc.org

Preface

Green chemistry is an emerging interdisciplinary field which has links to engineering, biochemistry, physics, material science and other areas. Modern free radical chemistry has flourished and developed into an independent, useful and powerful science linked to many aspects of the solving and understanding of chemical and life challenges at a molecular level. The aim of the two branches of science is to shift the paradigm and show the scientific growth and advances of the field of free radical green chemistry and the innovative and creative solutions adopted by scientists to date to address common deficiencies and limitations of conventional synthesis (such as poor yield, poor atom efficiency, E factor) in a move towards benign, biocompatible and highly efficient transformations, allowing scale-up from molecular to industrial scale, sustainable chemistry and development, green manufacturing, and process intensification through modern and innovative equipment and methods.

This book addresses and showcases the challenges and exciting modern innovative solutions and technologies in moving towards a common goal of the scientists: sustainable development through innovative free radical green chemistry as a powerful control as well as summarizing fundamental science relevant to the individual topics.

The potential environmental and health hazards of both industrial chemicals and chemical-based consumer products should be treated as intrinsic properties and should be minimized during the design. First of all the molecular level understanding of the relationships between structure and toxicity of chemicals is the only approach to design a less or, ideally, non-toxic product.

The book links together the pool of knowledge and concepts of chemists, engineers, material scientists, biochemists and biomedical experts as well as undergraduate and postgraduate students and directs them into thinking outside the conventional chemistry of today towards benign-by-design research and the challenges of tomorrow.

The target audience is chemists who are attempting to develop new processes and products which both meet the demands of society and have lower financial and environmental impact. The book will also make an excellent advanced

Streamlining Free Radical Green Chemistry
V. Tamara Perchyonok, Ioannis Lykakis and Al Postigo
© V. Tamara Perchyonok, Ioannis Lykakis and Al Postigo 2012
Published by the Royal Society of Chemistry, www.rsc.org

undergraduate textbook, with key reference materials at the end of each chapter.

The scope of the book is very broad and diverse as it covers all the aspects of efficient free radical green transformation, especially the initiation of the free radical transformations using conventional and contemporary means in organic and alternative media, the propagation step in the light of diastereoselective and enantioselective synthesis, and termination as an efficient and creative step in the overall radical transformation.

This book aims to fill the gap between scientific curiosity, academic research and industry with the powers and scope of a synthetically useful transformation being extended to solving problems at the interface of contemporary radical chemistry. Some transformations can arguably be described as being "more green than others", and some require significantly more investment if the technique is to replace conventional operation. In all cases, such adoption of the technology will only really take place if a real advantage can be shown. However, as each area develops and matures, all of them will no doubt hold important niches and advantages in the synthetic radical chemistry of the future. The practical and scope of chemical, bio-chemical and industrial application will make this book a unique highly practical text in a hot and rapidly developing field.

Acknowledgements

Writing this book was like following a free radical cycle, which would not be possible without the constant support, inspiration and belief from very fundamental people in my life—my parents Faina and Lazar, my close and dear family—and it is to them that I dedicate this book. Thank you for being a compass of this interesting and challenging journey.

I would like to express my gratitude to the co-authors of this book, Dr Ioannis N Lykakis and Dr Al Postigo, for a productive collaboration.

I am also indebted to enthusiastic mentors, colleagues and research associates who continuously stimulate my interest in science, research and general search for new challenges and who are too numerous to mention. I have been privileged to meet a few exceptional individuals, especially Professor Theunis Oberholzer and Dr Roy George, whose belief, support and enthusiasm encouraged me to look outside the box and strive to make dreams a reality.

A special thank you goes to Dr Merlin Fox for his generous assistance, support and guidance in making this project into an exciting opportunity.

Dr V. Tamara Perchyonok

Streamlining Free Radical Green Chemistry
V. Tamara Perchyonok, Ioannis Lykakis and Al Postigo
© V. Tamara Perchyonok, Ioannis Lykakis and Al Postigo 2012
Published by the Royal Society of Chemistry, www.rsc.org

Contents

Streamlining Free Radical Green Chemistry
V. Tamara Perchyonok, Ioannis Lykakis and Al Postigo
© V. Tamara Perchyonok, Ioannis Lykakis and Al Postigo 2012
Published by the Royal Society of Chemistry, www.rsc.org

Chapter 26 Innovative Reactions Mediated by Zirconocene: Advantages and Scope

Chapter 27 Applications of Conventional Free Radicals and Advances in Total Synthesis: Radical Cascades in Bio-inspired Terpene Synthesis

CHAPTER 1

Development of Free Radical Green Chemistry and Technology: Journey through Times, Solvents, Causes, Effects and Assessments*

1.1 INTRODUCTION

Smart and efficient science is an essential tool in combating the shortcomings of conventional chemistry through the synergy of the atom efficiency and versatility of free radical chemistry with the safeguards of pollution prevention at the molecular level through the principles of green chemistry. The point is that free radical green chemistry is not a solution to all environmental problems but rather the most fundamental approach to preventing pollution through development, innovation and creative thinking and an "outside the box" approach to problem solving of real life at the molecular level.[1]

Green chemistry concerns the development of chemical technology and processes that are designed to be incapable of causing pollution. We humans have dealt with toxicity and pollution throughout our entire history, but only recently have we been armed with an understanding of its sources and consequences. Scheme 1.1 portrays how green chemistry can diminish the need for other approaches to environmental protection. Ideally, the application of green chemistry principles and practice renders regulation, control, clean up, and remediation unnecessary, and the resultant environmental benefit can be expressed in terms of economic impact.

During the 1990s, environmental protection forces have been enveloping the science of chemistry with ideas and examples of green chemistry, and Paul Anastas coined the term "green chemistry" to focus attention on an area of

* Chapter written by V. Tamara Perchyonok.
Streamlining Free Radical Green Chemistry
V. Tamara Perchyonok, Ioannis Lykakis and Al Postigo
© V. Tamara Perchyonok, Ioannis Lykakis and Al Postigo 2012
Published by the Royal Society of Chemistry, www.rsc.org

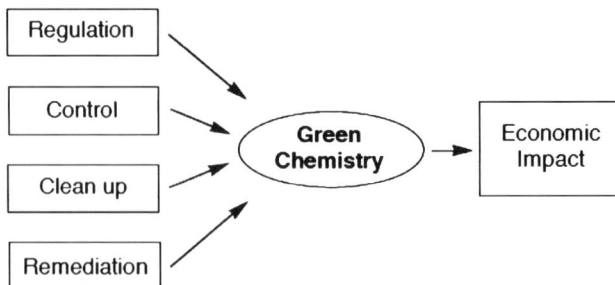

Scheme 1.1 Environmental protection activities that require the intervention of green chemistry to minimize their impact.

research and development which was undergoing rapid expansion and was increasingly characterized by the emergence of distinctive objectives and principles. The principles of green chemistry, as articulated by Anastas and Warner,[2] can guide chemists towards fulfilling their unique and vital role in achieving sustainable development. These principles are:

- It is better to prevent waste than treat or clean up waste after it is formed.
- Synthetic methods should be designed to maximize the incorporation of all materials used in the process into the final product.
- Wherever practicable, synthetic methodologies should be designed to use and generate substances that possess little or no toxicity to human health and the environment.
- Chemical products should be designed to preserve efficacy of function while reducing toxicity.
- The use of auxiliary substances (*e.g.*, solvents, separation agents, *etc.*) should be made unnecessary wherever possible, and innocuous when used.
- Energy requirements should be recognized for their environmental and economic impacts and should be minimized. Synthetic methods should be conducted at ambient temperature and pressure.
- A raw material or feedstock should be renewable, rather than depleting, wherever technically and economically practicable.
- Unnecessary derivatization (blocking group, protection/deprotection, temporary modification of physical/chemical processes) should be avoided whenever possible.
- Catalytic reagents (as selective as possible) are superior to stoichiometric reagents.
- Chemical products should be designed so that at the end of their function they do not persist in the environment but break down into innocuous degradation products.

- Analytical methodologies need to be further developed to allow for real-time in-processing monitoring and control prior to the formation of hazardous substances.
- Substances and the form of a substance used in a chemical process should be chosen so as to minimize the potential for chemical accidents, including releases, explosions, and fires.

These principles can motivate chemistry at all levels: research, reduction to practice, education, national and international policy, and public perception. Stated another way, green chemistry is about the redevelopment of chemistry to protect life itself. Such promise and intent hold enormous attraction for people. Just a few examples of recent initiatives and undertakings serve to illustrate the historical growth and incalculable potential of green chemistry:

- In the United States, green chemistry was an official focus area by the Environmental Protection Agency (EPA) at the beginning of the 1990s, and there was a great deal of activity in research, symposia, and education. In 1995, the United States launched the Presidential Green Chemistry Challenge Awards, which sought to provide visibility and recognition to those companies and academic researchers with outstanding achievements in green chemistry.
- In Italy, the Inter-university Consortium, Chemistry for the Environment (INCA), was established in 1993, with the aim to join together the academic groups dealing with chemistry and the environment; one of its focus areas is pollution prevention through research for cleaner reactions, products and processes. INCA organized its first meeting in Venice, "Processi Chimici Innovativie Tutela dell'Ambiente", in February 1993.
- In August 1996, IUPAC approved the formation of the Working Party on Green Chemistry under Commission III.2, which provided the beginnings of this work. The First International Green Chemistry Conference in Venice was held in September 1997 under the IUPAC sponsorship. The same year, the Green Chemistry Institute was founded.
- The European Commission has sustainability in its main research area, and the aims of green chemistry are present in the 4th and 5th Framework Programmes.

The growth of green chemistry has continued at an amazingly rapid pace in recent years, but it must be put into context. While there are numerous fine examples of how green chemistry is being used for the benefit of industry, the environment, and society, they constitute only a small fraction of the unrealized potential. The scientific tools and methodologies must be made available and used by the chemistry community for maximum benefit. The alternative of the costly and inefficient regulatory measures of the past, as a unilateral process, should no longer be an option. IUPAC is uniquely positioned to play a major role in supporting and advancing economic and environmental benefits through scientific innovation in green chemistry.[3–6]

1.2   THE MAJOR USE OF FREE RADICAL GREEN CHEMISTRY FROM THE BEGINNING

Economic considerations and environmental evaluations have pushed the chemical industry to adopt new eco-friendly technologies to survive in a market that becomes more demanding every day. Green chemistry will be one of the fields in which these sometimes conflicting forces will contend. Those companies that accept such a challenge and are the first to develop new, environmentally friendly technologies are most likely to gain the market, and to enjoy the support of their governments in promoting their initiatives.

The areas for the development of green chemistry have been identified as follows:

Use of alternative feedstocks. The use of feedstocks that are both renewable (rather than depleting) and less toxic to human health and the environment.

Use of innocuous reagents. The use of reagents that are inherently less hazardous and are catalytic whenever feasible.

Employing natural processes. Use of biosynthesis, biocatalysis, and biotech-based chemical transformations for efficiency and selectivity.

Use of alternative solvents. The design and utilization of solvents that have reduced potential for detriment to the environment and serve as alternatives to currently used volatile organic solvents, chlorinated solvents, and solvents that damage the natural environment.

Design of safer chemicals. The use of molecular structure design—and consideration of the principles of toxicity and mechanism of action—to minimize the intrinsic toxicity of the product while maintaining its efficacy of function.

Developing alternative reaction conditions. The design of reaction conditions that increase the selectivity of the product and allow for dematerialization of the product separation process.

Minimizing energy consumption. The design of chemical transformations that reduce the required energy input in terms of both mechanical and thermal inputs and the associated environmental impacts of excessive energy usage. This subdivision refers more specifically to organic synthesis and shows that the principles of green chemistry can be used for all aspects of chemical processes.

1.3   ALTERNATIVE FEEDSTOCKS

The synthesis and manufacture of any chemical substance begins with the selection of a starting material from which the final product will be built. In many cases the selection of a starting material can be the most significant factor in determining the impact of the synthesis on the environment. This selection not only needs to be assessed in terms of what the material's characteristics are, but also in terms of the upstream impacts of the origins of this starting material, as well as the implications for the rest of the process stemming from the choice of this particular starting material. If the substance itself does not pose any hazard to human health or the environment, for example, but the retrieval and/or isolation of the substance causes significant risk to either, then this factor must be taken into account in the selection. Or, if the use of the starting material means that a particularly toxic or otherwise hazardous reagent will be required in order to carry out the requisite synthetic transformation, then this factor also needs to be considered in the selection process.

1.3.1 Innocuous or More Innocuous

Certainly, a first-level assessment of any starting material must be whether or not the substance itself poses a hazard in the form of toxicity, accident potential, possible ecosystem damage, or any other form. These hazards can be evaluated as there are extensive data on properties ranging from dose response to flammability to ozone depleting potential. In the absence of empirical data, there are a wide range of structure–activity relationship models which can give reasonable approximations for the properties in question.

1.3.2 Renewable

In addition to the direct hazard associated with a particular chemical substance, the implications of using a renewable *versus* a depleting feedstock need to be included in the selection of that substance as a starting material in a synthetic transformation. The feasibility and benefits of using biobased instead of petroleum-based feedstocks, for example, is actively being researched in both academia and the chemical industry. For ensuring a high degree of product safety for consumers and the environment, renewable resources have often been shown to have advantages when compared with petrochemical raw materials and can therefore be regarded as being the preferred source of raw material.

1.3.3 LIGHT

Light is another emerging feedstock in a broad sense, a safe alternative to toxic catalysts in many synthetic transformations. Beside UV light, the most renewable and environmentally ideal energy source is sunlight. In this regard, the sentence (about one century ago) of Giacomo Ciamician, one of the founders of photochemistry must be quoted.[7] His sentence looks like a description by Jules Verne, but one concerning chemistry:

"On the arid lands there will spring up industrial colonies without smoke and without smokestacks; forests of glass tubes will extend over the plants and glass buildings will rise everywhere; inside of these will take place the photochemical processes that hitherto have been the guarded secret of the plants, but that will have been mastered by human industry which will know

how to make them bear even more abundant fruit than nature, for nature is not in a hurry but mankind is."

Although it appeared (and still is) futuristic, we now know that many of these former fictions can be realized and applied. To address such enormous tasks, photocatalytic systems that are able to operate effectively and efficiently not only under UV light but also under the most environmentally ideal energy source, sunlight, must be established.

1.3.4 Solve Other Environmental Problems

Selection of a starting material should be assessed not only for any hazards that the substance might or might not possess, but also for existing environmental problems that its use as a starting material might assist in solving. In many communities in the United States, for example, waste biomass has become a problem due to limited landfill capacity and other solid waste disposal issues. The utilization of waste biomass as a chemical feedstock in chemical manufacturing processes can alleviate if not completely eliminate such waste problems.

1.3.5 Biocatalysis

Biocatalysis harnesses the catalytic potential of enzymes to produce building blocks for the pharmaceutical and chemical industry. Located at the interface between fermentation processes and petroleum-based chemistry, biotransformation processes broaden the toolbox for bioconversion of organic compounds to functionalized products; a key technology to facilitate and smooth the necessary transition from petroleum-based chemistry to the use of renewable resources for the production of chemicals is represented by biocatalytic processes. While fermentations use the carbon source for *de novo* product synthesis, biocatalytic processes employ a different strategy. Precursor molecules are fed to the biocatalyst, which transforms them to the desired compound by a limited number of functionalizing steps (usually one). The carbon and energy required to produce the biocatalyst commonly come from a different, easily metabolizable carbon source such as a sugar. Here, the range of products is not limited by the metabolism of the biocatalyst: non-natural (xenobiotic) precursor molecules can be efficiently transformed because biocatalysts can transform non-natural as well as natural and renewable compounds.

The sequestering of carbon dioxide is another example of how the selection of a starting material can help solve an existing environmental problem. It is well known that carbon dioxide is among the most potent of "greenhouse gases" which have been found to contribute to global warming. The utilization of carbon dioxide in manufacturing processes in a way that does not allow its escape to the environment might not solve the global warming issue but can help reduce the amount of greenhouse gases in the atmosphere.

1.4 Benign Reagents/Synthetic Pathways

As in the selection of a starting material, the selection of a reagent must include an evaluation to identify the hazards associated with a particular reagent. This evaluation should include an analysis of the reagent itself, as well as an analysis of the synthetic transformation associated with the use of that reagent (*i.e.*, to determine product selectivity, reaction efficiency, separation needs, *etc.*). In addition, an investigation should be undertaken to determine if more alternative reagents are available that either are themselves more environmentally benign or are able to carrying out the necessary synthetic transformation in a more environmentally benign way.

1.4.1 Innocuous or More Innocuous

As in the selection of a starting material, the selection of a reagent should start with an evaluation of the reagent itself to determine if it possesses any hazardous properties. Inherent in this analysis is the question: is the transformation requiring the hazardous reagent necessary, or can the final target compound be obtained from an alternative pathway that uses a less hazardous reagent?

In order to answer this question, alternative reagents must be identified, and any hazardous properties which the alternatives possess must be compared with the hazardous properties associated with the reagent originally selected. One example of an innocuous reagent (which is produced from non-toxic intermediates) is dimethylcarbonate.[8]

1.4.2 Generates Less Waste

An important consideration and benefit associated with the use of a particular reagent is whether it is responsible for the generation of more or less waste than other reagents. The amount of waste either generated or eliminated, however, cannot be the only consideration. The type of any waste generated must also be assessed. Just as all chemical products are not equal in terms of their hazard, neither are chemical waste streams. Waste streams must therefore also be assessed for any hazard properties that they possess. In this regard, it is obvious that oxidation reactions involving oxygen and hydrogen peroxide will be of outstanding priority, as they produce water as a by-product.

Green oxidation reactions require the use of non-toxic solvents (water or CO_2) and mild reaction conditions. Oxidations using air as a reagent are difficult to control, being either too slow or too fast for industrial applications, and intrinsically non-selective when selectivity is very often a crucial parameter.

Hydrogen peroxide is a clean reagent, with water the only by-product formed, and a very high selectivity can be obtained. However, the use of hydrogen peroxide for fine chemical production is currently limited by its poor reactivity and its tendency to undergo radical decomposition easily. Therefore, there is a great effort to develop systems able to selectively activate oxygen and hydrogen peroxide for oxidative transformations.

In this context, both homogeneous and heterogeneous catalysis play a key role. Oxidation reactions are critical to pharmaceutical, petrochemical, and agricultural industries. Here are several examples of how environmentally benign oxidants such as molecular oxygen, hydrogen peroxide or nitrous oxide can be activated on heterogeneous catalysts. Direct oxidation of isoprenol, b-picoline and benzene are chosen as examples for continuous gas-phase processes, and oxidation of cyclopentanone, limonene, pinene, and propylene as examples for semi-continuous or batch-wise processes in the liquid phase.

1.4.3   Selective

Utilizing a reagent that is more selective means that more of the starting material is going to be converted into the desired product. High product selectivity does not always translate into high product yield (and less waste generated), however. Both high selectivity and high conversion must be achieved in order for a synthetic transformation to generate little or no waste. Utilizing highly selective reagents can mean that separation, isolation, and purification of the product will be significantly less difficult. Since a substantial portion of the burden on the environment that chemical manufacturing processes incur often results from separation and purification processes, highly selective reagents are very desirable in green chemistry.

1.4.4   Catalytic

If a catalyst is necessary, it should be used in the actual "catalytic amount". In fact, if a reagent can be utilized and yet not consumed in the process, it will require less material to continuously effect the transformation. This implies that the catalysis has to be as efficient (not only effective) as possible, involving a high turnover number. Other criteria that should be used in the selection of a reagent need to be balanced against one another in order to optimize the utility of the reagent and minimize the environmental impact. A large number of industrial processes are based on the use of inorganic or minerals acids. While many of these processes are catalytic, some (*e.g.*, acylation using $AlCl_3$) require stoichiometric amounts of Lewis acid. The final isolation of the product necessitates neutralization steps to remove the acid, resulting in enormous quantities of hazardous waste, with the cost of disposal of this waste often outweighing the value of the product.

1.5   BIOMASS: UTILIZATION AND SUSTAINABILITY

Biomass represents an abundant carbon-neutral renewable resource for the production of bioenergy and biomaterials, and its enhanced use would address several societal needs. Advances in genetics, biotechnology, process chemistry, and engineering are leading to a new manufacturing concept for converting renewable biomass to valuable fuels and products, generally referred to as the biorefinery. The integration of agro-energy crops and biorefinery manufacturing technologies offers the potential for the development of sustainable biopower and biomaterials that will lead to a new manufacturing paradigm.

First, the term "biomass" in general is extremely broad, covering everything from apricot pits to rice husks to the sawdust from a lumber mill. We should not be surprised that each different form of biomass will show different chemical characteristics. Second, the chemical reactions between different components of a single type of biomass are not well understood. Chemical components found in plants—mostly cellulose, hemicellulose and lignin—all undergo different reactions during pyrolysis. These reactions combine in complex ways that go beyond the simple superposition of their individual characteristics. Even if we only used a single source for all our biomass energy, the complex interplay between the component parts within this source would not be fully understood. An integrated biorefinery is an approach that optimizes the use of biomass for the production of biofuels, bioenergy, and biomaterials for both short- and long-term sustainability. The demands of future biorefineries will stimulate further advances in agriculture in which tailored perennial plants and trees will provide increasing amounts of bio-resources, as highlighted in the "Billion-Ton" report.[9] The advances in plant science will certainly be influenced by societal policies, land use practices, accelerated plant domestication programs, and research funding to develop this vision. Nonetheless, given humanity's dependence on diminishing non-renewable energy resources, this is a challenge that must be addressed—and we need to get on with it!

1.6 GREEN CHEMICAL SYNTHESES AND PROCESSES

Green chemistry is enjoying significant adoption by industry around the world and widespread activity from the research community.[10] One reason for this is that not only does green chemistry address the fundamental scientific challenges of protecting human health and the environment at the molecular level, but it accomplishes this in an economically beneficial way for industry.[11] One measure of this is the fact that while there is not a single regulation requiring industry to engage in the specific practices or methodologies of green chemistry,[12] there are nevertheless plentiful examples of excellent green chemistry techniques being commercialized.

Green chemistry, defined as the design of chemical products and processes that reduce or eliminate the use and generation of hazardous substances,[13] has been referred to as pollution prevention at the molecular level. This emerging area recognizes that during the design phase of any chemical synthesis, product, or process, minimized hazard must be viewed as a performance criterion. Moreover, the hazard must also be viewed as a physical/chemical property which it is possible to manipulate and control at the molecular level.

This book presents a number of the innovations that have been developed recently in the emerging area of free radical green chemistry and highlights the cutting edge science and engineering in this field being conducted in industry, academia and government.

1.7 BASIC RADICAL CHEMISTRY: STRUCTURE, REACTIONS AND RATES

1.7.1 General Aspects of Synthesis with Radicals: Advantages and Traditions

Radicals are species with at least one unpaired electron which, in contrast to organic anions and cations, react easily with themselves in bond-forming reactions. In the liquid phase most of these reactions occur with diffusion-controlled rates. Radical–radical reactions can be slowed down only if radicals are stabilized by electronic effects (stable radicals) or shielded by steric effects (persistent radicals). However, these effects are not strong enough to prevent diffusion-controlled recombination of, for example, benzyl radicals of *tert*-butyl radicals.[14] Only in extreme cases, *e.g.*, for the radical or di-*t*-butylmethyl radical, are recombination rates low.[15] While the recombination rate of the triphenylmethyl radical is reduced due to both steric and radical stabilizing effects, the steric effect alone slows down the recombination of the di-*tert*-butyl methyl radical. Since neither of the radicals have C–H bonds to the radical centre, a disproportionation reaction, in which the hydrogen atom is transferred, cannot occur.

1.7.2 Reactions Between Radicals

The fact that reactions between radicals are in most cases very fast could lead to a conclusion that direct radical combination is the most synthetically useful reaction mode.[16] This, however, is not the case because direct radical–radical reactions have several disadvantages:

- In the recombination reactions, the radical character is destroyed so that one has to work with at least an equivalent amount of radical initiators.
- The diffusion-controlled rates in radical–radical reactions give rise to low selectivity which cannot be influenced by reaction conditions.
- The concentration of radicals is so low that reactions with non-radicals, like solvents, which are present in high concentrations are very hard to prevent.

1.7.3 Reaction Between a Radical and a Non-radical

Nevertheless, there are several classes of synthetically useful reactions involving free radicals reacting with non-radicals.[17] It possesses the following advantages:

- The radical character is not destroyed during the reaction; therefore, one can work with catalytic amounts of radical initiators.
- Most of the reactions are not diffusion-controlled, and the selectivities can be influenced by variation of the substituents.
- The concentration of the non-radicals can be easily controlled.

In most cases, in order to use reactions between radicals and non-radicals for synthesis, chain reactions have to be encouraged and established. For the successful use of radical chains two fundamental conditions have to be met:

- The selectivities of the radicals involved in the chain have to differ from each other.
- The reaction between radicals and non-radicals must be faster than the radical combination reaction.

In practice these rules can be best illustrated by a chain reaction that has gained increasing synthetic application over the years and become one of the fundamental pillars of free radical chemistry.[18] In this chain reaction, alkyl halides and alkenes react in the presence of tributyltin hydride to give products.

For a successful application of the tin method, alkyl radicals must attack alkenes to form adduct radicals. Trapping the newly formed radical yields a formation of the products and tributyltin radicals, which react with alkyl halides to give back educts as radicals. The tin method can be synthetically useful only if these reactions are faster than all other possible reactions of formed radicals. Therefore, the radicals in the chain must meet certain selectivity and reactivity prerequisites.

1.7.4 Reactivity and Selectivity

As previously mentioned, radicals can undergo a number of different and competitive reactions. These processes have different rates of reaction and if one reaction proceeds at a much faster rate than all the rest we have a selective and high-yielding process. Alternatively, if a variety of reactions proceed at similar rates, the radical will react unselectively to produce a number of different products. The rates of these reactions can vary enormously and, for example, the rate constants of an abstraction reaction can vary by a factor of at least 10 000. The key factors that influence radical reactivity include enthalpy, entropy, steric effects, stereoelectronic effects, polarity and redox potential (Figure 1.1, from ref. 19).

1.7.5 Enthalpy: In Brief

Radical reactions will generally proceed so as to convert a radical into a more stable radical or non-radical product, whereas combination reactions leading to non-radical products have similar (diffusion-controlled) reaction rates to those of radical/non-radical reactions.

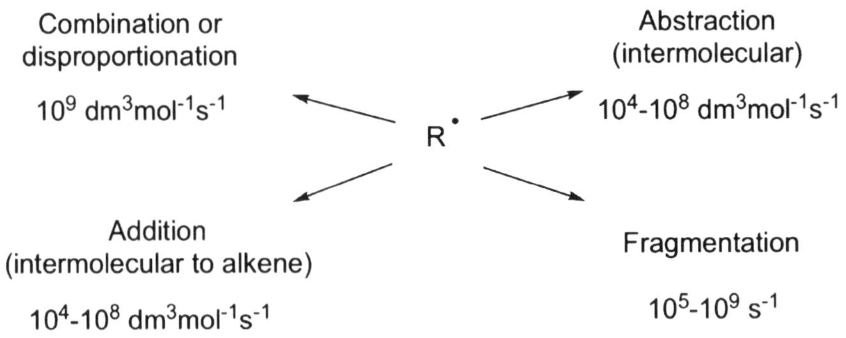

Combination or
disproportionation

10^9 dm^{3}mol^{-1}s^{-1}

Abstraction
(intermolecular)

10^4-10^8 dm^{3}mol^{-1}s^{-1}

R$^\bullet$

Addition
(intermolecular to alkene)

10^4-10^8 dm^{3}mol^{-1}s^{-1}

Fragmentation

10^5-10^9 s^{-1}

Figure 1.1   Factors that influence radical reactivity.

   Reactions can vary enormously. As a guide, the more reactive the radical reactant and the more stable the product, the faster the reaction rate. In practical terms, this explains why the reactive phenyl radical abstracts a chlorine atom much more quickly from carbon tetrachloride than does the *tert*-butyl radical (in solution at 25 °C). Reaction with the phenyl radical is favoured because the carbon–chlorine bond in chlorobenzene is stronger than that in 2-chloro-2-methyl propane. We can therefore predict whether a radical reaction will take place by considering the energies of the bonds that are broken and those that are formed. This will provide an approximate enthalpy change (ΔH^o) for the reaction; if energy is released, the reaction is exothermic ($\Delta H^o < 0$); if energy is absorbed, the reaction is endothermic ($\Delta H^o > 0$). Exothermic reactions result in the formation of strong bonds and these can proceed rapidly (often spontaneously), whereas endothermic reactions (which lead to products with weaker bonds) are generally very slow (Figure 1.2, ref. 20 and 21).

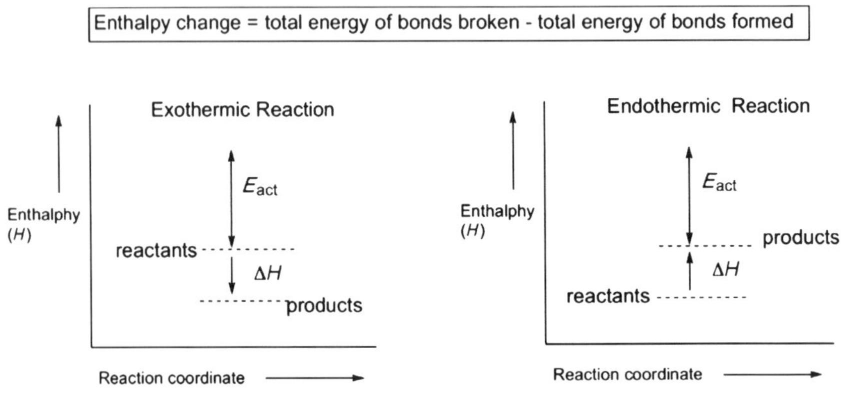

Enthalpy change = total energy of bonds broken - total energy of bonds formed

Exothermic Reaction

Enthalpy
(H)

E_{act}

reactants

ΔH

products

Reaction coordinate

Endothermic Reaction

Enthalpy
(H)

E_{act}

products

ΔH

reactants

Reaction coordinate

Figure 1.2   Enthalpy changes and chemical transformations.

1.7.6 Entropy

As you can probably guess, the enthalpy of a reaction gives only an approximate guide to selectivity in radical reactions. This is because the Gibbs free energy equation (see eqn (1.1)) shows that temperature and entropy are also important factors in determining the outcomes of a reaction. For a reaction to be thermodynamically favoured $\Delta G^{\ominus}$ should be negative; this occurs when there is a negative enthalpy change ($\Delta H^{\ominus}$), a positive entropy change ($\Delta S^{\ominus}$) and a high temperature (T). Therefore, reactions that produce an increase in entropy (or disorder) by increasing the number of molecular species on going from reactants to products are favoured. This explains why some endothermic reactions (positive $\Delta H^{\ominus}$) with an increase in entropy (positive $\Delta S^{\ominus}$) do not proceed spontaneously at room temperature. Higher temperatures are required to increase the $T\Delta S^{\ominus}$ contribution (above that of $\Delta H^{\ominus}$) to give a negative value of $\Delta G^{\ominus}$.

$$G = H - TS \quad \text{where } G \text{ is negative when } TS > H \qquad (1.1)$$

It's probably appropriate to link a few common concepts of a radical reaction such as radical initiators, reaction types and the driving force for reactions in the context of entropy to make the link between the physical organic and free radical chemistry even stronger.

The decomposition of peroxides (RO–OR) and azo-compounds (R–N=N–R) to form radicals is favoured by an increase in entropy. One peroxide molecule decomposes to give two alkoxyl (RO$^{\bullet}$) radicals, whereas an azoalkane will form three species: two carbon-centred radicals (R$^{\bullet}$) and nitrogen gas. The formation of the gas is particularly favourable because of the greater freedom of motion in gases (compared with liquids or solids), resulting in an increase in disorder.

The driving force for a number of fragmentation reactions, which are often endothermic, is an increase in entropy. The example of such transformation is decarbonylation reaction of acyl radicals [RC(=O)$^{\bullet}$] is generally endothermic, but can proceed with reasonable rates (10^{4}–10^{7} s^{-1} at 25 °C). In addition to these reactions being exothermic, they are also favoured by entropy, and both of these factors contribute to give a fast and irreversible decarboxylation reaction.

For radical rearrangement reactions, a single reactant leads to a single product with nearly the same mass and an almost identical structure. Although cyclization reactions can be categorized as rearrangements, they generally show a decrease in (rotational) entropy, as the bond rotation or the number of conformations in the cyclic product is not as great as that for the open-chain starting materials. The origin of favoured *versus* unfavoured radical cyclizations can be explained using these fundamental principles and is applicable to a broad range of cyclizations involving small, medium and large ring formations. Another important class of free radical transformations is the intermolecular hydrogen atom abstractions (S$_{H}$i) which are also controlled by entropy, and the linear geometry required in the transition state can be achieved in 1,5- and higher abstraction reactions. However, only 1,5- and 1,6- atom abstractions are commonly observed because, with longer chains, there is a greater loss of entropy in the transition state.

A much greater loss of entropy occurs in intermolecular reactions when two reactants collide to form one product. Although the system has now become more ordered, for radical–radical reactions this does not have a major impact on the rate of reaction because combination and disproportionation reactions are very exothermic. The reactions of radicals and non-radicals appear to be greatly influenced by entropic effects. These reactions are less exothermic (or entropy-favoured) because a radical rather than a non-radical product is formed. Entropy now becomes much more important and the rates of these intermolecular reactions are considerably slower. The entropy factor explains why intermolecular radical additions can be up to 10^5 times slower than related intramolecular cyclizations.

1.7.7 Steric Effects

A negative value for ΔG^o tells us that a reaction can take place, but the rate of the reaction can be determined by the enthalpy of activation ($\Delta^{\ddagger}H$). This is not only a measure of the difference in bond energy between the starting materials and the transition states, but also the difference in bond strain. The more strained the transition state, the higher the value of $\Delta^{\ddagger}H$ and the slower the reaction. Therefore, even though two reactions can be thermodynamically favoured and have similar (negative) values for ΔH^o they will not have the same rate if $\Delta^{\ddagger}H$ is different. It's a reasonable observation that radical reactions usually have low values for $\Delta^{\ddagger}H$ because the majority of radicals are reactive. However, this statement does not hold true for persistent or long-lived radicals that have very bulky substituents surrounding the radical centre. The reaction of these sterically hindered radicals would require a very strained transition state with a high enthalpy of activations and this is disfavoured (Figure 1.3).

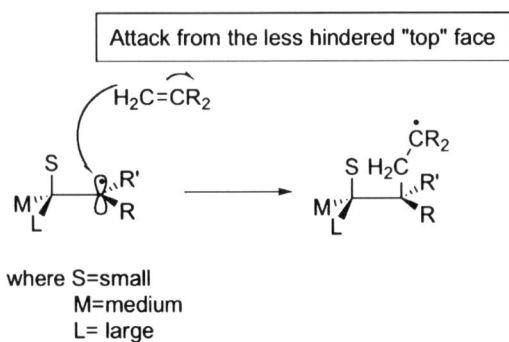

Figure 1.3   Steric effects and radical transformations.

Steric effects are used to explain the regioselectivity of the addition of radicals to alkenes. The radical preferentially attacks the less substituted (less hindered) end of the double bond to give a less strained transition state with lower $\Delta^{\ddagger}H$ in *anti*-Markovnikov-type reactions. If we introduce substituents onto the double bond, the rate of addition is lowered because of greater steric interaction. Thus, for example, the methyl radical will add three times more quickly to ethene (CH_2CH_2) than to the di-substituted alkene (*E*)-2-butene.

Steric effects can also explain why carbocyclic radicals (which are not able to undergo free rotation about the C–C bonds) can add to alkenes stereoselectively. The introduction of an adjacent chiral centre can make the two "faces" of the radical non-equivalent, and this can lead to the alkene preferentially adding to the less hindered face.

1.7.8 Stereoelectronic Effects

As previously mentioned, radicals with electron-withdrawing or electron-donating substituents can be stabilized by interaction of the singly occupied orbital with an n-, π- or σ*-orbital. For effective stabilization, the interacting orbitals must overlay efficiently, and this will depend on their geometry, or position in space. Similarly, for a radical to react, the singly occupied orbital must be able to overlap with either another of its own orbitals (for intramolecular reactions) or another radical or non-radical orbital in a different molecule. For reaction of a radical with a different molecule, there is often no restriction in the orbital geometries and they can rotate freely so as to combine with the maximum overlap (Figure 1.4).

For an SH2 reaction, a radical and non-radical can therefore orientate themselves to give a linear transition state which maximizes the interaction of the radical orbitals and the vacant σ*-orbital of the bond which is broken. These may, however, be difficult for very bulky molecules, where steric hindrance can prevent the orbitals from becoming close enough for efficient overlap.

For intramolecular radical reactions, there is often a restriction in the orbital geometries. The structure of the starting material dictates how far the orbitals are apart and their relative position; if they are situated a long way apart or held rigidly on different sides of a ring, they will find it hard to interact. When the radical bonds are not in close proximity to the C–H σ*-orbital,

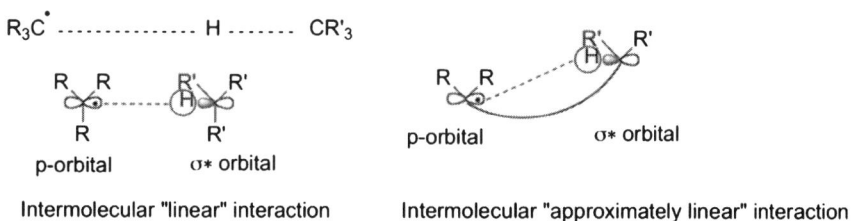

Figure 1.4 Stereoelectronic effects and radical transformations.

intermolecular hydrogen atom abstraction can compete with the intramolecular processes. Even when the orbitals are very close together, they may not interact because their geometry can prevent orbital overlap. This so-called electronic strain explains why 1,3- and 1,4-hydrogen atom transfers are very slow. The short chain length restricts the orbital positions and so the radical orbital cannot attack the vacant σ*-orbital at an angle of approximately 180 °C to give the most efficient overlap.

Stereoelectronic factors have been used to explain why the 5-*exo* mode of cyclization is faster than the competitive 6-*endo* cyclization. For the hex-5-en-1-yl radical, the rate of 5-*exo* cyclization is around is around 60 times faster than that for the 6-*endo* reaction at 25 °C. This is surprising as the 6-*endo* product is a secondary radical and (thermodynamically) more stable than the primary radical formed on 5-*exo* cyclization. In addition, the five-membered cyclopentane ring is more strained than the cyclohexane ring. The cyclization is therefore under kinetic control, and this is because the singly occupied orbital of the radical attacks the alkene at an angle close to 107 °C and so can overlap more favourably with the alkene π*-orbital at the 'internal' carbon atom. The carbon chain will position the radical's p-orbital closer to the internal (rather then the external) alkene carbon, and the energy calculations have confirmed that the smaller ring is formed because the chair-like transition states are less strained and lower in energy (Figure 1.5).

Radical fragmentation reactions are also stereoelectronically controlled. For the β-elimination or fragmentation of a carbon–carbon centred radical, the radical p-orbital and the σ*-orbital of the C–X bond that is broken must lie in the same plane so that the double bond can be easily formed. This is not a problem for molecules that can rotate freely around the central carbon–carbon

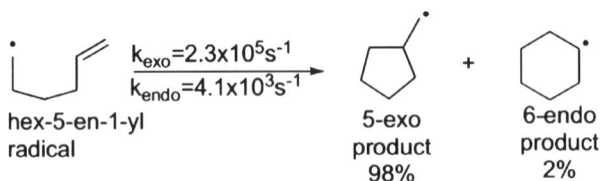

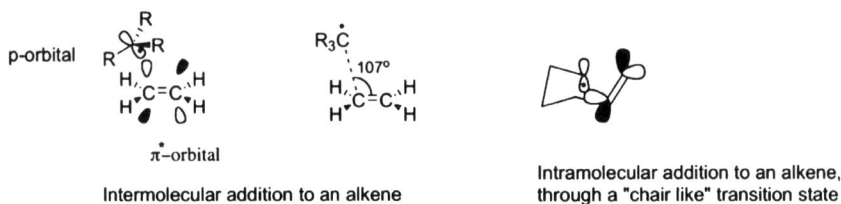

Figure 1.5   5-Exo *vs.* 6-endo transformations and radical clocks.

bond, as rotation can position the radical and the C–X bond in the correct orientation. The cyclopropylmethyl radical, for example, can undergo rotation around the exocyclic carbon carbon bond to align the p- and σ*-orbitals, and very rapid β-elimination occurs to open the strained three-membered ring. In contrast, the cyclopropyl radical cannot undergo ring opening because the carbon–carbon bonds within the ring cannot rotate, and so the p- and σ-orbitals lie at 90° (or orthogonal) to one another. This is in spite of the fact that the three-membered ring is very strained and ring opening is thermodynamically favoured.

1.7.9 Polarity

For reactions to occur, the interacting frontier orbitals must not only interact efficiently, but also have similar energies. The occupied frontier orbital for a radical is called the singly occupied molecular orbital (SOMO), we have previously discussed that radicals bearing different substituents have different SOMO energies. Radicals adjacent to electron-donating groups interact with a filled orbital to give a high-energy SOMO, while radicals next to electron-withdrawing groups interact with an unfilled orbital to give a low-energy SOMO. The SOMO energies lie somewhere between that of the highest occupied molecular orbital (HOMO) and that of the lowest unoccupied molecular orbital (LUMO) of a non-radical. Therefore, for the reaction of a radical with a non-radical, we need to consider the SOMO–HOMO and SOMO–LUMO interactions. In both cases, the interaction will lead to a decrease in energy and the formation of a bond; a SOMO–HOMO interaction places two of the three electrons in a low-energy bonding orbital, whereas a SOMO–LUMO interaction places the single electron in a low-energy bonding orbital. The energy of the SOMO will determine whether interactions with the HOMO or the (higher energy) LUMO predominates. Electrophilic radicals (with a low-energy SOMO) will be closer in energy to the HOMO, and therefore the SOMO–HOMO interactions will predominate. In comparison, nucleophilic radicals (with a high-energy SOMO) will be closer in energy to the LUMO, and therefore the SOMO–LUMO interaction will predominate (Figure 1.6).

1.8 SOLVENT EFFECT AND FREE RADICAL TRANSFORMATIONS: GENERAL UNDERSTANDING

The past 80 years have seen an incredible growth in our understanding of free radical reactions in homogeneous gaseous and liquid systems. Fifty years ago, so far as the vast majority of chemists were concerned, radicals were overly reactive species of no practical value or interest since all radical-mediated reactions were presumed to give gunk and tars. This negative view of radical reactions persisted even long after the successful commercialization of the free radical polymerization of vinyl monomers. Half a century ago the prevailing view of an organic or physical organic chemist about any of their colleagues

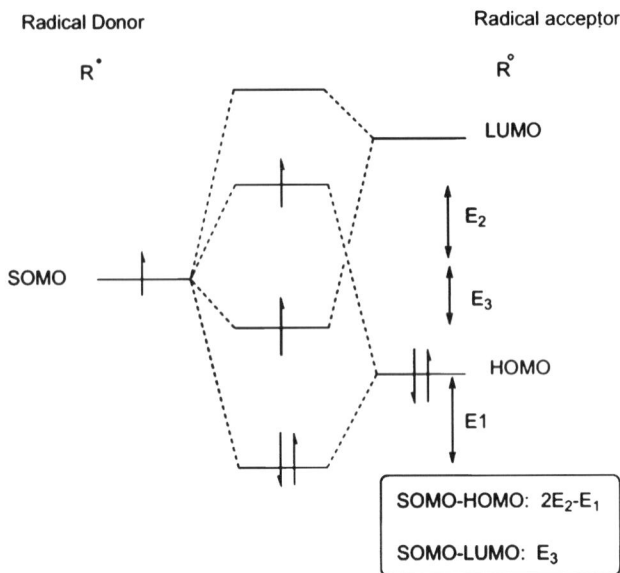

Figure 1.6   Polarity and radical donor/acceptor SOMO/LUMO interactions.

who expressed an interest in research into free radical reactions can best be summarized by a joking remark once made by the father of physical organic chemistry:

"Homolysis, even between consenting adults, is grounds for instant dismissal from this Department." C. K. Ingold, *ca.* 1955.[22]

The idea that radical chain reactions might actually be useful in organic synthesis was regarded with scorn, and the idea that radical chemistry might be important in living organisms was dismissed as utter nonsense. These fallacious views about radical chemistry were only slowly discarded by "mainstream" chemists, biochemists, and the medical profession. This change in outlook is largely due to the perseverance of a few (a very few) physical organic chemists who devoted much of their lives to careful kinetic and reaction product studies of complex chain reactions, particularly polymerizations, chlorinations, and autoxidations. Careful rate measurements under controlled conditions using pure materials and homogeneous systems not only proved to be reproducible, but also yielded overall kinetic rate equations from which overall reaction mechanisms could be deduced and rate-limiting elementary reactions identified. In all free radical chain reactions in homogenous systems the overall rate was found to depend on the rate of chain initiation, R_i or $R_i^{1/2}$, and on the ratio of the rate constants for the rate-controlling step of chain propagation, k_p, and for chain termination, $2k_i$ (or its square root, $\sqrt{2}k_i^{1/2}$).

Further understanding of radical chemistry evolved as physical organic chemists turned their attention to measuring the absolute rate constants for individual elementary reactions in homogeneous systems. First came the (accidentally discovered) rotating sector technique which allowed the rate constants for propagation and termination to be measured for a chain reaction which could be photochemically initiated and which exhibited bimolecular chain termination, *i.e.*, which obeyed the overall kinetic rate law, $k_p(R_i/2k_i)^{1/2}$. A couple of decades later, electron spin resonance (ESR) spectroscopy started to be employed to measure the rates of bimolecular radical–radical self-reactions ($R^{\bullet} + R^{\bullet}$ = products; rate constant = $2k_i$). This new ESR technique quickly displaced the rotating sector method because it could be applied both to radicals which did and to those which did not participate in the terminating step of radical chain reactions. Later, ESR methods were developed to measure the rates of unimolecular radical reactions and the rates of radical–molecule reactions.

The measurement of the rate constants for radical–molecule reactions by direct time-resolved monitoring of the decay of the radical or growth of a product, generally using UV-visible absorption spectroscopy, was first exploited by radiation chemists. The chemistry of the radiolysis of water was well understood to give hydroxyl radicals, hydrogen atoms, and the solvated electron. Pulse radiolysis of aqueous solutions was used to provide specific organic and inorganic radicals, which were formed by reaction of the hydroxyl radicals (oxidizing) or the solvated electron (reducing) with particular solutes. Reactions of these radicals with water-soluble substrates were monitored in real time and yielded a wealth of kinetic data. Although water is nature's favorite solvent it is certainly not the solvent of choice for most organic chemists. The pulse radiolysis kinetic data were therefore of little or no value to synthetic organic chemists, no matter how interesting it might be to bio-chemists and radiation biologists. The development of high powered, pulsed UV lasers some 15–20 years ago opened the door for physical organic chemists to obtain kinetic data for virtually any radical in virtually any homogeneous system. The technique became known as laser flash photolysis (LFP) and it quickly displaced kinetic ESR spectroscopy as the method of choice for free radical kineticists. The loss of a reactant or the appearance of a product is monitored in real time, generally by UV-visible absorption spectroscopy, but occasionally by other techniques such as time-resolved infrared spectroscopy. The explosion of free radical kinetic data "useful" to synthetic organic chemists is documented in the many volumes of essential reference texts and the "usefulness" of these kinetic data has been pointed out very nicely by Curran in 1988.[23]

"Since many of the relevant (radical) rate constants required for synthetic planning are known, a chemist can evaluate the possibilities for the success of a given (radical) reaction under a specific set of conditions. This ability to plan is a great asset of radical reactions."

One of the main limitations which retarded the practical application of "known" radical kinetics to "unknown" organic chemical systems was the worry about potential solvent effects on the rates of radical reactions. That there are substantial solvent effects on at least some radical reactions has been known since the early work on free radical chlorination by Russell in 1957 (ref. 24) and on alkoxyl radical chemistry by Walling in 1963.[25] In the last few years, solvent effects have been extensively investigated at the National Research Council Canada (NRC) by Ingold and co-workers.[22] This fundamental work can be summarized briefly as follows:

(i) There are generally no solvent effects on the rates of hydrogen atom abstraction from a C–H bond.

(ii) There may be solvent effects on the rates of unimolecular radical scission reactions but these are generally fairly small.

(iii) There are large solvent effects on the rates of hydrogen atom abstraction from O–H bonds (and to a lesser extent from N–H bonds).

Point (i) means that radical reactivities, in so far as hydrogen abstraction (and addition) reactions are concerned, are not influenced by the solvent. Point (iii) means that O–H containing substrates *(e.g.,* phenols, hydroperoxides, *etc.)* have their reactivities towards radicals influenced by the solvent. This influence was shown to be due to deactivation of the substrate, XOH, when it can act as a hydrogen bond donor to a hydrogen bond accepting (HBA) solvent, S. To a first approximation, the hydrogen bonded XOH, *i.e.,* XOHS, is unreactive towards radicals, with all of the hydrogen atom abstraction from the substrate actually occurring from "free" XOH, *i.e.,* non-hydrogen bonded XOH. This is a kinetic solvent effect (KSE) on the molecular reactant, XOH. Therefore, it was predicted that the KSE should be dependent on the ability of XOH to participate as a hydrogen bond donor to HBA solvents but should be independent of the reactivity or nature of the radical which does the hydrogen atom abstraction. This prediction, which is the first new and unifying principle for organic free radical chemistry to have been proposed in the last forty years, has been dramatically confirmed for hydrogen abstraction from phenol by the highly reactive cumyloxyl radical, $PhCMeO^{\bullet}$, and by the very unreactive diphenyl picrylhydrazyl radical (DPPH), $Ph_2NN^{\bullet}C_6H_2(NO_3)_3$. In the same solvent, the cumyloxyl radical abstracts the phenolic hydrogen atom from phenol ten billion (10^{10}) times faster than DPPH. However, a change in solvent, which reduces the rate of the cumyloxyl radical's reaction by a factor of 100 *(e.g.,* as occurs on changing from CCl_4 as solvent to ethyl acetate as solvent), also reduces the rate of the DPPH reaction by the same factor of 100.

Thus, it is now, at last, possible to predict the rate constant for hydrogen abstraction from phenol by any radical, $Z^{\bullet}$, in any solvent provided only that one single rate constant in one solvent has been measured for the $Z^{\bullet} + PhOH$ reaction. This new principle allows the quantitative prediction of free radical kinetics for hydroxylic substrates. Combined with the knowledge that there are generally no (or only very small) solvent effects on hydrogen, this made the measurement of the kinetics of free radical reactions in homogeneous solution a subject no longer worthy of further basic scientific research. Even as late as 1994 it looked as though the subject of homogeneous free radical kinetics was infinite (*i.e.*, it lay in a flat universe) with a measurement being required (or, at least, worth a paper) for all radical–molecule reactions in all solvents. In 1995 it became obvious that homogeneous free radical kinetics was a very finite subject (it lies in a very sharply curved universe). A few hundred, or at most a few thousand, rate constants for radical–molecule reactions in homogeneous solution are all that are ever likely to require measurement—and many of these rate constants have already been measured! Any and all other radical–molecule kinetic data which might be required in any solvent will be easy to predict with relatively high precision (probably generally better than a factor of 2–3).

If there really are no more worthwhile scientific goals for free radical kineticists in homogeneous systems, what direction should we expect this discipline to take in the 21st century? Obviously, forefront kinetic research on free radicals will abandon studies in homogeneous systems. Instead, it will tackle the currently daunting challenge of making reproducible rate measurements in heterogeneous systems and will concentrate on those systems which are of great interest and importance to organic chemistry, to biochemistry and to medicinal chemistry. One of the challenges is that many very important heterogeneous systems are themselves inherently somewhat irreproducible, *e.g.,* cells in culture, isolated organs and whole animals. Indeed, even such an apparently simple reaction as the autoxidation of linoleic acid dispersed in water in sodium dodecylsulfate micelles and inhibited by various phenolic antioxidants has been found to give different relative antioxidant activities in different laboratories! This particular problem of kinetic irreproducibility should be a simple matter to resolve. Nevertheless, it does serve to illustrate the kinds of difficulties the measurement of radical kinetics in heterogeneous systems is likely to encounter before inter-laboratory consensus as to the "truth" is achieved.

Undoubtedly, novel instrumentation and new applications for existing instruments will play a major role in heterogeneous radical kinetics. For example, there is a need to be able to monitor the movement of neutral and charged free radicals across the lipid/aqueous interface of biomimetic systems, *e.g.,* phospholipid vesicles, and biological supramolecular assemblies, *e.g.,* low density lipoprotein (LDL) particles. Similarly, there is a need to monitor the rates and products of radical-induced chemical change within single phospholipid vesicles and single LDL particles because so much valuable information becomes "scrambled" (averaged) when all the lipid particles are not only assumed to be identical but also are assumed to talk to one another (chemically speaking) as readily as if the overall systems were actually homogeneous.

In short, physical organic chemistry in the 21st century (insofar as free radical kinetics is concerned) is that it will continue (as in the past) to be applied to solve mechanistic questions regarding chemical processes of scientific interest and of fundamental importance in all areas of organic chemistry, biochemistry, and medicinal chemistry. The free radical kineticists of the 21st century will use new instruments, and old instruments in novel ways. They will develop theoretical models for even the most complex heterogeneous systems and subject these models to rigorous experimental verification. They will ensure that their branch of physical organic chemistry remains vibrant in itself and relevant and important to 21st century chemists. The challenges are exciting, the goals are worthy, and the value of the scientific knowledge to be gained is incalculable. Let's get started!

1.9   WHY WATER AS A SOLVENT? REASONS AND ADVANTAGES

Until recently, the use of water as solvent for organic reactions was mainly restricted to simple hydrolysis reactions.[26] Accordingly, most reagents and catalysts in organic synthesis have been imperiously developed for use in anhydrous, organic reaction media.[27] Why should we now spend time "rediscovering" reactions for use in water that already work well in familiar organic solvents such as THF, toluene, or methylene chloride? Because there are many potential advantages of replacing these and other unnatural solvents with water. The most obvious are the following:

- Cost. It does not get any cheaper than water!
- Safety. Most of the organic solvents used in the lab today are associated with risks: flammables, explosives, carcinogenics, *etc.*
- Environmental concerns. The chemical industry is a major contributor to environmental pollution. With increasing regulatory pressure focusing on organic solvents, the development of non-hazardous alternatives is of great importance.

It is, however, important that the benefits listed above are not gained at the expense of synthetic efficiency. Even a small decrease in yield, catalyst turnover, or selectivity of a reaction can lead to a substantial increase in cost and amount of waste generated. Fortunately, many theoretical and practical advantages of the use of water as solvent for organic synthesis do exist. These will be elaborated upon below but are briefly introduced here:

- Experimental procedures may be simplified since the isolation of organic products and recycling of water-soluble catalysts and other reagents could be made by simple phase separation.
- Laborious protecting-group strategies for functionalities containing acidic hydrogens may be reduced.
- Water-soluble compounds could be used in their "native" form without the need for hydrophobic derivatization, again eliminating tedious protection–deprotection steps from the synthetic route.
- As will be amply exemplified in this review, the unique solvating properties of water have been shown to have beneficial effects on many types of organic reactions in terms of both rate and selectivity.

1.9.1 Solubility of Organic Compounds in Water

Most chemical reactions are performed in solvents. The solvent provides a reaction medium in which reactants can be mixed over a very wide concentration range. In general, a good solvent should readily dissolve all or most of the participating reactants, should not interact adversely with the reaction, and should be easily separated during workup for facile isolation of products. On the basis of the knowledge of the chemical properties of the reactants, the chemist chooses a suitable solvent to meet these criteria. From this perspective, it is not surprising that water has found limited use as a solvent for organic reactions. In truth, the poor solubility of reactants and the deleterious effect on many organic transformations are the main obstacles to the use of water as reaction solvent. Nonetheless, the fact that many of the most desirable target molecules (*e.g.*, carbohydrates, peptides, nucleotides) and their synthetic analogues, as well as many alkaloids and important drugs, are readily soluble in water is inconsistent with the disproportionate bias toward the use of organic solvents for their preparation. It can be argued that our shortcomings as synthetic chemists prompt the use of exhaustive protecting-group strategies, thus limiting the possibility of using water as solvent because of the low solubility of the reactants. Moreover, with the limited number of organic transformations in water that is presently available to the synthetic chemist, intermediates soluble in organic solvents are preferred to those soluble in water, even if it means adding extra synthetic steps for derivatization. This may have particular relevance in carbohydrate chemistry.

Notwithstanding the above, many organic targets and their intermediates have very low solubility in water, which may lead to thwarting of reactions due to phase separation and inefficient mixing of reactants, although hetero-geneous mixtures may retain, at least partly, the positive influence of water, sometimes with the aid of sonication or microwave heating. A variety of strategies has been investigated in order to expand the scope of water-based organic synthesis to embrace highly hydrophobic reactants as well, and these will be briefly discussed below.[28]

1.9.2 Organic Co-solvents

One of the more efficient and versatile methods of increasing solubility—one that does not require modification of the solute—is to use an organic co-solvent. The co-solvent reduces the hydrogen-bond density of aqueous systems, so that it is less effective in squeezing out non-polar solutes from solution. Co-solvents can be structurally diverse, but they all carry hydrogen-bond donor and/or acceptor groups for aqueous solubility and a small hydrocarbon region that serves to disrupt the strong hydrogen-bond network of pure water, thereby increasing the solubility of non-polar reactants.[29] Some of the most commonly used co-solvents are the lower alcohols, DMF, acetone, and acetonitrile. The increase in solubility, however, comes at the cost of many of the properties that make water a unique solvent for synthesis such as high polarity, high cohesive energy density, and the hydrophobic effect. The significance of this erosion of the bulk properties of water depends on the nature of the chemical reaction. Reactions that involve charged or highly polar species will suffer more by a decrease in solvent polarity than reactions involving only uncharged species. Likewise, reactions with a negative activation volume (Diels–Alder, Claisen rearrangement) are expected to be adversely affected by the addition of co-solvent because of the concurrent decrease in cohesive energy density. Nevertheless, because of the efficiency and flexibility of co-solvents in solubilizing organic solutes, the major part of the development of reactions in aqueous media has been made with water–co-solvent mixtures.[30]

1.9.3 Ionic Derivatization (pH Control)

Adding a positive or negative charge to an ionizable solute generally brings about a substantial increase in its solubility in water. The adjustment of solution pH is therefore an efficient method of solubilizing weak electrolytes in aqueous media.[31] This approach, of course, changes the chemical nature of the reactant and may limit its use as a method of solubilization for synthetic purposes. For some types of reactions, however, the presence of a charged, highly polar moiety can have a very positive effect. For example, the reactions of diene-carboxylates, 2,4-sulfonates, or -ammonium salts with dienophiles in aqueous Diels–Alder reactions display significantly higher reaction rates over the corresponding neutral dienes.[32] It is generally accepted that the rate enhancement of the Diels–Alder reaction in water is at least partly due to the influence of the hydrophobic effect in this media. In making the diene amphiphilic, increased solubility comes with the added bonus of an enhanced hydrophobic effect and faster reaction. This is principally the same effect one would achieve by adding "salting-out" agents known to increase the hydrophobic effect, only in this case, the salt is the reactant itself.[33] When a buffered reaction is feasible it is one of the more efficient ways to keep organic molecules solubilized. A potential practical advantage of using pH control in organic reactions is that the product may be recovered from solution by precipitation upon suitable adjustment of pH or by extraction after addition of appropriate phase-transfer counterions.

1.9.4 Surfactants

An intriguing means of achieving aqueous solubility is by using surfactants. These are amphiphilic molecules, that is, they contain one distinctly polar and one distinctly non-polar region. In water, surfactants tend to orient themselves so that they minimize contacts between the non-polar region and the polar water molecules and when the concentration of surfactant monomer exceeds a certain critical value (critical micelle concentration), micellization occurs. Micelles are spherical arrangements of surfactant monomers with a highly hydrophobic interior and a polar, water-exposed surface. Organic solutes interact with micelles according to their polarity; non-polar solutes are buried in the interior of the micelle, moderately polar molecules locate themselves closer to the polar surface, while distinctly polar solutes will be found at the surface of the micelle. This compartmentalization of solutes is believed to be responsible for the observed catalytic or inhibitory influence on organic reactions in micellar media.[34]

1.9.5 Hydrophilic Auxiliaries

Another method to increase the solubility of aqueous organic reactions is by grafting hydrophilic groups onto insoluble reactants. This strategy has been only cautiously explored for synthetic purposes but has a pivotal role in medicinal chemistry and modern drug design because of the low water solubility of many drugs, which causes limited bioavailability and thus reduced therapeutic efficacy. One way of improving the solubility of drugs is by converting them into water-soluble prodrugs through covalent attachment of a hydrophilic auxiliary. Ideally, the attachment should be of a transient and reversible nature, allowing for release of the parent drug from the auxiliary upon distribution, either by enzymatic or chemical means. The size and nature of the solubilizing function ranges from small to medium-sized acidic and basic ionizable moieties (*e.g.*, carboxylic acids).[35]

1.9.6 Summary

Aqueous radical chemistry predominated in biological processes, and the development of the synthetic radical aqueous chemistry will aid in our understanding of the detailed mechanism of the chemistry of life, such as biocatalysis for example, and in turn will have applications in the evolving areas such as artificial biocatalysis, biomaterials and biotechnology in general. The main purpose of this book is to attempt to lift the greatest restriction on the implementation of aqueous radical synthesis, which is a misconception that might persist with many synthetic chemists regarding the inadequacy of aqueous radical chemistry as an equal partner in synthesis and chemistry in general.

1.10 CLASSICAL SYNTHESIS IN MODERN SOLVENTS

The reaction medium plays many crucial roles in the synthesis. Of these the most obvious are as reaction media and in product separation and purification.[36] In recent years there has been a dramatic increase in demand for development of new solvent systems which will provide particular advantages over existing methods. Part of this comes from the recognition that many solvents have undesirable properties, such as relative toxicity, harm to environment as well as problems with recyclability and reuse.

The development of "Green Organic Chemistry" is due to the recognition that environmentally friendly products and processes will be economical in the long term as they circumvent the need for treating 'end-of-the-pipe' pollutants and by-products generated by conventional synthesis. A new kind of chemical revolution, "green chemistry", is brewing—150 years after the first chemical revolution transformed modern life with a host of conveniences—which protects the environment, not by cleaning it up, but by inventing new chemistry and new chemical processes that prevent pollution. In essence, it prompts the chemical and pharmaceutical manufacturer to consider how human life is impacted after these chemicals are generated and introduced into their society.[37-41] By rethinking chemical design from the ground up, chemists are developing new ways to manufacture products that fuel the economy and lifestyles, without the damages to the environment that have become all too evident in recent years. This can be achieved through the proper choice of starting materials, atom economic methodologies with a minimum number of chemical steps, the appropriate use of greener solvents and reagents, and efficient strategies for product isolation and purification. Free radicals are ubiquitous, reactive chemical entities. Free radical reactions are an important class of synthetic reactions that have been traditionally performed in organic solvents. In recent years, the number of reports of free radical reactions that use water and alternative media such as supercritical CO_2, ionic liquids, fluorous solvents and the solid state has increased.[42,43] Reaction with radicals is one of the most useful methods for organic reactions in alternative media, because most of the organic radical species are stable in water and alternative media, and they do not react with water or unusual media. In addition, by harnessing free radical reactivity within the laboratory, biological processes can be studied and controlled, leading in turn to the prevention of disease and the development of new treatments for disease states mediated by free radicals. Whilst there have been several excellent reviews on carbon–carbon (C–C) bond formation and reactions of carbon–hydrogen (C–H) bonds in water, this book addresses synthetically useful C–H and C–C bond formations in alternative media (supercritical CO_2, ionic liquids, fluorous solvents, microwave mediated reactions and in solid state) *via* radical chain reactions (hydrogen atom transfer reactions, alkylations, radical cyclization and tandem radical reactions) within the past decade. There is a specific focus on C–H and C–C bond-forming reactions as they represent the major classes of the most useful and utilized free radical reactions. In addition, several important electron transfer processes as well as synthetically useful free radical non-chain reactions in alternative media will be discussed with a specific emphasis on mechanistic and application aspects of these transformations. All relevant stereo-aspects of the transformations in question will be discussed and presented.

1.10.1 Perfluorinated Solvents—a Novel Reaction Medium in Organic Chemistry: General Introduction

Most organic reactions are carried out in a solvent which has several important roles. At the molecular level, it breaks the crystal lattice of solid reagents, interacts with gaseous reagents and often lowers considerably the transition state of many reactions. Because of the intermolecular interactions between a solvent and organic reagents, it may not only enhance the reaction rate but also change the product distribution.[44] From the macroscopic point of view, the solvent also removes the excess heat produced during a reaction or allows a uniform supply of calories to the reagents. All these advantages of using a solvent for carrying out a reaction are also valuable for large-scale reactions. However, in this case the separation of the product from the solvent at the completion of the reaction may be costly and tedious, especially if other by-products have been formed or if an expensive catalyst has to be separated and recovered as well. Since the solubility of reagents in a solvent S is strongly dependent on the temperature, the question arises as to whether a solvent can be chosen in such a way that this solvent S solubilizes the reagents R^1, R^2, ...,R^n at the reaction temperature but not these reagents or more importantly the products P^1, P^2, ...,P^n at room temperature. This would allow a facile separation of the products from the reaction mixture simply by decantation. If the reaction is selective, *i.e.*, only one product is formed, no special workup conditions are required. Especially important are the variations of this concept when one regenerable reagent R^a or a catalyst C is selectively soluble in the solvent S at room temperature and reaction temperature whereas the starting materials R^1, ..., R^n and products P^1, ...,P^n are insoluble. This leads to a biphase system using two different solvents. S^1 solubilizes the organic reagents and products, whereas the second solvent S^2 solubilizes the regenerable reagent R^a or the catalyst C. Such a process is of industrial interest since, on the one hand, it allows a facile separation of the often costly catalyst and, on the other hand, a run of cyclic operations with the same catalyst solution is possible. Furthermore, the product phase S^1 should not be contaminated with the catalyst C or the regenerable reagent R^a, which is often a complex problem in industry (Scheme 1.2).

To be applicable to a broad range of substrates, the natures of the two solvents S^1 and S^2 have to be very different in physical properties and chemical behaviour so that low miscibility is observed. The use of water as solvent S^2 (S^1 being an organic solvent) has found many applications including industrial processes.[44] However, the incompatibility of many organometallic reagents with water and the difficulty of removing metal catalysts from water for disposal led to the search for alternative solvents.[45,46,47] These solvents display low chemical reactivity, low toxicity and low miscibility with many organic solvents. Furthermore, these solvents are inflammable and show a relatively low volatility. In this chapter, the basic physical and chemical properties will be indicated and the applications of perfluorinated solvents for the performance of organic reactions will be presented in detail as well as the reactions performed under conditions of a fluorous biphase system which has been extended to perfluorinated solvents.[47]

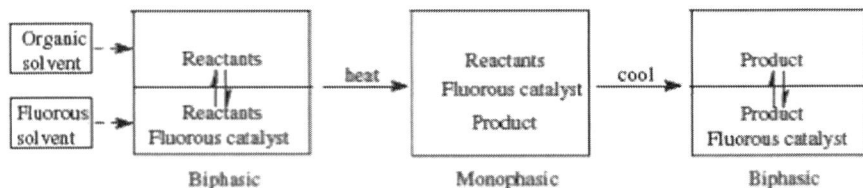

Scheme 1.2   Representation of a reaction performed in a bi-phase system.

1.10.2 Benzotrifluoride and Derivatives: Useful Solvents for Organic Synthesis and Fluorous Synthesis

Benzotrifluoride (BTF, trifluoromethylbenzene-a,a,a-trifluorotoluene, $C_6H_5CF_3$) and related compounds are introduced as new solvents for traditional organic synthesis and for fluorous synthesis. BTF is more environmentally friendly than many other organic solvents and is available in large quantities. BTF is relatively inert and is suitable for use as a solvent in a wide range of chemistry including ionic, transition-metal catalyzed and thermal reactions. It is especially useful for radical reactions, where it may replace benzene as the current solvent of choice for many common transformations. BTF and related solvents are also crucial components of fluorous synthesis since they can dissolve both standard organic molecules and highly fluorinated molecules.[48]

BTF belongs to an important group of trifluoromethyl-substituted aromatic compounds which have broad applications as intermediates or building blocks for crop protection chemicals, insecticides and pharmaceuticals, as well as dyes. Related higher boiling compounds that are produced in multimillion pound quantities include 4-chlorobenzotrifluoride (PCBTF) and 3,4-dichlorobenzotrifluoride (3,4-DCBTF). While these materials are manufactured as intermediates, BTF and its analogs are relatively inert and, hence, potentially useful as solvents for reactions and extractions, as well as for non-chemical applications such as solvents for coatings and cleaning of surfaces. The trifluoromethyl group on the aromatic ring is very stable to basic conditions even at elevated temperatures, and somewhat stable to aqueous acid conditions at moderate temperatures. BTF has recently become available at low price. Moreover, its lower toxicity and higher boiling point (which minimizes losses during evaporation) make it an ecologically suitable replacement for solvents like dichloromethane and benzene. Despite these favorable characteristics, BTF is relatively unknown as a solvent. However, recent evaluation by Curran,[49] and others' experience, is beginning to show that BTF is indeed suitable as solvent for many different reactions. In this book, we provide an overview of the features of BTF and related solvents and summarize their recent uses as reaction solvents in both traditional organic synthesis and new fluorous reactions. This information should prove helpful to others evaluating when and how to employ BTF as a solvent in organic synthesis.[49]

1.10.3 Reactions in Supercritical Carbon Dioxide (scCO₂) as a Novel Reaction Medium

Supercritical fluids (SCFs) have fascinated researchers ever since the existence of a critical temperature was first noted more than 175 years ago.[50] Although initial studies focused mainly on the physical properties of supercritical phases, their chemical reactivity was of interest from the beginning, too.[51] In fact, several well-established industrial processes for the production of bulk chemicals occur under temperatures and pressures beyond the critical data of the reaction mixture, the Haber–Bosch process and the high-pressure polymerization of ethylene being just the most outstanding examples. Supercritical fluid extraction (SFE) using supercritical carbon dioxide (scCO₂) is part of the current state-of-the-art for the industrial production of decaffeinated coffee and of hops aroma. In sharp contrast, the interest in the use of SCFs as reaction media for complex organic syntheses has sparked mainly during the last ten to fifteen years and it is just fair to refer to SCFs as "modern solvent systems" in this context.[52]

The unique physico-chemical properties of the supercritical state, as briefly outlined, make SCFs in general highly attractive reaction media for chemical synthesis. Probably the most important incentive for the use of SCFs in organic chemistry comes, however, from an increasing demand for environmentally and toxicologically benign processes for the production of high-value chemicals. Water and carbon dioxide are clearly the most attractive solvents for such applications of green chemistry.[53] In practical systems, the choice of an SCF for a synthetic application will also depend on the critical data, which must not be too drastic in order to keep equipment and process costs to a minimum. Chapter 2 of this book will therefore focus on the use of compressed (supercritical or liquefied) CO_2, because it best meets the criteria of ecological and economical constraints. The reader should be aware, however, of the rich and exciting chemistry which is carried out using other SCFs, especially in supercritical alkanes and in the more polar solvents $scCHF_3$ and scH_2O. We will further restrict our discussion mainly to applications of CO_2 in free radical organic syntheses using homogeneous or heterogeneous metal catalysts. Related areas like free-radical polymerization, the synthesis of inorganic or organometallic compounds, enzymatic reactions, and reactions that are used mainly as probes for near-critical phenomena are outside the scope of the present book.

1.10.4 Solvent-free Reactions as an Alternative: General Interest for Solvent-free Processes

For reasons of economy and pollution, solvent-free methods are of great interest in order to modernize classical procedures making them more clean, safe and easy to perform. Reactions on solid mineral supports, reactions without any solvent/support or catalyst, and solid/liquid phase transfer catalysis can be thus employed with noticeable increases in reactivity and selectivity. These methodologies can, moreover, be improved to take advantage of microwave activation as a beneficial alternative to conventional heating under safe and efficient conditions with large enhancements in yields and savings in time.

Nowadays, one of the main duties assigned to the organic chemist is to organize research in such a way that it preserves the environment and to develop procedures that are both environmentally and economically acceptable. One major objective is therefore to simplify and accommodate in a modern way the classical procedures with the aim of keeping pollution effects to a minimum, together with a reduction in energy and raw materials consumption.

Among the most promising ways to reach this goal, solvent-free techniques hold a strategic position as solvents are very often toxic, expensive, and problematic both to use and to remove. This is the main reason for the development of such modern technologies. These approaches can also enable experiments to be run without strong mineral acids (*i.e.* HCl, H_2SO_4 for instance) that can in turn cause corrosion, safety, manipulation and pollution problems as wastes.

1.10.4.1   *Reactivity*

An enhancement in kinetics can result from increasing concentrations in reactants when a diluting agent such as a solvent is avoided (see eqn (1.2)).

$$A + B \rightarrow product \qquad v = k[A][B] \qquad (1.2)$$

As concentrations in reactive species are optimal, reactivity is increased and only *mild conditions* are required. In several cases, reactions which are difficult (even impossible) using solvents are easily achieved under solvent-free conditions. Another unquestionable advantage lies in the fact that *higher temperatures*, when compared to classical conditions, can be used without the limitation imposed by solvent boiling points.

1.10.4.2   *Selectivity*

The layout of reacting systems can be increased when high concentrations and/or aggregation of charged species are involved. It can lead to some modifications in mechanisms resulting in a decrease in molecular dynamics and induce subsequent *special selectivities* (stereo-, regio- or enantioselectivity). Weak interactions can, for instance, appear (such as π-stacking) which are usually masked by solvents, inducing further effects on selectivity.

1.10.4.3   *Simplification of Experimental Procedures*

Firstly, complex apparatus is not needed and, for instance, reflux condensers are not required, which in turn allows the handling of a smaller quantity of material as there is no solvent. It also allows an operation to be carried out *with increased amounts of products in the same vessels.*

Washing and extraction steps are made easier or even suppressed. In the case of equilibrated reactions leading to light polar molecules (MeOH, EtOH or H_2O), the equilibrium can be easily shifted by a simple heating just above the boiling points or under reduced pressure.

1.10.4.4 Overall Benefits

Solvent-free techniques represent a clean, economical, efficient and safe procedure which can lead to substantial savings in money, time and products. They can be efficiently coupled to non-classical methods of activation that include ultrasound and microwaves.

1.11 METHODS OF GENERATING FREE RADICALS

The homolytic cleavage of covalent bonds produces radicals and, since this is an endothermic process, it requires the introduction of energy from the surroundings. Heat serves this purpose by collisional interconversion of kinetic energy into vibrational energy, and the temperature required for bond homolysis will be proportional to the bond dissociation energy. Absorption of light may also lead to radical species by intra- or intermolecular conversion of the increased electronic energy into vibrational energy. As expected, weaker covalent bonds dissociate into radicals more readily than stronger covalent bonds. The following table lists standard bond energies (Δ) for the C–C, C–O and C–H bonds commonly found in organic compounds, together with bond energies for some weaker bonds that have been found useful for generating radicals. Approximate homolysis temperatures at which half the bonds are cleaved in one hour are also given (Table 1.1).[54]

1.11.1 Thermal Cracking

At temperatures greater than 500 °C, and in the absence of oxygen, mixtures of high molecular weight alkanes break down into smaller alkane and alkene fragments. This cracking process is important in the refining of crude petroleum because of the demand for lower boiling gasoline fractions. Free

Table 1.1 Some Standard Bond Energies and Approximate Homolysis Temperatures.

Bond	Δ/kcal mol^{-1}	T/°C	Bond	Δ/kcal mol^{-1}	T/°C	Bond	Δ/kcal mol^{-1}	T/°C
C–C	85	670	O–O	34	160	O–Cl	49	280
C–H	99	850	N–N	39	230	C–I	51	350
C–O	84	680	S–S	55	440	C–Br	67	480

radicals, produced by the homolysis of C–C bonds, are known to be intermediates in these transformations. Studies of model alkanes have shown that highly substituted C–C bonds undergo homolysis more readily than do unbranched alkanes. In practice, catalysts are used to lower effective cracking temperatures.[55]

1.11.2   Homolysis of Peroxides and Azo Compounds

In contrast to stronger C–C and C–H bonds, the very weak O–O bonds of peroxides are cleaved at relatively low temperatures (80 to 150 °C). The resulting oxy radicals may then initiate other reactions, or may decompose to carbon radicals, depicted in the first two equations. Organic azo compounds (R–N=N–R) are also heat sensitive, decomposing to alkyl radicals and nitrogen. Azobisisobutyronitrile (AIBN) is the most widely used radical initiator of this kind, decomposing slightly faster than benzoyl peroxide at 70 to 80 °C. The thermodynamic stability of nitrogen provides an overall driving force for this decomposition, but its favourable rate undoubtedly reflects weaker than normal C–N bonds.[56]

1.11.3   Photolytic Bond Homolysis

Compounds having absorption bands in the visible or near-ultraviolet spectrum may be electronically excited to such a degree that weak covalent bonds undergo homolysis. Examples include the halogens Cl_2, Br_2 & I_2 (bond dissociation energies are 58, 46 & 36 kcal mol^{-1} respectively), alkyl hypochlorites, nitrite esters and ketones. The covalent bonds that undergo homolysis are coloured red, and the unpaired electrons in the resulting radicals are coloured pink. Ketones undergo n to π^* electronic excitation near 300 nm. The resulting excited state is a diradical in which one of the odd electrons is localized on the oxygen atom. Cleavage of an alkyl group may then take place.[57]

1.11.4   Electron Transfer

The action of inorganic oxidizing and reducing agents on organic compounds may involve electron transfers that produce radical or radical ionic species. Ferrous ion, for example, catalyzes the decomposition of hydrogen peroxide (Fenton's reagent) and organic peroxides. In some cases the radical intermediates formed in this manner are sufficiently stable to be studied in the absence of oxygen. The phenoxy radical formed in Würster's salt is another. The alkali metals lithium, sodium and potassium reduce the carbonyl group of ketones to a deep blue radical anion called a "ketyl". Subsequent chemical reactions of these useful intermediates are discussed elsewhere. A similar reduction of benzene and its derivatives also proceeds by way of radical anion intermediates.[59]

1.11.5 Hydrogen and Halogen Atom Abstraction

If free radical reactions are to be useful to organic chemists, methods for transferring the reactivity of the simple radicals generated by the previously described homolysis reactions to specific sites in substrate molecules must be devised. The most direct way of doing this is by an atom abstraction, as shown here.[59]

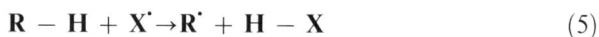

$$R - H + X^{\cdot} \rightarrow R^{\cdot} + H - X \tag{5}$$

Indeed, when X is Cl or Br, this is a key step in the alkane halogenation chain reaction. Hydrogen abstraction reactions of this kind are sensitive to the nature of both the attacking radical ($X^{\cdot}$) and the R–H bond. Thus the rate of reaction of 1° C–H with $Cl^{\cdot}$ is a thousand times faster than with $Br^{\cdot}$. However, the less reactive bromine atom shows much greater selectivity in discriminating between 1°, 2° and 3° C–H groups.

Certain C–H bonds are so susceptible to radical attack that they react with atmospheric oxygen (a diradical) to form peroxides. Typical groups that exhibit this trait are 3°-alkyl, 2° & 3°-benzyl and alkoxy groups in ethers.

$$R - H + O_2 \rightarrow R^{\bullet} + {}^{\bullet}O_2H \rightarrow R - O - O - H \tag{6}$$

The exceptional facility with which S–H and Sn–H react with alkyl radicals makes thiophenol and trialkyltin hydrides excellent radical quenching agents, when present in excess. At equimolar or lower concentrations they function well as radical transfer agents. Carbon halogen bonds, especially C–Br and C–I, are weaker than C–H bonds and react with alkyl and stannyl radicals to generate new alkyl radicals. This reaction has been put to practical use in a mild procedure for reducing alkyl halides to alkanes.

An important modification of this reduction is shown in the third example above. The use of equimolar amounts of tributyltin hydride in reactions presents certain problems, including the toxicity presented by organostannanes; the difficulty in separating non-polar stannanes, such as halides, bis(tributyltin) and bis(tributyltin) oxide from desired products; and the formation of tin oxides by reaction with moisture. To reduce these difficulties, a catalytic amount of the stannanes may be used together with enough $NaBH_4$ (or an equivalent reagent) to convert the tributyltin halides to the hydride. Indeed, the reduction is so facile that traces of peroxides in the reactants often initiate reaction without added AIBN.

1.11.6 The Configuration of Free Radicals

The configurational preferences of different reactive intermediates were noted in Section 1.7. Since the difference in energy between a planar radical and a rapidly inverting pyramidal radical is small, radicals generated at chiral centers generally lead to racemic products. However, unlike carbocation intermediates, which prefer to be planar, radicals tolerate being restricted to a pyramidal configuration. The initial formation of a carboxyl radical is followed by the loss of carbon dioxide to give a pyramidal bridgehead radical. This radical abstracts a chlorine atom from the solvent, yielding the bridgehead chloride as the major product. Although this is a 3°-alkyl halide, it does not undergo S_N1 solvolysis reactions because of the strain imposed on the carbocation intermediate by its pyramidal confinement.[60]

The concurrent formation of ester and dimeric cycloalkane products from acyl peroxides is common, and reflects a cage effect in homolysis reactions. When a pair of radicals is formed by homolysis, they are briefly held in proximity by the surrounding solvent molecules (the cage). Rapid decomposition to other radicals may occur, but until one or both of these radicals escapes the solvent cage a significant degree of coupling (recombination) may occur. Cage recombination of radicals may be sufficiently rapid to preserve the configuration of the generating species.

1.11.7   Elementary Reaction Steps between Radicals and Non-radicals: Reactions of Free Radicals

For many years organic chemists considered free radical reactions to have limited applications, and to be of little interest outside some fields of industrial chemistry. This view has changed markedly, and important examples of substitution, addition and elimination reactions proceeding by way of radical intermediates have been developed and used in the synthesis of complex molecules.

Since most free radical intermediates are very reactive and have short lifetimes, all the steps in a practical chain reaction sequence must be fast compared with possible competing reactions. This means that atom abstractions and radical additions should be exothermic, or only mildly endothermic. Both C–H and C–X abstractions satisfy this requirement, but the strong O–H bond does not. Also addition to C=C is energetically more favourable than addition to C=O; in addition, when radicals add to certain C=O functions they bond to the oxygen not the carbon. Another characteristic of radical reaction sequences is fragmentation with expulsion of a small stable molecule, such as CO_2, CO or N_2.

A radical chain is built up by different types of propagation steps all of which lead to the new radicals:

- Addition reactions
- Substitution (abstraction) reactions
- Elimination (fragmentation) reactions
- Rearrangement reactions
- Electron transfer reactions

As a rule of thumb there are two simple conventions that can be useful for fast radical chain reactions:

- Most chain-propagating steps are exothermic and one can use the strength of bonds that are broken and formed as a rough guide for the rate of the reaction (thermodynamic parameter).
- Because of the early transition states in fast radical reactions, frontier molecular orbitals theory can be utilized for these reactions.

1.12 SUSTAINABLE CHEMISTRY METRICS AND RADICAL CHEMISTRY: COMPARATIVE APPROACH

"Green chemistry" was formulated in the last decade of the 20th century as an answer to the social and scientific concerns about environmental problems related to the handling and use of potentially toxic and dangerous chemicals. The philosophy of green chemistry can be summarized in its "Twelve Principles",[61] which have been extended to the field of chemical engineering to formulate the so-called "Twelve Principles of Green Engineering".[62] Both academic and industrial processes can be benchmarked according to sustainability criteria.[63] The mission of green chemistry is to promote innovative chemical technologies that reduce or eliminate the use or generation of hazardous substances in the design, manufacture, and application of chemical products.[64] The mnemonics of the Twelve Principles of Green Chemistry and Green Engineering have been reported.[65,66]

- Prevent wastes
- Renewable materials
- Omit derivatization steps
- Degradable chemical products inputs
- Use of safe synthetic methods
- Catalytic reagents
- Temperature, pressure ambient
- In-process monitoring
- Very few auxiliary substances
- Environmental (E) factor, maximize feed in product
- Low toxicity of chemical products
- Yes, it is safe

- Inherently non-hazardous and safe
- Minimize material diversity
- Prevention instead of treatment
- Renewable materials and energy
- Output-led design
- Very simple
- Efficient use of mass, energy, space, and time
- Meet the need
- Easy to separate by design
- Networks for exchange local mass and energy
- Test the life cycle of the design
- Sustainability throughout product life cycle

Both afford a qualitative relationship between the concept of sustainability and chemical events. To quantify the sustainability of a chemical process over a number of years, several mass- and energy-based parameters have been introduced to evaluate the environmental impact of chemical reactions. With these and similar parameters, it is possible to achieve efficient and eco-friendly chemical reactions.

The field of green metrics is plagued by authors' continuous changing of the names of the same ideas, which gives the impression that there is yet another new metric to be defined, with a new insight. This has been a serious liability towards the acceptance of this topic as a useful concept in the routine practice of tracking the optimization of chemical reactions.

The aim of this section is to summarize the main parameters for analyzing chemical reactions and processes, in order to evaluate their undesirable environmental consequences. These parameters are useful for developing sustainable, new chemical processes and to improve the aforementioned reactions for the manufacture of chemical substances and the design of green products.[67]

1.12.1 Classical Metrics of Chemical Reactions

Implementing these principles requires a certain investment since the current, very inexpensive chemical processes must be redesigned. However, in times when certain raw materials become more expensive (for example, as the availability of transition metals becomes limited) and also the costs for energy increase, such an investment should be paid back as the optimized processes become less expensive than the unoptimized ones. The development of greener procedures can therefore be seen as an investment for the future, which also helps to ensure that the production complies with possible upcoming future legal regulations.

A typical chemical process generates products and wastes from raw materials such as substrates, solvents and reagents. If most of the reagents and the solvent can be recycled, the mass flow looks quite different (Figure 1.7).

Thus, the prevention of waste can be achieved if most of the reagents and the solvent are recyclable. For example, catalysts and reagents such as acids and bases that are bound to a solid phase can be filtered off, and can be regenerated (if needed) and reused in a subsequent run. In the production of chemical products on very large scale, heterogeneous catalysts and reagents can be kept stationary while substrates are continuously added and pass through to yield a product that is continuously removed (for example by distillation).

The mass efficiency of such processes can be judged by the E factor (environmental factor):

$$\text{E factor} = \frac{\text{mass of wastes}}{\text{mass of product}}.$$

Whereas the ideal E factor of 0 is almost achieved in petroleum refining, the production of bulk and fine chemicals gives E factors of between 1 and 50.

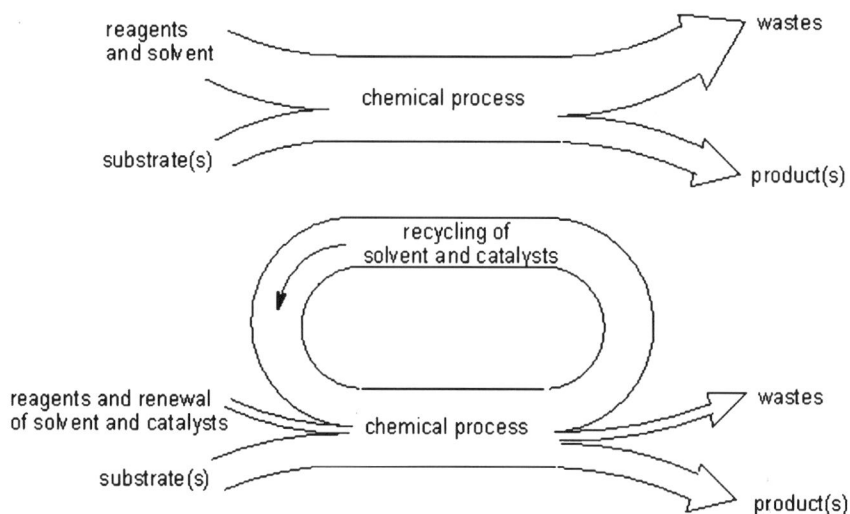

Figure 1.7 General diagram of a typical chemical process.

Typical E factors for the production of pharmaceuticals lie between 25 and 100. Note that water is not considered in this calculation, because this would lead to very high E factors. However, inorganic and organic wastes that are diluted in the aqueous stream must be included. Sometimes it is easier to calculate the E factor from a different viewpoint, since accounting for the losses and exact waste streams is difficult:

$$\text{E factor} = \frac{\text{mass of raw materials} - \text{mass of product}}{\text{mass of product}}.$$

In any event, the E factor and related factors do not account for any type of toxicity of the wastes. Such a correction factor (an "unfriendliness" quotient, Q) would be 1 if the waste has no impact on the environment, less than 1 if the waste can be recycled or used for another product, and greater than 1 if the wastes are toxic and hazardous. Such discussions are at a very preliminary stage, and E factors can be used directly for comparison purposes as this metric has already been widely adopted in the industry.

Another attempt to calculate the efficiency of chemical reactions that is also widely used is that of atom economy or efficiency. Atom efficiency is a highly theoretical value that does not incorporate any solvent, nor the actual chemical yield. An experimental atom efficiency can be calculated by multiplying the chemical yield by the theoretical atom efficiency. In any case, the discussion remains more qualitative than quantitative, and does not yet quantify the type of toxicity of the products and reagents used. Still, atom economy as a term can readily be used for a direct qualitative description of reactions. Considering specific reactions, the development of green methods is focused on two main aspects: the choice of solvent, and the development of catalyzed reactions. An additional advantage of such polymer-bound catalysts is the avoidance of toxic transition metal impurities, for example in pharmaceutical products.

A key point is still the choice of solvent, as this is the main component of a reaction system by volume (approx. 90%). Chlorinated solvents should be avoided, as many of these solvents are toxic and volatile, and are implicated in the destruction of the ozone layer. Alternative solvents include ionic liquids for example, which are non-volatile and can provide non-aqueous reaction media of varying polarity. Ionic liquids have significant potential, since if systems can be developed in which the products can be removed by extraction or distillation and the catalyst remains in the ionic liquid, theoretically both the solvent and the catalyst can be reused. The solvent of choice for green chemistry is water, which is a non-toxic liquid but with limited chemical compatibility. On the one hand, reactions such as free radical transformations are often even accelerated when run in an aqueous medium while, on the other hand, many reactants and reagents (including most organometallic compounds) are totally incompatible with water. There is thus a great need to develop newer methods and technologies that would make interesting products available through reactions in water or other aqueous media. Chemical reactions run under neat conditions (no solvent) and in a supercritical CO_2 medium can also be considered as green choices. Other possible improvements can be considered, such as for example the replacement of benzene by toluene (as a less toxic alternative), or the use of solvents that can be rapidly degraded by microorganisms. It is quite astonishing to consider the progress that has been made in the development of greener alternatives to traditional reactions.

1.12.2 How do Contemporary Free Radical Transformations Hold Up? Focus on Sustainability, Atom Efficiency and Advantages

1.12.2.1 *Straightforward Radical Organic Chemistry in Neat Conditions: Advantages of Reactions "on Water"*

Organic solvents represent the biggest pollution problem of many synthetic organic processes[68] and the development of efficient synthetic methodologies for organic reactions, in the absence of organic solvents, is an important challenge toward reducing the amount of waste.[69] An ideal organic reaction would proceed in neat substrate(s)[70] or an environmentally benign solvent, such as water.[71,72] The major problem of a wider use of water instead of organic solvents is the very low solubility of most organic compounds in aqueous media. The use of additives, such as surfactants, helps to increase the solubility of organic compounds;[73] however, the reaction protocols become more complicated. It was noted many years ago that certain organic reactions are greatly accelerated when performed in an aqueous suspension rather than solution.[74] These "on water" reactions have received a great deal of attention in recent years, yet their scope is still limited.[75,76] In particular, organic synthesis involving free radicals under the "on water" conditions remains largely unexplored.[77] Interestingly, it was often found that many organic reactions that proceed well "on water" also work well under solvent-free conditions, neat reactions being next in line or on par with reactions in aqueous suspensions.[78] In this book we highlight several selective and efficient radical organic reactions that take place in neat reactants and "on water" demonstrating the unique advantages of free radical chemistry and green chemistry in achieving interesting and unexpected synthetic aims. In general radical reactions "on water" lead to efficient atom economic synthetic approaches towards formation of carbon–carbon, carbon–hydrogen and carbon–oxygen bonds and can be classified as highly atom efficient transformations (AE = 80–100%)

1.12.2.2 Solvent Free Carbon–Carbon Bond Formation in Ball Mills

The simple use of a mortar and a pestle to mix reagents can be useful in order to perform organic reactions under solvent-free conditions.[79,80] However, for a more efficient and automated mixing more sophisticated instrumentation has been developed. In that context, ball mills appear to be interesting.[81] In inorganic chemistry and material sciences, ball milling is a well-established technique for the grinding of minerals and the preparation and modification of inorganic solids,[82] both at laboratory and industrial scales.[83] It allows the preparation of nanostructured alloys and the synthesis of new materials.[84,85] Various types of ball mills are known including (drum) ball mills, jet-mills, bead-mills, vibration ball mills, planetary ball mills and horizontal rotary ball mills.[86] All of these devices are based on the principle that a starting material is placed between two surfaces and crushed due to impact and/or frictional forces that are caused by collisions between these surfaces. The various mills differ in the method of how the motion causing these collisions is created. Besides the intensive grinding effect, the collisions lead to an energy transfer, which results in an increase of internal temperature and pressure. For achieving better control of these factors, some ball mills have cooling/ heating devices. In general, ball mills are able to produce materials with a particle size of approximately 100 nm. Several synthetically useful free radical transformations are discussed in the following chapters where the advantage and degree of control are highlighted by conducting reactions using mechanical milling techniques as an effective method for mixing reactants leading to various C–C bond-forming reactions. Often, cleaner reaction profiles and higher yields are obtained, and the use of harmful organic solvents can be avoided (except in the workup). Sometimes, running the reactions in a solid state also leads to new and unexpected products, not available from the liquid-phase reactions.

In terms of atom efficiency, solvent-free carbon–carbon bond formation in ball mills satisfies many of the principles of green chemistry and in all cases shows a high atomic economy with a minimum generation of by-products, owing to the incorporation of most moieties of the starting material into the final product.

1.12.2.3   Molecular Reactors and Radical Transformations: Cyclodextrins

Molecular reactors are miniature reaction vessels that control the assembly of reagents to affect the outcomes of chemical transformations at the molecular level. In many ways, they are analogous to the reaction vats used in the chemical industry, the flasks used in chemical laboratories, and even the cooking pots used in kitchens. In each case, containers are used to bring together the required ingredients. After the chemical reactions have taken place, sometimes as a result of stirring or heating, the products are removed and the containers may be reused. The unique aspect of molecular reactors is that they act at the molecular level, and it is implicit that this changes the outcomes of the reactions, to make them different from those that would result when using bulk reaction media, such as common solvents. Where molecular reactors act in this manner, without themselves being altered, by definition they are operating as catalysts. The naturally occurring cyclodextrins and their derivatives have been developed as miniature reaction vessels, to manipulate the outcomes of chemical transformations at the molecular level. Several fundamental free radical transformations have been evaluated in this book with particular focus on the interplay between free radical chemistry with modern molecular technology and green chemistry tastes.

In terms of atom efficiency, fundamental free radical reactions in molecular reactors including hydrogen transfer, carbon–carbon bond formations and deoxygenation reactions have proceeded smoothly in the above-mentioned media in good to excellent yields. The advantage of these molecular reactors or medium combinations lie in their affordability, low toxicity, the avoidance of using highly toxic "tin-based hydrogen donors", reusability and the broad range of synthetic, mechanistic and catalytic applications.

1.12.2.4   Streamlining Free Radical Chemistry through Modern Technology

An increasing emphasis has been placed on producing synthetic organic compounds faster and more efficiently than can be produced using conventional methodology. It is often observed that tried and true methods for performing synthesis, workup and purification on an individual or low number of reactions do not scale up well even to a modest level of parallel synthesis. As a result, novel methodologies or parallel and combinatorial chemical synthesis continue to be developed at a rapid pace.

In general, the specific structure of the target molecule(s) guides the selection of the synthetic methodology (*e.g.*, solid or solution phase) used in a given reaction sequence. The phrase "streamlined organic synthesis" has been coined as a descriptor for a methodology that lends itself to efficient application to parallel organic synthesis. The advantages of the past few years in microreactors have demonstrated the miniaturizing of chemistry which has significant advantages with respect to cost, safety, throughput, kinetics and scale-up. The use of chemical microreactors for a broad spectrum of important chemical transformation has illustrated the utility and benefits for both chemical discovery and the development of novel and unique applications. The main objective of this chapter is to bring together two important topics of synthetic free radical chemistry and its applications and advancements to "streamlined organic synthesis" on the interface of conventional and alternative materials, such as polymer supported reagents/reactants, use of ultraporous materials (PolyHIPEs) as well as continuous microflow systems of natural zeolite and commercial man-made microreactors. There is a specific focus on the major classes of the most useful and utilized class of free radical reactions in "Streamlined Free Radical Organic Synthesis". In addition, several important electron transfer processes as well as free radical, non-chain, synthetically useful reactions in a high throughput environment and their applications will be discussed with a specific emphasis on the mechanistic and application aspects of these transformations.

1.12.2.5 *Microwaves and Synthesis: the Radical Focus*

While fire is now rarely used in synthetic chemistry, it was not until Robert Bunsen invented the burner in 1855 that the energy from this heat source could be applied to a reaction vessel in a focused manner.[87,88] The Bunsen burner was later superseded by the isomantle, oil bath or hot plate as a source of applying heat to a chemical reaction. In the past few years, heating chemical reactions by microwave energy has been an increasingly popular theme in the scientific community. Since the first published reports on the use of microwave irradiation to carry out organic chemical transformations in 1986,[89] more than 3500 articles have been published in this fast-moving and exciting field, today generally referred to as microwave-assisted organic synthesis (MAOS).[90,91] In many of the published examples, microwave heating has been shown to dramatically reduce reaction times, increase product yields and enhance product purities by reducing unwanted side reactions compared to conventional heating methods. The advantages of this enabling technology have, more recently, also been exploited in the context of multistep total synthesis[92] and medicinal chemistry/drug discovery,[93] and have additionally penetrated related fields such as polymer synthesis,[68] material sciences,[94] nanotechnology[95] and biochemical processes.[96] The use of microwave irradiation in chemistry has thus become such a popular technique in the scientific community that it might be assumed that, in a few years, most chemists will probably use microwave energy to heat chemical reactions on a laboratory scale.

The statement that, in principle, any chemical reaction that requires heat can be performed under microwave conditions has today been generally accepted as a fact by the scientific community. The short reaction times provided by microwave synthesis make it ideal for rapid reaction scouting and optimization of reaction conditions, allowing very rapid progress through the "hypothesis–experiment–results" iterations, resulting in more decision points per unit time. In order to fully benefit from microwave synthesis one has to be prepared to fail in order to succeed. While failure could cost a few minutes, success would gain many hours or even days. The speed at which multiple variations of reaction conditions can be performed allows a morning discussion of "What should we try?" to become an after-lunch discussion of "What were the results?" Not surprisingly, therefore, many scientists, both in academia and in industry, have turned to microwave synthesis as a frontline methodology for their projects when their attempts to perform a particular reaction had failed, or when exceedingly long reaction times or high temperatures were required to complete a reaction. This practice is now slowly changing and, due to the growing availability of microwave reactors in many laboratories, routine synthetic transformations are now also being carried out by microwave heating.

One of the major drawbacks of this relatively new technology remains the equipment cost. While prices for dedicated microwave reactors for organic synthesis have come down considerably since their first introduction in the late 1990s, the current price range for microwave reactors is still many times higher than that of conventional heating equipment. As with any new technology, the current situation is bound to change over the next several years and less expensive equipment should become available. By then, microwave reactors will have truly become the "Bunsen burners of the twenty first century" and will be standard equipment in every chemical laboratory.

1.12.2.6 Sonochemistry and Free Radical Synthesis: Advantages and Progress

Since the early 1980s, sonochemistry has attracted considerable attention as an innovative method for accelerating organic and organometallic reactions in a wide variety of solvents and conditions, often accompanied by unexpected selectivity patterns. Some pioneers made the very significant observation that sonochemistry could further be subdivided into reactions susceptible of real chemical effects induced by cavitation and those mainly sensitive to the mechanical impact of bubble collapse.[97] Despite our modest understanding of cavitation and sonoluminescence, this non-classical high-energy technique is heavily applied in organic synthesis, the fine-tuned fabrication of micro- and nanosized particles, interfacial phenomena, and the environmental degradation of pollutants.[98] Sonochemistry also offers stimulating challenges such as the role of ultrasound in homogeneous reactions, in unusual media (low vapor pressure media, *e.g.* room temperature ionic liquids (RTILs) and gels), or the enhancement of certain key effects in controlling reactivity (*e.g.* hydrophobic interactions).[99]

Sonochemistry—a Time Line

- Broadly defined, this describes a subject in which sound energy is used to affect chemical processes. Non-electromagnetic radiation is used to accelerate and execute chemical reactions.
- The potential of sonochemistry was first identified by Loomis. Sonic waves of great intensity (100 to 500 kHz) were produced and led to "clear

accelerating effects" *e.g.*, explosion of NI$_3$ and "atomization" of glass fragments from container walls.

- Up until *ca.* 1940, a large amount of pioneering work in polymers and chemical processes took place.
- 1955–1970 ... "doldrum years" ... "period of neglect"
- Since the early 1970s, there has been an increase in research with ultrasound due to the general availability of commercial ultrasound equipment.
- National groups devoted to sonochemistry appeared in the UK, Germany, Romania, Japan and China.
- This presentation is mostly limited to the application of ultrasound to organic synthesis.

Chapter 15 in this book highlights a succinct overview of sonochemistry, intended for those not familiarized with this technique, and a series of current problems and applications found in free radical chemistry in recent years. The discussion of the current status of the methodology is certainly timely and will undoubtedly stimulate further applications and developments of the new field.

1.12.2.7 Black Light and Free Radical Transformations

It is well recognized that non-ionizing radiation can react photochemically with biological chromophores, producing end products that are toxic and/or mutagenic in mammalian cells. Most studies have concentrated on the role of UV irradiation due to its high energy, photo-reactivity, wide range of biological chromophores, specific cellular responses, and association with pathologies such as skin melanoma and cataracts.[100–104] However, the role of visible light has been less extensively investigated, even though studies have demonstrated that visible light can induce cellular dysfunction and cell death both *in vitro* and *in vivo*.[100–104] The blue region (400–500 nm) of the visible spectrum is likely to be particularly important because it has a relatively high energy, can penetrate tissue(s), and is associated with the occurrence of malignant melanoma in animal models.[104,105] Surprisingly, despite its potential for damage, the use of blue light blockers in sunscreens and spectacle lenses has, until recently, received only limited attention. Studies have shown that the irradiation of mammalian cells with visible light induces cellular damage primarily *via* reactive oxygen species (ROS).[106] ROS such as the hydroxyl radical, superoxide anion, and singlet oxygen can be produced when visible light excites cellular photosensitizers.[107,108] Although photosensitizers such as melanin and lipofuscin in pigmented cells and retinoids in photoreceptor cells have been identified, the identity and location of photosensitizers in non-pigmented cells remain largely unknown. However, a number of options exist, including flavin-containing oxidases, the cytochrome system, heme-containing proteins, and tryptophan-rich proteins. The interaction of these chromophores with light can generate ROS, which in turn can damage lipids, proteins, and DNA. Violet–blue light stimulated H$_2$O$_2$ production from peroxisomes and mitochondria in cultured 3T3 and CV1 mammalian cells.[109] Hydrogen peroxide production was enhanced by over-expression of flavin-containing oxidases, which suggests that violet–blue light initiates the photoreduction of flavins, which activate flavin-containing oxidases in mitochondria and peroxisomes, resulting in H_2O_2 production. Furthermore, the mechanism by which photosensitization leads to cellular dysfunction is unclear but may center on DNA damage. As reviewed by Goyns,[110] the role of DNA damage in aging mammals appears to be pivotal, and there is increasing evidence that oxidative damage is an important factor in producing mutations in genes, shortening telomeres, and damaging mitochondrial DNA.[110,111]

Photolysis is one of the most appropriate and widely used methods of generating nucleosidyl radicals, with numerous reports in the literature reported of sources and setups for its generation; however, most of the methods require specialized experimental setups, glassware and rather expensive light sources for the desired transformations. In order to overcome these problems the use of a black light which has a maximum peak wavelength at 352 nm has been reported in a broad range of free radical transformations in organic and aqueous media. This book summarizes the up-to-date collection of synthetically useful free radical transformations under the black light initiation conditions and demonstrates the advantages of the methodology and its bright future in free radical green chemistry.

1.12.2.8   Sunny Side of Radical Chemistry: Efficiency and Natural Abundance

Solar energy is the prerequisite for maintaining life on Earth, as recognized many thousands years ago when the sun had been an object of worship in many ancient civilizations. Sunlight-induced transformations on Earth are significantly older than life itself. It has been demonstrated that the sun produces light with a distribution similar to what would be expected from a 5525 K (5250 °C) black body. Roughly, we can say that half lies in the visible part of the electromagnetic spectrum and the other half in the near-infrared part. Recently, the total solar irradiance has been measured, confirming a solar constant value of 1366 W m^{-2} (ref. 112) as the average intensity hitting the Earth's atmosphere.[88] As a matter of fact, the energy from sunlight reaching our planet in 1 h (4.3×10^{20} J) is more than that consumed globally in 1 year (4.1×10^{20} J, 2001 data).[113] From the above, it can be concluded that solar radiation is the best energy source since it is cheap, clean and available throughout the entire world.[114] Unfortunately, the light flux reaching the Earth's surface is discontinuous depending on the weather conditions and not available at night. Nevertheless, such conditions were sufficient for nature to perform the photochemical reaction par excellence *viz.* chlorophyllian photosynthesis used by organisms and plants to fix atmospheric carbon dioxide and produce biomass (especially sugars).[115] The use of solar light for the synthesis of valuable compounds is a challenge since the solar spectrum must necessarily match that absorbed by starting materials. This is not trivial since many compounds (*e.g.* organics) do not absorb in the visible or in the limited UV portion region of the solar spectrum.

On the other hand, the sun was the only accessible light source in the earliest experiments of organic photochemistry since artificial sources such as UV lamps were not yet available. Moreover, the use of photochemical reactions mediated by sunlight is obviously an appealing green way for storing the solar energy into the newly formed chemical compounds.[116] As for the use in synthesis, solar radiation is clean and renewable and, furthermore, leaves no residues in the reaction mixture, in accord with the postulates of green or eco-sustainable chemistry.[117]

The storage of solar energy is one of the main challenges in the near future. A rather unexploited way to fulfil this goal is the solar light induced formation of new chemical bonds, *i.e.* the synthesis of chemicals. Solar photons can be considered *the* ideal green reagents since they are costless and leave no residue in the reaction mixture. In many cases the solar radiation could be successfully used in place of toxic or expensive chemical reagents to overcome the activation energy in organic synthesis. In this book, the emerging trends on the use of solar light for green synthesis are summarized, highlighting the advantages of this photochemical method.

1.13 CLASSICS AND CATALYSIS IN FREE RADICAL CHEMISTRY: REAGENTS, REACTANTS AND PROTOCOLS

'...most chemists have avoided radical reactions as messy, unpredictable, unpromising, and essentially mysterious...' Chryssostomos Chatgilialoglu[118]

For many years, radicals—molecules that contain a single unpaired electron—were considered too reactive to be used productively in organic synthesis. This myth has been dispelled and, somewhat ironically, it is now clear that radicals frequently offer higher levels of selectivity and predictability than analogous ionic reactions.[117] Even with increased understanding, dogma dictated that radicals could not participate in highly stereoselective reactions despite being simple organic species, subject to the same steric and electronic interactions as all other molecules. This too has proved incorrect as the last decade has seen tremendous progress in enantioselective radical reactions.[118] The majority of naturally occurring compounds are chiral and not super-imposable on their mirror images. One of the major challenges for organic chemists is to develop enantioselective reactions, *i.e.* reactions that can discriminate between mirror image enantiomers. The domination of enantio-selective transformations by metal-based reagents is coming to an end as it becomes clear that small, metal-free molecules, or organocatalysts,[119] can achieve complementary reactions without recourse to potentially toxic or expensive metals.

The recent introduction of radical intermediates into organocatalysis by Kim and MacMillan[120] and Sibi and Hasegawa[121] has attracted considerable attention and there is no doubt that the principles underpinning this methodology will have a major impact on organic synthesis. It is often overlooked that radical chemistry has always been conducive to organocatalysis, with many of the general characteristics of radicals being ideally suited for a synergistic relationship with organic-based catalysts. Radicals are largely impervious to the effects of water, display greater functional group tolerance than ionic reagents and operate over a wider pH range. It is the aim of this book to briefly outline the shared history of radicals and photocatalysis (metal-based homogenous catalysis, the complementary approach of radicals and transitional metal catalysis, asymmetric organometallic catalysis, organocatalysis using classical and alternative reagents for free radical transformations) and to speculate on the future directions of this profitable partnership.

1.14   RADICAL CASCADES AND FREE RADICAL GREEN CHEMISTRY

Cascade, domino, or tandem processes, which link together two or more transformations in one pot, are increasing in popularity because they lead to improvements in synthetic efficiency and decreases in environmental impact. Not only do these cascades contain choice mechanistic gems but they also deliver compact and elegant syntheses of complex natural products. Longer cascades require more functional groups precisely configured within carefully designed initial molecular architectures. Such "purposeful" molecules can be thought of as chemical algorithms. This book surveys the phenomenal range of unimolecular free radical cascades. A convenient system for classifying free radical cascades is described which is useful for evaluating and comparing cascades and aids the design of synthetic routes to polycyclic structures. Double cyclization cascades lead to cyclopentylcyclopentane or bicyclo[3.3.0]-octane derivatives. Precursors that contain a ring as a template have been used to control stereochemistry in syntheses of triquinanes and many related compounds. Of the cascades containing ring-cleavage steps, the most useful are the ring expansions which have opened up new synthetic routes to medium ring polycycles. The key design features of three-stage unimolecular free radical cascades that yielded steroid structures are linear arrays of radical acceptor units associated with methyl groups distributed at every fifth C-atom in the precursor polyenes. Ring cleavage is the reverse of cyclization. In special, symmetrical structures, therefore, this led to sequences that were reversible, thus launching endlessly repeating cascades supported by delightfully fluxional structures. The science of "programming" organic molecules to achieve particular target structures is maturing rapidly. Coordination and classification of the welter of information in this area is intended to facilitate design and hence to extend the range and complexity of attainable structures.

1.15 ARTIFICIAL ENZYMES IN FREE RADICAL SYNTHETIC CHEMISTRY: THE CHEMIST'S PERSPECTIVE

All living things in nature maintain their internal metabolic balances quite well when they are in healthy conditions. In other words, metabolic materials are in equilibrium in each living creature primarily in a collaboration of biological catalyses by numerous enzymes. Enzymes are sophisticated proteins having catalytic groups and often require specific cofactors or coenzymes for catalytic performance. If we look at enzymatic functions from physicochemical viewpoints rather than biological ones, catalytically active amino acid residues of enzyme proteins as well as coenzyme factors are buried in hydrophobic and water-lacking reaction sites which aer furnished by enzyme proteins and well separated from the bulk aqueous phase needed to attain thermodynamic stabilities. In consideration of such physicochemical roles of enzyme proteins, we are allowed to use man-made materials for construction of artificial enzymes that are capable of simulating catalytic functions demonstrated by enzyme proteins.[122,123]

Two types of such artificial enzymes or apoenzymes, macrocyclic compounds and molecular assemblies, are cited in this article as those which can provide specific microenvironments for substrate-binding and subsequent catalysis in aqueous media. Those micro-environmental properties are primarily due to a hydrophobic internal cavity and the internal domain of molecular assemblies in aqueous media, and other non-covalent intermolecular interactions (such as electrostatic, charge-transfer, and hydrogen-bonding modes) between a substrate and an apo-enzyme model are greatly enhanced in such microenvironments.[124] In this book, we primarily summarize recent studies on the functional simulation of holoenzymes requiring coenzyme factors, such as vitamin B_{12}, artificial methylmalonyl-CoA mutase and glutamate mutase as representative enzymes. Most of those coenzymes are soluble in aqueous media when separated from the corresponding apo-protein and, consequently, cannot be readily incorporated into hydrophobic microenvironments provided by the artificial systems. Modified proteins which are derived by mutagenic treatments of natural enzymes are often called artificial enzymes. It must be emphasized at this point that the artificial enzymes described here are not directly related to protein structures but are capable of carrying out functional simulation of enzymatic catalysis in the overall reaction schemes.

The field of artificial enzymes is a rapidly evolving subject. As the barrier between chemistry and biology becomes less distinct, a range of new methods, which combine expertise from both areas, is developing. In recognition of both the fact that the *de novo* design approach can be time-consuming, and that a tiny miscalculation will have a detrimental effect, a trend in all these recent techniques is the use of "selection approaches". The natural process of selection and amplification is after all, the way in which enzymes have evolved their sophisticated function and also the area is constantly evolving—the sky is the limit as far as crossing and eliminating the barrier between "traditional chemistry" and "traditional biological sciences" and developments and understanding of life science at the molecular level is concerned.

1.16   FUTURE CHALLENGES AND OPPORTUNITIES FOR THE CHEMICAL PROFESSION AND THE SCIENCE OF CHEMISTRY

The principles of green chemistry are also a substantial beginning for the chemical profession in trying to deal with the novel ethical context in which humanity has been placed by the unprecedented power afforded to it in the 20th century by science and technology. Thus, green chemistry is unusual in that it is here that what has hitherto been called "pure chemistry" must be integrated into the much broader questions associated with sustainability and *vice versa*. Although green chemistry will underpin many activities in environmental protection, Scheme 1.3 shows that its overall impact is potentially much wider.

Green chemistry has major contributions to make to the quality of life, human welfare, and sustainable development. However, before green chemistry can contribute fully to these areas, it must be integrated into the discipline of chemistry itself. This requirement presents a number of major challenges to the chemical profession:

- Chemists will need to integrate into pure chemistry the questions of why or why not, on environmental protection grounds, a particular technology should be abandoned, improved, or adopted. These questions must become as important in research and education, and made as concrete, as the ubiquitous questions associated with what comprises chemical technology and how it actually works.
- It is vital that green chemistry not become a fad, in which chemistry that is not really "green" gets paraded as such before the scientific community and the world.
- Certain of the largest sustainability issues, where chemists have so much to offer, will require new approaches that can only be built with long-term commitment. For example, finding efficient methods for converting solar to chemical energy is a large sustainability issue. The culture of present-

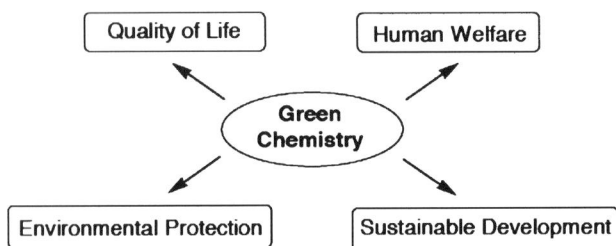

Scheme 1.3   Radical green chemistry in the context of the major drivers for new developments in modern chemistry.

day chemistry places too many short-term obstacles in the path of strategic research problems of this type. But chemists must solve such problems to achieve sustainability. Thus, the culture must adapt to recognize that certain sustainability problems will require novel approaches with inherently longer incubation times. Chemists must learn how better to evaluate and sustain research programs that, while they may not rapidly produce large numbers of publications, do offer reasonable promise of bringing, within the fullness of time, those critical advances that will genuinely promote the cause of sustainability.

- There is the very difficult issue of achieving a wholeness of scientific intelligence where more than the current specialized and professionalized expertise is paid due homage.

The science of chemistry cannot escape this growth and remain meaningful and important to humanity. Chemistry exerts an enormous influence on human action and is thus inextricably intertwined with the forces that guide human action, especially ethics and certain forms of passion. As such a wholeness is achieved, the power of scientific objectivity will be more openly directed by the action-orienting insights and passions that make us all human, such as our love of life and our desire to protect it. Such passions are neither vacuous nor disorienting. Rather, they are forces for great good that are fully capable of directing chemists toward research that really matters to each and every human being living and to come. Chemistry will have much more to offer, become more meaningful to humanity, increase in attractiveness as a career choice, grow to be more worthy of support, spawn large new economic developments, and progress to be more interesting and compelling if chemists work to define and follow their natural and unique role in achieving a virtuous civilization that sees broad validity within the community of living things for the claim to continuity of existence in an environment of natural genesis.

1.17 AN ENVIRONMENTALLY FRIENDLY ECONOMY FROM GREEN CHEMISTRY

Chemistry is the central and essential science necessary to deal with the very large questions of global sustainability. Chemists have proven over time that they possess the imagination, the technical ability, and the opportunities to produce the brilliant science that will be needed to decisively solve the technical problems of sustainability. Chemistry needs to begin work on sustainability problems with a force that befits their importance. The most obviously important research fields centre on finding alternatives to non-renewable (depleting) resources and pollution prevention through hazard reduction.

1.17.1 Renewable Energy Sources

In an energy-hungry world that has demonstrated an inability to engage in effective conservation measures on the scale necessary, the question of energy becomes paramount. The vast majority of energy production in the world is being generated through the use of finite depleting resources. These same finite resources (petroleum) are the basic feedstock for over 95% of all high value organic chemicals. It is clear that burning of fossil fuels is not sustainable, and the research of alternative sources is not proceeding at a rate necessary to address the problem in the necessary timeframe.

1.17.2   Renewable Feedstocks

It is undeniable that the fossilized carbon reserves of the planet, built up over millions of years, are being consumed at a rapid and ever-increasing pace. These reserves are consumed not only as energy sources, but also to produce feedstocks for the chemical industry. How are we going to achieve renewable feedstock alternatives? It is the role of creative chemists who will learn how to process renewable biomass into the chemical building blocks of future economies. In this area, chemists could have a great deal to offer also. Again, chemists need to develop a strategic plan and ensure that it is adequately backed.

1.17.3   Pollution Reduction

While progress is clearly being made in many arenas, it is still evident that much needs to be done to reduce the pollution pressure on the environment. Pollution largely results from imperfect applied chemistry. So who will reduce and eliminate this pressure, if not chemists? There are different kinds of pollutants. Persistent pollutants are more troubling than others because the environment does not have sufficiently rapid ways of dealing with them. By designing products and processes throughout the life cycle such that hazards are reduced, it will be to the benefit of sustainability, both environmentally and economically.

1.17.4   Interdisciplinary Approach

Authentic green chemistry is designed to eliminate pollution at its source with the intention that the new chemistry cannot imperil beneficial life forms. However, the earth's environment is perhaps the most complex chemical system we know about. The realization that a multitude of conundrums lie in the path of green chemistry does not constitute a reason not to proceed. Rather, it provides us with a clear warning to temper our natural enthusiasm for the good we might do with a healthy skepticism of our ability to foresee the future. We must be careful and aim this skepticism at avoiding environmental harm. Designers of the best green chemistry technologies will design for flexibility and deal with unknown perils in a prudent and correctable manner. Because of the complexity of the environment, green chemistry projects will require a certain level of multidisciplinary expertise reaching beyond the actual chemistry. This will be best achieved by collaborations among appropriate specialists.

1.18   CONCLUSION AND FUTURE DIRECTION

The potential environmental and health hazards of both industrial chemicals and chemical-based consumer products should be treated as intrinsic properties and should be minimized during the design. First of all the molecular level understanding of the relationships between structure and toxicity of chemicals is the only approach to design less toxic, or ideally non-toxic, products. The book links together the pool of knowledge and concepts of chemists, engineers, material scientists, biochemists and biomedical experts as well as undergraduate and postgraduate students and directs them into thinking outside the conventional chemistry of today towards "benign-by-design" research and the challenges of tomorrow.

The book as a whole also aims to identify the gap between the scientific curiosity, academic research and industry, with the power and scope of synthetically useful transformations being extended to solving problems on the interface of unconventional reaction media in radical chemistry. Ionic liquids, fluorous chemistry, solid state synthesis and supercritical CO_2 all have their merits and are fascinating media in which to carry out synthetic radical and non-radical transformations. Some can arguably be described as being "more green then others", and some require significantly more investment if the technique is to replace conventional operation. In all cases, such adoption of the technology will only take place if a real advantage can be shown. However, as each area develops and matures, all of them will no doubt hold important niches and advantages in the synthetic radical chemistry of the future. The practical and theoretical scope of chemical, biochemical and industrial application makes this book a unique highly practical text in this hot and rapidly developing field.

REFERENCES

1. J. Elkington, http://www.sustainability.co.uk/sustainability
2. P. T. Anastas and J. C. Warner, *Green Chemistry: Theory and Practice*, Oxford Science Publications, Oxford, 1998.
3. T. J. Collins, *Green Chemistry*, MacMillan, Encyclopedia of Chemistry, MacMillan Inc., New York, 1997.
4. (a) *Green Chemistry: Frontiers in Chemical Synthesis and Processes*, ed. P. T. Anastas and T. C. Williamson, Oxford University Press, Oxford, 1998; (b) *Benign by Design: Alternative Synthetic Design for Pollution Prevention*, ed. P. T. Anastas and C. A. Farris, ACS Symp. Ser. 577, Washington DC, 1994; (c) *Green Chemistry: Challenging Perspectives*, ed. P. Tundo and P. T. Anastas, Oxford University Press, Oxford, 2000.
5. OECD Environmental Health and Safety Publications, Series on Risk Management No. 10 (1999), "Proceedings of the OECD workshop on sustainable chemistry", Venice, 15–17 October 1998, pp. 204–205.
6. (a) B. M. Trost, *Science,* 1991, **254**, 1471; (b) B. M. Trost, *Angew. Chem., Int. Ed. Engl.*, 1995, **34**, 259.

7. G. Ciamician, *Science*, 1912, **36**, 385.

8. P. Tundo and M. Selva, Simplify gas-liquid phase transfer catalysis, *CHEMTECH*, 1995, **25**(5), 31–35.

9. L. A. Paquette, T. M. Mitzel, M. B. Isaac, C. F. Crasto and W. W. Schomer, *J. Org. Chem.*, 1997, **62**, 4293–4301.

10. R. Breslow, *Acc. Chem. Res.*, 1991, **24**, 159.

11. P. Tundo, *Continuous-Flow Methods in Organic Synthesis*, E. Horwood Pub., Chichester, UK, 1991.

12. H. Jonas, *The Imperative of Responsibility: In Search of Ethics for the Technological Age*, University of Chicago Press, Chicago, 1984.

13. C.-J. Li and T.-H. Chan, *Comprehensive Organic Reactions in Aqueous Media*, Wiley, Weinheim, 2nd edn, 2007.

14. D. C. Rideout and R. Breslow, *J. Am. Chem. Soc.*, 1980, **102**, 7816.

15. S. H. Yalkowsky, *Solubility and Solubilization in Aqueous Media*, Oxford University Press, New York, 1999.

16. U. Lindstrom, Stereoselective Organic Reactions in Water, *Chem. Rev.*, 2002, **102**, 2751–2772.

17. B. D. Anderson and K. P. Flora, in *The Practice of Medicinal Chemistry*, ed. C. G. Wermuth, Academic Press, London, 1996.

18. P. A. Grieco, K. Yoshida and P. Garner, *J. Org. Chem.*, 1983, **48**, 3137; J. F. W. Keana, A. P. Guzikowski, C. Morat, and J. J. Volwerk, *J. Org. Chem.*, 1983, **48**, 2661; P. A. Grieco, P. Galatsis and R. F. Spohn, *Tetrahedron*, 1986, **42**, 2847.

19. R. Breslow and C. J. Rizzo, *J. Am. Chem. Soc.*, 1991, **113**, 4340.

20. S. Tascioglu, *Tetrahedron*, 1996, **52**, 11113.

21. C. G. Wemuth, in *The Practice of Medicinal Chemistry*, ed. C. G. Wermuth, Academic Press, London, 1996.

22. K. U. Ingold, *Pure Appl. Chem.*, 1997, **69**, 241–243.

23. D. P Curran, Angew. Chem., Int. Ed., 1998, 37(9), 1175–1196.

24. G. A. Russell, Tetrahedron, 1959, 5, 101–102.

25. C. Walling, J. H. Cooley, A. A. Ponaras, E. J Racah and J. Elia, J. Am. Chem. Soc., 1966, 88(22), 5361–5363.

26. (*a*) P. Scrimin, in *Supramolecular Control of Structure and Reactivity*, ed. A. Hamilton, John Wiley & Sons, Chichester, UK, 1996, p. 101; (*b*) M. N. Khan, *Micellar Catalysis*, CRC Press, Boca Raton, USA, 2006.

27. (*a*) D. C. Rideout and R. Breslow, *J. Am. Chem. Soc.*, 1980, **102**, 7816; (*b*) R. Breslow, U. Maitra and D. C. Rideout, *Tetrahedron Lett.*, 1983, **24**, 1901; (*c*) J. B. N. F. Engberts, *Pure Appl. Chem.*, 1995, **67**, 823.

28. (*a*) S. Narayan, J. Muldoon, M. G. Finn, V. V. Fokin, H. C. Kolb and K. B. Sharpless, *Angew. Chem., Int. Ed.*, 2005, **44**, 3275; (*b*) B. K. Price and J. M. Tour, *J. Am. Chem. Soc.*, 2006, **128**, 12899; (*c*) A. El-Batta, C. Jiang, W. Zhao, R. Anness, A. L. Cooksy and M. Bergdahl, *J. Org. Chem.*, 2007, **72**, 5244; (*d*) D. Gonzalez-Cruz, D. Tejedo, P. de Armas and F. Garcia-Tellado, *Chem.–Eur. J.*, 2007, **13**, 4823; (*e*) M. C. Pirrung and K. Das Sarma, *J. Am. Chem. Soc.*, 2004, **126**, 444; (*f*) M. C. Pirrung

and K. Das Sarma, *Tetrahedron*, 2005, **61**, 11456; (*g*) A. R Extance, D. W. M. Benzies and J. J. Morrish, *QSAR Comb. Sci.*, 2006, **25**, 484; (*h*) I. Kanizsai, Z. Szakonyi, R. Sillanpaa and F. Fulop, *Tetrahedron Lett.*, 2006, **47**, 9113; (*i*) G. L. Turner, J. A. Morris and M. F. Greany, *Angew. Chem., Int. Ed.*, 2007, **46**, 7996; (*j*) P. G. Cozzi and L. Zoli, *Green Chem.*, 2007, **9**, 1292; (*k*) H.-B. Zhang, L. Liu, Y.-J. Chen, D. Wang and C.-J. Li, *Eur. J. Org. Chem.*, 2006, 869; (*l*) Y. Jung and R. A. Marcus, *J. Am. Chem. Soc.*, 2007, **129**, 5492.

29. For a very recent review see: (*a*) A. Chanda and V. V. Fokin, *Chem. Rev.*, 2009, **109**, 725. See also: (*b*) S. Otto and J. B. F. N. Engberts, *Org. Biomol. Chem.*, 2003, **1**, 2809; (*c*) M. C. Pirrung, *Chem.–Eur. J.*, 2006, **12**, 1312.

30. V. T. Perchyonok, I. N. Lykakis and K. L. Tuck, *Green Chem.*, 2008, **10**, 153.

31. For some selected examples of solvent-free organic reactions performed by grinding the reagents with a mortar and pestle, see: (*a*) F. Toda, K. Tanaka and A. Sekikawa, *J. Chem. Soc., Chem. Commun.*, 1987, 279–280; (*b*) F. Toda, K. Kiyoshige and M. Yagi, *Angew. Chem.*, 1989, **101**, 329–330 (*Angew. Chem., Int. Ed. Engl.*, 1989, **28**, 320–321); (*c*) F. Toda, K. Tanaka and S. Iwata, *J. Org. Chem.*, 1989, **54**, 3007–3009; (*d*) K. Tanaka, S. Kishigami and F. Toda, *J. Org. Chem.*, 1991, **56**, 4333–4334; (*e*) D. Braga, L. Maini, M. Polito, L. Mirolo and F. Grepioni, *Chem.–Eur. J.*, 2003, **9**, 4362–4370; (*f*) E. Y. Cheung, S. J. Kitchin, K. D. M. Harris, Y. Imai, N. Tajima and R. Kuroda, *J. Am. Chem. Soc.*, 2003, **125**, 14658–14659; (*g*) L. Rong, X. Li, H. Wang, D. Shi, S. Tu and Q. Zhuang, *Synth. Commun.*, 2007, **37**, 183–189.

32. For overviews on solvent-free organic reactions, see: (*a*) G. W. V. Cave, C. L. Raston and J. L. Scott, *Chem. Commun.*, 2001, 2159–2169; (*b*) K. Tanaka, *Solvent-Free Organic Synthesis*, Wiley-VCH, Weinheim, 2003; (*c*) P. J. Walsh, H. Li and C. A. de Parrodi, *Chem. Rev.*, 2007, **107**, 2503–2545.

33. Reviews: (*a*) G. Kaupp, M. R. Naimi-Jamal, H. Ren and H. Zoz, in *Advanced Technologies Based on Self-Propagating and Mechanochemical Reactions for Environmental Protection*, ed. G. Cao, F. Delogu and R. Orrx, Research Signpost, Kerala, 2003, pp. 83–100; (*b*) E. Gaffet, F. Bernard, J.-C. Niepce, F. Charlot, C. Gras, G. Le Caer, J.-L. Guichard, P. Delcroix, A. Mocellin and O. Tillement, *J. Mater. Chem.*, 1999, **9**, 305–314; (*c*) S. Kipp, V. Šepelák and K. D. Becker, *Chem. Unserer Zeit,* 2005, **39**, 384–392.

34. (*a*) C. Suryanarayana, *Mechanical Alloying and Milling*, CRC Press, New York, 2004; (*b*) A. L. Garay, A. Pichon and S. L. James, *Chem. Soc. Rev.*, 2007, **36**, 846–855.

35. G. Kaupp, *Cryst. Eng. Commun.*, 2006, **8**, 794–804, and references cited therein.

36. C. Reichardt, *Solvent Effects in Organic Chemistry*, VCH, Weinheim, 1978.

37. (*a*) *Applied Homogeneous Catalysis with Organometallic Compounds*, ed. B. Cornils and W. Herrmann, Wiley-VCH, Weinheim, 1996; (*b*) *Aqueous Phase Organometallic Catalysis* ed. B. Cornils and W. Herrmann, Wiley-VCH, Weinheim, 1998; (*c*) B. Cornils, *Angew. Chem., Int. Ed. Engl.*, 1995, **34**, 1574.

38. M. Vogt, PhD thesis, University of Aachen, 1991.

39. I. T. Horváth and J. Rábai, *Science*, 1994, **266**, 72.

40. (*a*) B. Cornils, *Angew. Chem., Int. Ed. Engl.*, 1997, **36**, 2057; (*b*) I. T. Horváth, *Acc. Chem. Res.*, 1998, **31**, 641.

41. E. P. Wesseler, R. Iltis and L. C. Clarc, *J. Fluorine Chem.*, 1977, **9**, 137.

42. (*a*) J. G. Riess, *New J. Chem.*, 1995, **19**, 893; (*b*) V. W. Sadtler, M. P. Krafft and J. G. Riess, *Angew. Chem., Int. Ed. Engl.*, 1996, **35**, 1976.

43. C. M. Sharts and H. R. Reese, *J. Fluorine Chem.*, 1978, **11**, 637.

44. (*a*) J. G. Riess and M. Le Blanc, *Angew. Chem., Int. Ed. Engl.*, 1978, **17**, 621; (*b*) J. G. Riess and M. Le Blanc, *Pure Appl. Chem.*, 1982, **54**, 2383.

45. D. W. Zhu, *Synthesis*, 1993, 953.

46. S. M. Pereira, P. Savage and G. W. Simpson, *Synth. Commun.*, 1995, **25**, 1023.

47. I. Klement and P. Knochel, *Synlett*, 1995, 1113.

48. I. Klement and P. Knochel, *Synlett*, 1996, 1005

49. H. C. Brown and V. H. Dodson, *J. Am. Chem. Soc.*, 1957, **79**, 2302.

50. A. Ogawa and D. P. Curran, *J. Org. Chem.*, 1997, **62**, 450.

51. A. K. Barbour, L. F. Belf and M. W. Buxton, in *Advances in Fluorine Chemistry*, ed. M. Stacey, J. C. Tatlow and A. G. Edwards, Butterworths, London, 1963, vol. 3, p. 81.

52. J. H. Brown, C. W. Suckling and W. B. Whalley, *J. Chem. Soc.*, 1949, S95.

53. I. Chen and S. Sarkanen, *Phytochem. Rev.*, 2003, **2**, 235.

54. S. Robota, US Pat. 3 859 372, 1975.

55. J. Baxamusa and S. Robota, US Pat. 4 183 873, 1980.

56. J. A. Darr and M. Poliakoff, *Chem. Rev.*, 1999, **99,** 495–541.

57. K. U. Ingold, *J. Am. Chem. Soc.*, 1992, **114**, 4589.

58. A. Watanabe, N. Noguchi, M. Takahashi and E. Niki, *Chem. Lett.*, 1999, 613.

59. H. Ohya and J. Yamauchi, *Electron Spin Resonance*, Kodansha, Tokyo, 1997.

60. J. R. Morton, K. F. Preston, P. J. Frusic, S. A. Hill and E. Wasserman, *J. Am. Chem. Soc.*, 1992, **114**, 5454.

61. M. Yoshida, F. Sultana, N. Uchiyama, T. Yamada and M. Iyoda, *Tetrahedron Lett.*, 1999, **40**, 735.

62. B. Giese, W. Damm, F. Wetterich and H. G. Zeitz, *Tetrahedron Lett.*, 1992, **33**, 1863.

63. M. E. Brik, *Tetrahedron Lett.*, 1995, **36**, 5519.

64. A. Mercier, Y. Berchadsky, Badrudin, S. Pietri and P. Tordo, *Tetrahedron Lett.*, 1991, **32**, 2125.
65. F. L. Moigne, A. Mercier and P. Tordo, *Tetrahedron Lett.*, 1991, **32**, 3841.
66. P. Braslau, H. Kuhn, L. C. Burrill, II, K. Lanham and C. J. Stenland, *Tetrahedron Lett.*, 1996, **37**, 7933.
67. J. M. Tronchet, *Tetrahedron Lett.*, 1991, **32**, 4129.
68. K. Awaga, T. Inabe, U. Nagashima, T. Nakamura, M. Matsumoto, Y. Kawabata and Y. Maruyama, *Chem. Lett.*, 1991, 1777.
69. A. Kamimura and N. Ono, *Bull. Chem. Soc. Jpn.*, 1988, **61**, 3629.
70. M. Ballestri, C. Chatgilialoglu, M. Lucarini and G. F. Pedulli, *J. Org. Chem.*, 1992, **57**, 948.
71. L. Grossi and P. C. Montevecchi, *Tetrahedron Lett.*, 1991, **32**, 5621.
72. Y. Miura, E. Yamano, A. Tanaka and Y. Ogo, *Chem. Lett.*, 1992, 1831.
73. J. E. Baldwin, *J. Chem. Soc., Chem. Commun.*, 1976, 734.
74. A. L. J. Beckwith and C. H. Schiesser, *Tetrahedron*, 1985, **41**, 3925.
75. J. H. Horner, F. N. Martinez, M. Newcomb, S. Hadida and D. P. Curran, *Tetrahedron Lett.*, 1997, **38**, 2783.
76. M. Newcomb, M. A. Filipkowski and C. C. Johnson, *Tetrahedron Lett.*, 1995, **36**, 3643.
77. E. W. Della, C. Kostakis and P. A. Smith, *Org. Lett.*, 1999, **1**, 363.
78. J. Hartung, *Eur. J. Org. Chem.*, 2001, 619.
79. W. R. Dolbier, Jr., A. Li, B. E. Smart and Z. Y. Yang, *J. Org. Chem.*, 1998, **63**, 5687.
80. A. Li, A. B. Shtarev, B. E. Smart, Z. Y. Yang, J. Lusztyk, K. U. Ingold, A. Bravo and W. R. Dolbier, Jr., *J. Org. Chem.*, 1999, **64**, 5993.
81. J. M. Baker and W. R. Dolbier, Jr., *J. Org. Chem.*, 2001, **66**, 2662.
82. A. Philippon, M. D. Castaing, A. L. J. Beckwith and B. Maillard, *J. Org. Chem.*, 1998, **63**, 6814.
83. A. L. J. Beckwith and B. P. Hay, *J. Am. Chem. Soc.*, 1989, **111**, 230.
84. A. L. J. Beckwith and B. P. Hay, *J. Am. Chem. Soc.*, 1989, **111**, 2674.
85. A. L. J. Beckwith and K. D. Raner, *J. Org. Chem.*, 1992, **57**, 4954.
86. A. L. J. Beckwith, *Tetrahedron*, 1981, **37**, 3073.
87. J. Lusztyk, B. Maillard, S. Deycard, D. A. Lindsay and K. U. Ingold, *J. Org. Chem.*, 1987, **52**, 3509.
88. D. P. Curran, J. Xu and E. Lazzarini, *J. Am. Chem. Soc.*, 1995, **117**, 6603.
89. D. P. Curran, J. Y. Xu and E. Lazzarini, *J. Chem. Soc., Perkin Trans. 1*, 1995, 3049.
90. D. P. Curran and M. Palovick, *Synlett*, 1992, 631.
91. C. Chatgilialoglu, C. Ferreri, M. Lucarini, A. Venturini and A. A. Zavitsas, *Chem.–Eur. J.*, 1997, **3**, 376.
92. D. J. Constable, A. D. Curzons and V. L. Cunningham, Metrics to green chemistry—which are the best? *Green Chem.*, 2002, **4**, 521–527.
93. C. Jiménez-González, D. J. C. Constable, A. D. Curzon and V. L. Cunningham, Developing GSK's green technology guidance: methodol-

ogy for case–scenario comparison of technologies, *Clean Technol. Environ. Policy,* 2002,`**4**, 44–53.

94. D. J. C. Constable, A. D. Curzons, L.M. Freitas dos Santos, G. R. Geen, J. Kitteringham, P. Smith, R. E. Hannah, M. A. McGuire, R. L. Webb, M. Yu, J. D. Hayler and J. E. Richardson, Green chemistry measures for processes research and development, *Green Chem.,* 2001, **3**, 7–9.

95. A. D. Curzons, D. J. C. Constable, D. N. Mortimer and V. L. Cunningham, So you think your process is green, how do you know?—Using principles of sustainability to determine what is green—a corporate perspective, *Green Chem.,* 2001, **3**, 1–6.

96. R. Sheldon, Catalytic reactions in ionic liquids, *Chem. Commun.,* 2001, 2399–2407.

97. W. T. Jiaang, H. C. Lin, K. H. Tang, L. B. Chang and Y. M. Tsai, *J. Org. Chem.*, 1999, **64**, 618.

98. P. Tauh and A. G. Fallis, *J. Org. Chem.*, 1999, **64**, 6960.

99. P. Tauh and A. G. Fallis, *J. Org. Chem.*, 1999, **64**, 6960.

100. B. Maillard, D. Forrest and K. U. Ingold, *J. Am. Chem. Soc.*, 1976, **98**, 7024.

101. J.-L. Luche, in *Advances in Sonochemistry*, ed. T. J. Mason, JAI Press Ltd, London, 1993, vol. 3, pp. 85–24.

102. (*a*) P. Cintas and J.-L. Luche, *Green Chem.*, 1999, **1**, 115–125; (*b*) T. J. Mason and P. Cintas, in *Handbook of Green Chemistry and Technology*, ed. J. Clark and D. Macquarrie, Blackwell, Oxford, 2002, pp. 372–396; (*c*) P. Cintas, in *Transition Metals for Organic Synthesis,* ed. M. Beller and C. Bolm, Wiley-VCH, Weinheim, 2nd edn, vol. 2, pp. 583–596.

103. (*a*) J. D. Oxley, R. Prozorov and K. S. Suslick, *J. Am. Chem. Soc.*, 2003, **125**, 11138–11139; (*b*) A. Tuulmets, S. Salmar and H. Hagu, *J. Phys. Chem. B*, 2003, **107**, 12891–12896; (*c*) S. Höfinger and F. Zerbetto, *Chem.–Eur. J.*, 2003, **9**, 566–569.

104. E. Paternò, Synthesis in organic chemistry by means of light. I. Introduction, *Gazz. Chim. Ital.*, 1909, **39**(1), 237–250 (*Synthesis in organic chemistry by means of light*, ed. M. D'Auria, Società Chimica Italiana, Rome, 2009).

105. C. A. Gueymard, The sun's total and spectral irradiance for solar energy applications and solar radiation models, *Sol. Energy*, 2004, **76**, 423–453.

106. R. M. Navarro, M. C. Sánchez-Sánchez, M. C. Alvarez-Galvan, F. del Valle and J. L. G. Fierro, Hydrogen production from renewable sources: biomass and photocatalytic opportunities, *Energy Environ. Sci.*, 2009, **2**, 35–54.

107. N. Armaroli and V. Balzani, The Future of Energy Supply: Challenges and Opportunities, *Angew. Chem., Int. Ed.*, 2007, **46**, 52–66.

108. V. Balzani, A. Credi and M. Venturi, Photochemical conversion of solar energy, *ChemSusChem*, 2008, **1**, 26–58.

109. For a basic discussion on storing solar radiation as chemical energy, including the isomerization of quadricyclane and related molecules, see:

H.-D. Scharf, J. Fleischhauer, H. Leismann, I. Ressler, W.-G. Schleker and R. Weitz, Criteria for the efficiency, stability and capacity of abiotic photochemical solar energy storage systems, *Angew. Chem., Int. Ed. Engl.*, 1979, **18**, 652–662; A. Behr, W. Keim, G. Thelen, H.-D. Scharf and I. Ressler, Solar energy storage with quadricyclane systems, *J. Chem. Technol. Biotechnol.*, 1982, **32**, 627–630.

110. A. Albini and M. Fagnoni, Green chemistry and photochemistry were born at the same time, *Green Chem.*, 2004, **6**, 1–6.

111. C. J. Adams, Neoteric solvents: An examination of their industrial attractiveness, *ACS, Symp. Ser.*, 2002, **818**, 15–29.

112. D. J. Constable, A. D. Curzons and V. L. Cunningham, Metrics to green chemistry—which are the best? *Green Chem.*, 2002, **4**, 521–527.

113. C. Chatgilialoglu, *Acc. Chem. Res.*, 1992, **25**, 188.

114. G. J. Rowlands, *Annu. Rep. Prog. Chem., Sect. B*, 2003, **99**, 3–20; 2004, **100**, 33–49; 2005, **101**, 17–32; 2006, **102**, 17–33; 2007, **103**, 18–34; S. Z. Zard, *Radical Reactions in Organic Synthesis*, OUP, Oxford, 2003.

115. G. Bar and A. F. Parsons, *Chem. Soc. Rev.*, 2003, **32**, 251–263; D. P. Curran, N. A. Porter and B. Giese, *Stereochemistry of Radical Reactions: Concepts, Guidelines, and Synthetic Applications*, VCH, Weinheim, 1995.

116. (a) M. P. Sibi, S. Manyem and J. Zimmerman, *Chem. Rev.*, 2003, **103**, 3263–3295; (b) J. Zimmerman and M. P. Sibi, *Top. Curr. Chem.*, 2006, **263**, 107–162.

117. (*a*) A. Dondoni and A. Massi, *Angew. Chem., Int. Ed.*, 2008, **47**; (*b*) H. Pellissier, *Tetrahedron*, 2007, **63**, 9267–9331.

118. C. Chatgilialoglu, *Chem.–Eur. J.*, 2008, **14**, 2310–2320.

119. C. Chatgilialoglu and V. I. Timokhin, *Adv. Organomet. Chem.*, 2008, **57**, 117–181.

120. C. Chatgilialoglu, *Acc. Chem. Res.*, 1992, **25**, 188–194.

121. H. Kim and D. W. C. MacMillan, *J. Am. Chem. Soc.*, 2008, **130**, 398–399; T. D. Beeson, A. Mastracchio, J. B. Hong, K. Ashton and D. W. C. MacMillan, *Science*, 2007, **316**, 582–585; H. Y. Jang, J. B. Hong and D. W. C. MacMillan, *J. Am. Chem. Soc.*, 2007, **129**, 7004–7005.

122. M. P. Sibi and M. Hasegawa, *J. Am. Chem. Soc.*, 2007, **129**, 4124–4125.

123. D. Voet and J. G. Voet, *Biochemistry*, J. Wiley & Sons, Hoboken, NJ, 2004; R. A. Copeland, *Enzymes, a practical introduction to structure, mechanism, and data analysis*, Wiley-VCH, New York, 2000.

124. R. Breslow, Biomimetic chemistry: centenary lecture, *Chem. Soc. Rev.*, 1972, **1**, 553–580.

125. (*a*) R. Breslow, Biomimetic chemistry, *Pure Appl. Chem.*, 1994, **66**, 1573–1582; (*b*) R. Breslow, Biomimetic chemistry and artificial enzymes: catalysis by design, *Acc. Chem. Res.*, 1995, **28**, 146–153.

Classical Synthetic Free Radical Transformations in Alternative Media: Supercritical CO$_2$, Ionic Liquids and Fluorous Media[*]

2.1 INTRODUCTION

Chemists are now moving away from volatile, environmentally harmful, and biologically incompatible organic solvents. With its low cost, ready availability, and capacity to remove environmentally unfriendly by-products, water (along with alternative biocompatible reaction media such as ionic liquids, fluorous solvents, or using a solid support) is an obvious replacement. Recent advances in free radical chemistry in water have expanded the versatility and flexibility of homolytic bond formations in aqueous media and alternative media from conventional synthesis to high-flow environments.[1–3] This chapter highlights the substantial progress which has been made in the last decade with respect to the reactive free radical species in alternative media, such as ionic liquids (supported and non-supported), fluorous solvents and supercritical CO$_2$.[4] It shows that, armed with an elementary knowledge of kinetics and some common sense, it is possible to harness radicals into tremendously powerful tools for solving synthetic problems and for a broad range of applications. The issue is a valuable and informative entry for growth and development of free radical green chemistry in aqueous and alternative media.

2.2 RADICALS IN SYNTHETIC CHEMISTRY IN THE NUTSHELL

The organic chemists are interested in reactions of compounds which yield products that may be isolated and purified. Radicals are species with at least

* Chapter written by V. Tamara Perchyonok.
Streamlining Free Radical Green Chemistry
V. Tamara Perchyonok, Ioannis Lykakis and Al Postigo
© V. Tamara Perchyonok, Ioannis Lykakis and Al Postigo 2012
Published by the Royal Society of Chemistry, www.rsc.org

one unpaired electron which, in contrast to organic anions and cations, react easily with themselves in bond-forming reactions. In the liquid phase most of these reactions occur with diffusion-controlled rates. Radical–radical reactions (Figure 2.1) can be slowed down only if radicals are stabilized by electronic effects (stable radicals) or shielded by steric effects (persistent radicals). However, these effects are not strong enough to prevent diffusion-controlled recombination of, for example, benzyl radicals or *tert*-butyl radicals.[5] Only in extreme cases, *e.g.* the radical or di-*tert*-butylmethyl radical, are recombination rates low.[6] While the recombination rates of the triphenylmethyl radical are reduced due to both steric and radical stabilizing effects, the steric effect alone slows down the recombination of the di-*tert*-butyl methyl radical. The disproportionation reaction, in which the hydrogen atom is transferred, cannot occur, since neither of the radicals have C–H bonds β to the radical centre.[1e,7]

2.3 REACTIONS BETWEEN RADICALS

The fact that reactions between radicals (Figure 2.2) are in most cases very fast could lead to a conclusion that direct radical combination is the most synthetically useful reaction mode.[8–16] This, however, is not the case because direct radical–radical reactions have several disadvantages:

- In the recombination reactions, the radical character is destroyed so that one has to work with at least an equivalent amount of radical initiators.
- The diffusion-controlled rates in radical–radical reactions give rise to low selectivity which cannot be influenced by reaction conditions.
- The concentrations of radicals are so low that reactions with non-radicals, like solvents, which are present in high concentrations, are very hard to prevent.

For reactions to occur, the interacting frontier orbitals must not only interact efficiently, but also have similar energies. The occupied frontier orbital

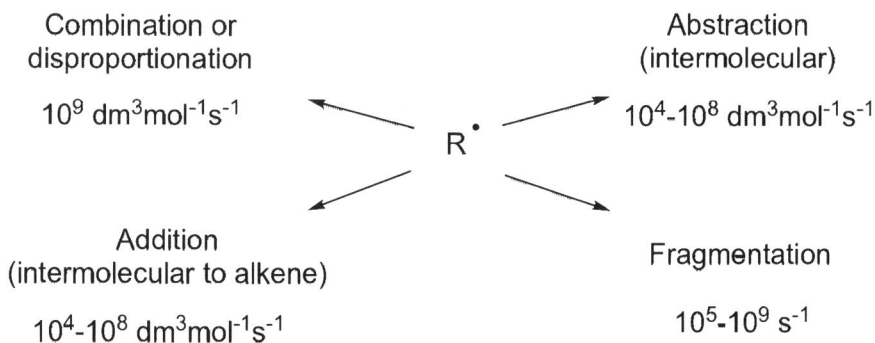

Combination or disproportionation
10^9 dm^{3}mol^{-1}s^{-1}

Abstraction (intermolecular)
10^4-10^8 dm^{3}mol^{-1}s^{-1}

R$^{\bullet}$

Addition (intermolecular to alkene)
10^4-10^8 dm^{3}mol^{-1}s^{-1}

Fragmentation
10^5-10^9 s^{-1}

Figure 2.1 Rates of fundamental radical reactions.

Unpaired Electron

Rad + A : B ⟶ Rad : A + B

Unpaired Electron

Reactant Radical   Substitution Product   Product Radical

Unpaired Electron

Rad +   ⟶   Rad

Unpaired Electron

Reactant Radical   Alkene   Addition product radical

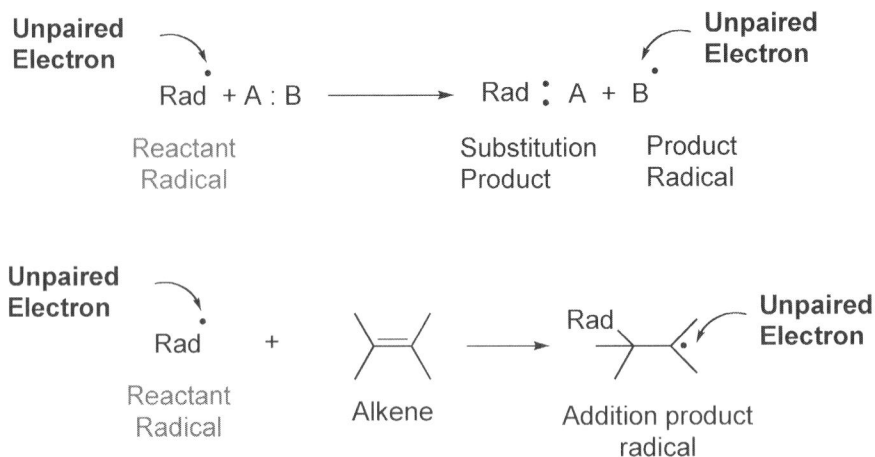

Figure 2.2   Radicals and reactions in general.

for a radical is called the singly occupied molecular orbital (SOMO); we have previously discussed that radicals bearing different substituents have different SOMO energies. Radicals adjacent to electron-donating groups interact with a filled orbital to give a high-energy SOMO, while radicals next to electron-withdrawing groups interact with an unfilled orbital to give a low-energy SOMO. The SOMO energies lie somewhere between those of the highest occupied molecular orbital (HOMO) and lowest unoccupied molecular orbital (LUMO) of the non-radical. Therefore, for the reaction of a radical with a non-radical, we need to consider the SOMO–HOMO and SOMO–LUMO interactions. In both cases, the interaction will lead to a decrease in energy and the formation of a bond; a SOMO–HOMO places two of the three electrons in a low-energy bonding orbital, whereas a SOMO–LUMO interaction places the single electron in a low-energy bonding orbital. The energy of the SOMO will determine whether interactions with the HOMO or the (higher energy) LUMO predominate (Figure 2.3). Electrophilic radicals (with a low-energy SOMO) will be closer in energy to the HOMO, and therefore the SOMO–HOMO interactions will predominate. In comparison, nucleophilic radicals (with a high-energy SOMO) will be closer in energy to the LUMO, and therefore the SOMO–LUMO interaction will predominate.

2.4   ELEMENTARY REACTION STEPS BETWEEN RADICALS AND NON-RADICALS

A radical chain is built up by different types of propagation steps, all of which lead to new radicals:

- Addition reactions
- Substitution (abstraction) reactions

Radical Donor   Radical acceptor

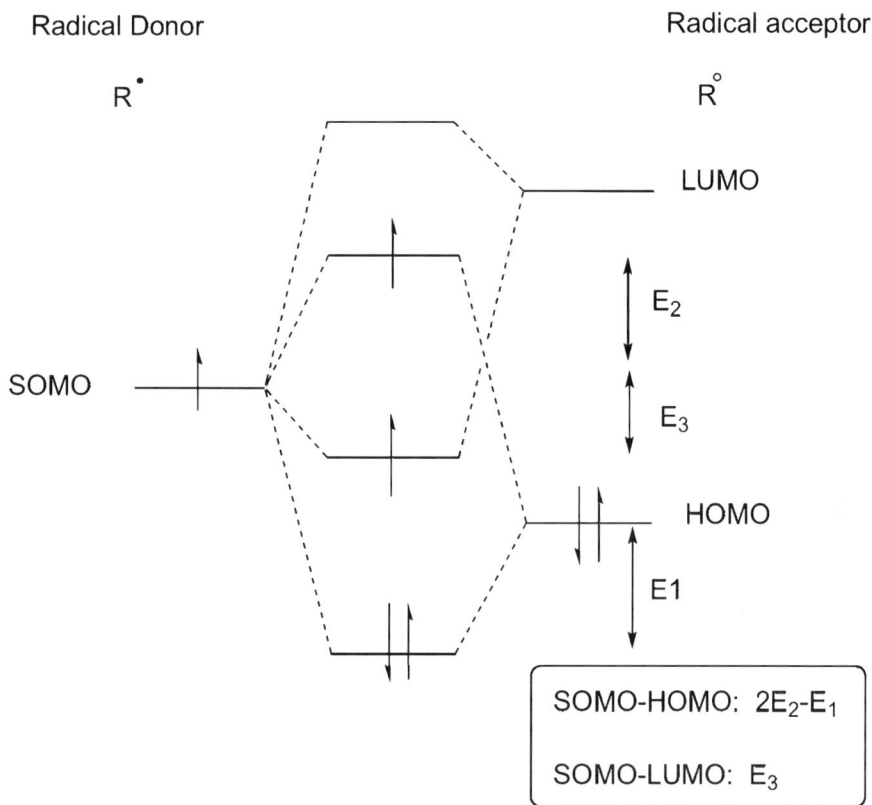

Figure 2.3 Orbital Interaction between a radical donor and a radical acceptor.

- Elimination (fragmentation) reactions
- Rearrangement reactions
- Electron transfer reactions

As a rule of thumb there are two simple conventions that can be useful for fast radical chain reactions:

- Most chain-propagating steps are exothermic and one can use the strength of bonds that are broken and formed as a rough guide for the rate of the reaction (thermodynamic parameter).
- Because of the early transition states in fast radical reactions, frontier molecular orbital theory can be utilized for these reactions.

2.4.1 Additions

Addition of alkyl radicals to alkenes is a useful C–C bond formation reaction in which a σ-CC bond is made from a π-CC bond in a very exothermic reaction. In contrast, the π-CO bonds of ketone and aldehyde are nearly as

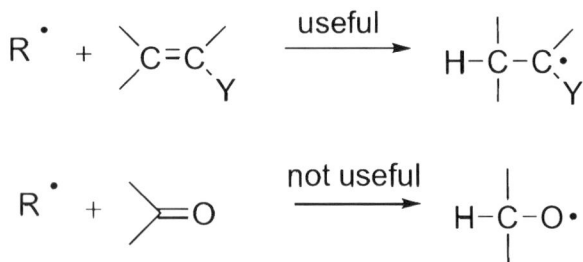

Figure 2.4   General addition of alkyl radicals to alkenes *versus* addition of alkyl radicals to aldehyde or ketone.

strong as σ-CC bonds. Therefore, ketones and aldehydes cannot be used as intermolecular traps in synthesis (Figure 2.4).

   The rate of addition of a radical to an alkene depends largely on the substituents on the radical and the alkenes. This substituent effect can be supported by the frontier orbital series. The singly occupied molecular orbital (SOMO) of the radical overlaps/interacts with the lowest unoccupied molecular orbital (LUMO) and/or the highest occupied molecular orbital (HOMO) of the CC-multiple bond. Radicals with a high-lying SOMO interact preferentially with the LUMO of the alkene (Figure 2.5).

   Electron-withdrawing substituents at the alkene, which lower the LUMO energy, increase the addition rate by reducing the SOMO–LUMO difference. Therefore, cyclohexyl radicals react almost 8500 times faster with acrolein than with 1-hexene.[4]

   The orbitals are crucial to the fact that a *tert*-butyl radical reacts faster than primary or secondary radicals with alkenes like vinylphosphinic esters

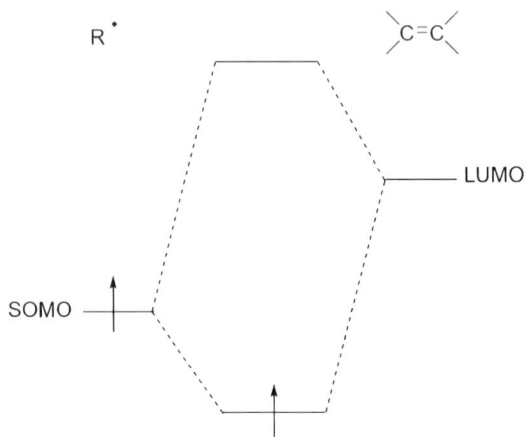

Figure 2.5   Orbital interaction between a nucleophilic radical and an electron-poor alkene.

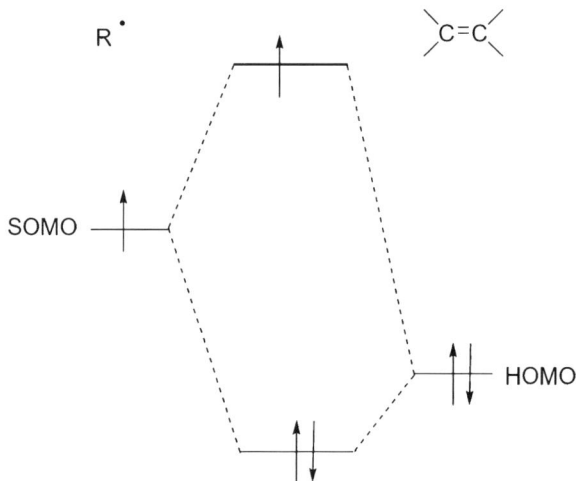

Figure 2.6 Orbital interaction between an electrophilic radical and an electron-rich alkene.

or acrylonitrile. Thus, the increase in the SOMO energy in going from primary to tertiary radicals has a larger effect on the rate than the decrease in the strength of the bonds that are formed. Alkyl, alkoxyalkyl, aminoalkyl, and other similar radicals are therefore nucleophiles. However, radicals with electron-withdrawing substituents at the radical centre have SOMO energies so low that the SOMO–HOMO interaction dominates (Figure 2.6).

These radicals react like electrophiles, that is, electron-donating substituents at the alkene increase the rate. The malonyl radical, for example, reacts with enamine 23 times faster than with acryl ester.

2.4.2 Substitution (Abstraction) Reactions

Since an O–H bond is much stronger than the C–H bond, a typical bimolecular reaction for the alkoxy radical is H-abstraction from the C–H bonds. However, the O–H bond of an alcohol is attacked too slowly for synthetic applications because the reaction is thermo-neutral (Figure 2.7).

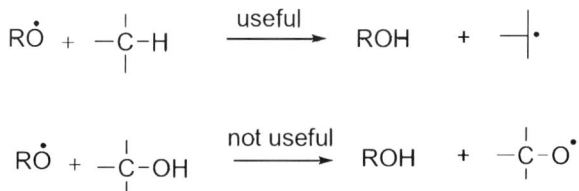

Figure 2.7 General methodology of hydrogen abstraction reaction from the C–H bonds *versus* O–H bonds.

Alkoxy radicals are electrophiles and they preferentially attack C–H bonds with high HOMO energies; for instance, the a-C–H bond of ethers and amines or the alkyl C–H bond of esters. In contrast, nucleophilic alkyl radicals abstract a hydrogen atom from the acyl group of esters because this C–H bond has a lower LUMO energy. These differences also account for the preferential abstraction of the β-hydrogen of propionic acid by the electrophilic chlorine atom and the abstraction of the α-hydrogen by the nucleophilic methyl radical.

2.4.3 Elimination Reactions

In elimination reactions, two molecules are formed from one. Thus, these reactions are favoured by activation entropies, and the free energy gain increases with increasing reaction temperatures. Therefore, even alkoxy radicals undergo fast β-elimination reactions, although the enthalpy differences between π-CO and σ-CC bonds are small. However, a C–OR bond α to a radical centre is cleaved too slowly to be of synthetic use, because less stable π-CC bonds are formed. Only with the weaker C–Br, C–SR or C–SnR$_3$ bonds are β-elimination reactions fast enough to be synthetically useful (Figure 2.8).

2.4.4 Rearrangement Reactions

Compared to their cationic counterparts, only a few radical rearrangements are fast enough to appear in synthesis. An example is the vinyl migration in which vinyl group migrates through a formation of the methyl cyclopropyl intermediate.

This reaction is a combination of an intramolecular addition and an elimination reaction (Figure 2.9). The addition of the nucleophilic radical to the electron-rich alkene is fast because the loss in entropy is much smaller than in intermolecular reactions. The β-C–C bond of the intermediate cleaves

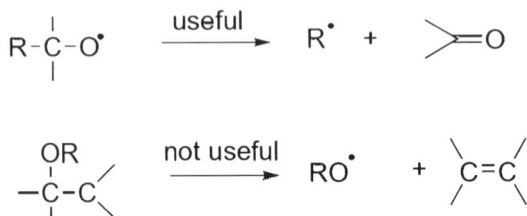

Figure 2.8 General methodology of the elimination reaction.

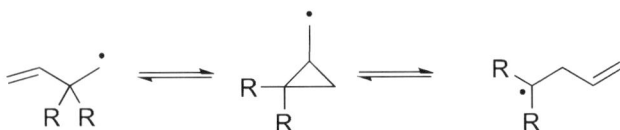

Figure 2.9 General methodology of a rearrangement reaction.

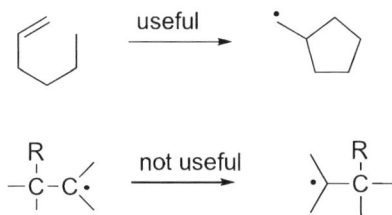

Figure 2.10   General methodology of a cascade intra-molecular addition/elimination reaction.

rapidly because of the ring strain, which is the driving force for the achieving of the desired outcome in the transformation. The much smaller ring strain in the methylcyclopentyl radical intermediate stops the rearrangement of the hexenyl radical at the cyclopentylmethyl radical step (Figure 2.10).

This is in accordance with the very slow 1,2-alkyl or hydrogen shift in radicals, which do not occur in synthesis because of the strong σ-bond (Figure 2.11).

2.4.5   Termination/Electron Transfer Reactions

The rate of electron transfer reactions depends on the difference of the redox potentials of educts and products. Since the alkyl radical possesses an unpaired electron in a non-bonding orbital, electron transfer reactions to many metals salts often occur with high rates. The higher the SOMO energies of the radicals, the faster the electron transfer.

With a brief historical background and general principles of free radical chemistry applicable to organic and aqueous media, we shall next turn our attention to the study of radical chemistry in aqueous and alternative media as it applies to organic synthesis and use the concepts that will emerge as the foundation upon which the rest of the material will build.

2.5   REACTIVITY AND SELECTIVITY

As was previously mentioned, radicals can undergo a number of different and competitive reactions. These processes have different rates of reaction and if one reaction proceeds at a much faster rate than all the rest we have selective and high-yielding processes. Alternatively, if a variety of reactions proceed at similar rate, the radical will react unselectively to produce a number of different products. The rates of these reactions can vary enormously and, for example, the rate constants of abstraction reaction can vary by a factor of at

Figure 2.11   General methodology and limitations of 1,2 alkyl or hydrogen shift

least 10 000. The key factors that influence radical reactivity include enthalpy, entropy, steric effects, stereo-electronic effects, polarity and redox potential (Figure 2.1).[6,16]

2.6 CHAIN *VS.* NON-CHAIN FREE RADICAL PROCESSES: REASONS, RELEVANCE AND OUTLOOK

The tremendous advantages and achievements in the field of free radical chain reactions were highlighted, with the desired end product being predetermined in the propagation steps. The faster the propagation step, the greater the observed efficiency: less initiator is needed, fewer unwanted side reactions can compete, and radical–radical interactions constituting the termination become negligible. In summary, the whole approach was aimed to reduce radical–radical interaction by keeping the steady-state concentration of the inter-mediate radical species as low as possible. Since the interactions are usually diffusion-controlled and therefore unselective, it might be seen (at first glance) as an impossible task to gain control of inter-radical reactions for synthetic purposes. However, this is not the case at all, due to the elegant and ingenious solution of this problem based on the persistent radical effect, also known as the Fischer–Ingold effect. This phenomenon, which is only recently being understood, underlies several reactions occurring in nature as well as novel synthetic applications recently discovered and elegantly applied in the special field of living radical polymerization and advanced material sciences, to name a few.

Following a brief look at the general principles of free radical chemistry applicable to organic and aqueous and alternative media under control media, we shall next turn our attention to the review of radical chemistry in alternative media, such as ionic liquids, supercritical CO_2 and perfluorinated solvents in conventional synthetically useful free radical transformations for organic synthesis. We will use the concepts that emerge as the foundation upon which our understanding of the detailed mechanism of the chemistry of life, such as biocatalysis for example, is built, and this in turn will have applications in evolving areas such as artificial biocatalysis, biomaterials and biotechnology in general. The main purpose of this chapter is to attempt to expand the implementation of radical synthesis in alternative media as an equal partner in synthesis and chemistry in general (Figure 2.12).

2.7 RADICAL REACTIONS IN SUPERCRITICAL FLUIDS

2.7.1 Radical Reactions and Supercritical CO_2: Is There a Hidden Advantage?

The use of supercritical fluids as reaction media offers the chemical and pharmaceutical industries the opportunity to replace conventional hazardous organic solvents and simultaneously optimize and control more precisely the effect of the solvent on reactions. Supercritical fluids, unlike conventional

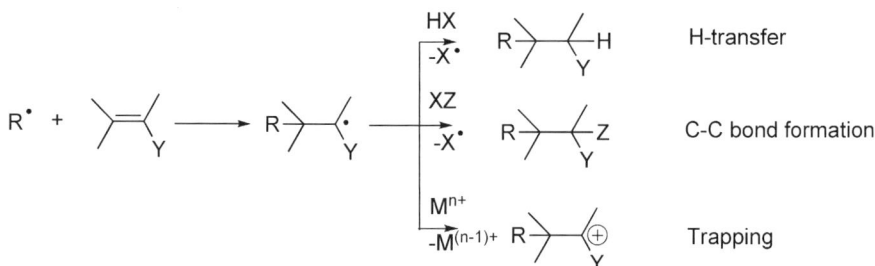

Figure 2.12   General summary of radical synthetic transformations in alternative reaction media to be discussed in this chapter.

liquid solvents, can be "pressure-tuned" to exhibit gas-like to liquid-like properties. Supercritical fluids have liquid-like local densities and solvent strength, which can be tuned by adjusting the pressure in the reactor while allowing for the control of the solubility of the reactants along with density-dependent properties such as the dielectric constant, viscosity and diffusivity. Additionally, solubility control through pressure can allow for easy separation of products and catalysts from the supercritical solvent.

Supercritical fluids are an attractive medium for chemical reactions because of their unique properties. Most of the important physical and transport properties which influence the kinetics of chemical reactions are intermediate between those of a liquid and a gas in a supercritical fluid medium. The reactants and the supercritical fluids frequently form a single supercritical fluid phase. Supercritical fluid reactions share many of the advantages of gas-phase reactions, including miscibility with other gases, low viscosity and high diffusivities, thereby providing enhanced heat transfer and the potential for a fast reaction. Supercritical fluids are especially attractive as a reaction medium for diffusion-controlled reagents (Figure 2.13).

Supercritical fluids (SCFs), and supercritical carbon dioxide (scCO$_2$) in particular, are emerging as some of the most promising sustainable technologies to have developed in recent years. They are not new, however, and can be

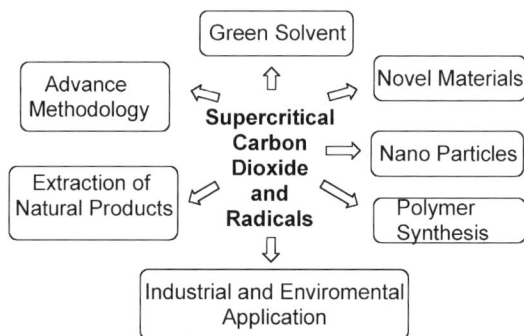

Figure 2.13   Summary of the application of supercritical CO$_2$ in radical chemistry.

traced back to original reports dated as long ago as the existence of the critical point of alcohol using equipment originally designed by Denys Papen in 1960.

A SCF is defined as a substance above its critical temperature (T_C) and critical pressure (P_C) (Figure 2.14). This definition should arguably include the clause "but below the pressure required for condensation into a solid"; however this is commonly omitted, as the pressure required for condensing a SCF into a solid is generally impracticably high. The critical point represents the highest temperature and pressure at which the substance can exist as a vapor and a liquid in equilibrium. The phenomenon can be easily explained with reference to the phase diagram for pure carbon dioxide. This shows the areas where carbon dioxide exists as a gas, liquid, solid or as a SCF. The curves represent the temperature and pressures where two phases coexist in equilibrium (at the triple point, the three phases co-exist). The gas–liquid coexistence curve is known as the boiling curve. If we move upwards along the boiling curve, increasing temperatures and pressure, then the liquid becomes less dense as the pressure rises. Eventually, the densities of the two phases converge and become identical, the distinction between gas and liquid disappears, and the boiling curve comes to an end at the critical point. The critical point of carbon dioxide occurs at a pressure of 73.8 bar and a temperature of 31.1 °C (Figure 2.14).

2.7.2 Radical Reactions in Supercritical Carbon Dioxide in Detail

As SCFs are low-density fluids with low viscosity resulting in high rates of diffusion, they offer potential benefits for radical processes. A significant proportion of the radical reactions studied have been in polymerization applications. Although radical reactions were some of the first synthetic processes to be considered in SCF, relatively few literature examples exist.

Many free radical halogenation reactions have historically been carried out in CCl_4, but $scCO_2$ is an attractive alternative. The reaction of molecular bromine with toluene and ethyl benzene forms the corresponding benzylic

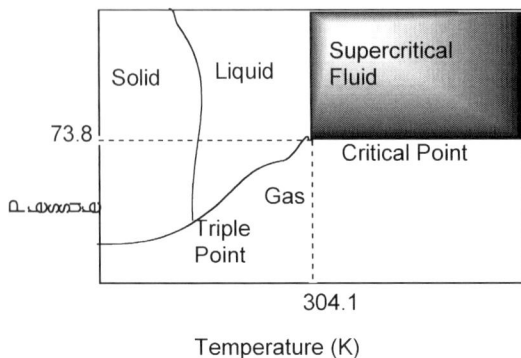

Figure 2.14   The phase diagram for pure carbon dioxide.

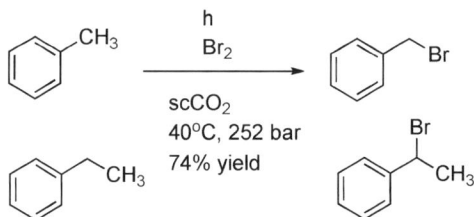

Scheme 2.1 Radical bromination in scCO$_2$.

bromides in good yields (Scheme 2.1) and Ziegler bromination using N-bromosuccinimide (NBS) was also successful.[17]

scCO$_2$ is also a practical medium for free radical carbonylation of organic halides to ketones and aldehydes. Using a silane-mediated carbonylation of an alkyl halide, alkene and CO using an azobisisobutyronitrile (AIBN) initiator gave yields comparable to those obtained in benzene. Related intramolecular reactions also proceeded efficiently and showed interesting pressure-dependent selectivity.

The use of tin hydride reagents in scCO$_2$ has also been reported. Both tributyl tin hydride and tris(perfluorohexylethyl)tin hydride were investigated, where tris(perfluorohexylethyl)tin hydride was miscible under reaction conditions whereas tributyl tin hydride was not (Scheme 2.2). These reagents are also being applied in fluorous-phase chemistry. Bromoadamantane was reduced by tris(perfluorohexylethyl)tin hydride (initiated by AIBN) under scCO$_2$ conditions to give a corresponding reduced product in 90% yield after 3 h. The workup for this reaction is particularly clean by partitioning between benzene and perfluorohexane. Surprisingly, despite the insolubility, tributyl tin hydride also facilitated reduction, with 1-bromoadamantane to adamantane to be isolated in 80% yield. Reactions of steroidal bromides, iodides and selenides with their respective hydrogen donors yielded the corresponding reduced products in high yields (85–98%).[18]

Several radical cyclization reactions were also studied (Scheme 2.3). Reduction of 1,1-diphenyl-6-bromo-1-hexene with tris(perfluorohexylethyl)tin hydride gave the 5-*exo* product in 87% yield. Similarly, reduction of aryl iodide with tris(perfluorohexylethyl)tin hydride gave quantitative conversion to the

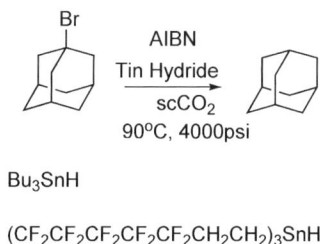

Bu$_3$SnH

(CF$_2$CF$_2$CF$_2$CF$_2$CF$_2$CH$_2$CH$_2$)$_3$SnH

Scheme 2.2 Tin hydride reduction in scCO$_2$.

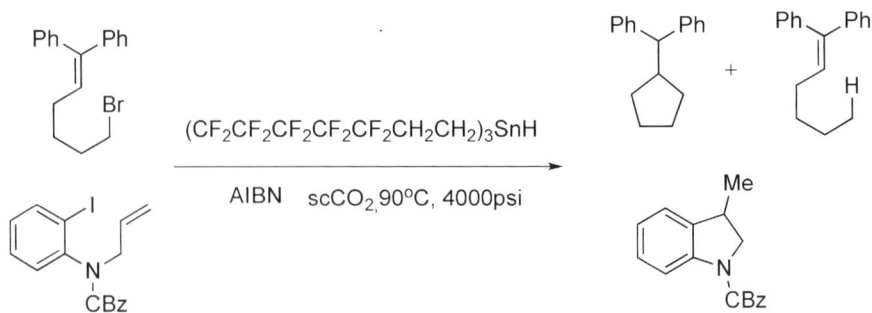

Scheme 2.3 Free radical cyclization reactions in $scCO_2$.

cyclized product. Interestingly no reaction was observed with tributyltin hydride in either case.

Free radical carbonylation reactions have extensively been investigated as a promising tool for the introduction of the carbonyl group because of their generality and scope of application in organic synthesis. Recently, Kishimoto and Ikariya applied the supercritical CO_2 as a reaction medium for the silane-mediated efficient free radical carbonylation of the organic halides to ketones.[19] The reductive carbonylation of 1-iodooctane and acrylonitrile with CO_2-soluble $(TMS)_3SiH$ in $scCO_2$ (CO 20 atm, 350 atm total pressure) containing AIBN as the radical initiator at 80 °C proceeded smoothly in 80% yield (Scheme 2.4). The radical carbonylative ring-closing reaction was expanded to the intramolecular radical/ carbonylation of 6-iodohexyl acrylate with 2 equivalents of $(TMS)_3SiH$ in $scCO_2$ including CO (CO 50 atm, total pressure 295 atm) at 80 °C for 2 h afforded the eleven-membered macrolide in 68% isolated yield. The yield of the macrolide is comparable to that achieved in benzene.

2.7.3 Future Directions

Using CO_2 as a solvent or a starting material for chemical synthesis and reactions has been a subject of extensive research since the early 1990s. Some of the developments in this area have already led to new processes in chemical separation and manufacturing. $scCO_2$ can replace environmentally hazardous benzene in broad range of free radical transformations. Carbon dioxide-based dry cleaning techniques and synthesis of fluoropolymers in $SF-CO_2$ are examples of industrial applications of these new supercritical fluid-based techniques. Demonstrations for the remediation of toxic metals in solid waste and the reprocessing of spent nuclear fuel in supercritical CO_2 have also been initiated recently. It is likely that research and development in CO_2-based technology for chemical separation and material processing will continue to expand in this decade towards industrial scale flow reactors and the development of novel and functional supercritical anti-solvents.

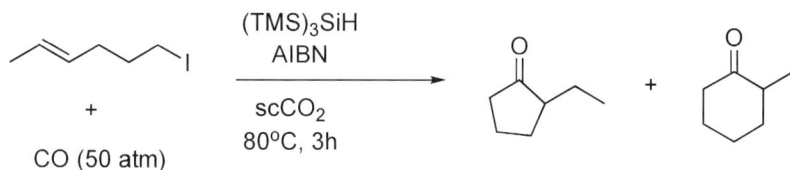

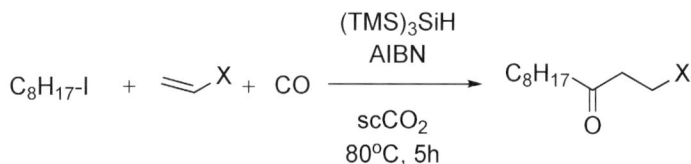

where X=CN, CO₂CH₃, COCH₃

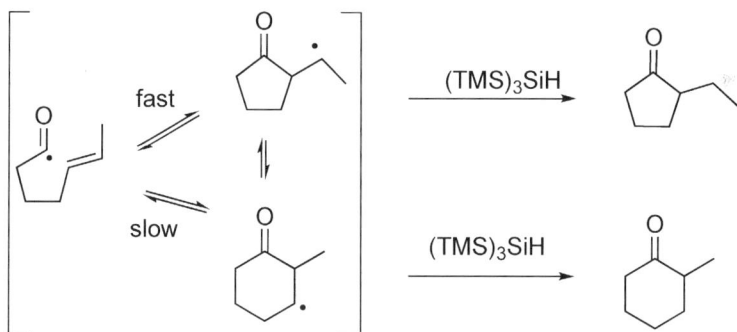

Mechanism:

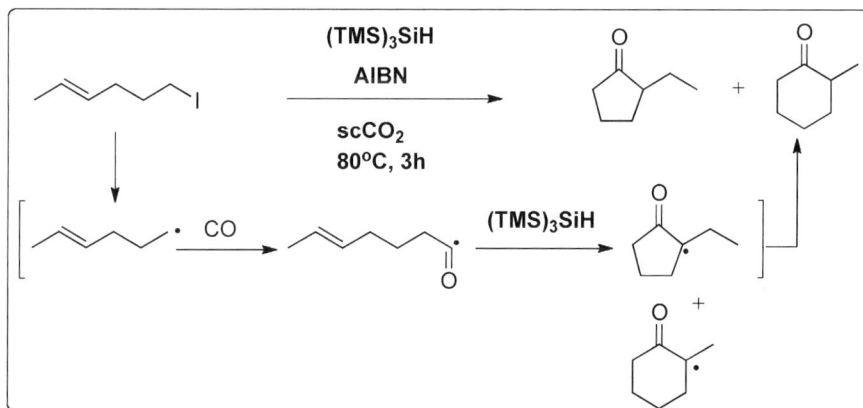

Scheme 2.4   Silane-mediated efficient free radical carbonylation of the organic halides to ketones in scCO₂.

2.8 RADICAL REACTIONS IN IONIC LIQUIDS

2.8.1 Ionic Liquids and Alternative Media: General Introduction

The ionic liquids are more closely related to conventionally used solvents and therefore chemistry and applications could be developed rapidly. A history of ionic liquids including properties, availability, costs, application, preparation and green potential can be found in several reviews. An increasing range of ionic liquids is now commercially available.[20,21]

The term "ionic liquid" is used to refer to a salt which exists in the liquid state at or around ambient temperature, hence the term "room temperature ionic liquids". The ionic liquid usually consists of an organic cation (often containing a nitrogen heterocycle) and an inorganic anion.

These room temperature ionic liquids exhibit properties that make them potentially useful reaction media for synthesis and catalysis. They are good solvents for a diverse range of chemical transformations due to their thermal stability and negligible vapour pressure as well as their ability to be used as *in situ* Lewis acids and basis and therefore catalyse the desired transformations, as well as being recyclable and having the potential to reuse the ionic liquids after transformation has been completed.

It should be appreciated, however, that the challenge here is not really whether a reaction can be carried out in an ionic liquid as they are very good solvents for many kinds of organic reactions, and its quite trivial to change from a conventional solvent to an ionic liquid. The important point is why you would want to change to an ionic liquid; does it allow you to do something that would otherwise be very difficult with a conventional solvent? Such considerations are particularly relevant for synthetic chemistry. It is beyond the scope of this review to describe this in detail; however, some practical examples will be presented and described.

2.8.2 Radical Chain Reactions in Ionic Liquids: Triethylborane-induced Radical Reactions

Some triethylborane-induced radical reactions were found to proceed in ionic liquids by Oshima *et al.*[21] The reactions include atom transfer radical cyclization reactions, hydrostannylation reactions of alkynes and atom transfer reactions (Scheme 2.5).

Benzyl bromoacetate participates in the bromine atom transfer reaction of 1-octene in 1-ethyl-3-methylimidazolium tetrafluoroborate to afford the corresponding adducts in excellent yields. The facile bromine atom transfer addition in the ionic liquid indicates that the ionic liquids may have a highly polar nature. Recycling the ionic liquids used in the investigations was attempted with varying degrees of success.[21]

$$\text{Ph}\overset{O}{\diagup}\text{O}\overset{\parallel}{\diagup}\text{Br} \xrightarrow[\substack{\text{C}_2\text{H}_5\text{-N}\overset{\oplus}{\diagdown}\text{N-C}_2\text{H}_5 \\ \text{BF}_4^{\ominus}}]{\text{Et}_3\text{B}} \text{Ph}\overset{O}{\diagup}\text{O}\overset{\parallel}{\diagdown}\underset{\text{Br}}{\diagdown}n\text{C}_6\text{H}_{13}$$

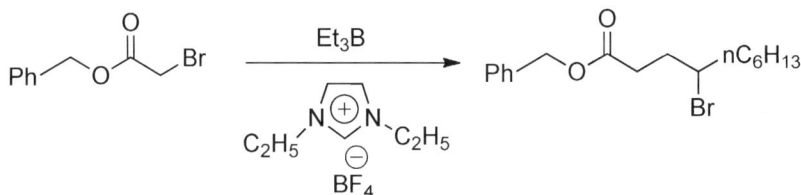

Proposed Mechanism:

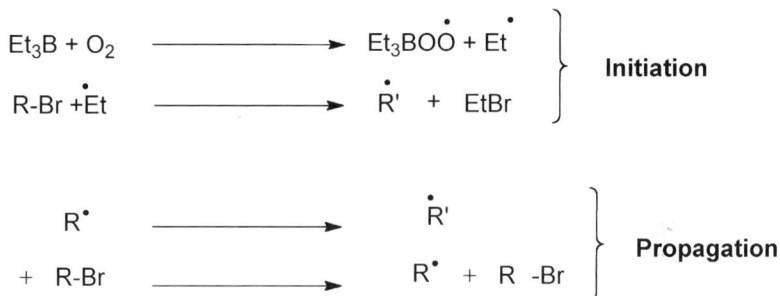

$$\text{Et}_3\text{B} + \text{O}_2 \longrightarrow \text{Et}_3\text{BO}\dot{\text{O}} + \dot{\text{Et}}$$

$$\text{R-Br} + \dot{\text{Et}} \longrightarrow \dot{\text{R'}} + \text{EtBr}$$

$\left.\vphantom{\begin{matrix}a\\b\end{matrix}}\right\}$ **Initiation**

$$\dot{\text{R}} \longrightarrow \dot{\text{R'}}$$

$$\dot{\text{R}} + \text{R-Br} \longrightarrow \dot{\text{R}} + \text{R -Br}$$

$\left.\vphantom{\begin{matrix}a\\b\end{matrix}}\right\}$ **Propagation**

Scheme 2.5   Bromine atom transfer reaction in ionic liquid.

2.8.3   Radical Additions of Thiols to Alkenes and Alkynes in Ionic Liquids

Nanni *et al.* have investigated the radical addition of thiols to double and triple carbon–carbon bonds examined in typical ionic liquids ([bmim][PF$_6$], [bmim][Tf$_2$N], and [bmim][BF$_4$]) under different temperature/initiator conditions (*i.e.* 80–100 °C / AIBN-VAZO®, r.t. / triethylborane, r.t. / AIBN / UV radiation, r.t. / photoinitiator / UV radiation).[22] All the addition products were usually obtained with high efficiencies and very good recyclability of the ionic liquid. In some cases, small but significant differences were noticed by changing the reaction medium from benzene to an ionic liquid (Scheme 2.6).[22e]

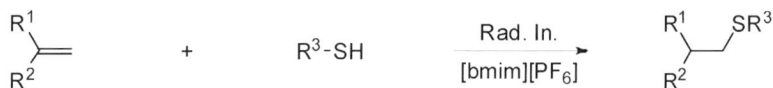

$$\underset{\text{R}^2}{\overset{\text{R}^1}{\diagup}}\!\!=\quad + \quad \text{R}^3\text{-SH} \xrightarrow[\text{[bmim][PF}_6]]{\text{Rad. In.}} \underset{\text{R}^2}{\overset{\text{R}^1}{\diagup}}\!\!\diagdown\text{SR}^3$$

a: R^1 = *n*-C$_6$H$_{13}$, R^2 = H   A: R^3 = Ph

b: R^1 = Ph, R^2 = H   B: R^3 = MeO(CO)CH$_2$

c: R^1 = *p*-Cl-C$_6$H$_4$, R^2 = H

d: R^1 = *p*-MeO-C$_6$H$_4$, R^2 = H

e: R^1,R^2 = *c*-C$_6$H$_{10}$

f: R^1 = *n*-BuO, R^2 = H

g: R^1 = MeC(O)O, R^2 = H

Scheme 2.6   Radical addition of thiol to alkene in ionic liquids.

This outcome suggests that hydrothiolation of alkynes is somewhat faster in ionic liquid with respect to aromatic solvents, in line with what was suggested above for hydrothiolation of alkenes. Of course, formation of the kinetic product is favored at lower temperatures, although kinetic control was never attained in any ionic liquid. By comparing the various results obtained at r.t., it seems that [bmim][PF$_6$] would be the best solvent for promoting formation of the *Z*-isomer, and hence the medium in which hydrothiolation of alkynes is faster.

Finally, Nanni's group briefly examined the possible use of ionic liquid solvents in click-chemistry reactions leading to biologically interesting molecules.[22c] The preliminary results include the addition of L-*N*-Fmoc-cysteine *tert*-butyl ester to 1-hexadecene and to an *O*-allyl glucoside, and hydrothiolation of phenylacetylene with L-cysteine ethyl ester hydrochloride (Scheme 2.7).

The first adduct (**1**), which has been recently synthesized in 42% yield through a radical thiol-ene, racemization-free procedure using thermal conditions, was chosen as a target compound for the recent interest in accessing hydrolysis-resistant non-natural *S*-alkylated cysteine derivatives. The reaction was carried out in [bmim][PF$_6$] under DMPA-promoted photolysis conditions with 3 equiv. of 1-hexadecene for 5 h and, after usual workup followed by chromatography, similarly gave adduct **1** but in somewhat lower yield (30%).

The second sulfide (**2**) was selected in view of the ever-growing importance of *O*-linked glycosides and their use for preparation of glyco- and/or peptido-mimetics. Also in this case the reaction was performed by DMPA-promoted photolysis of the peracetylated *O*-allyl glucoside (*ca.* 1 : 1 α/β mixture) in [bmim][PF$_6$] in the presence of orthogonally-protected L-*N*-Fmoc-cysteine *tert*-butyl ester (1 equiv.). After workup and chromatography, the target compound **2** was obtained in 50% yield.

The third reaction was simply a trial experiment to see whether mercapto-substituted aminoacids can add efficiently to an alkyne to give **3**-like vinyl

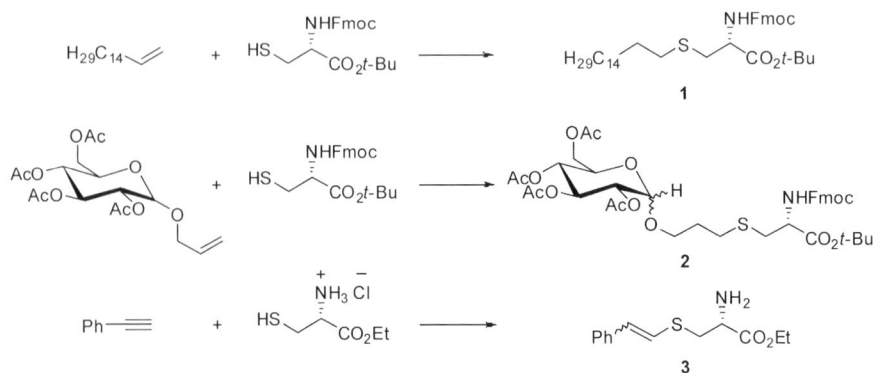

Scheme 2.7 Hydrothiolation as a tool for the synthesis of biologically interesting compounds.

sulfide adducts. The reaction between phenylacetylene and L-cysteine ethyl ester hydrochloride was carried out in [bmim][PF$_6$] by DMPA-promoted photolysis and yielded vinyl sulfide **3** in 78% yield (55:45 *E/Z* ratio). It is worth noting that this reaction gave **3** only to a slightly higher extent (86%) (55:45 *E/Z* ratio) when repeated in a traditional solvent such as DMF.

In conclusion, Nanni *et al.* reported that radical hydrothiolation of alkenes and alkynes can occur in ionic liquids with at least the same efficiency as in traditional solvents. The ionic liquids are compatible with the use of different radical initiation conditions, *i.e.* thermal decomposition of azo-initiators (80–100 °C), reaction of triethylborane with dioxygen (r.t.), and UV-photolysis at r.t. in the presence of either an azo-initiator (AIBN) or a photosensitizer (DMPA). The reaction products can be efficiently isolated from the ionic liquid by centrifuge-mediated extraction with diethyl ether, and the ionic liquid can be usually recycled up to 3 times without any significant change in yields and by-products under any reaction conditions. Some results show that the ionic liquids, if compared with traditional solvents, seem to favour the hydro-thiolation reaction, probably through stabilization of the transition state for hydrogen transfer from the starting thiol to the intermediate alkyl or vinyl radical. Although the results are merely preliminary and yields are not optimized, it seems that this protocol could be successfully applied to the synthesis of biologically interesting molecules through click-chemistry pro-cedures carried out in ionic liquids.

2.9 RADICAL NON-CHAIN REACTIONS IN IONIC LIQUIDS

2.9.1 Formation of Radicals by Oxidation with Transition Metal Salts: General Perspective

Transition metals in high oxidation states are often capable of extracting one electron from electron-rich organic substances. Ketones, esters, nitriles and various other "carbon acids" that can form enols, enolates and related struc-tures are by far the most commonly used substrates. Their oxidation can lead to a free radical, which then follows one or more of the pathways available. It is important to take into account the fact that the rate of radical production will depend on the exact structure of the substrate, its propensity to exist as the corresponding enol or enolate in the medium, the pH, the solvent, the temperature, and, of course, the redox potential of the metallic salt (which can be strongly affected by the nature of the ligand around the metal) and the exact mechanism by which electron transfer actually occurs (*i.e.* inner or outer sphere).

2.9.2 Oxidations involving Mn(III) in Ionic Liquids

Radical generation through oxidation of enolizable substrates using Mn(III) salts is by far the most common and the field is rapidly expanding. Mn(OAc)$_3 \cdot x$H$_2$O is the usual oxidant; it is actually a trimer made up of an oxo-centre triangle of Mn(III) ions bridged by acetate units, but for simplicity and convenience we shall use the simplified formula of Mn(OAc)$_3$ throughout the rest of the examples.[23] These reactions are considered to proceed *via* a free radical process where Mn(III) initiates the oxidation of the carbonyl compound and then the newly formed radical undergoes an intermolecular addition to the olefins to produce a new radical. The addition of a radical to the aromatic ring gives rise to a stabilized radical, which is then oxidized with Mn(III) to restore the aromaticity and yield the corresponding furan (Scheme 2.8).[23]

Many of the methods that were previously employed for Mn(OAc)$_3$-mediated radical reactions involved the use of acetic acid as a solvent. Because of the poor solubility of Mn(OAc)$_3$ in organic solvents and the need for high temperatures for many reactions, the use of acetic acid limited the range of substrates that could be employed. In order to improve this drawback, Parsons *et al.* investigated the elegant way of using ionic liquids to establish milder reaction conditions in Mn(OAc)$_3$ mediated reactions.[24] They showed that ionic liquids, such as 1-butyl-3-methylimidazolium tetrafluoroborate ([bmim][BF$_4$]), which is miscible with polar solvents (*e.g.* methanol, dichloromethane) could be used in Mn(OAc)$_3$-mediated radical reactions (Scheme 2.9).[24]

Cerium(IV) ammonium nitrate-mediated (CAN-mediated) oxidative radical reactions are carried out in the presence of ionic liquids, including 1-butyl-3-methylimidazolium tetrafluoroborate, for the first time. The presence of the ionic liquid not only increases the rate and yield of reactions in dichloromethane

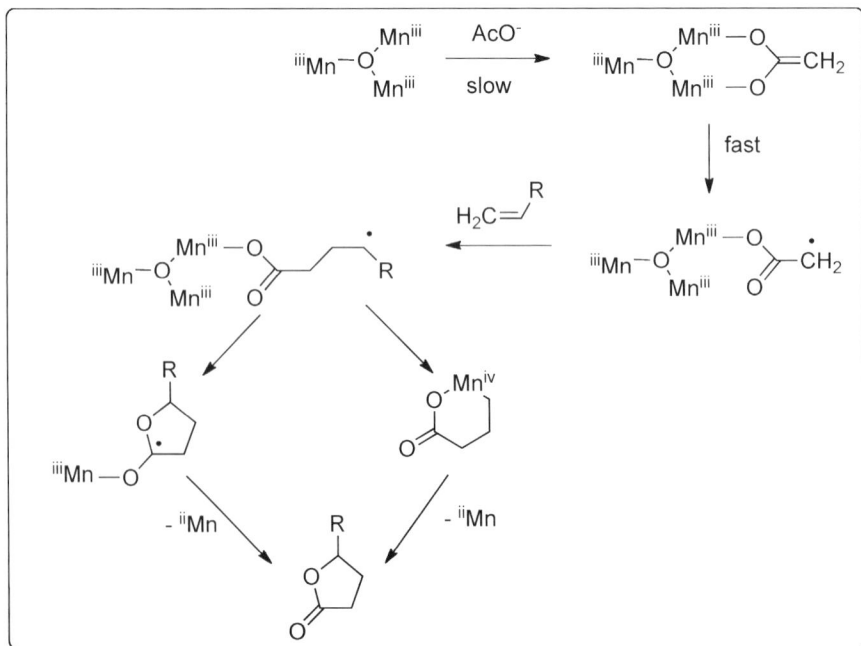

Scheme 2.8   Mechanism of Mn(OAc)$_3$-mediated radical oxidation reaction.

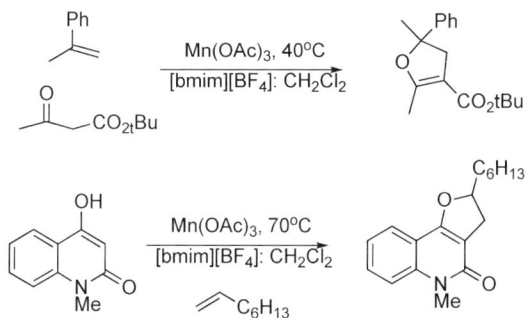

Scheme 2.9   Mn(III)-mediated reactions in ionic liquids.

but also extends the range of 1,3-dicarbonyl precursors, which can be utilized in these carbon–carbon bond-forming reactions (Scheme 2.10).[25]

Cerium(IV) ammonium nitrate (CAN) has been widely used to oxidize numerous organic compounds, including 1,3-dicarbonyls, to form radicals. The resulting dicarbonyl radicals can add intra- or inter-molecularly to a range of electron-rich alkenes to form radical adducts, which can be oxidized using a second equivalent of Ce(IV). The carbocations can then undergo nucleophilic attack or deprotonation to form, for example, nitrates or alkenes. Indeed, the use of CAN in intermolecular carbon–carbon bond-forming reactions has been shown to have advantages over the more commonly used manganese(III) acetate. These oxidative radical reactions are synthetically appealing as they permit the formation of functionalized products in one-pot reactions using inexpensive CAN.

The most common solvents employed in CAN oxidation reactions are acetonitrile or methanol. However, these polar solvents are not always ideal for non-polar substrates, and alcohols can react with intermediate carbocations to produce unwanted ethers. With the aim of extending the range of solvents which can be employed in CAN oxidative cyclizations, Parsons *et al.*

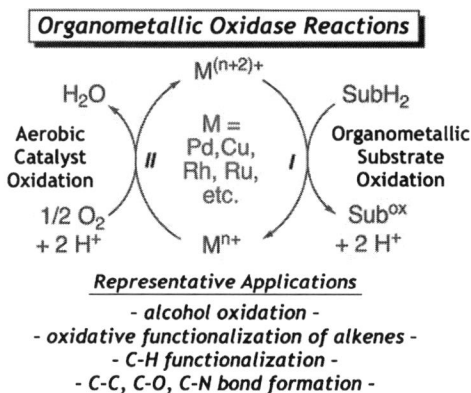

Scheme 2.10   General oxidation mechanism for CAN mediated radical reactions.

have investigated the novel use of cerium(IV) ammonium nitrate-mediated oxidative radical reactions in the presence of ionic liquids, including 1-butyl-3-methylimidazolium tetrafluoroborate. The presence of the ionic liquid not only increases the rate and yield of reactions in dichloromethane but also extends the range of 1,3-dicarbonyl precursors which can be utilized in these carbon–carbon bond-forming reactions (Scheme 2.11).[26]

Another important reaction which was explored in ionic liquids is the α-halogenation of carbonyl compounds. Ionic liquids (acetylmethylimidazolium halides) in combination with ceric ammonium nitrate promoted the halogenation of a wide variety of ketones and 1,3-ketoesters at the α-position. The ionic liquid acts here as a reagent as well as a reaction medium, and thus the reaction does not require any organic solvent or conventional halogenating agent. The reaction is completely stopped when the radical quencher TEMPO is used (Scheme 2.12).[27] The mechanism proposed for the transformation suggests that CAN plays a vital role as a one-electron oxidant. The preferential

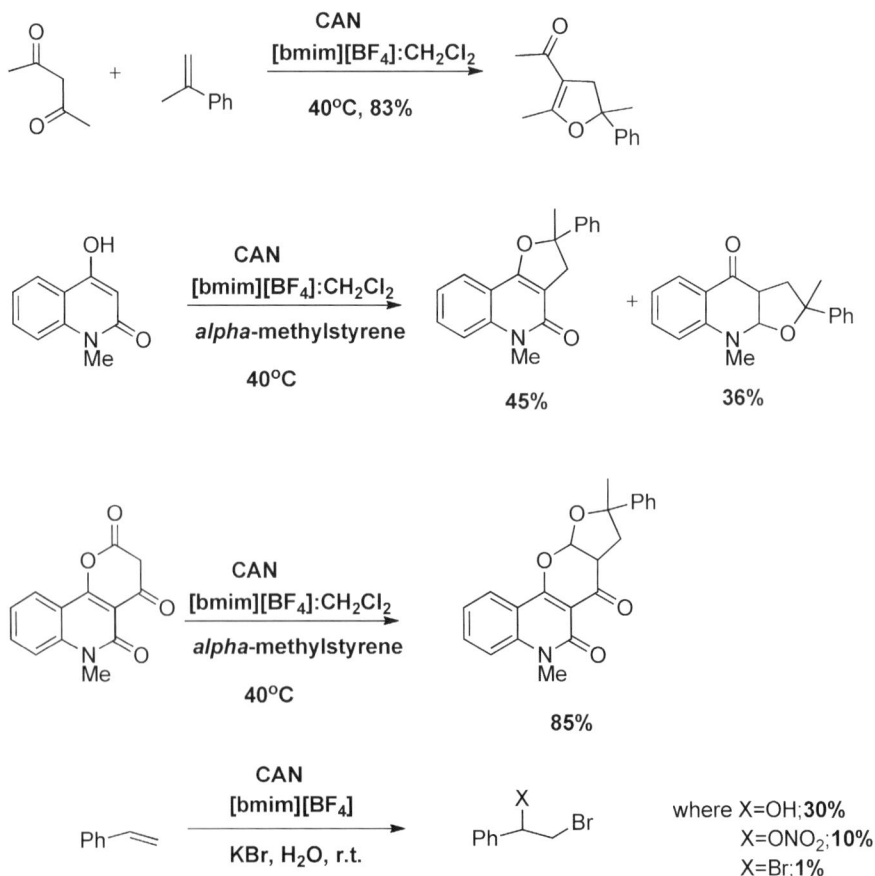

Scheme 2.11 CAN-mediated oxidative radical reactions in ionic liquids.

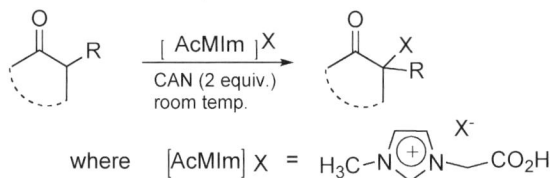

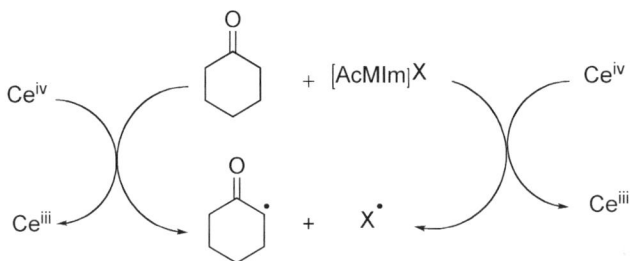

Scheme 2.12. Use of [AcMIm]X and CAN in the α-halogenation of carbonyl compounds and plausible mechanism for the α-halogenation.

halogenation in the non-substituted α-position of α-substituted ketone is also in accordance with various radical stabilities of newly formed radicals.

2.9.3 Supported Ionic Liquids: Versatile Reaction and Separation Media—the Latest Developments

As previously mentioned, ionic liquids have attracted growing interest as alternative reaction media to replace volatile organic solvents in catalysis. Their ionic nature, non-volatility and thermal stability make them highly suitable for biphasic ionic liquid–organic transition metal catalysis. The almost unlimited combination of cation–anion pairs further allows the synthesis of tailor-made ionic liquids that can stabilize catalytic species. However, traditional biphasic ionic liquid–organic systems require larger amounts of ionic liquids which make them unattractive based on economic considerations since ionic liquids are still expensive solvents, even though they are commercially available now. In addition, the high viscosity of ionic liquids can induce mass transfer limitations if the chemical reaction is fast, in which case the reaction takes place only within the narrow diffusion layer and not in the bulk of the ionic liquid catalyst solution. Therefore, only a minor part of the ionic liquid and the dissolved precious transition metal catalyst are utilized. It is for this reason that interest in supported ionic liquids has come to the fore due to its potential as a versatile reaction medium for a variety of synthetically useful transformations including the Kharasch reaction, *i.e.* solvent-free addition reaction of styrene and carbon tetrachloride to produce 1-(1,3,3-tetrachloropropyl)benzene (Scheme 2.13).[28]

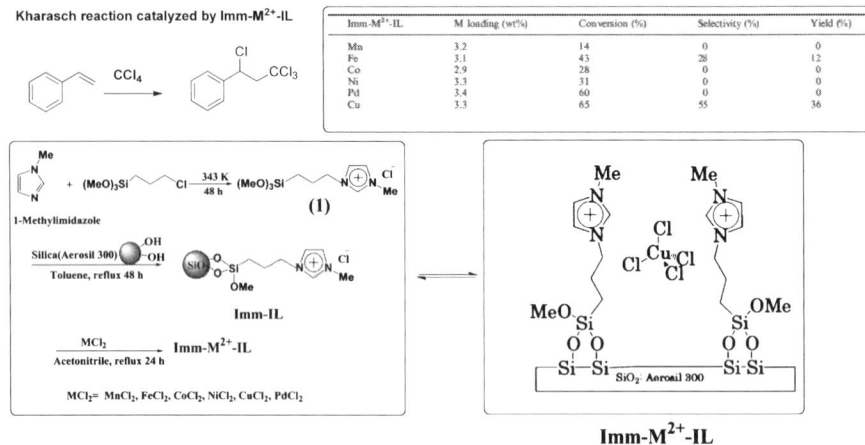

Imm-M²⁺-IL	M loading (wt%)	Conversion (%)	Selectivity (%)	Yield (%)
Mn	3.2	14	0	0
Fe	3.1	43	28	12
Co	2.9	28	0	0
Ni	3.3	31	0	0
Pd	3.4	60	0	0
Cu	3.3	65	55	36

Imm-M²⁺-IL

Scheme 2.13 Kharasch reaction catalyzed by Imm-M²⁺-IL.

Normally the reaction is carried out using homogeneous catalysts, thus making this approach one of the first examples using a heterogeneous metal complex system. Of the system examined, only Fe^{2+} and Cu^{2+} containing catalyst were found to be active, and when optimizing the copper(II) system a conversion of 98% and an excellent product selectivity of 95% could be obtained. However, during recycling experiments the yield decreased from 93% to 80%. The existence of the immobilized imidazolium group were found, by studies of analogous silica-supported $CuCl_2$ catalyst (*i.e.* without ionic liquid), to be essentially important as copper ions for the catalytic activity. The immobilization of the complex on the silica was assumed to restrict the conformation of reactants and thereby facilitate the high selectivity by favoring the formation of the chlorinated addition products rather than oligomerization of styrene.

2.9.4 Conclusions and Future Directions

Since its birth over a decade ago, the field of green chemistry has seen rapid expansion, with numerous innovative scientific breakthroughs associated with the production and utilization of chemical products. The concept and ideal of green chemistry now goes beyond chemistry and touches subjects ranging from energy to societal sustainability. The key notion of green chemistry is "efficiency", including material efficiency, energy efficiency, man-power efficiency, and property efficiency (*e.g.*, desired function *vs.* toxicity). Any "wastes" aside from these efficiencies are to be addressed through innovative green chemistry means. "Atom-economy" and minimization of auxiliary chemicals, such as protecting groups and solvents, form the pillar of material efficiency in chemical productions. By far, the largest amount of "auxiliary waste" in most chemical productions is associated with solvent usage. In a classical chemical process, solvents are used extensively for dissolving reactants, extracting and washing products, separating mixtures, cleaning reaction apparatus, and dispersing products for practical applications. While the invention of various exotic organic solvents has resulted in some remarkable advances in chemistry, the legacy of such solvents has led to various environmental and heath concerns. Consequently, as part of green chemistry efforts, a variety of cleaner solvents have been evaluated as replacements. However, an ideal and universal green solvent for all situations does not exist. Among the most widely explored greener solvents are ionic liquids, supercritical CO_2, and water. These solvents complement each other nicely both in properties and applications. Importantly, the study of green solvents goes far beyond just solvent replacement. The use of green solvents has led science to uncharted territories. For example, the study of ionic liquids made large-scale supported synthesis possible for the first time and the utilization of supercritical CO_2 has led to breakthroughs in microelectronics and nanotechnologies.

2.10 FLUOROUS CHEMISTRY AS AN ALTERNATIVE REACTION MEDIUM FOR FREE RADICAL TRANSFORMATIONS

2.10.1 Fluorous Separation Techniques: from "Liquid–Liquid" to "Solid–Liquid" and "Light Fluorous"

"What is fluorous chemistry?" That is a question still too common amongst the chemical community, which becomes more involved with supercritical media, ionic liquids and aqueous media as quite common for a broad range of free radical synthetically useful transformations.[29–40] "Fluorous" was the term coined for highly fluorinated (or perfluorinated) solvents, in an analogous way to "aqueous" for water-based systems. Fluorous solvents are immiscible with both organic and aqueous solvents, thus hexane (C_6H_{14}) and perfluorohexane (C_6F_{14}, commonly known as FC-72[TM]) are immiscible.

Fluorous chemistry has been developed as a broad-based technology platform that addresses many challenges in the modern drug discovery environment.[29] Fluorous technologies can be divided into two broad groups: fluorous tagging strategies and fluorous separation.[30] The advantage of fluorous tagging methods lies in the opportunity to combine a solution-phase reaction with phase tag-based separation. Perfluorinated (fluorous) chains such as C_6F_{13} and C_8F_{17} are employed as phase tags to facilitate the separation process. The fluorous chain is usually attached to the parent molecule through a $(CH_2)_m$ segment to insulate the reactive site from electron-withdrawing fluorines. A fluorous chain $C_nF_{2n+1}(CH_2)_m$ can be abbreviated to Rf_nh_m when it is presented in the reaction equation. In principle, any synthetic development in conventional solution-phase or in polymer-bound chemistry can be adapted to fluorous chemistry. Fluorous groups are used to tag reagents (including scavengers and catalysts) or to tag substrates. Fluorous reagents are commonly used for single- or short-step parallel synthesis, while fluorous-tagged substrates are more suitable for multistep parallel synthesis (Scheme 2.14).

Fluorous separation relies on the strong and selective affinity interaction between fluorous molecules and fluorous separation media.[32] The separation medium can be fluorous solvents which are immiscible with common organic solvents at room temperature and thus can be used for liquid–liquid extractions. A "heavy fluorous" tag (60% or more fluorine by molecular weight) is required to drive the fluorous molecule to a fluorous phase from non-fluorous phase. Perfluorinated alkanes such as FC-72 (perfluorohexanes) and highly fluorinated ethers such as HFC-7100 ($C_4F_{17}OCH_3$) are good solvents for fluorous extractions.

The important features which make fluorous solvents/synthesis so easy to integrate into synthetic applications is the synergy achieved through the solution-phase reactions and solid-phase separations as well as good combinatorial capabilities; it offers the following advantages:

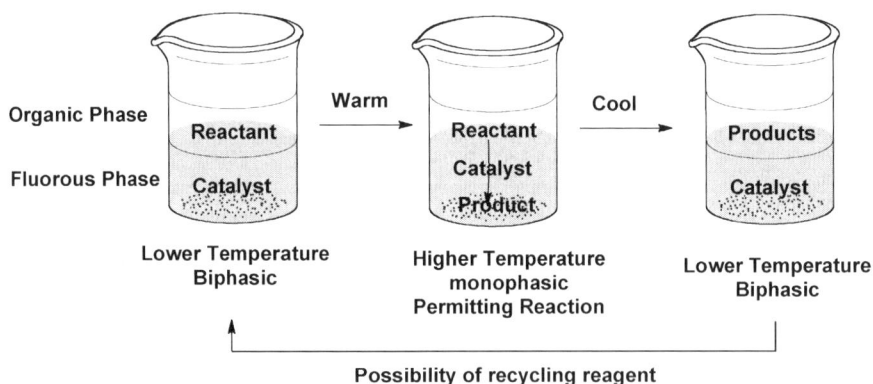

a. Tagging reagents/scavengers/catalysts

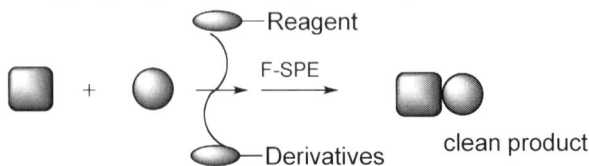

b. Tagging substrates for parallel synthesis

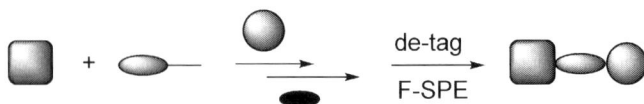

Scheme 2.14   Fluorous tagging strategies: (a) tagging reagent/scavengers/ catalyst; (b) tagging substrates for parallel synthesis.

- Possibility of following reactions by TLC, HPLC, IR and NMR.
- Ability to utilize the fluorous and non-fluorous separations to maximize the recover of reactants and products.
- Good solubility in the range of conventional organic solvents.
- More than one fluorous reagent can be used in a single reaction.
- Easily adaptable methods from the non-fluorous procedures.
- Recovery of fluorous materials after separation.

The most important characteristic that makes fluorous synthesis superior to solid-supported synthesis is the favourable reaction kinetics associated with the solution-phase reaction.

All chemical reactions are limited both by the efficiency of the transformation and also the ease of purification of the reaction mixture. Chemists have traditionally concentrated on the former of these two problems (the conversion of starting materials to products), to the detriment of the latter purification issue. In recent years, several new techniques have appeared to try to address this, including fluorous-phase chemistry. The aim of this chapter is to highlight the development of fluorous chemistry from what was often considered to be an expensive laboratory "curiosity" to a technique that is now an independent and valid scientific tool ready for adoption by a chemical community in general, especially in free radical-mediated synthesis: chain and non-chain reactions.

2.10.2   Fluorous Chemistry and Radicals—Combined Efforts to the Rescue

Fluorous methods have emerged as new and powerful techniques that have greatly influenced the way in which preparative organic chemistry is currently performed. In cooperation with the Curran group, who first reported on fluorous tin hydride technology (Scheme 2.15).[34]

2.10.3   Fluorous Radical Carbonylation Reactions: from Synthetic Approach to Practical Applications

Ryu developed "fluorous radical carbonylation" reactions.[35] The great advantage of fluorous radical carbonylation systems is the way in which they enable the easy separation of products and reagents by fluorous–organic liquid–liquid extraction or fluorous–solid-phase extraction, to ensure the reuse of tin reagents (Scheme 2.16).

As illustrated in Scheme 2.16, Ryu obtained comparable results when comparing conventional tin hydride with ethylene-spaced fluorous tin hydride. At an identical CO pressure, the use of tributyltin hydride afforded higher formylation/reduction ratios than ethylene-spaced fluorous tin hydride. This is consistent with the fact that the rate constant for primary alkyl radical trapping by fluorous tin hydride is about double that for tributyltin hydride at 20 °C. Thus, to obtain results identical to those for tributyltin hydride, a higher CO pressure and/or a higher dilution are required for the fluorous

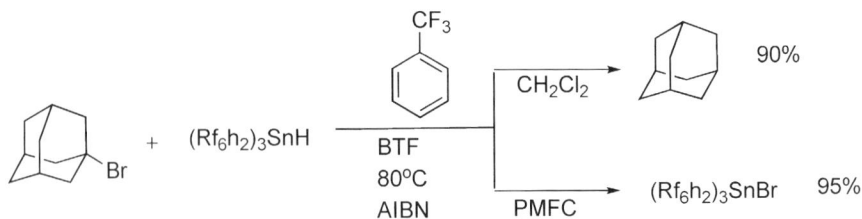

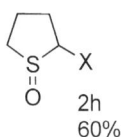 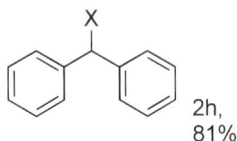

phenyl selenide   X=SePh ———→ X = H

2h
60%

2h,
81%

nitro compounds   X=NO₂ ———→ X = H

1-X-adamantane 24h, 88%   X-C(CH₂OMs)₃

Xanthates   X=OCS₂CH₃ ———→ X = H

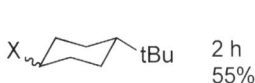 2 h
55%

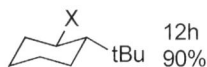 12h
90%

Scheme 2.15   Scope of free radical transformations in the presence of fluorous tin
            hydride and tributyl tin hydride.

analogue. Fluorous hydroxymethylation of organic halides using a catalytic
quantity of a fluorous tin hydride was also investigated and demonstrates a
successful future direction of research and development in the area. Kahne
and Gupta previously reported a similar hydroxymethylation, using a catalytic
amount of triphenylgermyl hydride and an excess amount of sodium cyanoborohydride.[36] Ryu *et al.* found that the reaction worked well even in the case
of fluorous tin hydride (Scheme 2.17).[37] Interestingly, this fluorous reagent,
as is usually the case with the related fluorous radical reactions, permits
simple purification through a three-phase (aqueous/organic/fluorous) extrac-
tive workup. After the workup, fluorous tin reagent is recovered from the
perfluorohexane layer and reused. Very recently, Ryu reported on the use of
F-626, 1H,1H,2H,2H-perfluorooctyl 1,3-dimethylbutyl ether, as a versatile
alternative solvent for both fluorous and non-fluorous reactions.[32]. As
illustrated in the second example of Scheme 2.18, the successful hydroxymethylation of 1-iodoadamantane was possible using this new solvent as a
fluorous/organic amphiphilic solvent instead of BTF (benzotrifluoride).

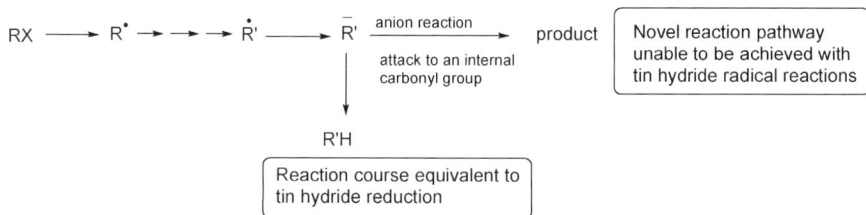

Mechanism:

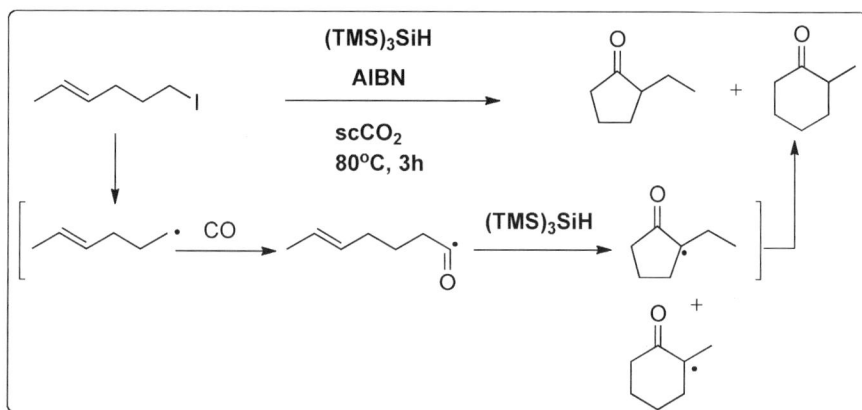

Scheme 2.16 Schematic diagram of tandem carbonylation reaction with hydride species in a one-electron reductive system.

Propylene-spaced fluorous allyltin and methallyltin proved particularly useful as mediators in the four-component coupling reactions, where alkyl halides, CO, alkenes, and allyltin are combined in a given sequence (Scheme 2.19).[39] Once the reaction was complete, the BTF (benzotrifluoride) was removed by vacuum-evaporation and the resulting oil was partitioned into acetonitrile and FC-72 (perfluorohexanes). Evaporation of the acetonitrile layer, followed by short column chromatography on silica gel provided us with the desired product. This study also revealed that the FC-72-layer contained fluorous tin compounds such as fluorous allyltin and tin iodide and quantified reproduce fluorous allyltin reagents by treating the tin residue with an ethereal solution of allyl and methallyl magnesium bromides that we were able to reuse without any appreciable loss in their activity.

Next, Ryu examined whether acyl radical cyclization is being preferred over one-electron reduction using zinc to form an acyl anion, and, if so, would the resulting radical species consisting of a cyclic ketone moiety undergo a further radical addition reaction.[40] This idea was tested with the reaction of 4-alkenyl iodide, CO, and acrylonitrile as a radical trap, as a three-component coupling reaction took place to produce cyclopentanone derivatives (Scheme 2.20).

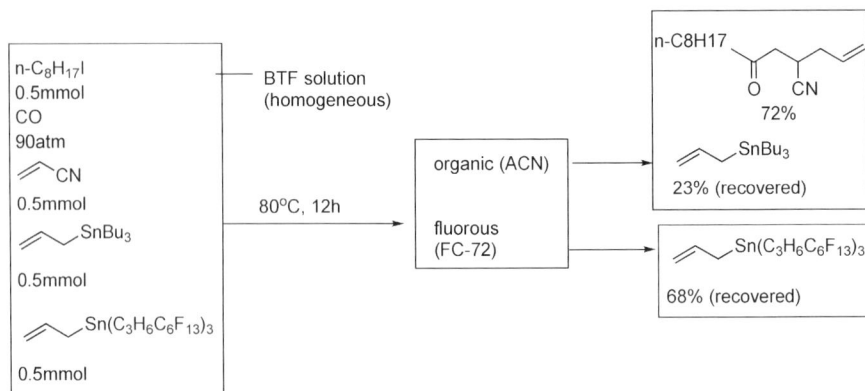

Scheme 2.17   Competition experiments between fluorous allyltin and allyltributyltin.

The observed stereochemical properties of the isolated products were identical to those obtained with the corresponding tin hydride-mediated reaction and thus support the theory that free radical generation, radical carbonylation, and acyl radical cyclization take place simultaneously in a zinc-induced system. The overall transformation is equivalent to a tin hydride-mediated system with an exception that the final step of the zinc/protic solvent system involves reduction to produce carbanions that are subsequently protonated (Scheme 2.21).[41]

The proposed mechanism is as follows and leads to the production of the bicyclic alcohol: (i) formation of an alkyl radical by one-electron reduction of the starting iodide; (ii) addition of the radical to CO; (iii) 5-*exo*-cyclization of the resulting acyl radical to give a tertiary alkyl radical; (iv) the addition of the tertiary radical to acrylonitrile; (v) one-electron reduction of the resulting a-cyano radical leading to an a-cyano anion; (vi) addition of the anion to an internal carbonyl group; and (vii) proton quenching of the alkoxy anion to give bicyclo[3.3.0]octan-1-ol. Ryu applied the zinc-induced dual [4+1] and

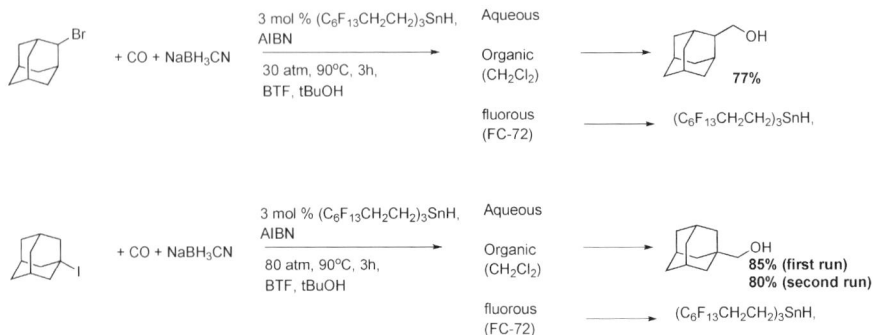

Scheme 2.18   Hydromethylation of RX using a catalytic amount of fluorous tin hydride.

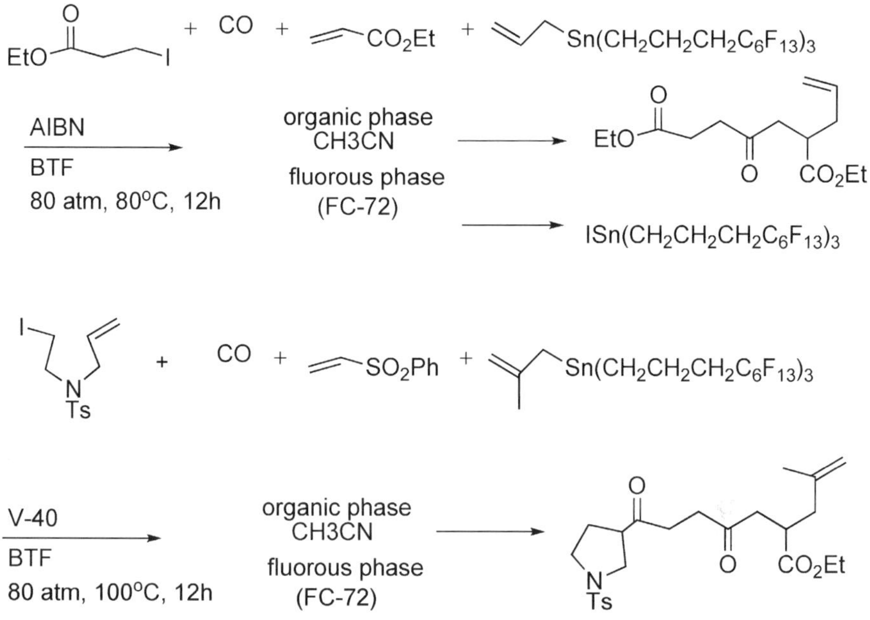

Scheme 2.19 Fluorous allyltin mediated four-component coupling reaction.

[3+2] annulation reaction to other alkenes, such as methyl acrylate and diethyl fumarate.[32] Also the construction of bicyclo[3.2.1]octanol skeletons was completed by using a zinc-induced cyclization. This system revealed that 6-*endo* cyclization is favored over 5-*exo* cyclization (Scheme 2.22). The reaction of 5-iodo-2-methylhex-1ene with CO and acrylonitrile in the presence of zinc led to bicyclo[3.2.1]octanol at a 51% yield with a 57/43 *exo/endo* ratio. Similarly,

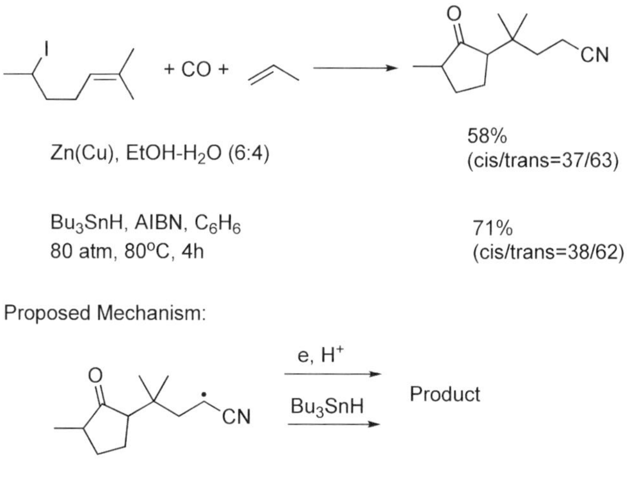

Zn(Cu), EtOH-H$_2$O (6:4)

58%
(cis/trans=37/63)

Bu$_3$SnH, AIBN, C$_6$H$_6$
80 atm, 80°C, 4h

71%
(cis/trans=38/62)

Proposed Mechanism:

Scheme 2.20 Three-component coupling reaction under two different sets of conditions.

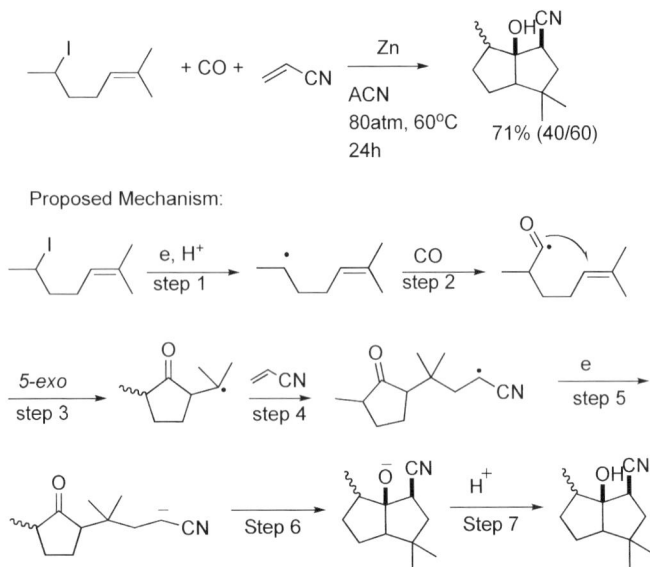

Scheme 2.21   Zn-induced dual annulation conducted in CAN.

trapping with methyl acrylate produced a 43% yield of the corresponding
bicyclo[3.2.1]octanol. Authors have demonstrated that radical carbonylation,
when combined with radical/anionic sequential reactions, provides a means
of producing bicyclo[3.3.0]octanols and bicyclo[3.2.1]octanols from readily
accessible starting materials. Using this "one-pot" procedure, we were able to
form four C–C bonds through three radical reactions and one anion reaction.

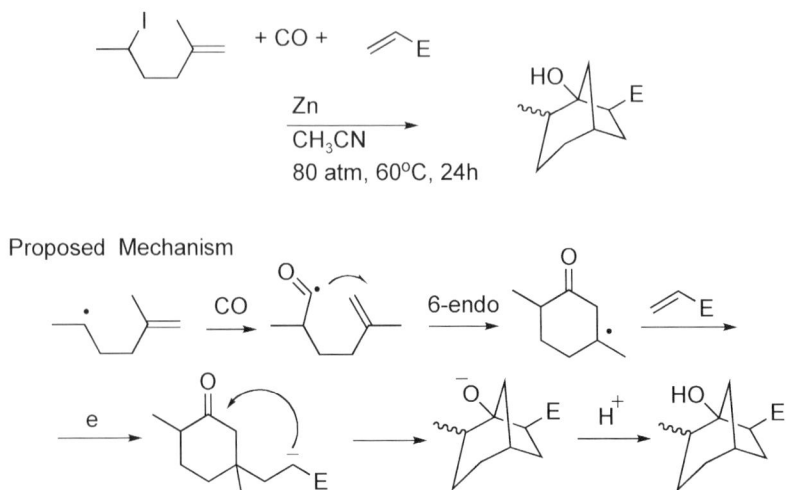

Scheme 2.22   Dual [5+1]/[3+2] annulation leading to bicyclo[3.2.1]octanols.

Ogawa *et al.* reported on the one-electron reduction of alkyl chlorides by SmI$_2$.[42] When they coupled this system with photo-irradiation, they observed a highly efficient reaction. When they carried out this under CO pressure, they obtained unsymmetrical ketones. Each product consisted of two molecules of alkyl chlorides and two molecules of carbon monoxide. Because their control experiments suggested that the dimerization of acyl anions is a likely key step, we can conclude that the one-electron reduction of an acyl radical to an acyl anion occurred quickly in this SmI$_2$ system. The rapid conversion to acyl anions is also supported by the absence of cyclized products, as in the case of 6-hexenyl chloride is used as a substrate (Scheme 2.23).

2.11 ISHII OXIDATION IN DETAIL

An innovation of the aerobic oxidation of hydrocarbons through catalytic carbon radical generation under mild conditions was achieved by using N-hydroxyphthalimide (NHPI) as a key compound. Alkanes were successfully oxidized with O$_2$ or air to valuable oxygen-containing compounds such as alcohols, ketones, and dicarboxylic acids.

Ishii and coworkers have developed an innovative strategy for the catalytic carbon radical generation from hydrocarbons by a phthalimide N-oxyl (PINO) radical generated *in situ* from N-hydroxyphthalimide (NHPI) (Scheme 2.24) and molecular oxygen in the presence or absence of a cobalt ion under mild

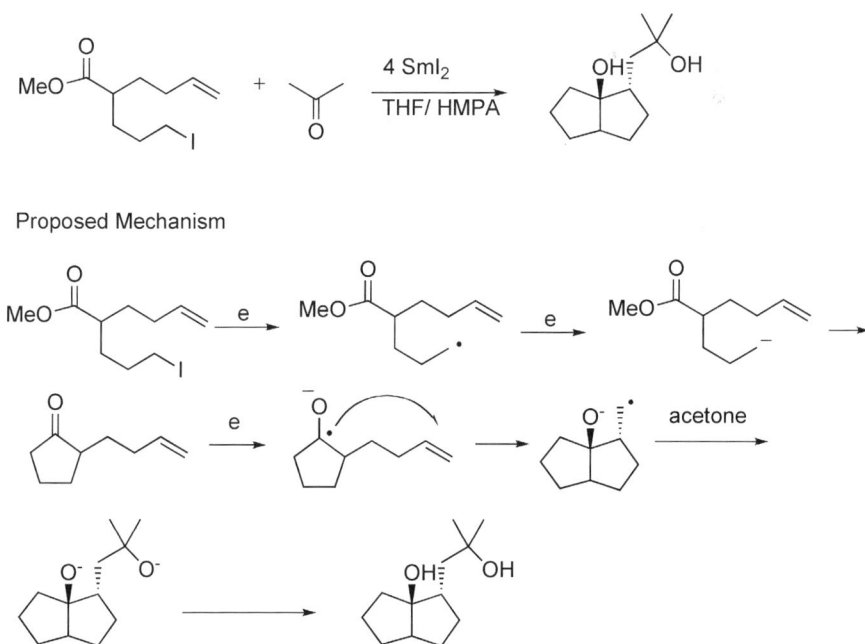

Scheme 2.23 Molander's tandem processes by species hybridization.

Electrolytic Oxidation of Alcohols mediated by NHPI

Epoxidation catalyzed by Mn-TPPCl using
hydroperoxides generated from NHPI, styrene, and O₂

Generation of PINO from NHPI under non-electrolytic
conditions

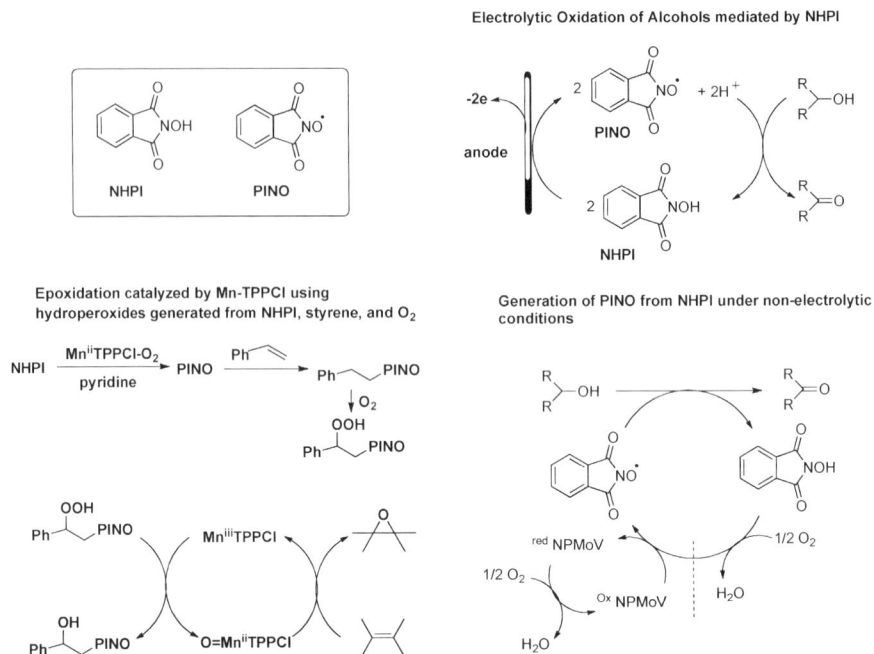

Scheme 2.24   Ishii oxidation in mechanistic detail: electrolytic oxidation, epoxidation
catalyzed by Mn-TPPCl using hydroperoxides generated from NHPI,
styrene, and O_2.

conditions. The carbon radical derived from a variety of hydrocarbons under the influence of molecular oxygen led to oxygenated products like alcohols, ketones, and carboxylic acids in good yields (Scheme 2.24).[43]

   The N-hydroxyphthalimide derivatives, F15- and F17-NHPI, bearing a long fluorinated alkyl chain, were prepared and their catalytic performances were compared with that of the parent compound, N-hydroxyphthalimide (NHPI). The oxidation of cyclohexane under 10 atm of air in the presence of fluorinated F15- or F17-NHPI, cobalt diacetate, and manganese diacetate without any solvent at 100 °C afforded a mixture of cyclohexanol and cyclohexanone (K/A oil) as major products along with a small amount of adipic acid. It was found that F15- and F17-NHPI exhibit higher catalytic activity than NHPI for the oxidation of cyclohexane without a solvent (Scheme 2.25). However, for the oxidation in acetic acid all of these catalysts afforded adipic acid as a major product in good yield and the catalytic activity of NHPI in acetic acid was almost the same as those of F15- and F17-NHPI. The oxidation by F15- and F17-NHPI catalysts in trifluorotoluene afforded K/A oil in high selectivity with little formation of adipic acid, while NHPI was a poor catalyst under these conditions, forming K/A oil as well as adipic acid in very low yields (Scheme 2.26). The oxidation in trifluorotoluene by F15- and F17-NHPI catalysts was considerably accelerated by the addition of a small amount of

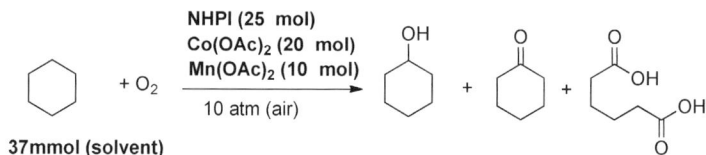

		OH	O	O OH / OH O
NHPI	6h 14h	622% 1060	468% 836	trace 5	TON = 10.9 19.0
F-NHPI	6h 14h	1278 2108	1485 2424	21 32	TON = 27.8 45.6

Scheme 2.25   Ishii's system for aerobic oxidation of cyclohexane.

zirconium(IV) acetylacetonate to the present catalytic system to afford selectively K/A oil, but no such effect was observed in the NHPI-catalyzed oxidation in trifluorotoluene (Scheme 2.27).[44]

2.12   FROM PHASE-SEPARATION TO PHASE-VANISHING METHODS BASED ON FLUOROUS-PHASE SCREEN: A SIMPLE WAY FOR THE EFFICIENT EXECUTION OF ORGANIC SYNTHESIS

Relying upon the unique properties of perfluorinated organic compounds, fluorous-phase chemistry has opened fresh ground in the combination of organic reactions and separation processes. Perfluorinated compounds are generally immiscible with most organic solvents and are denser than typical organic compounds. Intrigued by these unique properties of fluorous solvents, the groups of Ryu and Curran[34,35] have developed a synthetically convenient triphasic system comprising organic-phase-containing substrates, a fluorous phase, and a reagent phase (Scheme 2.28). In such a triphasic system, the fluorous layer functions as a liquid membrane bringing the two separate layers together. Thus, the reagent in the bottom layer diffuses slowly through the

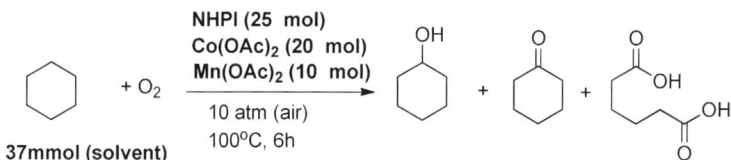

		OH	O	O OH / OH O	
F-NHPI	AcOH	2536%	408%	5120	TON = 82.6
F-NHPI	CF$_3$C$_6$H$_5$ (BTF)	1772	1504	36	33.1

Scheme 2.26   High selectivity for the oxidation of cyclohexane in BTF.

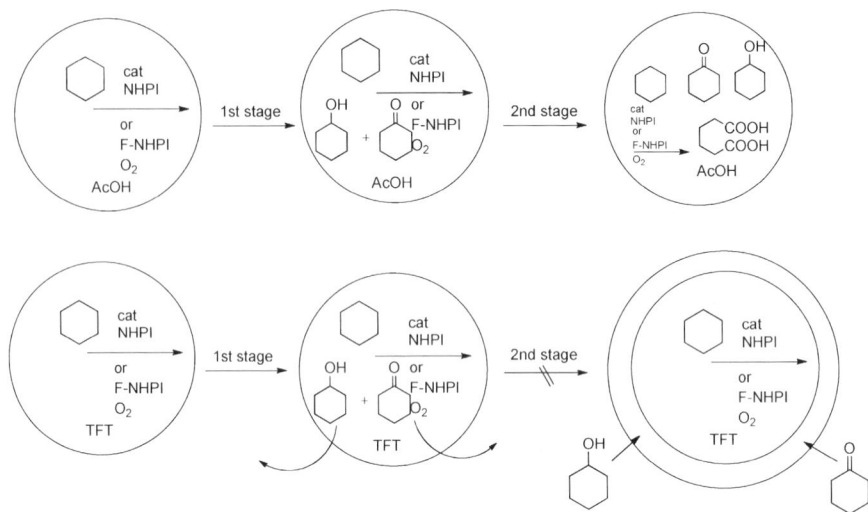

Scheme 2.27 Mechanistic aspects of oxidation and illustration of the selectivity observed for the aerobic oxidation of cyclohexane in AcOH and THF.

fluorous layer in the top layer. It encounters the substrate in the upper layer, then the reaction takes place. Eventually, the reagent phase disappears, leaving only two phases.

As a fluorous layer, typically FC-72 (perfluorohexane) is used for the triphasic phase-vanishing (PV) reaction. The more viscous and less volatile perfluorodecalin can also be used, but diffusion is slower. Recent studies reveal that less volatile and inexpensive perfluorinated polyether solvents, such as Galden HT-135, function as excellent fluorous phases in many cases. Inherent to the nature of very slow reagent addition like a "micro" syringe pump, control of heat evolution is also possible with thermal diffusion. This allows for the test-tube reaction to be performed without any cooling equipment. In many examples of exothermic reactions, the disappearance of the bottom

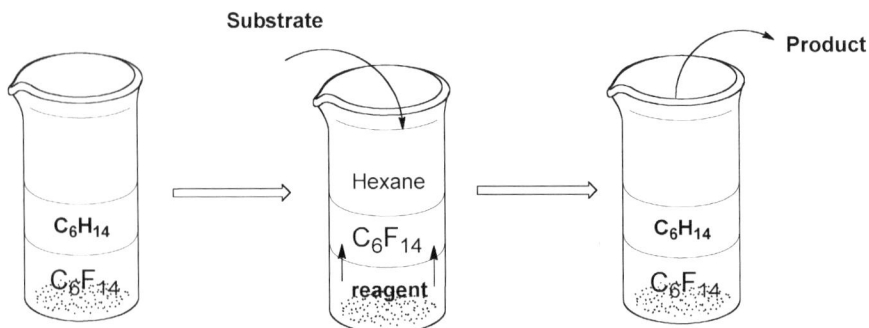

Scheme 2.28 General representation of the phase-vanishing method.

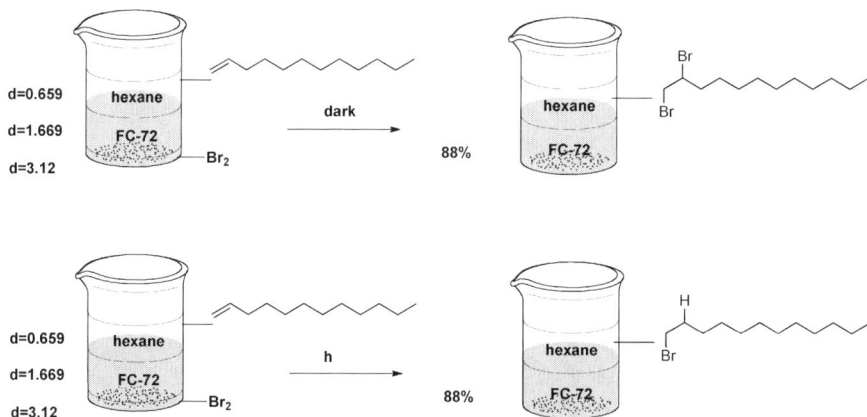

Scheme 2.29 Phase-vanishing bromination of alkenes.

phase is a useful sign of the end of the addition. The change from three layers to two layers is the origin of the name 'phase-vanishing (PV) method', and therefore, the concept of the PV method is not restricted to the fluorous phase. However, the superiority of fluorous media to water and some ionic liquids, which can also be used to constitute triphasic systems, has been demonstrated in many cases. Since the first report of Ryu and Curran in 2002, the PV method has demonstrated its utility in the bromination of alkenes under various conditions and is summarized in Scheme 2.29.

2.13   CONCLUSIONS AND FUTURE DIRECTIONS IN FLUOROUS CHEMISTRY

Over the past 15 years, fluorous chemistry has emerged as a real and viable environmentally attractive alternative to traditional reagents and catalytic systems. This section has attempted to highlight that these are no longer reasonable objections to raise: the costs of fluorous reagents and solvents are falling all the time and there are now specialist manufactures dealing solely with the production of fluorous reagents and compounds. Furthermore, large quantities of fluorous solvents are no longer required to effect reactions through the biphase system: there are light fluorous alternatives and, in addition, liquid–liquid extraction is a rapid method for the separation of both fluorous/organic and fluorous/fluorous compounds, without the need for fluorous solvents.

2.14   GENERAL CONCLUSION

The last 20 years have also seen the emergence of other so-called clean alternatives for the synthesis and catalysis, such as supercritical fluids and ionic liquids. It is becoming increasingly clear that no single system will, in its own right, ever be able to replace completely all conventional reagents and solvents as a truly environmentally friendly alternative. There are drawbacks associated with all three systems, but each method does, however, have its niche and role in the market and application. The great number of crossover applications continuously emerging means it is only matter of time until the use of fluorous technology, supercritical solvents and ionic liquids and aqueous media in synthesis will be widely adopted in chemical community.

REFERENCES

1. Selected reviews: (*a*) C.-J. Li, Organic reactions in aqueous media with a focus on carbon-carbon bond formations: a Decade update, *Chem. Rev.*, 2005, **105**, 3095–3166; (*b*) R. A. Sheldon, Green solvents for sustainable organic synthesis: state of the art, *Green Chem.*, 2005, **7**, 267–278; (*c*) C.-J. Li and L. Chen, Organic chemistry in water, *Chem. Soc. Rev.*, 2006, **35**, 68–82; (*d*) V. T. Perchyonok, I. N. Lykakis and K. L. Tuck, Recent advances in C–H bond formation in aqueous media: a mechanistic perspective, *Green Chem.*, 2008, **10**, 153–163; (*e*) V. T. Perchyonok and I. N. Lykakis, Recent advances in free radical chemistry of C–C bond formation in aqueous media: from mechanistic origins to applications, *Mini-Rev. Org. Chem.*, 2008, **5**, 19–32.
2. Microwave chemistry and radicals, review articles: A. Studer, Tin-free radical chemistry using the persistent radical effect: alkoxyamine isomerization, addition reactions and polymerizations, *Chem. Soc. Rev.*, 2004, **33**, 267–273 and references cited therein.
3. Ionic liquids—general reviews: R. D. Rogers and K. R. Seddon, Ionic liquids–solvents of the future? *Nature*, 2003, **302**(*5646*), 792 and references cited therein.
4. (*a*) S. Zhang, J. Chen, I. N. Lykakis and V. T. Perchyonok, Streamlining organic free radical synthesis through modern molecular technology: from polymer supported synthesis to microreactors and beyond, *Curr. Org. Synth.*, 2010, **7**, 177–188; (*b*) A. E. Johnson and V. T. Perchyonok, Recent advances in free radical chemistry in unconventional medium: ionic liquids, microwaves and solid state to the rescue, review article, *Curr. Org. Chem.*, 2009, **13**, 1705–1725.
5. For an excellent general reference book on free-radical chemistry see: B. Giese, *Radicals in Organic Synthesis: Formation of Carbon–Carbon Bonds*, Pergamon Press, Oxford, UK, 1986, and references cited therein.
6. V. T. Perchyonok, *Radical reactions in aqueous media*, RSC Green Chemistry Series, RSC Publishing, Cambridge, 1st edn, 2009.
7. V. T. Perchyonok and I. N. Lykakis, Radical reactions in aqueous media: origins, reason and applications, *Curr. Org. Chem.*, 2009, **13**, 573–598.
8. C. Chatgilialoglu, in *Organosilanes in Radical Chemistry*, John Wiley & Sons, Ltd, West Sussex, UK, 2004, and references cited therein.

9. S. Z. Zard, in *Radical Reactions in Organic Synthesis*, ed. R. G. Compton, S. G. Davies and J. Evans, Science, 2003, Oxford University Press, Oxford, UK.

10. (*a*) *Radicals in Organic Synthesis*, ed. P. Renaud and M. P. Sibi, Wiley-VCH, Weinheim, 2001, vol. 1; (*b*) Al. Postigo, C. Ferreri, M. L. Navacchia and C. Chatgilialoglu, The radical-based reduction with $(TMS)_3SiH$ 'on water', *Synlett*, 2005, **18**, 2854–2856; (*c*) Al. Postigo, S. Kopsov, C. Ferreri and C. Chatgilialoglu, Radical reactions in aqueous medium using $(Me_3Si)_3SiH$, *Org. Lett.*, 2007, **9**, 5159–5162; (*d*) C. Chatgilialoglu, M. Guerra and Q. G. Mulazzani, Model studies of DNA C5' radicals. Selective generation and reactivity of 2'-deoxyadenosin-5'-yl radical, *J. Am. Chem. Soc.*, 2003, **125**, 3839–3848; (*e*) L. B. Jimenez, S. Encinas, M. A. Miranda, M. L. Navacchia and C. Chatgilialoglu, The photochemistry of 8-bromo-2'-deoxyadenosine. A direct entry to cyclopurine lesions, *Photochem. Photobiol. Sci.*, 2004, **3**, 1042–1051; (*f*) C. Chatgilialoglu and C. Ferreri, Trans lipids: the free radical path, *Acc. Chem. Res.*, 2005, **38**, 441–448; (*g*) H. Sugiyama and Y. Xu, Photochemical Approach to Probing Different DNA Structures, *Angew. Chem., Int. Ed.*, 2006, **45**, 1354–1362.

11. Selected references on the use of organic hypophosphites as hydrogen donors in organic solvents: (*a*) A. F. Brigas and R. A. W. Johnstone, Mechanisms in heterogeneous liquid-phase catalytic-transfer reduction: the importance of hydrogen-donor concentration, *J. Chem. Soc., Chem. Commun.*, 1991, 1041–1042; (*b*) S. R. Graham, J. A. Murphy and A. R. Kennedy, Hypophosphite mediated carbon–carbon bond formation: total synthesis of epialboatrin and structural revision of alboatrin, *J. Chem. Soc., Perkin Trans. 1*, 1999, 3071–3073; (*c*) C. G. Martin, J. A. Murphy and C. R. Smith, Replacing tin in radical chemistry: *N*-ethylpiperidine hypophosphite in cyclisation reactions of aryl radicals, *Tetrahedron Lett.*, 2000, **41**, 1833–1836; (*d*) V. T. Perchyonok, K. L. Tuck, S. J. Langford and M. W. Hearn, On the scope of radical reactions in aqueous media utilizing quaternary ammonium salts of phosphinic acids as chiral and achiral hydrogen donors, *Tetrahedron Lett.*, 2008, **49**, 4777–4779.

12. D. H. Cho and D. O. Jang, Enantioselective radical addition reactions to the C=N bond utilizing chiral quaternary ammonium salts of hypophosphorous acid in aqueous media, *Chem. Commun.*, 2006, **9**, 5045–5047.

13. (a) U. M. Lindstrom, Stereoselective Organic Reactions in Water, *Chem. Rev.*, 2002, **102**, 2751–2772; (*b*) E. N. Jacobsen and M. S. Taylor, Asymmetric Catalysis by Chiral Hydrogen-Bond Donors, *Angew. Chem., Int. Ed.*, 2006, **45**, 1520–1543.

14. Selected publications for enantioselectivity in free radical transformations in the organic solvents: (a) D. Dakternieks, V. T. Perchyonok and C. H. Schiesser, Single enantiomer free-radical chemistry-Lewis acid-mediated reductions of racemic halides using chiral non-racemic stannanes, *Tetrahedron: Asymmetry*, 2003, **14**, 3057–3068; (b) M. P. Sibi and

K. Patil, Enantioselective H-Atom Transfer Reactions: A New Methodology for the Synthesis of β-Amino Acids, *Angew. Chem., Int. Ed.*, 2004, **43**, 1235–1238 and references cited therein.

15. Selected publications on the application of radical cascade reactions in total synthesis: (a) A. G. Myers and K. R. Condronski, Synthesis of (±)-7,8-Epoxy-4-basmen-6-one by a transannular cyclization strategy, *J. Am. Chem. Soc.*, 1993, **115**, 7926–7927; (b) A. G. Myers and K. R. Condronski, Synthesis of (±)-7,8-Epoxy-4-basmen-6-one by a transannular cyclization strategy, *J. Am. Chem. Soc.*, 1995, **117**, 3057–3083.

16. (a) A. F. Parson, *An Introduction to Free Radical Chemistry*, Blackwell Science; (b) *Supercritical Carbon Dioxide: Separation and Processes*, ACS Symp. Ser. 860, ed. A. S. Gopalan, C. M. Wai and H. K. Jacobs, and references cited therein.

17. B. Fletcher, N. K. Suleman and J. M. Tanko, Free radical chlorination of alkanes in supercritical carbon dioxide: The chlorine atom cage effect as a probe for enhanced cage effects in supercritical fluid solvents, *J. Am. Chem. Soc.*, 1998, **120**(*46*), 11839–11844.

18. For an interesting example of the use of supercritical CO_2, see: (a) S. Hadida, M. S. Super, E. J. Beckman and D. P. Curran, Radical reactions with alkyl and fluoroalkyl (fluorous) tin hydride reagents in supercritical CO_2, *J. Am. Chem. Soc.*, 1997, **119**, 7406–7407; (b) J. M. Tanko, Free-Radical Chemistry in Supercritical Carbon Dioxide, in *Green Chemistry using Liquid and Supercritical Carbon Dioxide*, ed. J. M. DeSimone and W. Tumas, Oxford University Press, New York, 2003, ch. 4, p. 64.

19. Y. Kishimoto and T. Ikariya, Supercritical Carbon dioxide as a reaction medium for silane-mediated free-radical carbonylation of alkyl halides, *J. Org. Chem.*, 2000, **65**, 7656–7659.

20. (a) *Ionic Liquids in Synthesis*, ed. P. Wasserscheid and T. Welton, Wiley-VCH, Weinheim, 2008; (b) V. I. Pârvulesku and C. Hardacre, Catalysis in ionic liquids, *Chem. Rev.*, 2007, **107**, 2615–2665; (c) J. B. Harper and M. N. Kobrak, Understanding organic processes in ionic liquids: achievements so far and challenges remaining, *Mini-Rev. Org. Chem.*, 2006, **3**, 253–269; (d) C. Chiappe, M. Malvaldi and C. S. Pomelli, Ionic liquids: solvation ability and polarity, *Pure Appl. Chem.*, 2009, **81**, 767–776.

21. H. Yorimitsu, T. Nakamura, H. Shinokubo, K. Oshima, K. Omoto and H. Fujimoto, Powerful solvent effect of water in radical reaction: triethylborane-induced atom-transfer radical cyclization in water, *J. Am. Chem. Soc.*, 2000, **122**, 11041–11047.

22. For some recent examples of 'green' radical reagents, see the references reported in: (a) J. C. Walton, Linking borane with *N*-heterocyclic carbenes: effective hydrogen-atom donors for radical reactions, *Angew. Chem., Int. Ed.*, 2009, **48**, 1726–1728; (b) G. Bencivenni, T. Lanza, R. Leardini, M. Minozzi, D. Nanni, P. Spagnolo and G. Zanardi, Tin-free generation of alkyl radicals from alkyl 4-pentynyl sulfides via homolytic

substitution at the sulfur atom, *Org. Lett.*, 2008, **10**, 1127–1130; (c) L. Benati, G. Bencivenni, R. Leardini, M. Minozzi, D. Nanni, R. Scialpi, P. Spagnolo and G. Zanardi, Reaction of azides with dichloroindium hydride: very mild production of amines and pyrrolidin-2-imines through possible indium-aminyl radicals, *Org. Lett.*, 2006, **8**, 2499–2502; (d) M. P. Healy, A. F. Parsons and J. G. T. Rawlinson, Consecutive approach to alkenes that combines radical addition of phosphorus hydrides with Horner−Wadsworth−Emmons-type reactions, *Org. Lett.*, 2005, **7**, 1597–1600; (e) L. Benati, R. Leardini, M. Minozzi, D. Nanni, R. Scialpi, P. Spagnolo, S. Strazzari and G. Zanardi, A novel tin-free procedure for alkyl radical reactions, *Angew. Chem., Int. Ed.*, 2004, **43**, 3598–3601; (f) A. Studer and S. Amrein, Tin hydride substitutes in reductive radical chain reactions, *Synthesis*, 2002, 835. See also: H. Togo, *Advanced Free Radical Reactions for Organic Synthesis*, Elsevier, Amsterdam, 2004, ch. 12, p. 247.

23. X.-Q. Pan, J.-P. Zou and W. Zhang, Manganese(III)-promoted reactions for formation of carbon–heteroatom bonds, *Mol. Diversity*, 2009, **13**, 421–438 and references cited therein.

24. (a) G. Bar, A. F. Parsons and C. B. Thomas, A radical approach to araliopsine and related quinoline alkaloids using manganese(III) acetate. *Tetrahedron Lett.*, 2000, **41**, 7751–7755; (b) G. Bar, A. F. Parsons and C. B. Thomas, Manganese(III) acetate mediated radical reactions leading to araliopsine and related quinoline alkaloids, *Tetrahedron*, 2001, **57**, 4719–4728; (c) G. Bar, A. F. Parsons and C. B. Thomas, Manganese(III) acetate mediated radical reactions in the presence of an ionic liquid, *Chem. Commun.*, 2001, 1350–1351.

25. (a) V. Nair, J. Mathew and J. Prabhakaran, Carbon-carbon bond forming reactions mediated by cerium(IV) reagents, *Chem. Soc. Rev.*, 1997, 127–132; (b) E. Baciocchi, A. B. Paolobelli and R. Ruzziconi, Cerium(IV) ammonium nitrate promoted oxidative cyclization of dimethyl 4-pentenylmalonate, *Tetrahedron*, 1992, **48**, 4617–4622; (c) A. D'Annibale, A. Pesce, S. Resta and C. Trogolo, Ceric ammonium nitrate promoted free radical cyclization reactions leading to β-lactams, *Tetrahedron Lett.*, 1997, **38**, 1829–1832.

26. G. Bar, F. Bini and A. F. Parsons, CAN-Mediated Oxidative Free Radical Reactions in an Ionic Liquid, *Synth. Commun.*, 2003, **33**, 213–222.

27. B. C. Ranu, L. Adak and S. Banerjee, Halogenation of Carbonyl Compounds by an Ionic Liquids, [AcMIm]X and Ceric Ammonium Nitrate (CAN), *Aust. J. Chem.*, 2007, **60**, 358–362.

28. A. Riisager, R. Fehrmann, M. Haumann and P. Wasserscheid, Supported Ionic Liquids: Versatile Reactions and Separation Media, *Top. Catal.*, 2006, **40**(*1–4*), 91–102.

29. I. T. Horvath and J. Rabai, Facile catalyst separation without water: fluorous biphase hydroformylation of olefins, *Science,* 1994, **266**, 72–75.

30. P. L. Nostro, Phase separation properties of fluorocarbon, hydrocarbons and their copolymers, *Adv. Colloid Interface Sci.,* 1995, **56**, 245–287.

31. L. P. Barthel-Rosa and J. A. Gladysz, Chemistry in Fluorous media: a user's guide to practical consideration in the application of fluorous catalyst and reagents, *Coord. Chem. Rev.,* 1999, **190–192**, 587–605 and references cited therein.

32. D. Clark, M. A. Ali, A. A. Clifford, A. Parratt, P. Rose, D. Schwinn, W. Bannwarth and C. M. Rayner, Reactions in Unusual Media, *Curr. Top. Med. Chem.,* 2004, **4**, 729–771.

33. M. S. Kharasch and F. R. Mayo, The peroxide effect in the addition of reagents to unsaturated compounds. The addition of hydrogen bromide to allyl bromide, *J. Am. Chem. Soc.,* 1933, **55**, 2468–2496.

34. D. P. Curran and S. Hadida, Tris(2-(perfluorohexyl)ethyl)tin Hydride: A new fluorous reagent for use in traditional organic synthesis and liquid phase combinatorial synthesis, *J. Am. Chem. Soc.*, 1996, **118**, 2531–2532.

35. I. Ryu, New approaches in radical carbonylation chemistry: Fluorous applications and designed tandem processes by species-hybridization with anion and transition metal species, *Chem. Rec.*, 2002, **2**, 249–258.

36. V. Gupta and D. Kahne, Direct introduction of CH_2OH by intermolecular trapping of CO, *Tetrahedron Lett.,* 1993, **34**, 591–594.

37. H. Matsubara, S. Yasuda, H. Sugiyama, I. Ryu, Y. Fujii and K. Kita, A new fluorous/organic amphilic ether solvent, F-626: execution of fluorous and high temperature classical reactions with convenient biphase workup to separate product from high boiling solvents, *Tetrahedron*, 2002, **58**, 4071–4076.

38. I. Ryu, T. Niguma, S. Minakata, M. Komatsu, Z. Luo and D. P. Curran, Radical carbonylation with fluorous allyltin reagents, *Tetrahedron Lett.,* 1999, **40**, 2367–2370.

39. S. Tsunoi, I. Ryu, H. Fukushima, M. Tanaka, M. Komatsu and N. Sonoda, Free-Radical Carbonylation Using a Zn(Cu) Induced Reduction System, *Synlett,* 1995, 1249–1251.

40. S. Tsunoi, I. Ryu, S. Yamasaki, M. Tanaka, N. Sonoda and M. Komatsu, Tandem annulations: a one operation construction of bicyclo[3.3.0]octan-1-ol and bicyclo[3,2,1]octan-1-ol skeleton by a three-component coupling reaction of alk-4-enyl iodides with CO and alkenes in the presence of zinc, *Chem. Commun.*, 1997, 1889–1890.

41. E. G. Hope and A. M. Stuart, Fluorous Biphase Catalyst, *J. Fluorine Chem.*, 1999, **100**(*1–2*), 75–83.

42. A. Ogawa, Y. Sumino, T. Nanke, S. Ohya, N. Sonoda and T. Hirao, *J. Am. Chem. Soc.*, 1997, **119**, 2745–2746.

43. Y. Ishii and T. Yanagida, Single molecule detection in life science, *Single Mol.*, 2000, **1**, 5–16.

44. S. K. Guha, Y. Obora, D. Ishihara, H. Matsubara, I. Ryu and Y. Ishii, Aerobic oxidation of cyclohexane using N-hydroxyphthalimide bearing fluoroalkyl chains, *Adv. Synth. Catal.,* 2008, **350**, 1323–1330.

CHAPTER 3

Solvent-Free Carbon–Carbon Bond Formations in Ball Mills and in the Solid State[*]

3.1 INTRODUCTION

The simple use of a mortar and a pestle to mix reagents can be useful in order to perform organic reactions under solvent-free conditions.[1,2] However, for a more efficient and automated mixing more sophisticated instrumentation has been developed. In that context, ball mills appear to be interesting.[3] In inorganic chemistry and material sciences, ball milling is a well-established technique for the grinding of minerals and the preparation and modification of inorganic solids,[4] at both laboratory and industrial scales.[5] It allows the preparation of nanostructured alloys and the synthesis of new materials.[3a,6]

Various types of ball mills are known including (drum) ball mills, jet-mills, bead-mills, vibration ball mills, planetary ball mills and horizontal rotary ball mills.[3] All of these devices are based on the principle that a starting material is placed between two surfaces and crushed due to impact and/or frictional forces that are caused by collisions between these surfaces. The various mills differ in terms of how the motion causing these collisions is created. Besides the intensive grinding effect, the collisions lead to an energy transfer which results in an increase of internal temperature and pressure. For achieving better control of these factors, some ball mills have cooling/ heating devices. In general, ball mills are able to produce materials with a particle size of <100 nm. Although in laboratories planetary and vibration ball mills are most commonly used, horizontal rotary ball mills are often found in industrial applications.

As all C–C bond formations that are discussed in this chapter have been performed in one of these three types of ball mill, they will be briefly described.

---

* Chapter written by V. Tamara Perchyonok.
Streamlining Free Radical Green Chemistry
V. Tamara Perchyonok, Ioannis Lykakis and Al Postigo
© V. Tamara Perchyonok, Ioannis Lykakis and Al Postigo 2012
Published by the Royal Society of Chemistry, www.rsc.org

A planetary ball mill contains a main disk, which can rotate at a high rotational speed and accommodates one to eight grinding bowls. These bowls hold a number of balls as the grinding medium and rotate around their own axes in opposite directions relative to the main disc. The rotational speeds are of the order of 100–1000 rpm. Vibration ball mills contain only one or two grinding chambers which accommodate one or more grinding balls, and the chambers can be shaken at a frequency of 10–60 Hz in three orthogonal directions. Some vibration ball mills have cooling/heating systems, which allow temperature control while grinding.

Other terms found in the literature describing the horizontal rotary ball milling technique are high-speed ball milling (HSM), high-speed vibrational ball milling (HSVM), shaker milling or high-energy ball milling. Horizontal rotary ball mills have the advantage that they can be operated at a high relative velocity of the grinding medium (up to 14 m s^{-1}) that cannot be reached by other types (up to 5 m s^{-1}).[7] Furthermore, they allow the use of closed circuits and controlled conditions such as vacuum or inert gas. In this type of ball mill a horizontally arranged rotor inside the grinding vessel accelerates the grinding medium. Thus, an advantage of this device is that it only moves reactants and balls, but not the large masses of the containers as the other mills. These ball mills are especially interesting for industrial applications being available with a grinding chamber capacity of 0.5–400 L and allowing rotation frequencies of up to 1800 rpm as well as control of the reaction temperature by cooling or heating devices. In all three cases, the grinding medium and the vessels are usually made of chemically inert and non-abrasive material such as hardened steel, tungsten carbide, zirconia, *etc.*

In chemical synthesis, ball milling modifies the conditions under which a chemical reaction commonly takes place. Generally, there are two possibilities. Either the modification is accomplished by changing the reactivity of the reagents (mechanical activation) or by inducing chemical reactions via mechanical breaking of molecular bonds during milling (mechanochemistry).[3b,8] Since none of the C–C coupling reactions dealt with in this review is based on mechanochemical processes, we will exclusively discuss transformations induced by mechanical activation.

Changes in reactivity during ball milling (compared to conventional methods) can mainly be ascribed to the more efficient mixing and the enormous increase of the reagents' P surfaces (P surface is defined as mill circuit product size). Both lead to a close contact between the starting materials on a (almost) molecular scale, leading to a reactivity increase. However, other factors such as the increased temperature and pressure can also lead to a change of reactivity. While milling, some extreme conditions occur on the external surfaces of two colliding bodies only for times on the order of microseconds. Although the ball milling process is extremely complicated and not fully understood yet,[9] a model developed by Urakaev and Boldyrev[10] permits one to estimate the magnitude of the extreme shifts in temperature and pressure during this short time scale. Thus factors such as the rotational frequency of the mill, the milling time, the radius of the vessel, size, weight and number of balls, weight and heat capacity of the material, *etc.* have to be taken into account in order to estimate the energy transfer onto the material, the local, temporary, extreme conditions and the overall rise in temperature and pressure during ball milling. According to this model, local temperatures of 400–1500 K and local pressures of some thousand atmospheres can characterize typical extreme conditions in the ball mill.[10]

Especially in the field of material science, ball milling has gained importance as these short-time extreme effects allow the generation of metastable phases under macroscopically very mild conditions. Many examples are known where materials that usually require high temperatures (*e.g.* >800 °C) and pressures to be formed can be synthesized with much less additional heating under ball milling conditions. As a result of these local effects in the absence of temperature control, one can expect that the average temperature in a ball mill typically rises by approximately 40–60 °C. Without taking the described complex processes into account, ball mills are commonly operated by controlling the average temperature through the application of cooling devices and the local effects roughly through the choice of the rotational speed, time of milling and pause periods, material sizes, weight and number of balls, *etc.* When using a ball mill without temperature control, breaks (pause periods) during the milling process can help to keep the temperature at a reasonable level.

Ball milling in organic chemistry is less common, and generally it has been applied with the goal to affect highly efficient mixing of reagents under solvent- free conditions. This issue is particularly important in reactions between solids.[11,12] In a more detailed description, solvent-free reactions can be categorized into three classes: (1) reactions between solids (solid-state reactions), (2) reactions between solids with intermediate local melting, and (3) reactions with at least one liquid reagent. The first type benefits the most from performance under ball milling conditions, which relates to the mechanism of solid-state molecular reactions, as recently elucidated by atomic force microscopy (AFM), scanning near-field optical microscopy (SNOM) and grazing incidence diffraction (GID) studies.[11c,12,13] Thus, the successful progression of three different stages is a requirement for a solid-state reaction to proceed. First, the crystal lattice must allow for long-range anisotropic molecular migration to form crystal-correlated surface features in a phase rebuilding step.[14] Second, the product lattice must form with a reasonable rate in a phase transformation step. Third, the crystal disintegration step (with formation of the product crystal structure) must provide a fresh surface of the reacting crystals. Thus, solid–solid reactions require continuous fresh contact between the reacting solids which can more efficiently be achieved by milling than by grinding or sonicating (especially for large-scale reactions).[12] Importantly, the three-step phase rebuilding solid-state mechanism avoids a decrease of reactivity by solvation (as it occurs in solution or melts) and additionally benefits from the crystal packing that leads to a regular self-assembled alignment and approach of the reacting molecules.[5] Consequently, the kinetics of solid state reactions are favorable and melt reactions suffer more often from side reactions. It is also noteworthy that all types of solvent-free reactions take advantage of the high concentrations of the reagents.

The aim of this chapter is to give an overview of the application of the ball milling technique for the performance of C–C bond-forming reactions[15] and especially to demonstrate the potential of the ball milling methodology in situations where it modifies behaviour in terms of selectivity and reactivity.

3.2 RADICAL ADDITIONS TO IMINES MEDIATED BY MN(III)

Over the past decades, manganese(III)-mediated free radical reactions have been extensively explored. Since they exhibit remarkable advantages over traditional peroxide or light-initiated processes, they have found widespread applications in organic synthesis.[16] In 2004, the first manganese(III)-mediated radical addition to imines.[17] In organic chemistry, manganese(III) acetate [Mn(OAc)$_3$] has commonly been used in the generation of carbon-centered radicals from various carbonyl compounds and their oxidative addition to alkenes.

However, since Mn(OAc)$_3$ is poorly soluble in organic solvents, harsh reaction conditions together with a tedious separation procedure are normally required.[16,17,18] In order to establish efficient and milder reaction conditions, solid-state radical reactions of 1,3-cyclohexanedione with *in situ* generated imines mediated by manganese(III) acetate under mechanical milling conditions were reported by Wang and coworkers.[16] The novel radical addition reaction to imines mediated by Mn(OAc)$_3$·2H$_2$O under solid-state conditions was reported for the first time. The high efficiency and good to excellent yield, no separation of the *in situ* generated imines, no use of any solvents in carrying out the reaction and thus the facile addition treatment make this method a potential alternative to a conventional methodology (Scheme 3.1).

3.3 SOLID-PHASE HOMOLYTIC SUBSTITUTION IN ACTION

Bowman *et al.* reported a solid-phase intramolecular aromatic homolytic substitution of benzoimidazole precursors.[17,18] The approach involves attachment of radical precursor to the resin *via* the radical leaving groups in the hemolytic aromatic substitution. When the radical reaction is complete, the leaving group, unaltered starting material and reduced uncyclized products remain attached to the resin, which facilitates easy separation of the cyclized products. Novel application of focused microwave irradiation in solid-phase radical reactions is advantageous and drastically reduces reaction times. Tributylgermanium hydride (Bu$_3$GeH) has been used to replace the toxic and troublesome tributyl tin hydride (Bu$_3$SnH) in the radical reactions (Scheme 3.2).

Relatively little effort has been made to alter the properties of a polymer to influence reaction outcomes. Polymers have been mostly used as phase anchors

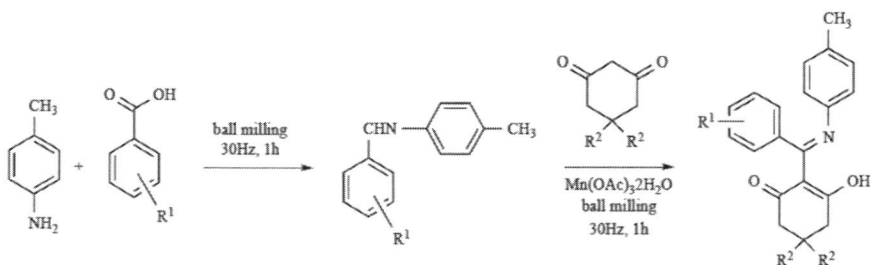

Proposed Mechanism:

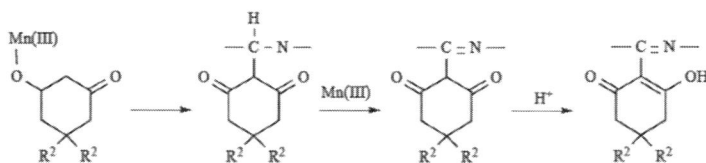

Scheme 3.1 Solid-state radical reactions of 1,3-cyclohexanedione and dimedone with *in situ* generated imine and proposed mechanistic pathway.

for a catalyst, reagent or ligands. In the future the influence of polymer solubility, polyvalency or other properties should also be utilized in the optimization and further development of the synthetically useful reactions and application.

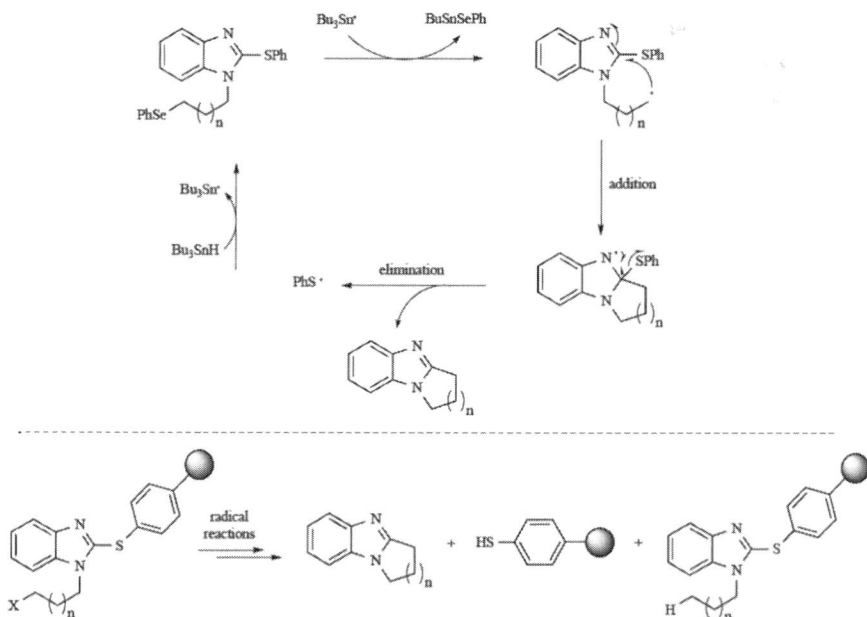

Scheme 3.2 Homolytic aromatic substitution and solid-phase radical protocol.

3.4 FUTURE DIRECTIONS

Ball milling has been applied to several classical free radical solvent-free carbon–carbon bond formations. In many cases, these transformations proved superior to the analogous reactions performed in solution. Ball milling is an environmentally benign versatile technique that can be performed at the kg scale and can be scaled up which gives it technical importance since environmentally friendly new processes are inevitable for the well-being of our environment.

REFERENCES

1. For some selected examples of solvent-free organic reactions performed by grinding the reagents with a mortar and pestle, see: (a) F. Toda, K. Tanaka and A. Sekikawa, *J. Chem. Soc., Chem. Commun.*, 1987, 279–280; (b) F. Toda, K. Kiyoshige and M. Yagi, *Angew. Chem.*, 1989, **101**, 329–330 (*Angew. Chem., Int. Ed. Engl.*, 1989, **28**, 320–321); (c) F. Toda, K. Tanaka and S. Iwata, *J. Org. Chem.*, 1989, **54**, 3007–3009; (d) K. Tanaka, S. Kishigami and F. Toda, *J. Org. Chem.*, 1991, **56**, 4333–4334; (e) D. Braga, L. Maini, M. Polito, L. Mirolo and F. Grepioni, *Chem.–Eur. J.*, 2003, **9**, 4362–4370; (f) E. Y. Cheung, S. J. Kitchin, K. D. M. Harris, Y. Imai, N. Tajima and R. Kuroda, *J. Am. Chem. Soc.*, 2003, **125**, 14658–14659; (g) L. Rong, X. Li, H. Wang, D. Shi, S. Tu and Q. Zhuang, *Synth. Commun.*, 2007, **37**, 183–189.

2. For overviews on solvent-free organic reactions, see: (a) G. W. V. Cave, C. L. Raston and J. L. Scott, *Chem. Commun.*, 2001, 2159–2169; (b) K. Tanaka, *Solvent-Free Organic Synthesis*, Wiley-VCH, Weinheim, 2003; (c) P. J. Walsh, H. Li and C. A. de Parrodi, *Chem. Rev.*, 2007, **107**, 2503–2545.

3. Reviews: (a) G. Kaupp, M. R. Naimi-Jamal, H. Ren and H. Zoz, in *Advanced Technologies Based on Self-Propagating and Mechanochemical Reactions for Environmental Protection*, ed. G. Cao, F. Delogu and R. Orrx, Research Signpost, Kerala, 2003, pp. 83–100; (b) E. Gaffet, F. Bernard, J.-C. Niepce, F. Charlot, C. Gras, G. Le Caer, J.-L. Guichard, P. Delcroix, A. Mocellin and O. Tillement, *J. Mater. Chem.*, 1999, **9**, 305–314; (c) S. Kipp, V. Šepelák and K. D. Becker, *Chem. Unserer Zeit*, 2005, **39**, 384–392.

4. (a) C. Suryanarayana, *Mechanical Alloying and Milling*, CRC Press, New York, 2004; (b) A. L. Garay, A. Pichon and S. L. James, *Chem. Soc. Rev.*, 2007, **36**, 846–855.

5. G. Kaupp, *Cryst. Eng. Commun.*, 2006, **8**, 794–804, and references cited therein.

6. For example, under ball milling conditions the normally immiscible metals iron and indium (or bismuth and copper) can react to give new alloys: N. Lyakhov, T. Grigorieva, A. Barinova, S. Lomayeva, E. Yelsukov and A. Ulyanov, *J. Mater. Sci.*, 2004, **39**, 5421–5423.

7. In the literature they are also called high-energy ball mills.

8. (a) In 1984, Heinicke defined the concept of mechanochemistry as follows: "Mechanochemistry is the branch of chemistry which studies the solid-state physical and chemical transformations, which are induced by the application of mechanical influence". G. Heinicke, *Tribochemistry*, Akademie Verlag, Berlin, 1984. For selected reviews on mechanochemistry, see: (b) J. F. Fernandez- Bertran, *Pure Appl. Chem.*, 1999, **71**, 581–586; (c) V. V. Boldyrev and K. Tkáčová, *J. Mater. Synth. Process.*, 2000, **8**, 121–132; (d) M. K. Beyer and H. Clausen-Schaumann, *Chem. Rev.*, 2005, **105**, 2921–2948.

9. For a macrokinetic approach of mechanical treatment by ball milling, see: F. Delogu, R. Orrx and G. Cao, *Chem. Eng. Sci.*, 2003, **58**, 815–821.

10. (a) F. K. Urakaev and V. V. Boldyrev, *Powder Technol.*, 2000, **107**, 93–107; (b) F. K. Urakaev and V. V. Boldyrev, *Powder Technol.*, 2000, **107**, 197–206; (c) for a qualitative summary of these calculations, see ref. 3*c*.

11. For organic solid state synthesis, see: (a) G. Kaupp, J. Schmeyers and J. Boy, *Chemosphere,* 2001, **43**, 55–61; (b) G. Rothenberg, A. P. Downie, C. L. Raston and J. L. Scott, *J. Am. Chem. Soc.*, 2001, **123**, 8701–8708; (c) G. Kaupp, *Cryst. Eng. Commun.*, 2003, **5**, 117–133.

12. For an overview of organic solid-state reactions, mostly with application of a ball mill, see: G. Kaupp, *Top. Curr. Chem.*, 2005, **254**, 95–183.

13. G. Kaupp, in *Making Crystals by Design*, ed. D. Braga and F. Grepioni, Wiley-VCH, Weinheim, 2006, ch. 2.

14. For an experimental proof that molecular crystals react to internal pressure, as imposed either chemically or mechanically, see: G. Kaupp, *Cryst. Eng. Commun.*, 2005, **7**, 402–410.

15. Some examples of non-C–C bond-forming reactions performed in a ball mill can be found in: (a) J. Schmeyers, F. Toda, J. Boy and G. Kaupp, *J. Chem. Soc., Perkin Trans. 2,* 1998, 989–993; (b) G. Kaupp, M. R. Naimi-Jamal and V. Stepanenko, *Chem.–Eur. J.,* 2003, **9**, 4156–4160; (c) S. A. Sikchi and P. G. Hultin, *J. Org. Chem.*, 2006, **71**, 5888–5891; (d) T. Rantanen, I. Schiffers and C. Bolm, *Org. Process Res. Dev.*, 2007, **11**, 592–997.

16. Z. Zhang, G.-W. Wang, C.-B. Miao, Y.-W. Dong and Y.-B. Shen, Solid-state radical reactions of 1,3-cyclohexanediones with *in situ* generated imines mediated by manganese(III) acetate under mechanical milling conditions, *Chem. Commun.*, **2004**, 1832–1834.

17. S. M. Allin, W. R. Bowman, R. Karim and S. S. Rahman, Aromatic homolytic substitution using solid phase synthesis, *Tetrahedron*, 2006, **62**, 4306–4316.

18. W. R. Bowman, S. L. Krintel and M. B. Schilling, *Synlett*, **2004**, 2004, 1215–1218.

Microwaves in Synthesis: How do Microwaves Promote the Reaction in Conventional and Alternative Media?[*]

4.1 INTRODUCTION

Microwave (MW)-enhanced chemistry is based on the efficiency of the interaction of molecules in a reaction mixture (substrates, catalyst and solvents) with electromagnetic waves generated by a "microwave dielectric effect".[1] This process mainly depends on the specific polarity of molecules. Since water is polar in nature, it has good potential to absorb microwaves and convert them to heat energy, thus accelerating the reactions in an aqueous medium as compared to results obtained using conventional heating. This can be explained by two key mechanisms: dipolar polarization and the ionic conduction of water molecules. Irradiation of a reaction mixture in an aqueous medium by MW results in the dipole orientation of water molecules and reactants in the electric field.

This causes two distinguishing effects:

- Specific microwave effect: the electrostatic polar effects which produce the dipole–dipole type interaction of the dipolar water molecules and reactants with the electric-field component of MW, resulting in energy stabilizations of an electrostatic nature. This concept of a specific MW (non-thermal) effect is controversial and the subject of debate among various chemists.
- Thermal effect: the dielectric heating that ensues from the tendency of dipoles (mostly water molecules in addition to reactants) to follow the inversion of alternating electric fields and induce energy dissipation in the form of heat through molecular friction and dielectric loss, which allows

* Chapter written by V. Tamara Perchyonok.
Streamlining Free Radical Green Chemistry
V. Tamara Perchyonok, Ioannis Lykakis and Al Postigo
© V. Tamara Perchyonok, Ioannis Lykakis and Al Postigo 2012
Published by the Royal Society of Chemistry, www.rsc.org

more regular repartition in reaction temperatures compared to conventional heating.

The demand for green and sustainable synthetic methods in the fields of healthcare and fine chemicals, combined with the pressure to produce these substances expeditiously and in an environmentally benign fashion, poses significant challenges to the synthetic chemical community. This objective can be achieved, in part, through the development of synthetic protocols using MW heating. However, the paucity of success in implementing these aqueous protocols in large-scale operations leaves the area wide open for a considerable amount of research.

4.2 MICROWAVE-ASSISTED FLUOROUS SYNTHESIS

As previously mentioned, organotin hydrides are popular reagents for free radical reactions. However, removal of their derivatives from the reaction mixtures is not always an easy task. Potential tin residues left in the final products limit the application of tin reagents in medicinal chemistry. One of the ways to address the issue is to use fluorous tin reagents. The Curran and Hallberg groups developed a heavy fluorous tin hydride reagent and used it under continuous microwave heating conditions for halide reduction and cyclization reactions (Scheme 4.1).[2] Reactions were conducted in benzotrifluoride (BTF) to increase the solubility of the tin reagent. Reaction mixtures were worked up by a three-phase extraction with FC-84, dichloromethane and water. The products were further purified by circular chromatography.

4.3 NITROXIDE-MEDIATED RADICAL CYCLIZATION AND INTRAMOLECULAR ADDITION REACTIONS IN MICROWAVES

4.3.1 The Persistent Radical Effect: General Introduction

In earlier parts of the review the tremendous advantages and achievements in the field of free radical chain reactions was highlighted, with the desired end product being predetermined in the propagation steps. The faster the propagation step, the greater the observed efficiency: less initiator is needed,

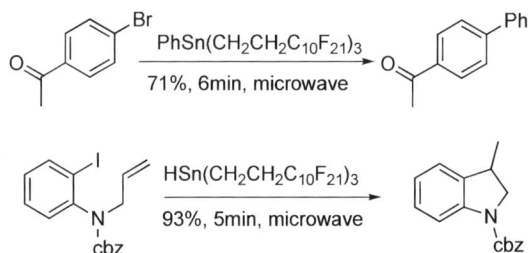

Scheme 4.1 Fluorous tin-promoted free radical reactions.

fewer unwanted side reactions can compete, and radical–radical interactions constituting the termination become negligible. In summary, the whole approach was aimed to reduce radical–radical interaction by keeping the steady-state concentration of the intermediate radical species as low as possible. Since the interactions are usually diffusion-controlled and therefore unselective, it might be seen (at first glance) as an impossible task to gain control of inter-radical reactions for synthetic purposes. However, this is not the case at all, due to the elegant and ingenious solution of this problem based on the persistent radical effect, also known as the Fischer–Ingold effect. This phenomenon, which is only recently being understood, underlies several reactions occurring in nature as well as novel synthetic applications recently discovered and elegantly applied in the special field of living radical polymerization.[3]

The Fischer–Ingold effect can be understood without the need for elaborate—but more rigorous—mathematical modelling. Let us consider a compound A which can be decomposed into 2 radicals $X^•$ and $Y^•$ which will then undergo recombination. If we neglected the eventual cage effect and assume that the rates of these recombinations are diffusion-controlled (and therefore comparable) then, statistically, from the three possible reactions shown by reactions (1)–(3) (see below), one would expect a yield of roughly 25% of each of the symmetrical dimers X–X and Y–Y, and a 50% yield of the cross product X–Y. There is therefore already some statistical selectivity in favour of the latter.

Now what will the effect on the selectivity be if $Y^•$ is a persistent radical, that is, if reaction (2) does not occur? We are now left with 2 reactions (1) and (3) as shown below and the first answer which comes to mind is that the selectivity in favour of X–Y would increase to around 66%, again on statistical grounds. In fact, the selectivity for the cross-product becomes almost total.

The reason is that we are not dealing with a hypothetical mathematical situation where all $X^•$ and $Y^•$ are generated instantly and react instantly.

In this real-life chemical transformation, compound A is being decomposed over a certain period of time (minutes, hours) which is vastly greater then the lifetime of the intermediate radicals (micro- to nanoseconds), whose concentration in the medium remains small throughout. Now, while the decomposition of A and the cross-coupling reaction (3) do not modify the relative concentration of $X^•$ and $Y^•$ (either one of each is produced or one of each is consumed), the formation of X–X in reaction (1) consumes, in contrast, only $X^•$ radicals.

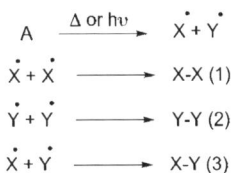

A $\xrightarrow{\Delta \text{ or } h\upsilon}$ $X^• + Y^•$	A $\xrightarrow{\Delta \text{ or } h\upsilon}$ $X^• + Y^•$
$X^• + X^• \longrightarrow$ X-X (1)	$X^• + X^• \longrightarrow$ X-X (1)
$Y^• + Y^• \longrightarrow$ Y-Y (2)	$X^• + Y^• \longrightarrow$ X-Y (3)
$X^• + Y^• \longrightarrow$ X-Y (3)	

Dimerization between 2 transient radicals   Dimerization between a transient and a persistent radical

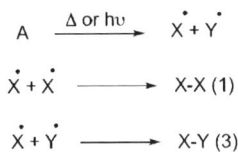

The consequence is that very soon after the radicals start being generated, the relative concentration becomes tilted greatly in favour of Y˙ radicals, even if the absolute, steady-state concentration of both X˙ and Y˙ remains small. Thus, every time a radical X˙ is created by the decomposition of A, its chances of capturing a radical Y˙ are much greater than capturing another X˙. Of course, Y˙ radicals, being by definition persistent, can only react with X˙. The formation of the cross-coupling product X–Y will therefore rapidly dominate and the selectivity, for all practical purposes, will be almost complete. The same result will qualitatively obtain if reaction (2) is (comparatively) slow or easily reversible under the experimental conditions.

About a decade ago, Studer *et al.* reported a first example of tin-free nitroxide-mediated radical cyclization reactions.[4] TEMPO-derived alkoximes were isomerized under thermal conditions to provide the corresponding 5-*exo* and 6-*endo* stylization products (Scheme 4.2). The products were obtained in moderate to good yields (54–63%).[5] The success of this transformation was fully guided by the persistent radical effect, with a specific emphasis on the stability of the C-centered radical formed upon thermal C–O bond photolysis. The isomerization worked well for alkoxyamines derived from stabilized radicals; however, for alkoxyamines leading to non-stable radicals, thermal isomerization fails. The C–O bond in these two alkoxyamines is too strong to be homolytically cleaved.

Studer has succeeded in performing thermal carboaminoxylations effectively in the microwave. For example, reaction of TEMPO-malonate with 1-octene (5 eq.) in DMF at 180 °C in a sealed tube was completed in 10 min to provide the corresponding addition product in 63% yield. Furthermore, alkoxyamine isomerization, which took 24 h applying classical heating, can be performed in 2.5 min under microwave conditions (Scheme 4.2).

More recently, Studer reported first results on microwave-induced hemolytic aromatic substitution followed by an ionic Horner–Wadsworth–Emmons (HWE) olefination for the preparation of α,β-unsaturated oxindoles (Scheme 4.3).[6]

Recently, the reactivity of phosphorous hydrides in radical additions reactions and have shown that the substituents on phosphorous are influencing the efficiencies of Horner–Wadsworth–Emmons type transformations under microwave irradiation conditions (Scheme 4.4).[7] Phosphorous hydrides with particularly weak P–H bonds are shown to undergo radical additions by microwave irradiation, in the absence of the conventional initiators. These

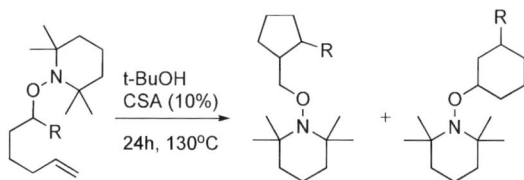

Scheme 4.2   Persistent radical effect mediated cascade reaction.

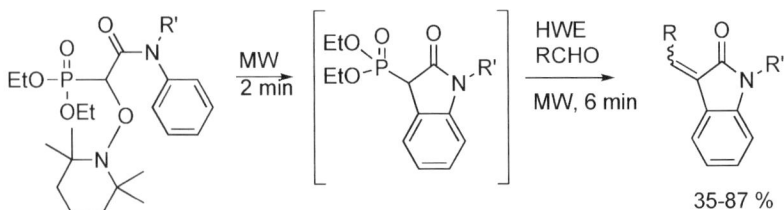

Scheme 4.3   Microwave-induced heating for the formation of oxindoles.

radical addition reactions produce phosphonothioates, phosphhinothioates and phosphane sulfides, which react in HWE-type reactions to afford substituted alkenes.

The scope and in-depth analysis of the reaction mechanism for the copper(II) carboxylate-promoted intramolecular carboamination of unactivated alkenes (Scheme 4.5).[8] The method provides access to N-functionalized pyrrolidines and piperidines. Both aromatic and aliphatic g- and s-alkenyl N-arylsulfona-mide undergo the oxidative cyclization reaction efficiently. The efficiency of the reaction was enchanced by the use of more organic soluble copper(II) salts, in particular copper(II) neodecanoate in particular. The reaction times were greatly reduced by the microwave heating. The reaction mechanism was established in the investigation and involves the N–C bond formation *via* intramolecular *syn* aminocupration and the C–C bond is formed *via* intramolecular addition of a primary carbon-centered radical to an aromatic ring.

4.4   RADICAL ADDITION TO C=N BONDS IN THE MICROWAVE

The microwave irradiation of 2-(aminoaryl)alkanone O-phenyl oximes and carbonyl compounds led to the generation of iminyl radicals, followed by imine ring closure to yield dihydroquinazolines or quinazolines when $ZnCl_2$ is included in the mixture (Scheme 4.6).[9] The advantage of the proposed methodology lies in the ability of the conducting imine to form condensation in the microwave leading to the possibility of a novel one-pot annulation reaction being performed in a green and efficient manner.

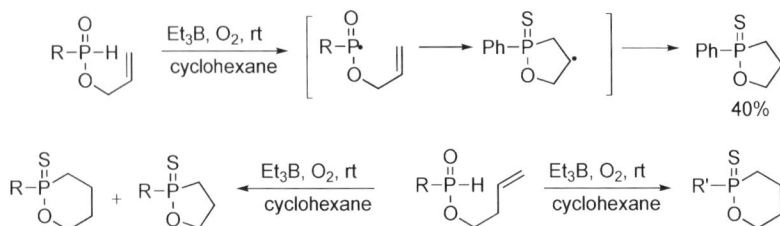

Scheme 4.4   Reaction of unsaturated phosphinothioates with triethylborane.

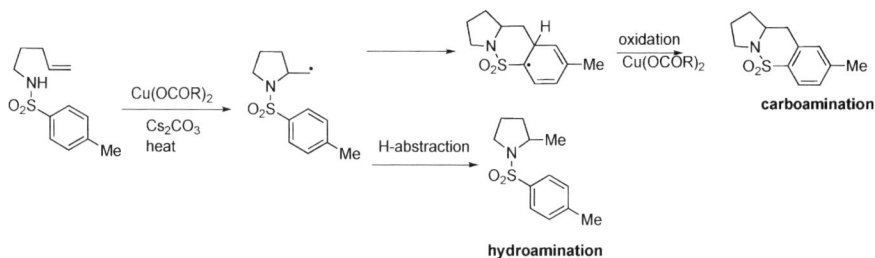

Scheme 4.5 Proposed pathway for formation of carboamination and hydroamination adducts.

The proposed mechanism is highlighted by the authors and involves coordination of the $ZnCl_2$ to the imine nitrogen prior to the cyclization and the subsequent ring closure which yields the amminium radical cation in a process similar to the iminium salt cyclization. The proton at position 2 of the heterocycle, adjacent to the radical cation, would have a considerably lower pK_a. Proton loss would then yield an intermediate in a process reminiscent of the Minisci reaction. The stabilized intermediate might transfer an electron to the starting oxime ether to give the corresponding cation or to convert to the final product on exposure to oxygen during work up. When the $ZnCl_2$ is not present, the C–H at position 2 is not acidic, and hence aromatization to a quinazoline does not proceed (Scheme 4.7).

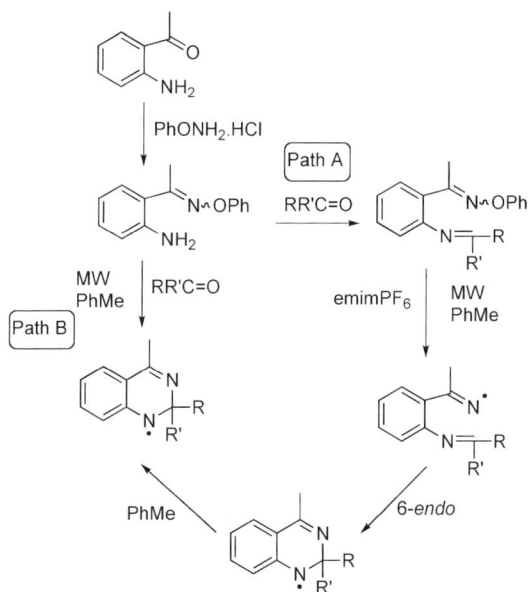

Scheme 4.6 Novel annulation of 2-aminoacetophenone O-phenyl oxime with carbonyl compounds.

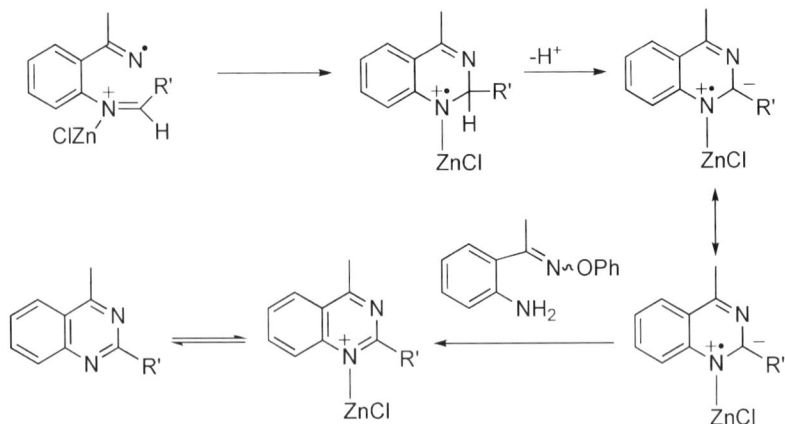

Scheme 4.7 Possible role of zinc chloride in quinazoline formation.

The overall process is a two-stage synthetic route from 2-aminoaryl ketone *via* O-phenyl oximes and then by a one-pot procedure with carbonyl compounds to dihydroquinazolines or quinazolines. The process is of wide scope and shown to be working well with alkyl, aryl and heterocyclic types of aldehydes. The reaction was shown to have several advantages over existing methods for quinazoline synthesis such as efficiency, neutral and mild reaction conditions and high yields.

A very important class of radical reactions is a Kharasch-type transformation, which involves the effective formation of carbon–carbon bonds. The reaction of interest is the Kharasch addition reaction of carbon tetrachloride to olefins which was developed through the involvement of a ruthenium catalyst (Scheme 4.8).[10] The procedure was shown to be straightforward and high yielding, even with some polymerizable substrates such as styrene and methyl methacrylate, an advantage over the conventional protocol.

Recently Kilburn reported another C–C bond-formation reaction with the use of two dialkylated isocyanides for the synthesis of indolizidines *via* a novel radical cyclization/N-alkylation/ring closing metathesis strategy in good to excellent yields under the condition of microwave irradiation (Scheme 4.9).[11] The results are exciting as they demonstrate for the first time that the simple isocyanides could be used as starting building blocks for the synthesis of functionalized and complex biologically relevant molecules with a high degree

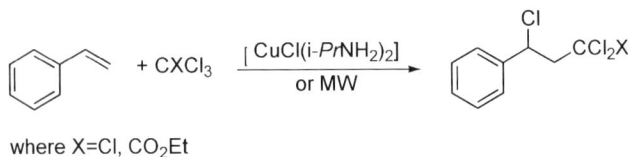

where X=Cl, CO$_2$Et

Scheme 4.8 Kharasch addition of carbon tetrachloride to olefins.

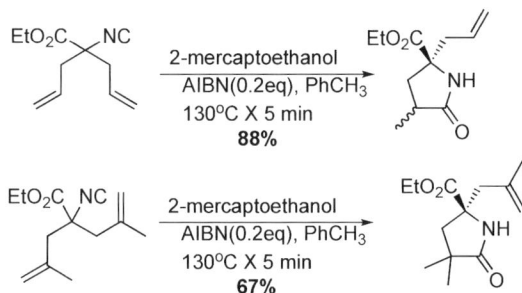

Scheme 4.9 Microwave-assisted radical cyclization of dialkylated isocyanides.

of stereospecificity under environmentally benign conditions such as micro-wave irradiations and tin-free radical chemistry.

4.5 MICROWAVE-ASSISTED GENERATION OF ALKOXYL RADICALS AND THEIR USE IN ADDITIONS, β-FRAGMENTATIONS AND REMOTE FUNCTIONALIZATION

Microwave irradiation of N-(alkoxy)thiazole-2(3H)-thiones in low-absorbing solvents afforded alkoxyl radicals. These radicals were identified by spin adduct formation, through EPR-spectroscopy and fingerprint-type selectivities in intramolecular additions (stereoselective synthesis of substituted tetrahy-drofurans), β-fragmentation/cleavage (formation of carbonyl compounds) and C, H-activation of aliphatic subunits, by d-selective hydrogen atom transfer.[12] The carbon-centered radicals formed from the oxygen-centered intermediates were trapped either by Bu$_3$SnH, L-cystein ethyl ester, the reduced form of glutathione (reductive trapping), or by the bromine atom donor BrCCl$_3$ (heteroatom functionalization) (Scheme 4.10).

The results suggested that microwave activation is superior to UV/Vis-photolysis and conductive heating for alkoxyl radicals generated from N-(alkoxy)thiazolethiones with significantly shorter reaction times as well

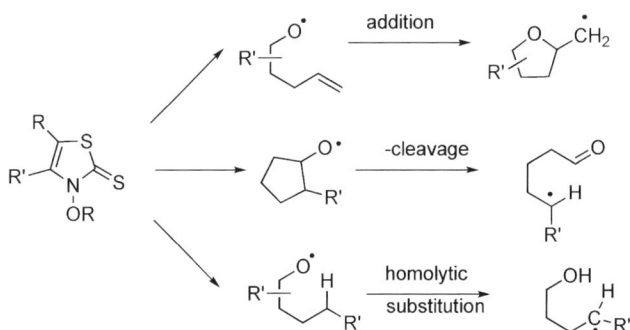

Scheme 4.10 The consequent alkoxyl radical chemistry upon microwave irradiation of N-(alkoxy)thiazole-2(3H)-thiones.

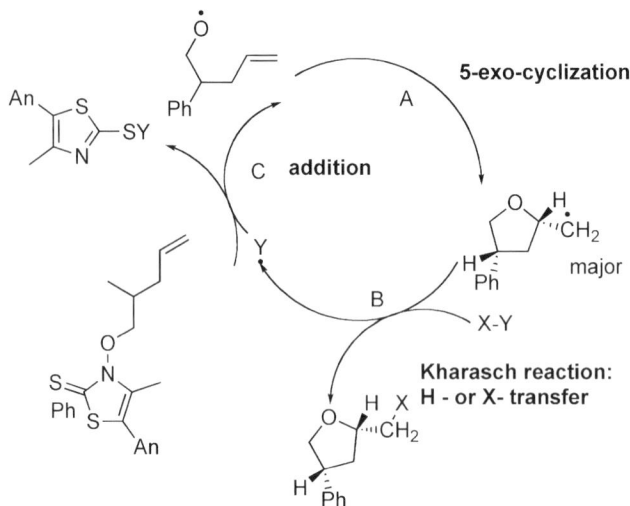

Scheme 4.11   Proposed mechanism for a transformation.

as an opportunity to significantly reduce the amount of trapping reagent. The proposed mechanism for the elementary reaction of N-(2-phenyl-4-pentenoxy)-thiazolethione through 5-*exo*-trig cyclization followed by hydrogen or bromine atom transfer respectively onto a cyclized radical, followed by the addition of a chain-propagating radical to a second molecule of thione, is presented in Scheme 4.11.

4.6   ATOM-TRANSFER REACTIONS AS EFFICIENT AND NOVEL BENZANNULATION REACTIONS IN THE MICROWAVE

Whereas the application of atom-transfer radical cyclization reactions (ATRC reactions) to the synthesis of g-butyrolactams and g-butyrolactones is now reasonably well established, the same cannot be said in the synthesis of medium-sized rings. Quayle *et al.* have developed a novel and efficient cyclization of tricloroacetates under ATRC conditions using the conventional and microwave irradiation conditions (Scheme 4.12).[12]

The reaction was thoroughly studied and generalized (by the application of a variety of readily available aryl trichloroacetates) to the predetermined experimental conditions. It was found that, in each case, benzannulation proceeded smoothly over 2 h at 200 °C to produce the respective naphthalene derivatives. It was confirmed that the copper catalyst appears to be intimately involved in the transformation; the reaction proceeds *via* the spirocyclic lactone, the product of 4-*exo* radical cyclization onto the aromatic ring. Once spirocyclization has taken place, presumably *via* the intermediacy of the radical, the lactone suffers a rapid loss of CO_2 to produce a corresponding vinyl chloride. Finally, double dehydrochlorination of vinyl chloride ultimately affords 1-chloronaphthalene.

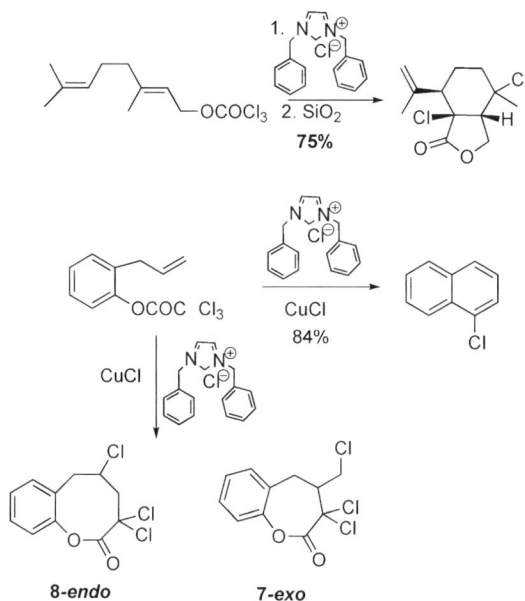

Scheme 4.12 Benzannulation sequences in microwaves.

4.7 CONCLUSION AND FUTURE DIRECTION

The fields of combinatorial and automated medicinal chemistry have been developed to meet the increasing requirement of new compounds for drug discovery; within these fields, speed is of the essence. The efficiency of microwave flash-heating chemistry in dramatically reducing reaction times (reduced from days and hours to minutes and seconds) has recently been proven in several different fields of organic chemistry. It is believed that the time saved by using focused microwaves is potentially important in traditional free radical organic synthesis but could be of even greater importance in high-speed combinatorial and medicinal chemistry.

REFERENCES

1. *Microwaves in Organic Synthesis*, ed. A. Loupy, Wiley, Weinheim, 2002; C. O. Kappe, *Angew. Chem., Int. Ed.*, 2004, **43**, 6250; C. O. Kappe and A. Stadler, *Microwaves in Organic and Medicinal Chemistry*, Wiley, Weinheim, 2005; P. Lidstrom, J. Tierney, B. Wathey and J. Wesman, *Tetrahedron*, 2001, **57**, 9225; B. L. Hayes, *Microwave Synthesis: Chemistry at the Speed of Light*, CEM Publishing, Matthews, NC, 2002.
2. D. P. Curran, Strategy-Level Separations in Organic Synthesis: From Planning to Practice, *Angew. Chem., Int. Ed. Engl.*, 1998, **37**, 1174; D. P. Curran and S. Hadida, Tris(2-(perfluorohexyl)ethyltin Hydride: A new

fluorous Reagent for Use in Traditional Organic Synthesis and Liquid Phase Combinatorial Synthesis, *J. Am. Chem. Soc.*, 1996, **118**, 2531; J. H. Horner, F. N. Martinez, M. Newcomb, S. Hadida and D. P. Curran, Rate constants for reactions of fluorous tin hydride reagent with primary alkyl radicals, *Tetrahedron Lett.*, 1997, **38**, 2783; K. Olofsson, S. Y. Kim, M. Larhed, D. P. Curran and A. Hallberg, High-speed, highly fluorous organic reactions, *J. Org. Chem.*, 1999, **64**, 4539.

3. H. Fischer, The Persistent Radical Effect: A Principle for Selective Radical Reactions and Living Radical Polymerization, *Chem. Rev.*, 2001, **101**, 3581.
4. A. Studer, The Persistent Radical Effect in Organic Synthesis, *Chem.–Eur. J.*, 2001, **7**, 1159.
5. A. Studer, Tin-Free Radical Cyclization Using the Persistent Radical Effect, *Angew. Chem., Int. Ed.*, 2000, **39**, 1108 and references cited therein.
6. C. M. Jessop, A. F. Parsons, A. Routledge and D. J. Irvine, Radical Additions of Phosphorous Hydrides: Tuning the reactivity of Phosphorous Hydride, the Use of Microwaves and Horner–Wadsworth–Emmons-Type Reactions, *Eur. J. Org. Chem.*, 2006, 1547–1554.
7. P. H. Fuller and S. R. Chemler, Copper(II) Carboxylate-Promoted Intramolecular Carboamination of Alkenes for the Synthesis of Polycyclic Lactams, *Org. Lett.*, 2007, **9**(*26*), 5477–5480.
8. F. Portela-Cubillo, J. S. Scott and J. C. Walton, Microwave-Assisted Syntheses of N-Heterocycles Using Alkenone-, Alkynone- and Aryl-carbonyl O-Phenyl Oximes: Formal Synthesis of Neocryptolepine, *J. Org. Chem.*, 2008, **73**(*14*), 5558–5565.
9. F. Nick, Y. Borguet, S. Delfosse, D. Bicchielli, L. Delaude, X. Sauvage and A. Demonceau, Microwave-Assisted Ruthenium-Catalyzed Reactions, *Aust. J. Chem.*, 2008, **62**(*3*), 184–207.
10. M. Lamberto, D. F. Corbet and J. D. Kilburn, Microwave assisted free radical cyclization of alkenyl and alkynyl isocyanides with thiols, *Tetrahedron Lett.*, 2003, **44**(*7*), 1347–1349.
11. J. Hartung, K. Daniel, T. Gottwald, A. Groß and N. Schneiders, Microwave-assisted generation of alkoxyl radicals and their use in additions, b-fragmentations and remote functionalization, *Org. Biomol. Chem.*, 2006, **4**, 2313–2322.
12. J. A. Bull, M. G. Hutchings and P. Quayle, A remarkable simple and efficient benzannulation reaction, *Angew. Chem., Int. Ed.*, 2007, **46**, 1869–1872.

CHAPTER 5

Asymmetric Free Radical Reductions Mediated by Chiral Stannanes, Germanes, and Silanes[*]

5.1 INTRODUCTION

Free radical technology had to come a long way before it gained its current well-deserved recognition among synthetic chemists. Traditionally, free radical chemistry has often been neglected for rational synthesis of fine chemicals because of the perception that radicals are uncontrollable and unselective. However, this negative view was revised with the adoption of tributyltin hydride in organic synthesis after it was recognized that reactions involving free radicals can indeed occur with a high degree of regioselectivity and diastereoselectivity. Nevertheless, the first examples of radical reactions that proceed with genuine enantiocontrol were reported only in the mid-1990s.[1-5]

The majority of applications that involve tributyltin hydrides are chain processes under reducing conditions and, from these, selective halogen abstractions from C–Hal bonds are by far the most prominent.[6-8] The general mechanism of such a reaction is outlined in Scheme 5.1. When the reaction has been initiated, the first step of the radical chain mechanism involves the abstraction of a halogen atom from the alkyl halide **A** by a triorganotin radical to give rise to a triorganotin halide and a planar (sp^2 configured) carbon radical **B**, which subsequently reacts with the triorganotin hydride producing the product **C** and a new triorganotin radical, which participates in a new cycle.

If the first reaction step affords a prochiral radical **B** then, in principle, the second step will produce a chiral product **C** (provided all three substituents of **B** are different and none are hydrogen) and the stereochemical outcome may be controlled by the stereochemistry of the organotin hydride. Optically

* Chapter written by V. Tamara Perchyonok.
Streamlining Free Radical Green Chemistry
V. Tamara Perchyonok, Ioannis Lykakis and Al Postigo
© V. Tamara Perchyonok, Ioannis Lykakis and Al Postigo 2012
Published by the Royal Society of Chemistry, www.rsc.org

117

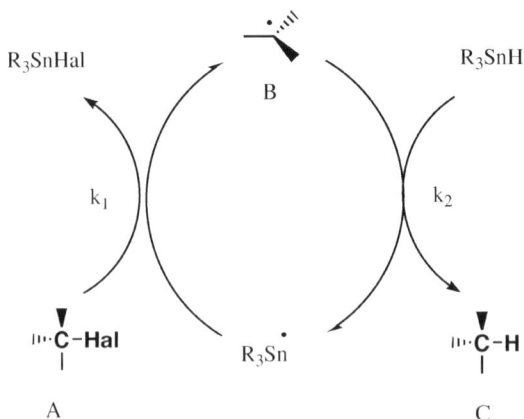

Scheme 5.1   General chiral stannane-mediated hydrogen transfer reaction.

inactive organotin hydrides (*e.g.*, Bu$_3$SnH) will attack the enantiotopic faces of the planar radical with equal probability, and the product will be racemic. However, when the organotin hydride is chiral, the transition states involved in the second reaction step are diastereotopic, and consequently the attack by the organotin hydride at the *Re*- and *Si*-faces of the planar radical will involve different activation energies (Figure 5.1). If the difference of the activation energies is sufficiently high and appropriate reaction conditions are applied (*e.g.*, low temperatures), effective discrimination of the enantiotopic faces of the planar radical by the chiral organotin hydride is possible and one enantiomer of **C** will be preferred over the other.

5.2   STOICHIOMETRIC FREE RADICAL REDUCTIONS

As a result of the limited configurational stability of optically active organotin compounds, in which the chirality is on the tin atom, most advances in enantioselective free radical reductions involve organostannanes where the elements of chirality are contained in the organic substituents. Selected early examples based on chiral 1,1′-binaphthyl-,[9,10] 2-[(1-dimethylaminoalkyl)phenyl]-,[11,12] and cholestannyl groups[13] are shown in Figure 5.2.

Feasibility studies of **1–4** in free radical reductions involving α-bromo ketones and esters demonstrated enantioselectivities of up to 51% ee. Although these results supply proof of concept, the aim to provide a widely applicable synthetic reagent required a significant improvement of ee values and access to less costly chiral-reducing agents. To address this problem, chiral organostannanes derived from menthol were developed.[14–24] Menthol, a naturally occurring terpene alcohol, provides a convenient source of chirality, where both enantiomers are inexpensive and readily available at the right scale. Up to three menthyl groups can be attached to a tin atom using the configurationally stable Grignard reagents derived from the corresponding menthyl chloride

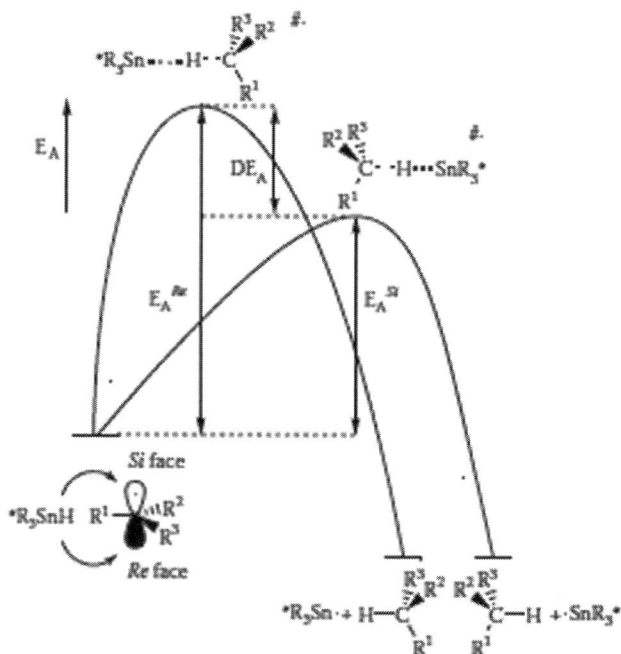

Figure 5.1 Diastereotopic transition states involved in the reaction of a prochiral carbon-centered radical with a chiral organotin hydride.

enantiomers.[25] The configurations of the respective menthyl groups are very similar to those of the parent alcohols, (−)- and (+)-menthol. This technology enables the cost-effective production of a variety of tailor-made menthyltin hydrides; a selection explored by Schiesser *et al.*[25] is depicted in Figure 5.3.

For α-bromoketones and related compounds, enhancement of the enantioselectivities has been achieved by the addition of simple Lewis acids (*e.g.*, BF3, Cp$_2$ZrCl$_2$, or magnesium salts, which presumably form Lewis acid–Lewis base complexes with the carbonyl functions of the substrates).[26,27] The stereochemical effect of some Lewis acid additives on the reduction of a racemic a-bromo esters, namely ethyl 2-bromo-2-phenylpropanoate, is summarized in Scheme 5.2.

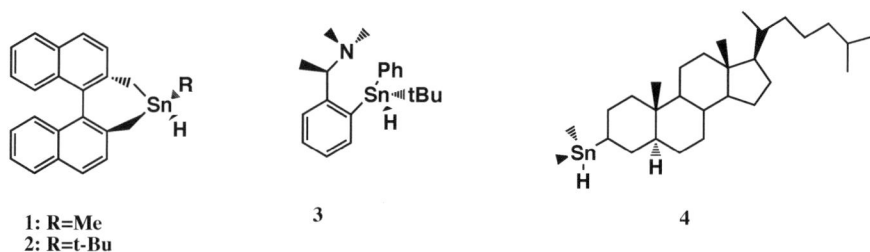

1: R=Me
2: R=t-Bu

3

4

Figure 5.2 Examples of C-centered optically active organotin hydrides.

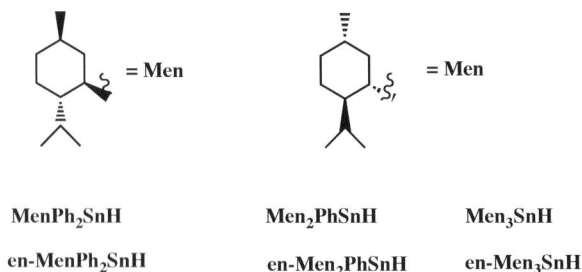

MenPh₂SnH

en-MenPh₂SnH

Men₂PhSnH

en-Men₂PhSnH

Men₃SnH

en-Men₃SnH

Figure 5.3   Examples of menthyltin hydrides evaluated by Schiesser *et al.* for enantioselective free radical reductions.

5.3   SCOPE AND LIMITATIONS

The versatility of organotin hydrides lies in their ability to reduce with a high degree of selectivity a variety of functional groups, including halides, sulfides, selenides, thionocarbonates, and dithiocarbonates; nitro groups; sulfoxide, sulfate, and sulfonyl moieties, and others.[6–8,28] Although the direct reduction of alcohols is not possible, appropriate derivatization to thiocarbonates or dithiocarbonates make them susceptible to radical transformations. These reactions are commonly referred to as Barton–McCombie reductions.[6–8,28] One of the most striking attributes of triorganotin hydrides, regardless of whether they are chiral, is their complementary nature to alternative reducing agents, such as NaBH₄, performing under polar reaction conditions. For instance, triorganotin hydrides selectively reduce halogen atoms in the presence of carbonyl functions or alcohol groups.[6–8,28] Triorganotin hydrides are generally applicable for sensitive molecules possessing a variety of functional groups that may decompose or epimerize under ionic reaction conditions. The use of triorganotin hydrides is particularly attractive where polar reactions suffer from steric hindrance that requires forcing reaction conditions. In these cases, free radical intermediates usually possess a greater accessibility to encumbered sites because they are neutral and only weakly solvated. Besides their application in radical reductions, triorganotin hydrides are also useful for initiating radical processes that involve selective additions to double or triple bonds (hydro-

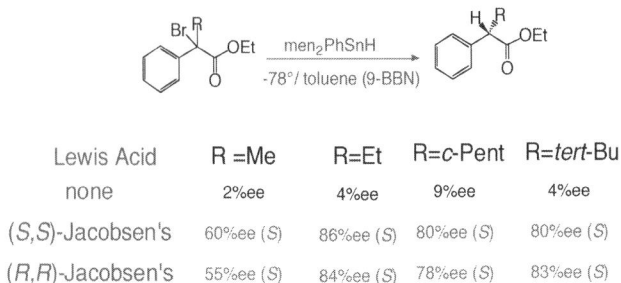

Lewis Acid	R = Me	R = Et	R = c-Pent	R = tert-Bu
none	2%ee	4%ee	9%ee	4%ee
(S,S)-Jacobsen's	60%ee (S)	86%ee (S)	80%ee (S)	80%ee (S)
(R,R)-Jacobsen's	55%ee (S)	84%ee (S)	78%ee (S)	83%ee (S)

Scheme 5.2   Lewis acid-mediated hydrogen transfer reactions.

stannylations) and proceed with the subsequent rearrangements of the radical intermediates. A variety of elegant tandem and cascade reactions have been reported that have been initiated by triorganotin hydrides.[28] Limitations to the applicability of triorganotin hydrides are encountered whenever functional groups susceptible to radical processes, such as nitro groups, may compete with the intended reaction or whenever radical inhibitors are present.

5.4 EXAMPLES RELEVANT TO THE FINE CHEMICAL INDUSTRY

Prime targets for free radical reductions are precursors possessing activated C–Hal bonds with a carbonyl functionality in the α-position. These cover a vast range of compound classes, such as α-halogenated ketones, esters, amino acids, α-hydroxy acids, *etc.* In general, these compounds are accessible to bromosuccinimide (NBS).[29] Thus, a two-step sequence involving the halogenation of a chiral (racemic) carbon atom situated in the α-position to a carbonyl function, followed by an enantioselective reaction, effectively provides a chemical deracemization alternative to classic resolution techniques. A selective free radical reduction produces the desired enantiomer exclusively. This illustrative example of a reaction sequence is the deracemization of methyl *N*-TFA phenylglycinate (Scheme 5.3)

Therefore, the methyl *N*-TFA phenylglycinate is brominated using NBS and subsequently reduced with either Men$_2$PhSnH (**6**) or *ent*-Men$_2$PhSnH (**6′**) in the presence of MgBr$_2$ to produce the same material, methyl *N*-TFA phenylglycinate, as either the *R*- or *S*-enantiomer with ee values greater than 99%.[29] The stereochemistry of the product can be easily selected by choosing the parity of the reducing agent (*e.g.*, **6** or **6′**). Compared to classic resolution technologies where the maximum yield of the desired enantiomer isolated from racemic mixture cannot exceed 50%, this chemical deracemization could provide yields close to 100%. Furthermore, this chemical deracemization can switch between the antipodes allowing the conversion of the (*S*)-enantiomer into the (*R*)-form and *vice versa*. In a similar way, both enantiomers of benzyl *N*-TFA *tert*-leucinate as well as ethyl esters of naproxen and ibuprofen, two non-steroidal anti-inflammatory drugs (NSAIDs), have been deracemized with enantioselectivities in excess of 96% ee (Figure 5.4).[29]

5.5 STRATEGIES FOR THE AVOIDANCE OF TIN WASTE

Despite the unique synthetic possibilities offered by organotin reagents, their use in the chemical industry has steadily declined in recent years as a result of

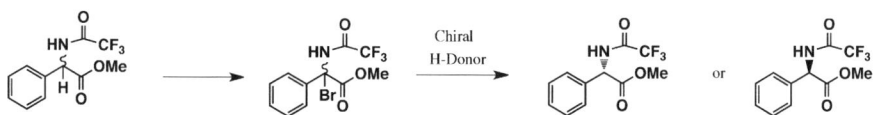

Scheme 5.3 Deracemization of methyl *N*-TFA-phenylglycinate.

Benzyl *N*-TFA *tert*-leucinate

NHTFA   Conditions: **6** or **6'** + MgBr₂

COOBz   (*R*) and (*S*)-form in 96 and 99% ee

Naproxen ethyl ester   Ibuprofen ethyl ester

COOEt   COOEt

MeO   *i*-Bu

Conditions: **6** or **6'** + MgBr₂   Conditions: **6** or **6'** + MgBr₂

(*R*) and (*S*)-form in ee's > 99%   (*R*) and (*S*)-form in 96% ee

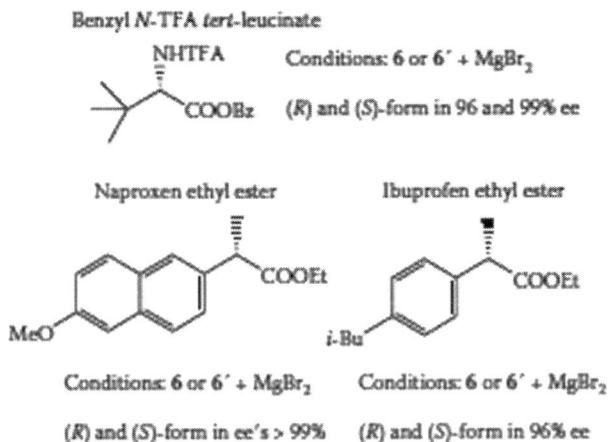

Figure 5.4 Deracemization of protected forms of *tert*-leucine, naproxen, and ibuprofen.

perceived toxicity and environmental concerns associated with disposal of tin wastes.[30,31] Consequently, appreciable efforts have been made to develop heavy metal-free alternatives based on silicon, germanium, or phosphorus chemistry. Other strategies have been applied to immobilize the tin reagents or to reduce their usage to less than stoichiometric amounts.

5.6 IMMOBILIZATION OF TIN REAGENTS

The immobilization of non-chiral organotin hydrides has been addressed in a number of works. Various solid supports (*e.g.*, silica, alumina, and polystyrene) and different ways to covalently bind organotin hydrides have been used.[32] Schiesser[32b,35b] has developed immobilized menthyltin hydrides attached to silica and polystyrene supports, as shown in Figure 5.5. Both systems have been developed to the proof-of-concept stage.

Although these systems are feasible for enantioselective free radical reductions, the enantioselectivities achieved are generally lower than those obtained in homogeneous reactions. As in the case for most immobilized reagents, the loading of active tin sites on the surface of the support materials is rather low,

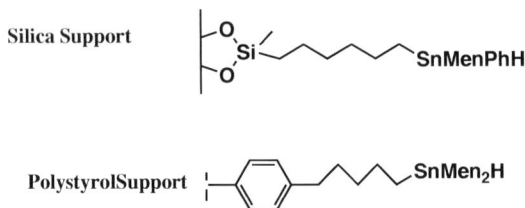

Silica Support

$$\begin{array}{c} -O \\ Si \\ -O \end{array}$$ SnMenPhH

PolystyrolSupport ◯ SnMen₂H

Figure 5.5 Immobilized menthyltin hydrides at silica and polystyrol supports.

especially for silica-supported systems. One associated problem is the relatively low concentration of radical donors, which reduces the radical delivery rate to levels that render the propagation of the radical chain difficult. Other drawbacks involve temperature control issues as a result of the heterogeneous nature of the reactions as well as leaching of the organotin moieties from the carrier support.

5.7 CATALYTIC REDUCTIONS IN TIN

An intriguing solution to the reduction of tin waste levels is the use of organostannanes (*e.g.*, Bu_3SnH) in catalytic amounts in the presence of a coreductant, such as $(MeHSiO)_n$ or $PhSiH_3$, which regenerates the organostannane but does not reduce the substrate directly.[33,34] Achiral systems were optimized to use as little as 5% of the stoichiometrically required amount of organostannane. Perchyonok[32b,35b] has developed a chiral version of this catalytic cycle for the use of its menthyltin hydrides (*e.g.*, compounds **5–7**) using borane as a coreductant. One advantage of this technology is that borane can cleave one phenyl group in phenyl-substituted organotin compounds to afford organotin hydrides.[35] Thus, the use of an excess of borane with a small amount of an appropriate triorganophenyltin compound (*e.g.*, Men_3SnPh) provides an effective catalytic system for (enantioselective) free radical reductions. It is important to note that borane does not undergo free radical reductions with C–Hal bonds of the substrate molecules. The basis of this technology is illustrated in Scheme 5.4.

Large amounts of borane can be prepared conveniently by the reaction of $NaBH_4$ with dimethyl sulfate. Excess of borane and associated by-products may be hydrolyzed after the reduction to yield the environmentally compatible boronic acid, $B(OH)_3$.

5.8 REDUCING AGENTS BASED ON GERMANIUM AND SILICON

Triorganotin hydrides are good reagents for free radical reductions because of the lability of the Sn–H bond and because of their ability to undergo homolytic

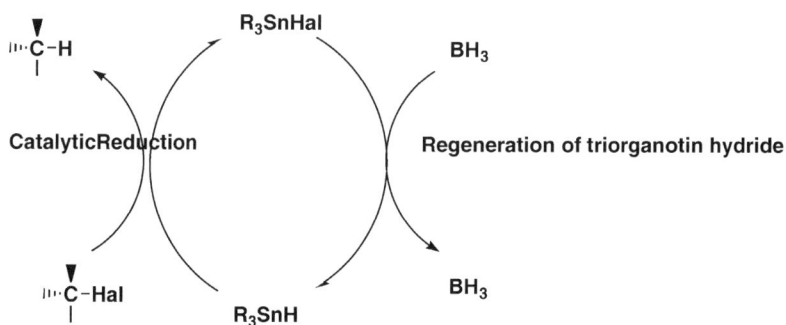

Scheme 5.4 Catalytic hydrogen atom transfer reaction and chiral organostannanes.

bond cleavages for the delivery of hydrogen radicals at acceptable rates. Compared to this, the M–H bonds in organosilanes and organogermanes, R_3MH (M = Si, Ge), are kinetically more stable and the hydrogen donor ability is much lower when R represents a simple alkyl or aryl group. Whereas organogermanes, R_3GeH (R = alkyl, aryl), may be considered as borderline cases, the hydrogen delivery rate toward common organic radicals is generally too slow with organosilanes, R_3SiH (R = alkyl, aryl).[36,37] However, the Si–H bond can be effectively activated by appropriate substituents, such as triorganosilyl- and organothio groups. Thus, tris(trimethylsilyl)silane, $(Me_3Si)_3SiH$, and tris(methylthio)silane, $(MeS)_3SiH$, undergo free radical reductions and are currently the most successful metal-free alternative to replace tributyltin hydride.[38,39] Organosilicon compounds are generally considered to be environmentally friendly because their ultimate degradation product is silica. Schiesser[32b,35b] has developed chiral organogermanes and organosilanes for the purpose of enantioselective free radical reductions; examples are depicted in Figure 5.6.

The chiral organogermanes, MenPh$_2$GeH (**8**) and *en*-MenPh$_2$GeH (**8′**) are again based on the availability of (+)- and (−)-menthol, whereas the chiral tetrasilanes, (*trans*-MyrMe$_2$Si)$_3$SiH (**9**) and *en*-(*trans*-MyrMe$_2$Si)$_3$SiH (**9′**) are derivatives of (+)- and (−)-β pinene **8** and **9** have been used successfully in free radical reductions of α-brominated methyl *N*-TFA phenylglycinate and benzyl *N*-TFA *tert*-leucinate, respectively, producing enantioselectivies of 99% ee in both cases. Besides environmental reasons, chiral organogermanes and organosilanes offer another advantage—namely, they produce excellent enantioselectivities even at comparatively high temperatures up to −20 °C. This fundamental difference from organostannanes, which generally provide best results at −78 °C, is attributed to the closer proximity of the organosilane with the radical substrate at the diastereotropic transition state of the hydrogen transfer, which allows a more effective transfer of chiral information.[40]

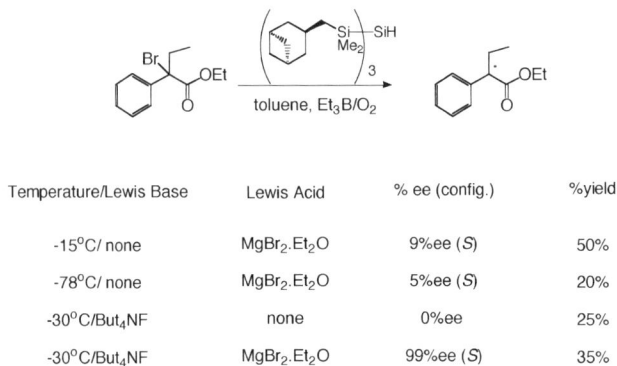

Temperature/Lewis Base	Lewis Acid	% ee (config.)	%yield
-15°C/ none	MgBr$_2$.Et$_2$O	9%ee (*S*)	50%
-78°C/ none	MgBr$_2$.Et$_2$O	5%ee (*S*)	20%
-30°C/But$_4$NF	none	0%ee	25%
-30°C/But$_4$NF	MgBr$_2$.Et$_2$O	99%ee (*S*)	35%

Figure 5.6   Heavy-metal-free chiral reducing agents based on Ge and Si chemistry.

5.9 SUMMARY

Free radical reductions mediated by chiral stannanes, germanes, and silanes may occur with enantioselectivities in excess of 99% ee. Owing to the involvement of radical intermediates and the mild reaction conditions, this process is applicable for a large variety of simple or even complex target molecules that are incompatible with asymmetric reductions that require ionic reaction conditions.

REFERENCES

1. D. P. Curran, N. A. Porter and B. Giese, Stereochemistry or Radical Reactions. Concepts, Guidelines, and Synthetic Applications, VCH, Weinheim, 1996.
2. M. P. Sibi and N. A. Porter, *Acc. Chem. Res.*, 1999, **32**, 163.
3. *Radicals in Organic Synthesis,* ed. P. Renaud and M. Sibi. Wiley-VCH, Weinheim, 2001.
4. G. Bar and A. F. Parsons, *Chem. Soc. Rev.*, 2003, **32**, 251.
5. M. P. Sibi, S. Manyem and J. Zimmerman, *Chem. Rev.*, 2003, **103**, 3263.
6. W. P. Neumann, *Synthesis*, 1987, 665.
7. D. P. Curran, *Synthesis,* 1988, 417.
8. D. P. Curran, *Synthesis,* 1988, 489.
9. D. Nanni and D. P. Curran, *Tetrahedron: Asymmetry*, 1996, **7**, 2417.
10. M. Blumenstein, K. Schwarzkopf and J. O. Metzger, *Angew. Chem., Int. Ed.*, 1997, **36**, 235.
11. K. Schwarzkopf, J. O. Metzger, W. Saak and S. Pohl, *Chem. Ber.*, 1997, **130**, 1539.
12. K. Schwarzkopf, M. Blumenstein, A. Hayen and J. O. Metzger, *Eur. J. Org. Chem.*, 1998, 177.
13. M. A. Skidmore and C. H. Schiesser, *Phosphorus, Sulfur Silicon Relat. Elem.*, 1999, **150–151**, 177.
14. H. Schumann and B. C. Wassermann, *J. Organomet. Chem.*, 1989, **365**, C1.
15. H. Schumann, B. C. Wassermann and F. E. Hahn, *Organometallics*, 1992, **11**, 2803.
16. H. Schumann, B. C. Wassermann and J. Pickardt, *Organometallics*, 1993, **12**, 3051.
17. J. C. Podesta and G. E. Radivoy, *Organometallics*, 1994, **13**, 3364.
18. J. C. Podesta, A. B. Chopa, G. E. Radivoy and C. A. Vitale, *J. Organomet. Chem.*, 1995, **494**, 11.
19. C. Lucas, C. C. Santini, M. Prinz, M.-A. Cordonnier, J.-M. Basset, M.-F. Connil and B. Jousseaume, *J. Organomet. Chem.*, 1996, **520**, 101.
20. C. A. Vitale and J. C. Podesta, *J. Chem. Soc., Perkin Trans. 1*, 1996, 2407.
21. A. de Mallmann, O. Lot, N. Perrier, F. Lefebvre, C. Santini and J. M. Basset, *Organometallics*, 1998, **17**, 1031.

22. D. Dakternieks, K. Dunn, D. J. Henry, C. H. Schiesser and E. R. T. Tiekink, *Organometallics*, 1999, **18**, 3342.
23. D. Dakternieks, K. Dunn, D. J. Henry, C. H. Schiesser and E. R. T. Tiekink, *J. Chem. Soc., Dalton Trans.*, 2000, 3693.
24. D. Dakternieks, K. Dunn, D. J. Henry, C. H. Schiesser and E. R. T. Tiekink, *J. Organomet. Chem.*, 2000, **605**, 209.
25. R. W. Hoffmann, *Chem. Soc. Rev.*, 2003, **32**, 225.
26. P. Renaud and M. Gerster, *Angew. Chem., Int. Ed.*, 1998, **37**, 2562.
27. D. Dakternieks, K. Dunn, V. T. Perchyonok and C. H. Schiesser, *Chem. Commun.*, 1999, 1665.
28. M. W. Carland and C. H. Schiesser, in *The Chemistry of Organic Germanium, Tin and Lead Compounds,* ed. Z. Rappoport, Wiley & Sons, Chichester, 2002, vol. 2, p. 1401.
29. D. Dakternieks, V. T. Perchyonok and C. H. Schiesser, *Tetrahedron: Asymmetry*, 2003, **14**, 3057.
30. P. A. Baguley and J. C. Walton, *Angew. Chem., Int. Ed.*, 1998, **37**, 3072.
31. A. Studer and S. Amrein, *Synthesis*, 2002, 835.
32. (a) K. Jurkschat and M. Mehring, in *The Chemistry of Organic Germanium, Tin and Lead Compounds,* ed. Z. Rappoport, Wiley & Sons, Chichester, 2002, vol. 2, p. 1543; (b) C. Schiesser, T. Perchyonok and D. Dakternieks, *PCT Int. Appl.*, 2004, WO 2004065335 A1 20040805.
33. R. M. Lopez, D. S. Hays and G. C. Fu, *J. Am. Chem. Soc.*, 1997, **119**, 6949.
34. D. S. Hays and G. C. Fu, *J. Org. Chem.*, 1998, **63**, 2796.
35. (a) M. B. Faraoni, L. C. Koll, S. D. Mandolesi, A. E. Zuniga and J. C. Podesta, *J. Organomet. Chem.*, 2000, **613**, 236; (b) C. Schiesser and T. Perchyonok, *PCT Int. Appl.*, 2004, WO 2004065334 A1 20040805.
36. C. Chatgilialoglu and M. Ballestri, *Organometallics*, 1995, **14**, 5017.
37. C. Chatgilialoglu, V. I. Timokhin and M. Ballestri, *J. Org. Chem.*, 1998, **63**, 1327.
38. C. Chatgilialoglu, *Chem. Rev.*, 1995, **95**, 1229.
39. C. Chatgilialoglu and C. H. Schiesser, in *The Chemistry of Organic Silicon Compounds*, ed. Z. Rappaport and Y. Apeloig, Wiley & Sons, New York, 2001, vol. 3, p. 341.
40. D. Dakternieks, D. J. Henry and C. H. Schiesser, *J. Chem. Soc., Perkin Trans 2*, 1998, 591.

CHAPTER 6

Organic Radical Reductions in Water: Water as a Hydrogen Atom Source[*]

6.1 INTRODUCTION

The chemistry of free radicals has undergone a massive renaissance over the past thirty years. The real burst in synthetic applications arose from the use of trialkyltin and triaryltin hydrides as radical-chain carriers. However, the toxicity of the tin reagents, coupled with the difficulty in separation of their by-products from the desired reaction products, meant that they were never acceptable to the pharmaceutical industry. Although improved methods of separation and operation have been developed, there is still a reluctance to use toxic tin reagents. This is unfortunate, since the properties of free radicals— *e.g.* lack of solvation (useful in assembling congested quaternary carbons) and predictable kinetics regardless of the reaction solvent (useful for predicting the relative speed of desired *vs.* side-reactions)—give them a unique advantage in certain synthetic manoeuvres.

In radical chain processes, the initial radicals are generated by some initiation. In organic solvents, the most popular initiator is 2′,2′-azobisisobutyronitrile (AIBN), with a half-life of 1 h at 81 °C, generating the incipient radicals that commence the radical chain reaction. Other azo-compounds are used from time to time as well as the thermal decomposition of di-*tert*-butyl peroxide depending on the reaction conditions. Triethylborane (Et$_3$B) in the presence of very small amounts of oxygen is an excellent initiator for low temperature reactions (down to −78 °C). Air-initiated reactions have also recently been reported in aqueous and neat mixtures. However, in water, the radical initiation varies, and several studies have been undertaken to assess the best methodology. In the last decade, Et$_3$B/ dioxygen has also been used as a radical initiator in water.

* Chapter written by V. Tamara Perchyonok and Al Postigo.
Streamlining Free Radical Green Chemistry
V. Tamara Perchyonok, Ioannis Lykakis and Al Postigo
© V. Tamara Perchyonok, Ioannis Lykakis and Al Postigo 2012
Published by the Royal Society of Chemistry, www.rsc.org

Following the progressive exploration of radical reactions in non-conventional media, water has been adopted as the solvent of choice as advantages such as rate acceleration, stereoselectivity, and regioselectivity have been observed when radical reactions are performed in this latter medium as compared to organic solvents. The majority of organic radical reactions employing Si-centred radicals take place in organic solvents, neat media, and on surface chemistry, and only recently has water been used as a convenient reaction medium for these radicals.[1,2]

6.2 WATER-SOLUBLE ORGANOSILANES AND SYNTHESIS

The development of novel water-soluble organosilane compounds and their applications to radical reactions in water is constantly being investigated, as a means of producing effective radical organic transformations in water, such as reductions of hydrophilic organic halides.

In the radical chain processes, the initial silyl radicals are generated by some initiation. The most popular initiator is $2',2'$-azobisisobutyronitrile (AIBN), with a half-life of 1 h at 81 °C, generating the incipient radicals that commence the radical chain reaction. Other azo-compounds are used from time to time[3] as well as the thermal decomposition of di-*tert*-butyl peroxide[4] depending on the reaction conditions. Triethyl borane (Et_3B) in the presence of very small amounts of oxygen is an excellent initiator for low temperature reactions (down to -78 °C). Air-initiated reactions have also recently been reported (eqn (6.1)).[5]

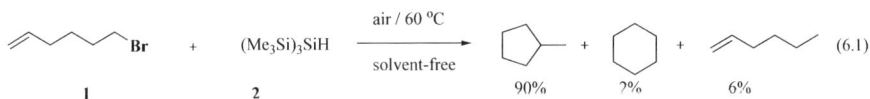

Togo and Yokoyama and coworkers[6] have dealt with the necessity of performing reduction reactions of water-soluble halides in water. They have shown that the reactivity of water-soluble arylsilanes, as reducing agents, is much higher than that of alkylsilanes, and that the reactivity of diarylsilanes was higher than that of monoarylsilanes and triarylsilanes. Thus, the solubility of bis[4-(2-methoxyethoxy)phenyl]silane **3** and bis{4-[2-(2-methoxyethoxy)ethoxy]phenyl}silane **4** were about 1.0×10^{-2} M and 1.5×10^{-2} M, respectively.

The reduction of potassium *o*-bromobenzoate and potassium *o*-iodobenzoate, which form the sp^2 carbon radicals, showed that the bromine atom abstraction is somewhat difficult, whereas the iodine atom of potassium *o*-iodobenzoate is easily removed by the silyl radical. These results and others obtained by the same authors showed that organosilanes **3** and **4** (Figure 6.1) can promote radical reductions of alkyl bromides, alkyl iodides, and aryl iodides, initiated by Et_3B, in aqueous media under aerobic conditions.

Radical cyclizations in water using organosilanes **4** and **5** have also been studied. The radical cyclization of potassium 7-bromo-2-heptenoate with **4** and

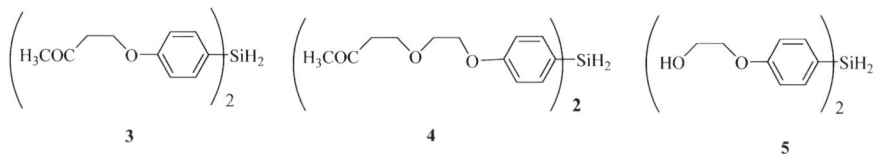

Figure 6.1 Water-soluble arylsilanes utilized in radical reductions of hydrophilic organic halides.

5 was carried out to afford 48 and 82% yields of cyclopentyl acetic acid, respectively, while no direct reduction product, *i.e.* 2-heptenoic acid, was formed. This clearly demonstrated the radical nature of the process.[6]

Water is also a choice of solvent for free radical polymerization. The high heat capacity of water allows effective transfer of the heat from polymerization. Compared to organic solvents, the high polarity of water distinguishes the miscibility of many monomers from polymers remarkably well. Today, aqueous free radical polymerization is applied in industry.[7]

Several interesting photoinitiators based on the silyl radical chemistry have been proposed as a means of effecting polymerization in aqueous suspensions.[8] Among these compounds, (4-tris(trimethylsilyl)silyloxy)benzophenone generates silyl radicals under light irradiation that produce high rates of polymerization. A water-soluble poly(methylphenylsilylene) derivative has been used as a photoinitiator of radical polymerization of hydrophilic vinyl monomers with great success.[9] In the following, we shall describe different radical triggering events that have recently been used for generating silicon-centered radicals in water that will be used throughout this chapter.

Recently, the hydrogen-atom donation reactions in water from diverse phosphorous compounds such as hypophosphorous acids, tetraalkylammonium hypophosphites, and reduction with SmI_2, $SmCl_2$, and titanium salts, have been thoroughly reviewed by Perchyonok and Lykakis.[10] Postigo and collaborators have reviewed the hydrogen donation reactions from silanes in water.[11]

6.3 *TRIS*(TRIMETHYLSILYL)SILANE IN WATER AND "ON WATER"

It is known that in organic solvents, *tris*(trimethylsilyl)silane, $(Me_3Si)_3SiH$, is an efficient reducing agent for organic halides. Also, the reported methodology of polarity-reversal catalysis is well documented in organic solvents. The thiol/silane couple shows not only an efficient synergy of radical production and regeneration, but could also provide for the use of an amphiphilic thiol, in order to enhance the radical reactivity at the interface. For the reduction of an organic halide (RX) by the couple $(Me_3Si)_3SiH/HOCH_2CH_2SH$ under radical conditions, the propagation steps depicted in Scheme 6.1 are expected. That is, the alkyl radicals abstract hydrogen from the thiol and the resulting thiyl

$$R^\bullet + HOCH_2CH_2SH \longrightarrow RH + HOCH_2CH_2S^\bullet$$

$$HOCH_2CH_2S^\bullet + (TMS)_3SiH \rightleftharpoons HOCH_2CH_2SH + (TMS)_3Si^\bullet$$

$$(TMS)_3Si^\bullet + RX \longrightarrow (TMS)_3SiX + R^\bullet$$

$$RX + (TMS)_3SiH \longrightarrow RH + (TMS)_3SiX$$

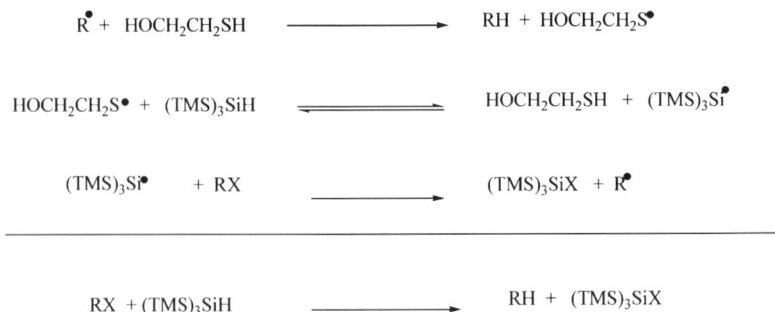

Scheme 6.1   Polarity-reversal catalysis of silanes with thiols.

radicals abstract hydrogen from the silane, so that the thiol is regenerated along with the chain-carrying silyl radical for a given RX.[12,13]

The proposal of $(Me_3Si)_3SiH$ in water is attractive from the point of view of its commercial availability. Recently, Postigo, Chatgilialoglu *et al.*[14] tested the reducing agent $(Me_3Si)_3SiH$ in water and observed its high stability in deaerated aqueous media and high temperatures. They subjected a series of organic halides to reduction with $(Me_3Si)_3SiH$ in water with different initiators, azo compounds and Et_3B. The initiators studied that afforded the best reduction yields with $(Me_3Si)_3SiH$ were the water soluble 2,2′-azobis(2-amidinopropane) dihydrochloride (AAPH) and 1,1′-azo*bis*(cyclohexanecarbonitrile) (ACCN, organic-solvent soluble). The half-life of ACCN at 100 °C is 2.33 h, while that of AAPH is *ca.* 1.1 h at 73 °C.

The reduction of hydrophilic 4-iodobutyric acid and hydrophobic 5-iodouracil, afforded the corresponding reduced products in yields >90%, with both initiators.

The reduction of hydrophilic (1*S*)-bromocamphor-10-sulfonic acid and 5-bromouridine were also considered under similar reaction conditions. Using the water-soluble AAPH initiator no reaction occurred for the camphor derivative, whereas 5-bromouridine afforded uridine in 82% yield (based on 17% converted substrate). However, when 3 mM ACCN is used as initiator, both substrates afforded 90% yields of the corresponding reduction products, although the conversion of the starting material was as low as 10%. By increasing the amount of ACCN, however, the disappearance of starting material increased in favor of reduction product.[14] In this work, the relevance of 2-mercaptoethanol in the reduction process was revealed. For the reduction of 5-bromouridine, the optimal ratio of substrate/2-mercaptoethanol was found to be 3–3.5.[14]

Other substrates of biological relevance bearing halogen atoms such as 8-bromoadenosine and 8-bromoguanosine were also subjected to reduction with $(Me_3Si)_3SiH/HOCH_2CH_2SH$ in water initiated by ACCN. Very high yields of reduced products were obtained under these reaction conditions (>80%).[15]

Later on, Postigo, Chatgilialoglu *et al.*[16] reported on two methods for the use of $(Me_3Si)_3SiH$ in water, depending on the hydrophilic or hydrophobic character of the substrates.

The reduction of water-insoluble organic substrates proceeded in a heterogeneous mixture of substrate, $(Me_3Si)_3SiH$, and ACCN , in water which is previously de-oxygenated with Ar and heated at 100 °C for 4 h. For water-soluble substrates, $HOCH_2CH_2SH$ is used. Thus, reduction of 5-bromo-nicotinic acid, 5′-iodo-5′-deoxyadenosine, and other hydrophilic halides do proceed by the couple $(Me_3Si)_3SiH/HOCH_2CH_2SH$, where the alkyl or aryl radicals (R) abstract hydrogen from the thiol in the water phase, and the resulting thiyl radicals migrate into the lipophilic dispersion of the silane and abstract a hydrogen atom, thus regenerating the thiol along with the chain-carrying silyl radical for a given RX (Scheme 6.1). The reaction of the silyl radical is expected to occur at the interface of the organic dispersion with the aqueous phase. It is worth mentioning that the reaction of thiyl radicals with silane is estimated to be exothermic by *ca.* -3.5 kcal mol^{-1}.[17]

The same reaction conditions were also applied to the radical cyclization of 1-allyloxy-2-iodobenzene derivative **6**, as shown in eqn (6.2), but in this case, 2-mercaptoethanol was not needed. The reaction afforded 85 % yield of cyclized product **7**.

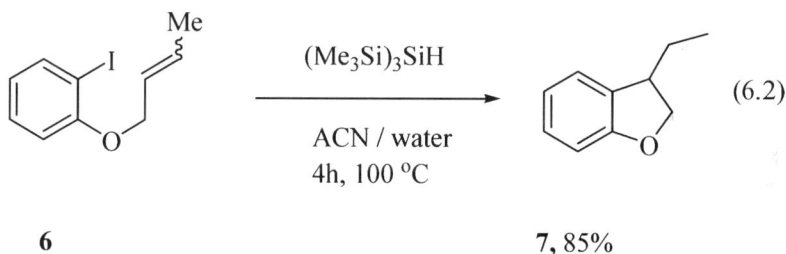

(6.2)

6   **7, 85%**

There is little knowledge about the reduction of *gem*-dichlorides by $(Me_3Si)_3SiH$ in organic solvents, and this knowledge was limited to some stereoselective examples.[18] Reduction of *gem*-dichlorides **8** and **9** in water (eqn (6.3)) with $(Me_3Si)_3SiH$ and ACCN under the usual experimental conditions (this time at 70 °C, 2 h) proceeded smoothly affording the corresponding monochloride derivatives, in quantitative yields, as a diastereoisomeric mixture of compounds **10** and **11** in a 1.7 : 1 ratio for both cases (**8** and **9**).

The diastereoselectivity outcome of this reaction is likely due to the influence of the substituents on the rate of the cyclopropyl radical and on the shielding of the two faces of the cyclopropyl ring.[18–20]

Another successful class of radical reductions in water has been obtained by the transformation of azides into primary amines under the same experimental conditions. The results are reported in Table 6.1.

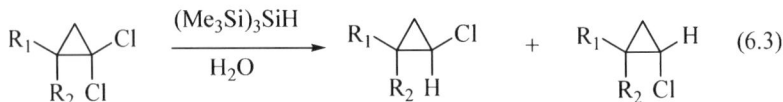

$$R_1 \underset{R_2 \ Cl}{\triangle} Cl \xrightarrow[\text{H}_2\text{O}]{(\text{Me}_3\text{Si})_3\text{SiH}} R_1 \underset{R_2 \ H}{\triangle} Cl \quad + \quad R_1 \underset{R_2 \ Cl}{\triangle} H \qquad (6.3)$$

8 $R_1 = C_5H_{11}$, $R_2 = H$ **10**   **11**
9 $R_1 = Me$, $R_2 = Ph$

Again, no reaction was observed in the absence of amphiphilic 2-mercaptoethanol. The mechanistic steps of this reaction are shown in Scheme 6.2, in analogy with the pathways reported for the radical reduction of aromatic azides with triethylsilane in toluene.[16,21,22]

De-oxygenation reactions in water (Barton–McCombie reaction) have also been attempted using $(\text{Me}_3\text{Si})_3\text{SiH}$ as reducing agent.[16] As observed in organic solvents, the reaction is independent of the type of the thiocarbonyl derivative (*e.g.* *O*-arylthiocarbonate, *O*-thiocarbamate, thiocarbonyl imidazole or xanthate). On the other hand, the water-soluble material does require the presence of the chain-carrier 2-mercaptoethanol.

There has been an increasing emphasis on the development of alternative hydrogen transfer agents due in large part to concerns about the toxicity of tin-containing compounds. Some of the more exciting advances in this direction involve the use of water and alcohols as safe "green" hydrogen atom transfer

Table 6.1 Reduction of water-soluble azides.

$$RN_3 \xrightarrow[\text{ACCN, H}_2\text{O}]{\substack{(\text{Me}_3\text{Si})_3\text{SiH} \\ \text{HOCH}_2\text{CH}_2\text{SH}}} RNH_2$$

RN_3	conversion (%)	RNH_2, (%)
12	70	99
13	>99	90
14	>99	95
15	>99	99

$$RN_3 + (Me_3Si)_3Si^\bullet \longrightarrow R\text{-}\overset{\bullet}{N}\text{-}Si(SiMe_3)_3 + N_2$$

$$R\text{-}\overset{\bullet}{N}\text{-}Si(SiMe_3)_3 + HOCH_2CH_2SH \longrightarrow R\text{-}NH\text{-}Si(SiMe_3)_3 + HOCH_2CH_2S^\bullet$$

$$HOCH_2CH_2S^\bullet + (Me_3Si)_3SiH \longrightarrow HOCH_2CH_2SH + (Me_3Si)_3Si^\bullet$$

$$R\text{-}NH\text{-}Si(SiMe_3)_3 + H_2O \longrightarrow RNH_2$$

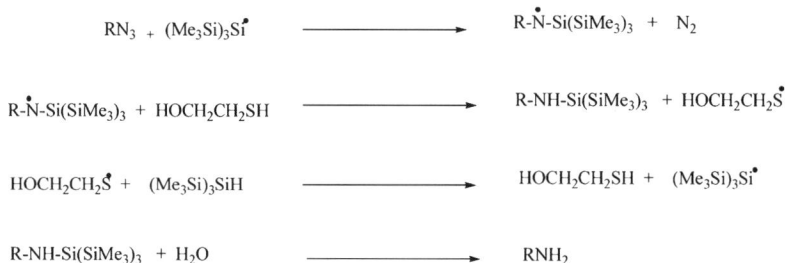

Scheme 6.2 Proposed reaction steps for the reduction of azides in water.

agents. The high O–H bond dissociation energies (BDEs) of alcohols (*ca.* 105 kcal mol^{-1}) and water (118 kcal mol^{-1}) suggest that hydrogen atom transfers from these sources to carbon-centered radicals will be too slow to be useful, but Lewis acid-complexed alcohols and water have much reduced O–H BDEs and can react with alkyl radicals rapidly.[23]

Renaud recently reported the reduction of B-alkylcatecholboranes into alkanes, and disclosed that alcohols and water, when complexed to an organo-borane species, can reduce alkyl radicals and replace toxic tin hydrides as a hydrogen atom source in radical reactions (Scheme 6.3, reaction (1)).[24]

A tentative mechanism is shown in Scheme 6.4. The alkyl radical is generated from the reaction between the B-alkylcatecholborane and a radical initiator (oxygen or an alkoxyl radical). The alkyl radical is then reduced by the complex **B** resulting from the coordination of the alcohol to the Lewis acidic MeO–BO$_2$C$_6$H$_4$. This complexation activates the the alcohol producing a substantial decrease of the O–H bond dissociation energy. The resulting radical **C** is closely related to the 'ate' complex **A** described in Scheme 6.1. It releases a molecule of MeO–BCat and a methoxyl radical which propagates the radical chain reaction. The scope and limitations of the reaction were investigated for a series of B-alkylcatecholboranes (Scheme 6.5). Good yields were found with primary (reactions (6) and (7)), secondary (reactions (8)–(11)) and tertiary alkyl radicals (reaction (12)). Benzyl radicals are not reduced with

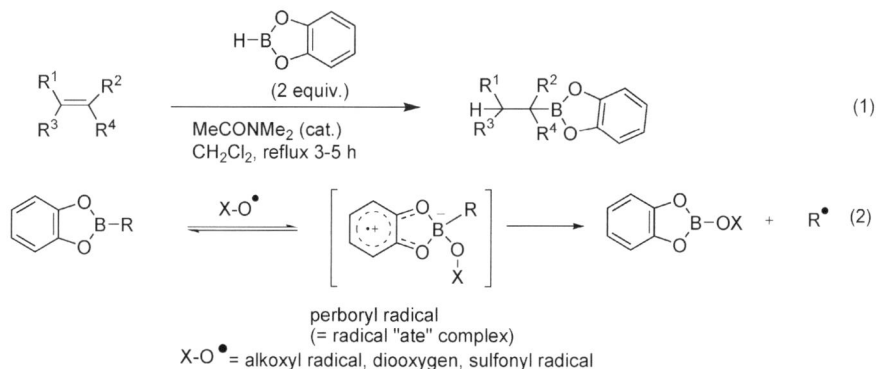

perboryl radical
(= radical "ate" complex)
X-O$^\bullet$ = alkoxyl radical, dioxygen, sulfonyl radical

Scheme 6.3 B-Alkylcatecholboranes.

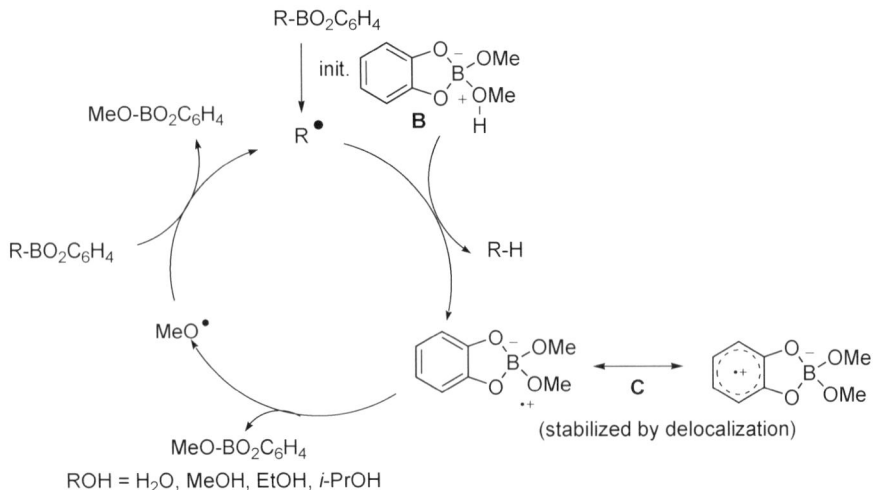

Scheme 6.4 Proposed mechanism for the reduction of B-alkylcatecholborane.

this system indicating that the O–H bond dissociation energy in complex **B** is not weak enough to reduce stabilized radicals.

6.4 TRIETHYLBORANE–WATER COMPLEX AS A REDUCING AGENT

A similar discovery was reported by Wood and coworkers while our manuscript was in preparation. They disclosed that the system H_2O/Et_3B can be

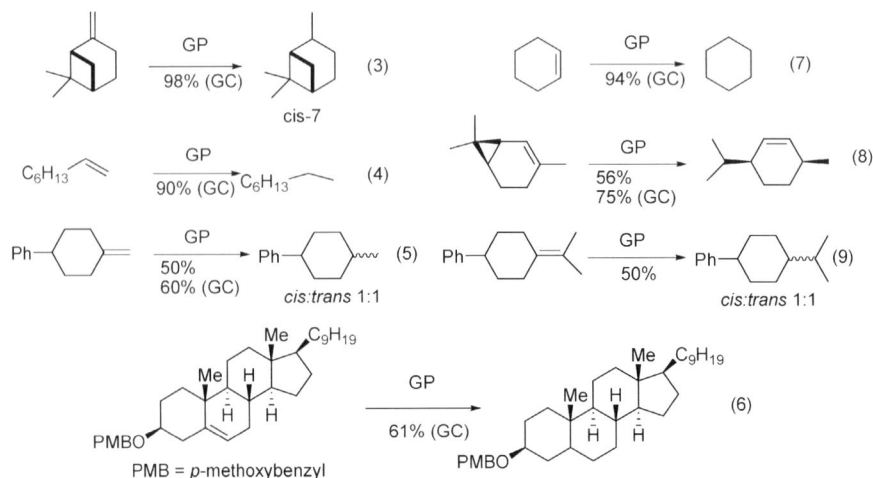

Scheme 6.5 Scope of transformation and general procedure (GP): (i) $C_6H_4O_2BH$ (2 equiv.), $MeCONMe_2$ (0.1 equiv.) in CH_2Cl_2; (ii) MeOH (4 equiv.), air (0.3 mol O_2), CH_2Cl_2, reflux, 90 min.

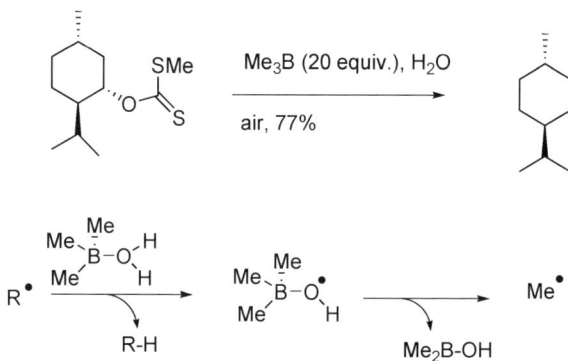

Scheme 6.6 Wood's modified Barton–McCombie reaction.

used as a source of hydrogen atom in the Barton–McCombie reaction (Scheme 6.6).[25] Computational studies confirmed the lowered dissociation energy of water complexed to the Me_3B (86 kcal mol^{-1}) compared to that of the free water (116 kcal mol^{-1}).

6.5 TITANIUM(III)–WATER AS A REDUCING AGENT

More recently, Oltra *et al.* observed that during the Cp_2TiCl-mediated epoxide ring opening, the reduced product was obtained when water was added to the reaction mixture (Scheme 6.7).[26] The mechanism of this reaction involves the aqua complex which may act as a hydrogen atom donor by single electron transfer to the oxygen atom resulting in a Ti(IV) complex. Density functional theory calculations showed again a marked decrease of the O–H hemolytic bond dissociation energy when water is coordinated to the $Cp_2Ti^{III}Cl$ complex (49 kcal mol^{-1}).

Newcomb and collaborators[27] have found that at room temperature the rate constants for reactions of triethylborane complexes of water and methanol as H-atom donors to alkyl radicals are only 2 orders of magnitude smaller than

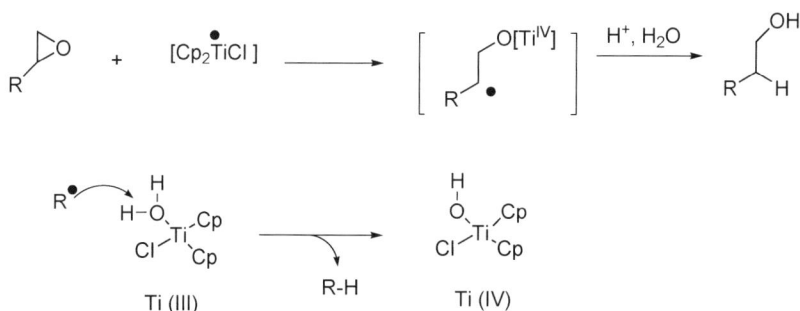

Scheme 6.7 $TiCp_2Cl$-water induced reductive opening of epoxides.

those for reactions of tin hydride reagents. These reactions are adequately fast at room temperature for some synthetic applications, and they are increasingly competitive with radical rearrangements at reduced temperatures. One anticipates increasing numbers of applications of borane complexes as radical reducing agents given the ease of removal of the boron-containing by-products, especially if a mixed reagent system can be developed that permits the use of common alkyl halides as radical precursors.[26]

Newcomb and collaborators[27] also determined the rate constants for reactions of $Cp_2Ti^{III}Cl$-complexed water and methanol with a secondary alkyl radicals. At ambient temperature in THF, the titanium(III) reagent complexed with deuterium oxide and water reacts with the secondary 1-dodecyl cyclobutyl radical with rate constants of 1.0×10 and $2.3 \times 10^4 \, M^{-1} \, s^{-1}$, respectively. In benzene containing 0.95 M methanol, the $Cp_2Ti^{III}Cl$–MeOH complex reacts with a rate constant of $7.5 \times 10^4 \, M^{-1} \, s^{-1}$.[27] The titanium(III) reagent apparently activates water and methanol more strongly than Et_3B with the result that the H-atom transfer reaction of the $Cp_2Ti^{III}Cl$–H_2O complex is 5 times as fast as the H-atom transfer reaction of Et_3B–H_2O at room temperature.[27]

The Arrhenius function for the reaction of $Cp_2Ti^{III}Cl$–H_2O had a normal entropic term; however, the unusually low entropic term for the borane–water complex leads to more efficient hydrogen-atom transfer trapping by this species at low temperatures. Radical reductions by H-atom transfer from water or alcohol complexes of $Cp_2Ti^{III}Cl$ will be useful when radicals are generated by reduction of epoxides or α,β-unsaturated ketones.[27]

It is noteworthy that the rate constants found in this work are about 1 order of magnitude smaller than those for radical reduction reactions of tin hydrides and similar to those for reactions of $((CH_3)_3Si)_3SiH$.[28] The reactions are fast enough to be used in many radical chain reaction sequences, and the kinetics illustrate a large degree of O–H bond activation made possible by complexation of simple hydroxylic compounds with a strong Lewis acid.

More recently, Cuerva, Cárdenas *et al.*[26] have demonstrated the initial assumption that titanocene(III)–water complexes are a unique class of hydrogen-atom transfer (HAT) reagents. They are able to reduce efficiently carbon-centered radicals of diverse nature. The success of this transformation is based on two key features: (a) an excellent binding capabilities of water toward titanocene(III) complexes and (b) a low activation energy for the HAT step.[29] Therefore, the observed reactivity can be explained in the framework of an unprecedented HAT reaction involving water.[29,30]

6.6 SUMMARY

The environmentally friendly methodology which involves the use of alcohols and water as sources of hydrogen atom under very mild conditions has been demonstrated. This trend in the current literature indicates that alcohols and water activated by Lewis acids may become valuable reducing agents in radical reactions offering substantial opportunities for the design of novel free radical reducing agents to replace toxic tin hydride derivatives in radical processes.

REFERENCES

1. T. Yorimitsu, H. Nakamura, K. Shinokubo, K. Oshima and H. Fujimoto, Powerful solvent effect of water in radical reaction: triethylborane-induced atom-transfer radical cyclization in water, *J. Am. Chem. Soc.*, 2000, **122**, 11041–11047.
2. H. Yorimitsu, H. Shinokubo and K. Oshima, Radical reaction by a combination of phosphinic acid and a base in aqueous media, *Bull. Chem. Soc. Jpn.*, 2001, **74**, 225–235.
3. H. Yasuda, Y. Uenoyama, O. Nobutta, S. Kobayashi and I. Ryu, Radical chain reactions using THP as a solvent, *Tetrahedron Lett.*, 2008, **49**, 367–370.
4. A. Naka, H. Ohnishi, J. Ohsita, J. Ikadai, A. Kunai and M. Ishikawa, Silicon-Carbon unsaturated compounds. 70. Thermolysis and photolysis of acylpolysilanes with mesitylacetylene, *Organometallics,* 2005, **24**, 5356–5363.
5. J. Wang, Z. Zhu, W. Huang, M. Deng and X. J. Zhou, Air-initiated hydrosilylation of unactivated alkynes and alkenes and dehalogenation of halohydrocarbons by tris(trimethylsilyl)silane under solvent-free conditions, *J. Organomet. Chem.*, 2008, **693**, 2188–2192.
6. O. Yamazaki, H. Togo, G. Nogami and M. Yokoyama, Radical addition of 2-iodoalkanamide or 2-iodoalkanoic acid to alkenols using a water-soluble radical initiator in water. A facile synthesis of γ-lactones, *Bull. Chem. Soc. Jpn.,* 1997, **70**, 2519–2523.
7. (a) S. Mecking, A. Held and F. M. Bauers, Aqueous catalytic polymerization of olefins, *Angew. Chem., Int. Ed.*, 2002, **41**, 544–561; (b) C.-J. Li and L. Chen, Organic chemistry in water, *Chem. Soc. Rev.*, 2006, **35**, 68–82.
8. J. Lalevée, N. Blanchard, M. El-Roz, B. Graff, X. Allonas and J. P. Fouassier, New photoinitiators based on the silyl radical chemistry: polymerization ability, ESR spin trapping and laser flash photolysis investigation, *Macromolecules,* 2008, **41**, 4180–4186.
9. I. Kminek, Y. Yagci and W. Schnabel, A water-soluble poly(methylphenylsilylene) derivative as a photoinitiator of radical polymerization of hydrophilic vinyl monomers, *Polymer Bull.*, 1992, **29**, 277–282.
10. V. T. Perchyonok and I. N. Lykakis, *Curr. Org. Chem.*, 2009, **13**, 573–598.
11. S. Barata-Vallejo, N. Sbarbati Nudelman and A. Postigo, *Curr. Org. Chem.*, 2011, **15**(7), 1–13.
12. B. P. Roberts, Polarity-reversal catalysis of hydrogen-atom abstraction reactions: Concepts and applications in organic chemistry, *Chem. Soc. Rev.,* 1999, **28**, 25–35.
13. C. Chatgilialoglu, *Organosilanes in Radical Chemistry*, Wiley-VCH, Chichester, 2004, pp. 1–142.

14. A. Postigo, C. Ferreri, M. L. Navacchia and C. Chatgilialoglu, The radical-based reduction with (TMS)$_3$SiH on water, *Synlett,* 2005, **2854**–2856.

15. C. Chatgilialoglu and V. I. Timokhin, Silyl radicals in chemical synthesis, *Adv. Organomet. Chem.*, 2008, **57**, 117–181; C. Chatgilialoglu, (Me$_3$Si)$_3$SiH: Twenty years after its discovery as a radical-based reducing agent, *Chem.–Eur. J.*, 2008, **14**, 2310–2320.

16. A. Postigo, S. Kopsov, C. Ferreri and C. Chatgilialoglu, Radical reactions in aqueous medium using (Me$_3$Si)$_3$SiH, *Org. Lett.,* 2007, **9**, 5159–5162.

17. C. Chatigilialoglu, The Tris(trimethylsilyl)silane/Thiol Reducing System: A Tool for Measuring Rate Constants for Reactions of Carbon-Centered Radicals with Thiols, *Helv. Chim. Acta,* 2006, **89**, 2387–2398.

18. (a) Y. Apeloig and M. Nakash, Reversal of stereoselectivity in the reduction of gem-dichlorides by tributyltin hydride and tris(trimethylsilyl)silane. Synthetic and mechanistic implications, *J. Am. Chem. Soc.,* 1994, **116**, 10781–10782; (b) Y. Liu, S. Yamazaki and S. Izuhara, Modification and Chemical Transformation of Si(111) Surface, *J. Organomet. Chem.*, 2006, **691**, 5809–5824.

19. A. Odedra, K. Geyer, T. Gusstafsson, R. Gimour and P. H. Seeberger, Safe, facile radical-based reduction and hydrosilylation reactions in a microreactor using tris(trimethylsilyl)silane, *Chem. Commun.*, 2008, 3025–3027.

20. C. Chatgilialoglu, C. Ferreri, Q. C. Mulazzani, M. Ballestri and L. Landi, *Cis–trans* Isomerization of Monounsaturated Fatty Acid Residues in Phospholipids by Thiyl Radicals, *J. Am. Chem. Soc.,* 2000, **122**, 4593–4601.

21. L. Benati, G. Bencivenni, R. Leardini, M. Minozzi, D. Nanni, R. Scialpi, P. Spagnolo and G. Zanardi, Radical Reduction of Aromatic Azides to Amines with Triethylsilane, *J. Org. Chem.*, 2006, **71**, 5822–5825.

22. Y. Liu, S. Yamazaki and S. Yamabe, Regioselective Hydrosilylations of Propiolate Esters with Tris(trimethylsilyl)silane, *J. Org. Chem.*, 2005, **70**, 556–561.

23. W. Tantawy and H. Zipse, *Eur. J. Org. Chem.*, 2007, 5817–5820.

24. D. Pozzi and P. Renaud, Alcohols and Water as Reducing Agents in Radical Reactions, *Chimia,* 2007, **61**, 151–154.

25. D. A. Spiegel, K. B. Wiberg, L. N. Schacherer, M. R. Madeiros and J. L. Wood, *J. Am. Chem. Soc.*, 2005, **127**, 12513.

26. J. M. Cuerva, A. C. Campagna, J. Justicia, A. Rosales, J. L. Oller-Lòpez, R. Robles, D. G. Càrdenas, E. Bunuel, E. J. Oltra, *Angew. Chem., Int. Ed.*, 2006, **45**, 5522.

27. J. Jin and M. Newcomb, Rate Constants for Reactions of Alkyl Radicals with Water and Methanol Complexes of Triethylborane, *J. Org. Chem.*, 2007, **72**, 5098–5103.

28. J. Jin and M. Newcomb, Rate Constants for Hydrogen Atom Transfer Reactions from Bis(cyclopentadienyl)titanium(III) Chloride-Complexed

Water and Methanol to an Alkyl Radical, *J. Org. Chem.*, 2008, **73**, 7901–7905.

29. C. Chatgilialoglu and M. Newcomb, Hydrogen donor abilities of the Group XIV hydrides, *Adv. Organomet. Chem.*, 1999, **44**, 67–112.

30. M. Paradas, A. G. Campaña, T. Jimenez, R. Robles, J. E. Oltra, E. Buñuel, J. Justicia, D. J. Cárdenas and J. M. Cuerva, Understanding the Exceptional Hydrogen-Atom Donor Characteristics of Water in Ti(III)-Mediated Free radical Chemistry, *J. Am. Chem. Soc.*, 2010, **132**, 12748–12756.

CHAPTER 7

Tin-Free Radical Reactions Mediated by Organoboron Compounds[*]

7.1 INTRODUCTION

The ability of organoboron compounds to participate in free radical reactions has been known since the earliest investigation of their chemistry.[1–3] For instance, the autoxidation of organoboranes (Scheme 7.1) has been proven to involve radical intermediates.[4,5] This reaction has led recently to the use of triethylborane as a universal radical initiator functioning under a very wide range of reaction conditions (temperature and solvent).[6,7]

Interestingly, homolytic substitution at boron does not proceed with carbon-centered radicals.[8] However, many different types of heteroatom-centered radicals, for example alkoxyl radicals, react efficiently with the organoboranes (Scheme 7.2).

This difference in reactivity is caused by the Lewis base character of the heteroatom-centered radicals. Indeed, the first step of the homolytic substitution is the formation of a Lewis acid–Lewis base complex between the borane and the radical. This complex can then undergo a β-fragmentation leading to the alkyl radical. This process is of particular interest for the development of radical chain reactions. The chapter of the use of organoboron compounds in radical chemistry will concentrate on applications where the organoborane is used as an initiator, as a direct source of carbon-centered radicals, as a chain-transfer reagent, as a radical-reducing agent as well as oxidation of alkyl trifluoroborates. The simple formation of carbon heteroatom bonds *via* a radical process is not treated in this chapter since it has been treated in previous review articles.[3,9]

* Chapter written by V. Tamara Perchyonok.
Streamlining Free Radical Green Chemistry
V. Tamara Perchyonok, Ioannis Lykakis and Al Postigo
© V. Tamara Perchyonok, Ioannis Lykakis and Al Postigo 2012
Published by the Royal Society of Chemistry, www.rsc.org

$$R_3B + O_2 \longrightarrow \left[R\overset{\underset{|}{R}}{\underset{\underset{|}{R}}{B}}\text{-}\overset{+\bullet}{O}\text{-}\overset{\bullet}{O} \right] \longrightarrow \left[Et_2BOO^{\bullet} + Et^{\bullet} \right] \longrightarrow \overset{\underset{|}{R}}{\underset{\underset{|}{R}}{B}}\text{-}O\text{-}\overset{\bullet}{O} + R^{\bullet}$$

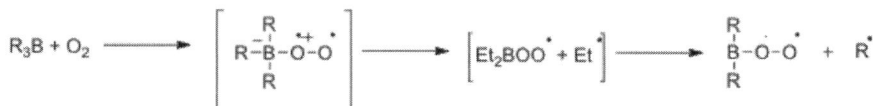

Scheme 7.1   Autoxidation of organoboranes.

7.2   ORGANOBORANES AS RADICAL INITIATORS

Utimoto, Oshima *et al.* were the first to apply the reaction of triethylborane with oxygen to initiate radical reactions.[6,10] Over classical initiators, the system Et_3B/O_2 offers the great advantage of being efficient even at low temperature (-78 °C). This aspect proved to be particularly important for the development of stereoselective radical reactions and for radical reactions involving thermally unstable adducts or products. Review articles describing the use of triethylborane as a radical initiator have appeared.[3,7] The majority of the reported examples involve tin reagents and will not be discussed here. A few selected examples involving tin-free chemistry will be presented.

7.3   IN REDUCTIVE PROCESSES

7.3.1   Reduction of Halides and Related Compounds

Triethylborane can initiate the formation of a silyl or germanyl radical from the related hydrides. For example, Evans and Roseman reported the cyclization of acyl radicals to vinylogous carbonates in the presence of (TMS)$_3$SiH. By using Et_3B/O_2 initiation rather than azobisisobutyronitrile (AIBN), the *cis*-oxepanones are obtained in higher stereoselectivity and yield since the decarbonylation of the intermediate acyl radical is suppressed (Scheme 7.3a).[11] In another striking example, a thermally unstable propargyl bromide cobalt complex cyclizes in the presence of Ph$_2$SiH$_2$ under Et_3B/O_2 initiation at 20 °C. A mixture of reduced and bromine atom transfer products are isolated (Scheme 7.3b).[12]

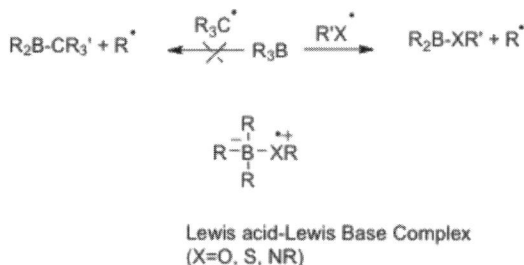

$$R_2B\text{-}CR_3' + R^{\bullet} \overset{R_3C^{\bullet}}{\underset{\times}{\longleftarrow}} R_3B \overset{R'X^{\bullet}}{\longrightarrow} R_2B\text{-}XR' + R^{\bullet}$$

$$R\overset{\underset{|}{R}}{\underset{\underset{|}{R}}{B}}\text{-}\overset{+}{X}R$$

Lewis acid-Lewis Base Complex
(X=O, S, NR)

Scheme 7.2   Reactivity of carbon- and heteroatom-centered radicals towards organoboranes.

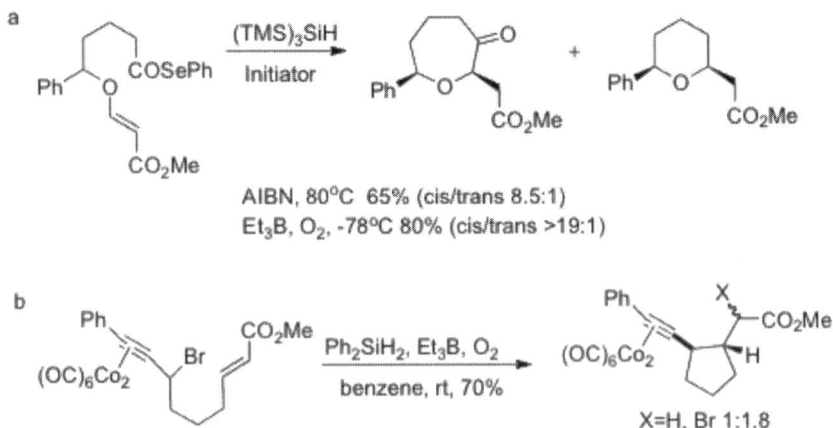

AIBN, 80°C 65% (cis/trans 8.5:1)
Et$_3$B, O$_2$, -78°C 80% (cis/trans >19:1)

X=H, Br 1:1.8

Scheme 7.3   Reductive cyclizations with silanes.

Interestingly, Et$_3$B/O$_2$ initiation can be performed in aqueous solution.[13] For instance, a wide range of aryl and alkyl halides are reduced in water by water-soluble organosilanes using Et$_3$B/O$_2$ initiation (Scheme 7.4).[14]

Recently, Gonzalez-Lopez de Turiso and Curran described a procedure using triethylborane for the synthesis of spirooxindoles and spirodihydroquinolones through intramolecular addition of aryl radicals at the *ipso* position of 4-alkoxy-substituted aromatic rings.[15] The key step for a formal synthesis of the vasopressin inhibitor SR121463A is described in Scheme 7.5. The initiation was performed with Et$_3$B in an open-to-air reaction vessel.

Germanes are also used for the reduction of various organic halides at ambient temperature under Et$_3$B/O$_2$ initiation. For example, tri-2-furylgermane-

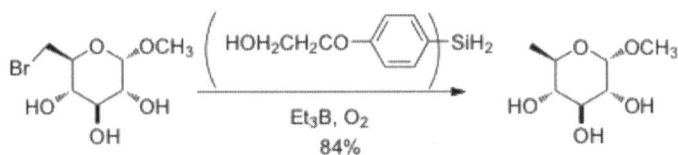

Et$_3$B, O$_2$
84%

Scheme 7.4   Silane-mediated reduction of a bromide in water.

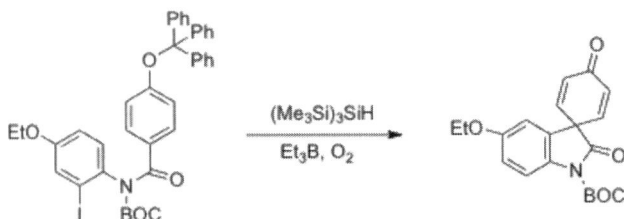

(Me$_3$Si)$_3$SiH

Et$_3$B, O$_2$

Scheme 7.5   Silane-mediated key cyclization for a formal synthesis of the vasopressin inhibitor SR121463A.

mediated radical cyclizations of aryl iodides proceed in good yields (Scheme 7.6a) and are also possible with $NaBH_4$ in the presence of a catalytic amount of triphenylgermane (Scheme 7.6b).[16]

Tin-free radical reduction by an organophosphite[17] and phosphinic acid can also be initiated by Et_3B/O_2. Radical cyclizations using phosphinic acid neutralized with sodium carbonate and Et_3B/O_2 as a radical initiator in aqueous ethanol were recently reported (Scheme 7.7a).[18] A similar stereo-selective cyclization of a β-alkoxyacrylate with 1-ethylpiperidinium hypopho-sphite (EPHP) at room temperature was described by Lee and Han (Scheme 7.7b).[19]

7.3.2 Reductive Addition of Heteroatom-centered Radicals to Alkynes and Alkenes

Tris(trimethylsilyl)silane,[20,21] thiols,[22] germanes[23–25] and gallium hydride[26] can be added easily to terminal alkynes in the presence of Et_3B/O_2. This process was extended to internal alkenes (Scheme 7.8a) as well as silyl enol ethers (Scheme 7.8b) by using tri-2-furylgermane. In this last case, basic or acidic treatment of the main *syn* β-siloxygermane furnishes the corresponding *E*- or *Z*-alkene, respectively.[24]

The addition of hypophosphites to alkenes under Et_3B initiation is also reported.[27] Piettre *et al.* described recently the addition of diethylthiophosphite to alkenes leading to the formation of thiophosphonates (Scheme 7.9a).[28]

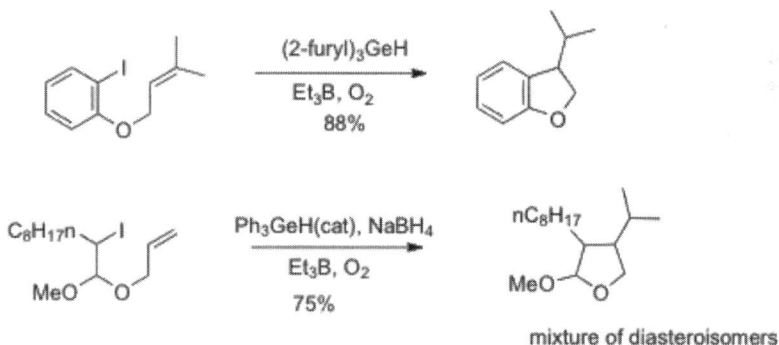

Scheme 7.6 Carbon-carbon bond formation mediated by Germanium hydrogen donor under various conditions.

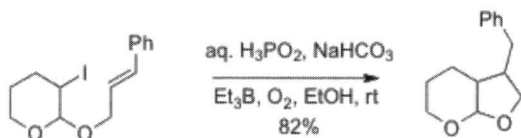

Scheme 7.7 Phosphinic acid-mediated radical cyclizations at room temperature.

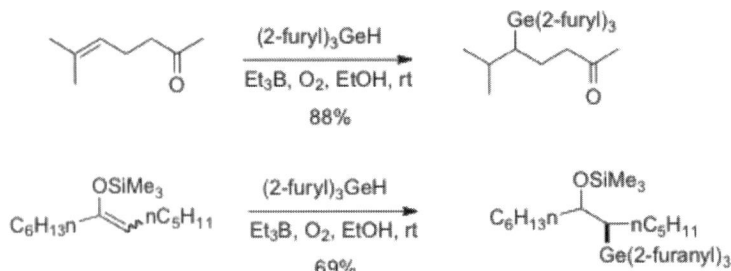

Scheme 7.8 Addition of triarylgermanes to alkenes and enol silanes.

Interestingly, this reaction can be used for cyclization of dienes and ring-opening of strained alkenes such as α-pinene (Scheme 7.9b). Parsons *et al.* prepared an alkenyl thiophosphonate by reaction of a chiral thiophosphite with phenylacetylene (Scheme 7.9c).[29]

7.3.3 In Fragmentation Processes

Oshima *et al.*[30] reported a radical alkenylation of α-halo carbonyl compounds under mild conditions by utilizing alkenylindium reagents. Using 0.5 equivalents of triethylborane as a radical initiator at ambient temperature, we demonstrated that this process affords the alkenylation products in high yield (Scheme 7.10a). The styrylation reaction showed retention of the stereochemistry from starting alkenylindium (Scheme 7.10b).

   An allylzirconium reagent, prepared from Cp$_2$ZrCl$_2$ and allylmagnesium chloride, can be used for the allylation of α-iodoesters.[31] The reaction of allyl zirconium reagent with α-halo esters and amides in presence of triethylborane provides a useful and efficient alternative for organotin chemistry (Scheme 7.11). This reaction has been extended to a three-component coupling process. Similar reactions with allylgallium reagent in water are also reported.[26]

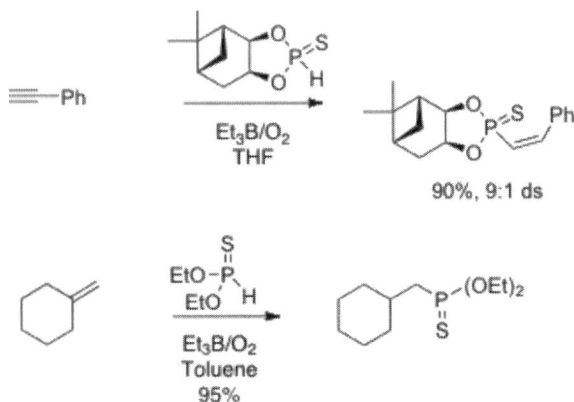

Scheme 7.9 Reductive processes with phosphorus-based reagents.

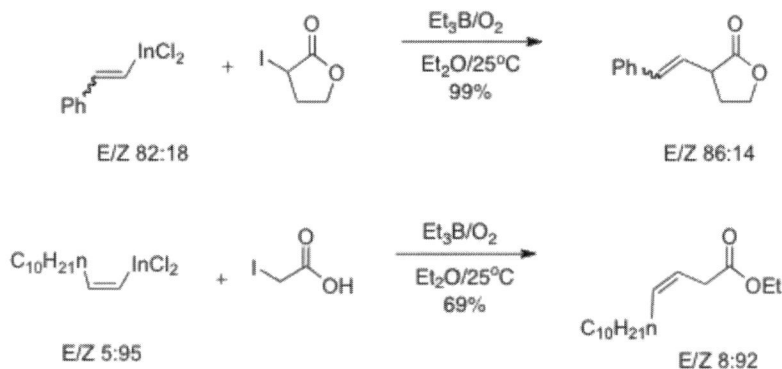

E/Z 82:18   E/Z 86:14

E/Z 5:95   E/Z 8:92

Scheme 7.10   Alkenylation reactions with alkenylindium derivatives.

Germyl enol ethers react with perfluoroalkyl iodides under Et_3B initiation to give α-perfluoroalkyl ketones. The intermediate radical adduct decomposes readily *via* β-elimination and provides the α-perfluoroalkyl ketone and a trialkylgermanyl radical as a chain carrier (Scheme 7.12).[32]

7.4   IN ATOM-TRANSFER PROCESSES

7.4.1   Iodine Atom Transfer

Triethylborane in combination with oxygen provides an efficient and useful system for iodine atom abstraction from alkyl iodide, and thus is a good initiator for iodine atom-transfer reactions.[13,33,34] Indeed, the ethyl radical, issued from the reaction of triethylborane with molecular oxygen, can abstract an iodine atom from the radical precursor to produce a radical $R^{\bullet}$ that enters into the chain process (Scheme 7.13). The iodine exchange is fast and efficient when $R^{\bullet}$ is more stable than the ethyl radical.

    Et_3B-induced addition of perfluoro alkyliodides,[35] α-iodoesters (Scheme 7.14a),[36] iodoamides,[37] α-iodonitriles,[36] and simple alkyl iodides[38] to alkenes

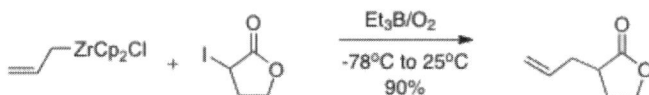

Scheme 7.11   Allylation with allylzirconium reagents.

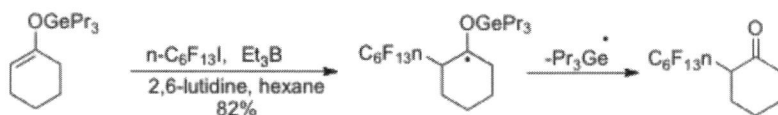

Scheme 7.12   Perfluoroalkylation of ketones *via* germyl enol ethers.

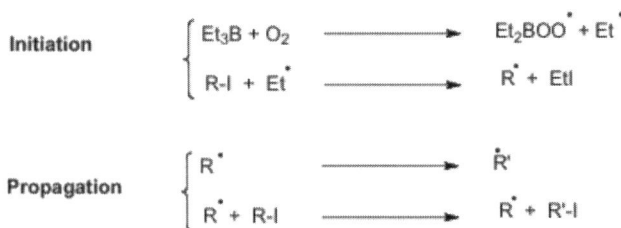

Initiation

$$Et_3B + O_2 \longrightarrow Et_2BOO^{\bullet} + Et^{\bullet}$$

$$R\text{-}I + Et^{\bullet} \longrightarrow R^{\bullet} + EtI$$

Propagation

$$R^{\bullet} \longrightarrow R'^{\bullet}$$

$$R'^{\bullet} + R\text{-}I \longrightarrow R^{\bullet} + R'\text{-}I$$

Scheme 7.13   Mechanism of the Et_3B-mediated iodine atom-transfer reaction.

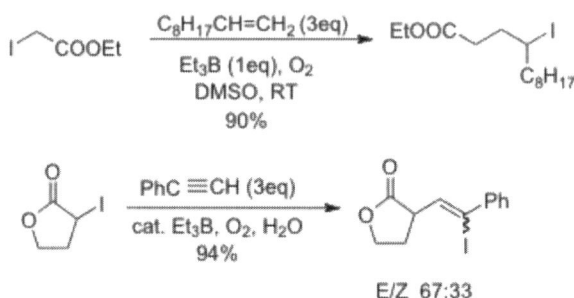

$$\underset{\text{COOEt}}{I} \xrightarrow[\substack{Et_3B \, (1eq), \, O_2 \\ DMSO, \, RT \\ 90\%}]{C_8H_{17}CH=CH_2 \, (3eq)} \underset{C_8H_{17}}{EtOOC}$$

$$\xrightarrow[\substack{\text{cat. } Et_3B, \, O_2, \, H_2O \\ 94\%}]{PhC{\equiv}CH \, (3eq)} \quad Ph$$

E/Z 67:33

Scheme 7.14   Intermolecular additions through iodine atom transfer.

and alkynes have been reported. Interestingly, these reactions were also performed with success in aqueous media[13,39,40] demonstrating the ability of Et_3B
to act as an initiator in water (Scheme 7.14b).[41]

Triethylborane is also an excellent initiator for intramolecular iodine atom-
transfer reactions. For example, cyclization of the propargyl α-iodoacetal
depicted in Scheme 7.15a gives the corresponding bicyclic vinyliodide in
high yield.[38] Allyl iodoacetamides (Scheme 7.15b) and allyl iodoacetates

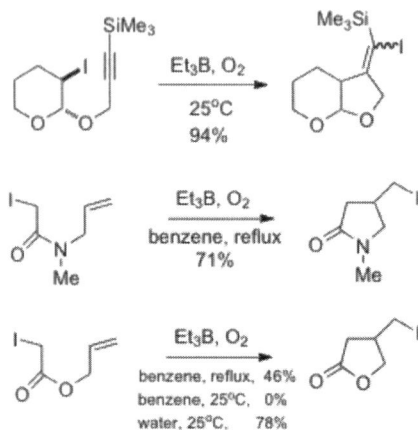

$$\xrightarrow[\substack{25°C \\ 94\%}]{Et_3B, \, O_2}$$

$$\xrightarrow[\substack{\text{benzene, reflux} \\ 71\%}]{Et_3B, \, O_2}$$

$$\xrightarrow[\substack{\text{benzene, reflux, } 46\% \\ \text{benzene, } 25°C, \quad 0\% \\ \text{water, } 25°C, \quad 78\%}]{Et_3B, \, O_2}$$

Scheme 7.15   Cyclizations through iodine atom transfer.

(Scheme 7.15c) cyclize cleanly under Et_3B/O_2 initiation. In the case of the ester, the reaction has to be run in refluxing benzene in order to allow *Z*/*E*-ester isomerization prior to cyclization.[42,43] No trace of cyclized product is detected when the reaction is carried out at room temperature. Interestingly, by running the same reaction in water, Oshima *et al.* obtained the desired lactone in 78% yield. It was suggested that water facilitates the *Z*/*E* isomerization. Efficient preparation of medium and large ring lactones in water have also been reported.[39,40]

Examples of tandem intermolecular addition–cyclization under iodine atom-transfer conditions are depicted in Scheme 7.16.[38,41]

Et_3B-induced radical cascade reactions with 1,5-enynes and 1,5-diynes have been applied to the synthesis of dioxatriquinanes and tricyclic glucoconjugates (Scheme 7.17).[44,45] Some of these elegant cascade cyclizations were also performed under mild conditions at −50 °C.

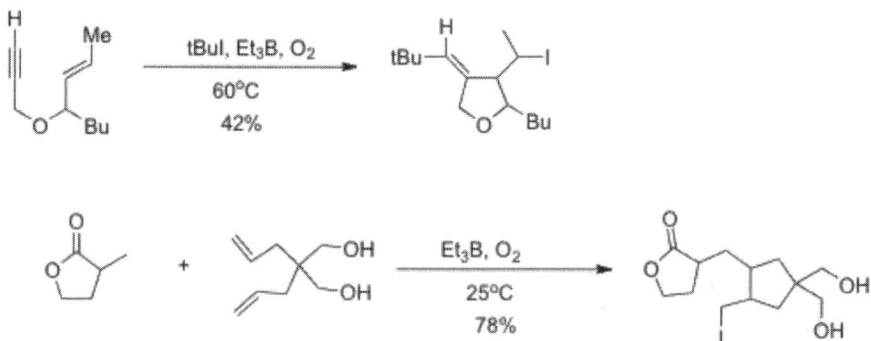

Scheme 7.16 Tandem intermolecular addition–cyclization reactions.

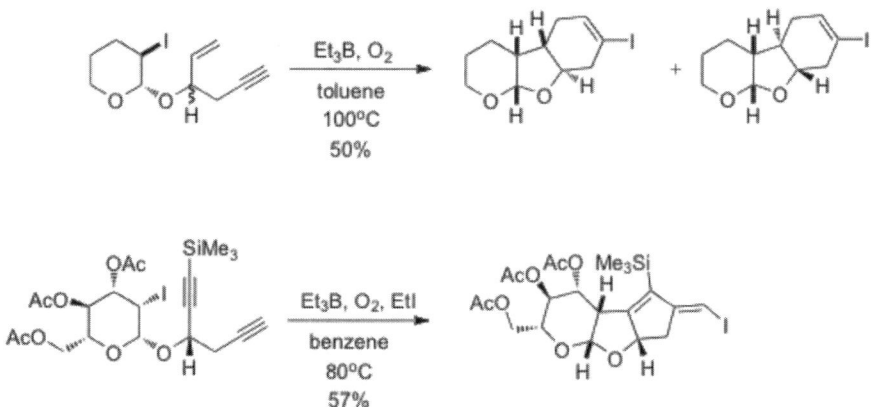

Scheme 7.17 Cascade cyclizations.

Using acyclic and cyclic *N*-tosylated iodomethylaziridines, Taguchi *et al.* investigated annulation reactions.[46] The reaction with electron-rich alkenes such as enol ethers proceeds smoothly as illustrated in Scheme 7.18.

Silyl enol ethers have also been used as a trap for electrophilic radicals derived from α-haloesters[36] or perfluoroalkyl iodides.[32] They afford the α-alkylated ketones after acidic treatment of the intermediate silyl enol ethers (Scheme 7.19a). Similarly, silyl ketene acetals are converted into α-perfluoroalkyl esters upon treatment with perfluoroalkyl iodides.[32,47] The Et_3B/O_2-mediated diastereoselective trifluoromethylation[48,49] (Scheme 7.19b) and (ethoxycarbonyl)difluoromethylation[50,51] of lithium enolates derived from *N*-acyloxazolidinones have also been achieved. More recently, Itoh and Mikami[52] succeeded in the trifluoromethylation of ketone enolates The mechanism of this transformation involves either a final iodine-transfer step or an electron-transfer process that gives back the trifluoromethyl radical.

7.4.2 Bromine Atom Transfer

Bromides are less reactive than the corresponding iodides in atom transfer processes. However, activated bromides such as diethyl bromomalonate[36] and bromomalonitrile[53] react with olefins under Et_3B/O_2 initiation. Kharasch-type reactions of bromotrichloromethane with alkenes are also initiated by Et_3B/O_2.[41] On the other hand, a remarkable Lewis acid effect was reported by Mero and Porter.[54] Atom-transfer reactions of an α-bromooxazolidinone amide with alkenes are strongly favored in the presence of Lewis acids such as $Sc(OTf)_3$ or $Yb(OTf)_3$; this reaction was successively applied to the diastereoselective

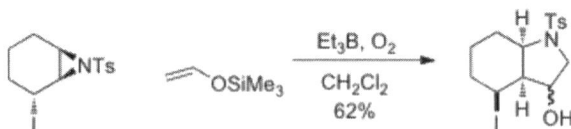

Scheme 7.18 Preparation of pyrrolidine derivatives *via* annulation with iodomethylaziridine derivatives.

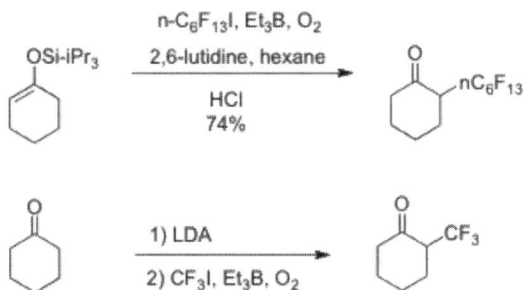

Scheme 7.19 Perfluoroalkylation of silyl enol ethers and lithium enolates *via* iodine atom transfer.

alkylation of chiral oxazolidinone derivatives (Scheme 7.20a).[54] More recently, Yorimitsu and Oshima reported that bromine atom transfers take place at room temperature in ionic liquid media (Scheme 7.20b).[55]

7.4.3 Chlorine Atom Transfer

Radical [3 + 2] annulation involving *N*-allyl-*N*-chlorotosylamide provides a route to pyrrolidine derivatives (Scheme 7.21).[56]

The reaction of *N*-chlorosulfonyl derivatives with enol ether initiated by Et₃B has been reported (Scheme 7.22).[57] The reaction mechanism involves a chlorine atom transfer followed by a Et₃N-promoted elimination of HCl to produce stable enol ethers.

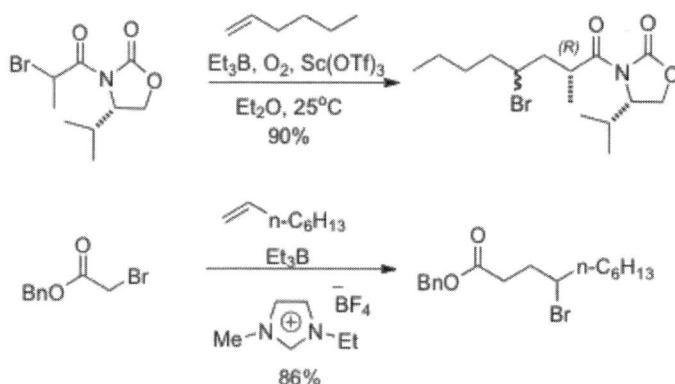

Scheme 7.20   Bromine atom transfer.

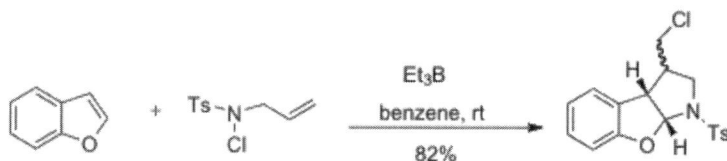

Scheme 7.21   Radical [3 + 2] annulation involving *N*-allyl-*N*-chlorotosylamide.

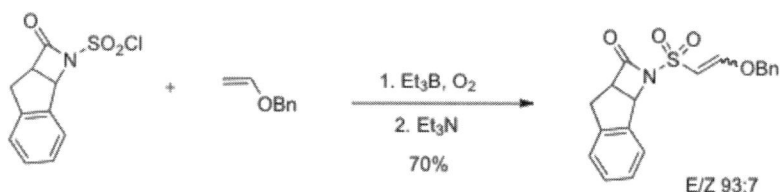

Scheme 7.22   Vinylation of β-lactamido *N*-sulfonyl chloride.

7.5 ORGANOBORON COMPOUNDS AS A SOURCE OF CARBON-CENTERED RADICALS

7.5.1 Conjugated Additions to Enones and Enals

One of the first synthetic applications of organoboranes in radical chemistry is the conjugate addition to enones (Scheme 7.23a) and enals reported by Brown and coworkers.[58–61] Addition reactions to β-substituted enones and enals are not spontaneous and initiation with the oxygen,[62] diacetyl peroxide,[63] or under irradiation[63] is necessary (Scheme 7.23b). A serious drawback of this strategy is that only one of the three alkyl groups is efficiently transferred, so the method is restricted to trialkylboranes derived from the hydroboration of easily available and cheap alkenes. To overcome this limitation *B*-alkylboracyclanes have been used but this approach was not successful for the generation of tertiary alkyl radicals.[64,65]

Brown proposed a mechanism where the enolate radical resulting from the radical addition reacts with the trialkylborane to give a boron enolate and a new alkyl radical that can propagate the chain (Scheme 7.24).[61] The formation of the intermediate boron enolate was confirmed by 1H NMR spectroscopy.[66,67] The role of the water present in the system is to hydrolyze the boron

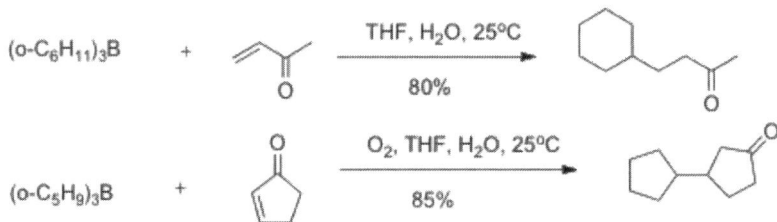

Scheme 7.23   The Brown conjugate addition.

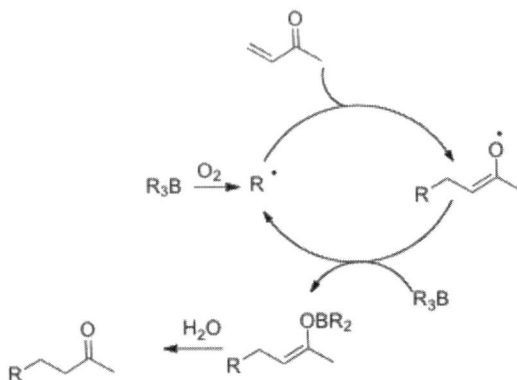

Scheme 7.24   Brown mechanism for the conjugate addition of organoboranes to methyl vinyl ketone.

enolate and to prevent its degradation by undesired free radical processes. This hydrolysis step is essential when alkynones[68] and acrylonitrile[58] are used as radical traps since the resulting allenes or keteneimines respectively react readily with radical species. Maillard, Walton *et al.* have shown, by ^{11}B NMR, ^{1}H NMR and IR spectroscopy, that triethylborane does complex with methyl vinyl ketone, acrolein and 3-methylbut-3-en-2-one.[69] They proposed that the reaction of triethylborane with these traps involves complexation of the trap by the Lewis acidic borane prior to conjugate addition.

The reaction between trialkylboranes and enones has found some interesting synthetic applications. An example is the preparation of prostaglandin precursors from *exo*-methylene cyclopentanone, generated *in situ* from a Mannich base. After dehydrogenation, a second conjugate addition of trioctylborane was used to introduce the ω-chain (Scheme 7.25).[70]

Several attempts to take advantage of the intermediate boron enolate to achieve a tandem conjugate addition–aldol reaction have been proposed.[71] Recently, Chandrasekhar *et al.*[72] reported the addition of triethylborane to methyl vinyl ketone followed by the *in situ* trapping of the enolate by aromatic aldehyde (Scheme 7.26).

Toru's group has investigated the stereoselectivity of the conjugate addition of trialkylboranes to 2-arylsulfinylcyclopentenones. Excellent stereocontrol is achieved with different alkyl radicals (Scheme 7.27).[73–76] In the acyclic series, the lack of diastereoselectivity in the addition step and a competitive Pummerer rearrangement have limited the synthetic potential of this reaction.[77]

A serious drawback of the trialkylborane approach is the requirement to use a 1:1 ratio of trialkylborane/radical trap to obtain good yields. Therefore, the method is restricted to trialkylboranes obtained by hydroboration of easily available and cheap alkenes. To overcome this limitation, *B*-alkylboracyclanes have been used.[64,65] According to Brown and Negishi, 3,3-dimethylborinane,

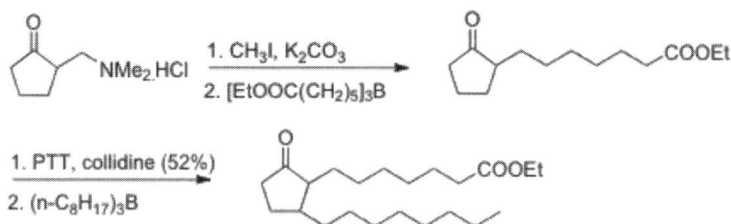

Scheme 7.25 Conjugate addition of a functionalized trialkylborane.

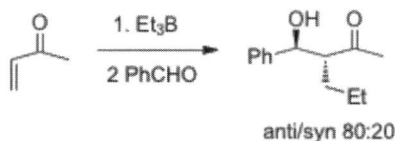

Scheme 7.26 Tandem conjugate addition–aldol reaction.

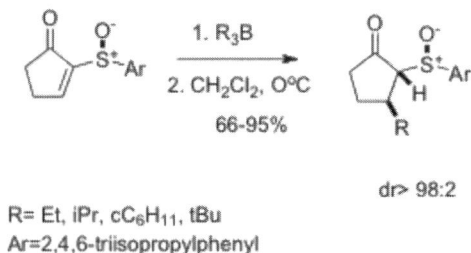

R= Et, iPr, cC₆H₁₁, tBu

$R = Et, iPr, cC_6H_{11}, tBu$
$Ar = 2,4,6\text{-triisopropylphenyl}$

Scheme 7.27 Stereoselective addition to 2-arylsulfinylcyclopentenones.

prepared from BH_3 and 2,4-dimethyl-1,4-pentadiene, is the most efficient reagent.[65] With this system, a selective cleavage of the boron-alkyl bond is possible for secondary and tertiary alkyl groups (Scheme 7.28). This method, referred to later as the Brown–Negishi reaction, is not suitable for primary alkyl radicals (yield <35%) or for radical traps substituted at the β-position. With these traps, the addition of extra oxygen is necessary to run the chain reaction and under these conditions the cleavage of the carbon–boron bond is no longer selective.

Recently, we have shown that similar results are obtained with cyclohex-yldiethylborane (easily prepared from Et_2BH and cyclohexene). The efficient addition to methyl vinyl ketone is possible (Scheme 7.29a). However, when cyclohexenone is used as radical trap, the addition of oxygen is necessary and a 3:1 mixture of products resulting from the addition of cyclohexyl and ethyl radicals is obtained (Scheme 7.29b).[78]

In order to circumvent the lack of selectivity in the cleavage of trialkyl-boranes, *B*-alkylcatecholboranes can be used as precursor of alkyl radicals. They are extremely sensitive towards oxygen and they react readily with alkoxyl radicals. It was clearly demonstrated by ESR that the perboryl radical

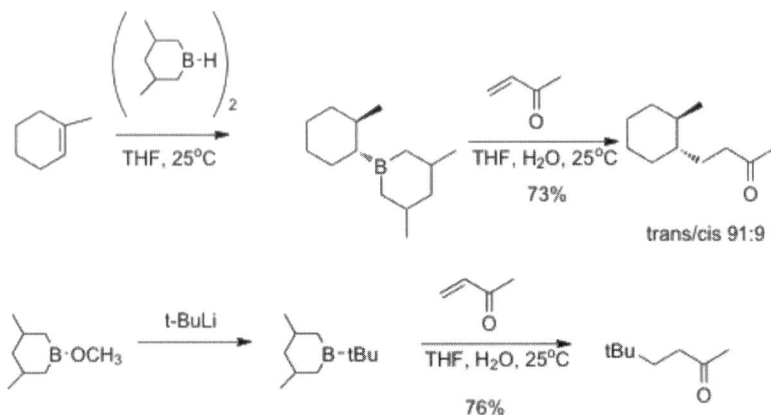

Scheme 7.28 Brown–Negishi reaction: selective formation of secondary and tertiary alkyl radicals.

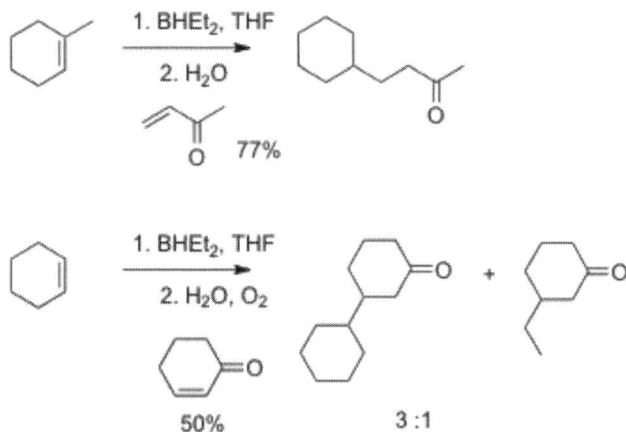

Scheme 7.29 Diethylborane-mediated conjugate addition of secondary alkyl radicals.

intermediate resulting from the complexation of *B*-methylcatecholborane with the alkoxyl radical is stabilized by delocalization onto the aromatic ring (Scheme 7.30).[79]

The observation of Davies and Roberts regarding the stability of the perboryl radical is at the origin of our own investigations about the use of *B*-alkylcatecholboranes as radical precursors. *B*-alkylcatecholboranes are expected to be more reactive than trialkylboranes and they are easily prepared from olefins *via* hydroboration with catecholborane with or without a catalyst. However, the most attractive feature of *B*-alkylcatecholboranes is the possibility of selectively generating one alkyl radical from an olefin, a possibility that trialkylboranes do not offer since no selective cleavage of the desired alkyl group is observed (*vide supra*). Indeed, reaction of *B*-alkylcatecholborane with a heteroatom-centered radical leads in an irreversible manner to the alkyl radical since cleavage of the "wrong" B–O bond is a reversible process that finally leads to the irreversible formation of an alkyl radical (Scheme 7.31).[9]

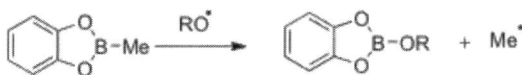

Scheme 7.30 Reaction of *B*-methylcatecholborane with alkoxyl radicals.

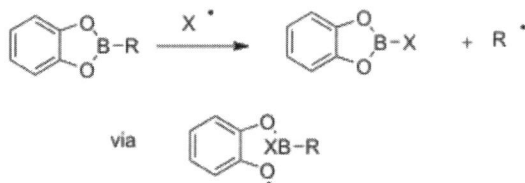

Scheme 7.31 Irreversible formation of alkyl radicals from *B*-alkylcatecholboranes.

A modified version of the Brown–Negishi reaction using *B*-alkylcatechol-boranes was reported (Scheme 7.32). This novel method is based on a simple one-pot procedure involving the hydroboration of various substituted alkenes with catecholborane, followed by treatment with catalytic amount of oxygen/DMPU/water and a radical trap. Efficient radical additions to α,β-unsaturated ketones and aldehydes have been reported. Primary alkyl radicals are efficiently generated by this procedure and the reaction has been applied to a 300 mmol scale synthesis of the γ-side chain of (–)-perturasinic acid (Scheme 7.32a).[80] The reaction was also applied to the radical addition to cyclohexenone (Scheme 7.32b) and to other β-substituted enones and enals as well as to cyclization (Scheme 7.32c) and annulation reactions.[78]

The reaction of trialkylboranes with 1,4-benzoquinones to give in quantitative yield 2-alkylhydroquinones was the first reaction of this type occurring without the assistance of a metal mediator.[81,82] An ionic mechanism was originally proposed but rapidly refuted since the reaction is inhibited by radical scavengers such as galvinoxyl and iodine.[83] This procedure is in many cases superior to the more widely use organometallic additions, for instance, when primary and secondary alkyl radicals have been used and afford the addition products in high yield (Scheme 7.33).[84]

The addition of *B*-alkylcatecholboranes to quinones has recently been investigated.[85] A good yield of the expected conjugate addition product is obtained with primary and most secondary radicals (Scheme 7.34a). However, hindered secondary radicals and tertiary alkyl radicals afford an unexpected product resulting from a radical addition to the oxygen atom of the quinone (Scheme 7.34b).

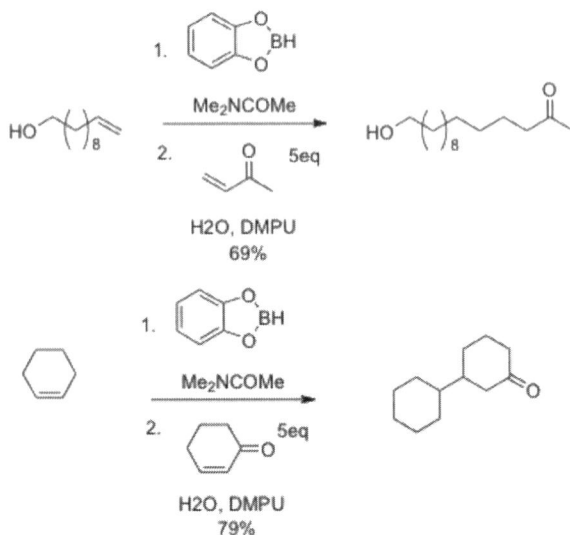

Scheme 7.32   *B*-Alkylcatecholborane-mediated addition of organoboranes to enones.

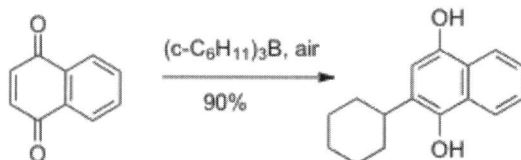

Scheme 7.33 Addition of alkyl radicals to quinones using trialkylboranes.

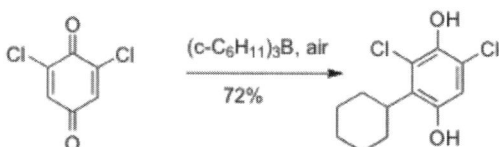

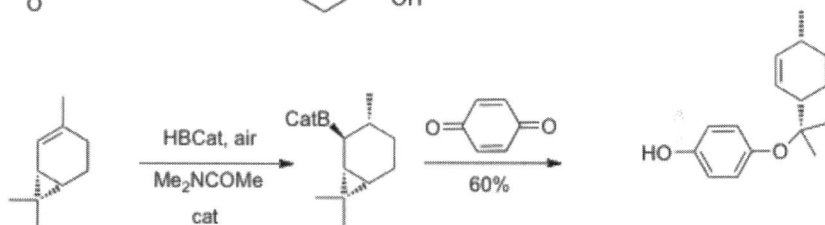

Scheme 7.34 Addition of *B*-alkylcatecholboranes to quinones: C- versus O-addition.

7.5.2 Conjugate Addition to Activated Olefins

The modified Brown–Negishi and the *B*-alkylcatecholborane conjugate additions described above are limited to enone and enal radical traps. Other radical traps such as unsaturated esters, amides and sulfones fail to react under these conditions. This failure was interpreted as a consequence of an inefficient reaction of the radical adduct and *B*-alkylcatecholboranes. This inefficiency is caused by the insufficient density of unpaired electrons on the oxygen atom of theses radicals relative to ketone–enolate and aldehyde–enolate radicals. The use of a chain-transfer reagent which is able to convert a carbon-centered radical into an oxygen-centered radical allows one to solve this problem. The Barton carbonate PTOC-OMe (PTOC = pyridine-2-thione-*N*-oxycarbonyl)[86,87] proved to be an excellent radical chain-transfer reagent according to Scheme 7.35a.[88] Interestingly, the same reagent proved to be an excellent initiator under irradiation with a standard 150 W tungsten lamp (Scheme 7.35b).

PTOC-OMe is a stable reagent easily obtained by the reaction of the commercially available sodium salt of *N*-hydroxypyridine-2-thione with methyl chloroformate. A related strategy has been developed by Dalko *et al.* using the Barton ester PTOC-Ph as a chain-transfer reagent.[89] In a preliminary study, *in situ* generated *B*-alkylcatecholboranes were allowed to react with PTOC-OMe under irradiation with a standard 150 W lamp. The *S*-pyridyl products coming from primary, secondary and tertiary alkyl radicals were

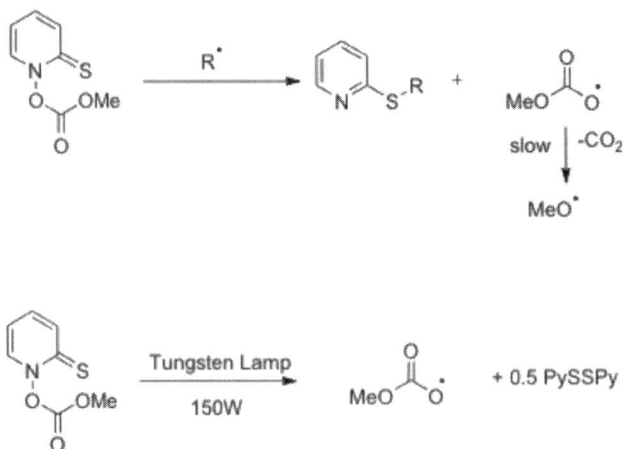

Scheme 7.35   Barton carbonate PTOC-OMe, a radical chain-transfer reagent able to convert a C-centered radical into an O-centered radical (a) and a radical initiator (b).

isolated in moderate to good yields.[88] Based on these initial results, a procedure for conjugate addition to various activated alkenes was developed. A one-pot procedure involving hydroboration of an alkene with catecholborane followed by irradiation in the presence of five equivalents of an activated alkene and three equivalents of the chain-transfer reagent PTOC-OMe was developed (Scheme 7.36).[88]

In contrast to the tin hydride-mediated reaction (Giese reaction),[90] no slow addition of the chain carrier is necessary. This is easily understandable from the reaction mechanism depicted in Scheme 7.37. In the Giese reaction, the tin hydride reduces the initial alkyl radical and the radical adduct at approximately the same rate. Therefore, in order to favour the product of conjugate addition, it is compulsory to work with a low concentration of tin hydride. In the catecholborane-mediated reaction, the initial radical reacts much more slowly than the radical adduct with the PTOC-OMe chain-transfer reagent. Indeed, a nucleophilic alkyl radical adds more slowly to the sulfur atom of a thiocarbonyl group than a radical having a marked electrophilic character such as the radical adduct.

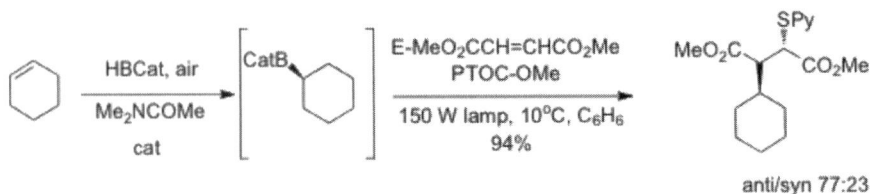

anti/syn 77:23

Scheme 7.36   PTOC-OMe mediated conjugate addition of *B*-alkylcatecholborane to activated alkenes.

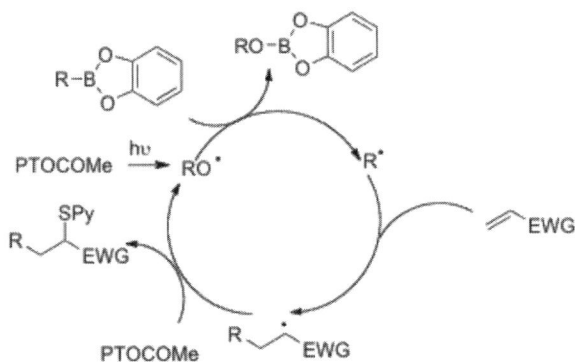

Scheme 7.37 Radical chain mechanism for the conjugate addition of *B* alkylca-techolboranes to activatedolefins (R = alkyl group; EWG = electron-withdrawing group; R–O$^{\bullet}$ = MeOC(O)O, MeO$^{\bullet}$).

B-alkylcatecholboranes, prepared by rhodium(I)-catalyzed hydroboration of alkenes, are suitable radical precursors for conjugate addition to activated olefins. This procedure is particularly useful for the control of the regio- and chemoselectivity of such tandem processes.[91] A one-pot enantioselective hydroboration–radical conjugate addition was successfully performed. For example, the reaction between norbornene and methyl methacrylate as a radical trap affords the product of conjugate addition in 68% yield and 85% ee (after desulfurization) using [Rh(COD)Cl]$_2$ and the chiral diphosphine (*S,S*)-BDPP as a catalyst for the hydroboration step (Scheme 7.38).

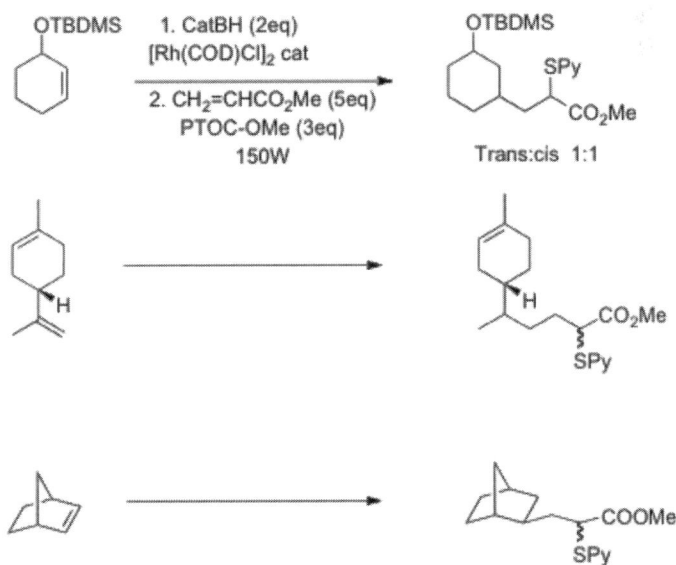

Scheme 7.38 Control of the enantioselectivity *via* rhodium(I)-catalyzed hydroboration.

The rhodium-catalyzed hydroboration has opened the way to cyclization reactions starting from dienes.[92] For instance, rhodium-catalyzed hydroboration of the terminal alkenyl group of an α,β-unsaturated lactone followed by reaction with the PTOC-OMe chain-transfer reagent afforded the bicyclic α-S-pyridyl lactone in 63% yield (Scheme 7.39). After oxidation of the sulfide with *m*-CPBA, thermal elimination of the sulfoxide afforded the corresponding α-methylene lactone in 65% yield. Interestingly, such bicyclic α-methylenelactones are substructures that can be found in many natural products such as mirabolide.[93]

7.5.3   Addition to Imine Derivatives

Intramolecular addition of trialkylboranes to imines and related compounds has been reported and the main results are in review articles.[94,95] The addition of ethyl radicals generated from Et_3B to aldimines affords the desired addition product in fair to good yield but low diaster control (Scheme 7.40a).[96] Similar reactions with aldoxime ethers,[97] aldehyde hydrazones,[97] and *N*-sulfonylaldimines[98] are reported. Radical addition to ketimines has been recently reported (Scheme 7.40b).[99] The addition of triethylborane to 2H-azirine-3-carboxylate derivatives is reported.[100] Very recently, Somfai *et al.* have extended this reaction to the addition of different alkyl radicals generated from trialkylboranes to a chiral ester of 2*H*-azirine- 3-carboxylate under Lewis acid activation with CuCl (Scheme 7.40c).[101]

   Naito *et al.* reported a radical addition to nitrone that occurs with high stereocontrol (Scheme 7.41).[102] Interestingly, most of the reactions reported with imines and related products such as oximes, oxime ethers, hydrazones and nitrones can be run in aqueous media.[103]

7.5.4   C–C Bond Formation *via* β-Fragmentation Processes

Brown and Suzuki have shown that treatment of trialkylboranes with ethenyl- (Scheme 7.42a) and ethynyloxiranes (Scheme 7.42b), in the presence of a catalytic amount of oxygen, affords the corresponding allylic or allenic

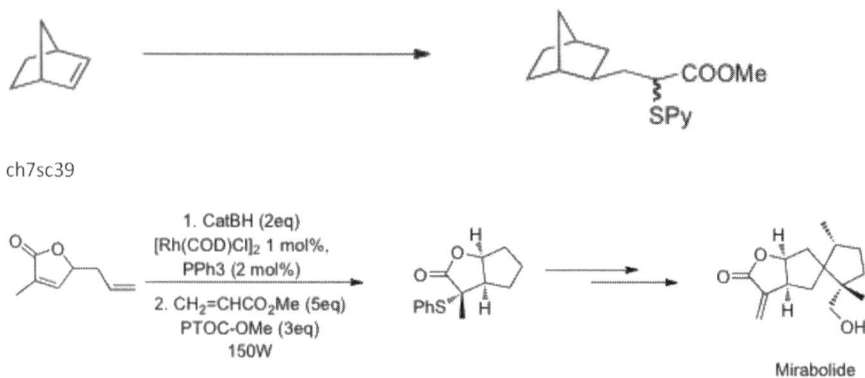

ch7sc39

Mirabolide

Scheme 7.39   Preparation of α-methylenelactone.

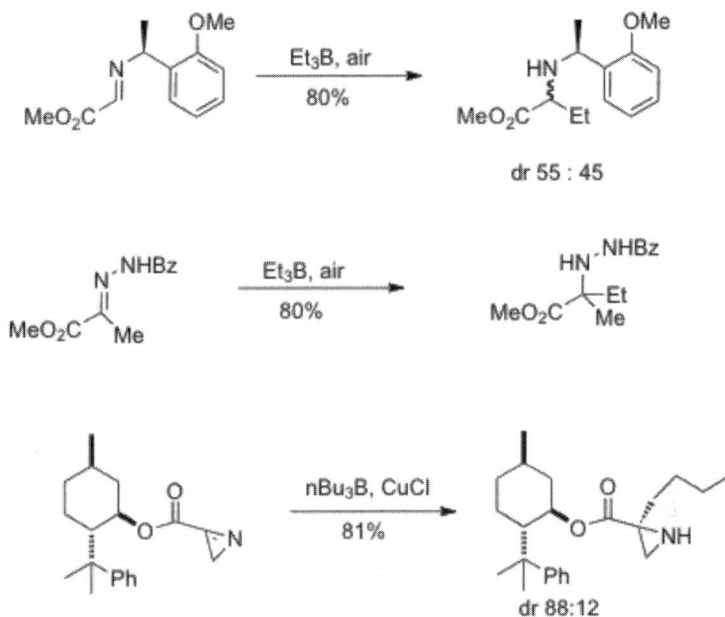

Scheme 7.40 Addition of trialkylboranes to aldimines, ketimines and related compounds.

alcohols.[61,104–106] The mechanism may involve the addition of alkyl radicals to the unsaturated system leading to 1-(oxiranyl)alkyl and 1-(oxiranyl)alkenyl radicals followed by rapid fragmentation to give alkoxyl radicals that finally complete the chain process by reacting with the trialkylborane.[104–106]

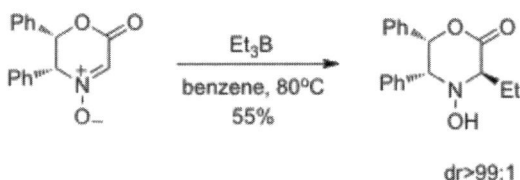

Scheme 7.41 Addition of triethylborane to nitrones.

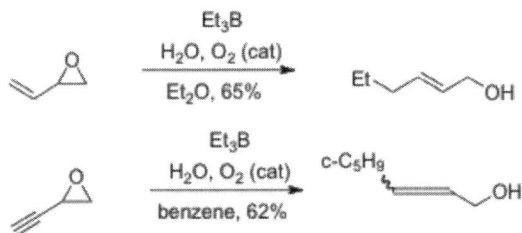

Scheme 7.42 Reaction of trialkylboranes with ethenyl- and ethynyloxiranes.

The free radical substitution of β-nitrostyrene (2-nitroethenylbenzene) by trialkylboranes involves a radical addition to the β-position (α to the nitro group) followed by fragmentation of $NO_2^•$ (Scheme 7.43). The reaction is *E* selective and works with a broad range of trialkylboranes, allowing the introduction of tertiary, secondary and allylic carbon moieties.[107]

Nozaki *et al.* reported the reaction of trialkylboranes with styryl sulfoxides and sulfones. Alkyl radicals generated from trialkylboranes add at the β-position of β-styryl sulfoxides and sulfones (α- to the sulfur atom). The resulting radicals fragment and deliver the β-styryl adducts.[108] Interestingly, the sulfoxides eliminate very rapidly leading to partially stereospecific substitution (Scheme 7.44). The radical nature of the process is demonstrated by the presence of a side product derived from the solvent (THF) by hydrogen atom abstraction.

Radical allylation of *B*-alkylcatecholboranes using easily available allylsulfones has been described,[109-111] By using phenylsulfones, the fragmentation produces a stable phenylsulfonyl radical that reacts with *B*-alkyl catecholborane to sustain the chain reaction (Scheme 7.45). Oxygen-centered radicals react efficiently with *B*-alkylcatecholboranes. Therefore, the easily available di-*tert*-butylhyponitrite was selected as an initiator due to its ability to furnish the *tert*-butoxyl radical at the refluxing temperature of dichloromethane. The thermal properties of this initiator allows one to run a one-pot hydroboration–radical reaction sequence by taking advantage of the very mild, efficient and cost effective hydroboration conditions developed by Garett and Fu.[112]

The desired products were obtained in satisfactory to excellent yields by using only 1.2 equivalents of the allylsulfones with primary, secondary and tertiary alkyl radicals. Many different types of allylic sulfones bearing an ester group, a sulfonyl group, and a bromine atom react equally well (Scheme 7.46).

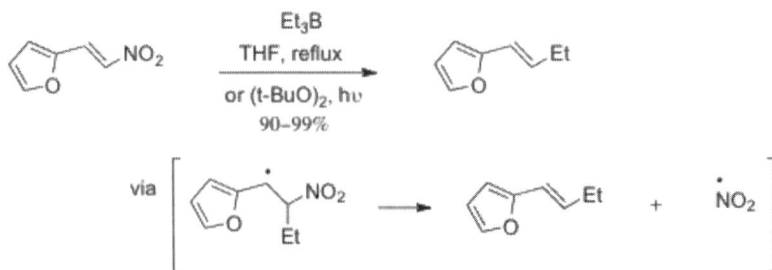

Scheme 7.43   Radical substitution of (2-nitrovinyl)arenes with Et_3B.

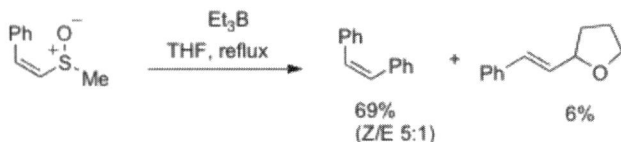

Scheme 7.44   Vinylation with styryl methyl sulfoxide.

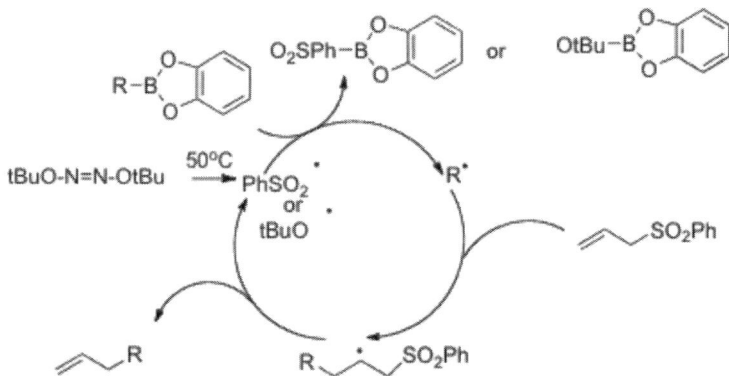

Scheme 7.45   Radical hydroallylation of alkenes.

The whole transformation represents formally a reductive allylation or hydroallylation of alkenes.

   Interestingly, this allylation process seems to be very general. For instance, the introduction of a dienyl moiety using penta-2,4-dienyl phenyl sulfone has been achieved (Scheme 7.47a). The modest yield (50%) for the conversion is

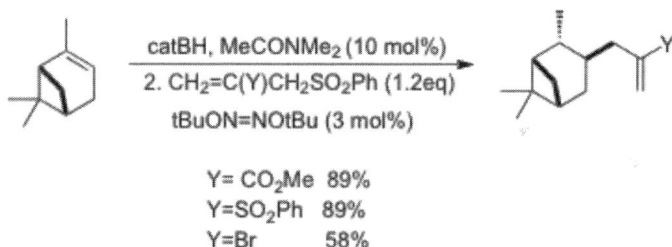

Y= CO₂Me 89%
Y=SO₂Ph 89%
Y=Br   58%

Scheme 7.46   Allylation of *B*-alkylcatecholboranes with allylsulfones.

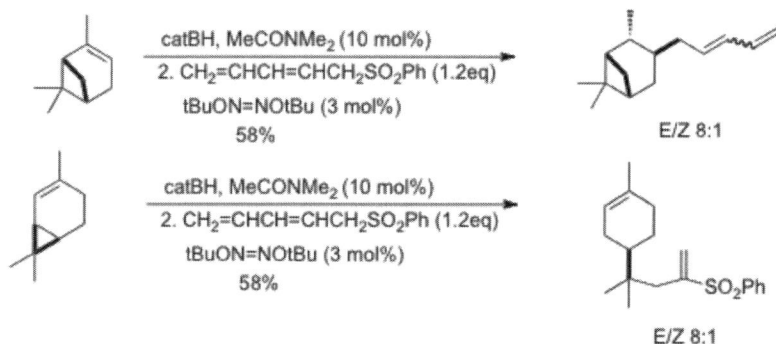

Scheme 7.47   Introduction of a dienyl moiety using penta-2,4-dienyl phenyl sulfone
(a) and hydroallylation of (+)-2-carene (b).

due to the instability of the dienyl sulfone which readily polymerizes. Finally, the radical nature of the process has been demonstrated by running an allylation reaction with (+)-2-carene (Scheme 7.47b).

With this radical probe, the intermediate cyclopropylmethyl radical undergoes ring opening to a homoallylic radical that is trapped by the allylic sulfone to afford the corresponding monocyclic compound in 58% yield. Radical coupling of *B*-alkylcatecholboranes, *in situ* generated from the corresponding alkenes, with ethyl 2-(benzenesulfonylamino)acrylate is reported (Scheme 7.48).[113]

This reaction represents an extension of the radical allylation of *B*-alkylcatecholboranes by allysulfones. This unique process allows one to prepare various α-ketoesters (alkylated pyruvates) in a straightforward manner. It also demonstrates the generality of the radical-mediated C–C bond formation starting from organoboranes and allylic benzenesulfonyl derivatives.

7.6 ORGANOBORANES AS CHAIN-TRANSFER REAGENTS

We have seen in the preceding chapters that trialkylboranes are useful radical initiators as well as an efficient source of alkyl radicals. Organoboranes can also be used as chain-transfer reagents. This approach is used when the direct reaction between the radical precursor and the radical trap cannot proceed (Scheme 7.49a). Alkyl radicals generated from the organoboranes are not involved in product formation, but they produce the radicals leading to

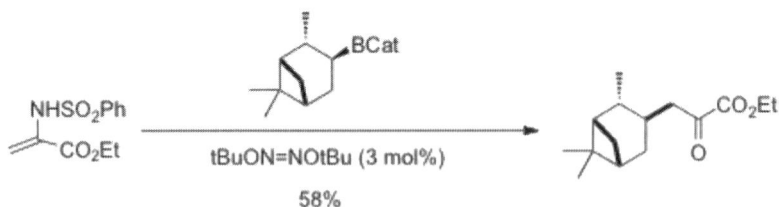

Scheme 7.48   Alkylation of ethyl pyruvate *via* reductive coupling of alkenes and ethyl 2-(benzenesulfonylamino)acrylate.

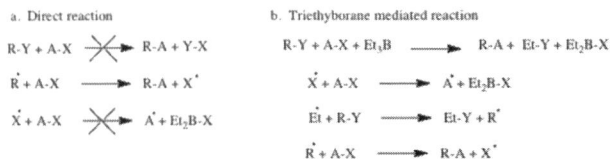

where R-Y = radical precursor, R-A = desired product, A-X = radical trap, X· = heteroatom centered radical, R· = radical involved in the formation.

Scheme 7.49   Triethylborane as a chain-transfer reagent for the conversion of R–Y to R–A.

products. For this purpose, an extra step such as an iodine atom transfer or a hydrogen abstraction is necessary. This point is schematically illustrated in Scheme 7.49 for a triethylborane mediated process (Scheme 7.49b). This reaction takes advantage of the high affinity of trialkylboranes for heteroatom-centered X$^{\bullet}$ radicals.

7.6.1   *Via* Iodine Atom Transfer

Radical addition of organoboranes to imines and related compounds is a promising alternative to the use of classical organometallic compound. However, this approach is limited to the few trialkylboranes that are easily available and cheap since only one of the three alkyl groups is transferred. By using a triethylborane as a chain-transfer reagent, the reaction could be extended to alkyl iodides as radical precursors. The groups of Bertrand[94,114] and Naito[95,97] both reported the use of triethylborane for the tin-free addition of alkyl iodides to imines. A typical example for a tentative of asymmetric addition to a glyoxylate imine is depicted in Scheme 7.50a. More recently, additions to isatin imines were reported (Scheme 7.50b) as well as addition to 2*H*-aziridine-3-carboxylates by Lemos *et al.*[100] and Somfai *et al.*[101] (Scheme 7.50c).

Enantioselective radical addition to *N*-acyl hydrazone using triethylborane as a chain-transfer reagent has been reported by Friestad and coworkers. Enantiomeric excesses up to 95% were obtained in the presence of copper(II)-bisoxazoline as a Lewis acid (Scheme 7.51).[115]

Tandem processes mediated by triethylborane involving conjugate addition to enones followed by an aldol reaction are reported (Scheme 7.52a). More

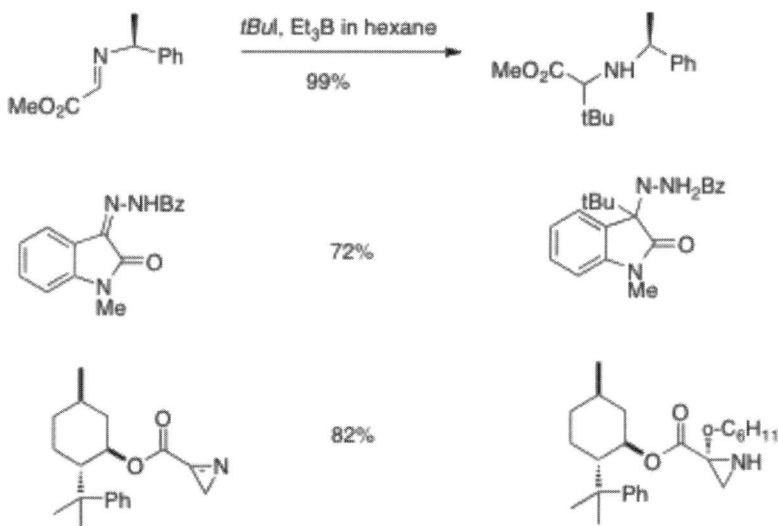

Scheme 7.50   Triethylborane-mediated addition of alkyliodides to imines.

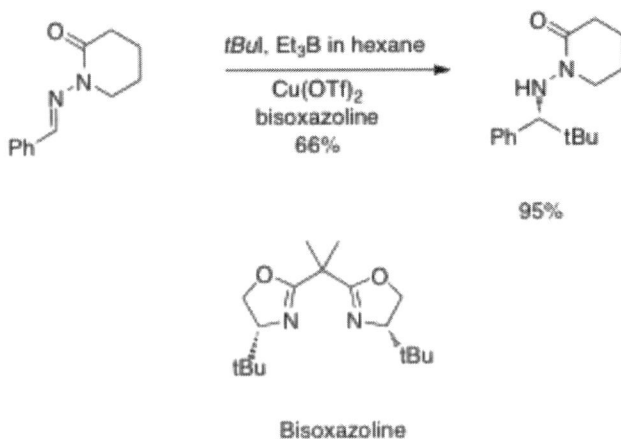

Bisoxazoline

Scheme 7.51   Enantioselective addition to *N*-acyl hydrazones.

recently, a tandem process involving addition of an isopropyl radical to an α,β-unsaturated oxime ether afforded an azaenolate intermediate that reacts with benzaldehyde in the presence of trimethylaluminum. The aldol product cyclizes to afford an isopropyl-substituted γ-butyroloactone in 61% overall yield (Scheme 7.52).[116] In these reactions, triethylborane is acting as a chain-transfer reagent that delivers a boron enolate or azaenolate necessary for the aldolization process.

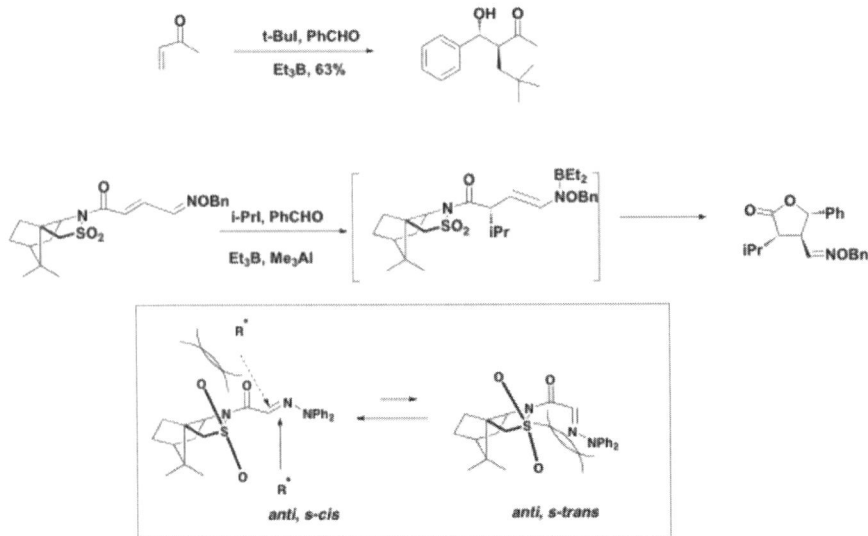

Scheme 7.52   Tandem radical addition aldol reaction.

Alkenylation of alkyl iodides with β-nitrostyrene derivatives has been reported (Scheme 7.53).[117]

The reaction is, however, not strictly a chain process since a stoichiometric amount of oxygen is necessary to run the reaction. The radical $NO_2^•$ is presumably not sufficiently reactive towards triethylborane to sustain the chain process. The radical carboazidation of alkenes has been achieved in water using triethylborane as an initiator.[118] This efficient process is complete in one hour at room temperature in an open-to-air reaction vessel (Scheme 7.54a). These new tin-free carboazidation conditions are environmentally friendly and allow one to run reactions with an excess of either the alkene or the radical precursor. They are also suitable for simple radical azidation of alkyl iodide as well as for more complex cascade reactions involving annulation processes (Scheme 7.54b). In both reactions (Scheme 7.54a and b), an excess of triethylborane (3 equivalents) is required to obtain a good yield. This may be an indication that the chain process, more precisely the reaction between the phenylsulfonyl radical and Et_3B, is not efficient.

7.6.2 *Via* Hydrogen Atom Transfer

All the examples presented under Section 7.6.1 used an iodine atom transfer to generate the desired radicals. Another approach involving abstraction of hydrogen atom is also reported. For instance, ethers and acetals undergo direct intermolecular addition to aldehydes under treatment with Et_3B/air

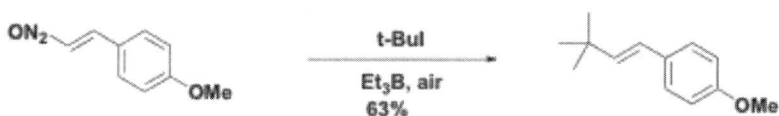

Scheme 7.53 Alkenylation of alkyl iodides.

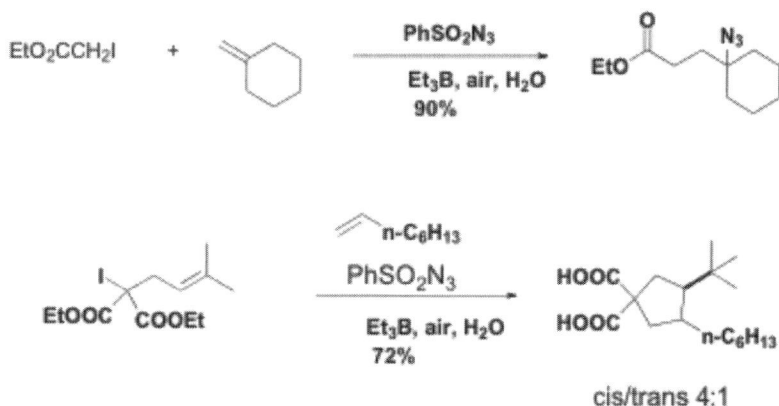

Scheme 7.54 Triethylborane-mediated carboazidation.

(Scheme 7.55a).[119] A plausible mechanism is depicted in Scheme 7.55 and involves radical addition of the 2-tetrahydrofuryl radical to the aldehyde followed by a rapid reaction of the alkoxyl radical with Et_3B. Triethylborane has a crucial role since by reacting with the alkoxyl radical it favours the formation of the condensation product relative to the β-fragmentation process (back reaction). A similar reaction with tertiary amines, amides and urea is also possible (Scheme 7.55b).[120]

7.7 ORGANOBORON COMPOUNDS AS RADICAL-REDUCING AGENTS

7.7.1 Complexes with Tertiary Amines

In pioneer work, Roberts investigated the use of amine–borane complexes as radical-reducing agents. This research led him to develop the concept of polarity-reversal catalysis.[121] He found that a slow hydrogen atom-abstraction step due to mismatched polarity can be replaced by two rapid steps with matched polarity. For example, the slow abstraction of an hydrogen atom from acetonitrile by a *tert*-butoxyl radical (mismatched polarity) is replaced by a rapid reduction of the *tert*-butoxyl radical with an amine–borane complex and by the abstraction of an hydrogen atom from acetonitrile by the amine–boryl radical.[122,123] Attempts to use this concept for the kinetic resolution of chiral esters and lactones by using chiral amine–borane complexes led to interesting enantioselectivities (Scheme 7.56).[124–126]

7.7.2 Complexes with Water and Alcohols

Wood *et al.*[127] reported an innovative development of the Barton–McCombie deoxygenation of alcohols allowed to work under tin-free conditions (Scheme 7.57). A trimethylborane–water complex proves to be an efficient reagent for the reduction of xanthates. The complexation of water by trimethylborane induces a strong decrease of the O–H bond dissociation energy from 116 kcal mol^{-1} (water) to 86 kcal mol^{-1} (Me$_3$B–water complex).

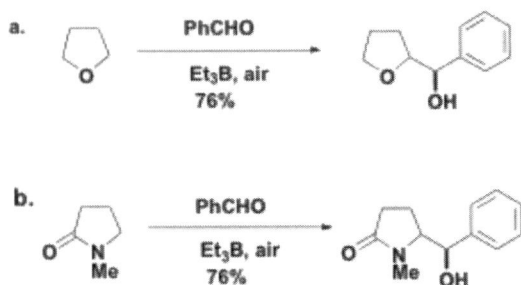

Scheme 7.55   Radical hydroxyalkylation of C–H bonds adjacent to oxygen and nitrogen.

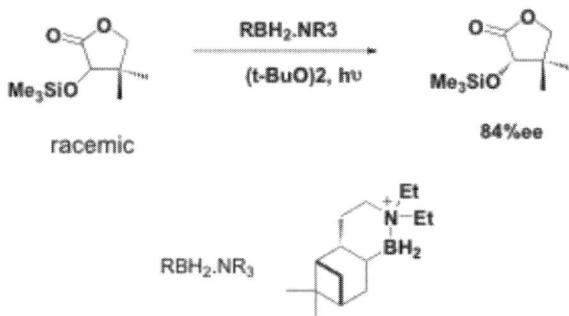

Scheme 7.56 Kinetic resolution of a γ-lactone.

Renaud's group made a similar observation when they reported a mild and efficient radical-mediated reduction of organoboranes (Scheme 7.58a).[128] An *in situ* generated *B*-methoxycatecholborane–methanol complex acts as a reducing agent. The radical nature of the process was demonstrated by using (+)-2-carene as a radical probe (Scheme 7.58b). Water, ethanol and trifluoroethanol can be used instead of MeOH with very similar efficiency.[129–138]

The reaction mechanism of this transformation is depicted in Scheme 7.59 and involves activation of the O–H bond of methanol by complexation with *B*-methoxycatecholborane. Interestingly, after fragmentation of the radical–ate complex the reduction leads to a methoxyl radical that reacts very efficiently with the *B*-alkylcatecholborane, warranting an efficient chain process.

7.8 CONCLUSIONS

The tremendous development of the use of radicals in organic synthesis and the necessity of avoiding the use of tin derivatives because of their toxicity has

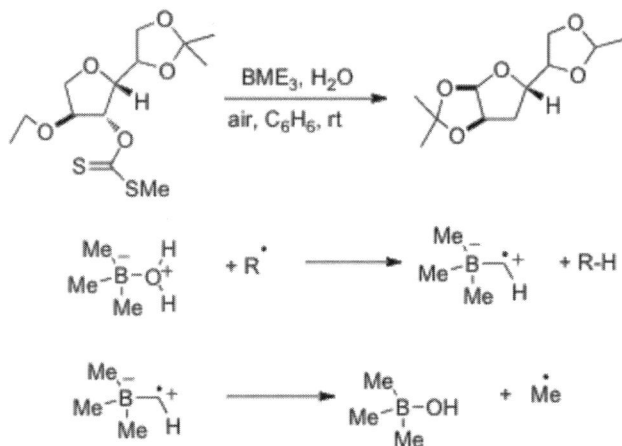

Scheme 7.57 Barton–McCombie deoxygenation with Me_3B–water complex.

1). CatBH (2eq)
MeCONMe2 (0.1eq)

2) MeOH (4eq)
Procedure A or B

Procedure A: air(30 mol% O2)
CH$_2$Cl$_2$, reflux, 90min, 80%
Procedure B:
tBuOOt-Bu (0.15eq)
C$_6$H$_4$Cl$_2$, microwave 300W, 140°C, 15 min, 74%

Proc. A
56%,
> 75% (GC)

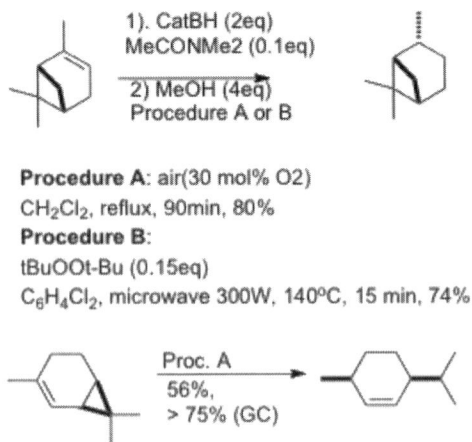

Scheme 7.58 Mild radical-mediated reduction of organoborane with methanol.

led to a revival of the radical chemistry of organoboranes. The use of triethylborane as an initiator for radical chain reactions is now part of the classical arsenal of organic chemists. The generation of more complex and functionalized radicals from organoboranes is of great interest since it allows one to consider olefins as a potential source of radicals. So far, the generation of radicals has not been extended to alkenyl and aryl radicals, but rapid progress is expected in this field. Interestingly, organoboranes could also play the role of chain-transfer reagents in radical processes. Due to the particularly rich reactivity of boron derivatives, the design of tandem processes involving radical and non-radical reactions is now possible. Finally, boron derivatives are promising reagents for activating water and alcohols and making them suitable reagents for the reduction of radicals. Spectacular developments in this particular field are expected in the near future.

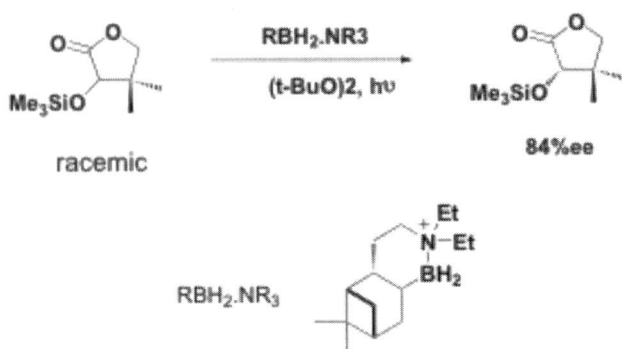

RBH$_2$.NR3

(t-BuO)2, hν

Me$_3$SiO   racemic

Me$_3$SiO   84%ee

RBH$_2$.NR$_3$

Et
N—Et
BH$_2$

Scheme 7.59 Chain mechanism of the reduction of organoboranes.

REFERENCES

1. H. C. Brown and M. M. Midland, *Angew. Chem., Int. Ed. Engl.*, 1972, **11**, 692.
2. A. Ghosez, B. Giese and H. Zipse, *Methoden Org. Chem. (Houben-Weyl)*, 1989, **E19a**, 753.
3. C. Ollivier and P. Renaud, *Chem. Rev.*, 2001, **101**, 3415.
4. A. G. Davies and B. P. Roberts, *Acc. Chem. Res.*, 1972, **5**, 387.
5. A. G. Davies, *Pure Appl. Chem.*, 1974, **39**, 497.
6. K. Nozaki, K. Oshima and K. Utimoto, *J. Am. Chem. Soc.*, 1987, **109**, 2547.
7. H. Yorimitsu and K. Oshima, in *Radicals in Organic Synthesis*, ed. P. Renaud and M. P. Sibi, Wiley, Weinheim, 2001, vol. 1, p. 11.
8. R. A. Batey and D. V. Smil, *Angew. Chem., Int. Ed.*, 1999, **38**, 1798.
9. A.-P. Schaffner and P. Renaud, *Eur. J. Org. Chem.*, 2004, 2291.
10. K. Miura, Y. Ichinose, K. Nozaki, K. Fugami, K. Oshima and K. Utimoto, *Bull. Chem. Soc. Jpn.*, 1989, **62**, 143.
11. P. A. Evans and J. D. Roseman, *J. Org. Chem.*, 1996, **61**, 2252.
12. K. L. Salazar and K. M. Nicholas, *Tetrahedron*, 2000, **56**, 2211.
13. H. Yorimitsu, H. Shinokubo and K. Oshima, *Synlett,* 2002, 674.
14. O. Yamazaki, H. Togo, G. Nogami and M. Yokoyama, *Bull. Chem. Soc. Jpn.*, 1997, **70**, 2519.
15. F. Gonzalez-Lopez de Turiso and D. P. Curran, *Org. Lett.*, 2005, **7**, 151.
16. T. Nakamura, H. Yorimitsu, H. Shinokubo and K. Oshima, *Synlett,* 1999, 1415.
17. D. H. R. Barton, S. I. Parekh and C.-L. Tse, *Tetrahedron Lett.*, 1993, **34**, 2733.
18. H. Yorimitsu, H. Shinokubo and K. Oshima, *Chem. Lett.*, 2000, 104.
19. E. Lee and H. O. Han, *Tetrahedron Lett.*, 2002, **43**, 7295.
20. K. Miura, K. Oshima and K. Utimoto, *Bull. Chem. Soc. Jpn.*, 1993, **66**, 2356.
21. K. Miura, K. Oshima and K. Utimoto, *Bull. Chem. Soc. Jpn.*, 1993, **66**, 2348.
22. Y. Ichinose, K. Nozaki, K. Wakamatsu, K. Oshima and K. Utimoto, *Tetrahedron Lett.*, 1987, **28**, 3709.
23. Y. Ichinose, K. Wakamatsu, K. Nozaki, J.-L. Birbaum, K. Oshima and K. Utimoto, *Chem. Lett.*, 1987, 1647.
24. S. Tanaka, T. Nakamura, H. Yorimitsu, H. Shinokubo and K. Oshima, *Org. Lett.*, 2000, **2**, 1911.
25. M. Taniguchi, K. Oshima and K. Utimoto, *Chem. Lett.*, 1993, 1751.
26. S. Usugi, H. Yorimitsu and K. Oshima, *Tetrahedron Lett.*, 2001, **42**, 4535.
27. S. Deprele and J.-L. Montchamp, *J. Org. Chem.*, 2001, **66**, 6745.
28. A. Gautier, G. Garipova, O. Dubert, H. Oulyadia and S. R. Piettre, *Tetrahedron Lett.*, 2001, **42**, 5673.

29. C. M. Jessop, A. F. Parsons, A. Routledge and D. J. Irvine, *Tetrahedron: Asymmetry*, 2003, **14**, 2849.

30. K. Takami, H. Yorimitsu and K. Oshima, *Org. Lett.,* 2004, **6**, 4555.

31. K. Hirano, K. Fujita, H. Shinokubo and K. Oshima, *Org. Lett.,* 2004, **6**, 593.

32. K. Miura, M. Tanigushi, K. Nozaki, K. Oshima and K. Utimoto, *Tetrahedron Lett.,* 1990, **31**, 6391.

33. D. P. Curran, M.-H. Chen, E. Spetzler, C. M. Seong and C.-T. Chang, *J. Am. Chem. Soc.,* 1989, **111**, 8872.

34. J. Byers, in *Radicals in Organic Synthesis*, ed. P. Renaud and M. P. Sibi, Wiley, Weinheim, 2001, vol. 1, p. 72.

35. Y. Takeyama, Y. Ichinose, K. Oshima and K. Utimoto, *Tetrahedron Lett.,* 1989, **30**, 3159.

36. E. Baciocchi and E. Muraglia, *Tetrahedron Lett.,* 1994, **35**, 2763.

37. Y. Tang and C. Li, *Org. Lett.,* 2004, **6**, 3229.

38. Y. Ichinose, S.-I. Matsunaga, K. Fugami, K. Oshima and K. Utimoto, *Tetrahedron Lett.,* 1989, **30**, 3155.

39. H. Yorimitsu, T. Nakamura, H. Shinokubo and K. Oshima, *J. Org. Chem.,* 1998, **63**, 8604.

40. H. Yorimitsu, T. Nakamura, H. Shinokubo, K. Oshima, K. Omoto and H. Fujimoto, *J. Am. Chem. Soc.,* 2000, **122**, 11041.

41. T. Nakamura, H. Yorimitsu, H. Shinokubo and K. Oshima, *Synlett,* 1998, 1351.

42. M. Ikeda, H. Teranishi, N. Iwamura and H. Ishibashi, *Heterocycles,* 1997, **45**, 863.

43. M. Ikeda, H. Teranishi, K. Nozaki and H. Ishibashi, *J. Chem. Soc., Perkin Trans. 1*, 1998, 1691.

44. U. Albrecht, R. Wartchow and H. M. R. Hoffmann, *Angew. Chem., Int. Ed.,* 1992, **31**, 910.

45. T. J. Woltering and H. M. R. Hoffmann, *Tetrahedron*, 1995, **51**, 7389.

46. O. Kitagawa, Y. Yamada, H. Fujiwara and T. Taguchi, *Angew. Chem., Int. Ed.,* 2001, **40**, 3865.

47. J. Sugimoto, K. Miura, K. Oshima and K. Utimoto, *Chem. Lett.,* 1991, 1319.

48. K. Iseki, Y. Nagai and Y. Kobayashi, *Tetrahedron Lett.,* 1993, **34**, 2169.

49. K. Iseki, Y. Nagai and Y. Kobayashi, *Tetrahedron: Asymmetry*, 1994, **5**, 961.

50. K. Iseki, D. Asada, M. Takahashi, T. Nagai and Y. Kobayashi, *Chem. Pharm. Bull.,* 1996, **44**, 1314.

51. K. Iseki, D. Asada, M. Takahashi, T. Nagai and Y. Kobayashi, *Tetrahedron Lett.,* 1994, **35**, 7399.

52. Y. Itoh and K. Mikami, *Org. Lett.*, 2005, **7**, 4883.

53. Y. Kita, A. Sano, T. Yamaguchi, M. Oka, K. Gotanda and M. Matsugi, *Tetrahedron Lett.,* 1997, **38**, 3549.

54. C. L. Mero and N. A. Porter, *J. Am. Chem. Soc.,* 1999, **121**, 5155.

55. H. Yorimitsu and K. Oshima, *Bull. Chem. Soc. Jpn.*, 2001, **74**, 225–235.
56. T. Tsuritani, H. Shinokubo and K. Oshima, *Org. Lett.*, 2001, **3**, 2709.
57. F. Montermini, E. Lacote and M. Malacria, *Org. Lett.*, 2004, **6**, 921.
58. A. Suzuki, A. Arase, H. Matsumoto, M. Itoh, H. C. Brown, M. M. Rogic and M. W. Rathke, *J. Am. Chem. Soc.*, 1967, **89**, 5709.
59. H. C. Brown, M. M. Rogic, M. W. Rathke and G. W. Kabalka, *J. Am. Chem. Soc.*, 1967, **89**, 5709.
60. H. C. Brown, G. W. Kabalka, M. W. Rathke and M. M. Rogic, *J. Am. Chem. Soc.*, 1968, **90**, 4165.
61. G. W. Kabalka, H. C. Brown, A. Suzuki, S. Honma, A. Arase and M. Itoh, *J. Am. Chem. Soc.*, 1970, **92**, 710.
62. H. C. Brown and G. W. Kabalka, *J. Am. Chem. Soc.*, 1970, **92**, 712.
63. H. C. Brown and G. W. Kabalka, *J. Am. Chem. Soc.*, 1970, **92**, 714.
64. A. Suzuki, S. Nozawa, M. Itoh, H. C. Brown, E. Negishi and S. K. Gupta, *J. Chem. Soc., Chem. Commun.*, 1969, 1009.
65. H. C. Brown and E. Negishi, *J. Am. Chem. Soc.*, 1971, **93**, 3777.
66. W. Fenzl, R. Köster and H.-J. Zimmermann, *Liebigs Ann. Chem.*, 1975, 2201.
67. T. Mukaiyama, K. Inomata and M. Muraki, *J. Am. Chem. Soc.*, 1973, **95**, 967.
68. A. Suzuki, S. Nozawa, M. Itoh, H. C. Brown, G. W. Kabalka and G. W. Holland, *J. Am. Chem. Soc.*, 1970, **92**, 3503.
69. V. Beraud, Y. Gnanou, J. C. Walton and B. Maillard, *Tetrahedron Lett.*, 2000, **41**, 1195.
70. O. Attanasi, G. Baccolini, I. Caglioti and G. Rosini, *Gazz. Chim. Ital.*, 1973, **103**, 31.
71. K. Nozaki, K. Oshima and K. Utimoto, *Bull. Chem. Soc. Jpn.*, 1991, **64**, 403.
72. S. Chandrasekhar, C. Narsihmulu, N. R. Reddy and M. S. Reddy, *Tetrahedron Lett.*, 2003, **44**, 2583.
73. T. Toru, Y. Watanabe, N. Mase, M. Tsusaka, T. Hayakawa and Y. Ueno, *Pure Appl. Chem.*, 1996, **68**, 711.
74. T. Toru, Y. Watanabe, M. Tsusaka and Y. Ueno, *J. Am. Chem. Soc.*, 1993, **115**, 10464.
75. N. Mase, Y. Watanabe and T. Toru, *J. Org. Chem.*, 1998, **63**, 3899.
76. N. Mase, Y. Watanabe, Y. Ueno and T. Toru, *J. Org. Chem.*, 1997, **62**, 7794.
77. N. Mase, Y. Watanabe, Y. Ueno and T. Toru, *J. Chem. Soc., Perkin Trans. 1*, 1998, 1613.
78. C. Ollivier and P. Renaud, *Chem.–Eur. J.*, 1999, **5**, 1468.
79. J. A. Baban, N. J. Goodchild and B. P. Roberts, *J. Chem. Soc., Perkin Trans. 2*, 1986, 157.
80. A. H. Forster, PhD Thesis, Université de Fribourg, Switzerland, Diss. No. 1242, 1999.
81. M. F. Hawthorne and M. Reintjes, *J. Am. Chem. Soc.*, 1964, **86**, 951.

82. Hawthorne MF, Reintjes M (1965) *J. Am. Chem. Soc.,* 87:4585
83. G. W. Kabalka, *J. Organomet. Chem.,* 1971, **33**, C25.
84. L. W. Bieber, P. J. Rolim Neto and R. M. Generino, *Tetrahedron Lett.,* 1999, **40**, 4473.
85. E. Kumli and P. Renaud, unpublished results.
86. M. Newcomb, M. U. Kumar, J. Boivin, E. Crépon and S. Z. Zard, *Tetrahedron Lett.,* 1991, **32**, 45.
87. A. L. J. Beckwith and I. G. E. Davison, *Tetrahedron Lett.,* 1991, **32**, 49.
88. C. Ollivier and P. Renaud, *Angew. Chem., Int. Ed.,* 2000, **39**, 925.
89. C. Cadot, J. Cossy, P. I. Dalko, *Chem. Commun.,* 2000, 1017.
90. B. Giese, *Angew. Chem., Int. Ed.,* 1983, **23**, 753.
91. P. Renaud, C. Ollivier and V. Weber, *J. Org. Chem.,* 2003, **68**, 5769.
92. B. Becattini, C. Ollivier and P. Renaud, *Synlett,* 2003, 1485.
93. F. Bohlmann, G. W. Ludwig, J. Jakupovic, R. M. King and H. Robinson, *Liebigs Ann. Chem.,* 1984, 228.
94. M. Bertrand, L. Feray and S. Gastaldi, *C. R. Chim.,* 2002, **5**, 623.
95. H. Miyabe, M. Ueda and T. Naito, *Synlett,* 2004, 1140.
96. M. P. Bertrand, S. Coantic, L. Feray, R. Nouguier and P. Perfetti, *Tetrahedron,* 2000, **56**, 3951.
97. H. Miyabe, R. Shibata, M. Sangawa, C, Ushiro and T. Naito, *Tetrahedron,* 1998, **54**, 11431.
98. H. Miyabe, M. Ueda and T. Naito, *Chem. Commun.,* 2000, 2059.
99. H. Miyabe, Y. Yamaoka and Y. Takemoto, *J. Org. Chem.,* 2005, **70**, 3324.
100. M. J. Alves, G. Fortes, E. Guimaraes and A. Lemos, *Synlett,* 2003, 1403.
101. E. Risberg, A. Fischer and P. Somfai, *Tetrahedron,* 2005, **61**, 8443.
102. M. Ueda, H. Miyabe, M. Teramachi, O. Miyata and T. Naito, *Chem. Commun.,* 2003, 426.
103. H. Miyabe, M. Ueda and T. Naito, *J. Org. Chem.,* 2000, **65**, 5043.
104. A. Suzuki, N. Miyaura, M. Itoh, H. C. Brown, G. W. Holland and E. Negishi, *J. Am. Chem. Soc.,* 1971, **93**, 2792.
105. A. Suzuki, N. Miyaura, M. Itoh, H. C. Brown and P. Jacob III, *Synthesis,* 1973, 305.
106. N. Miyaura, M. Itoh, N. Sasaki and A. Suzuki, *Synthesis,* 1975, 317.
107. C.-F. Yao, C.-M. Chu and J.-T. Liu, *J. Org. Chem.,* 1998, **63**, 719.
108. N. Miyamoto, D. Fukuoka, K. Utimoto and H. Nozaki, *Bull. Chem. Soc. Jpn.,* 1974, **47**, 503.
109. A.-P. Schaffner and P. Renaud, *Angew. Chem., Int. Ed.,* 2003, **42**, 2658.
110. A.-P. Schaffner, B. Becattini, C. Ollivier, V. Weber and P. Renaud, *Synthesis,* 2003, 2740.
111. V. Darmency, E. M. Scanlan, A.-P. Schaffner and P. Renaud, *Org. Synth.,* 2005, **83**, 24.
112. C. E. Garett and G. C. Fu, *J. Org. Chem.,* 1996, **61**, 3224.
113. V. Darmency and P. Renaud, *Chimia,* 2005, **59**, 109.
114. M. P. Bertrand, L. Feray, R. Nouguier and L. Stella, *Synlett,* 1998, 780.

115. G. K. Friestad, Y. Shen and E. L. Ruggles, *Angew. Chem., Int. Ed.*, 2003, **42**, 5061.

116. M. Ueda, H. Miyabe, H. Sugino, O. Miyata and T. Naito, *Angew. Chem., Int. Ed.*, 2005, **44**, 6190.

117. J.-T. Liu, Y.-J. Jang, Y.-K. Shih, S.-R. Hu, C.-M. Chu and C.-F. Yao, *J. Org. Chem.*, 2001, **66**, 6021.

118. P. Panchaud and P. Renaud, *J. Org. Chem.*, 2004, **69**, 3205.

119. T. Yoshimitsu, Y. Arano and H. Nagaoka, *J. Org. Chem.*, 2005, **70**, 2342.

120. T. Yoshimitsu, Y. Arano and H. Nagaoka, *J. Am. Chem. Soc.*, 2005, **127**, 11610.

121. B. P. Roberts, *Chem. Soc. Rev.*, 1999, **28**, 25.

122. V. Paul and B. P. Roberts, *Chem. Commun.*, 1987, 1322.

123. V. Paul and B. P. Roberts, *J. Chem. Soc., Perkin Trans. 2*, 1988, 1183.

124. H.-S. Dang, V. Diart and B. P. Roberts, *J. Chem. Soc., Perkin Trans. 1*, 1994, 1033.

125. P. L. H. Mok, B. P. Roberts and P. T. McKetty, *J. Chem. Soc., Perkin Trans. 2*, 1993, 665.

126. H.-S. Dang, V. Diart, B. P. Roberts and D. A. Tocher, *J. Chem. Soc., Perkin Trans. 2*, 1994, 1039.

127. D. A. Spiegel, K. B. Wiberg, L. N. Schacherer, M. R. Medeiros and J. L. Wood, *J. Am. Chem. Soc.*, 2005, **127**, 12513.

128. D. Pozzi, E. M. Scanlan and P. Renaud, *J. Am. Chem. Soc.*, 2005, **127**, 14204.

129. (a) *Radicals in Organic Synthesis*, ed. P. Renaud and M. P. Sibi, Wiley, Weinheim, 2001; (b) *Radicals in Synthesis I & II*, ed. A. Gansäuer, Springer, Berlin, 2006, vol. 263 & 264; (c) G. J. Rowlands, *Tetrahedron*, 2009, **65**, 8603–8655; (d) G. J. Rowlands, *Tetrahedron*, 2010, **66**, 1593–1636.

130. (a) P. A. Baguley and J. C. Walton, *Angew. Chem.*, 1998, **110**, 3272–3283 (*Angew. Chem., Int. Ed.*, 1998, **37**, 3072–3082); (b) A. Studer and S. Amrein, *Synthesis*, 2002, 835–849; (c) B. C. Gilbert and A. F. Parsons, *J. Chem. Soc., Perkin Trans. 2*, 2002, 367–397.

131. For very recent contributions, see: (a) J. W. Tucker, J. M. R. Narayanam, S.W. Krabbe and C. R. J. Stephenson, *Org. Lett.*, 2010, **12**, 368–371; (b) T. Taniguchi, N. Goto, A. Nishibata and H. Ishibashi, *Org. Lett.*, 2010, **12**, 112–115; (c) S.-H. Ueng, A. Solovyev, X. Yuan, S. J. Geib, L. Fensterbank, E. Lacôte, M. Malacria, M. Newcomb, J. C. Walton and D. P. Curran, *J. Am. Chem. Soc.*, 2009, **131**, 11256–11262, and references therein.

132. For a review, see: P. I. Dalko, *Tetrahedron*, 1995, **51**, 7579–7653.

133. For recent contributions, see: (a) M. P. Sibi and M. Hasegawa, *J. Am. Chem. Soc.*, 2007, **129**, 4124–4125; (b) M. Amatore, T. D. Beeson, S. P. Brown and D. W. C. McMillan, *Angew. Chem.*, 2009, **121**, 5223–5226 (*Angew. Chem., Int. Ed.*, 2009, **48**, 5121–5124); (c) M. P. DeMartino, K. Chen and P. S. Baran, *J. Am. Chem. Soc.*, 2008, **130**, 11546–11560; (d)

U. Jahn and E. Dinca, *Chem.–Eur. J.,* 2009, **15**, 58–62; (e) K. C. Nicolaou, R. Reingruber, D. Sarlah and S. Brse, *J. Am. Chem. Soc.,* 2009, **131**, 2086–2087; see also: (f) M. A. Ischay, M. E. Anzovino, J. Du and T. Yoon, *J. Am. Chem. Soc.*, 2008, **130**, 12886–12887; (g) J. M. R. Narayanam, J.W. Tucker and C. R. J. Stephenson, *J. Am. Chem. Soc.,* 2009, **131**, 8756–8757; (h) A. Wetzel, G. Pratsch, R. Kolb and M. R. Heinrich, *Chem.–Eur. J.,* 2010, **16**, 2547–2556.

134. For TEMPO-mediated oxidations, see: (a) M. S. Maji, T. Pfeifer and A. Studer, *Angew. Chem.,* 2008, **120**, 9690–9692 (*Angew. Chem., Int. Ed.,* 2008, **47**, 9547–9550); (b) T. Nagashima and D. P. Curran, *Synlett,* 1996, 330–332; (c) see also: M. Pouliot, P. Renaud, K. Schenk, A. Studer and T. Vogler *Angew. Chem.,* 2009, **121**, 6153 –6156 (*Angew. Chem., Int. Ed.,* 2009, **48**, 6037–6040), and references therein.

135. K. Tamao, J.-I. Yoshida, H. Yamamoto, T. Kakui, H. Matsumoto, M. Takahashi, A. Kurita, M. Murata and M. Kumada, *Organometallics*, 1982, **1**, 355–368.

136. (a) H. A. Stefani, R. Cella and A. S. Vieira, *Tetrahedron,* 2007, **63**, 3623–3658; (b) G. A. Molander and N. Ellis, *Acc. Chem. Res.*, 2007, **40**, 275–286; (c) S. Darses and J.-P. Gent, *Chem. Rev.*, 2008, **108**, 288–325.

137. Cu^{II} oxidation of boranes in halogenation reactions is known, see: (a) C. Ollivier and P. Renaud, *Chem. Rev.*, 2001, **101**, 3415–3434, and references therein, notably: (b) C. F. Lane, *J. Organomet. Chem.*, 1971, **31**, 421–431.

138. Electrochemical oxidation of polyphenyltrifluoroborate anions is known, see: (a) L. A. Shundrin, V. V. Bardin and H.-J. Z. Frohn, *Z. Anorg. Allg. Chem.*, 2004, **630**, 1253–1257; for PET oxidation of allyl- and benzyl-trifluoroborates, see: (b) Y. Nishigaichi, T. Orimi and A. Takuwa, *J. Organomet. Chem.*, 2009, **694**, 3837–3839; see also for other borates: (c) G. B. Schuster, *Pure Appl. Chem.*, 1990, **62**, 1565–1572; for the oxidation of arylboronic acids with Mn(OAc)$_3$, see: (d) A. Dickschat and A. Studer, *Org. Lett.,* 2010, **12**, 3972–3974, and references therein; see also: (e) I. B. Seiple, S. Su, R. A. Rodriguez, R. Gianatassio, Y. Fujiwara, A. L. Sobel and P. S. Baran, *J. Am. Chem. Soc.,* 2010, **132**, 13194–13196; for a Ritter reaction of trifluoroborates involving a two-electron oxidation, see: (f) C. Cazorla, E. Métay, B. Andrioletti and M. Lemaire, *Tetrahedron Lett.,.* 2009, **50**, 6855–6857.

CHAPTER 8

Thiols as Efficient Hydrogen Atom Donors in Free Radical Transformations in Aqueous Media[*]

8.1 INTRODUCTION

Following Nature's lead, the challenge for today's chemist is to move away from highly volatile and environmentally harmful organic solvents and towards friendly and biologically compatible media.[1] The obvious choice of solvent is water due to its abundance, cost effectiveness and biological compatibility. The potential usefulness of free radical reactions in water is demonstrated by ever-increasing studies over the last 5 years.[2] Free radical reactions are an important class of synthetic reactions that have been traditionally performed in organic solvents. In recent years, the number of reports of free radical reactions that use water has increased.[1-9] Water is an ideal solvent for free radical reactions as it possesses no reactive functional groups and strong O–H bonds that make hydrogen abstraction unlikely.[4,7,9] Whilst there have been several excellent reviews on carbon–carbon (C–C) bond formation and reactions of carbon–hydrogen (C–H) bonds in water,[4,6-9] this chapter addresses the efficient use of thiols in C–H and C–C bond formations in aqueous media *via* radical reactions (hydrogen-atom transfer (HAT), radical cyclization and deoxygenation) within recent decades. There is a specific focus on HAT reactions as they represent one of the major classes of the atom/group-transfer reactions. In HAT reactions, a hydrogen atom is transferred from a chain carrier to the newly formed radical centre, through the general mechanism involving radical initiation, radical propagation and radical termination steps. This reaction is important in biology, in the chemical industry and in the chemical laboratory.[1,3,10-15]

* Chapter written by Ioannis N. Lykakis and V. Tamara Perchyonok.
Streamlining Free Radical Green Chemistry
V. Tamara Perchyonok, Ioannis Lykakis and Al Postigo
© V. Tamara Perchyonok, Ioannis Lykakis and Al Postigo 2012
Published by the Royal Society of Chemistry, www.rsc.org

8.2 THE TRIS(TRIMETHYLSILYL)SILANE (TMS₃SIH)/THIOL SYSTEM IS AN EFFICIENT RADICAL HYDROGEN DONOR "ON WATER"

One of the drawbacks of radical reactions in aqueous media is the immiscibility of reagents, reactants or initiators. However, recently this viewpoint has changed and immiscibility has started to be considered advantageous. Chemical reactions "on water" are now given greater attention, since the reactivity in the suspension obtained by vigorous stirring of immiscible reactants seems to benefit from the enhanced contact surface of the resulting tiny drops, as well as from the unique molecular properties developed at the interface between water and the hydrophobic phase.[16] "Polarity-reversal catalysis" is ideal for free radical hydrogen-transfer reactions "on water".[17] The advantages of this methodology are based on the fact that the silane/thiol mixture not only enhances the radical production and regeneration, but is also flexible enough to accommodate the amphiphilic thiol which increases the radical reactivity at the interface (Scheme 8.1).

The reaction of thiyl radicals with silicon hydrides (Scheme 8.2) is a key step of the so called "*polarity-reversal catalysis*" that has been utilized extensively in the radical chain reductions of alkyl halides as well as in the hydrosilylation of olefins using a silane–thiol couple. The reaction is strongly endothermic and reversible. The rate constants k_{SH} and k_{SiH} were determined in cyclohexane at 60 °C, relative to $2k_T$ for the self-termination of the thiyl radicals, using the kinetic analysis of the thiol-catalysed reduction of 1-bromooctane and 1-chlorooctane by silane respectively.[18]

Recently, Chatgilialoglu and co-workers reported that, by employing a suitable thiol/silane mixture in the "polarity-reverse"-mediated HAT reaction,

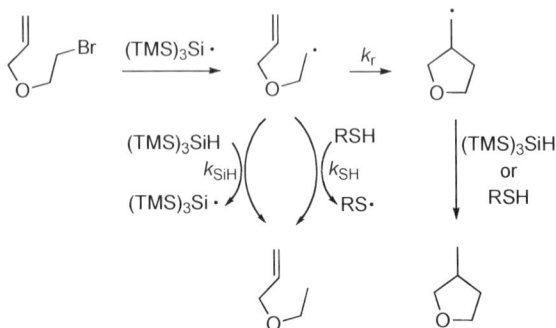

Scheme 8.1 Radical chain reduction/cyclization of 3-(2-bromoethoxy)prop-1-ene by the (TMS)₃SiH/thiol reducing system.

Scheme 8.2 Reaction between thiyl radicals and silanes.

the efficient synergy of radical production, regeneration and enchancement of the radical reactivity at the interface can be achieved.[17,19,20] For the reduction of the organic halide (RX) by the combination of $(TMS)_3SiH/HOCH_2CH_2SH$ under radical conditions, the propagation steps suggested in Scheme 8.3 are expected. That is, the alkyl radical abstracts hydrogen from the thiol and the resulting thiyl radicals abstract hydrogen from the silane, so that the thiol is regenerated along with the chain carrying silyl radical for a given RX (Scheme 8.3).[19]

The reduction of organohalides,[19] including bromonucleosides,[21] can be achieved in aqueous media using $(TMS)_3SiH$. This procedure employs 2-mercaptoethanol as the catalyst and either 1,1′-azobis(cyclohexanecarbonitrile) (ACCN) as the organic-solvent-soluble initiator or 2,2′-azobis(2-amidinopropane) dihydrochloride (AAPH) as the water-soluble radical initiator; the reduced compounds are obtained in yields between 75 and 100% (Table 8.1). As $(TMS)_3SiH$ does not react significantly with water, it is therefore a good reagent for radical-based reductions in aqueous media.

The extension and wide application of this work has come from the work of Chatgilialoglu *et al.* on the use of $(Me_3Si)_3SiH$ as a hydrogen donor suitable for water soluble and water-insoluble precursors, in the systems comprising substrate, silane and initiator (ACCN) mixed at 100 °C. A system was proposed that worked well for both hydrophilic and hydrophobic substrates, with the only variation that an amphiphilic thiol was also needed in the case of the water-soluble compounds (Scheme 8.4).[22]

8.3 THIOL/AZO INITIATOR SYSTEM IN *CIS–TRANS* ISOMERIZATION OF DOUBLE BONDS IN AQUEOUS MEDIA

Chatgilialoglu and co-workers have used the thiol/azo-initiator system for the isomerization of *cis* phospholipids to the corresponding *trans* compounds. Cascade reactions, as well as geometrical isomerization of the mono- and poly-unsaturated fatty acid double bonds can be achieved by using a combination of

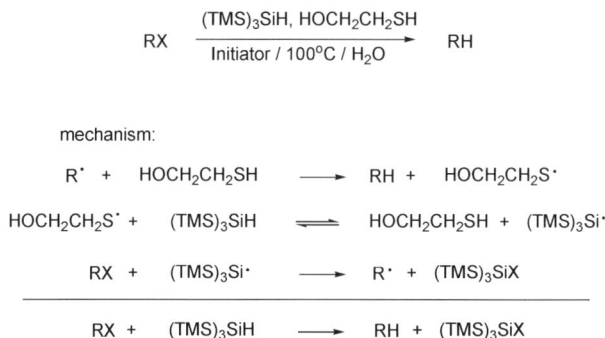

$$RX \xrightarrow[\text{Initiator / 100°C / H}_2\text{O}]{(TMS)_3SiH,\ HOCH_2CH_2SH} RH$$

mechanism:

$$R^\bullet + HOCH_2CH_2SH \longrightarrow RH + HOCH_2CH_2S^\bullet$$

$$HOCH_2CH_2S^\bullet + (TMS)_3SiH \rightleftharpoons HOCH_2CH_2SH + (TMS)_3Si^\bullet$$

$$RX + (TMS)_3Si^\bullet \longrightarrow R^\bullet + (TMS)_3SiX$$

$$RX + (TMS)_3SiH \longrightarrow RH + (TMS)_3SiX$$

Scheme 8.3 Proposed mechanism of polarity-reversal catalysis using the TMS_3SiH/thiol system.

Table 8.1 Reduction of organic halides (10 mM) with (TMS)$_3$SiH (12 mM) and HOCH$_2$CH$_2$SH (2.85 mM) in water.

Substrate	Initiator (mM)	Temperature /$^{\circ}C$	Yield (%)
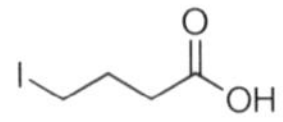	APPH (3) ACCN (3)	73 100	99 100
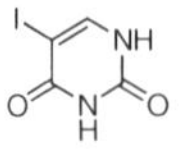	APPH (3)	73	90
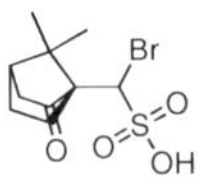	APPH (3) ACCN (3) ACCN (3)	73 100 100	— 90 95
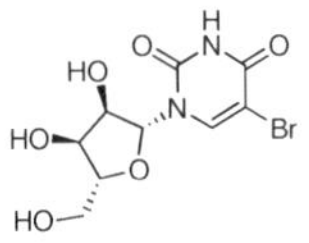	APPH (3) ACCN (3)	73 100	— 90
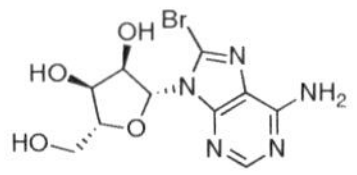	APPH (3) ACCN (3)	73 100	— 94
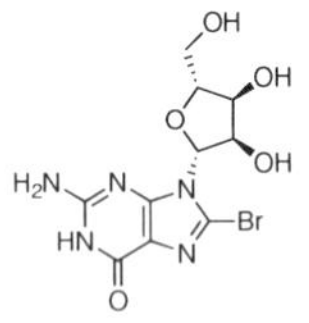	APPH (3) ACCN (3)	73 100	— 75

radical initiators and RSH as a hydrogen donor *in vitro* (Scheme 8.4). From the *in vitro* and *in vivo* studies it emerged that the thiyl radicals (RS•) are able to diffuse from the aqueous phase to the lipid phase and perform a reaction involving the membrane lipid double bonds thus forming *trans* fatty acids.[23,24] In these studies a catalytic amount of thiols and various hydrophilic and

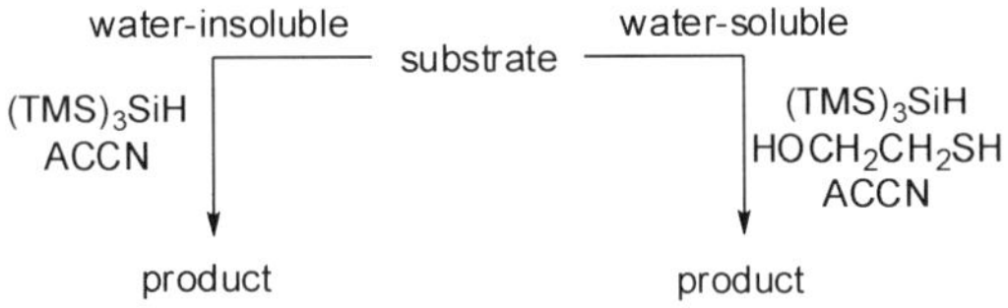

Scheme 8.4 On the use of TMS$_3$SiH in water and on water.

hydrophobic azo-initiators [such as AAPH, azobisisobutyronitrile (AIBN) and 2,2′-azobis(2,4-dimethylvaleronitrile) (AMVN)] were used at 37, 71 and 54 °C respectively, in aqueous media.[25] Thiyl radical addition to the *cis* and *trans* double bond of phospholipids in aqueous media is accompanied by isomerization, according to the mechanism proposed in Scheme 8.10. This addition is reversible and gives the *trans* geometric isomer in up to 80% relative yield. In all cases, HAT reactions take place at the initial pathways, between alcohol and alkyl radical intermediate (R•) forming the corresponding α-hydroxy carbon radical (•C–OH), which can abstract a H-atom from the thiol to generate the corresponding RS• (Scheme 8.5).

However, a variety of thiol compounds have been used for the generation of thiyl radicals, in a range of hydrophobic such as PhSH and/or 3-(2-mercaptoethyl)quinazoline-2,4(1*H*,3*H*)-dione (MECH),[26] and amphiphilic thiols such as 2-mercaptoethanol.[23] The mechanism of thiyl radical-catalyzed isomerization of unsaturated fatty acid residues in homogeneous solution and in liposomes was also studied by Sprinz and co-workers.[26] In that study, the effects of thiols on oleic and linoleic fatty acid residues using pulse radiolysis, γ-radiolysis and chemolysis (AAPH) to generate thiyl radicals was investigated. They proposed that the resulting isomeric equilibrium (*trans* fraction: 81%) does not depend on the structure of the thiyl radical or the organization of the lipids. Thiyl radicals were generated from MECH using pulse radiolysis and γ-radiolysis in aqueous and alcoholic solutions saturated with N_2O (Scheme 8.6).[26]

In addition, sulfides have been shown to catalyze the lipid isomerization processes[25] (with a series of hydrophobic and hydrophilic sulfides, such as $HOCH_2CH_2SCH_3$, $HOCH_2CH_2CH_2SCH_3$, S-methyl cysteine and S-methyl homocysteine) in the presence of AAPH as the initiator or under γ-radiation conditions. It was suggested that under γ-radiolysis the H• atoms react with

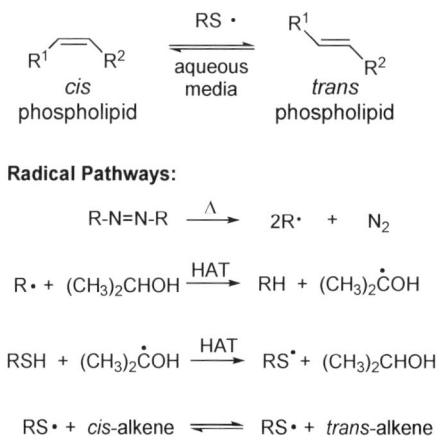

Scheme 8.5 Thiol/azo-initiator system catalyzes lipid double bonds: isomerization in aqueous media.

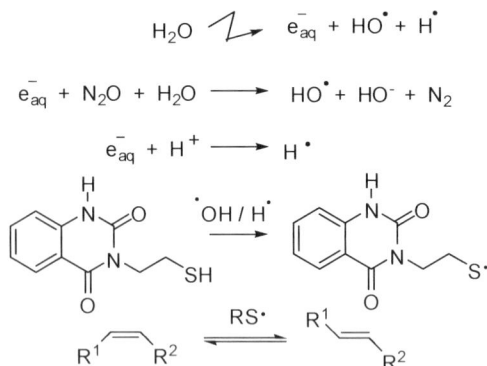

Scheme 8.6   Thiyl radical catalyzed isomerization of linoleate methyl ester.

sulfides to form a sulfuranyl species (Scheme 8.7) which in turn collapses to give directly or indirectly the MeS˙ radical (or HS˙ in the case of thiols). On the other hand the alkyl radicals R˙ or the ˙C–OH radical generated from the azo-initiator can abstract a hydrogen atom from positions that are activated by the neighboring groups forming the corresponding tertiary radical. These radicals are expected to generate MeS˙ radicals *via* β-elimination (Scheme 8.7). Similarly, thiols such as $HOCH_2CH_2SH$ and cysteine should form sulfhydryl radicals HS˙ (or S˙$^-$) radical, which are able to isomerize lipid double bonds efficiently in aqueous media.[27,28]

   A simple and high-yielding method to convert natural all-*cis* PUFA derivatives to the corresponding all-*trans* geometrical isomers was described recently.[29] The method is based on the thiyl radical-catalyzed *cis–trans*

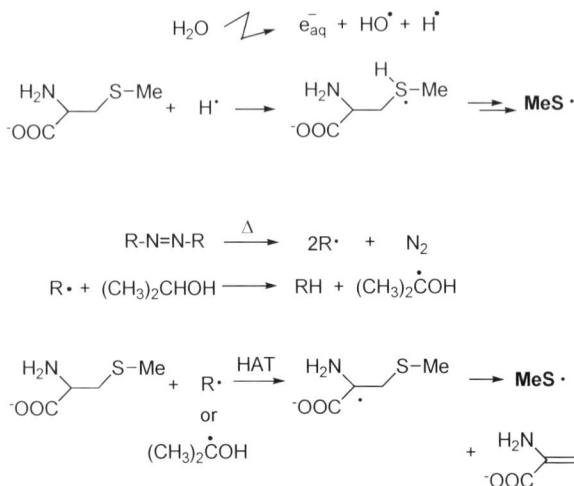

Scheme 8.7   Sulfide/azo-initiator system used in lipid double bonds isomerization in aqueous media.

isomerization. For example, thiyl radicals have been used for the synthesis of all-*trans* anandamide from the corresponding all-*trans* methyl arachidonate.[30] The latter was formed in good yield, up to 70%, from the corresponding arachidonic methyl ester irradiated under 50% equivalent of 2-mercaptoethanol in 2-propanol (Scheme 8.8).

Maleate esters can be also converted into fumarate esters in near quantitative yield through exposure to thiyl radicals generated in refluxing hexane by photolysis of diphenyl disulfide. When conditions are applied to dialkyl (hydroxyl alkyl)maleate esters, 3,2(5*H*)-furanones are formed in good yield. In all cases, the key reaction to overall process was the *E–Z* isomerization of the initial esters by addition–elimination of the RS$^\bullet$ radical to the double bond (Scheme 8.9).[31]

8.4 THIOLS IN PEPTIDES: DEGRADATION IN AQUEOUS MEDIA

The reaction of hydrogen atoms with peptides containing methionine (Met) residues such as Met-enkephalins,[32] RNase A[33–35] and amyloid β-peptides[36] have been also investigated. In these studies H$^\bullet$ atoms are generated by radiolytic conditions from the reaction of aqueous electrons with the dihydrogen phosphate anion ($H_2PO_4^-$), as shown in Scheme 8.10. For Met-enkephalin it was calculated that 30–40% of H$^\bullet$ should react with the Met residue.[32] The formation of diffusible thiyl radicals such as $CH_3S^\bullet$, derived from the reaction of the H$^\bullet$ with the Met moiety, was monitored with *trans* lipids as markers. Using this peptide–liposome model, it was proposed that $CH_3S^\bullet$ is able to migrate from the aqueous layer to the membrane compartment and thereby cause the geometrical isomerization of lipid double bonds.

Recently, the desulfurization reactions from zinc and cadmium complexes of a plant metallothionein under radical stress were studied.[37] Metallothioneins (MTs) are sulfur-rich proteins capable of binding metal ions to give metal

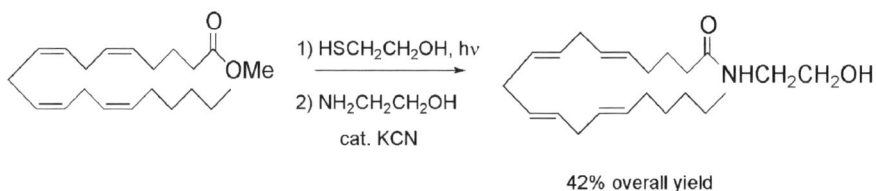

42% overall yield

Scheme 8.8 Mercaptoethanol in the *cis–trans* isomerization of methyl arachidonate.

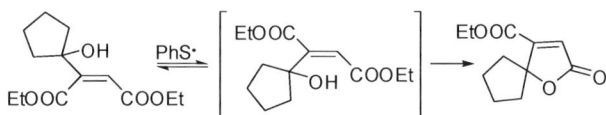

Scheme 8.9 *E–Z* isomerization by thiyl radical intermediate.

$$H_2O \xrightarrow{} e_{aq}^- + HO^{\bullet} + H^{\bullet}$$

$$e_{aq}^- + H_2PO_4^- \longrightarrow H^{\bullet} + H_2PO_4^{2-}$$

$$e_{aq}^- + H^+ \longrightarrow H^{\bullet}$$

$$X\text{-}SCH_3 + H^{\bullet} \rightleftharpoons CH_3S^{\bullet}$$

Peptide-containing
methionine

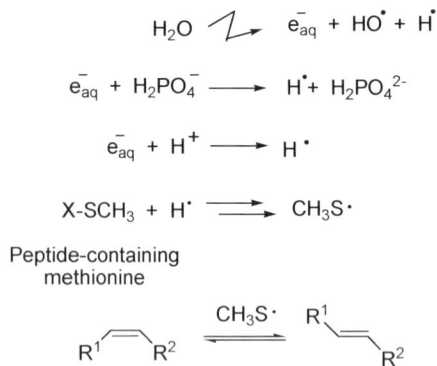

Scheme 8.10   Hydrogen-atom reaction with peptides containing Met-residues in aqueous media.

clusters. By using a biomimetic mode based on unsaturated lipid vesicle suspensions, the occurrence of tandem protein/lipid damage were shown under γ-irradiation conditions. It was suggested that under γ-radiolysis the H$^{\bullet}$ atoms and/or H$^+$ (e_{aq}^-) react with methionine residues which in turn collapse to give (directly or indirectly) the MeS$^{\bullet}$ radical, or react metal–sulfur complexes to give HS$^{\bullet}$/S$^{\bullet-}$ (Scheme 8.11). These thiyl radical species migrate to the lipids bilayer and induce *cis–trans* isomerization of unsaturated fatty acid residues.

8.5   THIOLS IN C–C BOND FORMATION IN WATER

Of the different types of sulfur-centered radicals used in organic synthesis, the sulfanyl radical (RS$^{\bullet}$) is one of the most attractive, and its ability to add to a multiple bond intramolecularly deserves special attraction. This strategy has been successfully employed in the synthesis of alkaloids and other bioactive heterocycles. Sulfanyl radicals trigger a tandem addition–cyclization protocol in linalool or citronelene derivatives for the efficient construction of the iridane monoterpene skeleton. Best results in yields and diastereoselectivity were

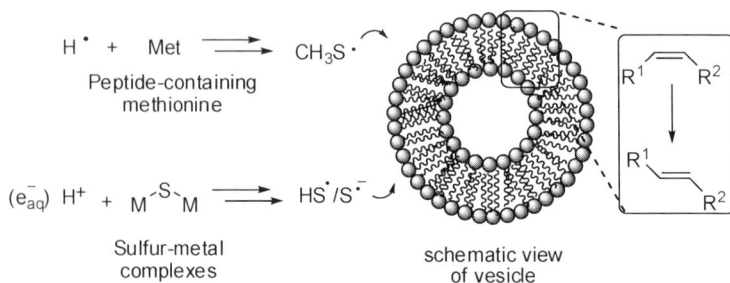

Scheme 8.11   Thiyl radicals induce the *cis–trans* isomerization of unsaturated fatty acids.

Table 8.2 Addition–cyclization reaction mediated by sulfanyl radicals.

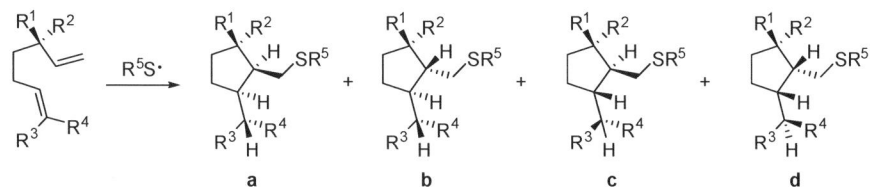

R^1	R^2	R^3	R^4	R^5	Yield (%)	Diastereomeric ratio **a**/**b**/**c**/**d**
OAc	CH_3	CH_3	CH_3	$PhCH_2CH_2$	87	46/26/17/11
OAc	CH_3	CH_3	CO_2Me	Ph	90	47/25/22/6
OAc	CH_3	CH_3	CO_2Me	$PhCH_2CH_2$	93	50/25/16/9
OAc	CH_3	CH_3	CO_2Me	t-Bu	95	48/23/20/19
OCOt-Bu	CH_3	CH_3	CO_2Me	Ph	87	46/20/17/17
CH_3	H	CH_3	CO_2Me	Ph	85	54/20/19/7
CH_3	H	CO_2Et	CO_2Et	Ph	96	58/21/14/7

obtained when phenylethylsulfanyl was used as radical initiator. Considering all the above and the different selectivity of the addition of sulfanyl radicals to double bonds reported elsewhere, Barrero and co-workers summarized that tandem processes of sulfanyl radical addition-cyclization of dienes possessing the basic structure of 3,7-dimethylocta-1,6-diene could be a suitable approach to the synthesis of natural iridoids.[38] According to this proposal, the key addition step should be both chemoselective towards the monosubstituted double bond and also regioselective to the less substituted extreme of the olefin (Table 8.2).

The radical-chain reductive alkylation of electron-rich alkenes mediated by silanes in the presence of thiols as polarity-reversal catalysts was also studied by Roberts and co-workers.[39] In the presence of a thiol catalyst, triphenylsilane mediates the reductive alkylation of electron-rich terminal alkenes by organic halides RHal *via* electrophilic carbon-centered radicals. Reactions were carried out in benzene or dioxane solvent using di-*tert*-butyl hyponitrite (TBHN) (at 60 °C) or dilauroyl peroxide (at 80 °C) as initiators with up to 60% isolated yield of the product $RCH_2CH_2R_1R_2$ (Scheme 8.12). In the presence of the thiol, the slow direct abstraction of hydrogen from the silane by the nucleophilic adduct radical is replaced by a cycle of more rapid polarity-matched reactions, in which hydrogen-atom transfer to the adduct radical from the thiol is followed by abstraction of hydrogen from the silane by the derived thiyl radical, to regenerate the catalyst. In the absence of thiol, negligible yields of reductive alkylation products were obtained. Initial reactions were carried out in the presence of di-*tert*-butyl hyponitrite ($t_{1/2} =$ *ca.* 55 min) which produces *tert*-butoxyl radicals that go on to abstract hydrogen from the silane and/or the thiol to afford chain-carrying silyl or thiyl radicals, shown in Scheme 8.12.

$$t\text{-BuON=NO}t\text{-Bu} \longrightarrow 2\ t\text{-BuO}^{\cdot} + N_2$$

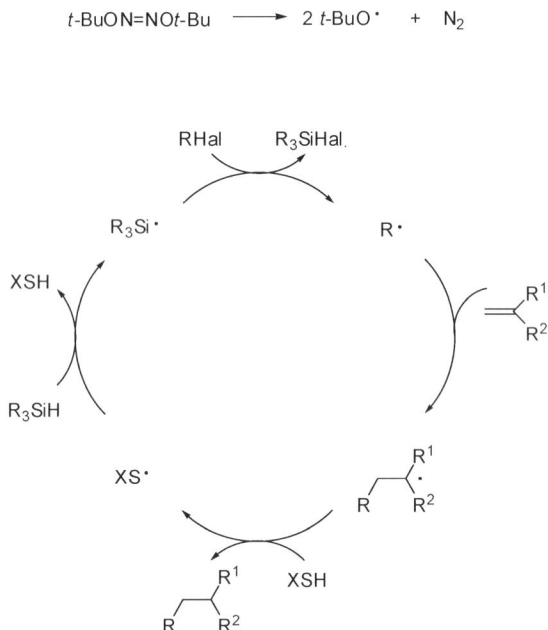

Scheme 8.12 Radical-chain reductive alkylation of electron-rich alkenes mediated by a silane/thiol catalytic system.

8.6 THIOL–ENE COUPLING AS A CLICK PROCESS FOR MATERIALS AND BIOORGANIC CHEMISTRY

One reaction that is emerging as an attractive click process is the century-old addition of thiols to alkenes, which is currently called thiol–ene coupling (TEC).[40] The photochemically/thermally-induced version of this reaction is known to proceed by a radical mechanism to give an anti-Markovnikov-type thioether.[41] The click status of this reaction is supported by it being highly efficient and orthogonal to a wide range of functional groups, as well as for being compatible with water and oxygen. Over the years, the TEC reaction has been extensively exploited in polymer chemistry,[42] and to synthesise several biomaterials for application in medicine, especially dentistry.[43] However, only recently has the click aspect of the TEC reaction been fully appreciated in the field of polymer science. Quite significantly a new term, "thio-click", was coined in a paper dealing with the modification of the backbone of poly[2-(3-butenyl)-2-oxazoline] by reaction with mercaptans.[44] In this way both hydrophobic fluoropolymers and water-soluble glycopolymers were prepared starting from the same readily available material. The products were well-defined polymers without giving side reactions such as thiyl radical coupling, which could lead to disulfide formation. In addition, the great potential of thiol–ene chemistry was exploited by Hawker and co-workers in the synthesis of poly(thioether) dendrimers.[45] In that work the key sulfur–carbon bond-forming reaction was used for the construction of both the dendritic backbone and the functionalization of the chain ends. Starting from a 2,4,6-triallyloxy-1,3,5-triazine core, the fourth generation dendrimer [G4]-OH48 was constructed through iterative photo-induced TEC and alkene generation in each cycle. After the transformation of [G4]-OH48 into the ene-functional dendrimer [G4]-ene48, the TEC reaction between this polyalkene substrate and monofunctionalized thiols enabled functionalization of the periphery of the dendrimer in an essentially complete manner. Substantial modifications of synthetic polymers with bioorganic molecules such as amino acids, peptides, and carbohydrates through TEC have been carried out by Schlaad and co-workers.[46]

One of the most practical and widely used routes for the synthesis of β-sulfido carbonyl compounds is the conjugate addition of thiols to α,β-unsaturated carbonyl compounds. Conversion of α,β-unsaturated carbonyl compounds to the corresponding β-sulfido carbonyl derivatives provides an elegant strategy for chemoselective protection of the olefinic double bond of conjugated enones due to the ease of generation of the double bond by removal of the sulfur moiety by copper(I)-induced and oxidative eliminations. Thus, thia-Michael addition is an important transformation and, apart from its versatile applications in synthetic organic chemistry, water[47] or ionic liquids[48] have been employed as solvents for the above reaction, with the presence of several reagents as catalysts. In contrast, the catalyst-free conjugate addition of thiols to α,β-unsaturated carbonyl compounds in water has recently been reported with the formation of the corresponding β-sulfido carbonyl compounds in short times and with excellent chemoselectivity.[49] Competitive dithiane/dithiolane formation, transesterification, and ester cleavage were not observed. It was proposed that water played a dual role in simultaneously activating the α,β-unsaturated carbonyl compound and the thiol. This new methodology constitutes an easy, highly efficient, and green synthesis of β-sulfido carbonyl compounds (Scheme 8.13).[49]

8.7 HYDROGEN SULFIDE IN OXIDATION AND/OR REDUCTION OF ORGANIC COMPOUNDS

An unusual reaction was presented from Barton and co-workers, in which cyclohexane is oxidised to cyclohexanol and cyclohexanone whilst at the same time hydrogen sulfide is converted to sulfur.[50] In that study, the only oxidant needed was the molecular oxygen. Fe(II) and O_2 furnish Fe(III) and superoxide with the latter to react with more Fe(II) to make the activated iron species and hence permit entry to Gif ketonization chemistry. The Fe(III) formed in the first step was reduced by H_2S to give Fe(II) and sulfur, with the parallel oxidation of saturated hydrocarbons to ketones and alcohols (Table 8.3).[50]

On the other hand, the reduction of allylic thiols to alkenes by hydrogen sulfide in aqueous solutions, a novel reaction that may explain reduction processes widely occurring in natural environments, has been proposed by

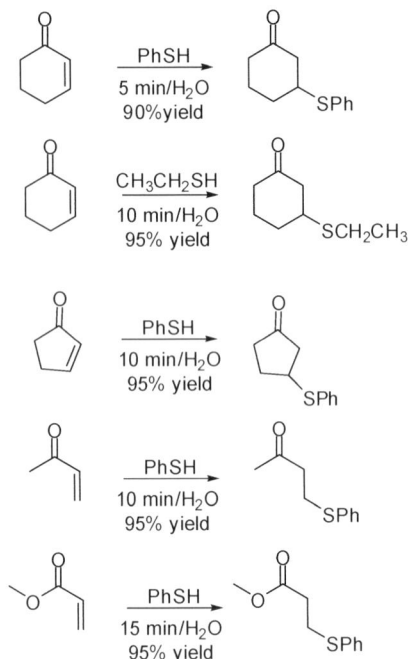

Scheme 8.13 Thia-Michael addition to enones in water.

Adam and co-workers.[51] Its mechanism has been studied and suggested to follow an SRN1-like pathway involving radical intermediates undergoing 1,4-hydrogen shifts. Hydrogen sulfide was used as an efficient hydrogen donor in water for this reduction process (Scheme 8.14).

8.8 THIYL RADICALS AND THE INFLUENCE OF ANTIOXIDANTS/VITAMINS

The effectiveness of the presence of the most common antioxidants with respect to the thiyl radical life time has been addressed.[52] The high reactivity of RS$^\bullet$ radical addition to all-*trans* retinol and β-carotene suggested that retinoid and carotenoid derivatives are the best trappers of thiyl radicals.[52,53] This result is based on the effect of the *cis–trans* isomerization of methyl oleate catalyzed by radiolytically generated HOCH$_2$CH$_2$S$^\bullet$ (RS$^\bullet$) radicals (Scheme 8.15).[52] The same experiment in the presence of 1 mM of all-*trans* retinol

Table 8.3 Synergistic oxidation of cyclohexane and hydrogen sulfide under Gif conditions.

Hydrocarbon	Time	Ketone/mmol	Alcohol/mmol	Yield (%)
Cyclohexane	9 h	3	4	36
Cyclodecane	6 h	5	5	32

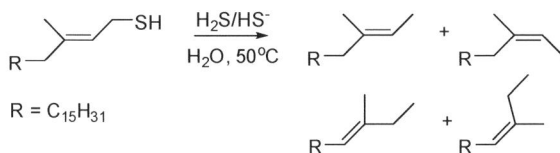

Scheme 8.14 Hydrogen sulfide in the reduction process of allylic thiols in water.

acetate shows an initial strong inhibition of the isomerization, although the *cis/trans* isomeric ratio reached the same equilibration point in a longer time, which means that the thiyl radicals are efficiently scavenged from the conjugated diene until it is consumed. Similar experiments by replacing the conjugated diene with α-tocopherol or ascorbic acid 6-*O*-palmitate showed that thiyl radicals are weakly scavenged by these two antioxidants.[52]

In addition, folic acid (FA-OH) was shown to scavenge thiyl radicals very efficiently.[54] In the reaction of thiyl radicals with folic acid, it has been observed that folic acid can not only scavenge thiyl radicals but can also repair these thiols at physiological pH. It is known that thiyl radicals react by two ways: either by addition to an olefinic bond or by abstraction of a hydrogen atom. The latter type of reaction is the only possible pathway in the presence of folic acid, and the interaction between the thiyl radical and folic acid can be explained as shown in Scheme 8.16.

The main product of the nitroxide reaction with thiyl radicals was identified as the corresponding amine.[55,56] For example, the reaction between gluta-thionyl radical (GS') and nitroxides has been suggested to protect cells from

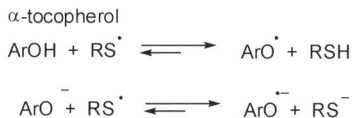

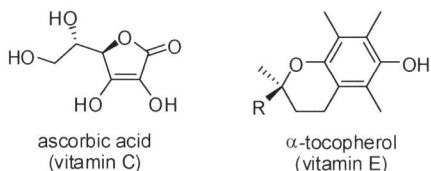

Scheme 8.15 Thiyl radicals trapped by several antioxidant vitamins.

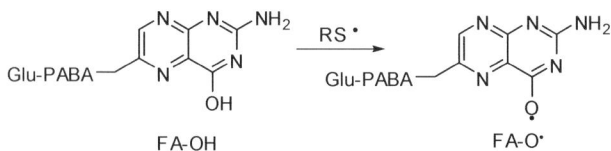

Scheme 8.16 Thiyl radicals trapped by folic acid.

GS˙ toxicity and also as a potential pathway of formation of secondary amines, which are known as major metabolites of nitroxides in cells.[55] The rate constant for the reaction of GS˙ in aqueous solutions with 4-((9-acridinecarbonyl)amino)-2,2,6,6-tetramethylpiperidine-1-oxyl (Ac-TPO) has been estimated to be in a range of 10^8–10^9 M^{-1} s^{-1}. In recent study, the reactions of piperidine and pyrrolidine nitroxides with thiyl radicals in water have been investigated to gain better understanding of the antioxidative activity of nitroxides. The rate constant for the reaction of those nitroxides with thiyl radicals has been determined to be approximately 5.7×10^8 M^{-1} s^{-1} at pH = 5–7, independent of the structure of the nitroxide and the thiyl radical.[57] It is proposed that this reaction yields an unstable adduct, which decomposes *via* a complex mechanism to yield the respective amine, sulfinic, and sulfonic acids. Under physiological conditions the proposed adduct is deprotonated and decomposes *via* heterolysis of the N–O bond, yielding the respective amine and sulfinic acid (Scheme 8.17). This mechanism accounts for the protective effect of nitroxides against thiyl radicals and also against reactive oxygen- and nitrogen-derived species in the presence of thiol.[57]

The reaction of carbon-centered radicals with thiols is important to a number of chemical as well as biological processes. For example, radical reactions on the sugar moieties of nucleic acids are of particular interest, because they play an important role in the damage and repair of these biomolecules. For this reason, a thorough study for the rate constant measurement of H-atom abstraction from thiols by C-centered radicals has been started by Rauk and co-workers. In that study, the rate of the reaction

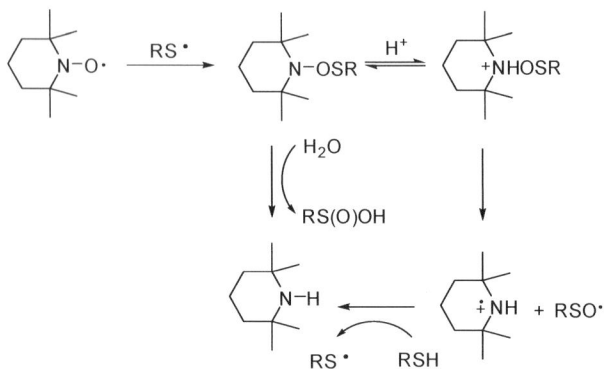

Scheme 8.17 Nitroxides as antioxidants in the presence of thiols.

between hydroxyradicals, methyl radical and primary C-centered radicals from *tert*-butyl alcohol and dithiothreitol was measured and found to be in a range of 10^7–10^8 M^{-1} s^{-1}.[58] All the experiments have been done by pulse radiolysis in water. Recently, Chatgilialoglu and co-workers focused on the measurement of the rate constants and the transition state geometry of the reactions between alkyl, alkoxyl and peroxyl radicals with thiols.[59] In that study, three different thiols have been used: alkanethiols (RSH) where the bond dissociation energy of SH is independent of the structure of alkyl attached to the sulphur atom, L-cysteine (HOOCCH(NH$_2$)CH$_2$SH) where the bond dissociation energy is slightly lower than those of other alkanethiols, and thiophenol (PhSH). They found that the measured rate constants ranged from 10^3 to 10^8 M^{-1} s^{-1}, with the highest value for the reaction between α-hydroxy alkyl radicals with thiols and the lowest one for the reaction between alkyl peroxy radicals with phenylthiol.

8.9 CONCLUSIONS

This chapter highlights the substantial progress which has been made in the last decade in shifting the paradigm in free radical chemistry from conventional tin-based hydrogen donors and organic solvents to environmentally benign and biocompatible reaction conditions and media applicable to the investigation of a broad range of synthetic and biological transformations.

REFERENCES

1. W. Wei, C. C. K. Keh and C.-J. Li, Water as a reaction medium for clean chemical processes, *Clean Technol. Environ. Policy*, 2005, **7**, 62–69.
2. D. Leca, L. Fensterbank, E. Lacote and M. Malacria, Recent advances in the use of phosphorus-centered radicals in organic chemistry, *Chem. Soc. Rev.*, 2005, **34**, 858–865, and references cited therein.
3. D. J. Adams, P. J. Dyson and S. J. Taverner, *Chemistry in Alternative Reaction Media*, Wiley-Interscience, Cambridge, 2004; W. M. Nelson, *Green Solvents for Chemistry, Perspectives and Practice*, Oxford University Press, New York, 2003.
4. U. M. Lindstrom, Stereoselective Organic Reactions in Water, *Chem. Rev.*, 2002, **102**, 2751–2772, and references cited therein.
5. D. Font, C. Jimeno, M. Pericas, and A. Miquel, Polystyrene-supported hydroxyproline: An insoluble, recyclable organocatalyst for the asymmetric aldol reaction in water, *Org. Lett.*, 2006, **8**, 4653–4655.
6. L. Chen and C.-J. Li, Catalyzed reactions of alkynes in water, *Adv. Synth. Catal.*, 2006, **348**, 1459–1484.
7. C.-J. Li, Organic reactions in aqueous media with a focus on carbon−carbon bond formations: A decade update, *Chem. Rev.*, 2005, **105**, 3095–3166, and references cited therein.

8. C. I. Herrerias, X. Yao, Z. Li and C.-J. Li, Reactions of C−H bonds in water, *Chem. Rev.*, 2007, **107**, 2546–2562, and references cited therein.

9. (a) C.-J. Li, Organic reactions in aqueous media—with a focus on carbon–carbon bond formation, *Chem. Rev.,* 1993, **93**, 2023–2035, and references cited therein; (b) V. T. Perchyonok, I. N. Lykakis and K. L. Tuck, Recent advances in C–H bond formation in aqueous media: a mechanistic perspective, *Green Chem.*, 2008, **10**, 153–163; (c) V. T. Perchyonok and I. N. Lykakis, Recent advances in free radical chemistry of C–C bond formation in aqueous media: From mechanistic origins to applications, *Mini-Rev. Org. Chem.*, 2008, **5**, 19–32.

10. Y. Xu and H. Sugiyama, Photochemical approach to probing different DNA structures, *Angew. Chem., Int. Ed.*, 2006, **45**, 1354–1362.

11. J. F. J. Coelho, M. Carreira, A. V. Popov, P. M. O. F. Goncalves and M. H. Gil, Thermal and mechanical characterization of poly(vinyl chloride)-b-poly(butyl acrylate)-b-poly(vinyl chloride) obtained by single electron transfer—degenerative chain transfer living radical polymerization in water, *Eur. Polym. J.*, 2006, **42**, 2313–2319.

12. M. K. Chaudhuri and S. Hussain, Boric acid catalyzed thia-Michael reactions in water or alcohols, *J. Mol. Catal. A: Chem.*, 2007, **269**, 214–217.

13. A. P. Brogan and T. J. Dickerson, Enamine-based aldol organocatalysis in water: Are they really "all wet"? *Angew. Chem., Int. Ed.*, 2006, **45**, 8100–8102.

14. H. Sugimoto, M. Tarumizu, H. Miyake and H. Tsukube, Bis(dithiolene) Molybdenum complex that promotes combined coupled electron-proton transfer and oxygen atom transfer reactions: A water-active model of the Arsenite oxidase Molybdenum center, *Eur. J. Inorg. Chem.*, 2006, 4494–4497.

15. Y. Gu, C. Ogawa, J. Kobayashi, Y. Mori and S. Kobayashi, A heterogeneous silica-supported Scandium/ionic liquid catalyst system for organic reactions in water, *Angew. Chem., Int. Ed.*, 2006, **45**, 7217–7220.

16. S. Narayan, J. Muldoon, M. G. Finn, V. V. Fokin, H. C. Kold and K. B. Sharpless, "On water": Unique reactivity of organic compounds in aqueous suspension, *Angew. Chem., Int. Ed.*, 2005, **44**, 3275–3279.

17. C. Chatgilialoglu, M. Guerra and Q. G. Mulazzani, Model studies of DNA C5' radicals. Selective generation and reactivity of 2'-deoxyadenosin-5'-yl radical, *J. Am. Chem. Soc.,* 2003, **125**, 3839–3848.

18. F. Villar, O. Andrey and P. Renaud, Diastereoselective radical cyclization of bromoacetals: Efficient synthesis of ($\pm$)-botryodiplodin, *Tetrahedron Lett.*, 1999, **40**, 3375–3378.

19. A. Postigo, C. Ferreri, M. L. Navacchia and C. Chatgilialoglu, The radical-based reduction with (TMS)3SiH 'On water', *Synlett,* 2005, 2854–2856.

20. L. B. Jimenez, S. Encinas, M. A. Miranda, M. L. Navacchia and C. Chatgilialoglu, The photochemistry of 8-bromo-2'-deoxyadenosine. A

direct entry to cyclopurine lesions, *Photochem. Photobiol. Sci.,* 2004, **3**, 1042–1046.

21. C. Chatgilialoglu, M. L. Navacchia and A. Postigo, A facile one-pot synthesis of 8-oxo-7,8-dihydro-(2′-deoxy)adenosine in water, *Tetrahedron Lett.,* 2006, **47**, 711–714.

22. C. Chatgilialoglu, A. Postigo, S. Kopsov and C. Ferreri, Radical reactions in aqueous medium using (Me₃Si)₃SiH, *Org. Lett.*, 2007, **9**, 5159–5162.

23. C. Chatgilialoglu and C. Ferreri, Trans lipids: The free radical path, *Acc. Chem. Res.,* 2005, **38**, 441–448.

24. C. Ferreri and C. Chatgilialoglu, Geometrical trans lipid isomers: A new target for lipidomics, *ChemBioChem,* 2005, **6**, 1722–1734.

25. (a) C. Chatgilialoglu, C. Ferreri, M. Ballestri, Q. G. Mulazzani and L. Landi, *Cis-trans* isomerization of monounsaturated fatty acid residues in phospholipids by thiyl radicals, *J. Am. Chem. Soc.,* 2000, **122**, 4593–4601; (b) C. Ferreri, C. Costantino, L. Perrotta, L. Landi, Q. G. Mulazzani and C. Chatgilialoglu, Cis-trans isomerization of polyunsaturated fatty acid residues in phospholipids catalyzed by thiyl radicals, *J. Am. Chem. Soc.,* 2001, **123**, 4459–4468; (c) C. Ferreri, A. Samadi, F. Sassatelli, L. Landi and C. Chatgilialoglu, Regioselective cis-trans isomerization of arachidonic double bonds by thiyl radicals: The influence of phospholipid supramolecular organization, *J. Am. Chem. Soc.,* 2004, **126**, 1063–1072.

26. H. Sprinz, J. Schwinn, S. Naumov and O. Brede, Mechanism of thiyl radical-catalyzed isomerization of unsaturated fatty acid residues in homogeneous solution and in liposomes, *Biochim. Biophys. Acta,* 2000, **1483**, 91–100.

27. C. Chatgilialoglu, C. Ferreri, I. N. Lykakis and P. Wardman, trans-Fatty acids and radical stress: What are the real culprits? *Bioorg. Med. Chem.,* 2006, **14**, 6144–6148.

28. I. N. Lykakis, C. Ferreri and C. Chatgilialoglu, The sulfhydryl radical (HS˙/S˙-): A contender for the isomerization of double bonds in membrane lipids, *Angew. Chem., Int. Ed.,* 2007, **46**, 1914–1916.

29. C. Ferreri, D. Anagnostopoulos, I. N. Lykakis, C. Chatgilialoglu and A. Siafaka-Kapadai, Synthesis of all-trans anandamide: A substrate for fatty acid amide hydrolase with dual effects on rabbit platelet activation, *Biorg. Med. Chem.,* 2008, **16**, 8359–8365.

30. D. Anagnostopoulos, C. Chatgilialoglu, C. Ferreri, A. Samadi and A. Siafaka-Kapadai, Synthesis of all-trans arachidonic acid and its effect on rabbit platelet aggregation, *Biorg. Med. Chem.,* 2005, **15**, 2766–2770.

31. D. C. Harrowven and J. C. Hannam, Thiyl radical induced isomerisations of maleate esters provide a convenient route to fumarates and furanones, *Tetrahedron,* 1999, **55**, 9341–9346.

32. O. Mozziconacci, K. Bobrowski, C. Ferreri and C. Chatgilialoglu, Reactions of hydrogen atoms with Met-Enkephalin and related peptides, *Chem.–Eur. J.,* 2007, **13**, 2029–2033.

33. C. Ferreri, I. Manco, M. R. Faraone-Mennella, A. Torreggiani, M. Tamba and C. Chatgilialoglu, A biomimetic model of tandem radical damage involving sulfur-containing proteins and unsaturated lipids, *ChemBio-Chem,* 2004, **5**, 1710–1712.

34. C. Ferreri, I. Manco, M. R. Faraone-Mennella, A. Torreggiani, M. Tamba, S. Manara and C. Chatgilialoglu, The reaction of hydrogen atoms with methionine residues: A model of reductive radical stress causing tandem protein–lipid damage, *ChemBioChem,* 2006, **7**, 1738–1744.

35. A. Torreggiani, M. Tamba, I. Manco, M. R. Faraone-Mennella, C. Ferreri and C. Chatgilialoglu, Investigation of radical-based damage of RNase A in aqueous solution and lipid vesicles, *Biopolymers,* 2006, **81**, 39–50.

36. V. Kadlcik, C. Sicard-Roselli, C. Houee-Levin, M. Kodicek, C. Ferreri and C. Chatgilialoglu, Reductive modification of a methionine residue in the amyloid-β peptide, *Angew. Chem., Int. Ed.,* 2006, **45**, 2595–2598.

37. A. Torregiani, J. Domenech, R. Orihuela, C. Ferreri, S. Atrian, M. Capdevila and C. Chatgilialoglu, Zinc and Cadmium complexes of a plant metallothionein under radical stress: Desulfurisation reactions associated with the formation of trans-lipids in model membranes, *Chem.–Eur. J.,* 2009, **15**, 6015–6024.

38. E. M. Sanchez, J. F. Arteaga, V. Domingo, J. F. Quilez del Moral, M. M. Herrador and A. F. Barrero, Tandem addition–cyclization mediated by sulfanyl radicals: a versatile strategy for iridoids synthesis, *Tetrahedron,* 2008, **64**, 5111–5118, and references cited therein.

39. H.-S. Dang, M. R. J. Elsegood, K.-M. Kim and B. P. Roberts, Radical-chain reductive alkylation of electron-rich alkenes mediated by silanes in the presence of thiols as polarity-reversal catalysts, *J. Chem. Soc., Perkin Trans 1,* 1999, 2061–2068.

40. A. Dondoni, The emergence of thiol–ene coupling as a click process for materials and bioorganic chemistry, *Angew. Chem., Int. Ed.,* 2008, **47**, 2–5, and references cited therein.

41. (a) F. R. Mayo and C. Walling, The peroxide effect in the addition of reagents to unsaturated compounds and in rearrangement reactions, *Chem. Rev.,* 1940, **27**, 351–412; (b) K. Griesbaum, Problems and possibilities of the free radical addition of thiols to unsaturated compounds, *Angew. Chem., Int. Ed. Engl.,* 1970, **9**, 273–287; (c) S. Z. Zard, *Radical Reactions in Organic Synthesis,* Oxford University Press, Oxford, 2003.

42. C. E. Hoyle, T. Y. Lee and T. Roper, Thiol-enes: Chemistry of the past with promise for the future, *J. Polym. Sci., Part A: Polym. Chem.,* 2004, **42**, 5301–5338.

43. J. A. Carioscia, H. Lu, J. W. Stanbury and C. N. Bowman, Thiol-ene oligomers as dental restorative materials, *Dent. Mater.,* 2005, **21**, 1137–1143.

44. A. Gress, A. Völkel and H. Schlaad, Thio-Click modification of Poly[2-(3-butenyl)-2-oxazoline]**,** *Macromolecules,* 2007, **40**, 7928–7933.

45. K. L. Killops, L. M. Campos and C. J. Hawker, Robust, efficient, and orthogonal synthesis of dendrimers via thiol-ene "click" chemistry, *J. Am. Chem. Soc.,* 2008, **130**, 5062–5064.

46. Y. Geng, D. E. Discher, J. Justynska and H. Schlaad, Dialkylphosphates as stereodirecting protecting groups in oligosaccharide synthesis, *Angew. Chem., Int. Ed.,* 2006, **45**, 7578–7581.

47. For examples in water in the presence of catalysts see: (a) B. N. Naidu, M. E. Sorenson, T. P. Connolly and Y. Ueda, Michael addition of amines and thiols to dehydroalanine amides: A remarkable rate acceleration in water, *J. Org. Chem.,* 2003, **68**, 10098–10102; (b) H. Firouzabadi, N. Iranpoor and A. Jafari, Micellar solution of sodium dodecyl sulfate (SDS) catalyzes facile michael addition of amines and thiols to α,β-unsaturated ketones in water under neutral conditions, *Adv. Synth. Catal.,* 2005, **347**, 655–661; (c) N S. Krishnaveni, K. Surendra and K. R. Rao, Study of the Michael addition of β-cyclodextrin–thiol complexes to conjugated alkenes in water, *Chem. Commun.,* 2005, 669–671.

48. For examples in ionic liquids see: (a) J. S. Yadav, B. V. S. Reddy and G. Baishya, Green protocol for conjugate addition of thiols to α,β-unsaturated ketones using a [Bmim]PF$_6$/H$_2$O system, *J. Org. Chem.,* 2003, **68**, 7098–7100; (b) B. C. Ranu and S. S. Dey, Catalysis by ionic liquid: a simple, green and efficient procedure for the Michael addition of thiols and thiophosphate to conjugated alkenes in ionic liquid, [pmIm]Br, *Tetrahedron,* 2004, **60**, 4183–4188.

49. G. L. Khatik, R. Kumar and A. K. Chakraborti, Catalyst-free conjugated addition of thiols to α,β-unsaturated carbonyl compounds in water, *Org. Lett.,* 2006, **8**, 2433–2436.

50. D. H. R. Barton, T. Li and J. MacKinnon, Synergistic oxidation of cyclohexane and hydrogen sulfide under Gif conditions, *Chem. Commun.,* 1997, 557–558.

51. Y. Hebting, P. Adam and P. Albrecht, Reductive desulfurization of allylic thiols by HS-/H2S in water gives clue to chemical reactions widespread in natural environments, *Org. Lett.,* 2003, **5**, 1571–1574.

52. C. Chatgilialoglu, L. Zambonin, A. Altieri, C. Ferreri, Q. G. Mulazzani and L. Landi, Geometrical isomerism of monounsaturated fatty acids: Thiyl radical catalysis and influence of antioxidant vitamins, *Free Radical Biol. Med.,* 2002, **33**, 1681–1692.

53. S. A. Everett, M. F. Dennis, K. B. Patel, S. Maddix, S. C. Kundu and R. L. Willson, Scavenging of nitrogen dioxide, thiyl, and sulfonyl free radicals by the nutritional antioxidant-caroten, *J. Biol. Chem.,* 1996, **271**, 3988–3994.

54. R. Jishi, S. Adhikari, B. S. Patro, S. Chattopadhyay and T. Mukherjee, Free radical scavenging behavior of folic acid: evidence for possible antioxidant activity, *Free Radical Biol. Med.,* 2001, **30**, 1390–1399.

55. G. G. Borisenko, I. Martin, Q. Zhao, A. A. Amoscato and V. E. Kagan, Nitroxides scavenge myeloperoxidase-catalyzed thiyl radicals in model systems and in cells, *J. Am. Chem. Soc.,* 2004, **126**, 9221–9232.

56. E. Damiani, P. Carloni, M. Iacussi, P. Stipa and L. Greci, Reactivity of sulfur-centered radicals with indolinonic and quinolinic aminoxyls, *Eur. J. Org. Chem.,* 1999, 2405–2412.

57. S. Goldstein, A. Samuni and G. Merenyi, Kinetics of the reaction between nitroxide and thiyl radicals: Nitroxides as Antioxidants in the Presence of Thiols, *J. Phys. Chem. A,* 2008, **112**, 8600–8605.

58. D. L. Reid, G. V. Shustov, D. A. Armstrong, A. Rauk, M. N. Schuchmann, M. S. Akhlaq and C. V. Sonntagz, H-atom abstraction from thiols by C-centered radicals. A theoretical and experimental study of reaction rates, *Phys. Chem. Chem. Phys.,* 2002, **4**, 2965–2974.

59. E. Denisov, C. Chatgilialoglu, A. Shestakov and T. Denisova, Rate constants and transition-state geometry of reactions of alkyl, alkoxyl, and peroxyl radicals with Thiols, *Int. J. Chem. Kinet.,* 2009, **41**, 284–293.

CHAPTER 9

Advances in the Use of Phosphorus-centered Radicals in Organic Synthesis in Conventional Flasks: Advantages, Reasons and Applications[*]

9.1 INTRODUCTION

This chapter aims to present recent contributions dealing with the organic chemistry of organophosphorus radicals. The first part briefly lays out the physical organic background of such intermediates. In a second part the use of organophosphorus radicals possessing a P–H bond that can undergo homolytic cleavage as alternative mediators is detailed. The third part is focused on radical additions of phosphorus-centered radicals to unsaturated compounds, an old reaction that is being rejuvenated. Lastly, radical eliminations of phosphorus-centered radicals are introduced in the fourth part. Most of the latter are relatively novel reactions, and have never been reviewed previously.

The organic chemistry of phosphorus is based on the rich array of stable compounds featuring a carbon–phosphorus bond. As a consequence, reactions involving organophosphorus radicals have a long history which has been told previously.[1] These last few years have witnessed a renewed awareness that P-centered radicals (especially those containing P–O bonds) could be of practical synthetic interest. In this chapter, we wish to present an overview of the recently published results in the field.

* Chapter written by V. Tamara Perchyonok.
Streamlining Free Radical Green Chemistry
V. Tamara Perchyonok, Ioannis Lykakis and Al Postigo
© V. Tamara Perchyonok, Ioannis Lykakis and Al Postigo 2012
Published by the Royal Society of Chemistry, www.rsc.org

9.2 PHYSICAL ORGANIC ASPECTS

The physical aspects underlying the reactivity of P-centered radicals are essential to synthetic chemists and industrialists, who need to know them in order to benefit from the full potentialities offered by phosphorus-containing compounds. Reductions, halogen-atom abstractions and additions are the most relevant transformations for synthesis. Several authors have determined key rate constants through various physical methods, among which are time-resolved electron spin resonance (ESR) spectroscopy and laser flash photolysis. Hydrogen abstraction to form phosphonyl radicals is comparable to that of phenyl-substituted silanes ($1.2 \times 10^5 \text{ M}^{-1} \text{ s}^{-1}$).[2] On the other hand, Turro showed that phosphinoyl radicals ($R_2P(O)^{\cdot}$) are roughly ten times more prone to reduction by thiophenol than acylphosphinoyl ones.[3] In any case, these rates are much lower than those of addition to double bonds (see below). Halogen abstraction is the most popular way to generate carbon radicals on an organic substrate, and thus a key feature of any synthetically relevant radical reaction.[4] Ingold *et al.* measured some rate constants for diethoxyphosphonyl radicals.[4] They follow the same trends as those of standard tin radicals, but are considerably lower. Turro *et al.* obtained similar results for acylphosphinoyl and phosphinoyl radicals.[5] The main characteristic of P-centered radicals is their high reactivity towards unsaturated compounds. Indeed, Ingold's diethoxyphosphonyl radicals add easily onto olefins (even hindered ones). Sumiyoshi, Schnabel and coworkers showed that phosphinoyl radicals also add rapidly onto double bonds,[6] albeit less than phosphonyl ones[7] ($\sim 10^6$–$10^7 \text{ M}^{-1} \text{ s}^{-1}$; see Figure 9.1).

Additional work by Schnabel, Kamachi *et al.* refined the understanding of the phosphinoyl radicals and confirmed that they added readily onto alkenes.[8] It also confirmed that additions onto electron-poor double bonds occurred faster than addition onto electron-rich ones. Phosphinoyl radicals are thus moderately nucleophilic. Acylphosphinoyl radicals add more slowly. In both cases though, steric effects are much more significant than polar ones.

A considerable amount of work has been spent on linking the structures and reactivities of the radicals. It had been shown early on that phosphinoyl and phosphonyl radicals were non-planar and as a result had a variable degree of s-character.[9,10] Phosphonyl radicals are more bent than phosphinoyl ones. Acylphosphinoyl radicals are further flattened.[5] In general, the more bent the radical, the faster its addition onto olefins (Figure 9.1). This trend has been

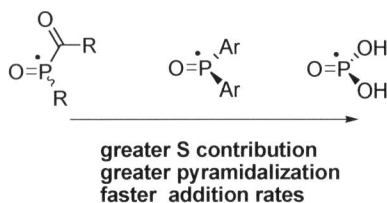

greater S contribution
greater pyramidalization
faster addition rates

Figure 9.1   Geometry of the P-radicals.

attributed to the relative accessibility of the localized spin to the trapping agent.[3,7]

9.3 USE OF P-CENTERED RADICALS AS MEDIATORS

Radical chemistry has relied on the use of tributyltin hydride (TBTH) as a mediator. While extremely useful, this compound is toxic and its by-products are difficult to remove from reaction mixtures. For these reasons, and because radical reactivities are complementary to other reactivities, the quest for alternative mediators has been extremely active. Barton, Jaszberenyi *et al.* re-examined some very ancient radical reductions involving hypophosphorous acid. Upon using organic salts of this acid, clean radical reduction, in particular azobisisobutyronitrile (AIBN)-initiated deoxygenation, was achieved cleanly in refluxing dioxane.[11] The best salt proved to be the N-ethylpiperidinium salt (EPHP). Dialkyl phosphonates were also examined, but they require initiation by peroxides and can lead to undesired by-products (Scheme 9.1).

This seminal contribution rapidly spawned several new contributions. In particular, Jang—who contributed to the initial work—showed that the sodium salt of hypophosphorous acid could reduce water-soluble organohalides in water.[12] His group also introduced dibutylphosphine[13] and diphenylphosphine[14,15] oxides as new reducing agents. Because they are not ionic, they are less hygroscopic than EPHP and thus could be used with water-sensitive substrates. In any case, the deoxygenation of hindered substrates was possible. A comparison of the yields to those obtained *via* Barton's method shows that the three mediators are complementary. Once these two main families of P-based mediators had been introduced, rapid progress arose (Scheme 9.2).

Murphy and Stoodley and their respective colleagues simultaneously reported that EPHP could trigger formation of carbon–carbon bonds either through a 6-*exo*-trig cyclization of an aryl radical obtained from an iodide (Murphy),[15] or through a 5-*exo*-dig cyclization of an alkyl α-keto radical obtained from a bromide (Stoodley).[16] In both cases, yields in densely functionalized products were quite high (Scheme 9.3).

Oshima *et al.* introduced deuterated hypophosphorous acid potassium salts to achieve radical deuteration.[17] Deuteration of hydrophobic substrates was

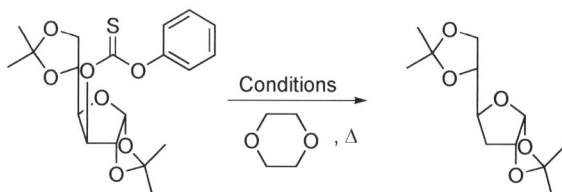

Conditions: EPHP, AIBN: 99%
$(OME)_2P(O)H$, Bz_2O_2: yield 92%

Scheme 9.1 Deoxygenation with EPHP.

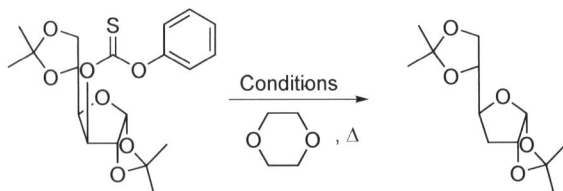

Conditions: Bu₂P(O)H, AIBN: yield 93%
Ph₂P(O)H, Bz₂O₂: yield 80%

Scheme 9.2   Deoxygenation with phosphine oxides.

possible, although the incorporation of deuterium was not optimal because of hydrogen-atom abstraction from either the solvent or the various additives used. As water is not prone to transfer a deuterium atom, less hydrophobic substrates led to deuteration with total incorporation (Scheme 9.4).

Kita *et al.* built on the previous studies to report EPHP-mediated cyclization of hydrophobic substrates in water (Scheme 9.5).[18]

This breakthrough was made possible by running the reaction in the presence of a water-soluble initiator (VA-061) and a surfactant (cetyl trimethylammonium bromide, CTAB). The authors explain this outstanding result by a micellar effect generated by CTAB. The organic ammonium probably contributes to the incorporation of the hypophosphoric acid in the micelles. By trapping hypophosphorous acid with a greasy tertiary amine, Jang *et al.* introduced a surfactant-type chain carrier and reported good yields for deoxygenations of alcohols in water, without additives.[19]

Tetraalkylammonium hypophosphites (TAHPs), recently reported by Jang and co-workers, were prepared by mixing the tetraalkylammonium hydroxide

Murphy

Where R = Ms or Tos
R' = Me

Stoodley

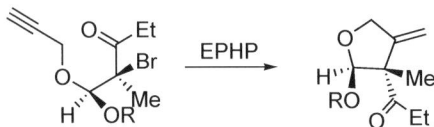

Where R = Me or Et

Scheme 9.3   Cyclizations mediated by EPHP.

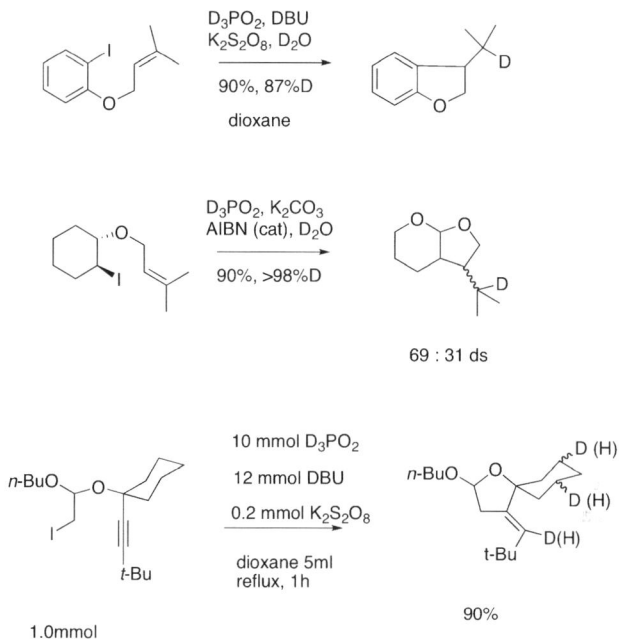

Scheme 9.4 Radical deuteration with phosphorous deuteride.

with aqueous H_3PO_4, both mild and efficient reagents for the radical deoxygenation of alcohols and the formation of carbon–carbon bonds in water without adding additives such as surfactants.[10] Cyclododecyl-S-methyl dithiocarbonate has been utilized as a model compound. The reaction of the xanthate with TAHP-1 in the presence of a water-soluble radical initiator, V-501 (4,4'-azobis(4-cyanovaleric acid)), in water yield 94% of the deoxygenated product after 5 h. Several other tetraalkylammonium hypophosphites were also used in the study and the results are summarized in Table 9.1.

This methodology was applied to the synthesis of 2',3'-didehydro-2',3'-dideoxynucleosides, potent anti-HIV agents. The bis-xanthate of the N^3-methyluridine derivative was subjected to the radical reaction conditions in water, giving the corresponding olefin in 82% yield. Similar results were also achieved in the case of adenosine derivative (76% yield). However, Kita and co-workers found that the combination of the water-soluble radical initiator [2,2'-azobis-(2-(2-imidazolin-2-yl)propane] (VA-061), water-soluble chain car-

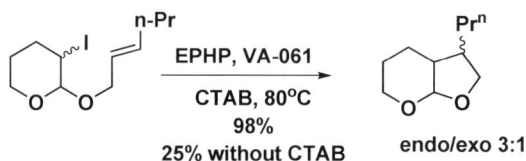

Scheme 9.5 Radical cyclization in water.

Table 9.1 Deoxygenation of O-cyclododecyl S-methyl dithiocarbonate with tetraalkylammonium hypophosphites in water.

Entry	TAPH (equiv.)	V-501 (equiv.)	Time /h	Yield (%)
1	TAPH-1(3)	0.5	5	94
2	TAPH-1(2)	0.5	8	84
3	TAPH-2(3)	0.75	7	90
4	TAPH-3(3)	0.75	8	62(29)
5	TAPH-4(3)	0.75	8	45(52)

rier 1-ethylpiperidine hypophosphite (EPHP) and surfactant cetyltrimethylammonium bromide (CTAB), resulted in a radical cyclization that occurred in water with a variety of hydrophobic substrates (Table 9.2).[16]

Although the results were not as high yielding in the case of TMS$_3$SiH in comparison to the optimised EPHP/surfactant/initiator combination, authors showed that tristrimethylsilyl silane (TMS$_3$SiH) could be used as a silicon-based hydrogen donor in water/aqueous medium for the first time. The importance of the observation opened the door for the application of TMS$_3$SiH as a environmentally benign and powerful hydrogen donor in organic solvents to hydrogen-atom transfer (HAT) "on water" and "in water".

Recently Perchyonok *et. al.* have reported further developments and applications of broad range of fundamental free radical reactions, such as hydrogen-atom transfer, radical deoxygenations and radical cyclizations utilizing quaternary ammonium salts of hypophosphorous acids as chiral and achiral hydrogen donors at room temperature and the results are

Table 9.2 Radical cyclization reactions using various initiators, hydrogen donors and surfactants.

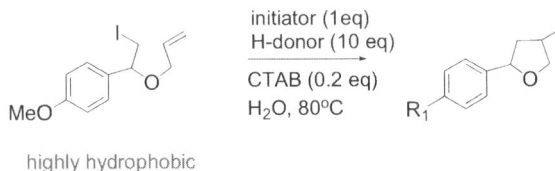

highly hydrophobic

Initiator EPHP	Yield (cis : trans)	H-donor/ V-061	Yield (cis : trans)	Additive EPHP V-061	Yield (cis : trans)
VA-061	98% (55 : 45)	None	None	None	64% (74 : 26)
AIBN	19% (57 : 43)	(TMS)$_3$SiH	94% (67 : 33)	CTAB	98% (55 : 45)
Et$_3$B	50% (67 : 43)	EPHP	98% (55 : 45)	SDS	98% (51 : 49)
V-501	72% (51 : 49)	H$_3$PO$_2$/ NaHCO$_3$	84% (78 : 22)	Triton X-100	98% (62 : 38)
VA-044	95% (50 : 50)	NaH$_2$PO$_2$	58% (78 : 22)	Et$_4$N$^+$Br$^-$	85% (65 : 35)

Table 9.3 Various free radical reactions at room temperature in the presence of chiral and achiral tetrasubstituted-ammonium hypophosphites in water using 0.5 eq. of Et$_3$B/air as the radical initiator.

Entry	QAHP	Radical Precursor	Product	Yield (%)
1	$(Bu_4N)^+H_2PO_2^-$			65
2	$(Bu_4N)^+H_2PO_2^-$			64
3	$(Bu_4N)^+H_2PO_2^-$			73
4	$(Bu_4N)^+H_2PO_2^-$			87
5	$(Bu_4N)^+H_2PO_2^-$			47
6	$(Bu_4N)^+H_2PO_2^-$			86
7	$(Bu_4N)^+H_2PO_2^-$			77
8	(S,S)-HMe$_2$NCHMePh$^+$ H$_2$PO$_2^-$			82
9	(S,S)-HMe$_2$NCHMePh$^+$ H$_2$PO$_2^-$			76

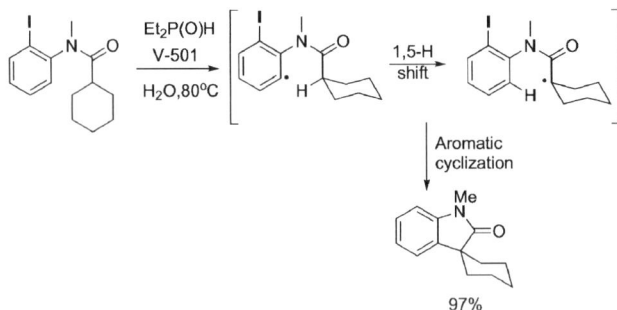

Scheme 9.6   DEPO-mediated arylation of lactams in water.

summarized in Table 9.3.[17] The same reactions were repeated at 80 °C under predetermined reaction conditions using either AIBN (10% wt) (organic-soluble radical initiator) or V-501 (10% wt) (a water-soluble radical initiator) and comparable results were obtained. The results have exceeded expectations as not only did the reactions proceed with good to excellent yields but they also showed a degree of stereoselectivity and enantioselectivity The results represent the first examples of enantioselectivity being observed in free radical hydrogen transfer reactions in aqueous media and work is currently in progress to explore this novel aspect of enantioselectivity in an aqueous environment. The advantages of these hypophosphite reagents (chiral and achiral hypophosphinates) lie in their affordability, low toxicity, avoidance of the use of highly toxic "tin-based hydrogen donors" and green reaction conditions.

Eventually, Murphy *et al.* introduced a water-soluble phosphine oxide which permits higher isolated yields than the corresponding reaction using EPHP, with no additional additive (Scheme 9.6).[20]

Upon using diethylphosphine oxide (DEPO), one can carry out sophisticated tin-free tandem radical reactions. Because DEPO is more lipophilic than hypophosphorous acid yet still water-soluble, it can facilitate the interaction between the water-soluble mediator and initiator and the lipophilic substrates without requiring a phase-transfer agent. Moreover, its pK_a is 6, thus ensuring that this almost neutral excess reagent can be extracted into base during workup.

One of the most impressive synthetic achievements of the P-based radical mediators is the deoxygenation of an erythromycin B derivative toward the industrial synthesis of ABT-229, a potent motilin receptor agonist. Clean deoxygenation was achieved on a 15 kg scale by using NaH_2PO_2 in an aqueous alcohol, and a phase transfer agent (Scheme 9.7).[21]

9.4   SYNTHETIC APPLICATIONS OF P-CENTERED RADICAL ADDITIONS

Additions of P-based radicals were the first reported reactions of those reactive intermediates.[1a] They suffered around two decades of relative neglect, but are once again being investigated actively.

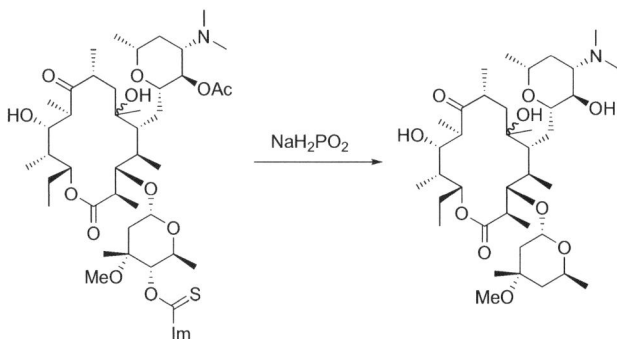

Scheme 9.7   Use of phosphorous acid in synthesis.

9.4.1   Phosphinyl Radicals

Because of the central role of phosphines as ligands for organometallic transformations, the radical addition of phosphines to olefins has been pursued, despite its being the oldest radical transformation involving phosphorus. Progress has been sought toward tandem reactions. Simpkins *et al.* have reported the domino preparation of bicyclic molecules triggered by initial addition of the diphenylphosphinyl radical to various unsaturated compounds (Scheme 9.8).[22]

Capretta *et al.* used phosphine to add to limonene, a chiral pool terpene. This different approach relied on the bidirectional functionalization of phosphine and yielded a strained bicyclic chiral phosphine in excellent yield (Scheme 9.8).[23]

Oshima *et al.* introduced very recently a highly elegant mild synthesis of vinylic diphosphines that provides an entry to organic compounds usable in material chemistry. In this process, the two phosphorus atoms were introduced

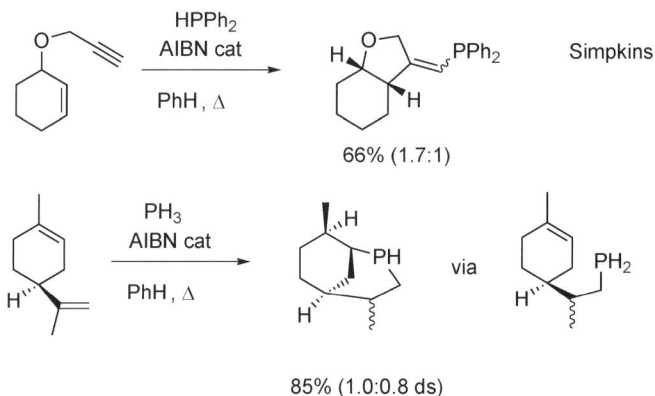

Scheme 9.8   Additions of phosphinyl radicals.

in one single pot through *in situ* formation of tetraphenyldiphosphine, homolytic cleavage of the P–P bond, highly chemoselective addition of the
diphenylphosphinyl radical to a terminal alkyne, and homolytic substitution
on the diphosphine to regenerate a diphenylphosphinyl radical (Scheme 9.9).[24]

9.4.2   Phosphonyl Radicals

By taking advantage of the rapid β-elimination of iodo radicals, Russell *et al.*
devised a vinylation of phosphonyl radicals derived from mercury compounds.[25] Probably because of the toxicity of mercuric salts, this method has
not been followed by synthetic applications. Piettre[26] and Motherwell *et al.*[27]
simultaneously used the very efficient addition of phosphonyl radicals to
alkenes to prepare 1,1-difluorophosphonates, which are believed to be isosteric
to phosphates and thus of high interest in pharmacology. Both authors'
disconnections relied on the addition of phosphonyl radicals to difluoro-olefins
(Scheme 9.10). By using difluoroenol ethers derived from sugars, Motherwell
and colleagues were able to achieve total regioselectivity.

   Phosphonyl radicals have subsequently been involved in tandem and
cascade processes. Observing that the phosphonyl radicals used as mediators
in the radical cyclization of dienes also led to the formation of phosphonate
by-products, Parsons *et al.* were able to cleanly prepare the corresponding
cyclic organophosphorus derivatives in good yields (Scheme 9.11).[28]

   Renaud *et al.* introduced a new tandem reaction based on the addition of
phosphonyl radicals, radical translocation and final cyclization onto the
vinylphosphonate. Various cyclopentane derivatives could be prepared
(Scheme 9.11).[29]

9.5   RADICALS FROM HYPOPHOSPHITES AND PHOSPHINATES

Hypophosphite hydrides can add to unsaturated compounds and thus lead to
the formation of the corresponding phosphinates. Deprèle and Montchamp
were able to carry out this process at room temperature thanks to the
triethylborane/air initiating system (Scheme 9.12).[30]

   Their study showed that formation of a radical from non-substituted hypophosphite salts or esters was much easier than from the mono-substituted
phosphinates; that H-abstraction from the alkylesters appears easier than from
the salts; and that the hypophosphite radicals were relatively electrophilic.

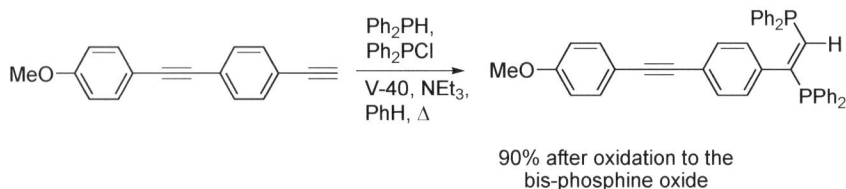

90% after oxidation to the
bis-phosphine oxide

Scheme 9.9   Diphosphinylation of alkynes.

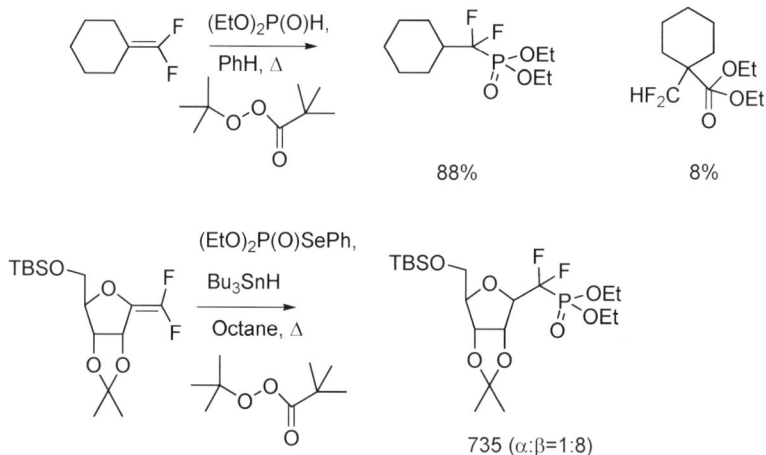

Scheme 9.10 Additions of phosphonyl radicals to gem-difluoro olefins.

Piettre has extended the scope of those radicals to the preparation of previously unreported a,a-difluorophosphinates. Radicals derived from the hypophosphorous acid sodium salt proved more reactive than both phosphonyl or phosphonothioyl radicals.[31]

Reding and Fukuyama designed a highly elegant synthesis of indoles by reacting hypophosphite salts with unsaturated thioanilides.[32] Initial regioselective addition of the P-centered radical onto the C=S bond generated a new stabilized carbon radical that could cyclize onto the double bond in the *ortho* position, thus giving birth to the carbon-skeleton of indoles. Aromatization of the compound generated the desired 2,3-substituted indoles. The author used this reaction as the key step toward the total synthesis of (±)-catharanthine, a presumed biological precursor of the anti-tumour alkaloids vinblastine and vincristine (Scheme 9.13).

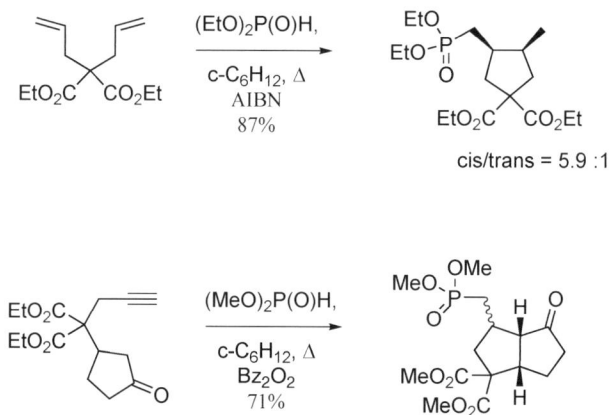

Scheme 9.11 Use of phosphonyl radicals for tin-free cascade processes.

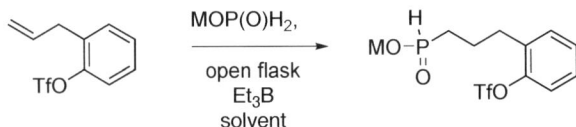

where M= Na, solvent=MeOH, yield=80%
n-Bu, solvent=cC$_6$H$_{12}$/n-BuOH, yield=59%

Scheme 9.12   Triethylborane/air-mediated formation of phosphinates.

9.6   PHOSPHINOYL RADICALS

Synthetic uses of additions of phosphinoyl radicals are relatively less abundant than what could have been expected when considering their popularity in polymer chemistry, spectroscopy, and as mediators (see above). Nonetheless, those radicals have been showed to be attractive addition partners. Taillades *et al.* showed that methanol was the best solvent to carry out radical addition of diphenylphosphine oxide onto olefins at room temperature,[33] while Parsons *et al.* reported the synthesis of a phosphorus-containing cyclic hydrazine through a radical tandem involving a hydrazone as the last radical acceptor.[28]

9.6.1   Thiophosphonyl and Other Sulfur-containing Radicals

As part of the work described previously, Piettre[26] and Motherwell *et al.*[27] examined the reactivity of thiophosphonyl radicals (Scheme 9.14).

Both authors observed that thiophosphonyl radicals led to higher yields. Motherwell attributed this improvement to the increased efficiency of the H-transfer step, due to a weaker P–H bond in thiophosphonates. This is consistent with further work by Piettre *et al.*, who carried out the same addition at room temperature using the ethyl radicals obtained from aerobic decomposition of triethylborane as the initiators.[34,35] In the same conditions, the corresponding phosphonates are left unchanged.

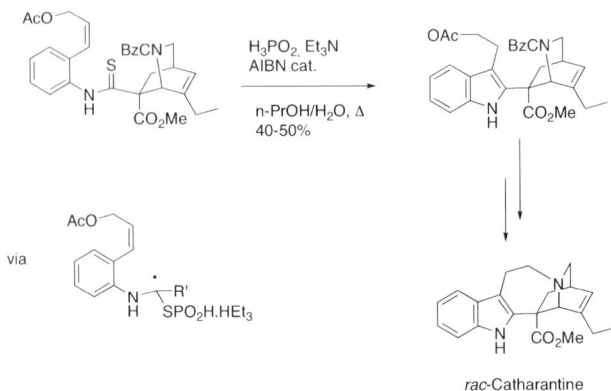

rac-Catharantine

Scheme 9.13   Formation of indoles from thioanilides.

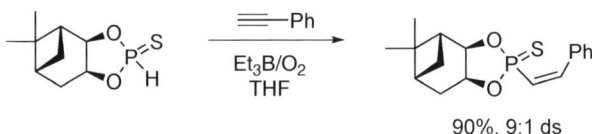

90%, 9:1 ds

Scheme 9.14 Addition of chiral thiophosphites.

Other sulfur-containing radicals were studied. The main results were published by Piettre *et al.*, who used the hemolytic cleavage of a P–Se bond introduced by Motherwell to access previously unknown *S,S*-dialkylphosphonodithioyl and phosphonotrithioyl radicals.[36] Those intermediates added smoothly onto olefins and were tentatively attributed a rather nucleophilic character. If confirmed, this latter result would be of high practical importance, since the nucleophilicity of the radical could be fine-tuned by simple choice of the substituents on phosphorus (Scheme 9.15).

Dingwall and Tuck had previously published an entry to phosphorus containing β-lactones, which were obtained by addition of various P-centered radicals to diketene. In particular, radicals produced from diphenylphosphine sulfide or thiophosphinates reacted smoothly and gave fair yields of the desired lactones (Scheme 9.16).[37]

9.7 ELIMINATION OF ORGANOPHOSPHORUS RADICALS

9.7.1 Phosphoranyl Radicals

Phosphoranyl radical reactivity (and notably elimination) has been extensively reviewed in the past.[1c,d] Chatgilialoglu *et al.* built on the affinity of silyl radicals for sulfur and selenium atoms to devise a radical method to reduce phosphine sulfides and selenides to phosphines.[38] Initial results show that the reactions proceed with retention of configuration. Zhang and Koreeda used the phosphoranyl radicals β-scission pathway to achieve versatile deoxygenation of alcohols (Scheme 9.17).[39]

The method works best with hindered alcohols, especially tertiary ones. Besides, the phosphonate by-product is easily removed from the desired compound. These two features ensure that this method is an attractive alternative to the Barton–McCombie deoxygenation of alcohols.

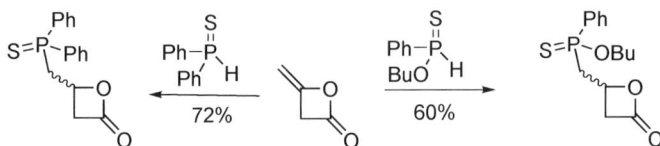

Scheme 9.15 Addition of phosphono di- and tri-thioates.

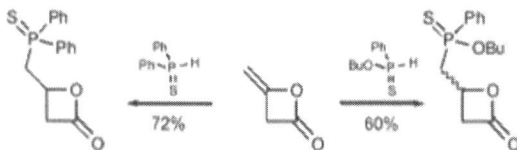

Scheme 9.16 Addition of thiophosphine oxides and thiophosphinates.

9.7.2 β-Elimination of P-centered Radicals

It has been known for quite a long time that the addition of phosphinyl radicals to olefins is reversible.[40] However, until recently, no work had addressed the β-elimination of P(v)-based radicals. Malacria *et al.* were first to report one such reaction, after they serendipitously observed that β-phosphinoyl radicals could indeed undergo elimination. In Malacria's initial report, this pathway accounted for a minor product of the reaction.[41] Malacria could nonetheless optimize the reaction, which proceeds at room temperature and is a formal radical vinylation (Scheme 9.18).[42]

En route to the synthesis of modified oligonucleotides through addition of hypophosphorous salts to suitable sugars, Piettre reported the unforeseen β-elimination of a phosphinoyl radical from a phosphinate (Scheme 19).[43]

This example demonstrates further that β-phosphinoyl radicals are prone to undergo elimination, even when the substituents are all alkyl groups.

Other oxidized P-centered radicals have since been shown to undergo β-elimination. Clive *et al.* used such a process to prepare biaryls *via ipso-*

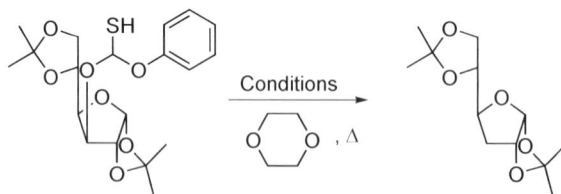

Conditions: EPHP, AIBN: 99%
$(OME)_2P(O)H, Bz_2O_2$: yield 92%

Scheme 9.17 Radical deoxygenation through phosphoranyl radicals.

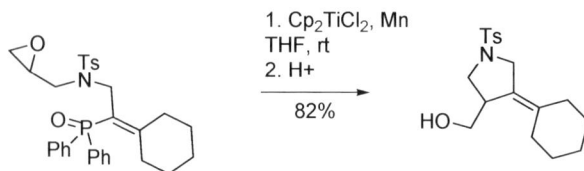

Scheme 9.18 Elimination of β-phosphinoyl radicals.

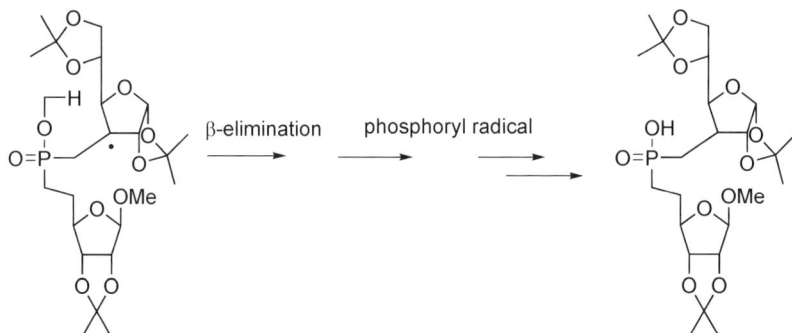

Scheme 9.19 Elimination of an alkyl β-phosphinoyl radical.

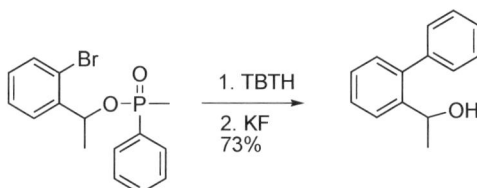

Scheme 9.20 Biaryl formation through *ipso*-substitution of arylphosphinates.

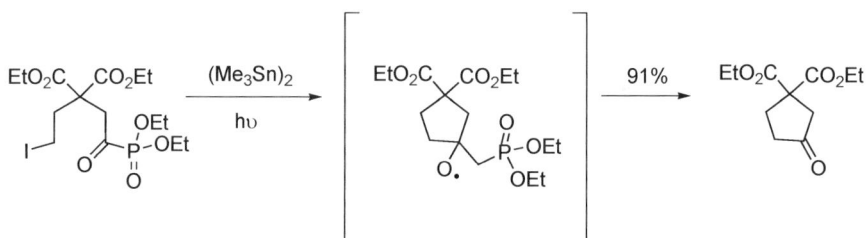

Scheme 9.21 Preparation of cyclic ketones *via* β-phosphonyl radical elimination.

substitution on arylphosphinates. Elimination from the intermediate radical is probably driven by rearomatization to the biarylic compound (Scheme 9.20).[44]

Eventually, Kim *et al.* introduced the β-elimination of a phosphonate radical from an alkoxy radical to prepare cyclic ketones (Scheme 9.21).[45] This reaction allows the intramolecular acylation of radicals, using an acylphosphonate as the key carbonylating agent. In that perspective, Kim's acylation and Malacria's vinylation (Scheme 9.18) are quite complementary.

9.8 CONCLUSION AND PERSPECTIVES

The growing awareness around green and sustainable methods stemming from legitimate concerns for the environment gives a strong and renewed relevance to the development of radical reactions involving phosphorus. Work in this field has moved from the understanding of physical aspects underlying the reactivities to optimization of the latter, and the use of phosphorus compounds in the total synthesis of complex molecules. Further progress should come from the introduction of more atom-efficient and stereoselective methods. Given the existing knowledge concerning phosphorus-containing compounds, progress will certainly also be achieved by integrating radical steps in more complex one-pot and domino processes.

REFERENCES

1. (a) F. W. Stacey and J. F. Harris, *Org. React.*, 1963, **13**, 150; (b) C. Walling and M. S. Pearson, *Top. Phosphorus Chem.*, 1966, **3**, 1; (c) W. G. Bentrude, *Acc. Chem. Res.*, 1982, **15**, 117; (d) W. G. Bentrude, in *The Chemistry of Organophosphorus Compounds*, ed. F. R. Hartley, Wiley, Chichester, 1990, vol. 1, pp. 531–566. For other tin-free radical mediators, see: P. A. Baguley and J. C. Walton, *Angew. Chem., Int. Ed.,* 1998, **37**, 3072–3082; or A. Studer and S. Amrein, *Synthesis*, 2002, 835.

2. C. Chatgilialoglu, V. I. Timokhin and M. Ballestri, *J. Org. Chem.*, 1998, **63**, 1327.

3. S. Jockusch and N. J. Turro, *J. Am. Chem. Soc.*, 1998, 120, 11773.

4. M. Anpo, R. Sutcliffe and K. U. Ingold, *J. Am. Chem. Soc.*, 1983, **105**, 3580.

5. G. W. Sluggett, P. F. McGarry, I. V. Koptyug and N. J. Turro, *J. Am. Chem. Soc.*, 1996, **118**, 7367.

6. T. Sumiyoshi, W. Schnabel, A. Henne and P. Lechtken, *Polymer*, 1985, **26**, 141.

7. T. Sumiyoshi and W. Schnabel, *Makromol. Chem.*, 1985, **186**, 1811.

8. A. Kajiwara, Y. Konishi, Y. Morishima, W. Schnabel, K. Kuwata and M. Kamachi, *Macromolecules*, 1993, **26**, 1656.

9. M. Geoffroy and E. A. C. Lucken, *Mol. Phys.*, 1971, **22**, 257.

10. C. M. L. Kerr, K. Webster and F. Williams, *J. Phys. Chem.*, 1975, **79**, 2650.

11. D. H. R. Barton, D. O. Jang and J. C. Jaszberenyi, *J. Org. Chem.*, 1993, **58**, 6838 and references therein.

12. D. O. Jang, *Tetrahedron Lett.*, 1996, **37**, 5367.

13. D. O. Jang, D. H. Cho and D. H. R. Barton, *Synlett*, 1998, 39.

14. D. O. Jang, D. H. Cho and J. Kim, *Synth. Commun.*, 1998, **28**, 3559.

15. C. Gonzalez Martin, J. A. Murphy and C. R. Smith, *Tetrahedron Lett.*, 2000, **41**, 1833.

16. R. McCague, R. G. Pritchard, R. J. Stoodley and D. S. Williamson, *Chem. Commun.*, 1998, 2691.

17. (a) H. Yorimitsu, H. Shinokubo and K. Oshima, *Bull. Chem. Soc. Jpn.*, 2001, **74**, 225. (b) V. T. Perchyonok, K. L. Tuck, S. J. Langford and M. W. Hearn, *Tetrahedron Lett.*, 2008, **49**(32), 4777–4779.

18. H. Nambu, G. Anilkumar, M. Matsugi and Y. Kita, *Tetrahedron*, 2002, **59**, 77.

19. D. H. Cho and D. O. Jang, *Tetrahedron Lett.*, 2005, **46**, 1799.
20. T. A. Khan, R. Tripoli, J. J. Crawford, C. G. Martin and J. A. Murphy, *Org. Lett.*, 2003, **5**, 2971 and references therein.
21. A. E. Graham, A. V. Thomas and R. Yang, *J. Org. Chem.*, 2000, **65**, 2583.
22. J. E. Brumwell, N. S. Simpkins and N. K. Terrett, *Tetrahedron*, 1994, **50**, 13533.
23. A. Robertson, C. Bradaric, C. S. Frampton, J. McNulty and A. Capretta, *Tetrahedron Lett.*, 2001, **42**, 2609.
24. A. Sato, H. Yorimitus and K. Oshima, *Angew. Chem., Int. Ed.*, 2005, **44**, 1694.
25. G. A. Russell, H. Tashtoush and P. Ngoviwatchai, *J. Am. Chem. Soc.*, 1984, **106**, 4622.
26. S. R. Piettre, *Tetrahedron Lett.*, 1996, **37**, 2233.
27. T. F. Herpin, W. B. Motherwell, B. P. Roberts, S. Roland and J.-M. Weibel, *Tetrahedron*, 1997, **53**, 15085.
28. C. M. Jessop, A. F. Parsons, A. Routledge and D. Irvine, *Tetrahedron Lett.*, 2003, **44**, 479.
29. F. Beaufils, F. Dénès and P. Renaud, *Angew. Chem., Int. Ed.*, 2005, **45**, 4292–4300.
30. S. Deprèle and J.-L. Montchamp, *J. Org. Chem.*, 2001, **66**, 6745.
31. A. Gautier, G. Garipova, C. Salcedo, S. Balieu and S. R. Piettre, *Angew. Chem., Int. Ed.*, 2004, **43**, 5963.
32. M. T. Reding and T. Fukuyama, *Org. Lett.*, 1999, **1**, 973.
33. P. Rey, J. Taillades, J. C. Rossi and G. Gros, *Tetrahedron Lett.*, 2003, **44**, 6169.
34. A. Gautier, G. Garipova, O. Dubert, H. Oulyadi and S. R. Piettre, *Tetrahedron Lett.*, 2001, **42**, 5673.
35. C. M. Jessop, A. F. Parsons, A. Routledge and D. J. Irvine, *Tetrahedron: Asymmetry*, 2003, **14**, 2849.
36. C. Lopin, G. Gouhier, A. Gautier and S. R. Piettre, *J. Org. Chem.*, 2003, **68**, 9916.
37. J. G. Dingwall and B. Tuck, *J. Chem. Soc., Perkin Trans. 1*, 1986, 2081.
38. R. Romeo, L. A. Wozniak and C. Chatgilialoglu, *Tetrahedron Lett.*, 2000, **41**, 9899.
39. L. Zhang and M. Koreeda, *J. Am. Chem. Soc.*, 2004, **126**, 13190.
40. J. Pellon, *J. Am. Chem. Soc.*, 1961, **83**, 1915.
41. S. Bogen, M. Gulea, L. Fensterbank and M. Malacria, *J. Org. Chem.*, 1999, **64**, 4920.
42. D. Leca, L. Fensterbank, E. Lacôte and M. Malacria, *Angew. Chem., Int. Ed.*, 2004, **43**, 4220.
43. O. Dubert, A. Gautier, E. Condamine and S. R. Piettre, *Org. Lett.*, 2002, **4**, 359.
44. D. L. J. Clive and S. Kang, *J. Org. Chem.*, 2001, **66**, 6083.
45. S. Kim, C. H. Cho and C. J. Lim, *J. Am. Chem. Soc.*, 2003, **125**, 9574.

CHAPTER 10

Metal-based Homogeneous Catalysis and Free Radical Synthesis: Advantages, Developments and Scope[*]

10.1 INTRODUCTION

Throughout this chapter, a thorough outlook on organic synthetic transformations accomplished through radical reactions employing different metallic species of the main-group metals and transition metals in water and aqueous media is provided. Thus, organic synthetic transformations achieved by the use of inorganic salts from diverse low-valent and high-valent metals such as U, Ti, V, Zn, Cu, Mn, Ce, Cr, Ga, Sn, In, Sm, Al, Mg, alone and in combination, allows the syntheses of important classes of organic compounds achieved by carbon–carbon coupling reactions, Mannich-type reactions, Reformatsky reactions, Barbier-type reactions, the syntheses of biologically active compounds, natural products, *etc.* The presentation intends to show the scope of organic radical transformations employing metal-radical species in environmentally friendly media.

10.2 METAL-MEDIATED REDUCTION AND OXIDATION REACTIONS IN WATER

Indium is currently used as a reducing agent in water for organic halides. A first systematic study on the dehalogenating power of indium was carried out by Ranu *et al.* This group provided an efficient and general methodology for the chemoselective reduction of α-halocarbonyl compounds and benzyl halides by indium metal in water under sonication. A wide range of structurally different α-iodo- and α-bromoketones and esters **1** underwent reduction, leading to the corresponding dehalogenated carbonyl compounds **2** (Scheme 10.1).[1]

---

* Chapter written by V. Tamara Perchyonok, Ioannis N. Lykakis and Al Postigo.
Streamlining Free Radical Green Chemistry
V. Tamara Perchyonok, Ioannis Lykakis and Al Postigo
© V. Tamara Perchyonok, Ioannis Lykakis and Al Postigo 2012
Published by the Royal Society of Chemistry, www.rsc.org

Brominated substrates were reduced more slowly than iodinated substrates. In fact, alkyl and aryl iodides remained inert although benzyl iodides and α-iodo-ketones were reduced. Selective deiodination was observed at the benzylic position *vs.* the aromatic carbon–iodine bond in the same substrate. The use of indium metal in aqueous medium was extended to the stereoselective reduction of aryl-substituted *gem*-dibromides to vinyl bromides. The reaction was performed in ethanol and saturated ammonium chloride solution under reflux, providing primarily the corresponding (*E*)-vinyl bromides (50 : 50 to 95 : 5 *E*/*Z* ratio) in high yields (70–95%). The compatibility with several sensitive functional groups (OMe, OBz, Cl, OTBDMS, and *o*-allyl) and the absence of over reduction processes are the main advantages of this methodology. However, thiophene- and furan-substituted *gem*-dibromides did not show any stereoselectivity, whereas low effectiveness was observed for alkyl-substituted *gem*-dibromides. The use of micellar solutions as reaction media has shown an enhancement in the reactivity of certain processes.[1]

Such is the case of the indium-mediated dehalogenation of α-halocarbonyl compounds **3** in water and in the presence of a catalytic amount of the surfactant sodium dodecyl sulfate (SDS).[2] These conditions were applied by Kim *et al.* to α-haloketones, esters, carboxylic acids, amides, and nitriles (Scheme 10.2).

For α-chlorocarbonyl compounds, the reaction was rather slow in comparison with that of the bromo derivatives, and a slightly higher temperature was required. In the absence of SDS, the reaction proceeded slowly and most of the starting materials were recovered unaltered after prolonged reaction times. The same group reported the efficient reductive conversion of 3-iodomethylcephalosporin into the corresponding 3-methylcephems by indium in an aqueous system.[3]

The capability of powdered zerovalent iron to dechlorinate DDT and related compounds at room temperature was investigated by Sayles *et al.*[4] Specifically, DDT (**5**), DDD [1,1-dichloro-2,2-bis-(*p*-chlorophenyl)ethane] (**6**), and DDE [1,1-dichloro-2,2- bis(*p*-chlorophenyl)ethylene] (**7**) were successfully dechlorinated by powdered zerovalent iron in buffered anaerobic aqueous solution at

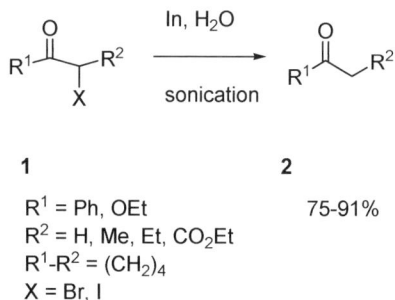

Scheme 10.1   Dehalogenation of α-halo carbonyl compounds.

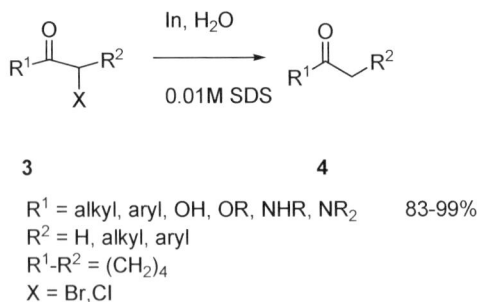

Scheme 10.2   Dehalogenation of α-halo carbonyl compounds.

20 °C, with or without the presence of nonionic surfactant Triton X-114 (Scheme 10.3). The rates of dechlorination of DDT and DDE were independent of the amount of iron, with or without surfactant, though rates with surfactant were much higher than without. A mechanistic model was constructed that quantitatively fits the observed kinetic data, indicating that the rate of dechlorination of the solid-phase reactants was limited by the rate of dissolution into the aqueous phase.

Granular iron metal was found to cause the reductive dechlorination of two important chloracetanilide herbicides, alachlor and metolachlor,[5] used for broadleaf weeds and annual grasses in domestic soybean and corn crops. The reaction was performed with granular cast iron in aqueous solutions at room temperature. A two-site, rate-limited sorption and first-order degradation model was applied to both batch data sets, with excellent agreement for alachlor and fair agreement for metolachlor.

The products of the reaction were chloride ions (84% mass balance for alachlor and 68% for metolachlor) and the corresponding dechlorinated acetanilides. The N-dealkylated acetanilide was a minor byproduct (9%) in the case of alachlor.

Atrazine (2-chloro-4-ethylamino-6-isopropylamino-1,3,5-triazine) **(8)** is a herbicide which has been used extensively in corn, sorghum, and sugarcane

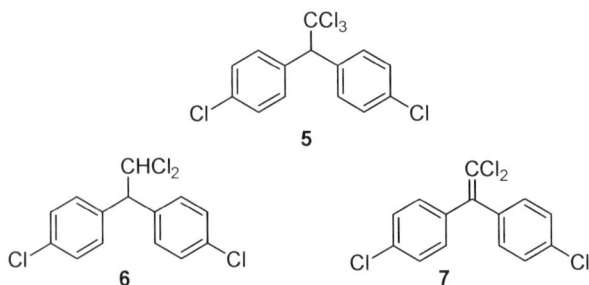

Scheme 10.3   Reductive dechlorination of DDT, DDD, and DDE by powdered Fe in aqueous solution.

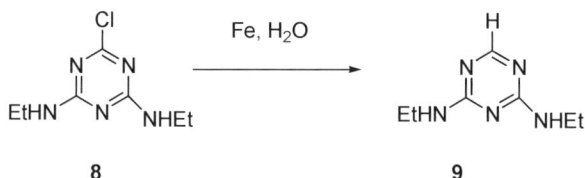

Scheme 10.4 Dechlorination of atrazine by Fe in water.

fields for the last 30 years,[6] with a long half-life in the environment (up to one year).[7] The possible water contamination, combined with the uncertainty of atrazine's carcinogenic and toxicological effects, has spurred interest in techniques that might more rapidly degrade atrazine and its metabolites. Batch aqueous experiments using fine-grained (100 mesh) zerovalent iron as an electron donor resulted in reductive dechlorination of atrazine to give 2-ethylamino-4-isopropylamino-1,3,5-triazine (**9**) (Scheme 10.4).[8] Identification of this compound initiated the development of analytical methods using HPLC, GC/MS, and HPLC/MS, to simultaneously quantify atrazine (**8**) and dechlorinated atrazine (**9**).

The dechlorination of atrazine (**8**) with metallic iron under low-oxygen conditions was studied at different reaction mixture pH values (2.0, 3.0, and 3.8).[9] The pH control was achieved by addition of sulfuric acid throughout the duration of the reaction. The lower the pH of the reaction, the faster the degradation of atrazine. The observed products of the degradation reaction were dechlorinated atrazine (**9**) and possibly hydroxyatrazine (2-ethylamino-4-isopropylamino-6-hydroxy-1,3,5-triazine). Triazine ring protonation was proposed to account, at least in part, for the observed effect of pH on atrazine by metallic iron.

Although the mechanisms of these reductions with zero-valent iron are not well elucidated, it appears that, generally, a two-electron transfer occurs either directly at the iron surface (by absorption of the organic halide) or through some intermediary (Scheme 10.5).[10] In a different mechanistic context, numerous studies have shown that dissociative adsorption of water takes place at clean iron metal surfaces, resulting in surface-bound hydroxyl, atomic oxygen, and atomic hydrogen ("nascent hydrogen").[11] The latter species can

$$RX + Fe + H^+ \longrightarrow RH + Fe^{+2} + X^- \quad (1)$$

$$Fe + 2H_2O \rightleftharpoons Fe^{+2} + H_2 + 2HO^- \quad (2)$$

$$RX + H_2 \longrightarrow RH + H^+ + X^- \quad (3)$$

$$RX + 2Fe^{2+} + H^+ \longrightarrow RH + 2Fe^{3+} + X^- \quad (4)$$

Scheme 10.5 Proposed reaction mechanism.

combine with itself, accounting for the formation of molecular hydrogen, or react with other compounds in the system, resulting in their hydrogenation (Scheme 10.5). A third possibility would be reduction by iron(II), resulting from corrosion of the metal (Scheme 10.5). A debate over the relative importance of these mechanisms[12] has gone on for many years, but the electron-transfer model is generally preferred.

Sugimoto, Tanji, *et al.* investigated the indium-mediated reduction of haloheteroaromatics in water.[13] The authors found that the deiodination of iodoheteroaromatics using indium in water was very effective. The proposed mechanism for the deiodination of iodoquinolines is depicted in Scheme 10.6.

When α- or γ-iodoquinoline is used as a substrate, the dihydroquinoline radical is generated smoothly since the radical is stabilized by an iodine atom. On the other hand, β-iodoquinoline reacts with indium in water more slowly than α- or γ-iodoquinoline because the dihydroquinoline radical is not stabilized by an iodine atom. Several haloheteroaromatics were successfully dehalogenated by indium metal in water, such as iodopyridines, and iodoquinoline derivatives.[13]

In an aqueous micelle system using *tert*-butyl hydroperoxide (TBHP) as an oxidant in the presence of O_2 (eqn (10.1)) it is possible to obtain oxygenated cycloalkanes from their hydrocarbon precursors.[14] The use of a surfactant was necessary to create the micelles, and no reaction occurred in its absence. The reaction gave a mixture of cyclohexanol, cyclohexanone, and *tert*-butylperoxycyclohexane in the case of cyclohexane and 2-cyclohexen-1-ol, and 2-cyclohexen-1-one and 3-(*tert*-butylperoxy) cyclohexene in the case of cyclohexene. The product ratio is dependent upon the amount of TBHP and starting material used. A radical mechanism, in which the favorable redox

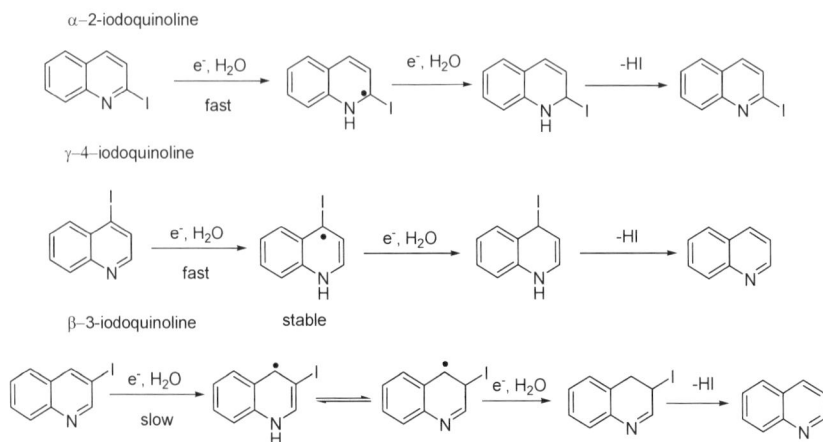

Scheme 10.6   Proposed deiodination mechanism of iodoquinolines in water mediated by indium metal.

chemistry of the iron complexes in the aqueous micelle system provided t-BuO$^{\bullet}$ and t-BuOO$^{\bullet}$ radicals as initiators (Haber–Weiss process), was proposed.

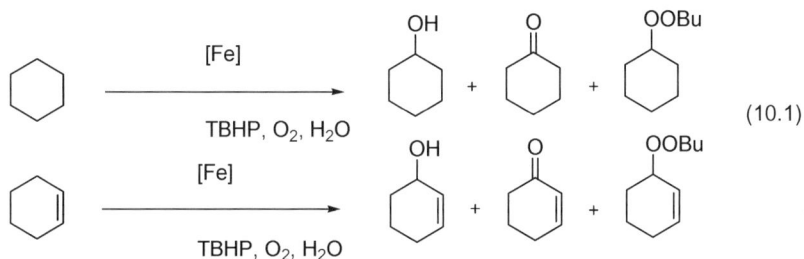

$$(10.1)$$

A simple and effective method for the transformation, under mild conditions and in aqueous medium, of various cycloalkanes (cyclopentane, cyclohexane, methylcyclohexane, *cis*- and *trans*-1,2-dimethylcyclohexane, cycloheptane, cyclooctane and adamantane) into the corresponding cycloalkanecarboxylic acids bearing one more carbon atom has been achieved by Pombeiro *et al.*[15] This method is characterized by a single-pot, low-temperature hydrocarboxylation reaction of the cycloalkane with carbon monoxide, water and potassium peroxodisulfate in a water/acetonitrile medium, proceeding either in the absence or in the presence of a metal promoter (Scheme 10.7). The influence of various reaction parameters, such as type and amount of metal promoter, solvent composition, temperature, time, carbon monoxide pressure, oxidant and cycloalkane, is investigated, leading to an optimization of the cyclohexane and cyclopentane carboxylations. The highest efficiency is observed in the systems promoted by a tetracopper(II)triethanolaminate-derived complex, which also shows different bond and stereoselectivity parameters (compared to the metal- free systems) in the carboxylations of methylcyclohexane and stereoisomeric 1,2-dimethylcyclohexanes.

A free radical mechanism is proposed for the carboxylation of cyclohexane as a model substrate, involving the formation of an acyl radical, its oxidation and consequent hydroxylation by water. Relevant features of the present hydrocarboxylation method, besides the operation in aqueous medium, include the exceptional metal-free and acid-solvent-free reaction conditions, a rare hydroxylating role of water, substrate versatility, low temperatures (*ca.* 50 °C) and a rather high efficiency (up to 72% carboxylic acid yields based on cycloalkane).

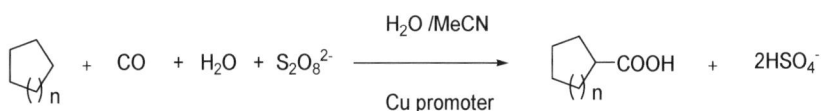

Scheme 10.7   Hydrocarboxylation of cycloalkanes to the corresponding cycloalkanecarboxylic acids in water/acetonitrile.

For both metal-free and copper-promoted carboxylations of cyclohexane, it involves the formation of a free cyclohexyl radical, which is generated by H atom abstraction from C_6H_{12} (reaction (1), Scheme 10.8) by the sulfate radical SO_4^- The latter is derived from the thermolytic and copper-promoted decomposition of $K_2S_2O_8$.

This involvement of cyclohexyl radical is confirmed by performing the carboxylations (both metal-free and copper-promoted) in the presence of the carbon-centered radical trap $CBrCl_3$ which results in the full suppression of cyclohexanecarboxylic acid formation and the appearance of cyclohexyl bromide as the main product. The radical pathway is also supported by the inhibiting effect of O_2, acting as a cyclohexyl trap to give the $C_6H_{11}COO^•$ peroxyl radical.

Subsequent carbonylation of the cyclohexyl radical by carbon monoxide results in the acyl radical $C_6H_{11}CO^•$ (reaction (2), Scheme 10.8) that upon oxidation by $S_2O_8^{2-}$ generates the acyl sulfate $C_6H_{11}C(O)OSO_3^-$ (reaction (3), Scheme 10.8). This is hydrolyzed by water (reaction (4), Scheme 10.8) furnishing the cyclohexanecarboxylic acid. In the copper-promoted process, an alternative route can occur, where the tetracopper(II)complex can behave as an oxidant of the acyl radical (reaction (5), Scheme 10.8).

This route involves the Cu(II)/Cu(I) redox couple and requires $K_2S_2O_8$ for regeneration (reaction (5)) of the Cu(II) form. The highest activity of copper(II) in comparison with the other tested metal compounds can be accounted for by its particular effectiveness in the oxidation of carbon-centered radicals. Hydrolysis of the thus formed acylation $C_6H_{11}CO^+$ ultimately leads to the $C_6H_{11}COOH$ product (reaction (6), Scheme 10.8), *via* protonated cyclohexanecarboxylic acid $C_6H_{11}C(OH)_2^+$ which is deprotonated by water, as supported by theoretical calculations on the corresponding species derived from the ethyl radical.

The hydroxylating role of water is played in both metal-free (3→4) and copper-promoted (5→6) pathways, as confirmed by experiments with $H_2^{18}O$ leading to $C_6H_{11}CO^{18}OH$ as the main product. Less favorable routes include the formation of unlabeled $C_6H_{11}COOH$, proceeding through the mixed anhydride $C_6H_{11}C(O)OSO_3H$ that is obtained by protonation of the acyl sulfate by HSO_4, or by coupling of $C_6H_{11}CO^+$ with HSO_4. This anhydride

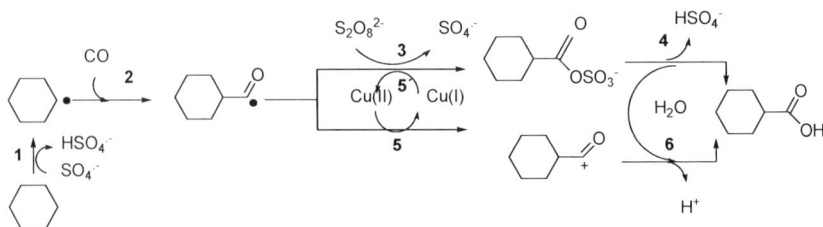

Scheme 10.8 Proposed simplified mechanism for the hydrocarboxylation of cyclohexane in water/acetonitrile.

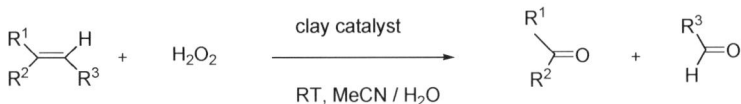

where R^1 = Ph
R^2 = H, Me, Ph
R^3 = H, NO_2, CHO, COOH, COOMe

Scheme 10.9 Oxidative cleavage of styrene in aqueous systems.

would undergo intramolecular H-transfer with elimination of SO_3, thus furnishing the $C_6H_{11}COOH$ product.

The oxidation of styrene and styryl derivatives can be accomplished by Fe-catalyzed oxidative cleavage in aqueous mixtures.[16] Oxidative cleavage of styrene yields benzaldehyde, as shown in Scheme 10.9.

A complete mechanistic interpretation of these results requires the consideration of two main pathways in which the O–O bond of hydrogen peroxide can be cleaved upon reaction with the catalyst (Scheme 10.10). Hydrogen peroxide typically reacts with a metal complex to form an initial metal–allylperoxo intermediate (a). The O–O bond of the coordinated peroxide can then cleave heterolytically to form a high-valent metal–oxo complex and water (b) or homolytically to form OH radicals and a metal hydroxide complex (c).

In the proposed mechanism (Scheme 10.11), the active oxidizing species (formed upon reaction of hydrogen peroxide with the catalyst) is described as high-valent Fe(v)O.[17] It can add onto the double bond leading to a carbon radical intermediate (proposed by Tuynman *et al.*[18]). This carbon radical intermediate is trapped by molecular oxygen followed by the abstraction of hydrogen or by the reaction between the carbon radical and activated hydrogen peroxide, which finally rearranges to give benzaldehyde as the sole product.

When $FeCl_3$ was used as a stoichiometric oxidation reagent and catalyst, homocoupling of 2-naphthols and substituted phenols successfully occurred in water.[19,20]

Kim and collaborators[21] undertook the reduction of nitroarene derivatives to anilines in the presence of 4 equiv. of indium and 0.4 equiv. of $InCl_3$ in THF/water (v/v = 5/1) at 50 °C (Table 10.1).

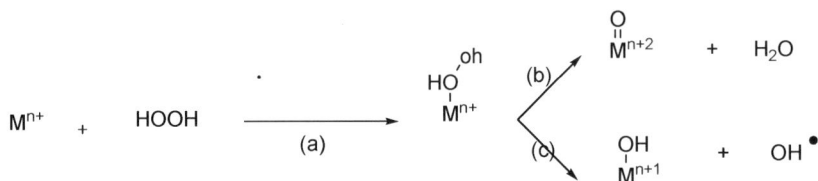

Scheme 10.10 Proposed reaction mechanism for the oxidation process.

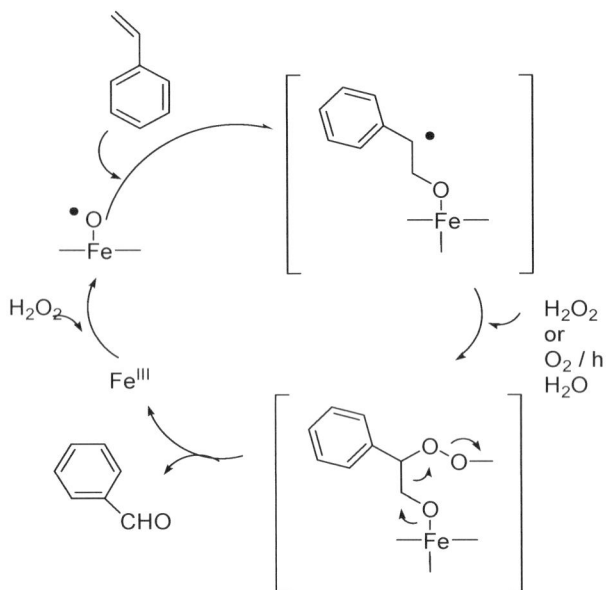

Scheme 10.11 Proposed mechanism for the oxidation of styrene derivatives to benzaldehyde derivatives.

Table 10.1 Reduction of nitroalkanes in water by SmI_2.

Entry	Starting material	Product	Yield (%)
1			99
2			92
3			95
4			60
5			96

Table 10.2   Reduction of nitroalkenes in water by SmI_2.

Entry	Starting material	Product	Yield (%)
1			60
2			52
3			47
4			22
5			45
6			75

Hilmersson and Ankner have accomplished the reduction of nitroalkanes and α,β-unsaturated nitroalkenes in water mediated by SmI_2/water/amine.[22]

Initial experiments revealed that addition of a dilute solution (0.1 M) of the nitro compound to a premixed THF solution of SmI_2 (0.1 M), isopropylamine (0.3 M) and water (0.6 M) gave a clean and almost quantitative conversion of aliphatic nitro compounds to the respective amines (Table 10.2). All the reactions were instantaneous.

As a result of the successful reduction of the nitro group, the possibility of reducing α,β-unsaturated nitroalkenes directly to amines using the SmI_2/H_2O/amine reagent was investigated. Gas chromatography (GC) analysis indicated clean and instantaneous conversion to saturated amines. However, the isolated yields after workup were only 22–75% (see Table 10.2). The dimethoxy derivative (entry 6) was isolated in fairly high chemical yields (75%). Again, the competing reduction of the aryl bromide was observed with the aryl bromide substrate (entry 4).

Pan and collaborators[23] developed an $InCl_3$-catalyzed reduction of anthrones and anthraquinones by using aluminum powder in aqueous media (Schemes 10.12 and 10.13). Alkyl and halide substituents can be present as R groups in the substrates. The yields of anthracene derivatives range from 72 to 92%.

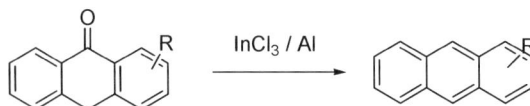

Scheme 10.12   Reduction of anthrone derivatives in water by InCl₃/Al.

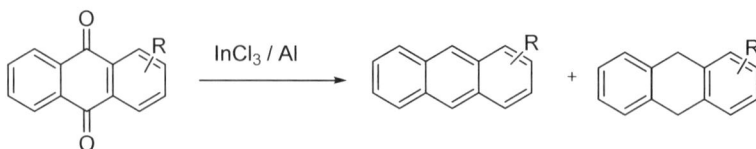

Scheme 10.13   Reduction of anthraquinones in water by InCl₃/Al.

The reaction of 1,4-disubstituted anthraquinones, in which R is H or C_2H_5, with InCl₃/Al gave different products. When the substituted anthraquinone 10 (1 mmol), indium chloride (0.2 mmol), Al powder (4 mmol), and AcOH (1 mL) were mixed in 50% aqueous alcohol and stirred at reflux for 11 h, it gave compound **11** in good yields. The new carbonyl groups in compound **11** were not reduced (Scheme 10.14).

Based on the results from experiments, proposed mechanisms were provided. It was thought that the mechanism of reduction of anthrones was similar to the reduction at metal surfaces involving ketyl radical anions (Scheme 10.15). Protonated anthrone obtained an electron to form the

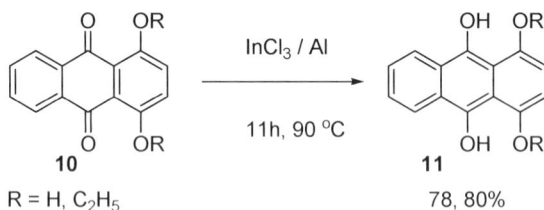

Scheme 10.14   Reduction of anthraquinone derivatives in water by InCl₃/Al.

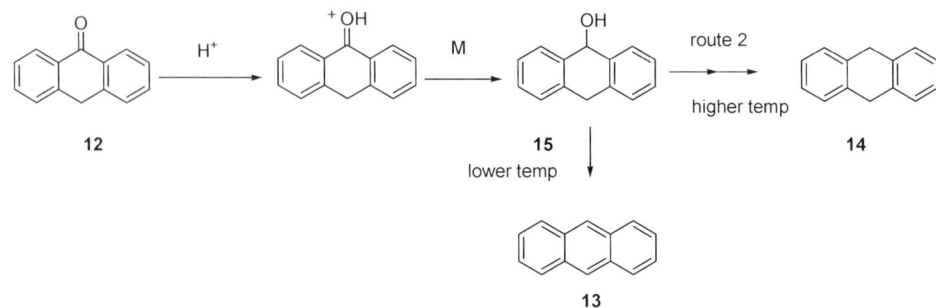

Scheme 10.15   Proposed reaction mechanism for the reduction of anthrone.

intermediate **15**. Intermediate **15** could react in two directions: anthracene **13** and 9,10-dihydroanthracene **14** were obtained, respectively. In these experiments, it was found that the intermediate **15** is easy to react along the route to **13**, which just requires room temperature. A higher temperature is required in the route to **14**. Therefore, when the temperature was reduced from 90 °C to rt, anthracene **13** was obtained as a single product in the InCl$_3$-catalyzed reduction of anthrone **12**.

For the reduction of 9,10-anthraquinone **16**, another possible mechanism was also proposed (Schemes 10.16 and 10.17). 9,10-Anthraquinone **16** was reduced to anthrone **12** firstly. Then anthrone formed from 9,10-anthraquinone reacted in the way described in Scheme 10.15. A higher temperature is required during the reduction of compound **16** to compound **12**, which is in agreement with experimental findings. At higher temperature, anthracene together with dihydroanthracene was obtained from anthrone. So anthracene could not be obtained as single product in InCl$_3$-catalyzed reduction of 9,10-anthraquinone **16**. In order to further verify the proposed mechanism, InCl$_3$-catalyzed reduction process of 9,10-anthraquinone **16** was monitored by ESI-MS. Finally, the mechanism of InCl$_3$-catalyzed reduction of 1,4-disubstituted anthraquinones (R ¼ H, C$_2$H$_5$) (Scheme 10.14) was proposed (Scheme 10.17). Firstly, compound **16** was hydrolyzed to 1,4-dihydroxy-9,10-anthraquinone.

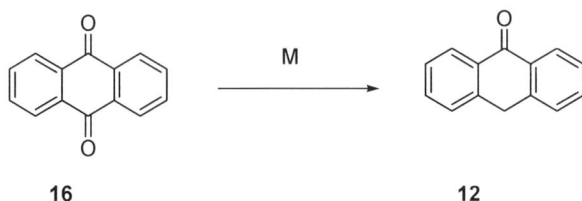

Scheme 10.16   Reduction of anthraquinone.

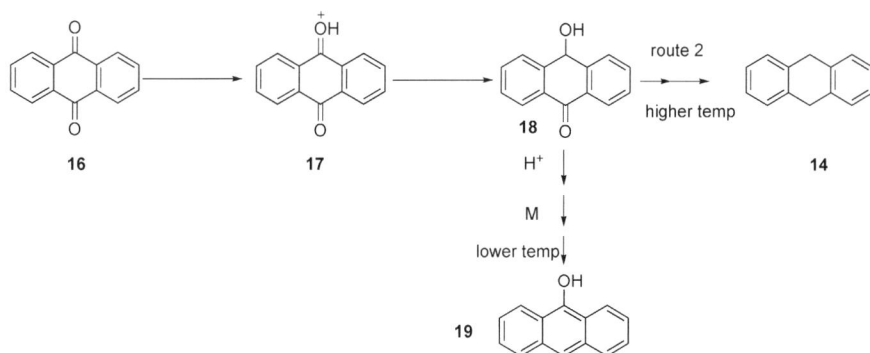

Scheme 10.17   Proposed reaction mechanism for the reduction of anthraquinone derivatives.

Then compound **17** was reduced to intermediate **18** by two possible routes. Finally, the product **19** was formed through the intermediate **18**.

Procter and collaborators have accomplished the reduction of lactones and cyclic 1,3-diesters in water mediated by SmI_2.[24] The authors reported on the first reduction of lactones to diols using SmI_2 in H_2O. The reagent system is selective for the reduction of lactones over esters; furthermore, it displays complete ring size-selectivity in that only 6-membered lactones are converted to the corresponding diols (Scheme 10.18). Experimental and computational studies suggest the selectivity originates from the initial electron-transfer to the lactone carbonyl and that anomeric stabilization of the radical-anion formed is an important factor in determining reactivity. In addition to the selectivity of the reagent system, SmI_2 is commercially available (or convenient to prepare), easy to handle, operates at ambient temperature, and does not require toxic cosolvents or additives, making the transformation an attractive addition to the portfolio of reductions.

Lactones can also be used in reductive carbon–carbon bond formation through cyclization of the radicals formed by one electron reduction, generating cyclic ketones (or ketals) often with high diastereoselectivity. The cyclizations constitute the A series of competition experiments has been carried out to illustrate further the chemoselectivity observed with the SmI_2/H_2O reagent system.

Mixtures of lactones were prepared and treated with SmI_2/H_2O. In all cases, no reduction products arising from 5-, 7- and 8-membered lactones were observed while 6-membered lactones were reduced smoothly (Scheme 10.18).

Modified SmI_2 reagent systems employing additives (HMPA, DMPU, LiBr) were also ineffective for the reduction of other lactones. A possible mechanism for the transformation is given in Scheme 10.19.

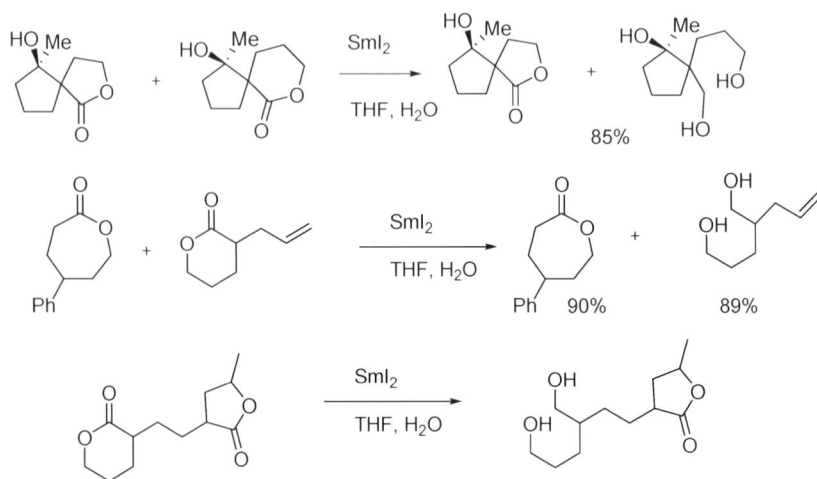

Scheme 10.18 Reduction of 6-membered lactones.

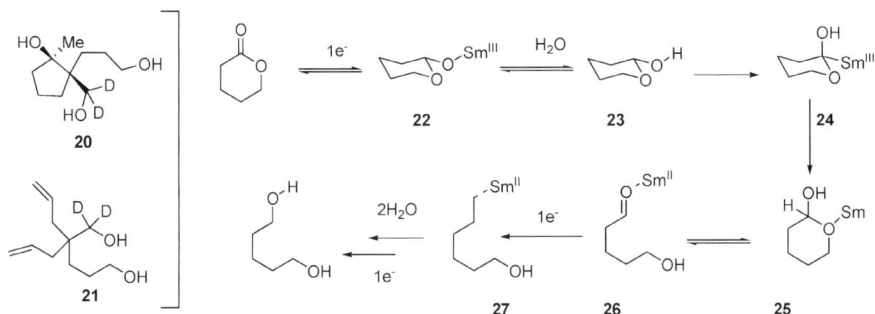

Scheme 10.19 Reduction mechanism of lactones.

Activation of the lactone by coordination to Sm(II) and electron-transfer generates radical anion **22** that is then protonated. A second electron transfer generates carbanion **24** that is quenched by the H_2O cosolvent. Lactol **25** is in equilibrium with hydroxy aldehyde **26** and is reduced by a third electron-transfer from Sm(II) to give a ketyl radical anion **27**. A final electron-transfer from Sm(II) gives an organosamarium that is protonated by H_2O. The amount of SmI_2 (approximately 7 equiv.) required experimentally is consistent with the amount predicted by the proposed mechanism (4 equiv.).

For 6-membered lactones, the authors believe that reduction generates a radical anion intermediate **22** (Scheme 10.19) that is stabilized by interaction with the lone-pairs on both the endocyclic and exocyclic oxygens. Such interactions are known to be more pronounced in 6-membered rings than in other, conformationally more labile, ring systems. It appears that the greater stability of the radical anion **22**, compared to analogous radicals formed from the reduction of 5-, 7- and 8-membered lactones, promotes the initial reduction step. This hypothesis is supported by the observation that 2-oxabicyclo[2,2,2]-octan-3-one **28** (Scheme 10.20), from which an intermediate radical-anion would be unable to adopt the chair conformation necessary for optimal stabilization, is not reduced by SmI_2/H_2O (Scheme 10.20).

Calculations suggest that the first electron-transfer to the lactone carbonyl is endothermic (100 kJ mol^{-1}) in all cases. The relative reaction energy of this step for 6-membered lactones, however, is calculated to be 116 kJ mol^{-1}, about 25–26 kJ mol^{-1} lower than those involving 5- and 7-membered rings. The relative reaction energy for the first electron-transfer to bicyclic lactone **28** is calculated

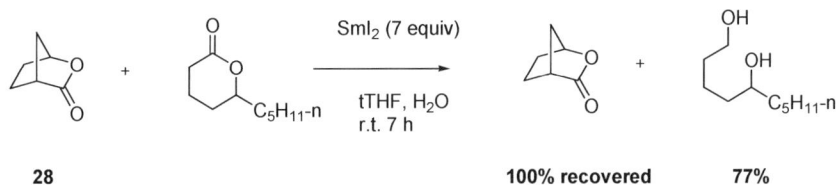

28   **100% recovered**   **77%**

Scheme 10.20 Reduction of lactones by SmI_2 in water.

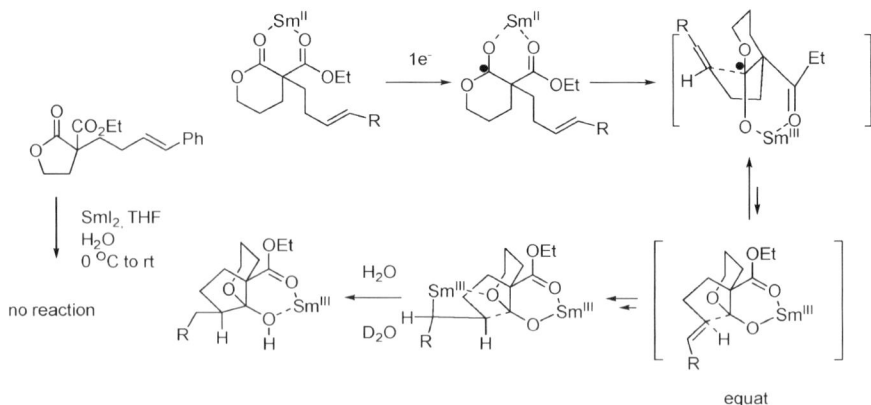

Scheme 10.21   Activation of the lactone by coordination to Sm(II) and electrostatic stabilization of the product radical-anion by coordination to Sm(III).

to be 147.4 kJ mol^{-1}. The second electron transfer is lower in energy and similar for all systems, agreeing with kinetic studies showing that the first electron-transfer is the rate-determining step. The calculated lowest energy conformation of the radical anion derived from a 6-membered anion suggests that the radical does indeed adopt a pseudoaxial orientation apparently enjoying stabilization by an anomeric effect. Activation of the lactone by coordination to Sm(II) and electrostatic stabilization of the product radical-anion by coordination to Sm(III) is likely to render these reductions more favorable than the calculated, relative reaction energies suggest (Scheme 10.21).

The same authors[24] also accomplished the reduction of cyclic 1,3-diesters employing SmI_2/H_2O, as shown in Table 10.3. The mechanism proposed also involves radical-ion intermediates as shown in Scheme 10.22.

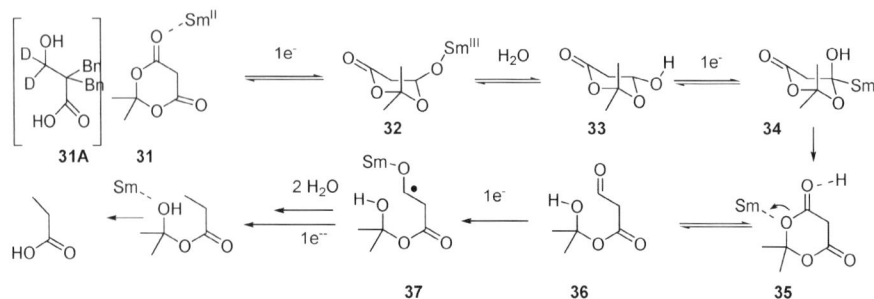

Scheme 10.22   Mechanism proposed for the reduction of lactones.

Table 10.3   Reduction of lactones.

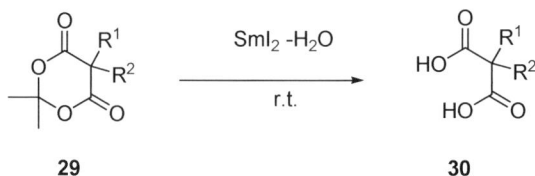

| | | | | |
| 29 | | | 30 | |
R^1	R^2	R^1	R^2	*Yield (%)* **30**
Bn	Bn	Bn	Bn	88
	$-(CH_2)_4-$		$-(CH_2)_4-$	81
H	Bn	H	Bn	68
H	4-MeOC$_6$H$_4$	H	4-MeOC$_6$H$_4$	78
H	4-BrC$_6$H$_4$	H	4-BrC$_6$H$_4$	77
H	*i*-Bu	H	*i*-Bu	94
Me	Bn	Me	Bn	98
H	Ph	H	Ph	72
	$=CHiPr$	H	*i*-Bu	87
	$-(CH_2CH_2)-$	H	Et	75

The reduction of **31A** with SmI$_2$/D$_2$O gave **31-D** (see Scheme 22) suggesting that anions are generated and protonated by H$_2$O during a series of electron transfer steps. A possible mechanism for the transformation is given in Scheme 10.22. Activation of the ester carbonyl by coordination to Sm(II) and electron transfer generates radical anion **32** that is then protonated. A second electron transfer generates carbanion **34** that is quenched by H$_2$O. Hemiacetal **35** is in equilibrium with aldehyde **36**, which is reduced by a third electron transfer from Sm(II) to give a ketyl-radical anion **37**. A final electron transfer from Sm(II) gives an organosamarium that is protonated. The amount of SmI$_2$ (approximately 7 equiv.) required experimentally is consistent with the amount predicted by the proposed mechanism (4 equiv.) (Scheme 10.22).

Hilmersson and collaborators[25] also investigated the mechanistic details of reduction of organic halides mediated by the SmI$_2$/H$_2$O/amine system. The kinetics of the SmI$_2$/H$_2$O/amine-mediated reduction of 1-chlorodecane was the subject of this study undertaken by Hilmersson, and studied in detail. The rate of reaction was found to be first order in amine and 1-chlorodecane, second order in SmI$_2$, and zero order in H$_2$O. Initial rate studies of more than 20 different amines show a correlation between the base strength (pK_{BH+}) of the amine and the logarithm of the observed initial rate, in agreement with the Brønsted catalysis rate law. To obtain the activation parameters, the rate constant for the reduction was determined at different temperatures. Additionally, the ^{13}C kinetic isotope effects (KIE) were determined for the reduction of 1-iododecane and 1-bromodecane. Primary ^{13}C KIEs (k_{12}/k_{13}, 20 °C) of 1.037 $\pm$ 0.007 and 1.062 $\pm$ 0.015, respectively, were determined for these reductions. This shows that cleavage of the carbon–halide bond occurs in the rate-determining step.

10.3 METAL-RADICAL-MEDIATED CARBON–CARBON BOND FORMATION REACTIONS IN WATER

10.3.1 Metal-mediated Radical Cyclizations in Water

Strategies involving tandem radical reactions or radical annulations offer the advantage of multiple carbon–carbon bond formations in a single operation. Thus a number of extensive investigations to this effect were reported in recent years.[26] However, the aqueous-medium tandem construction of carbon–carbon bonds has not been widely explored and therefore tandem radical reactions in aqueous media have been a subject of recent interest.[27]

Naito *et al.*[28] investigated the indium-mediated reaction of substrates having two different radical acceptors. At first, the tandem addition-cyclization-trap reaction (ACTR) of substrate **38** having acrylate and olefin moieties was examined (eqn (10.2)). To a suspension of **38** in water were added *i*-PrI (2 × 5 equiv.) and indium (2 equiv.), and then the reaction mixture was stirred at 20 °C for 2 h. The reaction proceeded smoothly affording the desired cyclic product **39a** in 63% yield as a *trans/cis* mixture in 3:2.1 ratio, along with 13% yield of the addition product **40a**. The preferential formation of cyclic products **39a–c** could be explained by a radical mechanism (Scheme 10.23).

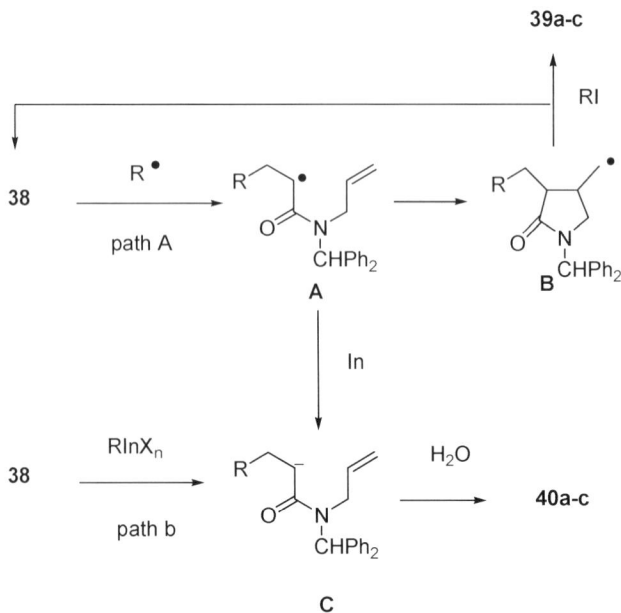

Scheme 10.23   Indium-mediated tandem addition–cyclization–trap reaction.

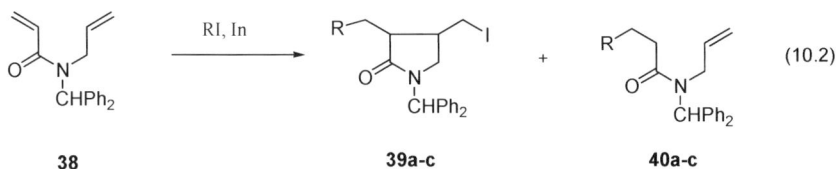

(10.2)

38   **39a-c**   **40a-c**

a: R =*i*-Pr, **b**: R =*c*-Pentyl, **c**: R=*t-Bu*

The indium-mediated reaction was initiated by single electron transfer (SET) to RI (alkyl iodide) with generation of an alkyl radical which then attacked the electrophilic acrylate moiety of **38** to form the carbonyl-stabilized radical A (path A, Scheme 10.23). The cyclic products **39a–c** were obtained *via* intramolecular reaction of radical A with the olefin moiety followed by iodine atom-transfer reaction from RI to the intermediate primary radical B. Although there are many examples of anions adding to isolated double bonds, these reactions have been limited to lithium-mediated reactions.[29]

Sulfonamides (electron-deficient alkenes) such as **41** (eqn (10.3)) have also been examined in indium-mediated tandem radical reactions.[28] As expected, sulfonamide **41** exhibited good reactivity to afford moderate and good yields of the desired cyclic products **42a–c** without the formation of other by-products.

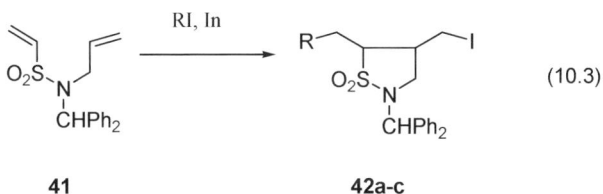

(10.3)

41   **42a-c**

a: R = *i*-Pr, **b** : R = *c*-pentyl, **c** : R = *t*-Bu

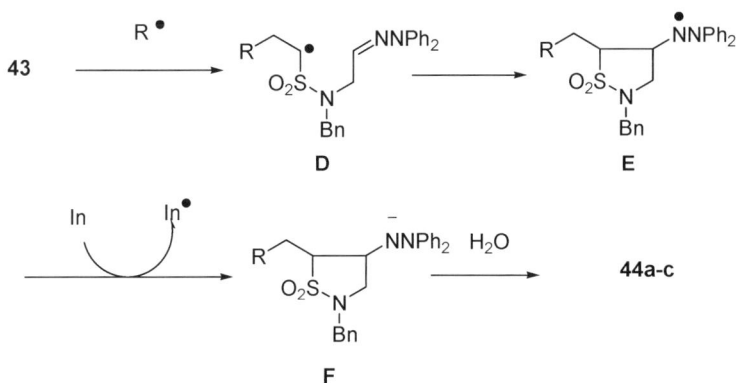

Scheme 10.24   Proposed reaction pathway for the indium-mediated radical addition–cyclization of hydrazones in water.

The indium-mediated tandem reaction of **41** with *i*-PrI in water afforded selectively the cyclic product **42a** in 81% yield as a *trans/cis* mixture in 1 : 1.4 ratio, with no detection of simple addition products. Thus indium was found to be a highly promising radical initiator in aqueous media. Hydrazones connected with a vinyl sulfonamide group such as **43** have also been investigated in the tandem radical addition–cyclization reaction of imines (eqn (10.4), Scheme 10.24). The radical reaction of **43** does not proceed *via* a catalytic radical cycle such as iodine atom-transfer; thus a large amount of indium was required for a successful reaction to take place (Scheme 10.24).

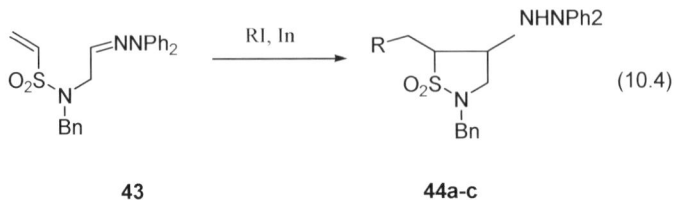

(10.4)

43   **44a-c**

a: R = *i*-Pr, **b** : R = *c*-pentyl, **c** : R = *t*-Bu

The tandem reaction of hydrazone **43** with isopropyl radical was carried out in water/methanol for 5 h by using *i*-PrI (2 × 5 equiv.) and indium (10 equiv.). As expected, the reaction proceeded smoothly to render the *iso*propylated product **98a** in 93% yield as a *trans/cis* mixture in a 1 : 1.2 ratio, without the formation of the simple addition product. The biphasic reaction of **43** in water–CH$_2$Cl$_2$ also proceeded effectively to afford 94% yield of **44a**. A cyclopentyl radical and a bulky *tert*-butyl radical worked well to give the cyclic product **44b**, and **44c** in 86% and 42% yields, respectively. The stereochemical outcome for the cyclization of hydrazone **43** is almost the same as that in the case of olefin **41** (eqn (10.3)) in which *cis* products were the major products.

An intrinsic drawback of indium is the need for almost stoichiometric amounts of this relatively expensive metal, or as seen above, a large excess. In response to the cost factor, various combinations containing catalytic amounts of indium and a secondary cheaper metal (such as Al, Zn, Sn, or Mn) have been developed, but these protocols are limited to allylation of carbonyl compounds (*vide infra*). Sakuma and Togo reported the Zn-mediated formation of cyclopropanes **47** *via* rare 3-*exo-trig* cyclizations of substrates **45** and **46** (Scheme 10.25).[30]

Geminal dialkyl substitution was required. The zinc presumably functions as a single-electron reductant both in forming the initial alkyl radical and in reducing the incipient α-carbonyl or sulfonyl radical faster than the potential fragmentation can occur. This method compares well to similar reactions promoted by SmI$_2$, as the latter reagent is air-sensitive.

Mangeney *et al.* performed a regio- and stereoselective 6-*exo* intramolecular RCA (Radical Conjugate Addition) of 1,4-dihydropyridine **48** under Luche conditions[31–35] (Scheme 10.26).[36,37] The product was transformed into both (K)-lupinine and (C)-epi-lupinine **49**.

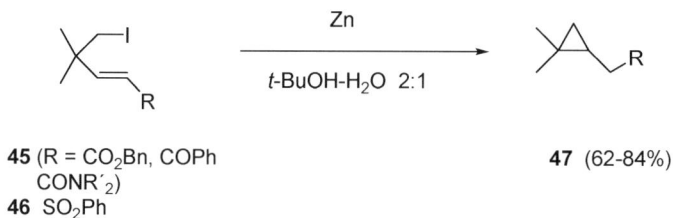

45 (R = CO$_2$Bn, COPh
CONR'$_2$)
46 SO$_2$Ph

47 (62-84%)

Scheme 10.25 Formation of cyclopropanes *via* rare 3-*exo-trig* cyclizations of β-iodo-alkenyl substrates.

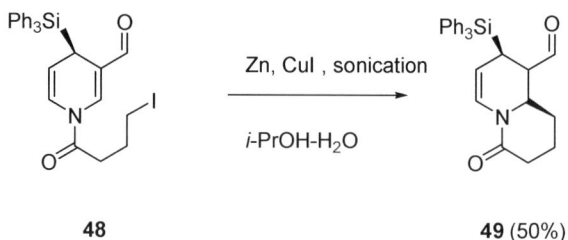

48

49 (50%)

Scheme 10.26 6-*exo* intramolecular RCA of 1,4-dihydropyridine derivatives under Luche conditions.

In 1990, Marshall and co-workers published that, upon treatment with AgNO$_3$ or AgBF$_4$, allenals (**50**, R^1 = H) and allenones (**50**, R^1 = CH$_3$, alkyl) afford furans (**51**) (Scheme 10.27).[38] These authors have developed this methodology and published many applications, always trying to improve the experimental conditions. The best set of conditions can be AgNO$_3$/CaCO$_3$/acetone/water[39] or 10% AgNO$_3$ on silica gel and hexane.[40] The following

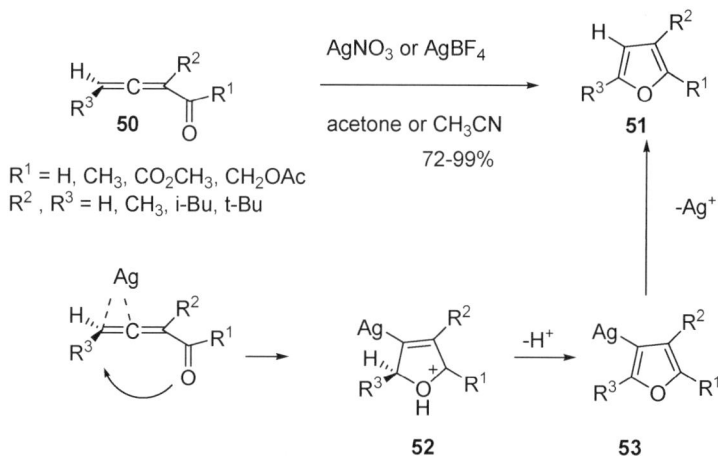

R^1 = H, CH$_3$, CO$_2$CH$_3$, CH$_2$OAc
R^2, R^3 = H, CH$_3$, i-Bu, t-Bu

50

AgNO$_3$ or AgBF$_4$

acetone or CH$_3$CN
72-99%

51

-Ag$^+$

52

-H$^+$

53

Ag

Scheme 10.27 Proposed reaction mechanism for the Ag-mediated synthesis of furans.

reaction pathway has been proposed: the process is initiated by coordination of Ag(I) with the allenyl π-system. Attack by the carbonyl oxygen would lead to the oxo-cation **52**. Ensuing proton loss from cation **52** would result in the Ag(I)-furan intermediate **53**. This could undergo direct protonolysis with loss of Ag(I) to afford furan product **51**. Deuterium incorporation experiments support this mechanism.[41] It is important to point out that the choice of the transition-metal catalyst is crucial to form this kind of substituted furans, because, under similar conditions, allenic ketones delivered different products when catalyzed by Pd(II) or Hg(II).[42]

Among various types of radical reactions, radical cyclizations in the *5-exo-trig* and *6-exo-trig* manners are the most powerful and versatile methods for the construction of 5- and 6-membered ring systems. Recently, a two-atom carbocyclic enlargement based on an indium-mediated Barbier-type reaction in water was reported.[43] A series of different ring-sized α-iodomethyl cyclic β-keto esters in a mixture of *tert*-amyl alcohol (TAA) and water was examined (eqn (10.5)).[44]

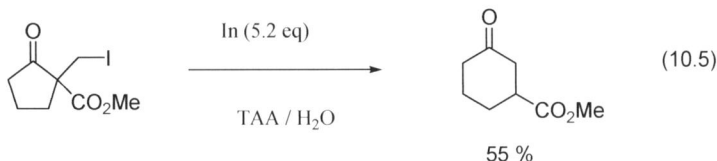

$$ \text{(10.5)} $$

55 %

The same ring expansions of α-iodomethyl cyclic β-keto esters with zinc powder, instead of indium powder were examined in a mixture of TAA (*tert*-amyl alcohol) and water (1:1). In these cases, the yields were surprisingly much increased. The presence of water in these reactions was found to be essential for an effective and high yield of the ring-expanded products. Moreover, both bromomethyl and iodomethyl cyclic β-keto esters can be used for the ring-expansion reaction to provide 6-membered, 8-membered, 9-membered, 13-membered, and 16-membered products in yields ranging from 60% up to 87%.[44] The reactions were extremely clean and operationally simple for isolation of products. A plausible reaction mechanism is depicted in Scheme 10.28.

The reaction is initiated by the first single electron transfer from metal (indium or zinc) to an α-halomethyl cyclic β-keto ester to form the corresponding methyl radical derivative, followed by *3-exo-trig* cyclization and its β-cleavage. In this mechanism, only ring-expansion products are formed since there is no hydrogen donor such as a tin hydride or silicon hydride.[44]

Recently, Li and Cao[45] demonstrated the efficiency of the *p*-methoxybenzenediazonium tetrafluoroborate–TiCl₃ couple in promoting/initiating the halogen atom-transfer radical addition (ATRA) reaction and the iodine atom-transfer radical cyclization (ATRC) reaction as an entry to heterocycles such as lactones and lactams, as shown in Table 10.4.

The active species in the *p*-methoxybenzenediazonium tetrafluoroborate/TiCl₃ chain process is the aryl radical. Initiation relies on the fact that the aryl

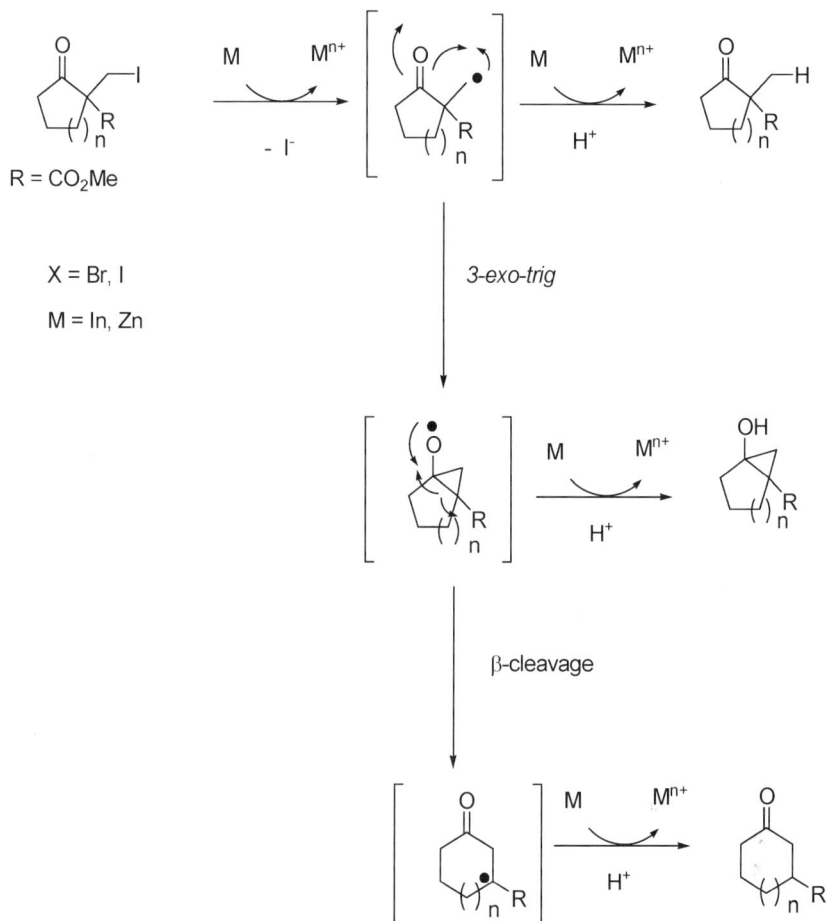

Scheme 10.28 Proposed reaction mechanism for the metal-mediated ring expansion of α-halomethyl cyclic β-keto esters.

radical is generated selectively, and it abstracts an iodine atom from the substrate rather than adding to the C=C bond. This is because the rate constant for the iodine atom abstraction of a phenyl radical from an alkyl iodide is close to the diffusion-controlled limit ($>10^9$ M^{-1} s^{-1}) which is about 100 times faster than the rate of phenyl radical addition to a monosubstituted alkene (*ca.* 3 × 10^7 M^{-1} s^{-1}). More importantly, the rate constant for the iodine atom-transfer from the substrate to the adduct radical is around 2.7 × 10^7 M^{-1} s^{-1}, at least one order of magnitude higher than that for the trapping of the adduct radical by the diazonium ion *p*-methoxybenzenediazonium tetrafluoroborate. This allows the iodine atom-transfer chain process to evolve smoothly without the intervention of a termination step.[45]

The reaction of trialkylboranes with 1,4-benzoquinones to give 2-alkylhydroquinones in quantitative yields was the first reaction of this type occurring

Table 10.4 *p*-Methoxybenzenediazonium tetrafluoroborate–TiCl$_3$-ediated iodine atom-transfer radical cyclization.

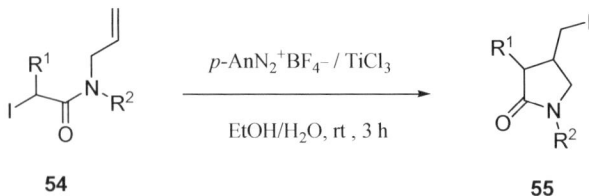

R^1	R^2	Product (yield, %)
H	allyl	99
H	Ts	96
H	Ms	86
H	Me	40
Me	allyl	96
Me	Ts	91

without the assistance of a metal mediator.[46] The reaction is inhibited by a radical scavenger such as galvinoxyl and iodine.[47]

Togo *et al.*[48] undertook the In-mediated cyclopropanation of 2,2-disubstituted 1,3-diiodopropanes and 1,3-dibromopropanes in dioxane solution of 20% water and THF solution of 20% water. However, the cyclopropanation of 2,2-disubstituted 1,3-dichloropropanes with indium powder could only be accomplished in ionic liquids.

As regards the reaction mechanism, when 2,2-disubstituted-1,3-diiodopropanes were treated with In powder (2.4 equiv.) in THF solution of 20% H$_2$O and dioxane solution of 20% H$_2$O, only 1,1-disubstituted cyclopropanes were obtained in high yields without the formation of 2,2-disubstituted 1-iodopropanes and 2,2-disubstituted propanes, which could be formed through the reactions of the corresponding carbanions with H$_2$O. This result suggests that the present cyclization reaction of 2,2-disubstituted 1,3-dihalopropane with In powder may proceed in the radical *3-exo-tet* manner, as shown in Scheme 10.29.

A rapid stereoselective route to the *trans* hydrindane ring system was achieved by Khan *et al.* using tin-, indium-, and ruthenium-based reagents starting from tetrabromo norbornyl derivatives.[49]

Kim and collaborators[50] reported on the syntheses of 2,1-benzisoxazol derivatives in aqueous media employing indium and 2-nitrobenzaldehydes, in the presence of 2-bromo-2-nitro-propane (BNP) in a methanol : water (v/v = 1 : 2) mixture (Scheme 10.30).

An interesting indium-mediated (Barbier-type) intramolecular allylation-ring expansion reaction in water has been reported by Haberman and collaborators, according to Scheme 10.31.

Usually, the reaction in an aqueous solution completed much faster than in methanol. The optimum condition was obtained when 2 equiv. of BNP and 5

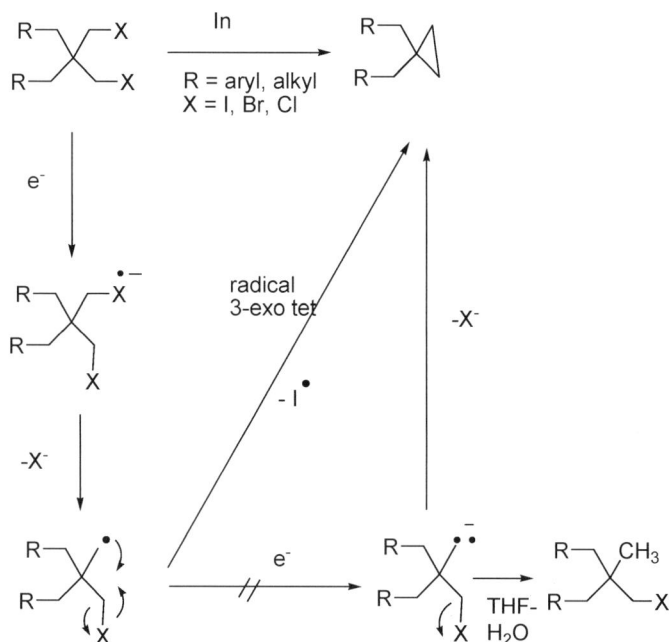

Scheme 10.29 Proposed reaction mechanism for the cyclopropanation of 2,2-disubstituted-1,3-dihalopropanes.

equiv. of indium were applied in methanol/water (v : v = 1 : 2) with 2-nitrobenzaldehyde at 50 °C and the reaction time was diminished dramatically compared to the previous zinc-mediated reaction. It produced almost a quantitative yield of desired 2,1-benzisoxazole within 10 min. The role of BNP is to be an electron acceptor due to its low-lying antibonding π-orbital and the

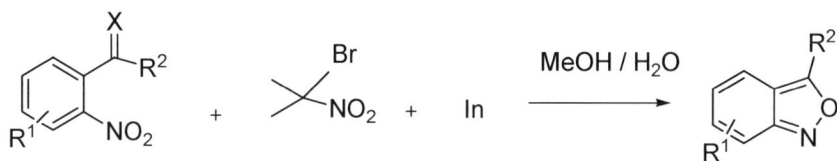

Scheme 10.30 Reactions of 2-bromo-2-nitropropane with nitrobenzaldehydes in water.

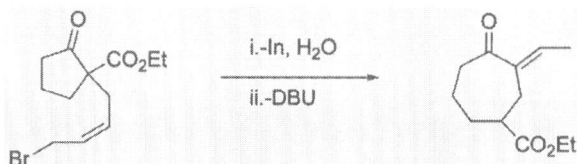

Scheme 10.31 Indium-mediated intramolecular allylation–ring expansion reaction in water.

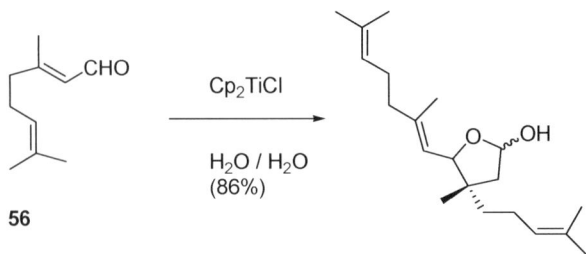

Scheme 10.32 Ti(III)/H$_2$O-promoted homocoupling of geranial **56**.

utility of BNP has been described by Russell *et al.*[51] Furthermore, addition of di-*tert*-butyl nitroxide or *m*-dinitrobenzene has shown strong inhibitory effects. Di-*tert*-butyl nitroxide is a known radical scavenger and *m*-dinitrobenzene is known to quench radical anion intermediates. The reactions of nitrobenzaldehyde /BNP/indium in MeOH in the presence of 10 mol% of di-*tert*-butyl nitroxide or *m*-dinitrobenzene had shown about 1 h initial retardation for each and completion time.

In the presence of water, titanocene(III) complexes were reported[52] to promote a stereoselective carbon–carbon bond-forming reaction that provides γ-lactols by radical coupling between aldehydes and conjugated alkenals. The method is useful for both intermolecular reactions and cyclizations. The relative stereochemistry of the products can be predicted with confidence with the aid of model Ti-coordinated intermediates. The procedure can be carried out enantioselectively using chiral titanocene catalysts.

Thus, the Ti(III) homocoupling of geranial was achieved (Scheme 10.32), and that of neral **57** was obtained according to these techniques (Scheme 10.33).

The radical cyclization in water of 1-((*E*)-but-2-enyloxy)-2-iodobenzene (**58**) to afford 3-ethyl-2,3-dihydrobenzofuran (**59**) in 85% yield (eqn (10.6)) has been reported, when (Me$_3$Si)$_3$SiH and an azo initiator is employed.[53] The reaction afforded 85% yield of cyclized product **59**.

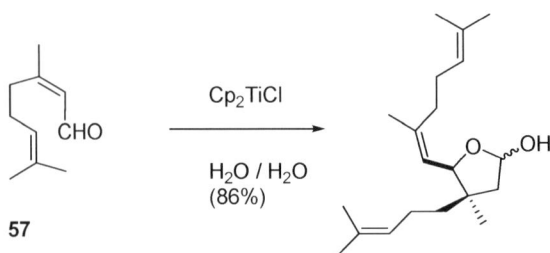

Scheme 10.33 Ti(III)/H$_2$O-promoted homocoupling of neral **57**.

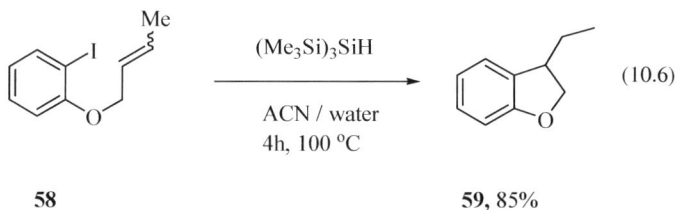

(10.6)

58   **59**, 85%

In another account, 6-bromo-3,3,4,4,5,5,6,6-octafluoro-1-hexene[54] (12 mM) **60** was subjected to reaction (24 h) with $(Me_3Si)_3SiH$ (8 mM) and dioxygen in water (5 mL), and obtained the *exo-trig* cyclization product 1,1,2,2,3,3,4,4-octafluoro-5-methylcyclopentane **61** (eqn (10.7)) in 76% yield (isolated).[55]

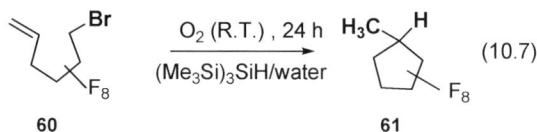

(10.7)

60   **61**

Though the measurement of the rate constant for cyclization in the heterogeneous water system is difficult to obtain, the cyclohexane cyclized product has not been observed in water under the reaction conditions reported. No uncyclized-reduced product is either observed.[55]

Analogously, cyclization of 5-bromo-1,1,2,3,3,4,4,5,5-nonafluoro-pent-1-ene (12 mM) **62** in water triggered by $(Me_3Si)_3SiH$ (8 mM)/dioxygen leads to nonafluorocyclopentane, the *exo-trig* cyclization product in 68% yield (isolated). No reduced product could be isolated from the reaction mixture. The reaction carried out in benzene-d_6 does not lead to cyclization product (eqn (10.8)).[55]

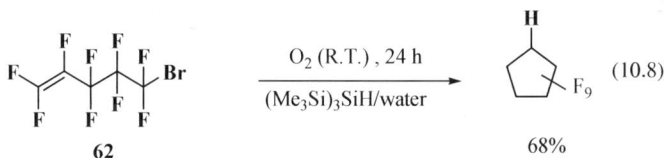

(10.8)

62   68%

When 3,3,4,4-tetrafluoro-1,5-hexadiene (40 mM) **63** is allowed to react (24 h) in water with $(Me_3Si)_3SiH$ (5 mM)/dioxygen and C_2F_5I (10 mM), product **64** is obtained in 61% yield (the 4-*exo* cyclization product), based on C_2F_5I (eqn (10.9)).

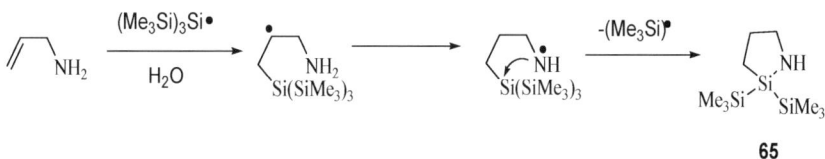

65

Scheme 10.34 Proposed mechanism for formation of cyclic product **65** from the hydrosilylation reaction of allyl amine in water.[57]

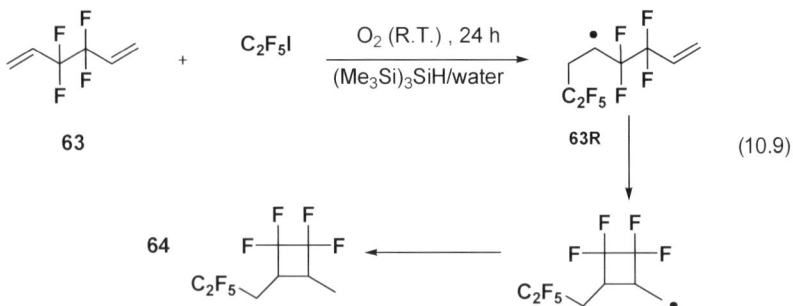

(10.9)

Though cyclobutanation is an uphill process due to considerable strain in ring formation, Dolbier found that fluorinated cyclobutanes appear to be less strained than their hydrocarbon counterparts and that fluorinated 4-pentenyl radicals (eqn (10.9)) cyclize both in a favored kinetic and thermodynamic manner.[55]

The question why 4-pentenyl radical **63R** (eqn (10.9)) prefers to cyclize in an *exo* fashion to render the four-membered ring compound **64** in water is not yet clear. Fluorinated 4-pentenyl radical derived from **62** (eqn (10.8)) cyclizes in an *endo* fashion to yield the five-membered ring nonafluorocyclopentane (eqn (10.8)).

These observations support the notion that both electron-rich and electron-poor alkenes are suitable substrates for radical perfluoroalkylation reactions in water. Notoriously, radical cyclization reactions of fluorinated alkenes have not been carried out before in water or aqueous heterogeneous mixtures.

Notably, the hydrosilylation product derived from allylic alcohol, only affords an open chain product (99% isolated yield), as opposed to allylamine where a cyclic product is observed. This difference could be related to the difference in the nucleophilicity of oxygen- and nitrogen-centered radicals in water, as opposed to organic solvents (Scheme 10.34).[56]

10.3.2 Reformatzky Reactions in Water

The recent interest in aqueous medium metal-mediated carbon–carbon bond formation led to the continuing search for more reactive and selective metal species for such reactions.

The Reformatsky reaction (eqn (10.10)) between a 2-halo ester and a carbonyl compound in the presence of Zn was the first example of a large number of now commonly used carbon–carbon bond-forming reactions: the addition of organometallic reagents to the carbonyl group. For nearly a century, these reactions were believed to require strictly anhydrous and oxygen-free conditions. Only during the past decade have chemists witnessed numerous examples of one-step reactions between organic halides, a reactive metal, and an electrophilic substrate commonly a carbonyl compound, which proceeded not only in wet solvents, but sometimes in water or salt solutions.[58] Many of these procedures gave addition products in preparative yields, comparable or superior to those obtained with preformed organometallic reagents under anhydrous conditions.

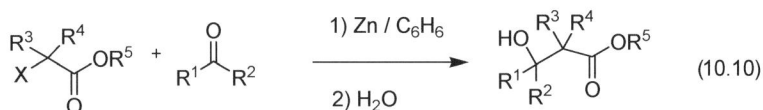

$$(10.10)$$

Bieber *et al.*[59] demonstrated that the Reformatsky reaction can be carried out in water with a wide range of carbonyl substrates including saturated and unsaturated aldehydes and ketones where the previously indium-promoted reaction in water was reported to be ineffective (*vide infra*).[60] Preparatively interesting yields comparable to those of the classical procedure in anhydrous solvents, can be obtained from substituted benzaldehydes with ethyl bromoacetate and from aromatic and unsaturated aldehydes with ethyl 2-bromo*iso*butyrate. From the mechanistic point of view, the reaction was inhibited by galvinoxyl and hydroquinone, which is consistent with a radical mechanism of two single electron transfer (SET) processes proposed by Chan *et al.*[60] In Scheme 10.35, an alternative radical chain mechanism was postulated by the authors, which does not involve hydrogen abstraction.[59] When Zn reacts with **66** (Scheme 10.35), it produces directly the Reformatsky reagent **67**. This will react with water to form ethyl acetate or with a benzoyl radical to form benzoate and radical **68** which adds to the aldehyde **69**, giving the oxyl radical **71**. Reduction of the intermediate **71** by another molecule of

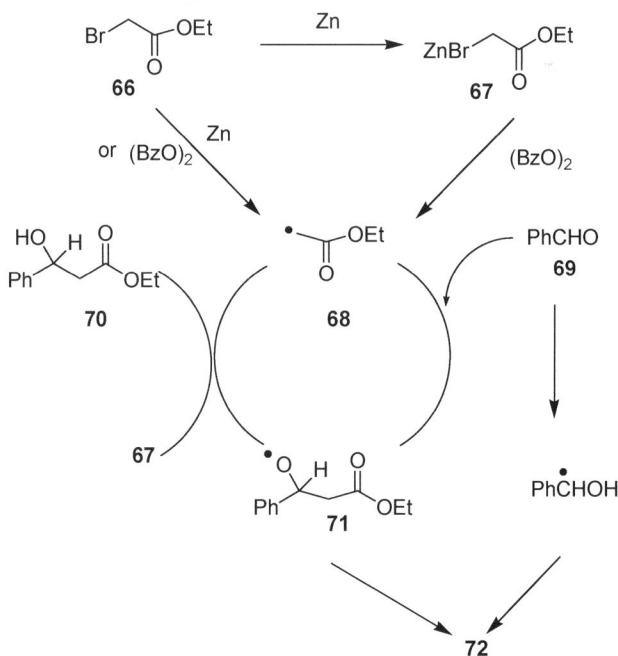

Scheme 10.35 Proposed mechanism for the Reformatsky reaction in water.

Reformatsky reagent **67** produces the final adduct **70** and a new radical **68** to continue the chain. Alternatively, the initial radical **68** may be produced by bromine abstraction from **66**, either by a phenyl radical or on the zinc surface.

10.3.3 Alkylation of Carbonyl Compounds, Imine Derivatives and Electron-deficient Alkenes in Water

An efficient Barbier–Grignard-type alkylation of aldehydes in water in the presence of CuI, Zn, and catalytic InCl in dilute aqueous sodium oxalate affords alkylated alcohols in good yields.[61] According to eqn (10.11), a series of alkyl halides can be used to afford alkylated alcohols in fairly good yields.

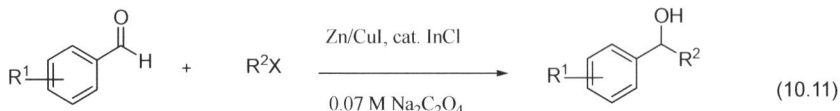

$$(10.11)$$

R^1 = 4-CN	R^2 = cyclohexyl iodide	71%
	*iso*propyl iodide	85%
	tert-butyl iodide	30%

Hammond *et al.*[62] have recently synthesized *gem*-difluorohomopropargyl alcohols from *gem*-difluorohomopropargyl bromides **73** using indium and a catalytic amount of Eu(Otf)$_3$ (5 mol%) as a water tolerant Lewis acid. Later on, the same authors employed a combination of Zn and catalytic amounts of indium and iodine (eqn (10.12)).

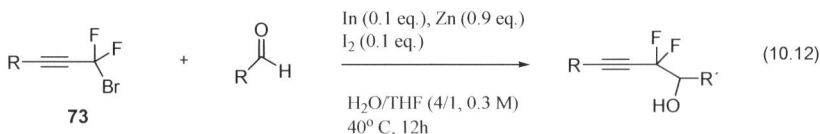

$$(10.12)$$

Only fluorinated propargyl alcohols were observed as products under the reaction conditions. The reaction is highly regioselective as the corresponding fluoroallenylalcohols were not detected.[62]

An interesting alkylative amination of aldehydes has been reported by Porta and collaborators.[63] The strategy consists of an aqueous acidic TiCl$_3$ solution that promotes alkylative amination of aldehydes in a one-pot reaction involving up to four components, according to the stoichiometry of Scheme 10.36.

The reactions were carried out by adding the phenyldiazonium salt (2.5-3.75 mmol), as the fluoroborate (method I) or as the chloride (method II), portionwise over 2 h at 20 °C to a solution containing 74 (2.5 mmol), 75 (3.75 mmol), 76 (7.5 mmol), and TiCl$_3$ (5.0-6.5 mmol of the 15% commercially aqueous acidic solution) in 15 mL of glacial acetic acid under N$_2$ atmosphere (Scheme 36).

The mechanism proposed (Scheme 10.37) involves the generation of phenyl radical by a redox process (i), which abstracts an iodine atom from RI to

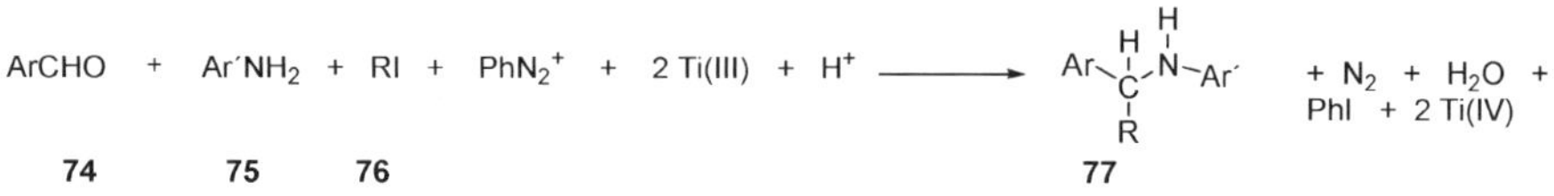

$$\text{ArCHO} + \text{Ar'NH}_2 + \text{RI} + \text{PhN}_2^+ + 2\,\text{Ti(III)} + \text{H}^+ \longrightarrow \text{Ar}\underset{R}{\overset{H}{\underset{|}{\overset{|}{C}}}}\text{N}\text{-Ar'} + \text{N}_2 + \text{H}_2\text{O} + \text{PhI} + 2\,\text{Ti(IV)}$$

74   75   76   77

Scheme 10.36 Alkylative amination of aldehydes with Ti(III).

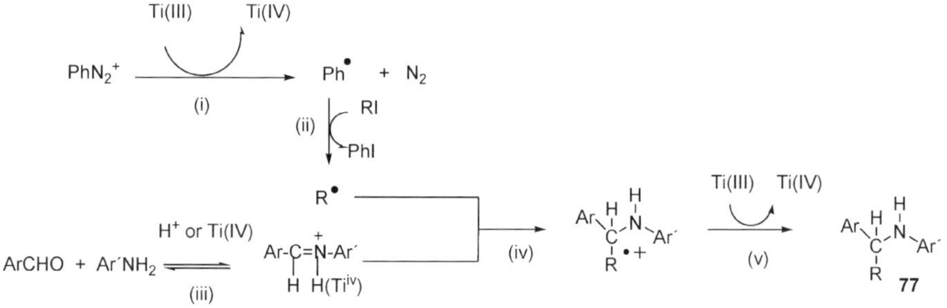

Scheme 10.37 Proposed mechanism for the alkylative amination of aldehydes.

generate the alkyl radical which adds to the benzaldehyde-amine adduct, protonated imine derivative, to generate an aminyl radical cation (iv) which upon further reduction by Ti(III) generates the alkylative amination products **15**.

Bieber and collaborators[64] have made benzylic chlorides react in aqueous dibasic potassium phosphate under silver catalysis with aromatic aldehydes in the presence of zinc dust to give 1,2-diaryl alcohols in moderate to good yields (Scheme 10.38). Dimerization to bibenzyls and reduction of the halide are important side reactions. A wide range of substituted aromatic and heteroaromatic aldehydes and of substituted benzylic chlorides can be used. Aliphatic aldehydes and ketones are unreactive. A mechanism of two SETs on the metal surface is discussed.

Among the many synthetic methods available for the synthesis of amines, the addition of organometallic reagents to imines provides one of the most straightforward methods to amine production. Loh *et al.*[65] reported on an efficient method for the alkylation of a wide variety of imines *via* a one-pot condensation of aldehyde, amine (including aliphatic and chiral amines), and alkyl iodides using indium–copper in aqueous media. These authors demonstrated that the combination of In/CuI/InCl$_3$, was an efficient system for

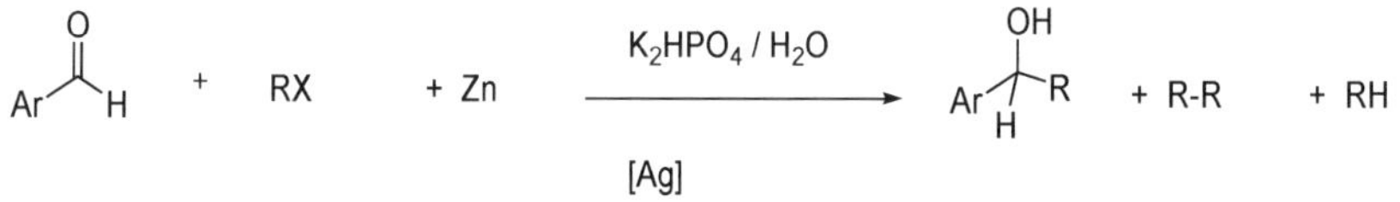

Scheme 10.38 Alkylation of aldehydes with dibasic potassium phosphate under silver catalysis.

the activation of amine-alkylation in water, to generate the corresponding products in high yields (eqn (10.13)).

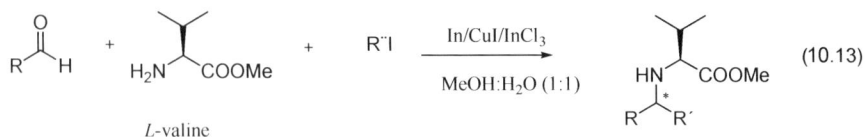

L-valine   (10.13)

Among the several metals screened, indium proved to be the best for this reaction, following the order for activation of the imine alkylation reaction: In $> Zn > Al > Sn$.[65]

It was worthwhile noting that the same reactions carried out in organic solvents such as MeOH, THF, CH_2Cl_2, DMF, DMSO, and hexane afforded the desired product in much lower yields. Even aliphatic amines, such as benzylamine could also react efficiently with different aldehydes and secondary alkyl iodides to furnish the desired products in fairly good yields. As shown in eqn (10.13), enantiomerically-enriched amino compounds were also obtained. The one-pot reaction employing various aldehydes and alkyl iodides condensed efficiently with L-valine methyl ester to generate the desired products in good yields and good diastereoselectivities. It was also worthwhile noting that even aliphatic aldehydes (cinnamaldehydes and nonyl aldehyde) were also good substrates for these reactions. A proposed reaction mechanism is shown in Scheme 10.39.

The reaction was initiated by a single electron transfer from indium-copper to alkyl iodide to generate an alkyl radical b (Scheme 10.39). This radical attacked the imine to furnish a radical intermediate c. Subsequent indium-promoted reduction of intermediate c and the quenching of the generated

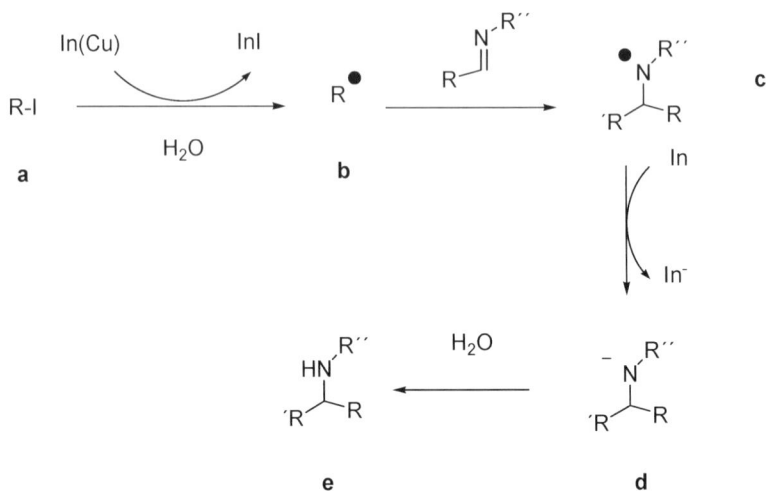

Scheme 10.39   Proposed reaction mechanism for the alkylation of imines in aqueous media.

amino anion d in the presence of water, afforded the desired product e (Scheme 10.39).

As depicted above, the carbon–nitrogen double bond could be considered a radical acceptor, and therefore several radical addition reactions have been reported in organic solvents.[66] On the other hand, it has been shown that imine derivatives such as oxime ethers, hydrazones, and nitrones are excellent water-resistant radical acceptors for the aqueous-medium reactions using Et_3B as a radical initiator.[67]

The reaction of glyoxylic oxime ether **78A** (Scheme 10.40), with *i*-PrI (5 equiv.) in H_2O–CH_2Cl_2 (4:1, v/v) and indium (7 equiv.) afforded the *iso*propylated product **79** in 76% yield without formation of significant by-products.[68]

It is noteworthy that no reaction of **78A** occurred in the absence of water. This result suggests that water would be important for the activation of indium and for the proton-donor to the resulting amide anion. In the presence of galvinoxyl free radical (radical scavenger) the reaction did not proceed, purporting that a free radical mechanism based on the single electron transfer (SET) process from indium is operative. The indium-mediated alkyl radical addition to glyoxylic hydrazone **18** afforded α-aminoacids **19** (eqn (10.14)).[69]

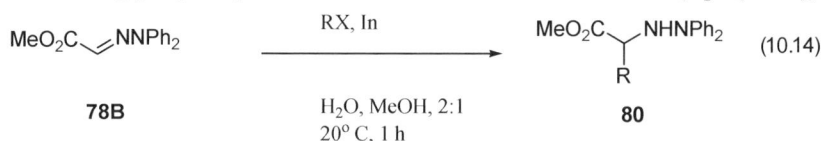

$$MeO_2C\diagdown_{\diagup}NNPh_2 \xrightarrow[\substack{H_2O,\ MeOH,\ 2:1 \\ 20°\ C,\ 1\ h}]{RX,\ In} MeO_2C\diagdown_{\underset{R}{\diagup}}NHNPh_2 \qquad (10.14)$$

78B   **80**

Integration of multi-step chemical reactions into one-pot reactions is of great significance as an environmentally benign method. The indium-mediated reaction leading to the one-pot synthesis of α-aminoacid derivatives was therefore considered (eqn (10.15)).

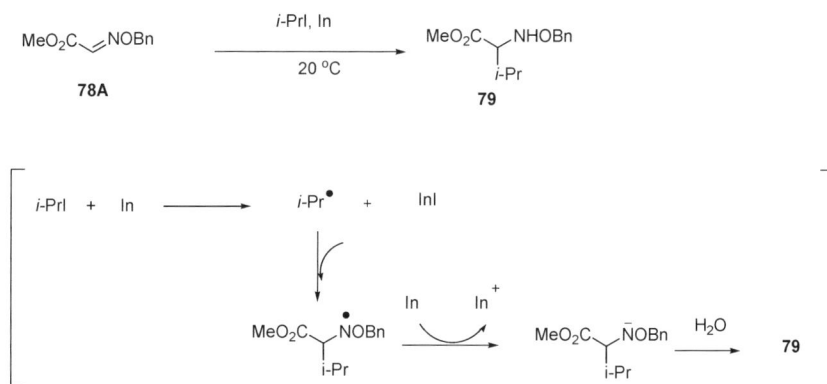

Scheme 10.40 In-mediated reaction of oximes with alkyl halides in water.

1) Ph$_2$NNH$_2$ HCl, MeOH, 2 h

2) RI, In, H$_2$O, 1 h

$$MeO_2C \overset{OMe}{\underset{OH}{\diagdown}} \longrightarrow MeO_2C \overset{NHNPh_2}{\underset{R}{\diagdown}} \quad (10.15)$$

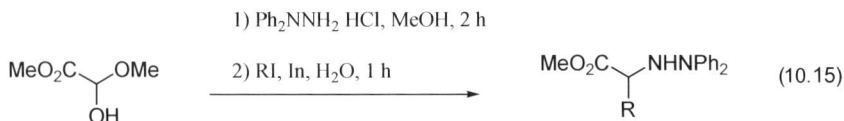

More recently, Loh and coworkers[69] have attempted the indium–copper-mediated Barbier-type alkylation of nitrones in water to furnish amines and hydroxylamines. Among the different metals investigated, indium and zinc were observed to be effective metals for the activation of the alkylation reaction in water to obtain the corresponding amines (eqn (10.16)).

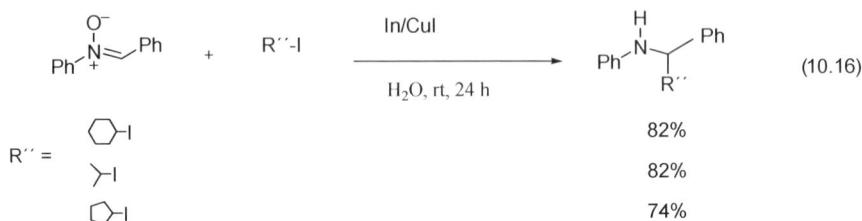

$$\underset{Ph}{\overset{O^-}{\underset{+}{N}}}{=}\diagup Ph \quad + \quad R''\text{-}I \quad \xrightarrow[\text{H}_2\text{O, rt, 24 h}]{\text{In/CuI}} \quad Ph{-}\overset{H}{\underset{R''}{N}}{\diagdown}\overset{Ph}{} \quad (10.16)$$

R'' =

⬡–I   82%

Ж–I   82%

⬠–I   74%

Ueda proposed a zinc-mediated diastereoselective addition of alkyl iodides in water[70] (Scheme 10.41).

To test the utility of indium as a single-electron transfer radical initiator, the indium-mediated alkyl radical addition to electron-deficient C=C bonds (Scheme 10.42) was considered.

To a solution of phenylvinyl sulfone **20** and RI (5 equiv.) in MeOH were added indium (7 equiv.) and H$_2$O, and the reaction mixture was stirred at 20 °C for 30 min. As expected, **81** exhibits a good reactivity to render the desired alkylated products **82a–d** in good yields with no detection of by-products such as reduced products. The reaction proceeded by a SET process from indium as shown in Scheme 10.42.

The alkylation of sulfones can also be carried out in the absence of metals and in water. Jang *et al.* reported the use of N-ethylpiperidine hypophosphite (EPHP) as a substitute for n-Bu$_3$SnH in radical conjugate additions (RCAs) to phenyl vinyl sulfone (Scheme 10.43).[71]

α,β-Unsaturated esters and ketones could also be used as acceptors with reduced yields. Significantly, use of the initiator 4,4-azobis(4-cyanovaleric acid) (ABCVA) in conjunction with cetyltrimethylammonium bromide (CTAB) allowed performance of the reaction in water (Scheme 10.44).[72]

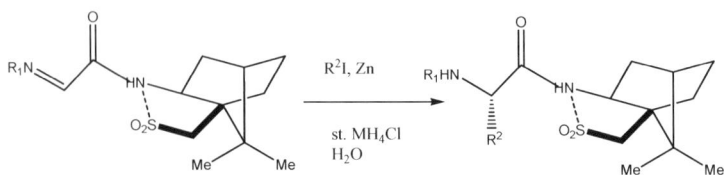

Scheme 10.41 Zn-mediated alkylation of imine derivatives.

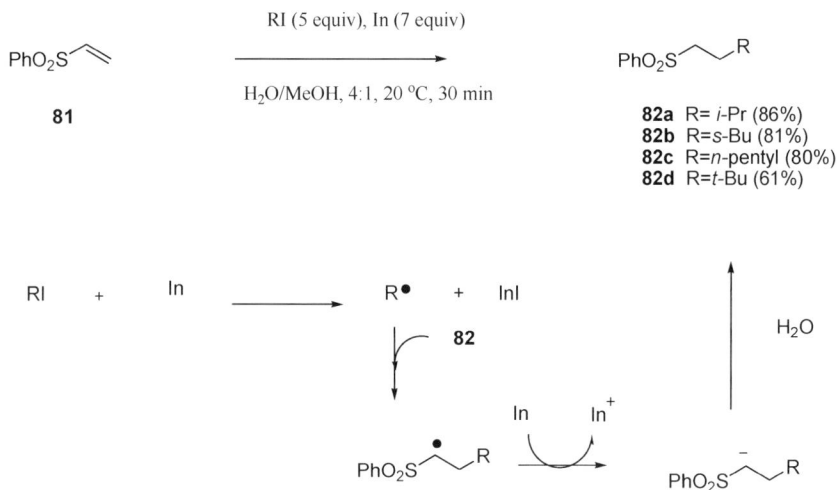

Scheme 10.42 Indium-mediated alkyl radical addition to electron-deficient C=C bond in water.

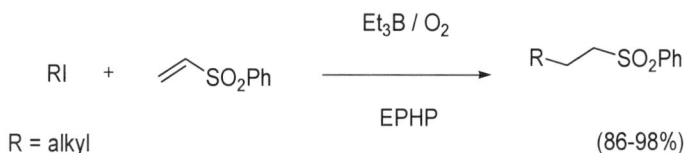

Scheme 10.43 Alkylation reactions of sulfones.

The surfactant could be omitted if tetraalkylammonium hypophosphites were employed instead of EPHP.[73]

Mouriño used the Luche conditions with acrylate, enolate, and vinyl sulfone acceptors to synthesize vitamin D_3 analogues (Scheme 10.45).[74–76]

In conjunction with studies of stereoselective Luche-type intermolecular RCAs,[77–79] Sarandeses and Perez Sestelo later employed chiral acceptors **22**, allowing introduction of a stereocenter at C-24 (Scheme 10.46).[80–82] In a synthesis of C-18-modified vitamin D_3 analogues, Sarandeses performed an inter-molecular RCA of hindered neopentylic iodide **84** to methyl acrylate (Scheme 10.47).[83]

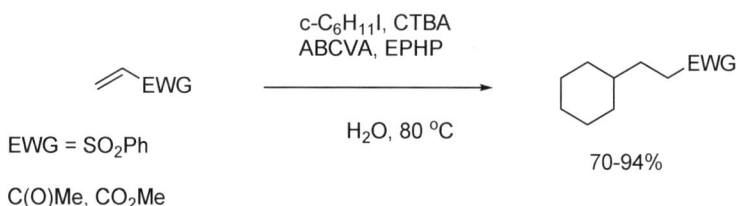

Scheme 10.44 Alkylation of sulfone derivatives.

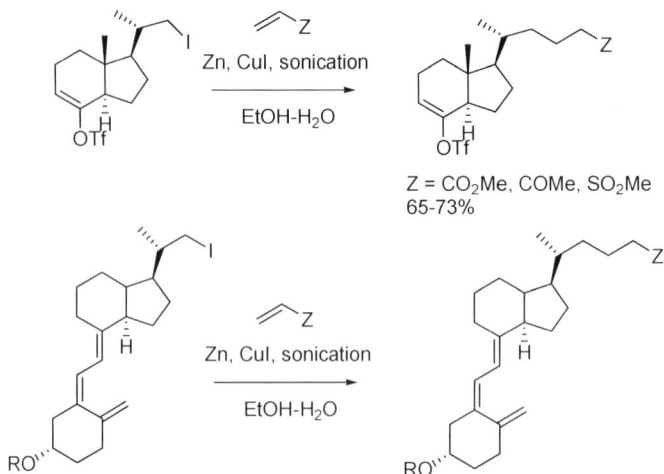

Scheme 10.45 Synthesis of vitamin D$_3$ analogues.

With a plethora of protective groups available for various types of functional groups, it is rather surprising that no practical protective groups have been developed for double bonds. Epoxidation can be used as a means of protecting double bonds; however, the successful implementation of this strategy would largely depend on the effective deoxygenation of epoxides back to alkenes.

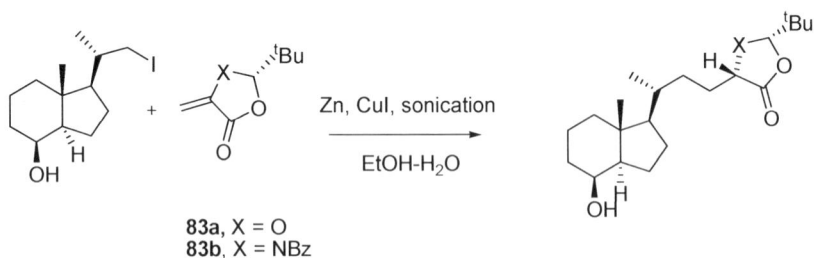

83a, X = O
83b, X = NBz

Scheme 10.46 Zn-mediated RCA on methylene lactones in EtOH-water.

84   **85**, 40%

Scheme 10.47 Zn-induced intermolecular RCA of hindered neopentylic iodide **84** to methyl acrylate **85**.

In(OAc)$_3$ 0.2 equiv
PhSiH$_3$

RI + [87] $\xrightarrow{\text{2,6-lutidine, 0.5 equiv} \atop \text{air, H}_2\text{O, EtOH, r.t.}}$ R⌒E

R = alkyl, aryl
E = electron withdrawing group

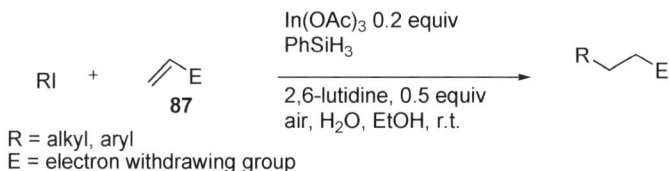

Scheme 10.48 In–silane-mediated addition of organic iodides to electron-deficient alkenes.

Although indium metal has been used for so many reduction reactions and carbon–carbon bond-forming reactions, it has not been exploited in the reduction of epoxides to form alkenes mediated by electron transfer from indium.

Miura, Hosomi *et al.* have recently developed a convenient In(III)-catalyzed intermolecular radical addition of organic iodides to electron-deficient alkenes.[84] In the presence of phenylsilane and catalytic amounts of indium(III) acetate, organic iodides add to electron deficient alkenes, according to Scheme 10.48. The mechanism for this useful catalytic reaction is illustrated in Scheme 10.49.

The first step is the formation of (AcO)$_2$InH by hydride transfer from PhSiH$_3$ to In(OAc)$_3$. The indium hydride undergoes H-abstraction by dioxygen from air to give (AcO)$_2$-In⋅. The active species abstracts halogen from a halide **86** (R–X) to generate the corresponding carbon radical R⋅ and (AcO)$_2$InX. The addition of R⋅ to an alkene **87** followed by H-abstraction from indium hydrides (In–H) gives the corresponding adduct **88** with regeneration of indium radicals (In⋅). The indium salt formed, (AcO)$_2$InX, is converted into In–H by the reaction with PhSiH$_3$ in ethanol/water. The formation of **89** is the result of direct H-abstraction of R⋅ from In–H. The

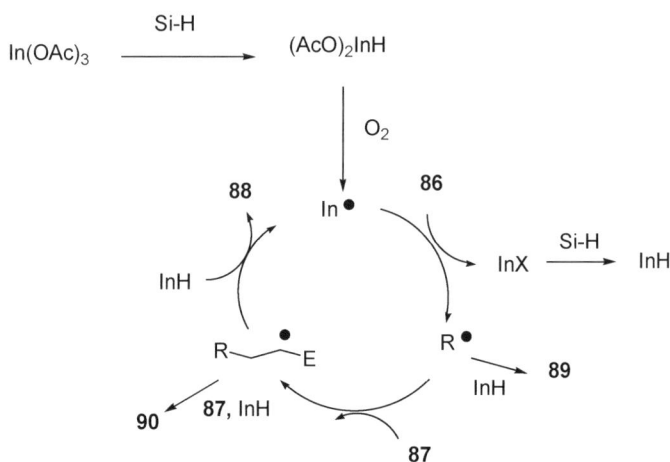

In(OAc)$_3$ $\xrightarrow{\text{Si-H}}$ (AcO)$_2$InH

Scheme 10.49 Proposed mechanism for the In–silane-mediated alkylation of alkenes.

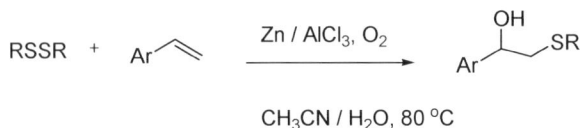

Scheme 10.50   Radical thioalkylation of styrene derivatives in aqueous media mediated by Zn/AlCl3 initiated by dioxygen.

successive addition of R· to two molecules of **87** forms the adduct **90**. The present system enables proper control of the concentration of In–H to avoid these side reactions.

Movassagh and Navidi[85] sought novel applications of zinc thiolates and zinc selenolates in chemical reactions. They investigated a convenient, catalyst-free method for the anti-Markovnikov addition of thiols to styrenes at room temperature in water. They have examined a new methodology for the synthesis of β-hydroxysulfides *via* the anti-Markovnikov addition of thiolate anions, generated *in situ* by reductive cleavage of diaryl disulfides in the presence of Zn/AlCl$_3$ to styrenes in aqueous acetonitrile in the presence of oxygen (Scheme 10.50).

The experiments were initially conducted with styrene and diphenyl disulfide, as a model reaction by varying the molar ratios, solvents, and temperatures under ambient atmosphere. The authors found that the reactants were converted readily to the corresponding β-hydroxysulfide using the Zn/AlCl$_3$ system with a molar ratio of disulfide/AlCl$_3$/Zn/styrene = 0.5 : 1 : 3.5 : 1.2 in acetonitrile/water (4 : 1) at 80 °C. The formation of β-hydroxysulfides may be explained as follows: the oxygen may complex with the styrene assisted by hydrogen bonding with the water hydroxyls, and this would be followed by nucleophilic attack by zinc thiolate, (RS)$_2$Zn, prepared *via* reductive cleavage of the disulfide with Zn/AlCl$_3$.

10.3.4 Allylation of Carbonyl Compounds and Imine-derivatives in Water

In the past two decades, the one-pot Barbier procedure for coupling allyl-halides with carbonyl compounds has gained renewed interest. Contrary to the Grignard reaction, the Barbier procedure does not require strictly anhydrous solvents but can be performed very efficiently in aqueous media. In fact, the allylation of aldehydes and ketones under the Barbier conditions usually occurs faster and gives rise to higher yields when water is used as a (co)solvent.[86] Allylation reactions of carbonyl compounds using Zn,[87] Bi,[88] Sn,[89] Mg,[90] Mn,[86] Sb,[91] Pb,[92] Hg,[93] and In[94] in aqueous media have been reported (Scheme 10.51).

Magnesium-mediated Barbier–Grignard-type alkylation of aldehydes with allyl halides was investigated by Zhang and Li.[95] It was found that the magnesium-mediated allylation of aldehydes with allyl bromide and iodide proceeded effectively in aqueous 0.1 M HCl or 0.1 M NH$_4$Cl. Aromatic aldehydes reacted chemoselectively in the presence of aliphatic aldehydes. A

Scheme 10.51 General metal-mediated allylation of carbonyl compounds in aqueous systems.

variety of aldehydes were tested with this alkylation method, according to eqn (10.17).

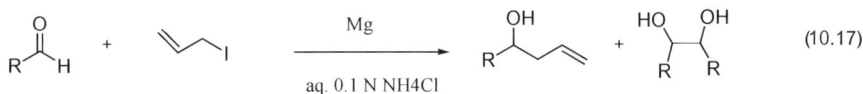

(10.17)

The allylation of aromatic aldehydes bearing halogen atoms proceeded without any problems. The allylation of hydroxylated aldehydes also afforded the allylation products in good yields. Reaction of 4-hydroxybenzaldehyde under the standard conditions led to the formation of the allylation product.

The mechanism of the classical magnesium-mediated Barbier and Grignard reactions have been studied intensively by several groups.[96] It is generally believed that the radicals on the metal surface are involved in the organomagnesium reagent formation. For the Barbier allylation of carbonyl compounds with magnesium in anhydrous solvent, it is assumed that the reaction of allyl bromide on the metal surface generates an organometallic intermediate that is in equilibrium with the charge-separated form and the radical form, as proposed by Alexander,[97] as shown in Scheme 10.52. The two forms will also lead to either the protonation of the carbanion (overall reduction of the halide) or Wurtz-type coupling, whereas the intermediate reacts with aldehydes through the usual six- membered ring mechanism. The radical intermediate could lead to the formation of 1,6-hexadiene, pinacol product and benzyl alcohol.

For the rationalization of the pinacol formation, the authors[95] postulate two potential pathways competing with each other generating either the pinacol-coupling product (path a, Scheme 10.53) or the benzyl alcohol product (path b, Scheme 10.53). The same authors observed that upon increasing the steric hindrance around the carbonyl group, a destabilization in the transition state responsible for the formation of the pinacol product (path a), would result in an increase in the formation of the benzyl product. Thus, they observed that no pinacol product was formed from the magnesium-mediated reaction with 2,6-dichlorobenzaldehyde, and a 74 % yield of the reduction product was encountered in this example (Scheme 10.53).

Madsen *et al.*[98] reported on a theoretical study of the Barbier-type allylation of aldehydes mediated by magnesium. They concluded that a radical anion was involved in the selectivity-determining event.

Other metals were used from time to time to mediate in the coupling reactions to construct new carbon–carbon bonds. Manganese was shown to be very effective for mediating aqueous medium carbonyl allylations and pinacol coupling reactions.

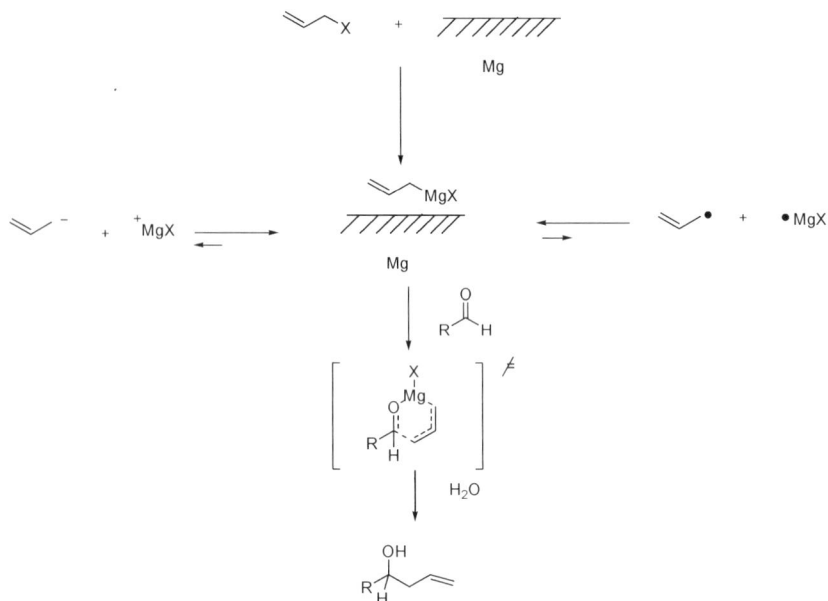

Scheme 10.52 Mg-surface mediated electron transfer mechanism of allylation of carbonyl compounds.

Li *et al.*[99] reported an unprecedented metal-mediated carbonyl addition between aromatic and aliphatic aldehydes. The allylation of aldehydes mediated by manganese in water in the presence of a catalytic amount of copper showed exclusive selectivity toward aromatic aldehydes. Manganese was found to be equally selective in promoting pinacol-coupling reactions of aryl aldehydes. When 3 equivalents of allyl chloride and manganese mediator were used, the isolated yield of the allylation product was 83% (eqn (10.18)).

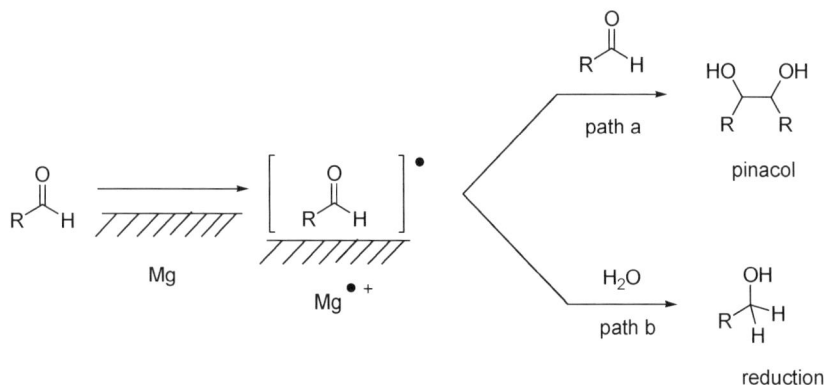

Scheme 10.53 Pinacol formation and reduction: competing paths.

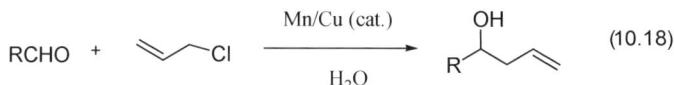

$$\text{RCHO} + \text{(allyl chloride)} \xrightarrow[\text{H}_2\text{O}]{\text{Mn/Cu (cat.)}} \text{(product)} \quad (10.18)$$

It was found[99] that various aromatic aldehydes were allylated efficiently by allyl chloride and manganese in water. It is worth mentioning that aromatic aldehydes bearing halogen atoms were allylated without any problems. The allylation of hydroxylated aldehydes was equally successful. On the other hand aliphatic aldehydes were inert under the reaction conditions. Such an unusual reactivity difference between an aromatic aldehyde and an aliphatic aldehyde suggested to the authors[99] the possibility of an unprecedented chemoselectivity. When competitive studies were carried out involving both aromatic and aliphatic aldehydes (eqn (10.19)), a single allylation of benzaldehyde was generated when a mixture of heptaldehyde and benzaldehyde was reacted with allyl chloride. Such a selectivity appeared unique when aqueous methodologies mediated by other metals such as Zn, Sn and In all generated a 1 : 1 mixture of allylation products of both aldehydes (*vide infra*).

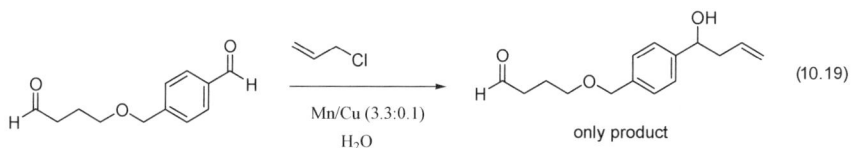

only product   (10.19)

Li and Chan[91] have recently reported that fluoride salts are equally effective in activating antimony in aqueous media to mediate in the coupling of allyl bromide with aldehydes to afford the corresponding homoallylic alcohols. 1 M concentrations of NaF and KF were found to be equally effective as RbF and CsF or 2 M KF. The reaction proceeded well with either aromatic or aliphatic aldehydes. The allylation of α,β-unsaturated aldehydes as represented by *trans*-cinnamaldehyde occurred in a regiospecific manner and furnished solely the 1,2-addition product. Furthermore, electron-donating or -withdrawing groups on the aromatic ring did not seem to affect the reaction significantly either in the yield of the product or the rate of the reaction. With this metal, activated antimony, no alcohols or pinacols were detected as side products of the reactions, as has been shown previously for the Mn and Mg (*vide supra*) cases. Even the nitro substituent on the aromatic ring of the aldehyde was not reduced under the reaction conditions obtaining the corresponding allylated alcohol from *p*-nitrobenzaldehyde (usually the nitro group is sensitive to reduction by metals and cannot be allylated under Barbier conditions).[100] In this sense, the authors argued[91] that the use of a fluoride salt as an activating agent is superior to the use of Al, Fe, or NaBH$_4$, reported previously. Efforts to allylate ketones failed by this Sb-mediated methodology. From the mechanistic point of view, the reaction proceeds between the allylmetal species and the aldehyde.[98]

In 1983, Nokami *et al.* first reported on the successful coupling of allyl bromide (**91**) with carbonyl compound (**92**) mediated by tin to give the

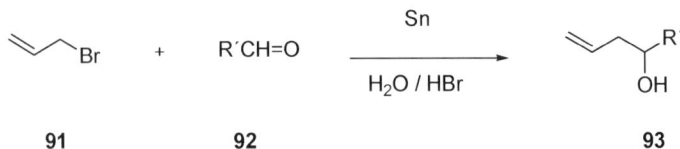

Scheme 10.54   Tin-mediated allylation of carbonyl compounds in water.

homoallylic alcohol (**93**) in water (Scheme 10.54).[101] However, the reaction requires a catalytic amount of hydrobromic acid.

Later on, the addition of metallic aluminum powder or foil was found to improve the yield of the product dramatically.[102] On the other hand, Wu and co-workers found that a higher temperature can be used to replace aluminum.[103] Alternatively, Einhorn and Luche found that the reaction can be performed in the absence of aluminum or hydrobromic acid by the use of ultrasonic irradiation together with saturated aqueous NH_4Cl/THF solution.[104] Various mechanisms have been proposed for the aqueous Barbier reactions, including the intermediacy of a radical,[105] radical anion,[106] and an allylmetal species.[107] In the latter case, it has been presumed that diallyltin dibromide is the organometallic intermediate in the tin-mediated allylation reactions; however, no experimental proof has been offered.[108,109]

Chan and co-workers[110] found that allylation of carbonyl compounds can easily be accomplished in water using diallyl tin dibromide to obtain homoallyl alcohols in good yields. Table 10.5 summarizes some examples of carbonyl substrates that can be allylated through this latter reagent.

Although an allyltin bromide and diallyltin bromide species have been postulated as intermediates in these reactions, these results do not, however, eliminate the possibility of a parallel process of metal surface-mediated radical or radical anion reactions. Nevertheless, the understanding that the reaction can proceed through an organotin intermediate has useful synthetic applications.

However, allylation reactions based on allylstannanes have serious drawbacks in the synthesis of biologically active compounds, because the inherent toxicity of organotin derivatives and the difficulty of removing residual tin compounds often prove fatal.

Bian and collaborators[111] have accomplished the allylation reaction of aromatic aldehydes and ketones with tin dichloride in water. The allylation reactions of aromatic aldehydes and ketones were carried out in 31–86% yield using a $SnCl_2$–H_2O system under ultrasound irradiation at rt for 5 h. The reactions in the same system gave homoallyl alcohols in 21–84% yield with stirring at rt for 24 h (Scheme 10.55). Compared with traditional stirring methods, ultrasonic irradiation is more convenient and efficient.

In Table 10.6, the aromatic aldehydes and ketones studied are summarized. It is observed that by ultrasound irradiation the allylation products are obtained in yields ranging from 73 to 86%, except for acetophenone, wich renders a low yield of the respective allylation product.

Table 10.5 Allylation reactions of carbonyl compounds with diallyltin dibromide in water.

Entry	Carbonyl compound	Product	Yield (%)
1	Benzaldehyde		95
2	Heptaldehyde		99
3	4-nitrobenzaldehyde		97
4	Cyclohexylaldehyde		95
5	OHCCHO		95
6	Cinnamaldehyde		99
7	4-chlorobenzaldehyde		99

Chan and Yang[112] investigated the nature of the allylation reaction of carbonyl compounds with indium in water (Scheme 10.56). The authors postulate a series of intermediates, among which those in Scheme 10.57 were considered.

However, through a series of experiments,[112] it was concluded that indeed the allyl indium intermediate contained indium(I), and the structure of the

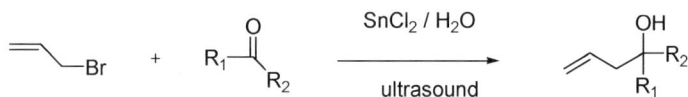

Scheme 10.55 Tin-mediated allylation of carbonyl compounds triggered by ultrasound.

Table 10.6   Allylation reactions of aldehydes with $SnCl_2/H_2O$.

Entry	Substrate	Yield (%, stirring)	Yield (%, ultrasound)
1	Benzaldehyde	84	86
2	4-Chlorobenzaldehyde	77	76
3	Furfural	58	59
4	Cinnamaldehyde	71	73
5	$3,4\text{-}(CH_2O)C_6H_3CHO$	84	79
6	$4\text{-}CH_3OC_6H_5CHO$	–	–
7	$C_6H_5COCH_3$	21	31

Scheme 10.56   Allylation of carbonyl compounds with indium metal in water.

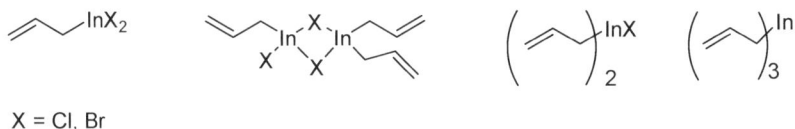

X = Cl, Br

Scheme 10.57   Intermediates proposed in the indium-mediated allylation of carbonyl compounds.

intermediate supported by experiments was that shown in eqn 10.20.

         (10.20)

Welch and collaborators[113] explored the generality of the indium-mediated allylation reaction with various difluoroacetyltrialkylsilanes in a water and THF mixture.

Desired homoallylic alcohols were synthesized in good yields without enol silyl ether formation. Substituents on silicon have no effect on the product formation. However, in reaction with a substituted allyl bromide such as 4-bromo-2-methyl-2-butene, the desired homoallylic alcohol was not formed, rather, the acylsilane was recovered (Table 10.7).

The regio- and stereoselectivity of metal-mediated allylation reactions in aqueous media is well understood.[114] In general, regioselectivity is governed by the substituent on the allyl halide. The formation of γ-adducts was exclusively observed under some conditions; however, allyl halides bearing bulky γ-substituents such as 1-bromo-4,4-dimethyl-2-pentene and (3-bromopropenyl)-trimethylsilane resulted in the formation of α-adducts. In contrast, diastereoselectivity is affected by the steric and chelating effect of substituents. These reactions have also been attempted with zinc.

Table 10.7 Indium-mediated allylation reaction of difluoroacetyltrialkylsilane in aqueous mixtures.

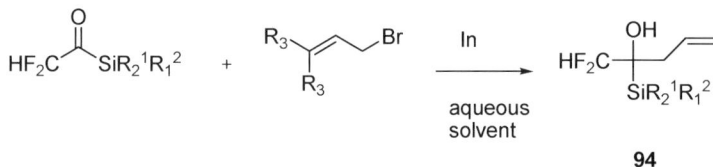

Entry	R^1	R^2	R^3	Solvent	% **94**
1	Phenyl	*t*-Butyl	H	THF–water (1:1)	97
2	Ethyl	Ethyl	H	THF–water (1:1)	83
3	Isopropyl	Isopropyl	H	THF–water (1:1)	85
4	Ethyl	Ethyl	CH_3	THF–water (1:1)	–
5	Ethyl	Ethyl	CH_3	THF–water (1:1) NH_4Cl	–

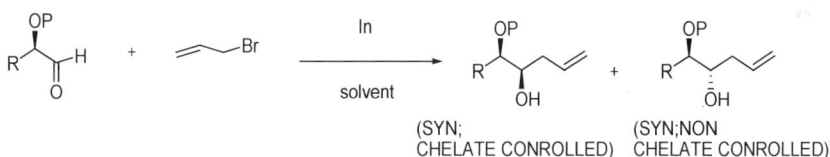

Scheme 10.58 In-mediated allylation of α-oxygenated aldehydes.

Paquette, Loh *et al.*[115] have investigated the stereochemical course of indium-promoted allylations to α- and β-oxy aldehydes in solvents ranging from anhydrous THF to pure H_2O. The free hydroxyl derivatives react with excellent diastereofacial control to give significantly heightened levels of *syn*-1,2-diols and *anti*-1,3-diols (Scheme 10.58). Relative reactivities were determined in the α-series, and the hydroxyl aldehydes proved to be the most reactive substrates. This reactivity ordering suggests that the selectivity stems from chelated intermediates. The rate acceleration observed in water can be heightened by initial acidification. Indeed, the indium-promoted allylation reaction mixtures become increasingly acidic on their own. Preliminary attention has been given to salt effects, and tetraethylammonium bromide was found to exhibit a positive synergistic effect on product distribution. Finally, mechanistic considerations are presented in order to allow for assessment of the status of these unprecedented developments at this stage of advancement of the field.

The results obtained with **95** (entry 1, Table 10.8) and **96** (entry 2, Table 10.8) provide important calibration points for non-chelate-controlled behavior. In both instances, the *anti*-product is favored. Presumably because the basicity of the *te*rt-butyldimethylsiloxy substituent falls below that of the benzyloxy, the *anti*-percentages reach a maximum for **95**.

It is noteworthy that in every example allylations performed in either H_2O or H_2O–THF (1:1) proceeded at appreciably more rapid rates than in THF

Table 10.8   Indium-mediated allylations of α-oxygenated aldehydes.

Entry	Aldehyde	Reaction time	Syn	Anti	Yield (%)
1	**95**	3.5 h	1	3.9	90
2	**96**	3 h	1	1.2	92
3	**97**	24–30 h	2.3	1	90–95

alone. For **95**, the diastereoselectivity realized is constant whether H_2O is present or not. Interestingly, the *anti*-preference in the case of **96** decreases by a factor of about 3 in pure H_2O. Product yields were found to be consistently high. Hemiacetal **97** must, of course, undergo ring opening prior to condensation with the allylindium reagent. Beyond that, it is not clear that carbon–carbon bond formation materializes prior to, or only after, the loss of formaldehyde.

pH considerations. The rate acceleration noted above for allylations promoted in water could, as for example with D-arabinose, be attributed to the improved solubility of the substrate in the aqueous medium. However, the phenomenon persists when the solubilities of the reagents are lower in H_2O than in THF. This behavior could be explained by attributing enhanced stability to the allylindium reagent in THF. Under these circumstances, reactivity toward an incoming aldehyde carbonyl would be reduced and condensation would proceed more slowly. Indeed, an increase in reaction rate has been observed by Araki *et al.*[116] when progressing from benzene to THF. In this latter study, it was noted that the pH of all allylations performed in water or aqueous THF dropped significantly as the reactions progressed. This aspect of indium-promoted condensations does not seem to have been previously recognized and was therefore explored more fully in order to elucidate the accompanying advantages or disadvantages. As indicated in Table 10.8, aldehydes **95–97** were closely scrutinized.

Table 10.9 Effect of pH on the rate and diastereoselectivity of allylindium additions in water at 25 °C.

Entry	Aldehyde	pH	Reaction time/h	Syn product	Anti product	Yield (%)
1	95	a	3.5	1	3.9	90
2	95	7	5.5	1	3.0	85
3	95	4	4	1	3.0	80
4	96	a	3	1	1.2	92
5	96	7	12.5	1	1.4	84
6	96	4	4	1	1.5	86
7	97	a	24–30	2.3	1	90–95
8	97	7	48	2.0	1	80–87
9	97	4	0.5	10.0	1	85–88

*a*pH not controlled.

When the pH was maintained at 7 by controlled infusion of sodium hydroxide solution, an increase in reaction times became necessary to achieve complete allylation (entries 1 and 3). This phenomenon could be due in part to increased dilution due to addition of the aqueous base and not constitute a manifestation of the pH itself. Reactions allowed to proceed without pH control were accompanied by a progressive development in acidity to a point below pH 4. When the allylations were initiated at a preset pH of 4, the transformations took place at notably accelerated rates (Table 10.9).

Noteworthily, the benzyl- and silyl-protected substrates exhibit the same diastereoselectivities at all ranges of pH tested. These findings dispel any concerns that product distribution might be dependent upon pH to the point where the *syn/anti*-ratios would vary as the reaction progressed. This feature of indium catalysis in water requires that care be exercised when acid-sensitive reactants are involved. Consequently, these considerations must be taken into account when designing alternative applications of this chemistry.

β-Oxy-substituted aldehydes. A free hydroxyl substituent β- to a carbonyl group was considered to be exploitable for 1,3-asymmetric induction during condensation with allylindium reagents in water. Aldehydes **98** were selected because of their structural simplicity, similarity to **95–97**, and varied basicity at the β-oxygen. If chelation were to gain importance and nucleophilic attack were to occur from the less hindered diastereotopic π-face of the aldehyde carbonyl, then anti-adduct **99** (Scheme 10.59) would result.

Homoallylic alcohols **99** are the Felkin–Anh (non-chelation-controlled) products. The diastereomeric ratios of **99** and **100** resulting from exposure of

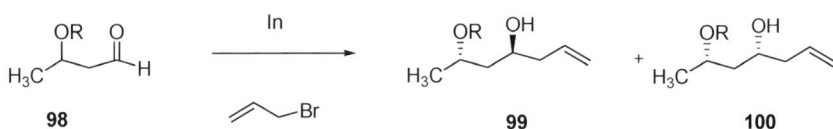

Scheme 10.59 In-mediated allylation of β-oxygenated aldehydes.

Table 10.10   Indium-promoted C-allylation (with allyl bromide) of β-oxygenated aldehydes in water.

Entry	Aldehyde	Reaction time/h	Syn	Anti	Yield (%)
1	OH O ... H **102**	2	1	8.2	77
2	OBn O ... H **103**	2.5	1	1	80
3	OTBSO ... O, H **104**	3.5	1	1	84
4	H$_3$CO ... O, H **105**	2.7	1	4	78

98 to allyl bromide and indium in water are compiled in Table 10.10. The product distributions are seen to correlate closely with the β-alkoxy series. Thus, α,β-methoxyl substituent is capable of modest levels of chelation control, irrespective of whether the allylation is performed in an anhydrous or aqueous medium. Since a benzyloxy or a *tert*-butyldimethyl-siloxy group results in production of totally stereo random mixtures of **99** and **100**, there is no evidence for transient structural rigidification prior to nucleophilic attack in these examples. This crossover could reflect the operation of a steric effect. The unprotected hydroxyl derivative **98** exhibits the most pronounced face selectivity as anticipated. In fact, the 8.5 : 1 ratio of **99** to **100** (Scheme 10.59) compares quite favorably with the product distributions exhibited by the β-hydroxy aldehydes **103** and **104** (Table 10.10). Clearly the free β-OH group is capable of chelation control in water, finding it possible to coordinate to the indium ion despite its preexisting solvation by water molecules.

Mechanistic considerations. Additions of the allylindium reagent to α- and β-hydroxy aldehydes in water have been demonstrated to be highly stereoselective and synthetically useful operations. The corresponding methoxy and methoxymethyl (MOM) derivatives exhibit comparable properties, although to a demonstrably lessened degree. The free hydroxyl derivatives represent the more reactive substrates in either series, this reactivity ordering conforming expectedly to chelation-controlled addition. This ability of the indium cation to lock the carbonyl substrate conformationally prior to nucleophilic attack is indicative that coordination to the substrate can indeed overcome the H$_2$O

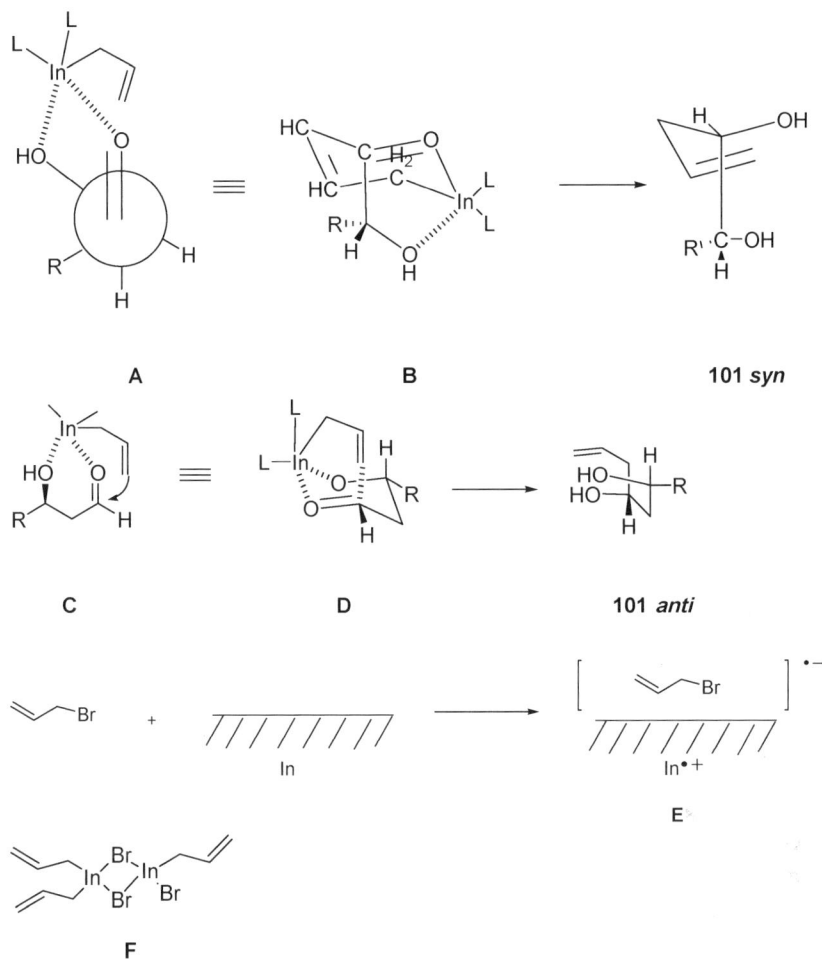

Figure 10.1 Cram's model for the formation of product *61-syn*. **B** Conformation adopted for product *61-syn*. **C** and **D**, conformations for the formation of the *anti* product *61*. **E** radical ion intermediates proposed. **F** allyl indium intermediate proposed.

solvation forces, especially when the neighboring functionality is an unprotected hydroxyl substituent.

The sense of asymmetric induction in the α-series, *viz.* a strong kinetic preference for formation of the *syn*-diol, is consistent with operation of the classic Cram model as in A (Figure 10.1). Once complexation occurs, the allyl group is transferred to the carbonyl carbon from the less hindered π-surface opposite to that occupied from the R group. In B (Figure 10.1), the chelation pathway is seen to be capable of adoption of a chair conformation which concisely accommodates the favored formation of the *syn*-diol. The reversal in stereoselectivity in going from **102** to **105** in the same aqueous environment

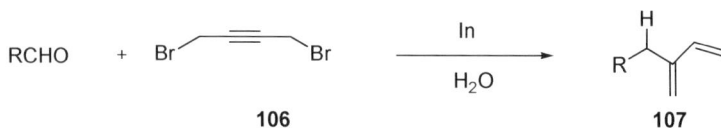

Scheme 10.60 In-mediated 1,3-butadien-2-ylation of carbonyl compounds in water.

is the classical test for chelation. For the β-chelate reactions, the factors which influence product formation appear to be the same. When C forms (Figure 10.1), intramolecular attack is guided to occur *syn* to the preexisting hydroxyl. This reaction trajectory leads preferentially to the *anti*-diol, provided that a chair-like transition state approximating D (Figure 10.1) is followed. Importantly, it is one single allylindium that chelates and reacts. Although similar working models have been advanced in explanation of the mode of addition of titanium[117] or borane reagents,[118] this behavior is distinct from other chelation-controlled reactions where the reacting reagent is different from the chelating agent. This may well be an argument that the indium-mediated reaction takes place on the metal surface.

The present studies have demonstrated a direct kinetic link between stereo-selectivity and the presence of a neighboring hydroxyl group. While this relationship has been extensively discussed,[119,120] the support of this concept is not universal. Several experimental and theoretical reports have appeared supporting the notion that π-complexation is not a kinetically important event.[121,122] Clearly, additional studies of this entire question would be welcomed.

The precise mechanism of indium-promoted reactions remains unclear. In the mid-1980s, the involvement of radical pairs was advanced in explanation of tin-promoted allylations.[123] Subsequent recourse to radical clock experiments demonstrated unambiguously that radicals could not be involved.[124] Single electron transfer process similar to that advanced by Chan have been proposed.[125] According to this reaction profile, the allyl bromide approaches the surface of the indium metal where the SET process generates the reactive radical anion/indium radical cation pair E. These conditions operate, of

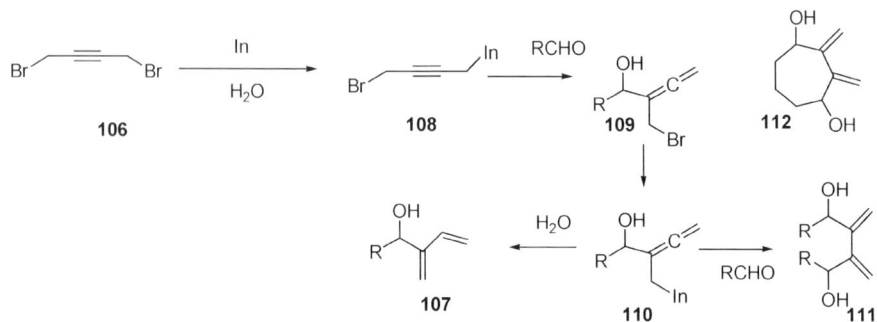

Scheme 10.61 Proposed mechanism for the In-mediated 1,3-butadien-2-ylation of carbonyl compounds in water.

Table 10.11 Indium-mediated 1,3-butadien-2-ylation of carbonyl compounds in water.

Entry	Aldehyde	Product	Yield (%)
1	Benzaldehyde		53
2	4-Methoxybenzaldehyde		55
3	Cinnalmaldehyde		60
4	Heptaldehyde		64
5	Cyclohexylaldehyde		64
6			67
7			68

course, only when indium metal is present as a reactant. Acyclic diastereo-facial control is presently recognized to occur in a wide range of reactions.[126,127] Suffice it to indicate at this point that the preformation of allylindium reagents may well bypass the involvement of E, suggesting an alternative pathway involving the more conventional species F can also operate.[128,129] Proper selection of reaction conditions could alter the precise pathway at work.

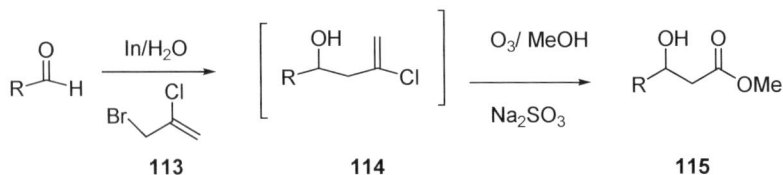

Scheme 10.62   In-mediated allylation of aldehydes with 3-bromo-2-chloro-1-propene
113.

Chan *et al.* reported[130] that indium can effectively mediate the coupling
between 1,4-dibromobut-2-yne (**106**) and carbonyl compounds in aqueous
media to give regioselective 1,3-butadien-2-ylmethanols **107** in good yields
(Scheme 10.60).

The reaction is likely to proceed *via* an organoindium intermediate **108**
which reacts with the aldehyde to give adduct **109** (Scheme 10.61). Further
reaction of the bromide with indium can lead to another organoindium
intermediate **110**, which is quenched by water to give 1,3-butadienyl-2-
methanol **111**. Reaction of **111** with another molecule of aldehyde to give di-
adduct **112** was not observed, presumably because of steric hindrance.
However, the authors were able to show that with glutaric dialdehyde
intramolecular trapping of intermediate **110** was possible and the cyclic di-
adduct **112** was obtained in 40% isolated yield.[130] With this methodology, a
series of dienes was synthesized, and the yields are reported in Table 10.11.

When benzaldehyde is treated with 3-bromo-2-chloro-1-propene **113** and
indium in water gave the corresponding allylation product **114** (Scheme
10.62).[131] ^{1}H NMR measurement of the crude product indicated, virtually, a
quantitative reaction. Isolation with chromatography provided 78% of the
allylation product **114**. Upon ozonolysis, compound **114** was converted to the
corresponding β-hydroxyl methyl ester **115** in 82% yield (Scheme 10.62).

Other aldehydes reacted similarly, and the results are summarized in
Table 10.12. Attachment of various functionalities on the aromatic ring provides an equivalent or better overall yields of the product (entries 2, 4, 5 and 7,
Table 10.12). Aliphatic aldehydes were similarly transformed to the corresponding γ-hydroxyl esters (entries 3 and 6, Table 10.12). The use of a mixture
of water and THF as solvent for the indium mediated allylation did not affect
the reaction result (entry 4, Table 10.12). It is worth mentioning that compounds with a free hydroxyl group (entry 8, Table 10.12) can be converted
directly to the corresponding γ-hydroxyl ester. In entry 9, Table 10.12, even
though the aldehyde existed in its cyclized hemiacetal form, the compound was
transformed to the desired compound without any difficulty.

Oshima and collaborators[132] have performed the radical allylation of α-halo
carbonyl compounds with allylgallium in water (Scheme 10.63).

The origin of the favorable solvent effect is not clear at this stage. Similar
phenomena were reported on the atom-transfer radical reaction of α-iodo
carbonyl compounds in aqueous media, where the high cohesive energy density

Table 10.12 In-mediated allylation of aldehydes with 3-bromo-2-chloro-1-propene 113

Entry	Substrate	Product (%)
1	Benzaldehyde	78
2	3-Bromobenzaldehyde	91
3	Heptaldehyde	58
4	4-Chlorobenzaldehyde	82
5	4-Methylbenzaldehyde	91
6	Cyclohexylaldehyde	
7	4-Methoxybenzaldehyde	77
8	4-Benzylic carbaldehyde	
9	2-Hydroxy-tetrahydropyrene	84

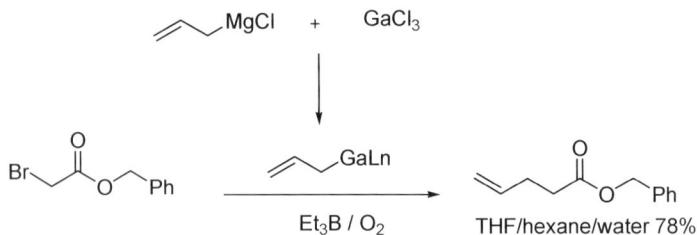

Scheme 10.63   Radical allylation of α-halo carbonyl compounds with allylgallium in water.

of water causes a reduction of the volume of an organic molecule. In the present case, the addition step could be accelerated because the addition necessarily accompanies the decrease of the total volume of the reactants. It is also probable that the structure of the allylgallium species would change and that the addition of water could increase the reactivity of allylgallium. Allylgallium dichloride is likely to be transformed into allylgallium hydroxide that is possibly more reactive for radical allylation.

   Various combinations of α-carbonyl compounds and allylic gallium reagents were examined (Table 10.13). More reactive α-iodo carbonyl compounds gave better results compared with their bromo analogs. 2-Halopropanoate or 2-halopropanamide also reacted with the allylgallium reagent to give 2-methyl-4-

Table 10.13   Radical allylation of α-halocarbonyl compounds with allyl gallium in water.

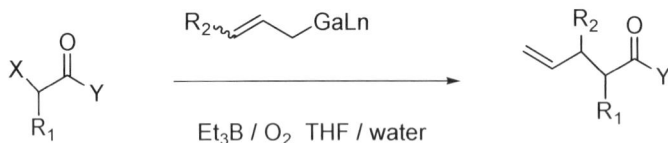

Entry	X	Y	R^1	R^2	Time/h	Yield (%)
1	Br	OCH$_2$Ph	H	H	2	78
2	Br	OCH$_2$Ph	Me	H	2	63
3	Br	NMe$_2$	Me	H	2	64
4	I	OCH$_2$Ph	H	H	0.5	89
5	I	OCH$_2$Ph	Me	H	0.5	81
6	I	NHCH$_2$CHCHC$_{10}$H$_{21}$	H	H	1	87
7	I	OCH$_2$CHCHC$_3$H$_7$	H	H	0.5	95
8	I	O(CH$_2$)$_6$Cl	H	H	1	85
9	I	O(CH$_2$)$_2$ OCH$_2$CHCH$_2$	H	H	0.5	64
10	I	OCHPhCH$_2$CHCH$_2$	H	H	2	71
11	I	OCH(C$_3$H$_7$)COC$_3$H$_7$	H	H	1	84
12	I	OCH$_2$Ph	H	Me	0.5	60
13	I	NHCH$_2$CHCHC$_{10}$H$_{21}$	H	Me	2	85
14	I	O(CH$_2$)$_6$Cl	H	Me	2	46
15	I	OCH$_2$Ph	H	Me	1	50
16	Br	NMe$_2$	Me	Me	7	65

pentenoate or 2-methyl-4-pentenamide (entries 2, 3 and 5, Table 10.13). In contrast, 2-bromo-2-methylpropanoate did not give the anticipated product, and the starting material was recovered unchanged, probably due to the steric hindrance around the carbon-centered radical. Interestingly, allylation was effective for the substrates having a terminal carbon–carbon double bond (entries 9 and 10, Table 10.13).

An electron-deficient (alkoxycarbonyl)methyl radical reacted faster with the highly electron-rich alkene moiety of the allylgallium species than with the olefinic parts of the substrate and of the product. Allylation of the ketone moiety was not observed.

Chan and collaborators[133] reported on a simple synthesis of 1,3-butadienes from carbonyl compounds in aqueous medium according to Figure 10.2. A typical experimental procedure consists of a mixture of the carbonyl compound **121** (1 mmol), 1,3-dichloropropene (**117**, X = Cl; 1 mmol), and zinc powder (2 mmol) in 10 mL of water was heated to 35 °C with vigorous stirring for 3–4 h. The reaction mixture was cooled and quenched with ether. The organic product was isolated from the ether phase and purified by flash column chromatography to give the corresponding 1,3-butadiene **121**. The reaction has a number of interesting features. First, it is important to note that, in the reaction of benzaldehyde, the yield of 1-phenylbutadiene was quite satisfactory under these conditions in aqueous medium but failed to proceed at all in diethyl ether or other organic solvents normally used for organometallic reactions. Second, the reaction seems to proceed with both aldehydes and ketones. With cinnamaldehyde, the corresponding triene was obtained. Third, the butadienes were formed stereoselectively and, in the case of aldehydes, exclusively as the *E* isomers. Furthermore, unprotected hydroxyl compounds such as glyceraldehyde and 5-hydroxypentanal underwent the diene conversion without difficulty. On the other hand, the yield of **121** was modest at best in all cases, in spite of efforts to vary the reaction temperature, time, amount of metal, *etc.* The poor yield was traced to the formation of the homoallylic

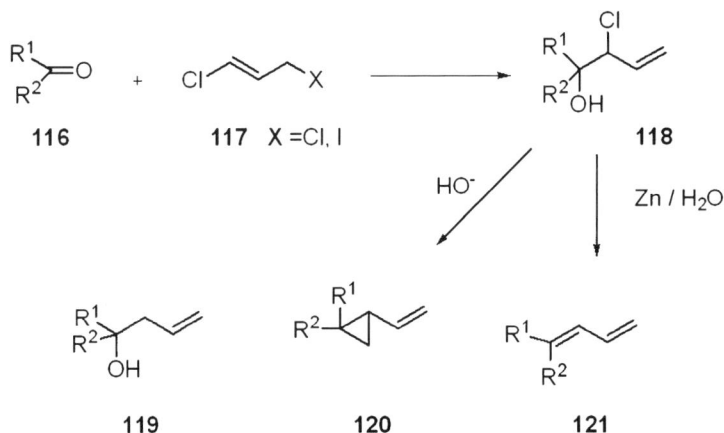

Figure 10.2   Formation of product **118–121**.

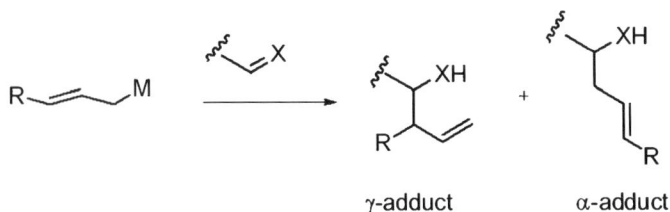

γ-adduct   α-adduct

Figure 10.3   Regioselectivity in allylation reactions.

alcohol **119**, which must have been formed by the zinc-mediated reduction of the intermediate chlorohydrin **118**.

The reaction of allylmetals with electrophiles has been developed as an important carbon–carbon bond-forming method. In general, γ-adducts (branched products) are obtained in the allylation of aldehydes and imines using allylmetals. The selective synthesis of α-adducts (linear products) has been a subject of current interest (Figure 10.3).

Recently, the preparation of allylindium reagents *via* transient organopalladium intermediates has been studied by Araki *et al.*[134] Allylation reactions of electron-deficient imines with allylic alcohol derivatives in the presence of a catalytic amount of palladium(0) complex and indium(I) iodide was studied by Takemoto *et al.*[135] The reversibility of allylation was observed in the reaction of glyoxylic oxime ether having camphorsultam. γ-Adducts were observed with high regioselectivity in water.

The reaction of **122** with allylic acetate **123a** in the presence of Pd(PPh₃)₄ and indium(I) iodide in the presence of water, affords the γ-adduct *syn*-**124a** in 90% yield as a single diastereomer (Scheme 10.64, Table 10.14).

In comparison with the reaction of glyoxylic oxime ether, the authors investigated the allylation of N-sulfonylimine under similar reaction conditions (Scheme 10.65).[135]

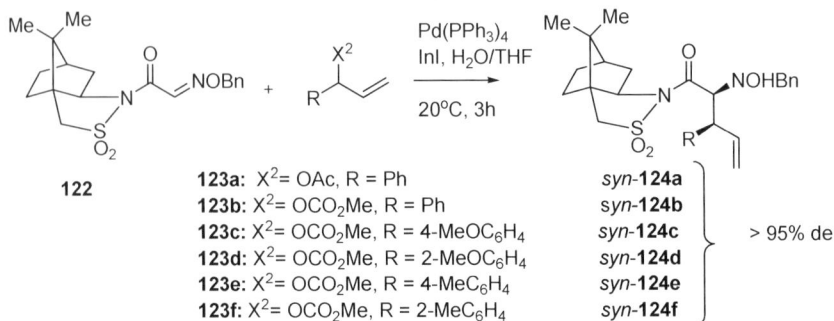

123a: X^2= OAc, R = Ph	*syn*-**124a**
123b: X^2= OCO₂Me, R = Ph	*syn*-**124b**
123c: X^2= OCO₂Me, R = 4-MeOC₆H₄	*syn*-**124c**
123d: X^2= OCO₂Me, R = 2-MeOC₆H₄	*syn*-**124d**
123e: X^2= OCO₂Me, R = 4-MeC₆H₄	*syn*-**124e**
123f: X^2= OCO₂Me, R = 2-MeC₆H₄	*syn*-**124f**

> 95% de

Scheme 10.64   The reaction of imine derivatives with allylic acetates in the presence of Pd(PPh₃)₄ and indium(I) iodide in water, to afford the γ-adduct *syn-124*a.

Table 10.14   Reaction of 122 with allyl alcohol derivatives 123a–f.

Entry	Reagent	Solvent	Time/h	Product	Yield (%)
1	123c	H$_2$O–THF	3	Syn-124c	90
2	123d	H$_2$O–THF	3	Syn-124d	81
3	123e	H$_2$O–THF	3	Syn-124e	71
4	123f	H$_2$O–THF	3	Syn-124f	72
5	123g	H$_2$O–THF	3	Syn-124g	66

Although the reaction of **125** proceeded smoothly, the formation of α-adducts were obtained even under anhydrous THF. These observations indicate that the allylation of N-sulfonylimine **125** was not a reversible process due to extra stabilization of indium-bonding adduct by electron withdrawing N-sulfonyl group. The bulky γ,γ-dimethylallyl acetate **58c** and carbonate **58d** were less effective for the allylation reaction of N-sulfonylimine **55**. The reaction of **55** with α,α,-dimethylallyl acetate **128c** gave the γ-adduct **126d** in 36% yield.

In another account, Takemoto *et al.* successfully attempted the propargylation reaction of hydrazones and glyoxylic oxime ethers in water mediated by Pd-indium iodide, demonstrating the role of water in directing the diastereoselectivity.[136]

Alcaide, Almendros and collaborators have performed the synthesis of 2-azetidinone allenol derivatives in high yields through the indium-mediated reaction in water (Scheme 10.66, Table 10.15).[137]

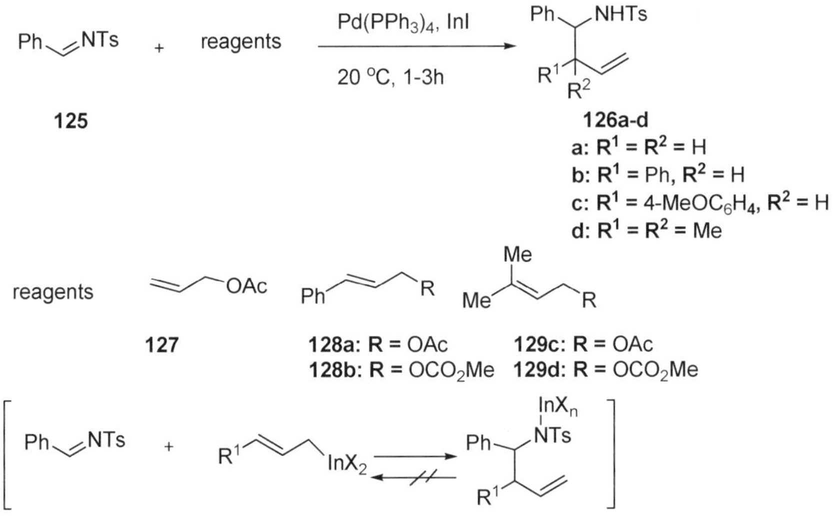

Scheme 10.65   Allylation of N-sulfonylimine with allylic acetate derivatives in the presence of Pd(PPh$_3$)$_4$ and indium(I) iodide in water.

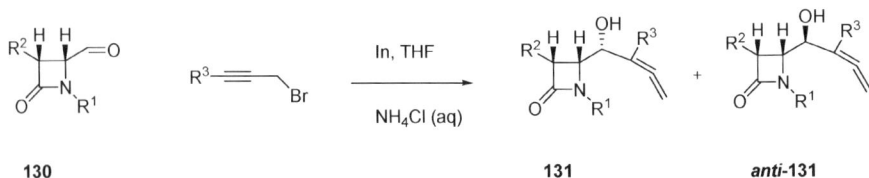

130   **131**   *anti*-**131**

Scheme 10.66   Synthesis of 2-azetidinone-tethered intermediates.

2-Azetidinone-tethered allenols **131a–k** (Table 10.15) were obtained by a metal-mediated Barbier-type carbonylallenylation reaction of β-lactam aldehydes **130a–f** in aqueous media by using our previously described methodologies (Scheme 10.66, Table 10.15).

It is observed that the yields of lactam **131** range from 50 to 89% yields, with the prevalence of the *syn* isomer. In several cases, the *syn* isomer is obtained exclusively.

10.3.5   Radical Conjugate Additions to α,β-Unsaturated Carbonyl Compounds in Water

Conjugate addition of alkyl groups to α,β-unsaturated carbonyl compounds is a versatile synthetic method for the construction of C–C bonds. Among the various methods available, the most commonly employed strategies involve the use of organometallic species such as Grignard reagents (RMgX) or organolithium (RLi) reagents. However, the use of these highly reactive organometallic reagents can lead to undesired side reactions such as hydrolysis, Wurtz coupling, β-elimination of the organometallic reagent, and the reduction of carbonyl compounds. Also, 1,2-addition of the alkyl group to the carbonyl group can compete with the 1,4-conjugate addition reaction. If this reaction could be developed to take place in water without the above-mentioned side reactions, it would greatly aid organic chemists.

Table 10.15   Regio- and stereoselective allenylation of 4-oxaazetidine-2-carbaldehydes 109 in water.

Aldehyde	R^1	R^2	R^3	Product	Syn : anti	Yield (%)
(+)-**130a**	Allyl	MeO	Me	(+)-**131a**	95 : 5	75
(+)-**130a**	Allyl	MeO	Ph	(+)-**131b**	100 : 0	64
(+)-**130b**	3-Methyl-but-2-enyl	MeO	Me	(+)-**131c**	85 : 15	68
(+)-**130b**	3-Methyl-but-2-enyl	MeO	Ph	(+)-**131d**	100 : 0	61
(+)-**130c**	3-Methyl-but-2-enyl	MeO	Ph	(+)-**131e**	100 : 0	51
(+)-**130d**	Methallyl	PhO	Me	(+)-**131f**	90 : 10	79
(+)-**130d**	Methallyl	PhO	Ph	(+)-**131g**	100 : 0	58
(+)-**130e**	Methallyl	Vinyl	Me	(+)-**131h**	10 : 90	60
(+)-**130e**	PMP	Vinyl	Ph	(+)-**131i**	70 : 30	89
(+)-**130f**	PMP	Isopropyl	Me	(+)-**131j**	10 : 90	73
(+)-**130f**	PMP	Isopropyl	Ph	(+)-**131k**	65 : 35	63

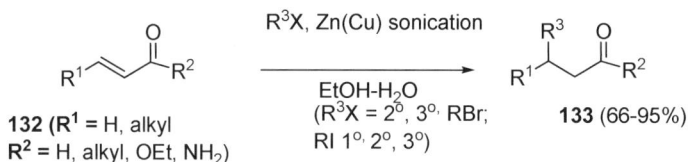

132 (R¹ = H, alkyl
R² = H, alkyl, OEt, NH₂)

133 (66-95%)

Scheme 10.67   Intermolecular addition of alkyl halides to α,β-unsaturated aldehydes, ketones and esters.

Luche found that the combination of Zn–Cu couple and sonication mediates in the intermolecular addition of alkyl radicals to α,β-unsaturated aldehydes, ketones, esters, and amides in aqueous EtOH (Scheme 10.67).[31–33]

These reactions were typically most efficient with tertiary and secondary radicals. Mechanistic studies suggested that the intermediate α-carbonyl radical is reduced to an enolate and subsequently protonated.[34]

Sarandeses and Perez Sestelo have employed the Luche protocol in diastereoselective RCA reactions. Methylene–dioxolanone **134** and γ,-dioxo-lanyl-α,β-unsaturated ester **136** serve as effective chiral acceptors for the stereoselective synthesis of α- and γ-hydroxy acid derivatives **135** and **137**, respectively (Scheme 10.68).[77,78]

Similar RCAs conducted with N-Cbz methyleneoxazolidinone **138** provided α-amino acid derivatives **139** with very good yields and dr's (diastereoselec-tivities) (Scheme 10.69).[79]

In the course of a synthesis of sinefungin analogs, Fourrey *et al.* discovered that sonication was not required to promote the RCA of ribose derivative **140** to dehydroalanine **142** under modified Luche conditions (Scheme 10.70).[35,138,139] Rather, vigorous stirring was sufficient. Similar reactions could also be induced by a Zn–Fe couple.[140]

Crich *et al.* demonstrated that alkyl radicals generated *via* the reductive mercury method underwent intermolecular conjugate addition to dehydroa-

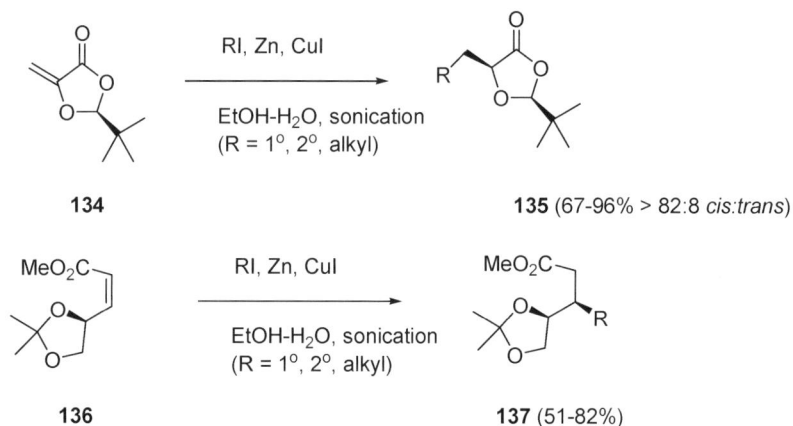

134

135 (67-96% > 82:8 *cis:trans*)

136

137 (51-82%)

Scheme 10.68   Diastereoselective radical conjugate addition.

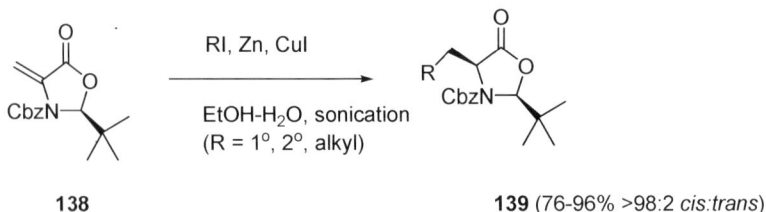

138   **139** (76-96% >98:2 *cis:trans*)

Scheme 10.69   Diastereoselective radical conjugate addition of N-Cbz methyleneoxazolidinone with alkyl iodides in aqueous media.

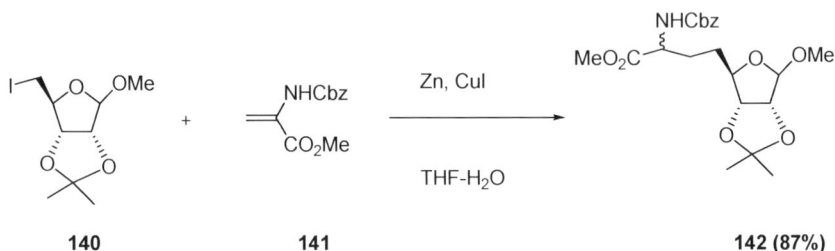

140   **141**   **142 (87%)**

Scheme 10.70   Radical conjugate addition with a ribose derivative.

lanine **143** (Scheme 10.71).[141] Primary, secondary, and tertiary alkylmercury
bromides and chlorides could all be used in this reaction.

   Crich extended this process to the intermolecular RCA of alkyl radicals to
dehydroalanine-containing dipeptides **145** (Table 10.16).[142] The stereocenter
present in each substrate exerted little influence over the hydrogen atom
abstraction, as the products **146** (Table 10.16) were obtained in low de
(diastereomeric excess). Tripeptides were also viable substrates in this reaction,
producing RCA adducts in good yield (87–88%) and poor de (3–15%).

   Yim and Vidal showed that the Crich method could be employed in the
solid-phase synthesis of α-amino acids by anchoring the dehydroalanine
radical acceptor to Wang resin (Scheme 10.72).[143] Cleavage of the N-acetyl
amino acid from the resin was accomplished by acid treatment.

   The pioneering works by Luche, Li, Naito and others have shown that it is
possible to carry out alkyl additions to conjugated systems in water. Unfortunately, in most cases, the use of harsh reaction conditions such as ultra-
sonication, inert atmospheres, cosolvent systems, and the narrow substrate
scope limit their applicability to complex molecule synthesis. Therefore, the

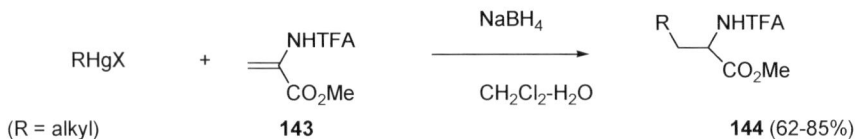

(R = alkyl)   **143**   **144** (62-85%)

Scheme 10.71   Synthesis of dehydroalanine derivatives *via* RCA.

Table 10.16 Alkylation of peptides in aqueous mixtures.

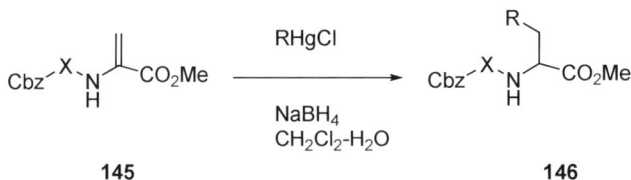

Entry	X	R	Yield (%)	de (%)
1	L-Val	c-C_6H_{11}	71	11
2	L-Phe	c-C_6H_{11}	64	5
3	L-Cys(Z)	c-C_6H_{11}	70	6
4	L-Ser	c-C_6H_{11}	42	1
5	L-Pro	i-Pr	98	1

development of more general and practical methods for alkyl addition to α,β-unsaturated carbonyl compounds under mild conditions is highly desirable.

More recently, Loh and collaborators[144] have attempted the alkylation reaction of carbonyl compounds in water using unactivated alkyl halides and In/CuI/I$_2$ or In/AgI/I$_2$ system. From their results, it became apparent that the use of organic solvents inhibited the occurrence of the Barbier–Grignard-type alkylation reaction. In contrast to the work reported by Li and coworkers[145] it was noteworthy that even aliphatic aldehydes could also react efficiently with alkyl iodides to furnish the alkylated products in good yields.

Lately, Loh and coworkers developed alkylation reactions of unactivated alkyl iodides to α,β-unsaturated carbonyl compounds in water (including a chiral version) using indium/copper in water.[115] In addition, the formation of symmetrical *vic*-diarylalkanes was observed when aryl-substituted alkenes were used as the substrate.

Initial studies focused on the reaction of α,β-unsaturated ester and cyclohexyl iodide under different reaction conditions.

As shown in Table 10.17, it was found that the combination of In/CuI/InCl$_3$ (6:3:0.1) was an efficient system for activation of the conjugate addition reaction of **149** in water (see Table 10.17). The reaction proceeded smoothly at room temperature to generate the corresponding adduct **150** in 80% yield (entry 1). It is important to note that, without the use of CuI, the reaction

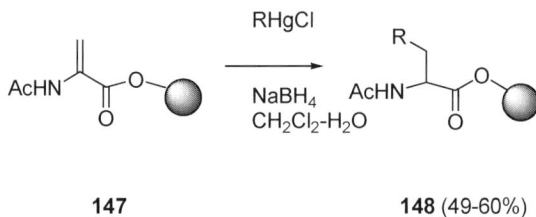

Scheme 10.72 Solid-state synthesis of ?-amino acids.

Table 10.17   Reactions of a α,β-unsaturated ester with cyclohexyliodide
under different conditions.

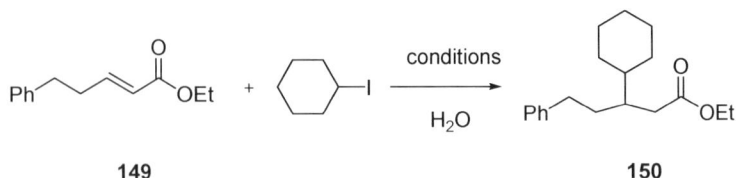

Entry	Conditions	Yield (%)
1	In/CuI/InCl$_3$	80
2	In/InCl$_3$	<20
3	In/CuI	54
4	CuI/InCl$_3$	0
5	In/CuBr/InCl$_3$	68
6	In/CuCl/InCl$_3$	<50
7	In/AgI/InCl$_3$	48

proceeded sluggishly to give the desired product in poor yield (entry 2).
Without the addition of InCl$_3$, the yield of the product decreased to 54% (entry
3). In addition, it was found that the use of the metal (*i.e.*, indium) was also
indispensable (entry 4). Among the several metals screened, indium proved to
be the best for this reaction. The following order was apparent for activation
of the conjugate alkylation: In > Zn > Al > Sn. Other copper and silver
compounds such as CuBr, CuCl, and AgI were also investigated, but all gave
the products in lower yields in comparison to CuI (entries 5–7). It was
worthwhile to note that the reactions proceeded more efficiently in water
than in organic solvents such as MeOH,THF,CH$_2$Cl$_2$,DMF,DMSO,and
hexane. Furthermore, the reactions were carried out without an inert atmosphere and ultra-sonication was unnecessary.

The reaction was extended to various α,β-unsaturated esters and enones, as
shown in Tables 10.18 and 10.19.

A plausible reaction mechanism is proposed (Scheme 10.73). The reaction is
possibly initiated by a single-electron transfer from indium/copper to alkyl
iodide a (Scheme 10.73) to generate an alkyl radical b. This radical can attack
the α,β-unsaturated carbonyl compound *via* 1,4-conjugate addition to furnish
a radical intermediate c. Subsequent indium-promoted reduction of intermediate c and quenching of the generated anion d in the presence of water
affords the expected product e. This method works with a wide variety of α,β-
unsaturated carbonyl compounds. The mild reaction conditions, moderate to
good yields, and the simplicity of the reaction procedure make this method an
attractive alternative to conventional methods using highly reactive organometallic reagents in anhydrous conditions.

Numerous and useful indium-mediated allylation reactions of carbonyl
compounds have been reported.[146] However, the corresponding reaction of
imine derivatives has not been widely studied because of the lower electro-

Table 10.18 Radical conjugate additions of alkyl iodides to different α,β-unsaturated esters in water, employing In/CuI/InCl₃.

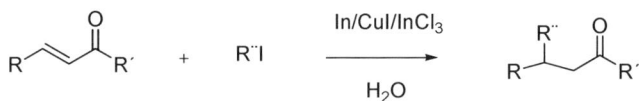

Entry	*α,β-unsaturated ester*	*Alkyl iodide*	*Yield (%)*
1	Ph, OEt	Cyclohexyl iodide	80
2	Ph, OEt	Cyclopentyl iodide	84
3	Ph, OEt	Isopropyl iodide	70
4	Ph, OEt	2-Iodobutane	73
5	Ph, OEt	Cyclohexyl iodide	70
6	Ph, OEt	Cyclohexyl iodide	75
7	OEt	Cyclohexyl iodide	61
8	C₅H₁₃, OEt	Cyclohexyl iodide	84
9	OEt	Cyclohexyl iodide	76
10	Ph-O, OEt	Cyclohexyl iodide	46

philicity of carbon–nitrogen double bonds. Therefore, the development of indium-mediated reactions of imines in aqueous media has been a subject of recent interest. Chan *et al.*[106,147] reported on the first studies of indium-mediated allylation of N-sulfonylimines in aqueous media. This is likely an area for further development and new discoveries. Then, it was demonstrated

Table 10.19 Radical conjugate additions of alkyl iodides to different α,β-unsaturated enones in water, employing In/CuI/InCl$_3$.

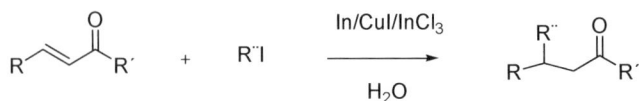

Entry	Enone	Alkyl iodide	Yield (%)
1		Cyclohexyl iodide	85
2		Cyclopentyl iodide	78
3		Isopropyl iodide	73
4		Hexyl iodide	45
5		Cyclohexyl iodide	53
6		Cyclohexyl iodide	81
7		Cyclohexyl iodide	83
8		Cyclohexyl iodide	74
9		Cyclohexyl iodide	65

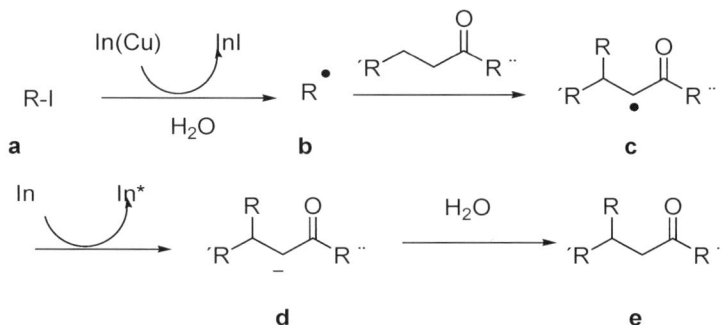

Scheme 10.73 Proposed mechanism for the RCA of α,β-unsaturated compounds with alkyl iodides employing In/CuI/InCl$_3$.

that trialkylboranes are excellent reagents for conjugate addition to vinyl ketones (eqn (10.21)), acrolein, α-bromoacrolein and quinones.

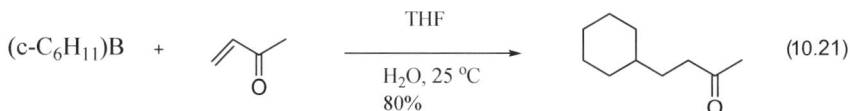

$$(c\text{-}C_6H_{11})B \quad + \quad \xrightarrow[\substack{H_2O,\ 25\ ^\circ C \\ 80\%}]{THF} \qquad\qquad (10.21)$$

Various attempts to extend this reaction to β-substituted-α,β-unsaturated carbonyl compounds such as *trans*-3-penten-2-one, mesityl oxide, 2-cyclohen-1-one, and *trans*-crotonaldehyde were unsuccessful unless radical initiators were used.

Fleming and Gudipati[148] have utilized a silica-supported zinc–copper matrix for promoting conjugate additions of alkyl iodides to alkenenitriles in water. Acyclic and cyclic nitriles react with functionalized alkyl iodides, overcoming the previous difficulty of performing conjugate additions to disubstituted alkenenitriles with non-stabilized carbon nucleophiles. Conjugate additions with ω-chloroalkyl iodides generate cyclic nitriles primed for cyclization, collectively providing one of the few annulation methods for cyclic alkenenitriles (Scheme 10.74).

10.3.6 Synthesis of α,β-Unsaturated Ketones

Indium- and zinc-mediated carbon–carbon bond-forming reactions in aqueous media have been of great importance from both economic and environmental points of view. Indium and zinc are stable under air, and it is much easier to run their reactions than those with SmI$_2$, and the toxicity is quite low.

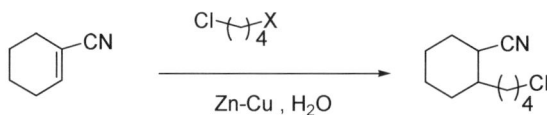

Scheme 10.74 Conjugate additions of alkyl halides to alkenenitriles in water.

Moreover, the metal species can be easily removed from the reaction mixture by simple filtration and washing with water, unlike tributyltin hydride.

Electron-transfer reactions mediated by indium have attracted the attention of synthetic chemists due to its low first ionization potential at 5.79 eV, which is lower than many reducing metals such as aluminum (5.98 eV), tin (7.34 eV), magnesium (7.65 eV), zinc (9.39 eV), and close to that of alkali metals such as sodium (5.12 eV) and lithium (5.39 eV). The second ionization potential for indium is much higher (18.86 eV). Having such low first ionization potential makes indium attractive for conducting reduction reactions. This is particularly so because it is so much easier to handle than alkali metals, for example, the metal remains unaffected by air or oxygen at ordinary temperatures and is practically unaffected by water even at high temperatures, and very resistant to alkaline conditions. As indium has a low toxicity, has found considerable utility in dental alloys. It has also to be pointed out that a favorable experimental feature of indium-mediated radical reactions is that the reactions proceed in the absence of toxic tin hydride, providing the carbon–carbon bond-forming method in aqueous media.

Indium has shown great potential for a number of carbon–carbon bond-forming reactions such as Reformatsky, Barbier-type alkylation, allylation, and propargylation of carbonyl compounds. This is largely due to the fact that a highly reactive metal, such as indium, is required to break the non-activated carbon–halogen bond (as well as to react with the carbonyl once the organometallic intermediate is formed). However, even if the desired intermediate is successfully generated, various competing side reactions may occur when utilizing a highly reactive metal, for example, the reduction of water, the reduction of starting materials, the hydrolysis of the organometallic intermediate, and pinacol-coupling (*vide infra*).[145]

The indium-mediated reaction of benzaldehydes and methyl vinyl ketones proceeded smoothly in the presence of $InCl_3$ in aqueous media to form β,γ-unsaturated ketones.[149] Thus benzaldehydes were reacted with methyl vinyl ketone in the presence of indium powder and $InCl_3$ in a solvent mixture of THF and H_2O at ambient temperature for 6–8 h. Addition of NH_4Cl to the reaction mixture afforded the desired β,γ-unsaturated ketones in good to moderate yields (eqn (10.22)).

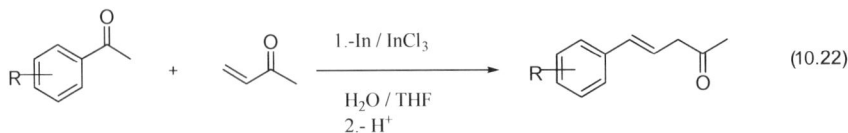

$$(10.22)$$

Generally, the yields of products are not affected by the nature of the substituents on the phenyl ring. The reaction also proceeded with heteroaromatic aldehydes. With the absence of In, or $InCl_3$ the reaction did not occur. When other Lewis acids such as $SnCl_4$, $FeCl_3$, and $CuCl_2$ were used instead of $InCl_3$, low yields of β,γ-unsaturated ketones resulted (20–38%). When an aldehyde reacted with ethyl vinyl ketone instead of methyl vinyl ketone as a

Michael acceptor, a coupling product was produced in 62% yield. The reaction conditions were extended to other Michael acceptors such as acrolein, acrylonitrile, ethyl acrylate, and acrylic acid; however, the reactions did not proceed.[149]

The reaction mechanism was postulated to be a radical mechanism involving the radical anion intermediate of methyl vinyl ketone formed from indium (Scheme 10.74).

The reaction intermediate undergoes radical cyclopropanation and addition to benzaldehyde. Upon addition of butylated hydroxytoluene (BHT) a rate retardation effect was observed.[149] When the reaction was followed by [1] H NMR in a 1:1 mixture of THF-d$_8$ and D$_2$O, the cyclopropanyl proton signals were observed at δ 1.2–0.5 as multiplet. Quenching the reaction mixture with DCl in D$_2$O after an appropriate reaction time and examination of the CDCl$_3$-extracted products by ^{1}H NMR showed the signal of the 5-phenyl-4-penten-2-one together with peaks of some methyl vinyl ketone (MVK) decomposed compounds.

Murphy and collaborators,[150] however, developed a facile and an environment-friendly protocol for the deoxygenation of epoxides with good radical-stabilizing groups adjacent to the oxirane ring, using indium metal and indium(I) chloride or ammonium chloride in alcohol/water mixtures (eqn (10.23)).

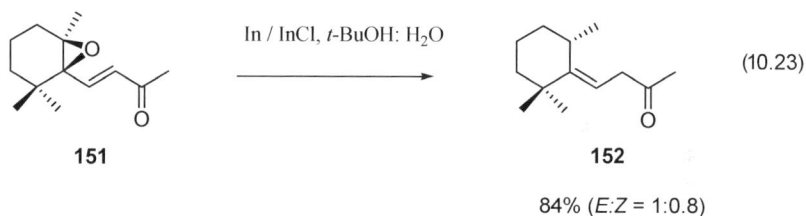

$$ (10.23) $$

151   **152**

84% (*E:Z* = 1:0.8)

Oxirane **151** underwent smooth deoxygenation to afford the alkene **152** in an excellent 81% yield. The reaction sequence is depicted in Scheme 10.75.

The formation of alkene **152** can be explained *via* the reduction of the expected dienone intermediate **153** by indium metal (Scheme 10.76).[150]

10.3.7 Metal-mediated Mannich-type Reactions in Water

A convenient method that allows the easy, mild and efficient synthesis of a large number of differently substituted propargylamines characterized by high atom economy, low environmental impact and use of non-toxic solvents and reagents has been developed by Bieber and da Silva.[151] The possibility of reacting unprotected primary amines and alkynols allows further reactions on these sites. As shown in Scheme 10.77, the reaction involves a three component procedure between terminal alkynes **157**, formaldehyde and secondary amines.

Thus, it was possible to obtain propargylamines **159–163** in pure water. Other propargylamines can also be obtained in DMSO–water mixtures, in yields ranging from 70 to quantitative.

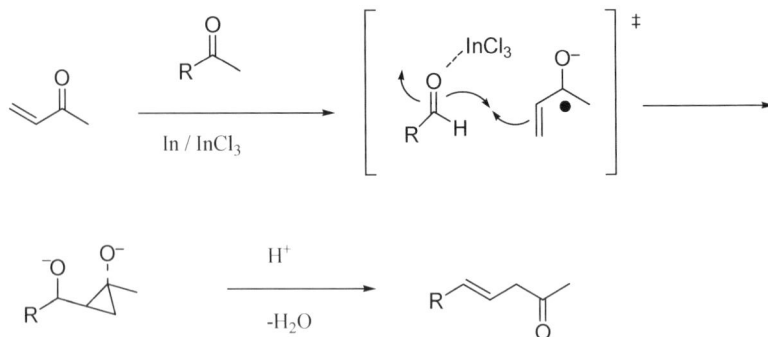

Scheme 10.75 Possible reaction mechanism for the In-mediated synthesis of β,γ-unsaturated ketones.

In view of the need for a copper catalyst and the exceptionally mild reaction conditions, a radical intermediate should also be considered. A radical addition of phenyl, acetyl and alkyl radicals to iminium ions has been postulated before in the Zn-Barbier reaction[152] and in the TiCl₃-promoted reaction with diazonium salts or *tert*-butyl hydroperoxide.[153]

10.3.8 Pinacol and Other Coupling Reactions in Water

Since the first report of the reaction of acetone with sodium in 1858,[154] various low-valent metals such as Al,[155] Sm,[156] V,[157] Mg,[158] Zn,[159] Mn,[160] Sn,[161] Ti,[162] Ce,[163] Te,[164] U,[165] Cr,[166] Ga,[167] and In[168] have been used to promote this reductive coupling reaction. Among these methods, some require an absolutely anhydrous system under inert atmosphere, and some reagents and solvents are costly, moisture-sensitive, and toxic. In order to find environmen-

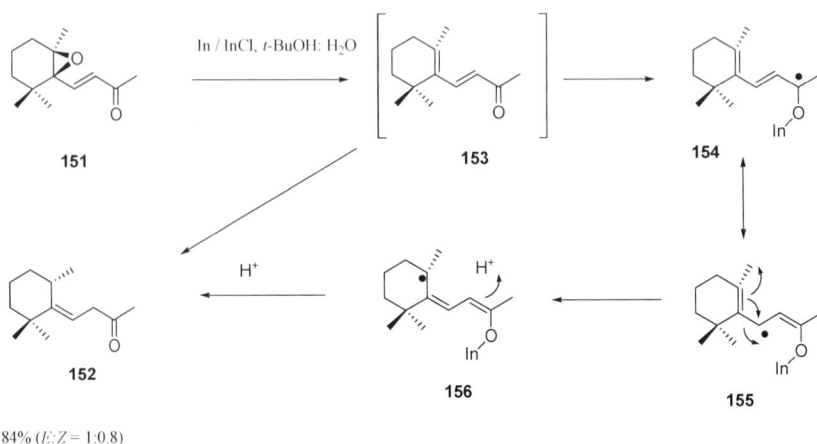

Scheme 10.76 Plausible mechanism for the Indium-mediated deoxygenation of epoxides in aqueous media.

$$R\!=\!\!=\ +\ CH_2O\ +\ HNR'R'' \xrightarrow[\text{[Cu]}]{H_2O} R\!=\!\!=\!-CH_2NR'R''$$

157	158	159-163

R= CH$_2$OH   R′, R″=(CH$_2$)$_4$ 159   70%

R= CH$_2$OH   R′,R″ =(CH$_2$)$_5$ 160   96%

R= CMe$_2$OH R′, R″ =Et, Et   161   87%

R= CMe$_2$OH R′,R″= (CH$_2$)$_4$ 162   80%

R=CMe$_2$OH R′,R″= (CH$_2$)$_5$ 163   82%

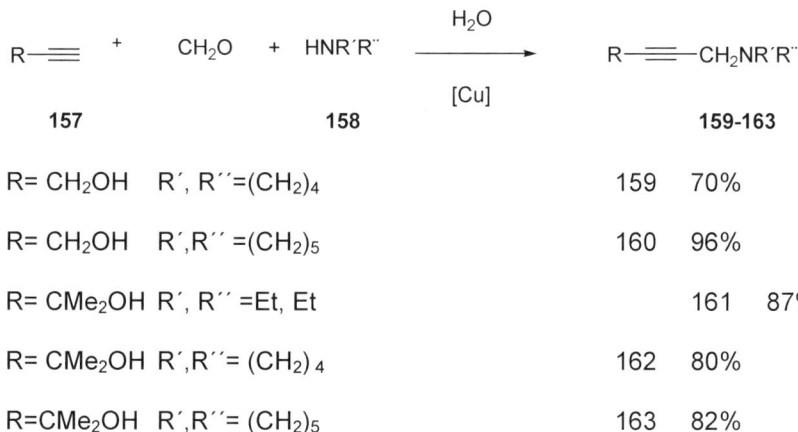

Scheme 10.77   Mannich-type reaction for the synthesis of propargylamines in water, mediated by CuI.

tally friendly conditions, it is very attractive to develop a new convenient method for the pinacol coupling by utilizing less toxic reagents and solvents. During past decades, great efforts have been devoted by chemists to explore environmentally benign systems for the pinacol reaction. Different catalysts/ co-catalysts in aqueous media (including TiCl$_3$, VCl$_3$/Al, Mn/HOAc, Al/MF, *etc.*) have been reported with promising results.

As is well known the reductive (pinacol) coupling of carbonyl compounds is a useful method for the creation of carbon–carbon bonds with 1,2-difunctionality (eqn (10.24)). Although detailed mechanistic studies of pinacol coupling are lacking, the reactions are generally considered to involve the generation and reaction of the substrate ketyl (radical anion) with either the neutral substrate or another ketyl species.

$$2\ \underset{R\quad R'}{\overset{O}{\|}}\ +\ 2e^-\ +\ 2H^+ \longrightarrow \underset{R\quad O}{\overset{OH\ \ R}{|}}\!R' \qquad (10.24)$$

The pinacol coupling reaction (*vide supra*) is a fundamental reaction in organic chemistry. The pinacol coupling reaction in water mediated by Ti(III) and other metals such as Zn–Cu have also been found to promote pinacol formation under ultrasonic radiation conditions in aqueous acetone. When benzaldehyde was reacted with manganese in the presence of a catalytic amount of acetic acid in water, the corresponding pinacol coupling product was obtained smoothly. Other aryl aldehydes were coupled similarly. On the other hand, aryl and aliphatic ketones appeared to be inert under the same reaction conditions, and only the reduced product was obtained with aliphatic aldehydes.[99]

Pan, Wu, and collaborators reported on the pinacol coupling of aromatic aldehydes and ketones using InCl$_3$/Al in aqueous media.[169]

Table 10.20 Pinacol coupling of substituted benzophenones by using InCl₃/Al.

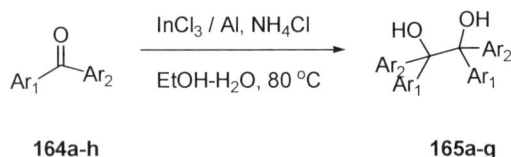

$$Ar_1\text{-CO-}Ar_2 \xrightarrow[\text{EtOH-H}_2\text{O, 80 }^\circ\text{C}]{\text{InCl}_3 \text{ / Al, NH}_4\text{Cl}} Ar_2(HO)(Ar_1)C\text{-}C(OH)(Ar_2)(Ar_1)$$

164a-h   **165a-g**

Entry	Ar_1	Ar_2	Time/h	165
A	Ph	Ph	5	165a
B	4-CH₃C₆H₄	Ph	5	165b
C	4-FC₆H₄	4-FC₆H₄	5	165c
D	4-MeOC₆H₄	4-FC₆H₄	11	165d
E	4-MeOC₆H₄	Ph	11	165e
F	4-ClC₆H₄	Ph	3	Mixture
G	4-C₆H₅C₆H₄	Ph	3	Mixture
F	4-ClC₆H₄	Ph	5	165f
G	4-C₆H₅C₆H₄	Ph	5	165g
H	4-ClC₆H₄	Ph	5	–

Under the optimized conditions (Table 10.20, entry 1), various substituted benzophenones were investigated to give a series of 1,1,2,2-tetraaryl substituted diols (**165a–h**) (Table 10.20).

From these results, the authors found that **165a–h** were obtained in excellent yields without the formation of by-products due to reduction of the carbonyls to the corresponding alcohols (Table 10.20, entries A–C). Benzophenones bearing electron-donating groups (**164d** and **164e**) *para* to ketones were reduced to the corresponding pinacols in the moderate yields even after 11 h (Table 10.20, entries D and E). Unfortunately, the reactions of benzophenones bearing electron-withdrawing groups (**164f** and **164g**) *para* to ketones with InCl₃/Al reagent at 80 °C for 3 h gave miscellaneous products, in which most starting materials had been consumed (Table 10.20, entries F and G). On the other hand, the lower temperature gave pinacols in moderate yields (Table 10.20, entries F and G) along with a small quantity of the corresponding alcohols. Unexpectedly, benzophenone bearing a chlorine group (**164h**) *ortho* to ketone only gave the corresponding alcohol as the main product.

The coupling of water-soluble acetonitrile derivatives has been developed by Holtz, Pinhas, and coworkers using a Fenton reagent (Scheme 10.78).[170]

$$2\ R\text{-CH}_2\text{-CN} \xrightarrow[\text{30\% H}_2\text{O}_2]{\text{FeSO}_4, \text{H}_2\text{SO}_4} 2\ R\text{-}\overset{\bullet}{C}\text{H-CN} \longrightarrow (R)(CN)CH\text{-}CH(R)(CN)$$

Scheme 10.78 Pinacol coupling by Fenton reagent.

For this reaction, it does not matter if a free radical or an iron–oxo complex is formed. What matters is that the Fenton chemistry generates a radical[171,172] or radical-equivalent[173] that can remove a hydrogen atom from the alkyl chain of the alkyl nitrile, and then two of these "alkyl radicals" can couple. This type of coupling reaction was first mentioned about 50 years ago,[174] but unfortunately, the yields were low and the regiochemistry was not investigated in detail. In this paper, the authors discuss the coupling of acetonitrile and other water-soluble alkyl nitriles.[175] Reaction yields are improved, and in addition, the authors have investigated the regiochemistry of the coupling reaction. This regiochemistry is not only important from a synthetic perspective, but also tells the energetics of hydrogen-atom removal from various positions on the alkyl chain.

Since the cyanomethyl radical is the only radical that can be obtained from acetonitrile, succinonitrile is the only dinitrile product. However, propionitrile can form two radicals: a resonance-stabilized secondary radical (**166**) formed by abstraction of an α-hydrogen atom or a primary radical (**167**) by abstraction of α-hydrogen atom. The products formed by all possible combinations of these two radicals are illustrated in Scheme 10.79.

Stereoisomers of 2,3-dimethylsuccinonitrile (**169**) (D,L-pair and a *meso* compound) are formed by the coupling of two molecules of the secondary radical (**166**), while two primary radicals (**167**) combine to form adiponitrile (**170**). Cross-coupling of these two radicals produces 2-methylglutamonitrile (**168**). The observed ratio for dinitrile isomers 168:169:170 is 50:45:5. After correcting for the number of abstractable hydrogens at each position, it was determined that the resonance-stabilized secondary radical (**166**) and the primary radical (**167**) form in a 3.5:1 ratio rather than the statistical ratio of 2:3.

*Iso*butyronitrile can also form two radicals: a resonance-stabilized tertiary radical (**171**) formed by abstraction of an α-hydrogen atom or a primary

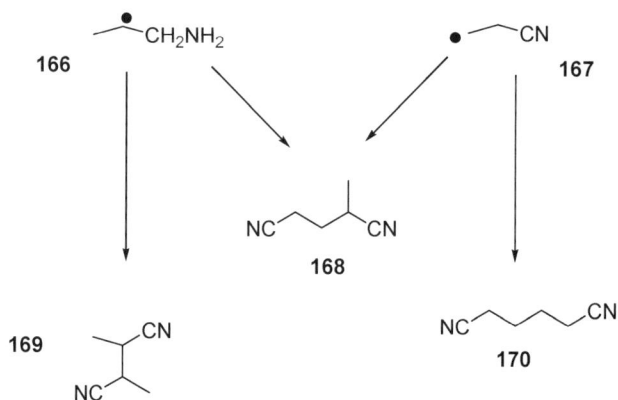

Scheme 10.79 Possible combination products from pinacol coupling of propionitrile.

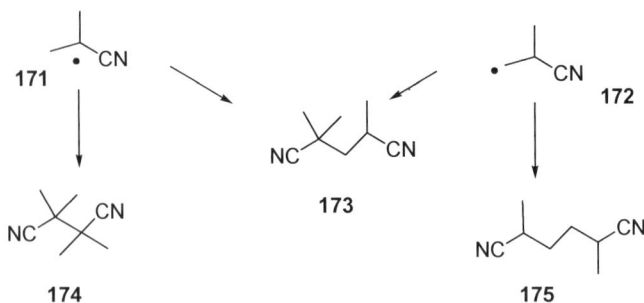

Scheme 10.80   Possible combination products from pinacol coupling of *iso*butyronitrile.

radical (**172**) by abstraction of α-hydrogen atom. The products formed by all
possible combinations of these two radicals are illustrated in Scheme 10.80.

Stereoisomers of 2,5-dimethyladiponitrile (**175**) (*d,l*-pair and a *meso* compound) are formed by the coupling of two molecules of the primary radical
(**172**), while two tertiary radicals (**171**) combine to form 2,2,3,3-tetramethy-
succinonitrile (**174**). Cross-coupling of these two radicals produces 2,2,4-
trimethylglutamonitrile (**173**).

The observed ratio for dinitrile isomers **173**:**174**:**175** is 26:29:45. After
correcting for the number of abstractable hydrogens at each position, it was
determined that the resonance-stabilized tertiary radical (**172**) and the primary
radical (**171**) form in a 4.3:1 ratio rather than the statistical ratio of 1:6.

Since the cyanomethyl radical is easily reduced to the cyano-methyl anion by
iron(II), with a stoichiometric amount of iron, the lower yield is not surprising.
To improve the yield, the concentration of iron(II) should be kept low. Thus
for large scale preparation of succinonitrile *via* Fenton reagent, the reaction
should be catalytic in iron(II), keeping its concentration as low as possible.
Iron(0) is an attractive candidate as a reducing agent because iron(II) is the
only product of the oxidation–reduction reaction.

Although the use of iron(0) creates a heterogeneous reaction which leads to
greater variability in the product yield, the increased production of succinonitrile indicates iron(0) has an effect on the coupling reaction by reducing
iron(III) to iron(II).

Loh and collaborators[176] have very recently reported an efficient pinacol
cross-coupling reaction of aldehydes and α,β-unsaturated ketones using Zn/
InCl$_3$ in aqueous media. The 1,2-diols **176** were thus obtained in moderate to
good yields, with up to 93.7% diastereoselectivity (eqn (10.25)).

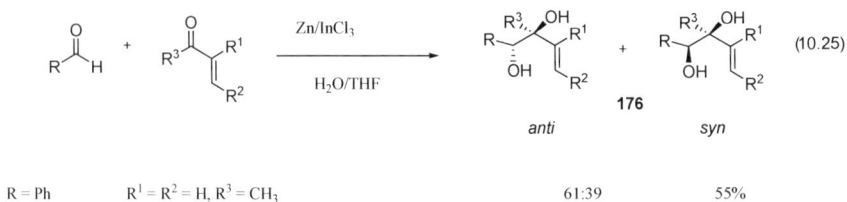

R = Ph   R^1 = R^2 = H, R^3 = CH$_3$ 61:39   55%

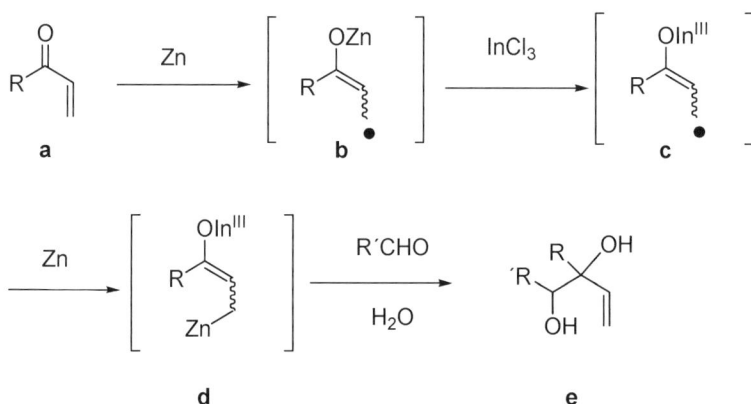

Scheme 10.81 Proposed mechanism for the pinacol cross coupling of enones and aldehydes.

A possible reaction mechanism is shown in Scheme 10.81. The reaction is initiated by a single electron transfer from zinc to the α,β-unsaturated ketone to form a radical enolate anion b. Fast trapping of the oxygenmetal bond in the radical enolate anion b by $InCl_3$ affords the γ-In(III)-substituted allylic radical c. The radical c is further reduced by zinc to furnish the corresponding allylic zinc species d. Finally, coupling of the γ-In(III)-substituted allylic zinc species d with an aldehyde followed by quenching of the resulting 1,2-diolate with water generates the desired product e.

Kalyanam and Rao[177] reported a novel reductive coupling of aldimines brought about by indium to vicinal diamines as described in eqn (10.26). The reaction occurs in aqueous ethanol. While NH_4I is not essential for the reaction, the reaction is accelerated by its presence. The reaction fails completely in CH_3CN, DMF and DMF containing small quantities of water. Indium used in the reaction was in the form of small rods made from a sheet of indium of about 1 mm thickness. It may be mentioned that optically pure derivatives of **178** have considerable potential in asymmetric synthesis.

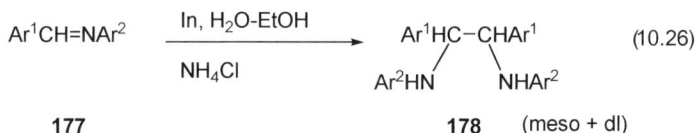

$$Ar^1CH=NAr^2 \xrightarrow[\text{NH}_4\text{Cl}]{\text{In, H}_2\text{O-EtOH}} \underset{Ar^2HN \quad NHAr^2}{Ar^1HC-CHAr^1} \qquad (10.26)$$

177   **178**   (meso + dl)

Some important aspects of the reaction are to be noted. The reaction occurs in aqueous medium and does not require exclusion of oxygen or anhydrous conditions as required by other reagents mentioned in the literature for effecting the same transformation as in eqn (10.26). The complete absence of the side product in the crude reaction mixture, $ArCH_2$–NHAr (eqn (10.26)) (a resultant of unimolecular reduction as happens in some of the methods employed for this transformation) is noteworthy. Further, the reaction is brought about by indium rods and does not require fine indium powder. The product **178** is invariably a 1 : 1 mixture of D,L and *meso* isomers indicating fast coupling of any putative radical intermediates. The reaction fails in the case of substrates, Ph(CH₃)C=NPh and PhCH=NCH₂Ph, showing selectivity for the coupling of aldimines obtained from aromatic aldehydes and aromatic amines.

Nair *et al.* have reported on the indium/indium trichloride-mediated pinacol cross coupling of reaction of aldehydes and chalcones in aqueous media to obtain substituted but-3-ene-1,2-diols.[178]

In an initial experiment, 3,4-dichlorobenzaldehyde was treated with 4-methylbenzylidene acetophenone in the presence of indium and indium trichloride in aqueous THF at room temperature to yield an isomeric mixture of 1-(3,4-dichlorophenyl)-2-phenyl-4-*p*-tolylbut- 3-ene-1,2-diols **179** in 66% yield (Scheme 10.82).

Similar results were obtained with other chalcones and aldehydes (Table 10.21). With benzylidene acetone and aldehydes, only one *trans*-isomer was formed whereas with other α.β-unsaturated ketones and aldehydes a mixture of *syn*- and *anti-trans* isomers was obtained.

Halterman and collaborators have achieved the pinacol coupling of benzaldehydes in water mediated by CrCl₂.[179] The pinacol coupling of benzaldehyde (0.25 M or 1.25 M) in water was catalyzed by 5–25 mol % CrCl₂ in the presence of Zn-dust or Al-dust at 20 °C. In all cases at most 50% of the pinacol coupling product (1,2-diphenyl-1,2-ethanediol) was obtained with the major product (benzyl alcohol) being formed by a competitive 2e⁻ reduction of the carbonyl. The D,L- to *meso* -diastereoselectivity of the pinacol products ranged from 0.6 : 1 to nearly 1 : 1 (Figure 10.4).

According to the catalytic scheme depicted in Figure 10.5, Cr(II) can initially reduce the aldehyde (step a) to form radical intermediate B .The reactive carbon site in B can combine with a second aldehyde unit (either before or after its reduction) as in step b to form coupled product C. Hydrolysis to release the 1,2-diol and reduction of the chromium species back to a lower valent metal as in step c completes the desired catalytic cycle. However, intermediate B can be further reduced by a second electron from coordinated chromium or by an electron from an external metal as in step d. This competitive side reaction can lead to the formation of the undesired reduced benzyl alkoxide D that can hydrolyze to form benzyl alcohol. In terms of the

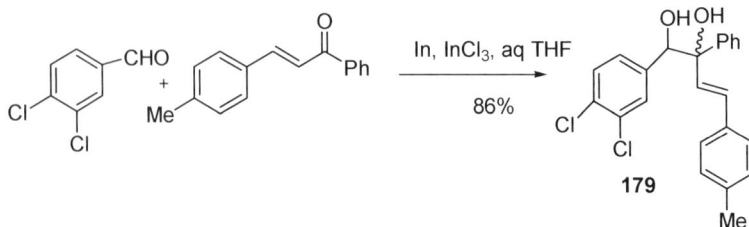

Scheme 10.82   Indium/indium trichloride-mediated pinacol cross coupling reaction of aldehydes and enones in aqueous media.

Table 10.21 Synthesis of substituted but-3-ene-1,2-diols.

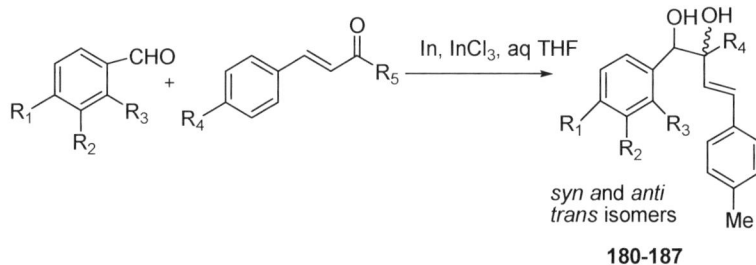

Entry	Substituents	Products	Yield (%)	Ratio
1	$R^1 = R^2 = Cl$, $R^3 = R^4 = H$, $R^5 = Ph$	**180**	56	1:2
2	$R^1 = R^2 = Cl$, $R^3 = R^4 = H$, $R^5 = Me$	**181**	42	0:1
3	$R^1 = Cl$, $R^2 = R^3 = H$, $R^4 = Me$, $R^5 = Ph$	**182**	60	1:2
4	$R^1 = Cl$, $R^2 = R^3 = R^4 = H$, $R^5 = Ph$	**183**	61	0:6.2
5	$R^1 = Cl$, $R^2 = R^3 = R^4 = H$, $R^5 = Me$	**184**	46	0:1
6	$R^1 = R^2 = H$, $R^3 = Cl$, $R^4 = Me$, $R^5 = Ph$	**185**	56	0:6.1
7	$R^1 = R^2 = R^4 = H$, $R^3 = Cl$, $R^5 = Ph$	**186**	42	0:4.1
8	$R^1 = R^2 = R^4 = H$, $R^3 = Cl$ $R^5 = Me$	**187**	56	0:1

chemical selectivity for the pinacol coupling versus reduction to form benzyl alcohol, the benzyl alcohol was always the major product under all conditions studied. Under most conditions, the ratio of pinacol to benzyl alcohol varied from 1:1 to 1:2. We noted that at higher temperature with a higher ratio of starting chromium catalyst, a lower selectivity for the pinacol coupling was obtained.

The catalytic reactions produced both diastereomeric pinacol products D,L-**188** and *meso*-**188** with the *meso*-product favored in ratios from 0.6:1 to nearly 1:1. The stereoselectivity in the presence of aluminum as the stoichiometric reductant was similar to when zinc-dust was used.

Biaryl coupling of 2-naphthols and substituted phenols was efficiently promoted by a supported Ru catalyst using O_2 as an oxidant in water (eqn (10.27)).[174] The supported catalyst can be reused seven times without losing catalytic activity. The big advantages of this method are that an environmentally friendly oxidant (O_2) and solvent (H_2O) can be used. The studies on the

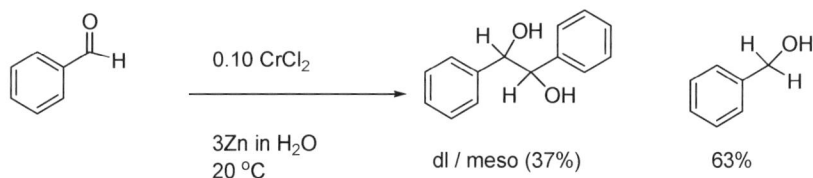

Figure 10.4 Pinacol coupling of benzaldehyde by $CrCl_2/Zn$ catalyst in water.

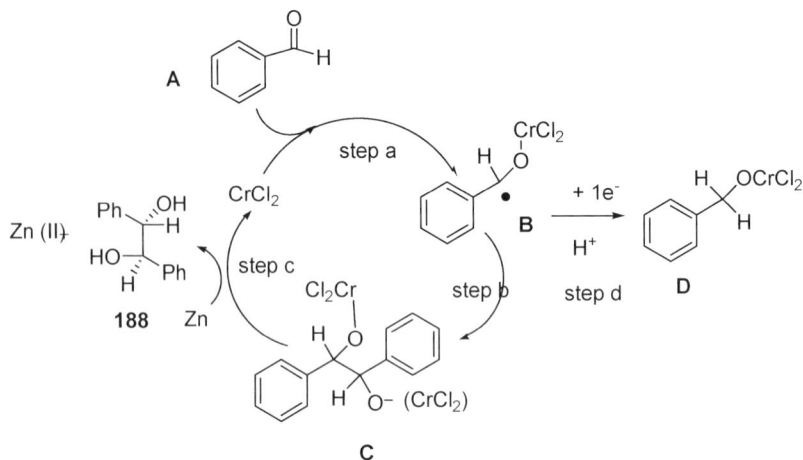

Figure 10.5 Catalytic cycle for Cr(II)-Zn-catalyzed pinacol reaction with competitive benzaldehyde reduction.

mechanism behind the reaction showed that the Ru-catalyzed biaryl coupling reaction proceeds through the radical coupling mechanism.

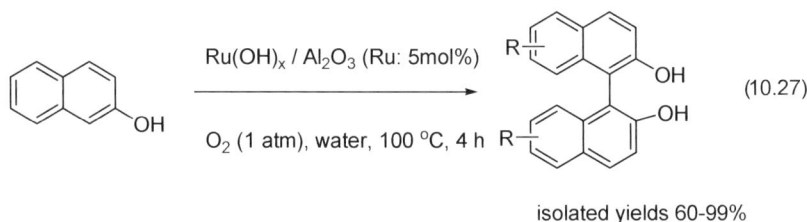

$$(10.27)$$

isolated yields 60-99%

Benzidine derivatives were obtained *via* oxidative coupling of *N,N*-dialkylarylamines using CuBr as a catalyst and H_2O_2 as an oxidant in water.[180] When CAN was used as oxidant, homocoupling of *N,N*-dialkylarylamines was also effectively promoted using water as solvent (eqn (10.28)).[181] A rationale for the mechanism of this coupling reaction is proposed *via* dimerization of diradical cations. Unlike homocoupling of 2-naphthols and substituted phenols which gave *ortho* products to the OH group, *para*-substituted products were selectively formed for *N,N*-dialkylarylamine substrates.

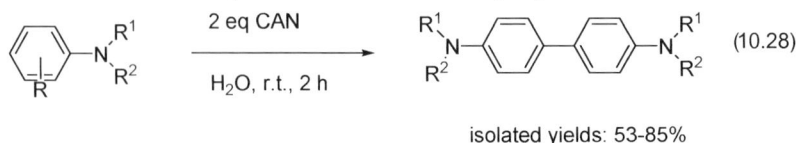

$$(10.28)$$

isolated yields: 53-85%

10.4   CONCLUSION AND FUTURE DIRECTION

Radical atom transfers reactions are classical examples of atom efficiency in organic synthesis devoted to waste minimization. For radical rearrangement reactions (intramolecular cyclization reactions), a single reactant leads to a single product with nearly the same mass and an almost identical structure. All these factors support the good prospects that place radical chemistry in water under the umbrella of green chemistry.

ACKNOWLEDGEMENTS

Thanks are given to Conicet, Argentina (National Council of Scientific and Technical Research), and to Agencia Nacional Científica y Técnica for financial support.

REFERENCES

1. (a) F. Alonso, I. P. Beletskaya and M. Yus, *Chem. Rev.*, 2002, **102**, 4009; for a review, see: S. Taciolu, *Tetrahedron*, 1996, **52**, 11113; (b) B. C. Ranu, P. Dutta, A. Sarkar, *Tetrahedron Lett.*, 1998, **39**, 9557; (c) B. C. Ranu, S. Samanta, A. Das, *Tetrahedron Lett.* 2002, **43**, 5993; (d) B. C. Ranu, J. Dutta, S. K. Guchhait, *Org. Lett.*, 2001, 3, 2603; (e) B. C. Ranu, K. Chattopadhyay, S. Banerjee, *J. Org. Chem.*, 2006, **71**, 423; (f) B. C. Ranu, S. Samanta, *J. Org. Chem.*, 2003, **68**, 7130; (g) B. C. Ranu, K. Chattopadhyay, L. Adak, *Org. Lett.*, 2007, **9**, 4595.

2. L. Park, G. Keum, S. B. Kang, K. S. Kim and Y. Kim, *J. Chem. Soc., Perkin Trans. 1,* 2000, 4462.

3. H. Chae, C. Sangwon, G. Keum, S. B. Kang, A. N. Pae and Y. Kim, *Tetrahedron Lett.,* 2000, **41**, 3899.

4. G. D. Sayles, G. You, M. Wang and M. J. Kupferle, *Environ. Sci. Technol.,* 1997, **31**, 3448.

5. G. R. Eykholt and D. T. Davenport, *Environ. Sci. Technol.,* 1998, **32**, 1482.

6. D. C. Bridges, in *Triazine Herbicides: Risk Assessment*, ed. L. G. Ballantine, J. E. McFarland and D. Hackett, *ACS Symp. Ser. 683*, American Chemical Society, Washington, DC, 1998, p. 24.

7. L. Wackett, Atrazine Pathway Map. *The University of Minnesota Biocatalysis/Biodegradation Database.*

8. S. J. Monson, L. Ma, D. A. Cassada and R. F. Spalding, *Anal. Chim. Acta,* 1998, **373**, 153.

9. T. Dombek, E. Dolan, J.Schultz and D. Klarup, *Environ. Pollut.,* 2001, **111**, 21.

10. E. J. Weber, *Environ. Sci. Technol.,* 1996, **30**, 716.

11. W.-H. Hung, J. Schwartz and S. L.Bernasek, *Surf. Sci.,* 1991, **248**, 332.

12. (a) L. J. Matheson and P. G. Tratnyek, *Environ. Sci. Technol.,* 1994, **28**, 2045; (b) T. L. Johnson, M. W. Scherer and P. G .Tratnyek, *Environ. Sci. Technol.,* 1996, **30**, 2634; (c) P. G. Tratnyek, T. L. Johnson, M. M. Scherer and G. R. Eykholt, *Ground Water Monit. Rem.,* 1997, **17**, 109; (d) P. G. Tratnyek and M. M. Scherer, *Proc. Natl. Conf. Environ. Eng.,* 1998,

110; (e) M. M. Scherer, B. A. Balko, D. A. Gallagher and P. G. Tratnyek, *Environ. Sci. Technol.,* 1998, **32**, 3026; (f) T. L Johnson, F. William, Y. A. Gorby and P. G. Tratnyek, *J. Contam. Hydrol.,* 1998, **29**, 379.

13. N. Hirasawa, Y. Takahashi, E. Fukuda, O .Sugimoto and K-i. Tanji, *Tetrahedron Lett.*, 2008, **49**, 1492.

14. A. Rabion, R. M. Buchanan, J.-L. Seris and R. H. Fish, *J. Mol. Catal. A: Chem.,* 1997, **116**, 43.

15. M. V. Kirillova, A. M. Kirillov and A. J. L. Pombeiro, *Adv. Synth. Catal.*, 2009, **351**, 2936.

16. A. Dhakshinamoorthy and K. Pitchumani, *Tetrahedron,* 2006, **62**, 9911.

17. (a) W. Nam, H. J. Han, S.-Y. Oh, Y. J. Lee, M.-H. Choi, S.-Y. Han, C. Kim, S. K. Woo and W. Shin, *J. Am. Chem. Soc.*, 2000, **122**, 8677 and references cited therein; (b) D. Mansuy, J. Leclaire, M. Fontecave and P. Dansette, *Tetrahedron,* 1984, **40**, 2847.

18. A. Tuynman, J. L. Spelberg, I. M. Kooter, H. E. Schoemaker and R. Wever, *J. Biol. Chem.*, 2000, **275**, 3025.

19. M. Matsushita, K. Kamata, K. Yamaguchi and N. Mizuno, *J. Am. Chem. Soc.,* 2005, **127**, 6632.

20. P. J. Wallis, K. J. Booth, A. F. Patti and J. L. Scott, *Green Chem.,* 2006, **8**, 333.

21. J. S. Kim, J. H. Jan, J. J. Lee, Y. M. Jun, B. M. Lee and B. H. Kim, *Tetrahedron Lett.,* 2008, **49**, 3733.

22. T. Ankner and G. Hilmersson, *Tetrahedron Lett.*, 2007, **48**, 5707.

23. C. Wang, J. Wan, Z. Zheng and Y. Pan, *Tetrahedron,* 2007, **63**, 5071.

24. (a) D. Parmar, L. A. Duffy, D. V. Sadasivam, H. Matsubara, P. A. Bradly, R. A. Flowers II and D. J. Procter, *J. Am. Chem. Soc.*, 2009, **131**, 15467; (b) G. Guazelli, S. De Grazia, K. D. Collins, H. Matsubara, M. Spain and D. J. Procter, *J. Am. Chem. Soc.*, 2008, **131**, 7214.

25. A. Dahlén and G. Hilmersson, *J. Am. Chem. Soc.*, 2005, **127**, 8340.

26. A. Dermican and P. J. Parsons, *Eur. J. Org. Chem.,* 2003, 1729.

27. (a) H. Miyabe, M. Ueda, K. Fujii, T.Goto and T. Naito, *J. Org. Chem.,* 2003, **68**, 5618; (b) H. Miyabe, K. Fujii, T. Goto and T. Naito, *Org. Lett.,* 2000, **2**, 4071; (c) H. Miyabe, K. Fujii, H. Tanaka and T. Naito, *Chem. Commun.,* 2001, 831.

28. M. Ueda, H. Miyabe, A. Nishimura, O. Miyata, Y. Takemoto and T. Naito, *Org. Lett.*, 2003, **5**, 3835.

29. W. F. Bailey and M. W. Carson, *J. Org. Chem.,* 1998, **63**, 361.

30. D. Sakuma and H. Togo, *Synlett,* 2004, 2501.

31. C. Petrier, C. Dupuy and J. L. Luche, *Tetrahedron Lett.*, 1986, **27**, 3149.

32. J. L. Luche and C. Allavena, *Tetrahedron Lett.*, 1988, **29**, 5369.

33. C. Dupuy, C. Petrier, L. A. Sarandeses and J. L. Luche, *Synth. Commun.,* 1991, **21**, 643.

34. J. L. Luche, C. Allavena, C. Petrier and C. Dupuy, *Tetrahedron Lett.,* 1988, **29**, 5373.

35. P. Blanchard, M. S. El Kortbi, J.-L. Fourrey and M. Robert-Gero, *Tetrahedron Lett.*, 1992, **33**, 3319.
36. S. Raussou, N. Urbain, P. Mangeney, A. Alexakis and N. Platzer, *Tetrahedron Lett.*, 1996, **37**, 1599.
37. P. Mangeney, L. Hamon, S.R aussou, N. Urbain and A. Alexakis, *Tetrahedron*, 1998, **54**, 10349.
38. J. A. Marshall amd E. D. Robinson, *J. Org. Chem.*, 1990, **55**, 3450.
39. J. A. Marshall and X. J. Wang, *J. Org. Chem.*, 1991, **56**, 960.
40. J. A. Marshall and C. A. Sehon, *J. Org. Chem.*, 1995, **60**, 5966.
41. J. A. Marshall and G. S. Bartley, *J. Org. Chem.*, 1994, **59**, 7169.
42. A. S. K. Hashmi, L. Schwarz and J. W. Bats, *J. Prakt. Chem.*, 2000, **342**, 40.
43. P. Panchaud and P. Renaud, *J. Org. Chem.*, 2004, **69**, 3205.
44. M. Sugi, D. Sakuma and H. Togo, *J. Org. Chem.*, 2003, **68**, 7629.
45. L. Cao and C. Li, *Tetrahedron Lett.*, 2008, **49**, 7380.
46. M. F. Hawthorne and M. Reintjes, *J. Am. Chem. Soc.*, 1965, **87**, 4585.
47. G. W. Kabalka, *J. Organomet. Chem.*, 1971, **33**, C25–C28
48. Y. Tsuchiya, Y. Izumisawa and H. Toho, *Tetrahedron*, 2009, **65**, 7533.
49. F. A. Khan, F. Satapathy, J. Dash and G. Savitha, *J. Org. Chem.*, 2004, **69,** 5295.
50. B. H. Kim, Y. Jim, Y. M. Jum, R. Han, W. Baik and B. M. Lee, *Tetrahedron Lett.*, 2000, **41**, 2137.
51. (a) G. A. Russell, M. Jawdosiuk and M. Makosza, *J. Am. Chem. Soc.*, 1979, **101**, 2355; (b) G. A. Russell and A. R. Metcalfe, *J. Am. Chem. Soc.*, 1979, **101**, 2359; (c) G. A. Russell and B. Mydryk, *J. Org. Chem.*, 1982, **47**, 1879; (d) G. A. Russell and W. Baik, *J. Chem. Soc., Chem. Commun.*, 1988, 196.
52. R. E. Estévez, J. L. Oller-López, R. Robles, C. R. Melgarejo, A. Gansäuer, J. M. Cuerva and J. E. Oltra, *Org. Lett.*, 2006, **8**, 5433.
53. A. Postigo, S. Kopsov, C. Ferreri and C. Chatgilialoglu, *Org. Lett.*, 2007, **9**, 5159.
54. S. Barata-Vallejo, N. Nudelman and A. Postigo, *Curr. Org. Chem.*, 2011, **15**, 1826.
55. (a) *Organic Radical Reactions in Water and Alternative Media*, ed. A. Postigo, Nova Science Publications, Happaauge, New York, 2011; (b) X. X. Rong, H.-Q. Pan, W. R. Dolbier, Jr. *J. Am. Chem. Soc.*, 1994, **116**, 4521.
56. J. Calandra, A. Postigo, D. Russo, N. Sbarbati-Nudelman and J. J. Tereñas, *J. Phys. Org. Chem.*, 2010, **23**, 944.
57. S. Barata-Vallejo and A. Postigo, *J. Org. Chem.*, 2010, **75**, 6141.
58. C.-J. Li, *Tetrahedron*, 1996, **52**, 5643.
59. L. W. Bieber, I. Malvestiti and E. C. Storch, *J. Org. Chem.*, 1997, **62**, 9061.
60. T. H. Chan, C.-J. Li and Z. Y. Wei, *Can. J. Chem.*, 1994, **72**, 1181.
61. C. C. K. Keh, C. Wei and C.-J. Li, *J. Am. Chem. Soc.*, 2003, **125,** 4062.

62. (a) S. Arimitsu, J. M. Jacobsen and G. B. Hammond, *Tetrahedron Lett.*, 2007, **48**, 1625; (b) S. Arimitsu and G. B. Hammod, *J. Org. Chem.*, 2006, **71**, 8665.

63. R. Cannella, A. Clerici, N. Pastori, E. Regolini and O. Porta, *Org. Lett.*, 2005, **7**, 645.

64. L. W. Bieber, E. C. Storch, I. Malvestiti and M. F. da Silva, *Tetrahedron Lett.*, 1998, **39**, 9393.

65. Z.-L. Shen and T.-P. Loh, *Org. Lett.*, 2007, **9**, 5413.

66. G. K. Friestad, *Tetrahedron,* 2001, **57**, 5461.

67. (a) H. Miyabe, M. Ueda and T. Naito, *J. Org. Chem.*, 2000, **65**, 5043; (b) M. Miyabe, M. Ueda, A. Nishimura and T. Naito, *Tetrahedron,* 2004, **60**, 4227.

68. H. Miyabe, M. Ueda, A. Nishimura and T. Naito, *Org. Lett.*, 2002, **4**, 131.

69. Y.-S. Yang, Z.-L. Shen and T.-P. Loh, *Org. Lett.*, 2009, **11**, 1209.

70. M. Ueda, *Yakagaku Zasshi*, 2004, **124**, 311.

71. G. S. C. Srikanth and S. L. Castle, *Tetrahedron,* 2005, **61**, 10377; D. O. Jang, D. H. Cho and C.-M. Chung, *Synlett,* 2001, 1923.

72. D. O. Jang and D. H. Cho, *Synlett,* 2002, 1523.

73. D. H. Cho and D. O. Jang, *Tetrahedron Lett.,* 2005, **46**, 1799.

74. J. L. Mascareñas, J. Perez Sestelo, L. Castedo and A. Mouriño, *Tetrahedron Lett.*, 1991, **32**, 2813.

75. J. Perez Sestelo, J. L. Mascareñas, L. Castedo and A. Mouriño, *J. Org. Chem.*, 1993, **58**, 118.

76. J. Perez Sestelo, J. L. Mascareñas, L. Castedo and A. Mouriño, *Tetrahedron Lett.,* 1994, **35**, 275.

77. R. M. Suarez, J. Perez Sestelo and L. A. Sarandeses, *Synlett,* 2002, 1435.

78. R. M. Suarez, J. Perez Sestelo and L. A. Sarandeses, *Chem.–Eur. J.*, 2003, **9**, 4179.

79. R. M. Suarez, J. Perez Sestelo and L. A. Sarandeses, *Org. Biomol. Chem.*, 2004, **2**, 3584.

80. J. Perez Sestelo, I. Cornella, O. de Uña, A. Mouriño and L. A. Sarandeses, *Chem.–Eur. J.,* 2002, **8**, 2747.

81. J. Perez Sestelo, O. de Uña, A. Mouriño and L. A. Sarandeses, *Synlett,* 2002, 719.

82. I. Cornella, R. M. Suarez, A. Mouriño, J. Perez Sestelo and L. A. Sarandeses, *J. Steroid Biochem. Mol. Biol.*, 2004, 89.

83. I. Cornella, J. Perez Sestelo, A. Mouriño and L. A. Sarandeses, *J. Org. Chem.*, 2002, **67**, 4707.

84. K. Miura, M. Tomita, J. Ichikawa and A. Hosomi, *Org. Lett.,* 2008, **10**, 133–136.

85. B. Movassagh and M. Navidi, *Tetrahedron Lett.*, 2008, **49**, 6712.

86. C. J. Li, Y. Meng and X. H. Yi, *J. Org. Chem.*, 1998, **63**, 7498.

87. (a) C. Petrier and J.-L. Luche, *J. Org. Chem.*, 1985, **50**, 910; (b) C. J. Li and T. H. Chan, *Organometallics,* 1991, **10**, 2548.

88. M. Wada, H. Ohki and K.-Y. Akiba, *J. Chem. Soc., Chem. Commun.*, 1987, 708.

89. B. F. Bonini, M. Comes-Franchini, M. Fochi, G. Mazzanti, C. Nanni and A. Ricci, *Tetrahedron Lett.*, 1998, **39**, 6737.

90. T. Fukuma, S. Lock, N. Miyoshi and M. Wada, *Chem. Lett.*, 2002, 376.

91. L.-H. Li and T. H. Chan, *Tetrahedron Lett.,* 2000, **41**, 5009.

92. J. Y. Zhou, Y. Jia, G. F. Sun and S. H. Wu, *Synth. Commun.*, 1997, **27**, 1899.

93. T. H. Chan and Y. Yang, *Tetrahedron Lett.*, 1999, **40**, 3863.

94. P. Cintas, *Synlett,* 1995, 1087.

95. W.-C. Zhang and C.-J. Li, *J. Org. Chem.*, 1999, **64**, 3230.

96. (a) L. M. Lawrence amd G. M. Whitesides, *J. Am. Chem. Soc.*, 1980, **102**, 2493; (b) H. M. Walborsky and C. Zimmermann, *J. Am. Chem. Soc.* 1992, **114**, 4996.

97. E. R. Alexander, *Principles of Ionic Organic Reactions*, John Wiley & Sons, New York, 1950, p. 188.

98. J. Hygum Dam, P. Fristrup and R. Madsen, *J. Org. Chem.*, 2008, **73**, 3228.

99. C.-J. Li, Y. Meng and X.-H. Yi, *J. Org. Chem.*, 1997, **62**, 8632.

100. T. H. Chan and B. M. Isaac, *Pure Appl. Chem.*, 1996, **68**, 919.

101. J. Nokami, J. Otera, T. Sudo and R. Okawara, *Organometallics,* 1983, **2**, 191.

102. J. Nokami, S. Wakabayashi and R. Okawara, *Chem. Lett.,* 1984, 869.

103. S. H. Wu, B. Z. Huang, T. M. Zhu, D. Z. Yiao and Y. L. Chu, *Acta Chim. Sin.,* 1990, **48**, 372.

104. C. Einhorn and J. L. Luche, *J. Organomet. Chem.*, 1987, **322,** 177.

105. See: (a) M. Pereyre, J.-P. Quintard and A. Rahm, *Tin in Organic Synthesis*, Butterworth, London, 1987; (b) P. J. Smith, *Toxicological Data on Organotin Compounds*, International Tin Research Institute, London, 1978.

106. T. H. Chan, C. J. Li and Z. Y. Wei, *J. Chem. Soc., Chem. Commun.,* 1990, 505.

107. (a) E. Kim, D. M. Gordon, W. Schmid and G. M. Whitesides, *J. Org. Chem.,* 1993, **58,** 5500; (b) A. Kundu, S. Prabhakar, M. Vairamani and S. Roy, *Organometallics,* 1997, **16,** 4796.

108. Diallyltin dibromide has been found to be an efficient allylation reagent of carbonyl compounds in organic solvents. See: S. Kobayashi and K. Nishio, *Tetrahedron Lett.,* 1995, **36**, 6729, and references cited therein.

109. In ref. 101, Nokami *et al.* showed that diallyltin dibromide (**4**) reacted with benzaldehyde in an ether–water mixed solvent, but the intermediacy of **4** in the aqueous Barbier reaction was assumed but not demonstrated.

110. T. H. Chan, Y. Yang and C. J. Li, *J. Org. Chem.*, 1999, **64**, 4452.

111. Y.-J. Bian, W.-L. Xue and X.-G. Yu, *Ultrason. Sonochem.,* 2010, **17**, 580.

112. T. H. Chan and Y. Yang, *J. Am. Chem. Soc.*, 1999, **121**, 3228.

113. W. J. Chung, S. Higashiya, Y. Oba and J. T. Welch, *Tetrahedron,* 2003, **59**, 10031.

114. (a) J. Podlech and T. C. Maier, *Synthesis,* 2003, 633; (b) L. Paquette, *Synthesis,* 2003, 765.

115. (a) Z.-L. Shen, H.-L. Cheong and T.-P. Loh, *Tetrahedron Lett.,* 2009, **50**, 1051; (b) L. Paquette and T. M. Mitzel, *J. Am. Chem. Soc.,* 1996, **118**, 1931.

116. S. Araki, S.-J. Jin, Y. Idou and Y. Butsugan, *Bull. Chem. Soc. Jpn.,* 1992, **65**, 1736.

117. M. T. Reetz and A. Jung, *J. Am. Chem. Soc.,* 1983, **105**, 4833.

118. (a) K. Narasaka and H. C. Pai, *Chem. Lett.,* 1980, 1415; (b) K. Narasaka and H. C. Pai, *Tetrahedron,* 1984, **12**, 2233; (c) D. A. Evans, K. T. Chapman and E. M. Carreira, *J. Am. Chem. Soc.,* 1988, **110**, 3560.

119. (a) X. Chen, E. R. Hortelano, E. L. Eliel and S. V. Frye, *J. Am. Chem. Soc.,* 1990, **112**, 6130; X. Chen, E. R. Hortelano, E. L. Eliel and S. V. Frye, *J. Am. Chem. Soc.,* 1992, **114**, 1778; (b) S. V. Frye, E. L. Eliel and R. Cloux, *J. Am. Chem. Soc.,* 1987, **109**, 1862.

120. R. C. Corcoran and J. Ma, *J. Am. Chem. Soc.,* 1992, **114**, 4536.

121. S. Mori, M. Nakamura, E. Nakamura, N. Koga and K. Morukuma, *J. Am. Chem. Soc.,* 1995, **117**, 5055.

122. (a) T. Poll, J. O. Metter and G. Helmchen, *Angew. Chem., Int. Ed. Engl.,* 1985, **24**, 112; (b) W. E. Buhro, S. Georgiou, J. M. Fernández, A. T. Patton, C. E. Strouse and J. A. Gladysz, *Organometallics,* 1986, **5**, 956–965; (c) M. Arai, T. Kawasuji and E. Nakamura, *J. Org. Chem.,* 1993, **58**, 5121.

123. (a) C. Petrier and J. L. Luche, *J. Org. Chem.,* 1985, **50**, 910; (b) C. Einhorn and J. L. Luche, *J. Organomet. Chem.,* 1987, **322**, 177.

124. S. R. Wilson and M. E. Guazzaroni, *J. Org. Chem.,* 1989, **54**, 3087.

125. T.-H. Chan, C. J. Li, M. C. Lee and Z. Y. Wei, *Can. J. Chem.,* 1994, **72**, 1181.

126. C. Petrier and J. L. Luche, *J. Org. Chem.,* 1985, **50**, 910.

127. W. Smadja, *Synlett,* 1994, 1.

128. S. Araki, H. Ito and Y. Butsugan, *Synth. Commun.,* 1988, **18**, 453.

129. J. A. Marshall and K. W. Hinkle, *J. Org. Chem.,* 1995, **60**, 1920.

130. W. Lu, J. Ma, Y. Yang and T. H. Chan, *Org. Lett.,* 2000, **2**, 3469.

131. X.-H. Yi, Y. Meng and C.-J. Li, *Tetrahedron Lett.,* 1997, **38**, 4731.

132. S.-i. Usugi, H. Yorimitsu and K. Oshima, *Tetrahedron Lett.,* 2001, **42**, 4535.

133. T.-H. Chan and C.-J. Li, *Organometallics,* 1990, **9**, 2649.

134. S. Araki, T. Kamei, T. Hirashita, H. Yamamura and A. Kawai, *Org. Lett.,* 2000, **2**, 847.

135. H. Miyabe, Y. Yamaoka, T. Naito and Y. Takemoto, *J. Org. Chem.,* 2003, **68**, 6745.

136. H. Miyabe, Y. Yamoaka, T. Naito and Y. Takemoto, *J. Org. Chem.,* 2004, **69**, 1415.

137. (a) B. Alcaide, P. Almendros, C. Aragoncillo, M. C. Redondo and M. R. Torres, *Chem.–Eur. J.*, 2006, **12**, 1539; (b) B. Alcaide, P. Almendros, C. Aragoncillo and M. C. Redondo, *J. Org. Chem.*, 2007, **72**, 1604.

138. A. D. Da Silva, E. J. Maria, P. Blanchard, J.-L. Fourrey and M. Robert-Gero, *Nucleosides Nucleotides,* 1998, **17,** 2175.

139. E. J. Maria, A. D. Da Silva and J.-L. Fourrey, *Eur. J. Org. Chem.*, 2000, 627.

140. P. Blanchard, A. D. Da Silva, M. S. El Kortbi, J.-L. Fourrey and M. Robert-Gero, *J. Org. Chem.*, 1993, **58,** 6517.

141. D. Crich, J. W. Davies, G. Negrón and L. Quintero, *J. Chem. Res., Synop.*, 1988, 140.

142. D. Crich and J. W. Davies, *Tetrahedron,* 1989, **45**, 5641.

143. A.-M. Yim, Y. Vidal, P. Viallefont and J. Martinez, *Tetrahedron Lett.,* 1999, **40**, 4535.

144. Z.-L. Shen, Y.-L. Yeo and T.-P. Loh, *J. Org. Chem.*, 2008, **73**, 3922.

145. J. J. Gajewski, W. Bocian, N. L. Brichford and J. L. Henderson, *J. Org. Chem.*, 2002, **67**, 4236.

146. C. J. Li and T. H. Chan, *Tetrahedron,* 1999, **55**, 11149.

147. W. Lu and T. H. Chan, *J. Org. Chem.*, 2001, **66**, 3467.

148. F. F. Fleming and S. Gudipati, *Org. Lett.*, 2006, **8,** 1557.

149. S. Kang, T. S. Jang, G. Keum, S. B. Kang, H. So-Yeop and Y. Kim, *Org. Lett.*, 2000, **2**, 3615.

150. M. Mahesh, J. A. Murphy and H. P. Wessel, Novel deoxygenation reactions of epoxides by indium, *J. Org. Chem.,* 2005, **70**, 4118–4123.

151. L. W. Bieber and M. F. da Silva, *Tetrahedron Lett.*, 2004, **45**, 8281.

152. I. H. S. Estevam and L. W. Bieber, *Tetrahedron Lett.*, 2003, **44**, 667.

153. A. Clerici and O. Porta, *Gazz. Chim. Ital.*, 1992, **122**, 165.

154. R. Fittig, *Liebigs Ann.,* 1859, **110**, 23.

155. (a) S. Bhar and S. Guha, *Tetrahedron Lett.*, 2004, **45**, 3775–3777; (b) L. H. Li and T. H. Chan, *Org. Lett.* 2000, **2**, 1129; (c) D. A. Sahade, S. Mataka, T. Sawada, T. Tsukinoki and M. Tashiro, *Tetrahedron Lett.,* 1997, **38**, 3745; (d) J. M. Khurana and A. Sehgal, *J. Chem. Soc., Chem. Commun.*, 1994, 571.

156. (a) T. Ueda, N. Kanomata and H. Machida, *Org. Lett.*, 2005, **7**, 2365; (b) H. C. Aspinall, N. Greeves and C. Valla, *Org. Lett.*, 2005, **7**, 1919; (c) S. Matsukawa and Y. Hinakubo, *Org. Lett.,* 2003, **5**, 1221; (d) L. Wang and Y. M. Zhang, *Tetrahedron Lett.*, 1998, **39**, 5257; (e) G. A. Molander and C. Kenny, *J. Am. Chem. Soc.*, 1989, **111**, 8236; (f) J. L. Namy, J. Souppe and H. B. Kagan, *Tetrahedron Lett.,* 1983, **24,** 765.

157. (a) X. L. Xu and T. Hirao, *J. Org. Chem.*, 2005, **70**, 8594; (b) T. Hirao, H. Takeuchi, A. Ogawa and H. Sakurai, *Synlett,* 2000, 1658; (c) T. Hirao, B. Hatano, Y. Imamoto and A. Ogawa, *J. Org. Chem.,* 1999, **64**, 7665; (d) B. Kammermeier, G. Beck, H. Jendralla and D. Jacobi, *Angew. Chem., Int. Ed. Engl.*, 1994, **33**, 685; (e) R. Annunziata, M. Benaglia, M. Cinquini, F. Cozzi and P. Giaroni, *J. Org. Chem.,* 1992, **57**, 782; (f)

D. J. Kempf, T. J. Sowin, E. M. Doherty, S. M. Hannick, L. Codavoci, R. F. Henry, B. E. Green, S. G. Spanton and D. W. Norbeck, *J. Org. Chem.*, 1992, **57**, 5692; (g) J. H. Freudenberger, A. W. Konradi and S. F. Pederson, *J. Am. Chem. Soc.*, 1989, **111**, 8014.

158. S. T. Handy and D. A. Omune, *Org. Lett.*, 2005, **7**, 1553.

159. (a) T. Mukaiyama, N. Yoshimura, K. Igarashi and A. Kagayama, *Tetrahedron,* 2001, **57**, 2499; (b) T.-Y. Li, W. Cui, J. G. Liu and J. Z. Zhao, *Chem. Commun.*, 2000, 139; (c) T. Tsukinoki, T. Awaji, I. Hashimoto, S. Mataka and M. Tashiro, *Chem. Lett.,* 1997, 235; (d) K. Tanaka, S. Kishigami and F. Toda, *J. Org. Chem.*, 1990, **55**, 2981.

160. K. Takai, R. Morita, H. Matsushita and C. Toratsu, *Chirality,* 2003, **15**, 17.

161. S. H. David and C. F. Gregory, *J. Am. Chem. Soc.,* 1995, **117**, 7283.

162. Y. Yamamoto, R. Hattori and K. Itoh, *Chem. Commun.*, 1999, 825.

163. (a) U. Groth and M. Jeske, *Synlett,* 2001, 129; (b) U. Groth and M. Jeske, *Angew. Chem., Int. Ed.,* 2000, **39**, 574.

164. R. H. Khan, R. K. Mathur and A. C. Ghosh, *Synth. Commun.*, 1997, **27**, 2193.

165. M. Ephritikhine, O. Maury, C. Villiers, M. Lance and M. Nierlich, *J. Chem. Soc., Dalton Trans.,* 1998, 3021.

166. A. Svatos and W. Boland, *Synlett,* 1998, 126.

167. Z. Y. Wang, S. Z. Yuan, Z. G. Zha and Z. D. Zhang, *Chin. J. Chem.*, 2003, **21**, 1231.

168. (a) T. Ohe, T. Ohse, K. Mori, S. Ohtaka and S. Uemura, *Bull. Chem. Soc. Jpn.*, 2003, **76**, 1823; (b) V. Nair, S. Ros, C. N. Jayan and N. P. Rath, *Tetrahedron Lett.*, 2002, **43**, 8967–8969; (c) K. Mori, S. Ohtaka and S. Uemura, *Bull. Chem. Soc. Jpn.*, 2001, **74**, 1497; (d) H. J. Lim, G. Keum, S. B. Kang, B. Y. Chung and Y. Kim, *Tetrahedron Lett.*, 1998, **39**, 4367.

169. C. Wang, Y. Pan and A. Wu, *Tetrahedron,* 2007, **63**, 429.

170. C. L. Keller, J. D. Dalessandro, R. P. Holtz and A. R. Pinhas, *J. Org. Chem.*, 2008, **73**, 3616.

171. C. Walling, *Acc. Chem. Res.,* 1998, **31**, 155.

172. (a) P. A. MacFaul, D. D. M. Wayner and K. U. Ingold, *Acc. Chem. Res.,* 1998, **31**, 159; (b) S. Goldstein and D. Meyerstein, *Acc. Chem. Res.,* 1999, **32**, 547.

173. D. T. Sawyer, A. Sobkowiak and T. Matsushita, *Acc. Chem. Res.,* 1996, **29**, 409.

174. (a) C. D. D. Hoffman, E. L. Jenner and R. D. Lipscomb, *J. Am. Chem. Soc.,* 1958, **80**, 2864; also see (b) C. Walling and G. M. El-Taliawi, *J. Am. Chem. Soc.,* 1973, **95**, 844.

175. Attempted radical-coupling reactions of phenylacetonitrile were unsuccessful due to its insolubility in water.

176. Y.-S. Yang, Z.-L. Shen and T.-P. Loh, *Org. Lett.*, 2009, **11**, 2213.

177. N. Kalyanam and V. Rao, *Tetrahedron Lett.*, 1993, **34**, 1647.

178. V. Nair, S. Ros, C. N. Jayan and N. P. Rath, *Tetrahedron Lett.*, 2002, **43**, 8967.
179. R. L. Halterman, J. P. Porterfield, S. Mekala, *Tetrahedron Lett.*, 2009, **50**, 7172.
180. Y. Jiang, C. Xi and X. Yang, *Synlett*, 2005, 1381.
181. C. Xi, Y. Jiang and X. Yang, *Tetrahedron Lett.*, 2005, **46**, 3909.

CHAPTER 11

Radicals and Transition-metal Catalysis: a Complementary Solution to Increase Reactivity and Selectivity in Organic Chemistry*

11.1 INTRODUCTION

The perception of radical reactions being wild and hard to control is linked to the short lifetime of radicals and the widely varying kinetics of the individual radical reaction steps.[1] Thanks to the ground-breaking studies by Ingold, Beckwith, Fischer, and Newcomb, the kinetics of radical reactions are known today.[2,3] Free radical chemistry represents an attractive alternative to its ionic counterpart with several advantages including high functional group tolerance and the use of mild reaction conditions.[4] Another advantage of applying free radicals is their central position among reactive intermediates, since they can be easily reduced to carbanions or oxidized to carbocations. A critical point remains, however, that radical reactions require a stoichiometric amount of a chain carrier, oxidant, or reductant. Transition-metal catalysis offers an excellent and complementary potential to solve synthetic problems.[5] As the fundamental reactivity patterns are known, reactivity can be tuned by the proper choice of substrate, metals, and ligands utilized. Cross-coupling reactions, in particular, enjoy a wide popularity, although optimization can be quite cumbersome. A major disadvantage of a number of coupling reactions is that they are often sluggish and thus require quite harsh conditions to accomplish the transformation. Recently, a novel strategy has emerged that combines the advantages of transition-metal catalysis and free radicals in organic chemistry, and it is proving to be very useful for the development of new efficient synthetic methodology. Although radicals were recognized early

* Chapter written by V. Tamara Perchyonok.
Streamlining Free Radical Green Chemistry
V. Tamara Perchyonok, Ioannis Lykakis and Al Postigo
© V. Tamara Perchyonok, Ioannis Lykakis and Al Postigo 2012
Published by the Royal Society of Chemistry, www.rsc.org

on to be involved in various transition-metal-catalyzed processes including palladium-catalyzed reactions,[6] Kharasch-type reactions,[7] and several nickel-catalyzed cross-coupling reactions,[8] an initial study in merging transition-metal catalysis with radical chemistry was first published in 2002 by Ryu and co-workers (Scheme 11.1).[9]

They reported a photolytic palladium-catalyzed cascade starting from homoallyl halides **1** and leading to diverse cyclopentanone derivatives **2** in good yields. The proposed mechanism involves the photolytic formation of a homoallyl radical **3** and a Pd^{II} species upon irradiation of the homoallyl halide in the presence of the Pd^0 catalyst. Radical **3** undergoes addition to CO to generate the acyl radical intermediate **4**. 5-*exo* cyclization followed by further carbonylation leads to radical **5**, which can couple with the Pd^{II} species to give the acylpalladium intermediate **6**. In the last step the palladium unit is displaced in the presence of a nucleophile to generate the final product. Pd^I complexes, though currently not very common, are isolable dimeric species, which have been used in Buchwald–Hartwig aminations[10] and enolate aryla-tions;[11] however, their role is most likely that of a precatalyst. Hor and coworkers summarized the use of Pd^I catalysts in Suzuki–Miyaura couplings

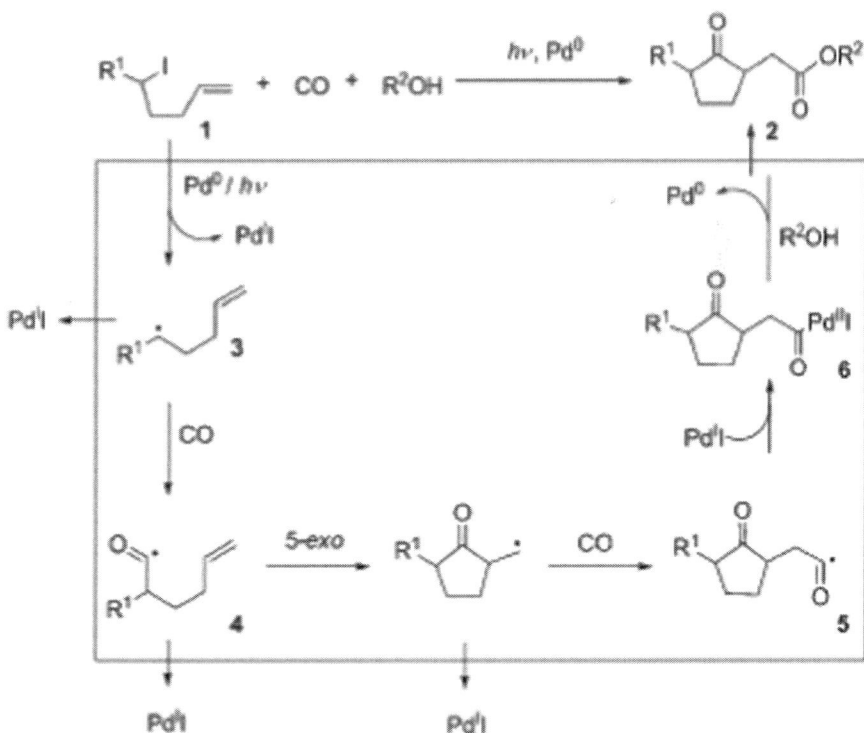

Scheme 11.1 Photolytic Pd-catalyzed radical carbonylation/ cyclization/ carbonyla-tion cascade. The arrows to the outside Pd(II) indicate that reversible combination may occur, thus modulating the lifetime of the radicals.

and mentioned a potential Pd^I/Pd^{III} manifold and thus a radical pathway in these reactions.[12]

Manolikakes and Knochel provided the first evidence for a Pd^I/Pd^{III} catalytic cycle involving a radical chain reaction in Kumada cross-coupling reactions of aryl Grignard reagents with aryl bromides (Scheme 11.2).[13]

The coupling was slow when the Grignard reagent was generated classically by the oxidative addition of magnesium. The reaction was significantly more facile and occurred at room temperature in only a few minutes when the aryl Grignard reagent was generated by iodine–magnesium exchange of aryl iodides with the iPrMgCl/LiCl complex and then added to a mixture of the aryl bromide and catalytic Pd(OAc)$_2$ and S-Phos (2-dicyclohexylphosphanyl-2′,6′-dimethoxybiphenyl) or PEPPSI ([1,3-bis(2,6-diisopropylphenyl)imidazol-2-ylidene](3-chloropyridyl) palladium(II) dichloride). This acceleration can be traced to the isopropyl iodide generated along with the Grignard reagent. Deliberate addition of other alkyl iodides displayed the same accelerating effect. This procedure was used to couple a wide variety of functionalized aryl and heteroaryl

Grignard compounds with various aryl bromides in excellent yields after only 5 min. Even the highly unstable ester-substituted organomagnesium compound **7** could be coupled with bromobenzenes **8** and **9** to give biphenyls **10** and **11** in 82 and 84% yield, respectively. Significantly, the cyclized product **13** was isolated from the coupling with aryl bromide **12**, whereas no cyclized product was observed with the *ortho*-alkenyl Grignard reagent **14** (Scheme 11.3).

Based on these observations, a radical catalysis mechanism involving a Pd^I/Pd^{III} system was proposed (Scheme 11.4). The initiation step is reaction of the Pd^0 catalyst **15** with the alkyl iodide (RI) to give an alkyl radical (RC) and the Pd^I species **16**. The latter abstracts bromine from aryl bromide (Ar$_1$Br) to release an aryl radical (Ar$_1$C), which is trapped by palladium(II) halide **17** to

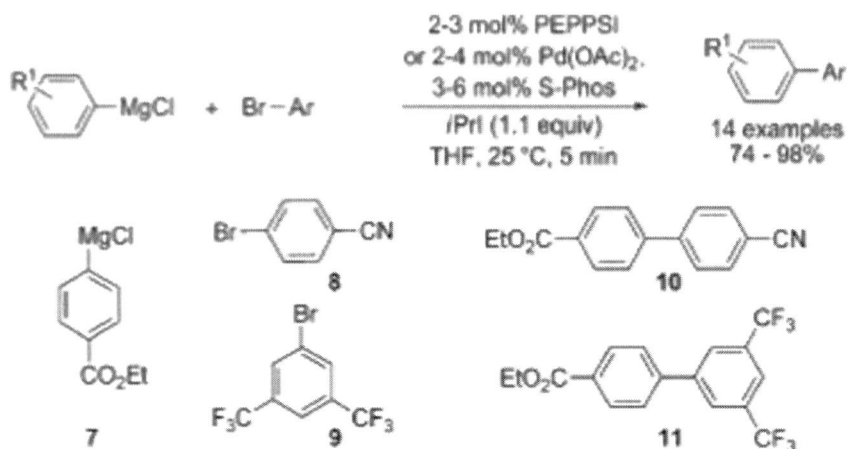

Scheme 11.2   Scope of the radical-catalyzed Kumada coupling.

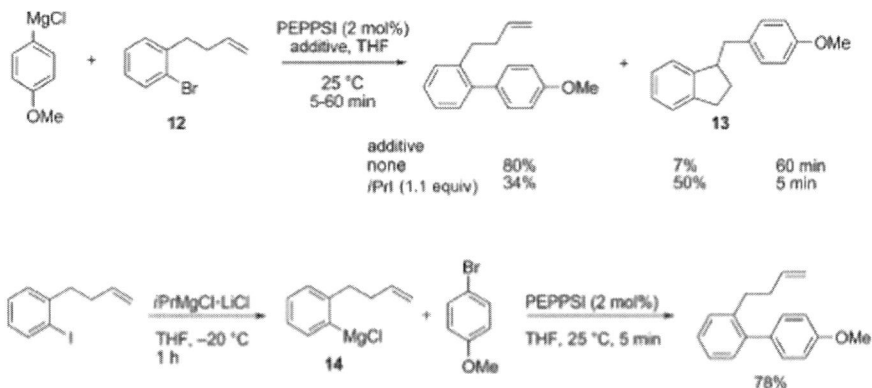

Scheme 11.3 Mechanistic experiments indicating radical catalysis.

give the Pd^{III} complex **18**, which undergoes transmetalation with the aryl Grignard reagent (Ar$_2$MgX) affording the diarylpalladium(III) halide **19**. Reductive elimination of **19** generates the cross-coupled product (Ar$_1$–Ar$_2$) and regenerates the LPd^{I}X radical chain carrier **20**.

The effect of radical catalysis in the Kumada reaction is highly beneficial from several points of view. The slow transition-metal-catalyzed reaction is

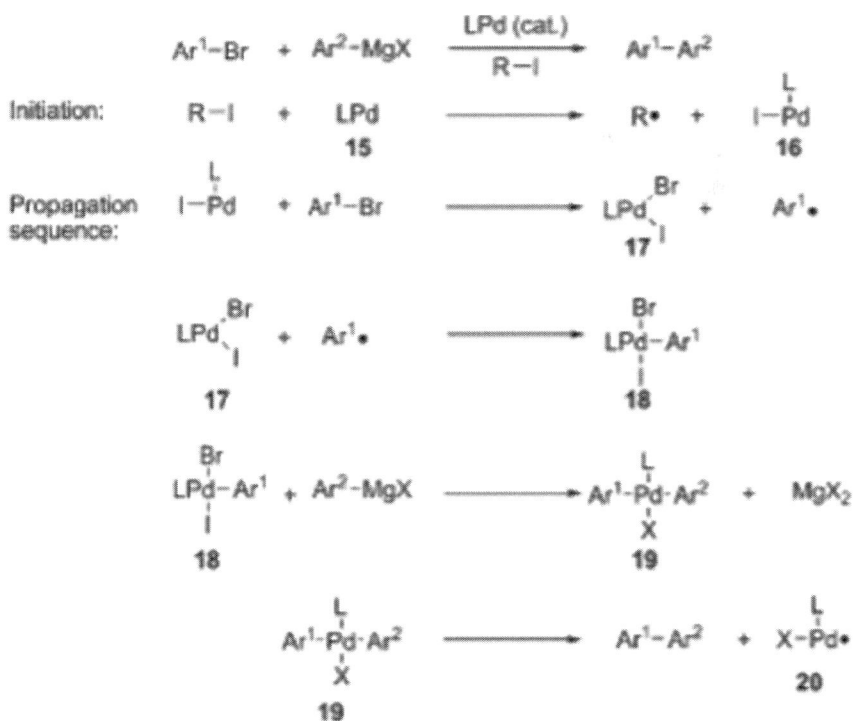

Scheme 11.4 Chain reaction of radical-catalyzed Kumada couplings.

accelerated dramatically, which allows the reaction to be conducted under very mild conditions and enhances the functional group tolerance of the organo-metallic reaction considerably. This obviates the need for transmetalation of the Grignard reagents to zinc or boron intermediates, which greatly improves the atom economy of the process. In addition, the ability of palladium (in its various oxidation states) to mediate the radical chain reaction process of the highly reactive aryl radicals is quite efficient, resulting in high yields of the cross-coupled products.[14]

Palladium is not the only metal suitable for radical catalysis. Recent studies highlight the versatility of other metal complexes and illustrate how textbook reactions, which historically often gave poor results under polar reaction conditions, can be vastly improved under free radical conditions. Oshima and coworkers have shown that a cobalt(II) salt, in combination with an N-heterocyclic carbene ligand (IMes·HCl), mediates the regioselective dehydrohalogenation of alkyl halides by dimethylphenylsilylmethylmagnesium chloride effectively and under mild conditions to give terminal alkenes almost exclusively (Scheme 11.5).[15]

In contrast, the corresponding ionic elimination produces a mixture of regio- and stereoisomers. A plausible mechanism for the reaction involves electron transfer from cobalt complex 21 to the alkyl halide, resulting in the formation of an alkyl radical. The Grignard reagent 23, which functions as a hydrogen

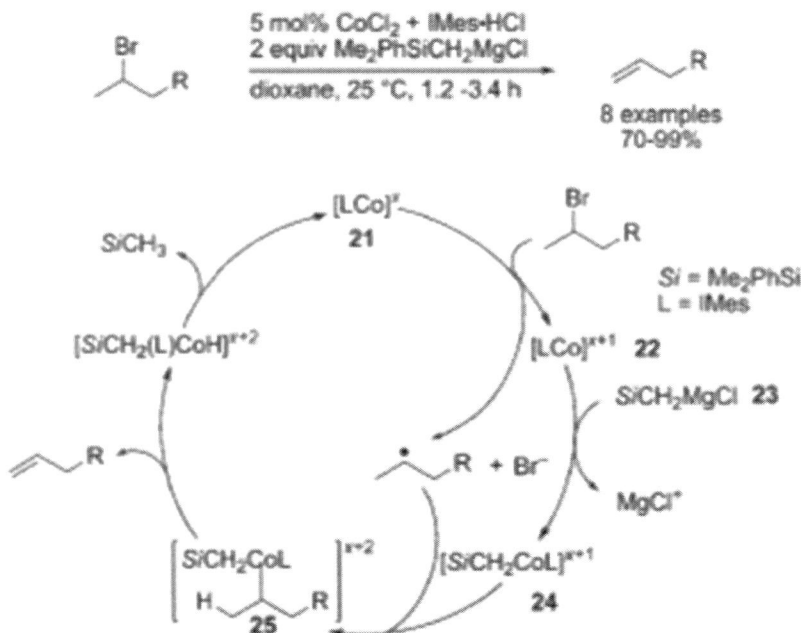

Scheme 11.5 Cobalt-catalyzed synthesis of terminal alkenes by radical-catalyzed dehydrobromination.

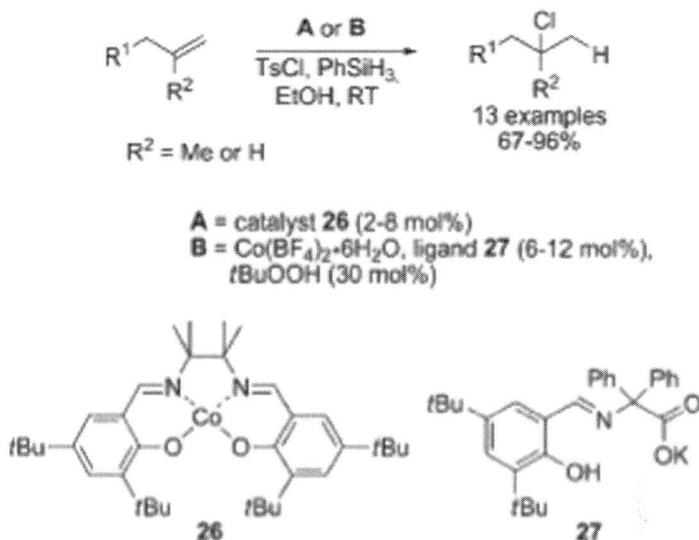

Scheme 11.6 Cobalt-catalyzed Markovnikov hydrochlorination of olefins.

acceptor, is transmetalated by cobalt complex **22** to give the intermediate **24**. The capture of the alkyl radical by cobalt complex **24** generates alkylcobalt complex **25**, which undergoes β-hydride elimination through a synperiplanar conformation to afford the 1-alkene selectively.

The reverse reaction, the polar Markovnikov addition of HCl to olefins is rarely utilized synthetically, despite the usefulness of alkyl chlorides. Carreira, Gaspar and coworkers reported a cobalt-catalyzed addition that is thought to involve radical intermediates (Scheme 11.6).[16]

Recently Gansäuer and co-workers developed a method for the catalytic reductive ring opening of epoxides (Scheme 11.7).[17] A synergistic system was used consisting of catalytic [Cp$_2$TiCl$_2$] to mediate the radical ring opening of the epoxide along with a rhodium hydride, derived from Wilkinson's catalyst in an atmosphere of H$_2$, to promote the hydrogen-atom transfer. In this way variously substituted epoxides can be cleaved to alcohols at ambient temperature in good yields. The reaction produces the less-substituted alcohol since the epoxide opening generates the most stable radical. The versatility of transition metals in radical chemistry is demonstrated with this catalytic system. Hydrogen-atom transfer from transition-metal hydrides represents an exciting alternative to classical hydrogen atom donors in radical chemistry. The low M–H bond strength (bond dissociation energy (BDE) of Rh^{III}–H is about 58 kcal mol^{-1})[18] indicates that such transfers could occur with rate constants of up to 10^9 m^{-1} s^{-1}.[19]

The method has been extended to the enantioselective opening of meso-epoxides with Ti complex **28**.

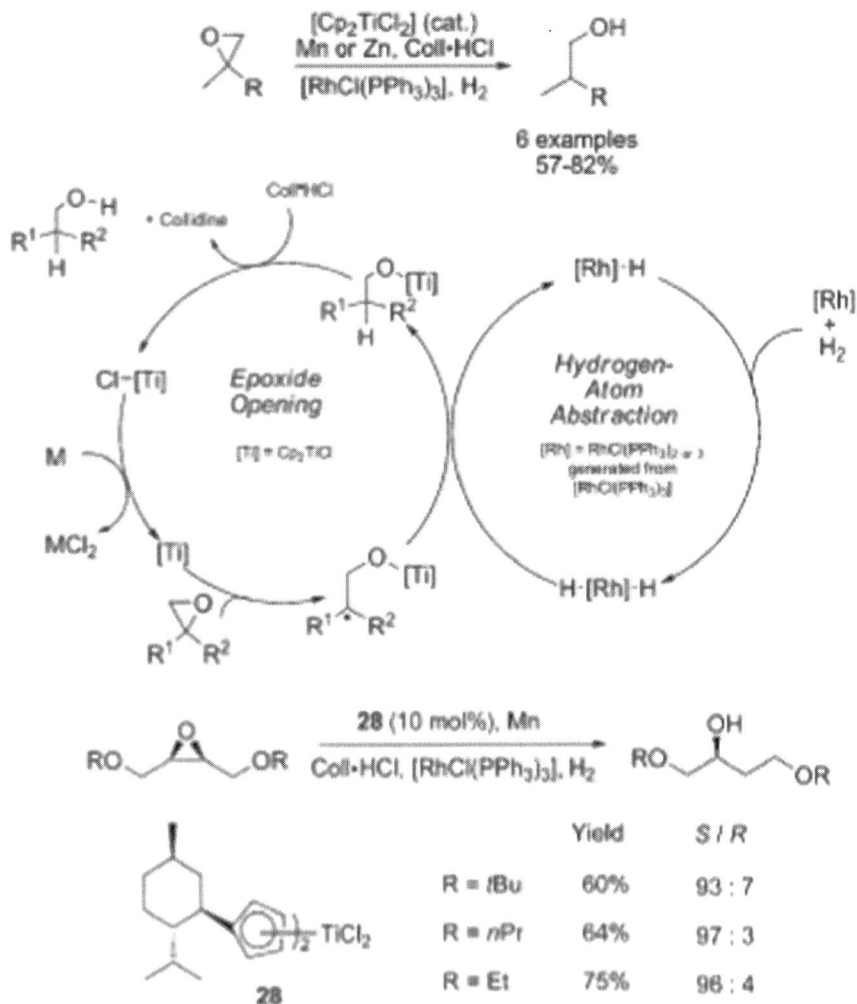

Scheme 11.7 Radical catalysis with a bi-metallic system for the ring opening of epoxides.

11.2 RADICAL CYCLIZATIONS TERMINATED BY IR-CATALYZED HYDROGEN-ATOM TRANSFER

Radical cyclizations are highly useful for the synthesis of complex molecular architectures due to their high selectivity and compatibility with densely functionalized substrates.[1] For synthetic applications, a catalytic, sustainable methodology without stoichiometric amounts of toxic and expensive substances, especially hydrogen-atom donors, is highly desirable. To this end, H_2O,[2] alcohols,[3] and H_2 are especially attractive.[4] A first example of a H_2-mediated, Cr-catalyzed radical cyclization was recently reported by Norton.[5] Unfortunately, the remarkable threefold task of the catalysts H_2 activation, radical generation, and radical reduction *via* hydrogen-atom transfer (HAT) limits the substrate scope of the reaction.

A conceptually different and unprecedented approach to such cyclizations is the coupling of catalytic cycles for radical generation and for hydrogen activation and HAT. In that manner, more general reactions become available. However, potential pitfalls for such methodology are numerous. Most notably, the kinetics of radical generation, cyclization, H_2 activation, and HAT have to be precisely adjusted to preclude undesired side reactions such as hydrogenation of radical acceptors or reduction of radical intermediates by a HAT before cyclization (Scheme 11.8).

Gansäuer *et al.* chose Vaska's complex, $IrCl(CO)(PPh_3)_2$ (**1**), as HAT catalyst to address these issues.[6] It forms a stable product of oxidative addition to H_2 without free coordination sites and hence displays low activity in the hydrogenation reactions of radical acceptors, especially alkynes. Since premature reduction of intermediate radicals must be avoided, the steric shielding of the hydrido ligands in $[IrH_2Cl(CO)(PPh_3)_2]$ seems favorable. The use of **1** also imposes limitations for a catalytic system for radical generation. No H_2 activation or ligand exchange with **1** should occur. The cyclization must take place before undesired HAT to intermediate radicals. Finally, the radical-generating agent must not intercept the final radical before the catalytic HAT. A titanocene-catalyzed reductive epoxide opening[7] is an attractive method for these purposes. The titanocene complexes involved do not bind phosphines and CO and do not activate H_2. The presence of the desired and undesired

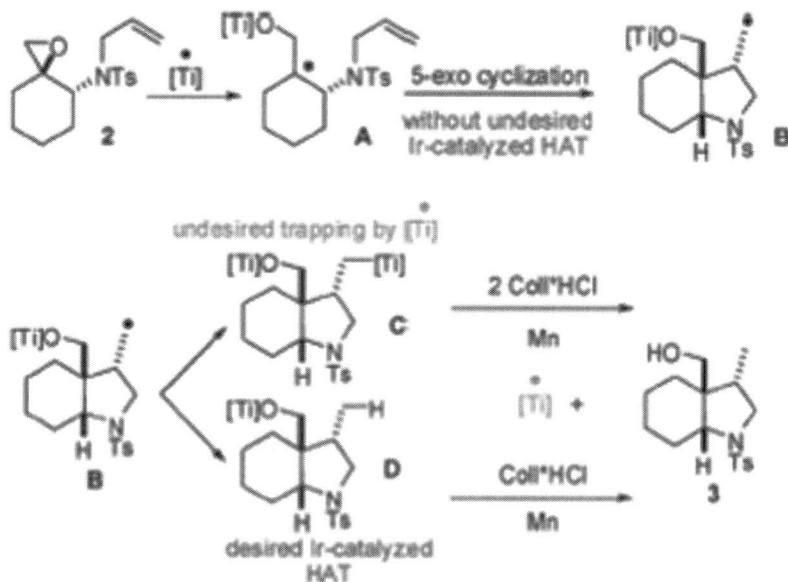

Scheme 11.8 Concept of the catalytic radical cyclization of **2** terminated by an Ir-catalyzed HAT.

pathways can be readily distinguished experimentally. If **A** is intercepted by HAT, an undesired product will be formed. In the absence of **1** or in the case of an inefficient HAT, **B** will be trapped by Cp$_2$TiCl to give **C**. From **C**, liberation of **3** and regeneration of Cp$_2$TiCl$_2$ requires the protonation of the Ti–O and Ti–C bonds by at least 2 equiv. of Coll·HCl. In the case of trapping of **B** by HAT, the formation of **3** and regeneration of Cp$_2$TiCl$_2$ from **D** requires only the protonation of the Ti–O bond by 1 equiv. of Coll·HCl. Subsequently, Cp$_2$TiCl is re-formed by reduction with Mn. Thus, the amount of Coll·HCl added to the reaction provides a simple experimental means for studying the efficiency of the Ir-catalyzed HAT.

Scheme 11.9 summarizes the results of the catalytic cyclizations of **2** and **4** in the absence and in the presence of **1**. In the absence of **1**, the yields of **3** and **5** are below 50%, even with 1.5 equiv. of Coll·HCl.[8] The remainder consists of starting material. In the presence of **1** and H$_2$ (4 atm), the isolated yields increase by more than a factor of 2. This indicates an efficient coupling of catalytic radical cyclization and Ir-catalyzed HAT. No products of undesired HAT to radicals of type **A** were observed. Even when taking into account that both radical and HAT reagent are present in catalytic amounts, and therefore a bimolecular trapping is disfavored, this is still amazing, because reactions of radicals with M–H bonds are exothermic and can have rate constants much higher than those of 5-*exo* cyclizations.[9] As a reason for the chemoselectivity of the HAT, we suggest that the steric shielding of the hydrido ligands in [IrH$_2$Cl(CO)(PPh$_3$)$_2$] by PPh$_3$ retards the trapping of radicals of type **A**. This could be especially relevant for the tertiary radicals employed here. In the case of **4**, the situation is more complex. After the cyclization, a highly reactive vinyl radical is generated that is not trapped Cp$_2$TiCl. Instead, its high reactivity results either in a reduction by the Ir-catalyzed HAT or by a HAT from THF.

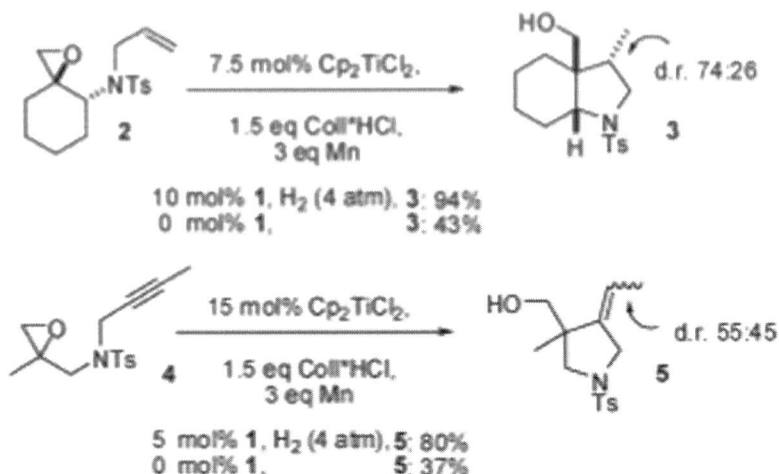

Scheme 11.9 Study of the competing pathways of trapping of radicals formed by 5-*exo* cyclization for **2** and **4**.

Table 11.1 Radical cyclizations terminated by Ir-catalyzed HAT (1.5 equiv. of Coll·HCl, 3 equiv. of Mn, 0.1 M THF, 4 atm H_2).

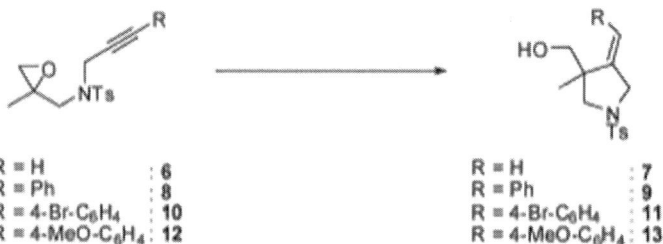

R = H	: 6	R = H	: 7
R = Ph	: 8	R = Ph	: 9
R = 4-Br-C_6H_4	: 10	R = 4-Br-C_6H_4	: 11
R = 4-MeO-C_6H_4	: 12	R = 4-MeO-C_6H_4	: 13

Entry	Substrate	Conditions	Product
1	6	15 mol% Cp_2TiCl_2, 5 mol% 1	7, 72%
2	6	15 mol% Kagan's complex, 5 mol% 1	7, 77%[a]
3	8	15 mol% Cp_2TiCl_2, 5 mol% 1	9, 76%[b]
4	10	15 mol% Cp_2TiCl_2, 5 mol% 1	11, 70%[c]
5	12	15 mol% Cp_2TiCl_2, 5 mol% 1	13, 70%[d]

[a]er = 73:37. [b]dr = 96:4. [c]dr = 91:9. [d]dr = 89:11.

In the latter case, a tetrahydrofuranyl radical is generated that can be either trapped by Cp_2TiCl or reduced by an Ir-catalyzed HAT. The 80% isolated yield of **5** obtained in the presence of **1** (37% without **1**) demonstrates that the coupling of the catalytic cycles is not affected by the nature of the final HAT step. The identical diastereoselectivity of the formation of **4** with or without **1** may suggest an initial HAT from THF. Hydrogenation catalysts, such as

Table 11.2 Radical cyclizations to carbocycles terminated by Ir-catalyzed HAT (1.5 equiv. of Coll·HCl, 3 equiv. of Mn, 0.1 M THF, 4 atm H_2).

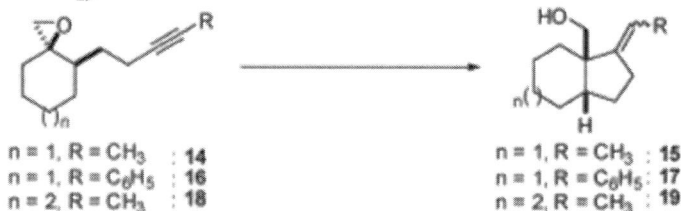

n = 1, R = CH_3	: 14	n = 1, R = CH_3	: 15
n = 1, R = C_6H_5	: 16	n = 1, R = C_6H_5	: 17
n = 2, R = CH_3	: 18	n = 2, R = CH_3	: 19

Entry	Substrate	Conditions	Product[a]
1	14[b]	10 mol% Cp_2TiCl_2, 1 mol% 1	15, 60%[c]
2	14[b]	10 mol% Cp_2TiCl_2, 5 mol% 1	15, 71%[c]
3	14[b]	15 mol% Cp_2TiCl_2, 5 mol% 1	15, 88%[c]
4	16[d]	15 mol% Cp_2TiCl_2, 5 mol% 1	17, 68%[e]
5	18[f]	15 mol% Cp_2TiCl_2, 5 mol% 1	19, 91%[g]

[a]All products can be diastereoconvergently hydrogenated to the *trans* products with Crabtree's catalyst.[12] [b]dr = 96:4. [c]dr = 63:37. [d]dr = 99:1. [e]dr = 85:15. [f]dr = 98:2. [g]dr = 67:33.

Wilkinson's catalyst, RhCl(PPh$_3$)$_3$,[10] are not useful, because hydrogenation, especially of alkynes, competes with the desired coupling of the catalytic cycles.

Table 11.1 summarizes further examples of the synthesis of pyrrolidines. For all substrates investigated, satisfactory yields of the desired products could be obtained. Gratifyingly, the aryl-substituted olefins are formed in much higher selectivity [(*E*):(*Z*) 89 : 11 to 96 : 4, see ESI for details] than **5**. Presumably, this is due to enhanced steric interactions by aryl substitution. Kagan's complex,[11] bearing two bulky menthyl substituents at the cyclopentadienyl ligands, gave only a slightly higher yield (entry 2, Table 11.1) than Cp$_2$TiCl$_2$. This indicates that interactions between the two metal complexes can be neglected and provides yet another hint that the two catalytic cycles operate independently. Reactions leading to carbocyclic products can also proceed in high yields.

Table 11.2 summarizes the development of efficient reaction conditions. For **1**, a catalyst loading of 5 mol % (entry 2) is sufficient, whereas a reduction to 1 mol % (entry 1) leads to unsatisfactory yields. For Cp$_2$TiCl$_2$, a catalyst loading of 15 mol% is adequate for high yields (entries 3 and 5). In summary, Gansäuer *et al.* have devised a system of coupled catalytic cycles for sustainable radical cyclizations terminated by Ir-catalyzed HAT with H$_2$ as terminal reductant. It is essential that the HAT catalyst, Vaska's complex, is not a hydrogenation catalyst.

It must be mentioned that MacMillan's work on "SOMO catalysis"[20] is similar in principle to the chemistry highlighted here. It is worthwhile to compare MacMillan's "SOMO catalysis" with "radical catalysis". Both concepts refer to the combination of a radical reaction for the key bond formation (or breaking) with a catalytic polar process that controls the radical process in a favorable way. The generation and fate of the radical is determined by the quantity and nature of the transition-metal complex in "radical catalysis", and by the quantity of organocatalyst used to generate the enamine intermediate in "SOMO catalysis". In the latter method, the stereoselectivity of the process is, of course, also efficiently controlled. Thus, in both processes the catalyst secures the desired low radical concentration during the overall process, guaranteeing that undesired side reactions of the reactive radicals are minimal.

11.3 CONCLUSION

Radical reactions initiated and controlled by transition-metal catalysis show great promise in mediating a variety of reactions in excellent yields and under mild conditions. By utilizing transition-metal complexes in catalytic amounts for the generation and transformation of radicals, these reactions have distinct advantages over standard methods of radical chemistry and transition-metal catalysis. In addition, radical catalysis can greatly improve reactions that perform poorly under polar conditions. Thus transition-metal-"tamed" radicals represent powerful and versatile intermediates in organic chemistry.

REFERENCES

1. B. Quiclet-Sire and S. Z. Zard, *Pure Appl. Chem.*, 1997, **69**, 645.
2. *Landolt–Börnstein: Numerical Data and Functional Relationships in Science and Technology*, New Series, Group II, ed. H. Fischer, Springer, Berlin, 1983, vol. 13.
3. M. Newcomb, *Tetrahedron*, 1993, **49**, 1151.
4. S. Z. Zard, *Radical Reactions in Organic Synthesis*, Oxford University Press, Oxford, 2003.
5. (a) *Metal-Catalyzed Cross-Coupling Reactions*, ed. A. de Meijere and F. Diederich, Wiley-VCH, Weinheim, 2nd edn, 2004; (b) *Transition Metals for Organic Synthesis*, ed. M. Beller and C. Bolm, Wiley-VCH, Weinheim, 2004.
6. (a) Q. Y. Chen, Z. Y. Yang, C. X. Zhao and Z. M. Qui, *J. Chem. Soc., Perkin Trans. 1*, 1988, 563; (b) D. P. Curran and C.-T. Chang, *Tetrahedron Lett.*, 1990, **31**, 933; (c) T. Ishiyama, S. Abe, N. Miyaura and A. Suzuki, *Chem. Lett.*, 1992, 691; (d) H. Stadtmüller, A. Vaupel, C. E. Tucker, T. Stüdemann and P. Knochel, *Chem.–Eur. J.*, 1996, **2**, 1204–1220.
7. Reviews: (a) A. J. Clark, *Chem. Soc. Rev.*, 2002, **31**, 1; (b) T. Pintauer and K. Matyjaszewski, *Chem. Soc. Rev.*, 2008, **37**, 1087; (c) H. Matsumoto, T. Motegi, T. Nakano and Y. Nagai, *J. Organomet. Chem.*, 1979, **174**, 157, and references therein; (d) M. Kameyama, N. Kamigata and M. Kobayashi, *J. Org. Chem.*, 1987, **52**, 3312, and references therein.
8. A. Rudolph and M. Lautens, *Angew. Chem.*, 2009, **121**, 2694 (*Angew. Chem., Int. Ed.*, 2009, **48**, 2656) and references therein.
9. (a) I. Ryu, S. Kreimerman, F. Araki, S. Nishitani, Y. Oderaotoshi, S. Minakata and M. Komatsu, *J. Am. Chem. Soc.*, 2002, **124**, 3812; for applications see: (b) T. Fukuyama, S. Nishitani, T. Inouye, K. Morimoto and I. Ryu, *Org. Lett.*, 2006, **8**, 1383.
10. M. W. Hooper, M. Utsunomiya and J. F. Hartwig, *J. Org. Chem.*, 2003, **68**, 2861.
11. Leading reference: T. Hama and J. F. Hartwig, *Org. Lett.*, 2008, **10**, 1545.
12. Z. Q. Weng, S. H. Teo and T. S. A. Hor, *Acc. Chem. Res.*, 2007, **40**, 676.
13. G. Manolikakes and P. Knochel, *Angew. Chem.*, 2009, **121**, 211 (*Angew. Chem., Int. Ed.*, 2009, **48**, 205).
14. Generally the coupling of free aryl radicals is not synthetically useful. For a recent oxidative homocoupling of aryl Grignard reagents mediated or catalyzed by the free radical TEMPO, see: M. S. Maji, T. Pfeifer and A. Studer, *Angew. Chem.*, 2008, **120**, 9690 (*Angew. Chem., Int. Ed.*, 2008, **47**, 9547). The mechanism has, however, not been elucidated.
15. T. Kobayashi, H. Ohmiya, H. Yorimitsu and K. J. Oshima, *J. Am. Chem. Soc.*, 2008, **130**, 11276.
16. (a) B. Gaspar and E. M. Carreira, *Angew. Chem.*, 2008, **120**, 5842 (*Angew. Chem., Int. Ed.*, 2008, **47**, 5758). Though no proof exists for the presence of radicals in this example, the proposed mechanism parallels the one proposed and studied in detail for the hydrohydrazination and hydro-

azidation reactions. See: J. Waser, B. Gaspar, H. Nambu and E. M. Carreira, *J. Am. Chem. Soc.,* 2006, **128**, 11693.

17. A. Gansäuer, C.-A. Fan and F. J. Piestert, *J. Am. Chem. Soc.,* 2008, **130**, 6916.

18. R. S. Drago, J. G. Miller, M. A. Hoseton, R. D. Farris and M. J. Desmond, *J. Am. Chem. Soc.,* 1991, **113**, 4888.

19. A. Bakac and L. M. Thomas, *Inorg. Chem.*, 1996, **35**, 5880.

20. (a) T. D. Beeson, A. Mastracchio, J. B. Hong, K. Ashton and D. W. C. MacMillan, *Science*, 2007, **316**, 582; (b) D. A. Nicewicz and D. W. C. MacMillan, *Science*, 2008, **322**, 77; For a recent highlight: (c) P. Melchiorre, *Angew. Chem.,* 2009, **121**, 1386 (*Angew. Chem., Int. Ed.,* 2009, **48**, 1360).

Reagent Control in Transition-metal-initiated Radical Reactions*

12.1 INTRODUCTION

Over the past decades radical chemistry has developed into an important and integral part of organic chemistry. Although the first example of an organic radical (**1**) was observed as early as 1900 by Gomberg,[1] the pace of development was rather slow over the next couple of decades and radicals were rarely used in synthesis. The development of efficient chain reactions constituted an important breakthrough in the application of radical chemistry in organic synthesis.[2] An important and very attractive feature of these reactions is their high degree of functional group tolerance. Since radicals are usually stable under protic conditions, alcohols or even water can, in principle, be used as solvents in radical chemistry. Consequently, protic functional groups do not need protection. As soon as the underlying principles of the kinetic and thermodynamic behavior of free radicals were firmly established, efficient synthetic applications became feasible.[3] The use of chain reactions has resulted in a number of very impressive total syntheses of natural products. The characteristic features of free radicals can by now be deduced from electron spin resonance (ESR) data.[5] The course of some radical reactions can be understood by theoretical means.[6] However, because the crucial intermediates are free radicals, no influence of the ligand sphere of the reagent generating the radical on the selectivities of the reaction is usually observed. These transformations are, therefore, classical examples of substrate-controlled reactions. An alternative approach to radical chemistry is constituted by controlling the course of the radical reaction by a suitably designed reagent both during radical generation and the ensuing transformation of the metal-bound radical. This concept of reagent control has been applied with excellent

* Chapter written by V. Tamara Perchyonok.
Streamlining Free Radical Green Chemistry
V. Tamara Perchyonok, Ioannis Lykakis and Al Postigo
Published by the Royal Society of Chemistry, www.rsc.org

success in organometallic chemistry and in catalysis.[7] Until recently, use of this otherwise very successful approach to radical chemistry has been rare.[8] The purpose of this chapter is to describe exactly these novel emerging concepts in C–C and C–H bond-forming reactions. Examples where the element of stereocontrol depends on chiral auxiliaries on the starting material will not be dealt with here. The existing excellent reviews and book chapters on this topic should be consulted by the interested reader.[3b,9] Metal-initiated reactions leading to transformations of free radicals will not be treated because no metal-bound radicals are obtained. Thus, vitamin B_{12}- initiated reactions[10] and cobaloxime chemistry[11] will not be discussed here. The recently described living radical polymerizations initiated by well-designed metal complexes[12] are also thought to proceed *via* chain reactions of free radicals. The selectivity-determining step of allylic oxidations catalyzed by chiral copper complexes is thought to proceed through an organocopper(III) reagent.[13] Thus, these oxidations will not be treated here. C–H activation by manganese porphyrins,[14] DNA cleavage by metal complexes, *e.g.*, bleomycin,[15] and DNA foot-printing[16] are not included because the radicals formed have not been used in C–C bond-forming reactions. In principle, reagent control can be exercised at different stages in a radical reaction. (a) The first step in the series of transformations of a radical reaction is constituted by the generation of the radical from a suitable precursor. The usual selectivities of this generation, *e.g.*, by electron transfer, can be controlled by the electron-transfer reagent and its ligand sphere. Clearly, the radical precursor needs to have a functional group that allows for binding of the reagent in close proximity of the newly generated radical prior to its formation. (b) In the subsequent transformation of the radical, the selectivities of the reaction, *e.g.*, addition to carbon–carbon multiple bonds, should, in principle, also be amenable to reagent control if the metal complex remains bound to the radical. Here one is, of course, not dealing with the chemistry of intermediates usually described as free radicals but with metal-bound radicals. (c) In the case of a free radical reaction, reagent control is possible if the radical or the radical trap is complexed by a carefully designed reagent. The stereochemical course of the following transformation is thus amenable to reagent control by the metal and its ligand. Although at the stage the reagent-controlled radical reaction is completed, an additional attractive feature of the desired process becomes immediately apparent. If the reagent determining the course of the overall transformation can be cleaved off the reaction product and recycled, a catalytic reaction emerges. This is obviously of great economic advantage if the reagent contains an expensive metal or a ligand that has to be synthesized in a multistep sequence.

Metal-initiated radical reactions with suitable radical precursors allow for reagent control in both of the above-mentioned two points, a and b. Although these radical reactions have been applied to demanding synthetic reactions with great success for some time,[17] attempts to influence the usual selectivities by ligand variations have appeared in the literature only recently. These emerging novel concepts and reactions will be the subject of this section. The

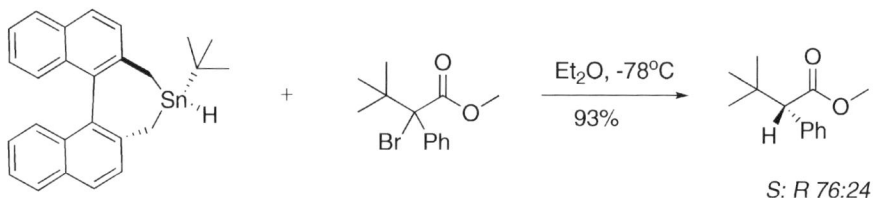

Scheme 12.1 Enantioselective radical reaction in a nutshell.

focus will be on the use of carbonyl- and epoxide-containing molecules as radical precursors for reagent control in radical chemistry. The concepts outlined as point c were reviewed by Renaud[18] in late 1998 and by Sibi and Porter[19] in early 1999. Diastereoselective and enantioselective reactions in these fields have been realized during the last 10 years. Therefore, this subject will not be covered comprehensively in this article and only recent examples will be discussed here.

Chiral, non-racemic stannanes have been used to enantioselectively reduce radicals by Schumann *et al.*[20] and later independently by Nanni and Curran[21] and Metzger *et al.* (Scheme 12.1).[22]

An intriguing extension of this work has been reported by Schiesser *et al.* only recently.[23] It was demonstrated that the performance of chiral, non-racemic stannanes in these transformations can be significantly improved in the presence of enantiomerically pure Lewis acids for substrates containing carbonyl groups. Obviously this concept of double stereocontrol is very promising for future applications. Since the first reports on the activation of prochiral radical traps by enantioselective Lewis acid catalysis by Porter, Sibi and colleagues (Scheme 12.2), this exiting field has developed at a rapid pace.[24] The interested reader is referred to the review by these authors.[19]

12.2 REAGENT CONTROL IN TRANSITION-METAL-INITIATED RADICAL REACTIONS

Interesting classes of radical sources in transition-metal-initiated reactions are aldehydes and ketones. Carbonyl compounds are good ligands for Lewis acidic metal complexes, and thus reagent control during the formation of ketyl anions seems possible.[8] Moreover, the ketyl anions formed during electron transfer enables binding of the metal ions or complexes *via* oxygen. Control of

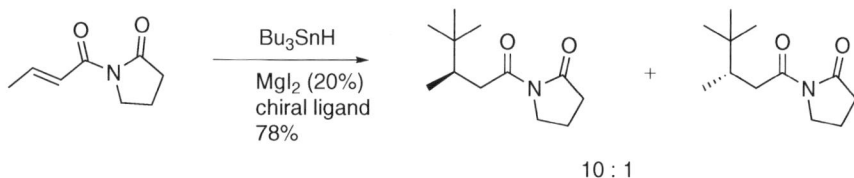

Scheme 12.2 Enantioselective alkylation and Lewis Acid.

the following radical transformations can, therefore, be achieved. Ketyl radicals are, of course, interesting intermediates in organic synthesis. They can either dimerize to give 1,2-diols in pinacol coupling[25] or add to carbon–carbon multiple bonds in inter- or intramolecular reactions.[26] In these classes of transformations, reagent control can be exercised in directing the diastereo- and enantioselectivity of the products derived from the ketyl anions by variation of the ligands and the metal ions of the electron-transfer reagent. Another intriguing class of radical precursors that is to date used rarely and is thus probably less well established than carbonyl compounds are epoxides.[8] The epoxide oxygen is well-suited for complexation by a metal complex. Therefore, the regio- and stereoselectivity of epoxide opening *via* electron transfer can be influenced by the ligands. The initial product of the opening *via* electron transfer is a α-metal oxy radical. These radicals that are bound to the metal *via* the oxygen can participate in the usual reactions of carbon-centered radicals, *e.g.*, hydrogen-atom abstraction, cyclizations, and intermolecular addition reactions to activated olefins. The course of these transformations should be amenable to reagent control if a properly chosen metal complex is utilized. Therefore, these metal-bound radicals constitute an interesting class of intermediates for a number of synthetically useful transformations. An example of a reaction of reagent control in radical chemistry without using carbonyl compounds or epoxides was reported by Kamigata *et al.*[27] It was demonstrated that the addition of sulfonyl chlorides to styrene and phenylpropene is catalyzed by chiral ruthenium complexes (Scheme 12.3). The enantioselectivities obtained were low. A reaction mechanism involving a radical redox transfer chain has been proposed. The exact reason for the stereochemical induction is still unclear, however.

12.3 CARBONYL COMPOUNDS AS RADICAL SOURCES: PINACOL COUPLINGS

12.3.1 Stoichiometric Reagent-controlled Couplings

The reductive coupling of two carbonyl compounds, the pinacol coupling, is probably the most direct way of forming the C–C bond of 1,2-diols.[25] Since the first report of the reaction of acetone with sodium in 1858 by Fittig from Göttingen[28] (Scheme 12.4), considerable effort has been devoted to the development of milder and more selective ways to achieve this important transformation.

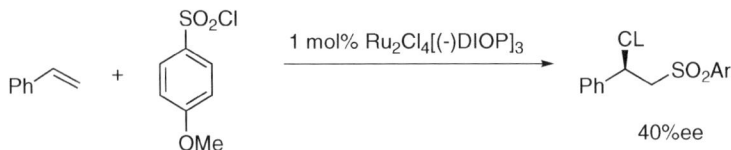

Scheme 12.3 Catalysis and chiral ruthenium complexes.

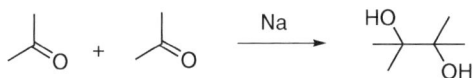

Scheme 12.4 First reported reaction of acetone with Na.

Pinacol couplings have been used as key steps in a number of elegant syntheses of natural products.[29] A complete coverage of modern methods is beyond the scope of this chapter. The emerging catalytic variations will be emphasized. Stoichiometric titanium-based complexes have turned out to be excellent reagents for the pinacol coupling of aromatic and R,α-unsaturated aldehydes.[30] Raubenheimer and Seebach reported that titanium trichloride generated *in situ* from titanium tetrachloride and butyllithium was an excellent reagent for highly diastereoselective couplings of aromatic aldehydes to racemic C_2-symmetrical 1,2-diols.[31] Aliphatic aldehydes and aromatic ketones were not affected. Thus, the chemoselectivity of this mild reagent is high. Later, Porta *et al.* showed (Scheme 12.5) that commercial titanium trichloride in THF/CH_2Cl_2 solution was also an efficient reagent for this transformation.[32]

Diastereoselectivities were similar to the titanium(III) reagent generated *in situ*. Unfortunately, the addition of a tartrate-derived ligand did not induce significant enantioselectivities (ee < 5). It remains to be seen whether this conceptually simple and attractive approach is to be successful in enantioselective synthesis. In 1987, Inanaga and Handa disclosed their results on the pinacol coupling of aromatic and R,α-unsaturated aldehydes.[33] They found that reduction of titanocene dichloride with Grignard reagents lead to a green trinuclear titanium(III) reagent that was formulated as $(Cp_2TiCl)_2MgCl_2$. This complex coupled the aldehydes in high yield and with high diastereoselectivities to give the racemic C_2-symmetrical 1,2-diols (Scheme 12.6).

The mechanistic rationale offered for the high diastereoselectivity of this process is shown in Figure 12.1.

Both ketyl anions are coordinated in a trinuclear complex consisting of two titanocene(IV) units and a central magnesium ion. Each ketyl anion is bound to

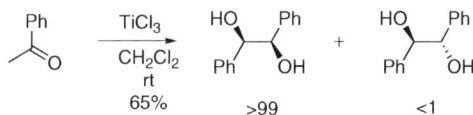

Scheme 12.5 Pinacol coupling and commercial titanium trichloride.

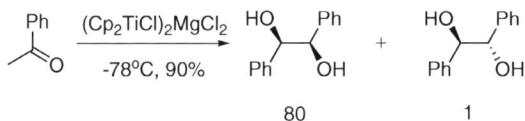

Scheme 12.6 Pinacol formation mediated by $(Cp_2TiCl)_2MgCl_2$.

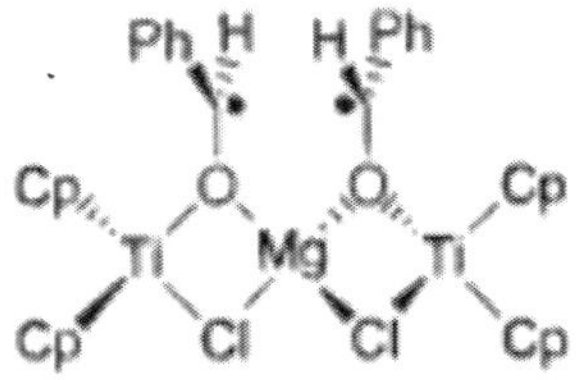

Figure 12.1   Possible decisive intermediate in titanocene-mediated pinacol couplings.

one titanium atom and to magnesium. This arrangement results in the depicted orientation of the phenyl groups minimizing steric interactions. In this case, aliphatic aldehydes and ketones were unreactive. Clearly, this highly ordered trinuclear complex should allow for control of diastereo- and enantioselectivity of the pinacol coupling if the cyclopentadienyl ligands are chosen properly.[34] Titanocene(III)-initiated pinacol couplings were later reinvestigated by Barden and Schwartz.[35] It was found that reduction of Cp_2TiCl_2 with aluminum powder[34] led, after washing with diethyl ether, to the dimeric $(Cp_2TiCl)_2$ as the active reagent for the highly diastereoselective coupling of activated aldehydes. Interestingly, the coupling could be performed in the presence of water without significant loss in diastereoselectivity. However, more than 50 equiv. of NaCl had to be added to the reaction mixture to conserve the high selectivities. This observation is nevertheless an intriguing manifestation of the stability of radicals in a protic environment. A common feature of all titanium(III) reagents reported to date is their high chemoselectivity. Obviously simple titanium(III) reagents are incapable of transferring an electron to non-activated carbonyl compounds, *e.g.*, simple aldehydes and ketones. To achieve this goal, titanium(II) reagents have been developed by Mukaiyama *et al.* In this manner, ketones can be coupled to the 1,2-diols in moderate to excellent diastereoselectivities without significant formation of the deoxygenation products.[36] An important achievement in this area is Matsubara's observation that the addition of chelating diamines greatly improves the performance of the titanium(II) reagent.[37] It was not established whether the amine led to an acceleration of the reaction or simply tamed Lewis-acid species initiating side reactions. The goal of an enantioselective coupling was also pursued by addition of diamines and amino alcohols to the reaction mixture by Matsubara (Scheme 12.7). The enantioselectivity obtained (44% ee) so far is not yet fully satisfying.

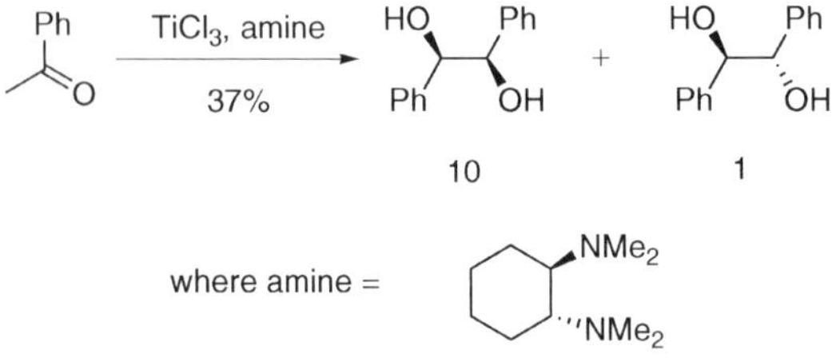

Scheme 12.7   Enantioselective pinacol formation mediated by $TiCl_2$/chiral diamine.

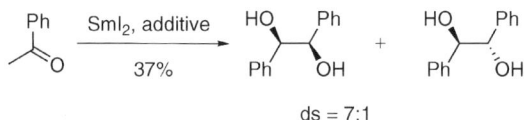

ds = 7:1

where additive is

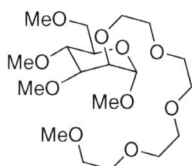

Scheme 12.8 Pinacol and SmI_2/chiral ligand transformation.

However, this simple and efficient concept still seems very promising. Interesting modifications of the selectivity of samarium diiodide-based reagents have been reported by Skrydstrup *et al.*[38] The addition of chelating ligands allowed for distinct improvements in the diastereoselectivity of the pinacol coupling (Scheme 12.8).

Samarium diiodide has also been employed in stereo-controlled pinacol cyclizations.[39] Examples are shown in Schemes 12.9 and 12.10.

The origin of stereocontrol in the first example is thought to be the formation of a nine-membered cyclic ketyl radical. The other transformations also proceed under chelation control. Low-valent vanadium reagents constitute a very interesting class of reagents in pinacol couplings. Pedersen demonstrated as early as 1989 that crossed pinacol couplings are readily achieved using $[V_2Cl_3-(THF)_6]_2[Zn_2Cl_6]$ (ref. 40) with substrates allowing for chelation.[41] Excellent diastereoselectivities were sometimes achieved (Scheme 12.11).

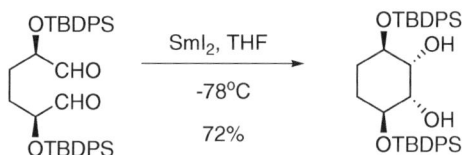

Scheme 12.9 Reagent control of stereoselctive pinacol transformation mediated by SmI_2.

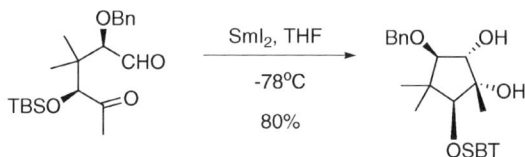

Scheme 12.10 Stereo-control in SmI_2 mediated pinacol cyclization.

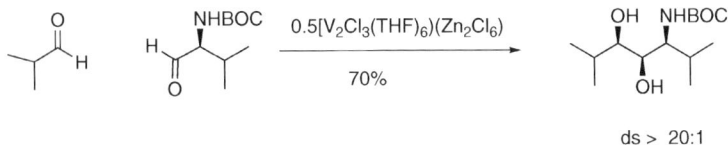

ds > 20:1

Scheme 12.11   Vanadium mediated diastereoselective pinacol coupling.

This important methodology can also be applied to pharmacologically important R-amino aldehydes.[42] Although variations of the ligand sphere have not yet been reported, reagent control is exercised in binding the substrates in a chelation-controlled manner.[43] Intriguing applications of low-valent niobium complexes[44] have been found in the reductive coupling of imines and in the crossed pinacol coupling of imines and aldehydes by Pedersen in one of the earliest examples of highly selective couplings in 1987 (Scheme 12.12).[45]

12.3.2   From Stoichiometric to Catalytic Pinacol Couplings

12.3.2.1   *Chlorosilanes as Mediators*

Although some of the reagents discussed above allow for excellent results, they all suffer from the principal drawback of having to be employed in stoichiometric amounts. This is especially disadvantageous when more complex and expensive reagents are to be used to obtain reagent control. Clearly a catalytic reaction would circumvent this problem and result in a more efficient use of resources. In 1995, Fürstner and Hupperts reported their McMurry reaction (Scheme 12.13) as catalytic in titanium and introduced a novel concept for conducting catalyzed redox reactions.[46]

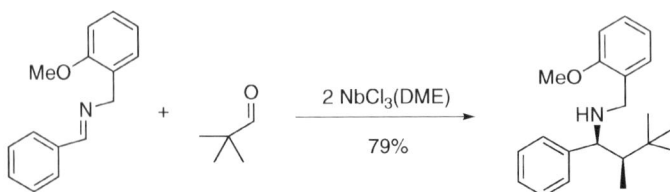

Scheme 12.12   Low-valent Niobium mediated stereospecific pinacol coupling of imines and aldehydes.

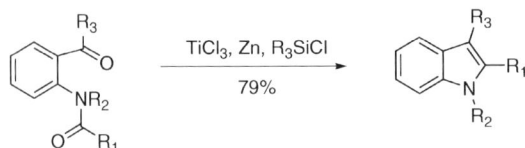

Scheme 12.13   Ti-mediated catalyzed redox reaction.

ds = 63:30:7

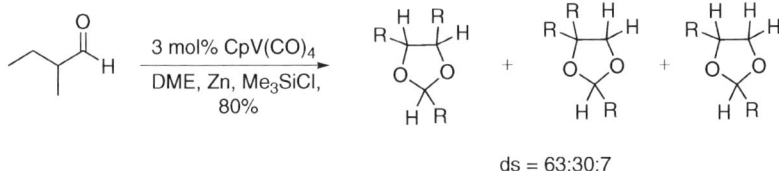

Scheme 12.14 Vanadium-catalyzed pinacol coupling and diastereoselectivity.

Independently, Hirao *et al.* developed a vanadium-catalyzed pinacol coupling based on the same considerations.[47] However, in this reaction the initially formed diolate was cleaved off the vanadium catalyst by formation of a dioxolane. Thus, a third of the aldehyde was consumed for catalyst regeneration. The role of Me₃SiCl seems to be the activation of the aldehyde for dioxolane formation. The diastereoselectivities of the pinacol coupling were moderate (Scheme 12.14).

No further investigations concerning ligand variations were undertaken. More recently, this concept[48] has also been applied to the synthetically very useful Nozaki–Hiyama reactions by Fürstner and Shi.[49] The essential novel step in both catalytic cycles consists of the removal of oxo or alkoxides from metal complexes for the regeneration of metal chloride species that could be reduced to the redox-active reagent *in situ*. Metal chloride formation was achieved by adding chlorosilanes to the reaction mixture. The redox-active reagent was simply recovered by reduction with zinc or manganese dust. A catalytic enantioselective variation of the Nozaki–Hiyama reaction was very recently reported by Cozzi[50] (Scheme 12.15).

Ephritikhine *et al.* demonstrated that cleavage of the metal oxo species can also be achieved by adding aluminum trichloride instead of chlorotrimethylsilane.[51,52] The catalysts used in these reactions were titanium trichloride and uranium tetrachloride. In 1996, Endo *et al.* reported the samarium-catalyzed

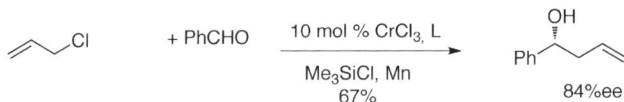

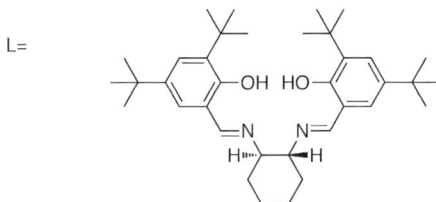

Scheme 12.15 Catalytic enantioselective variation of the Nozaki-Hiyama reaction.

PhCHO $\xrightarrow[\substack{1.5\ MgBr_2,\ Zn \\ 2.5\ Me_3SiCl \\ 90\%}]{3mol\%\ Cp_2TiCl_2}$

OH

Ph—⟨⟩—Ph

OH

de= 95:5

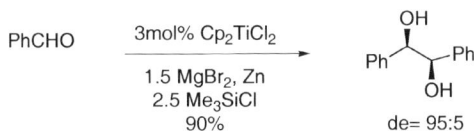

Scheme 12.16   Titanocene-catalyzed pinacol coupling of aromatic aldehydes.

pinacol coupling of aldehydes and ketones using chlorotrimethylsilane.[53] Diastereoselectivities in the intermolecular reaction were moderate, and no attempts were made to change the ligands on samarium to alter the course of the reaction. Another study bearing promise for a reagent-controlled catalytic pinacol coupling was the titanocene-catalyzed pinacol coupling of aromatic aldehydes as shown in Scheme 12.16.[54]

It turned out to be essential that the aldehyde, Me$_3$SiCl, and MgBr$_2$ were added slowly to a mixture of Cp$_2$TiCl$_2$ and zinc dust in THF. In this manner, the uncatalyzed coupling of aromatic aldehydes in the presence of ClSiMe$_3$ was suppressed.[55] MgBr$_2$ was necessary to obtain a tight tri-nuclear complex ensuring high diastereoselectivity.[33,34] It was established that silylation was the slowest step in the catalytic cycle. Without the catalyst, the reaction was slower and yielded the 1,2-diols with substantially reduced diastereoselectivity. A drawback of these conditions was that aldehydes that are reduced only slowly for steric reasons, *e.g.*, *o*-tolyl aldehyde, or electronic reasons, *e.g.*, anisaldehyde, were not coupled by the catalyst but by the stoichiometric reductive system. Although acetophenone was transformed to product under the catalytic conditions, this was not due to the titanocene reagent. In the absence of the catalyst, essentially the same results were obtained. Thus, a chlorotrimethylsilane-initiated coupling was taking place.[55] If the proper reaction conditions and substrates, *i.e.*, unhindered aromatic aldehydes, were chosen, the catalytic system delivered the reaction products in good yields and with reasonable diastereoselectivity (90 : 10 up to 95 : 5). THF constituted the best solvent. Later, Dunlap and Nicholas reported a similar reaction using manganese dust as the stoichiometric reductant.[56] With the titanocene reagents, control of diastereoselectivity is readily accomplished by variation of the ligand sphere. Using Brintzinger's *ansa*-titanocene[57] (Figure 12.2) in racemic form as catalyst lead to a distinct improvement of diastereoselectivity compared to titanocene dichloride with zinc as reductant.[58]

Dunlap and Nicholas reported the first catalytic enantioselective pinacol coupling using enantiomerically pure *ansa*-metallocene as a catalyst with manganese as the reductant.[56] Although reasonably high levels of enantiose-

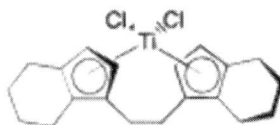

Figure 12.2   Brintzinger's *ansa*-metallocene.

lection were observed (ee 60% for benzaldehyde), the reaction suffered from a decrease in diastereoselectivity (8 : 1 *vs.* 13 : 1) compared to Cp$_2$TiCl$_2$ (Scheme 12.17).

This effect was even more pronounced in the presence of zinc dust.[59] The trinuclear complex responsible for the binding of both ketyl radicals probably could not be formed with the enantiomerically pure *ansa*-metallocene. It remains to be seen if this problem can be circumvented by employing different titanocene complexes as catalysts. However, titanocene(III) complexes constitute the catalysts allowing for the highest enantioselectivities in pinacol couplings so far. Hirao *et al.* reported the pinacol coupling of aliphatic aldehydes using the titanocene dichloride/ClSiMe$_3$/zinc dust system.[60] Diastereoselectivities were usually substantially lower than for aromatic aldehydes. The authors claim that the titanium(III) reagent transfers an electron to an aldehyde activated by the strong Lewis acid ClSiMe$_3$. The silyl-bound neutral ketyl radicals then coupled with low selectivity. Thus, the system does not offer obvious potential for controlling selectivity by the ligands of the metal. A similar observation has been made by Svatos and Boland using chromium(II) compounds as electron-transfer catalysts.[61] They found that using bulkier chlorosilanes leads to greatly enhanced selectivity, albeit at the expense of lower yields. These reports clearly demonstrate the major disadvantage of chlorosilane-mediated catalytic pinacol couplings. The high Lewis acidity of the employed silanes allows for activation of the carbonyl-containing substrates toward electron transfer and thus a background reaction through silyl bound ketyl radicals. Therefore, the reaction conditions have to be carefully controlled. A milder way of cleaving the metal–oxygen bonds prior to *in situ* reduction of the catalyst is therefore desirable. A solution to this problem will be discussed in the next section. Three other titanium-catalyzed pinacol couplings based on silylation reactions to recycle the catalyst have recently been described. Nelson *et al.* developed a catalyst based on titanium trichloride.[62] Interestingly, protic additives, *e.g.*, *tert*-butyl alcohol, and donor ligands, *e.g.*, 1,3-diethyl-1,3-diphenyl urea, showed distinct improvements in yield and selectivity. Salen ligands have been employed by Cozzi *et al.* in titanium-catalyzed reactions.[63] Aromatic and aliphatic aldehydes could be coupled (Scheme 12.18).

Diastereoselectivities were excellent in many cases. A single example of a chiral ligand was reported resulting in low enantioselectivity (10% ee).

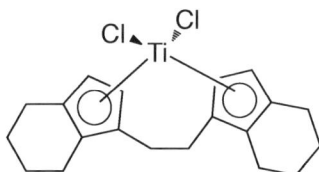

Scheme 12.17 Catalytic enantioselective pinacol coupling using enantiomerically pure ansametallocene as a catalyst with manganese as a co-reductant.

PhCHO →(10mol% chiral ligand / Mn / Me₃SiCl / no yield given)

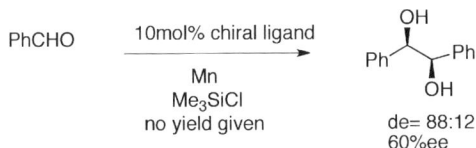

de= 88:12
60%ee

Scheme 12.18 Aromatic and aliphatic aldehydes and titanium catalyzed reaction.

Therefore, the titanocene catalysts still seem to bear greater promise for enantioselective catalysis. The intriguing catalyst phenyl titanocene(III) (**3**) was employed by Itoh *et al.*[64] (Scheme 12.19).

However, for benzaldehyde, diastereoselectivities (71:29) are somewhat lower than with titanocene chloride. Phenyltitanocene is a more active electron-transfer reagent than titanocene(III) chloride. The phenyl group constitutes a less electron-withdrawing substituent on titanium than chloride. Thus, aliphatic dialdehydes could be coupled intramolecularly. This attractive novel approach toward selective catalysis offers an additional element of reagent control in radical reactions through variation of the aryl group attached to titanium. Hirao *et al.* further optimized their vanadium catalysts for the coupling of aldehydes and imines.[65] Varying the ligands on vanadium and changing the solvent lead to dramatic improvements in the diastereoselectivity of the reaction (Scheme 12.20). It turned out that vanadocene dichloride in THF constituted by far the most efficient system for catalysis. This catalytic system has also been used by Hirao *et al.* for coupling imines.[66] No enantioselective reactions using vanadocenes have been reported so far.

12.4 PROTONATION OF METAL–OXYGEN BONDS IN CATALYTIC RADICAL REACTIONS

As indicated above, silylation is not always an ideal way to regenerate the redox active complexes *in situ*. The problem to be solved is the cleavage of

PhCHO →(3mol% chiral ligand / Zn / Me₃SiCl / 89%)

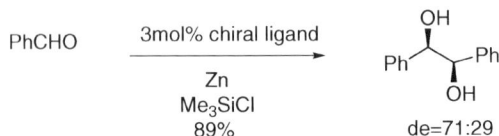

de=71:29

Scheme 12.19 Pinacol coupling and phenyl titanocene.

PhCHO →(3mol% Cp₂VCl₂ / Zn / Me₃SiCl,THF / 66%)

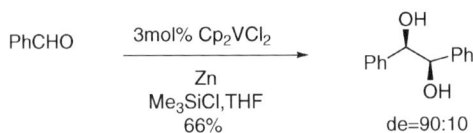

de=90:10

Scheme 12.20 Vanadium mediated pinacol coupling.

metal–oxygen bonds to yield metal halides by an oxophilic reagent without activating carbonyl compounds toward electron transfer. It has to be kept in mind that the reactive intermediates dealt with in pinacol couplings are radicals. Thus, any reagent employed should be tailor-made to account for the stability of radicals. Protonation is, in principle, the simplest way of cleaving metal oxides and alkoxides. Protonation also seems to be well-suited in radical reactions. Radicals are usually stable under protic conditions, and alcohols and even water are suitable solvents for radical reactions.[3] The reason for this stability is the low tendency for homolytic cleavage of O–H bonds. It should not be forgotten, however, that even addition reactions to carbonyl compounds where classical carbanionic species have until now not been ruled out as intermediates, *e.g.*, magnesium compounds under Barbier conditions, can be performed in aqueous acidic media.[67] Bearing these general considerations in mind, we decided to screen buffered forms of hydrochloric acid as mediators in titanocene-catalyzed pinacol couplings. Neat hydrochloric acid as the ultimate proton source is desirable because titanocene dichloride is readily reduced by zinc or manganese dust to the corresponding titanium(III) reagents.[34] Except for the bromide and iodide, this is not readily achieved with other ions. Chloride constitutes the most convenient choice among the halides. The catalytic cycle is depicted in Figure 12.3.

Some features of the acid and the stoichiometric reductant, *i.e.*, the metal powder, to be employed to achieve catalytic turnover become immediately apparent.

- The acid must be strong enough to protonate a metal–oxygen bond. The acid's pK_a in water should thus be lower than that of typical alcohols

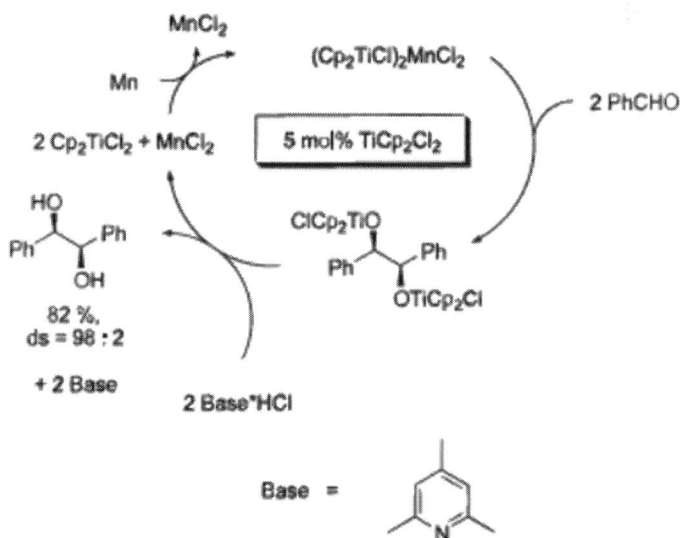

Figure 12.3 Titanocene-catalyzed pinacol coupling employing protic conditions.

(CH$_3$OH, 15.5; *t*-BuOH, 19.2).[68] To ensure complete protonation, the pK_a should, therefore, be at most 12.5. Also, protonation should occur fast to exclude any undesired side reactions.

- Neither the stoichiometric reductant, *i.e.*, the metal powder, nor the active titanium(III) reagent may be oxidized by the acid. Therefore, neat hydrochloric acid cannot be employed directly.
- The corresponding base must not complex and deactivate any titanium species in the catalytic cycle.
- The employed acid should not activate the aldehyde strongly toward electron transfer by protonation.
- The metal salt formed during the reduction of titanocene dichloride should not act as a Lewis acid, initiating uncatalyzed electron transfer to the carbonyl compound.

Another feature of the catalytic cycle is that free 1,2-diols are formed as compared to the silyl ethers obtained in silylations. The free diol can, in principle, act as a ligand for titanium, and care has to be taken to avoid product inhibition. With these considerations in mind, we decided to use pyridine hydrochlorides as acids. Both the pK_a and steric demand of the acid and the corresponding base can be readily altered by variation of the substituents.[68] 2,4,6-Collidine hydrochloride was an appropriate acid to achieve catalytic turnover.[69] Manganese as a stoichiometric reductant was distinctly superior to zinc with respect to both yield and diastereoselectivity of the coupling. It should be noted that under catalytic conditions, the diastereoselectivity is almost the same as in the stoichiometric parent system.[33] Product inhibition did not seem to pose serious problems. Under the optimized conditions, hardly any benzyl alcohol was formed. The stoichiometric reductive system is, therefore, exceptionally mild. *o*-Tolyl and *p*-anisaldehyde gave the desired product in good yields and with excellent diastereoselectivities. Both aldehydes reacted with distinctly lower selectivity when Me$_3$SiCl was used as the mediator for catalysis. The chemoselectivity of the system was high. Neither aliphatic aldehydes nor aromatic ketones were affected under the reaction conditions. These findings indicate the almost complete absence of uncatalyzed electron transfer from manganese to the aldehydes. Our mild stoichiometric reducing agent is, thus, clearly superior to the systems employing zinc or manganese and chlorotrimethylsilane. It remains to be seen if this reagent combination will be of value in other catalytic reactions. Although no attempts have been made to investigate ligand variations, this area of research certainly remains promising.

12.5 CARBONYL COMPOUNDS AS RADICAL PRECURSORS: ADDITIONS OF KETYL RADICALS TO C–C AND C–X BONDS

A very productive part of radical chemistry during the last 20 years has been addition reactions of ketyl radicals to olefins. It is fair to say that the rapid development of this field is due to the introduction of samarium diiodide as an

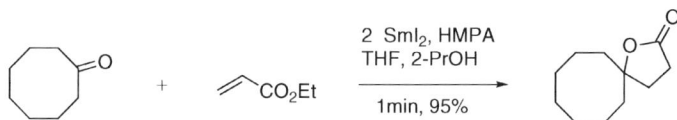

Scheme 12.21 SmI_2 and C-C bond formation.

electron-transfer reagent to organic synthesis by Namy *et al.* in 1980.[70] Application of low-valent complexes in ketyl anion chemistry of samarium has resulted in a number of impressive applications in synthesis.[17,71] Among the best acceptors in the coupling reactions are R,α-unsaturated esters.[72] Both aldehydes and ketones can be used as carbonyl partners in these reactions (Scheme 12.21).

Interestingly, even formaldehyde in aqueous solutions constitutes a good precursor for the ketylradicals.[73] The products of these transformations, γ-lactones and γ-hydroxyesters, are a common structural motif in natural product synthesis and are valuable synthetic intermediates. Addition of simple aldehydes to crotonates has been reported to be highly diastereoselective.[72b] However, the results are somewhat confusing because in closely related systems substantially different selectivities have been observed[72c,d] (Scheme 12.22).

The reactivity of samarium diiodide can be dramatically increased if hexamethylphosphoric acid triamide (HMPA) is added as ligand for samarium to the reaction mixture.[72c,73] Lewis acid cocatalysis has been reported as an efficient means to accelerate ketyl additions reaction by Inanaga *et al.*[74] The diastereoselectivity of the addition reactions of ketyl radicals seems to be governed by the preferred configuration of the ketyl radical as deduced from theoretical studies[75,76] (Scheme 12.23).

One would expect reagent control by addition of ligands to be an interesting means of influencing the selectivity. The high oxophilicity of samarium allows for excellent results in chelation-controlled ketyl olefin couplings.[74d] The diastereoselectivity can be very high if the properties of the reagent and the

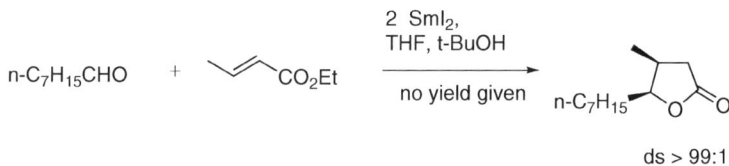

ds > 99:1

Scheme 12.22 Synthesis of gamma-lactones and SmI_2.

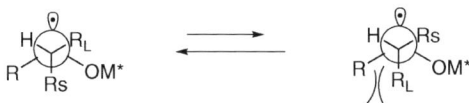

Scheme 12.23 Reasons for stereoselective outcomes.

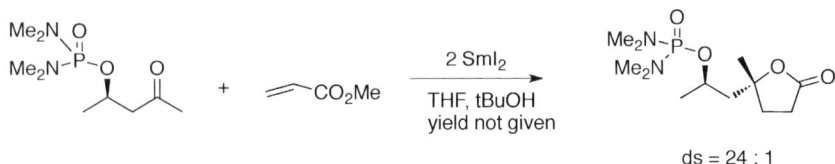

Scheme 12.24   Chelation-controlled ketyl-olefin coupling.

substrate are suitably adjusted. An instructive example is shown in Scheme 12.24.

An eight-membered chelate containg the $[P(O)(NMe_2)_2]$ group has been postulated as the decisive intermediate in the coupling of the protected hydroxy ketone and acrylic acid ethyl ester.[63d] Simple hydroxy ketones yielded excellent results also by forming a five-membered chelate[67a,b] (Scheme 12.25). Thus, the matching of the steric and electronic features of both substrate and the reagent samarium diiodide, a typical scenario of reagent control, can lead to excellent results.

An important class of ketyl couplings is cyclization reactions. With simple substrates and reactions giving bicyclic products, low to reasonable diastereoselectivities can be obtained[72b,77] (Scheme 12.26).

Chair-like transition structures have been proposed, minimizing interactions with the samarium complex and the pseudo-axial substituent of the olefin. Olefin geometry can be an important factor[77] (Scheme 12.27).

Substrates allowing for chelation usually react with exceedingly high diastereoselectivity.[78] For these reactions, convincing transition-state models have been proposed. It should be noted, though, that proper choice and positioning of the chelating group can be essential for optimizing the steric interactions in the respective transition states.[79] In this manner, even four-membered rings can be obtained in reasonable yields[80] (Scheme 12.28).

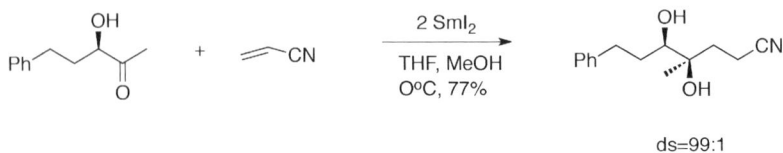

Scheme 12.25 SmI_2 and five membered chelate model.

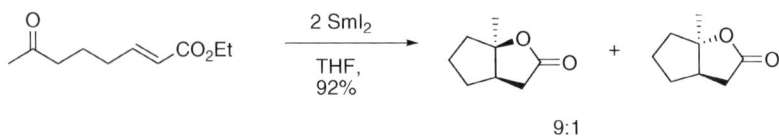

Scheme 12.26   Ketyl coupling and cyclization reaction.

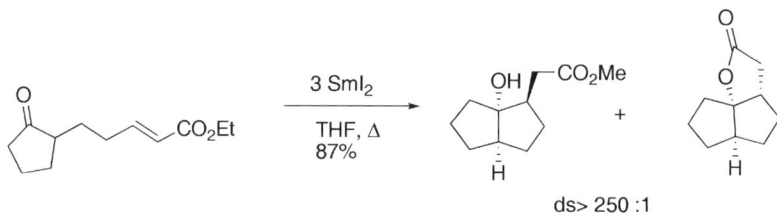

ds> 250 :1

Scheme 12.27 SmI$_2$ mediated stereospecific cyclization reactions.

Seven-[73a] and eight-membered rings[81] and bicyclic and bridged systems[81] are readily accessible using this methodology (Scheme 12.29).

Two efficient systems for the intermolecular addition of ketyl radicals to acrylates and for ketyl cyclizations employing chiral auxiliaries have recently been described by Fukuzawa *et al.*[82] and Molander *et al.*[83] Although the use of auxiliaries does not fit the subject of reagent control in a strict sense, these examples show aspects of the coordination chemistry of samarium. Fukuzawa demonstrated that (1*R*,2*S*)-*N*-methyl ephedrine is an excellent chiral auxiliary for the addition of a wide variety of ketyl radicals to acrylates (Scheme 12.30).

The ephedrine is vastly superior to esters containing no groups for binding samarium, *e.g.*, menthol. It has been postulated that the auxiliary enables binding of both the acrylate and the ketyl radical to samarium in a sterically well-defined manner. In this manner, excellent stereoselectivity is achieved. This assumption is supported by the observation that addition of hexam-

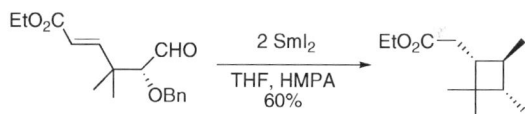

Scheme 12.28 SmI$_2$ mediated stereospecific cyclization towards four membered ring.

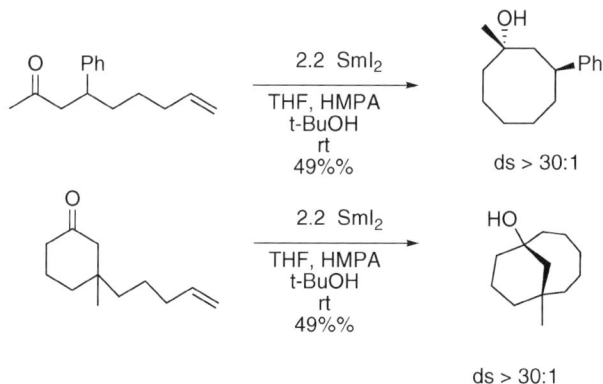

Scheme 12.29 Formation of seven and eight membered rings and SmI$_2$.

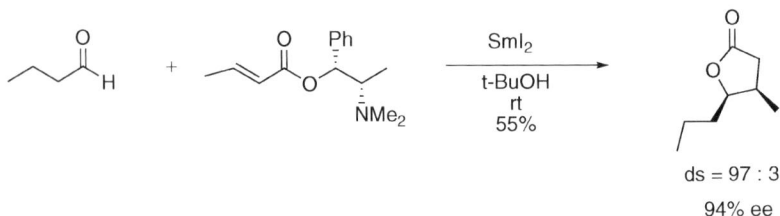

ds = 97 : 3

94% ee

Scheme 12.30   Synthesis of ephedrine derivatives using SmI$_2$.

ethylphosphoramide (HMPA) results in complete loss of selectivity. No structures of the chelates were proposed. Molander *et al.* reported chiral auxiliaries for cyclization reactions based on tartaric acid amides.[83] A typical example with the best auxiliary is shown in Scheme 12.31.

The improved donor ability of the amide compared to ester groups is crucial for the success of the reaction. Both relative stereoselectivity and stereo-induction from the chiral auxiliary are usually excellent. It should be noted that the sense of relative stereoselectivity is reversed compared to samarium diiodide-induced cyclization containing no auxiliary ligand system. The effect of the auxiliary acting as ligand for the reactivity of samarium is therefore dramatic.

Recently, the first example of chiral ligand control in intermolecular ketyl additions was described by Mikami and Yamaoka.[84] This report constitutes a very interesting example of an enantioselective addition of ketyl radicals to acrylates. A number of ligands containing the highly polar P–O bond were examined. This approach is based on the beneficial effects of HMPA on samarium chemistry. By using "chiral HMPA", the coordination sphere of samarium was modified to allow for enantioselective synthesis. The best ligand described was the oxide of the well established (*R*)-BINAP (Scheme 12.32).

Enantiomeric excesses of up to 89% were obtained. However, diastereoselectivities were modest in some cases. This was due to an essentially non-stereoselective protonation of the obtained enolates. Using different protic acids could result in further improvements. This promising approach is very interesting in connection with the recently developed reactions catalytic in

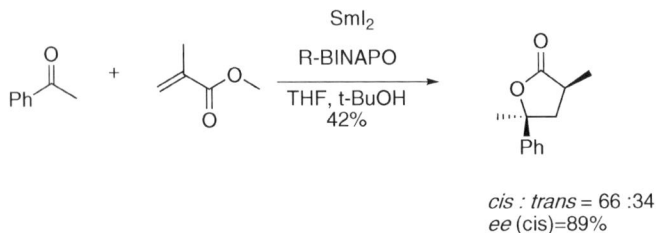

cis : *trans* = 66 :34
ee (cis)=89%

Scheme 12.31   Cyclization mediated by SmI$_2$.

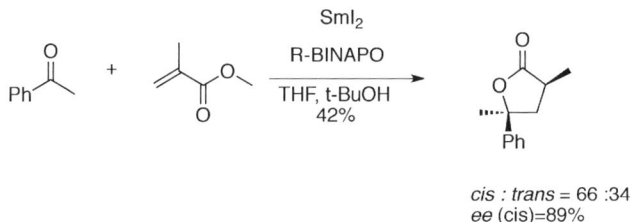

Scheme 12.32 Enantioselective radical addition under chiral ligand control.

samarium.[85] Another application of ketyl cyclizations is the addition to nitriles, hydrazones, and oximes. These transformations are possible using samarium diiodide[86–88] (Scheme 12.33).

As in the addition to carboncarbon multiple bonds, diastereoselectivities are high, although the yields are not always as satisfying. Itoh *et al.* described an intriguing titanocene-based system for the addition of ketones to nitriles.[89] As in the pinacol coupling, the phenyl ligand on titanium is essential for the ability of the reagent to transfer electrons (Scheme 12.34). The authors convincingly demonstrated that the reagent is also necessary to activate the nitrile group by complexation toward attack by the ketyl radical. Both the cyclopentadienyl and the phenyl ligands offer potential for enantioselective synthesis.

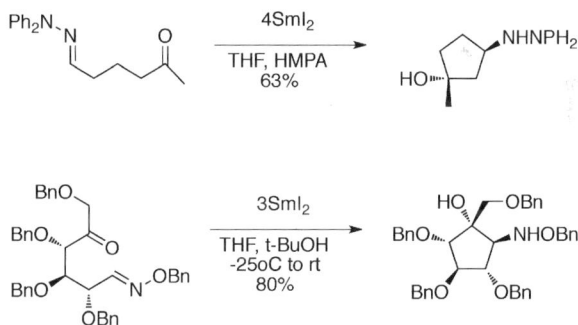

Scheme 12.33 Addition to nitriles, hydrozones and oximes via ketyl cyclization mediated by SmI_2.

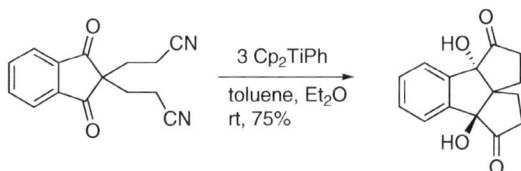

Scheme 12.34 Titanocene-based system for the addition of ketones to nitriles.

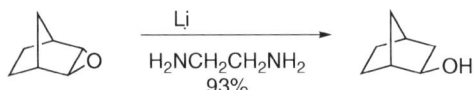

Scheme 12.35 Early epoxide ring opening via electron transfer.

12.6 EPOXIDES AS RADICAL PRECURSORS

12.6.1 Stoichiometric Reagents

Probably the first examples of epoxide opening *via* electron transfer were reported by Birch in 1950, followed by others.[90] An example of this type of transformation is shown in Scheme 12.35.

It was originally proposed that the reactions proceed through a nucleophilic opening of the epoxide *via* a solvated electron to yield the radical anion, an α-lithium oxyradical. However, it seems more likely that, as in the formation of ketyl anions, an electron is transferred to the epoxide with concomitant opening of the radical anion. Clearly the use of solvated electrons does not readily allow for a reagent-controlled course of the reaction and, thus, structurally more complex electron-transfer reagents are desirable. The first convincing evidence that α-metaloxy radicals can be obtained from epoxides *via* electron transfer emerged from investigations by Kochi *et al.* in 1968.[91] Deoxygenation of cyclohexene oxide and styrene oxide in the presence of chromium(II) reagents yielded cyclohexene and styrene in high yields. This finding was explained by the presence of a long-lived intermediate trapped by a second equivalent of the chromium(II) reagent. The resulting species, a α-metaloxy metal compound, fragmented to give the olefin. To achieve C–C bond formation or reduction of the α-metaloxy radical with hydrogen-atom donors, it is essential that the subsequent reaction of the α-metaloxy radical is faster than the trapping with a second equivalent of the electron-transfer reagent and concomitant α-elimination. Thus, highly active electron donors such as SmI$_2$ were, as yet, not suitable for this purpose and led to clean deoxygenation.[92] As expected, this elimination yielded mixtures of the (*E*)- and (*Z*)-olefins (Scheme 12.36). This observation can be an intermediate (**33–37**) that is long-lived enough to rotate around the C–C bond before being trapped by a second equivalent of SmI$_2$.

An important discovery was made in 1986 by Bartmann.[93] He demonstrated that epoxides could be reductively opened to α-lithio lithiumoxy compounds at low temperatures by radical anions of aromatic compounds, *e.g.*, of biphenyl.

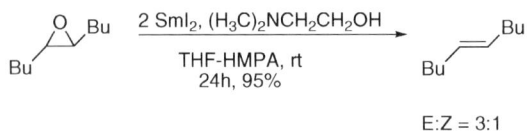

E:Z = 3:1

Scheme 12.36 Clean deoxygenation and SmI$_2$.

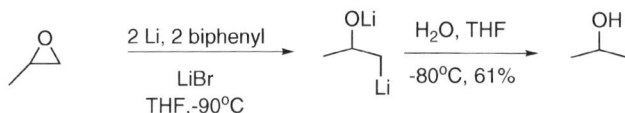

Scheme 12.37 Epoxide and organolithiums.

These organometallic intermediates generally did not decompose instantaneously due to the low tendency to eliminate the ionic "O_2^-" group. Epoxides are opened to give the less substituted organolithium compound. These species could be trapped with reactive electrophiles, *e.g.*, protons, allylic halides, and aldehydes, as shown in Scheme 12.37.

However, the low thermal stability of these intermediates and their high reactivity makes them somewhat difficult to use in the synthesis of complex organic molecules. Interesting applications have nevertheless been reported by Cohen *et al.*[94] and Yus *et al.*[95] Especially attractive substrates employed in these studies are "Sharpless epoxides"[96] (Scheme 12.38).

In this manner, a number of 1,3-diols can be readily obtained. The mechanism of these transformations has been studied in some detail by Cohen *et al.*[97] Calculations suggest that an electron is transferred from lithium to the epoxide to yield the epoxide radical anion as shown in Scheme 12.38. This unstable intermediate fragments to yield a α-lithio oxy radical that is subsequently trapped by a second equivalent of the aromatic radical anion. The crucial α-lithio lithiumoxy compound is obtained. The reason for the formation of the less stable radical during these transformations is explained by the formation of the more stable higher substituted alkoxide. The difference in stability between a secondary and primary alkoxide is obviously greater than the difference in stability of a primary and secondary radical! Interestingly the lithium ion was postulated to have very little influence on the course of the reaction. Therefore, reagent control is unfortunately difficult to achieve in these reactions also.

Important steps toward reagent-controlled epoxide openings were achieved between 1988 and 1994 when Nugent and RajanBabu discovered that titanocene(III) complexes are useful stoichiometric reagents for the reductive opening of epoxides with or without deoxygenation.[98] Obviously the reduced redox potential of the titanium(III) reagent compared to samarium diiodide combined with the higher steric demand of the cyclopentadienyl ligands can, if desired, prevent trapping of the α-metaloxy radicals with the titanium reagent under properly chosen conditions. The usual reactivity of carbon-centered radicals toward radical traps can be exploited in synthetically useful reactions. Therefore, the cyclopentadienyl ligands of titanium determine the chemos-

Scheme 12.38 Mechanistic aspects.

electivity of the reaction by tuning the redox properties and the steric demand of the metal complex. Reagent control is also exercised in the formation of the higher substituted radical, *i.e.*, the regioselectivity of epoxide opening. It seems that after complexation of the epoxide by the titanocene(III) reagent, the resulting adduct, presumably the radical anion of the epoxide bound to a titanocene(IV) species,[97] avoids substantial unfavorable steric interactions between the metal complex and the bulky substituent on the epoxide in opening the epoxide. Thus, the higher substituted α-titanoxy radical is formed. This selectivity is complementary to the above-mentioned Bartmann opening with aromatic radical anions, although one can also imagine a reversible epoxide opening with the titanocene(III) reagent leading to the more stable radical (Figure 12.4).

This typical Curtin–Hammett scenario[99] seems unlikely. If an equilibrium existed, the ratio of the products formed should depend on the radical trap employed. This is, however, not the case. *tert*-Butyl acrylate, acrylonitrile, and 1,4-cyclohexadiene give the same ratio of products derived from decene oxide.[98] Thus, it seems that the regioselectivity of epoxide opening is determined by the interaction of the metal's ligand sphere with the substrate, the typical scenario of reagent control. Epoxides can be readily deoxygenated in the presence of low-valent titanocene reagents under extremely mild conditions. Schobert,[100] and independently Nugent and RajanBabu,[98] provided convincing evidence that α-metaloxy radicals are indeed intermediates in these reactions. Both *cis*- and *trans*-5,6-epoxy decane yield the same 27:73 mixture of (*E*)- and (*Z*)-5-decene as products.[98] Both deoxygenations are thought to proceed *via* the same long-lived α-titanoxy radical that can rotate freely around the adjacent carbon–carbon bond to yield the same mixture of α-metal metaloxy species. After elimination, both (*E*)- and (*Z*)-5-decenes are obtained. This very mild deoxygenation procedure[98] has been used

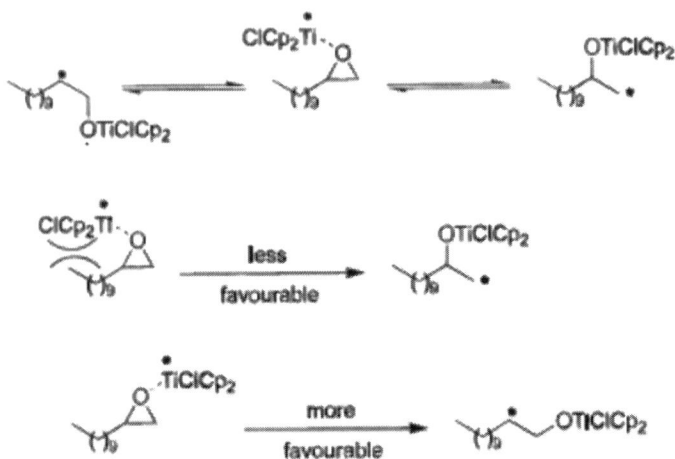

Figure 12.4   Possible mechanisms of titanocene-initiated epoxide opening.

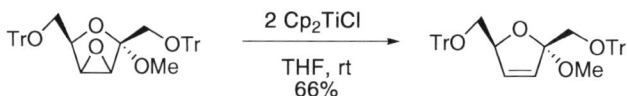

Scheme 12.39 Mild deoxygenation procedure.

in the synthesis of a number of highly acid-sensitive products (Scheme 12.39) that are not readily accessible *via* different methods.

The method is especially useful in the synthesis of deoxy sugar derivatives since the corresponding epoxides are readily accessible. An elegant application of the deoxygenation reaction is the synthesis of enantiomerically pure allylic alcohols from "Sharpless epoxides" as demonstrated in Scheme 12.40.[101]

It should be noted that the disadvantage of the loss of one-half the allylic alcohol as in the kinetic resolutions of allylic alcohols is not a problem when this protocol is employed. Considering these results, it is obvious that titanocene(III) chloride is a superior reagent compared to samarium diiodide in terms of chemoselectivity and yields for the deoxygenation of epoxides. An interesting and preparatively important extension of the deoxygenation emphasizing the radical character of the pivotal intermediates is the reduction of the α-metaloxy radical with hydrogen-atom donors, *e.g.*, 1,4-cyclohexadiene or *tert*-butyl thiol.[99] This useful transformation has a number of attractive features. Epoxides are opened with high regioselectivity opposite to that of S_N2 reactions to yield the corresponding alcohols. It is highly chemoselective, *e.g.*, ketones, tosylates, and halides are not reduced, and it is applicable in the synthesis of complex and sensitive molecules. The use of properly functionalized "Sharpless epoxides"[96] as substrates allows for an efficient synthesis of 1,2- and 1,3-diols. Tuning of the alcohol's protecting group allows the choice between a chelating and non-chelating binding mode of the epoxide by the titanium reagent. In this manner, it is possible to open the epoxide and obtain either the 1,2- or 1,3-diol with a high level of regioselectivity (Scheme 12.41).

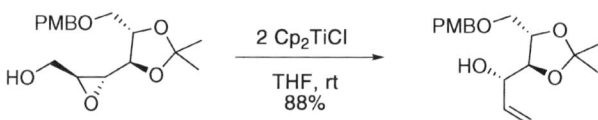

Scheme 12.40 Deoxygenation of enantiomerically pure allylic alcohols.

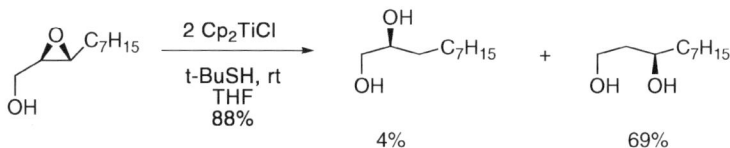

Scheme 12.41 Opening of epoxide and formation of 1,2- and 1,3-diols.

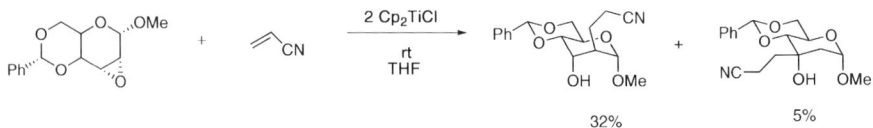

Scheme 12.42   Diastereoselectivity in carbohydrate system and titanium.

As in the deoxygenation reactions, the synthesis of sensitive molecules has been demonstrated. Deoxy sugars are an interesting class of compounds readily accessible by this methodology. Efficient C–C bond-forming reactions, being even more important than the formation of C–H bonds, become available too. The first class of these reactions to be discussed here is intermolecular addition to R,α-unsaturated carbonyl compounds.[98] After reductive epoxide opening, the resulting radical readily adds to esters of acrylic and methacrylic acid. The resulting compounds, α-hydroxyesters, can be lactonized, thus allowing a convenient entry to the synthesis of α-lactones from epoxides in a single step. Yields are usually high. The corresponding esters can be readily obtained by using *tert*-butyl acrylate. Unfortunately α-substitution of the ester is not tolerated. Diastereoselectivities in carbohydrate systems are the same as those in related systems using a free radical methodology (Scheme 12.42).

Using "Sharpless epoxides" as substrates, derivatives of 1,3-diols incorporating the additional ester group are readily obtained. This intriguing approach to hydroxyesters and lactones still offers synthetic potential. Acrylonitrile and methacrylnitrile are also useful radical traps in these reactions. The corresponding hydroxynitriles are valuable intermediates in organic synthesis. Arguably one of the synthetically most important applications of radicals is the 5-*exo*-cyclization reaction.[102,103] Suitably unsaturated epoxides are good substrates for titanocene(III)-initiated cyclization reactions as shown in Scheme 12.43.

The desired products are obtained in good to high yields, and diastereoselectivities are in the usual range for radical cyclizations.[104] Optically pure carbocyclic compounds can thus be readily obtained from carbohydrates. The resulting densely functionalized products are important intermediates for organic synthesis. In similar carbohydrate systems of free radicals studied by Giese *et al.*, comparable results were obtained.[105]

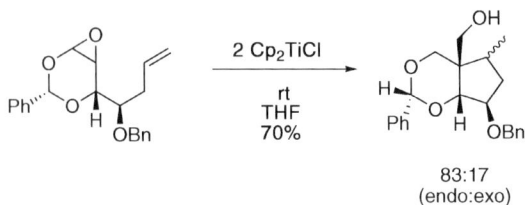

83:17
(endo:exo)

Scheme 12.43   Unsaturated oxides and titanocene (III)-initiated cyclization reaction.

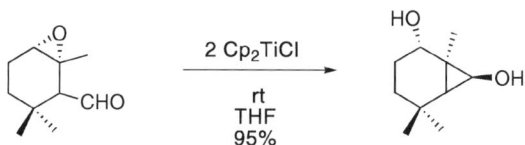

Scheme 12.44 Intramolecular addition to aldehydes and ketones.

Recently, the scope of the cyclization reactions was further increased by Fernández-Mateos *et al.* through intramolecular additions to aldehydes and ketones.[106] In the example shown in Scheme 12.44, a rare example of a highly efficient 3-*exo*-cyclization has been realized.

An intriguing aspect of titanocene-mediated reactions is the reductive trapping of the radical formed after the cyclization step by a second equivalent of the titanium(III) reagent. This is clearly advantageous compared to free radical cyclization reactions conducted in the presence of stannanes and silanes. In these latter cases, no further functionalization of the cyclization product can be achieved *in situ*. In the titanocene-initiated reductions, tandem reactions are readily possible. The nucleophilic titanium species obtained after reductive termination can be reacted with electrophiles other than protons, *e.g.*, iodine, to yield iodoalcohols. These compounds can be readily transformed to other useful products, *e.g.*, tetrahydrofuran derivatives (Scheme 12.45).

In principle, this approach combines the advantages of radical chemistry, *e.g.*, high functional group tolerance and mildness of the reaction conditions, with the advantages of organometallic chemistry, *e.g.*, determining the course of reactions by ligand variations. Unfortunately the applicability of these reactions is some what limited for practical use by the need to employ at least 2 equiv. of titanocene dichloride. This is especially disadvantageous for complexes that have to be synthesized in a number of steps and cannot be recycled.[107]

12.6.2 Titanocene-catalyzed Epoxide Openings

As outlined above and in the section on stoichiometric pinacol couplings, the main obstacle in investigating the influence of different ligands from simple cyclopentadienyl on the selectivities is the stoichiometric use of the titanocene complex. However, the improvement of reagent control in the regioselectivity of the opening of monosubstituted epoxides, the control of diastereoselectivity

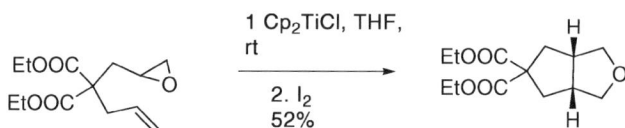

Scheme 12.45 Titanocene mediated reactions and reductive trapping of radicals.

of cyclization reactions, and the enantioselective opening of *meso*-epoxides are synthetically important goals. Clearly a catalytic reaction would be suited to circumvent the limitation of using stoichiometric amounts of titanocene complexes to achieve reagent control. Of course, the aim must be to develop a catalytic system preserving the advantages of the stoichiometric reagent. The planned catalytic cycle for the reductive opening in the presence of 1,4-cyclohexadiene as a hydrogen-atom donor is outlined in Figure 12.5.[108]

As in the titanocene-catalyzed protic pinacol coupling, the resulting titanocene alkoxide has to be cleaved to yield titanocene dichloride and to liberate the product of the reaction, the alcohol. It is crucial for the success of the catalytic reaction that the epoxide is not opened *via* S_N2 or S_N1 under the reaction conditions either by the employed acid or the metal salt MCl_2 formed during the reduction of Cp_2TiCl_2. As in the catalytic pinacol coupling, the base generated during protonation should not deactivate any titanium species by coordination and product inhibition must be avoided. Pyridine hydrochloride is known to open epoxides to the corresponding chlorohydrines as a mild protic acid in chloroform.[109] Thus, an acid with a higher pK_a in water than pyridine hydrochloride should be chosen. The acid should also be at least as strong as triethylamine hydrochloride in order to be able to quantitatively protonate alkoxides. 2,4,6-Collidine hydrochloride was a very useful acid to protonate titanocene alkoxides in combination with manganese as a reductant, as in the protic catalytic pinacol coupling. No significant amounts of by-products could be detected in the crude reaction mixture. Zinc performed distinctly inferiorly as a stoichiometric reductant. Presumably the zinc chloride formed during reduction of the titanocene dichloride complexed the epoxide

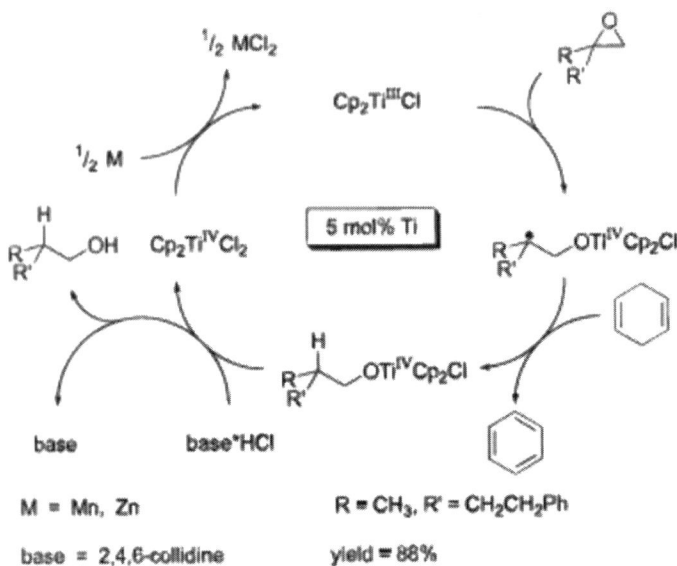

Figure 12.5   Titanocene-catalyzed reductive epoxide opening.

and liberated chlorohydrines *via* an S_N1 reaction. The catalytic system showed the same regioselectivity as the stoichiometric system. It should be noted, however, that in the case of 1-dodecene oxide this selectivity is somewhat higher in the catalytic transformation (94:6 *vs.* 88:12). This could be due to the 1- and 2-dodecanol formed during the course of the reaction. According to the general reasoning described for the selectivity of epoxide opening, this should lead to increased regioselectivity of the reaction. An important issue is the chemoselectivity of the catalytic epoxide opening. The stoichiometric reductive system has to be chosen carefully to ensure that electron transfer from the metal powder occurs exclusively to reduce titanocene dichloride. The high functional group tolerance of the stoichiometric reaction was preserved under the catalytic conditions. The mild acid 2,4,6-collidine hydrochloride is obviously not able to promote electron transfer from manganese to a variety of functional groups, *e.g.*, esters, nitriles, ketones, and even aliphatic aldehydes. Other easily reduced functional groups, *e.g.*, bromides, chlorides, and tosylates, are also perfectly stable. Our stoichiometric reductive system is mild and could be useful in other catalytic radical reactions, also. Preparatively more important than the catalytic reductive opening of epoxides are catalytic C–C bond forming reactions. Two important stoichiometric applications, cyclization reactions and intermolecular additions to R,α-unsaturated carbonyl compounds, have been reported to be successful using 2 equiv. of titanocene by Nugent and RajanBabu.[98] Recently the catalytic reaction was also developed. Intermolecular additions worked well under the conditions outlined in Figure 12.6.

The reaction exploits the stability of radicals and the instability of titanocene alkoxides and enolates under protic conditions. Once the enolate is formed, protonation liberates the reaction product with formation of titanocene dichloride. *In situ* reduction regenerates the redox-active titanocene(III) complex. The catalytic cycle is closed. Since the radicals formed after the intermolecular addition step can be trapped by the titanium(III) reagent or the stoichiometric reductant without concomitant elimination of the titanium oxo species, no hydrogen donor, *i.e.*, 1,4-cyclohexadiene, is necessary for the completion of the catalytic cycle. In intermolecular addition reactions, manganese as the stoichiometric reductant was by no means ideal.[110] Conversions were low even after prolonged reaction times. This is in contrast to the reductive opening yielding simple alcohols. With the additional ester group present, the product can chelate the titanium catalyst and initiate product inhibition. This problem was solved simply by using zinc dust as the stoichiometric reductant or by adding zinc chloride to the reaction mixture. In this manner, the catalyst can be reactivated. The stronger Lewis acid zinc dichloride chelates the product and liberates the catalyst. The same effect, although less pronounced, could be achieved by addition of excess collidine. The stability of the intermediate radical under the reaction conditions is crucial for the success of the reaction. Under the optimized conditions, the reaction can be run with as little as 1 mol% of the catalyst. For the use of titanocene

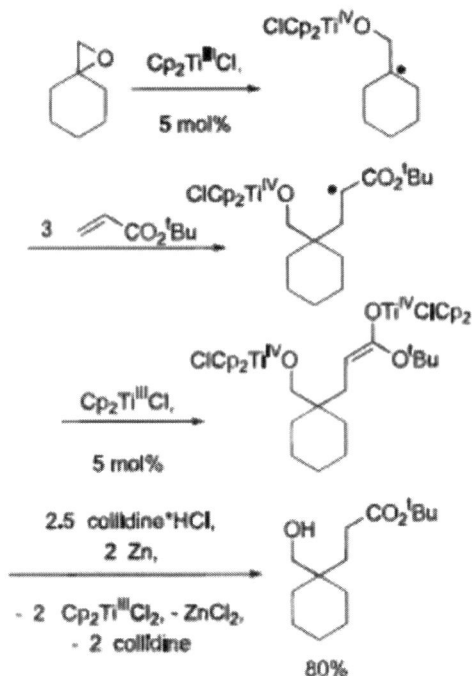

Figure 12.6 Titanocene-catalyzed intermolecular addition reactions.

complexes as catalysts in cyclization reactions, a similar concept led to an efficient reaction[111] (Figure 12.7).

   As in the stoichiometric reaction, the radical formed after the cyclization step is trapped by a titanocene(III) reagent. To achieve catalytic turnover, both titanium–carbon and titanium–oxygen bonds have to be cleaved while the reaction product and titanocene dichloride are liberated. Protonation constitutes an ideal means to achieve these goals. 2,4,6-Collidine hydrochloride represented a suitable acid in these reactions. The products can be isolated in good yields. Generally, diastereoselectivities were in the usual range for radical cyclizations. It should be noted that the diastereoselctivity in the formation of the [3.3.0] system is somewhat higher (98:2 *vs.* 90:10) than in the stoichiometric system. As in the reductive opening, this seems to be due to the presence of the product alcohol.

12.6.3 Catalytic Enantioselective Epoxide Openings

With the catalytic system described in the section above, the goal of enantioselective reagent-controlled radical reactions by variation of the cyclopentadienyl ligand was within reach. A good point to start with is the enantioselective opening of *meso*-epoxides *via* electron transfer. Many excellent examples of catalytic enantioselective openings of *meso*-epoxides by S_N2 reactions have recently been reported.[112] However, the S_N2 reactions are

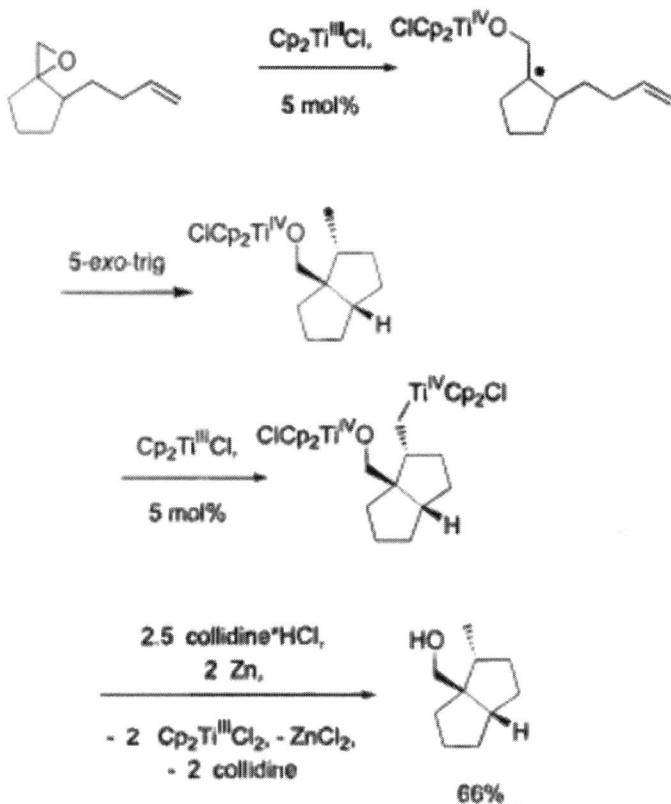

Figure 12.7 Titanocene-catalyzed cyclizations

conceptually different from the approach described here, because in S_N2 reactions the path of the incoming nucleophile has to be controlled. In the titanocene-catalyzed reaction, the intermediate radical has to be formed selectively. If an intermediate similar to the Bartmann opening is postulated here,[93,97] the selectivity determining interaction should be that of the epoxide radical anion with a titanocene(IV) complex as depicted in Figure 12.8.

According to the introductory remarks, reagent control is thus exercised in the radical forming step. Thus, two diastereomeric radicals are initially formed

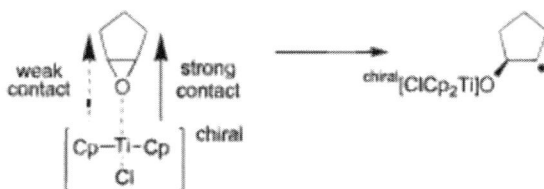

Figure 12.8 Plausible crucial intermediate in enantioselective epoxide openings.

due to the chirality of the titanocene complex. The diastereoselectivity of the following reaction may also be controlled by the ligand sphere of titanium. After protic cleavage of the titanium–oxygen bond, enantiomeric products are formed. This mechanistic reasoning allowed for the rational design of the cyclopentadienyl ligands.[113] To achieve efficient differentiation in the steric interaction of the catalyst with the *meso*-epoxide, the ligand should be able to interact with the substrate in regions distant from the initial binding site, the epoxy group. Thus, efficient chirality transfer from the periphery of the titanocene complex to regions of the substrate distant from the binding site of the catalyst has to be achieved. Inspection of the extensive literature on titanocene and cyclopentadienyl complexes[107] suggested ligands from terpenes as suitable for achieving this purpose.[114] In *ansa*-metallocenes that have been used in enantioselective catalysis with great success recently,[115] the chirality is centered around the metal. Chirality transfer to the periphery of these complexes is not obvious in studies of molecular models and the crystal-lographic structures. Epoxide **4** was chosen as a test substrate as shown in Scheme 12.46 because it is readily accessible from (*Z*)-butene diol in two steps and the absolute stereochemistry of the opening product can be established by synthesis of authentic samples from malic acid.

The results of the investigation of a number of titanocene complexes are shown in Figure 12.9.

Brintzinger's complex **2** shown in Figure 12.3 performed poorly concerning the enantioselectivity (56% ee) of the epoxide opening and the yield of product (55%) in the presence of 10 mol% catalyst. The titanocene complex[114] **5** obtained from (1*R*,2*S*,5*R*) menthol *via* tosylation, S$_N$2 reaction with sodium cyclopentadienide, and metallation performed somewhat better, although the axially positioned cyclopentadienyl group is not ideal. A satisfactory result was obtained with the ligand from *neo*-menthol **6** containing an equatorial cyclopentadienyl ligand.[116] The enantioselectivity of the opening reached synthetically useful levels (97:3), and the isolated yields were reasonable. Complex **7** with a ligand derived from phenyl menthone[117] performed well, giving an enantioselectivity of 96.5:3.5. Phenyl menthol[118] has already been extensively and successfully used as chiral auxiliary.[119] These results suggest that both **6** and **7**, after being reduced to the redox-active species, contain a chiral pocket well-suited for the steric differentiation of the enantiotopic groups of *meso*-epoxide **4**. The corresponding bis-*tert*-butyl ether epoxide constituted a more difficult example due to the increased steric demand of the bulky groups. Both catalysts performed distinctly worse. With **6**, an enantioselectivity of 92.5:7.5 was obtained, whereas **7** gave the lower value

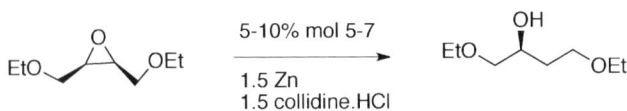

Scheme 12.46   Chirality transfer and epoxide opening.

5

6

7

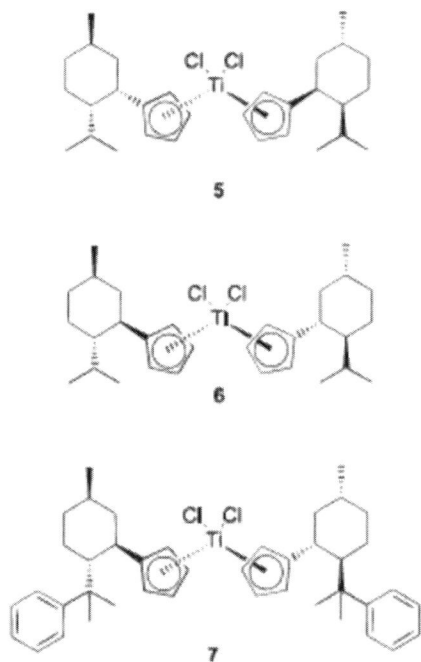

Figure 12.9   Titanocenes utilized in the enantioselective opening of *meso*-epoxides.

of 87.5:12.5. An interesting and demanding problem is the opening of cyclic *meso*epoxides, *e.g.*, cyclopentene oxide, and trapping of the resulting radical with an acrylate, *e.g.*, *tert*-butyl acrylate. Besides the enantioselectivity of epoxide opening, the diastereoselectivity of the C–C bond-forming step has to be controlled.[119] Complex **6** was the most selective catalyst, giving higher enantioselectivity (81% ee) while preserving high diastereoselectivity. Titanocene catalyst **7** gave a lower enantioselectivity (86.5:13.5) with cyclopentene oxide as the substrate. It should be noted that the diastereoselectivity of the addition reaction to *tert*-butyl acrylate (>97.5:<2.5) was substantially higher using **5**, **6**, and **7** than with Cp_2TiCl_2 as catalyst (86:14). Thus, the ligands derived from *neo*-menthol performed better than those derived from *neo*-phenyl menthol in all cases investigated. As for the opening of **4**, the Brintzinger complex **2** was not an efficient catalyst for the opening of cyclopentene oxide. Both chemical yield (24%) and enantioselectivity were low (29% ee). The diastereoselectivity of the addition to *tert*-butyl acrylate was rather low for this system (trans:cis = 90:10). The opening of cyclohexene and cycloheptene oxide with **7** as catalyst proceeded with somewhat higher enantioselectivity (91:9 in both cases). Diastereoselectivity of the addition reaction was lower than with cyclopentene oxide (81:19 for the cyclohexene oxide and 87:13 for cycloheptene oxide). In both cases this constituted an improvement compared to Cp_2TiCl_2 as the catalyst (about 60:40 and 70:30, respectively).

12.7   CONCLUSION AND FUTURE DIRECTION

Several fundamentals of reagent control in the reductive opening of *meso*-epoxides in stoichiometric and catalytic manner can be exercised both at the stage of radical generation and at the subsequent transformations of the formed radicals. Although some of the initial results are promising, further investigations have to establish whether reagent control in these reactions can be improved further to reach practically useful levels of stereoselection and catalytic activity in other simple cases and in natural product synthesis.

REFERENCES

1. M. Gomberg, *Chem. Ber.*, 1900, **33**, 3150.
2. (a) B. Giese, *Radicals in Organic Synthesis: Formation of Carbon–Carbon Bonds*, Pergamon Press, Oxford, 1986; (b) D. P. Curran, N. A. Porter and B. Giese, *Stereochemistry of Radical Reactions*, VCH, Weinheim, 1996; (c) J. Fossey, D. Lefort and J. Sorba, *Free Radicals in Organic Chemistry*, Wiley, New York, 1995; (d) T. Linker and M. Schmittel, *Radikale und Radikalionen in der Organischen Synthese*, Wiley-VCH, Weinheim, 1998.
3. (a) B. Giese, *Angew. Chem.*, 1983, **95**, 771 (*Angew. Chem., Int. Ed. Engl.*, 1983, **22**, 753); (b) B. Giese, *Angew. Chem.*, 1985, **97**, 555 (*Angew. Chem., Int. Ed. Engl.*, 1985, **24**, 553); (c) B. Giese and J. Dupuis, *Angew. Chem.*, 1983, **95**, 633 (*Angew. Chem., Int. Ed. Engl.*, 1983, **22**, 622); (d) J. Dupuis, B. Giese, D. Ruegge, H. Fischer, H.-G. Korth and R. Sustmann, *Angew. Chem.*, 1984, **96**, 887 (*Angew. Chem., Int. Ed. Engl.*, 1984, **23**, 896).
4. For some leading references, see: (a) P. Bazukis, O. O. S. Campos and M. L. F. Bazukis, *J. Org. Chem.*, 1976, **41**, 3261; (b) D. P. Curran and D. M. Rakiewicz, *J. Am. Chem. Soc.*, 1985, **107**, 1448; (c) D. P. Curran and M.-H. Chen, *Tetrahedron Lett.*, 1985, **26**, 4991; (d) S. L. Danishefsky and J. S. Panek, *J. Am. Chem. Soc.*, 1987, **109**, 917; (e) Y.-J. Chen and W.-Y. Lin, *Tetrahedron Lett.*, 1992, **33**, 1749.
5. For a comprehensive account, see: J. J. Brocks, H.-D. Beckhaus, A. L. J. Beckwith and C. Ruchardt, *J. Org. Chem.*, 1998, **63**, 1935 and references therein.
6. See, for example: S. J. Francisco and J. A. Montgomery, Jr, in *Energetics of Organic Free Radicals*, ed. J. A. M. Simoes, A. Greenberg and J. F. Liebmann, Blackie Academic & Professional, London, 1996.
7. (a) I. Ojima, *Catalytic Asymmetric Synthesis*, VCH, Weinheim, 1993; (b) R. Noyori, *Asymmetric Catalysis in Organic Synthesis*, Wiley, New York, 1994.
8. A. Gansäuer, *Synlett,* 1998, 801.
9. (a) N. A. Porter, B. Giese and D. P. Curran, *Acc. Chem. Res.*, 1991, **24**, 296; (b) W. Smadja, *Synlett,* 1994, 1.
10. For a review and recent examples, see: (a) G. Pattenden, *Chem. Soc. Rev.*, 1988, **17**, 361; (b) S. Busato and R. Scheffold, *Helv. Chim. Acta,* 1994, **77**,

92; (c) D. L. Zhou, P. Walder, R. Scheffold and L. Walder, *Helv. Chim. Acta,* 1992, **75**, 995.

11. For some recent work, see: (a) B. P. Branchaud and R. M. Slade, *Tetrahedron Lett.*, 1994, **35**, 4071; (b) R. M. Slade and B. P. Branchaud, *J. Org. Chem.*, 1998, **63**, 3544.

12. For some recent references, see: (a) T. Ando, M. Kamigaito and M. Sawamoto, *Tetrahedron,* 1997, **53**, 15445; (b) V. Percec, B. Barboiu and H.-J. Kim, *J. Am. Chem. Soc.*, 1998, **120**, 305; (c) F. Simal, A. Demonceau and A. F. Noels, *Angew. Chem.,* 1999, **111**, 559 (*Angew. Chem., Int. Ed.*, 1999, **38**, 538).

13. (a) M. B. Andrus, A. B. Argade, X. Chen and M. G. Pamment, *Tetrahedron Lett.*, 1995, **36**, 2945; (b) M. B. Andrus, D. Asgari and J. A. Scafani, *J. Org. Chem.*, 1997, **62**, 9365; (c) T. Kasuki, in *Comprehensive Asymmetric Catalysis*, ed. E. N. Jacobsen, A. Pfaltz and H. Yamamoto, Springer, Berlin, 1999, vol. 2, p. 791.

14. (a) M. D. Kaufman, P. A. Grieco and D. W. Bougie, *J. Am. Chem. Soc.*, 1993, **115**, 11648; (b) R. Breslow, X. J. Zhang and Y. Huang, *J. Am. Chem. Soc.*, 1997, **119**, 4535; (c) B. Meunier, in *Transition Metals for Organic Synthesis,* ed. M. Beller and C. Bolm, Wiley-VCH, Weinheim, 1998, vol. 2, p. 173.

15. For two examples, see: (a) P. G. Schultz, J. S. Taylor and P. B. Dervan, *J. Am. Chem. Soc.*, 1982, **104**, 6861; (b) P. G. Schultz and P. B. Dervan, *J. Am. Chem. Soc.*, 1983, **105**, 7748.

16. (a) J. Stubbe and J. W. Kozarich, *Chem. Rev.*, 1987, **87**, 1107; (b) C. A. Claussen and E. C. Long, *Chem Rev.*, 1999, **99**, 2797.

17. (a) G. A. Molander, *Chem. Rev.*, 1992, **92**, 29; (b) G. A. Molander and C. R. Harris, *Chem. Rev.*, 1996, **96**, 307; (c) G. A. Molander and C. R. Harris, *Tetrahedron,* 1998, **54**, 3321.

18. P. Renaud and M. Gerster, *Angew. Chem.*, 1998, **110**, 2704 (*Angew. Chem., Int. Ed.*, 1998, **37**, 2562).

19. M. P. Sibi and N. A. Porter, *Acc. Chem. Res.*, 1999, **32**, 163.

20. H. Schumann, B. Pachaly and B. C. Schutze, *J. Organomet. Chem.*, 1984, **265**, 145.

21. D. Nanni and D. P. Curran, *Tetrahedron: Asymmetry,* 1996, **7**, 2417.

22. (a) M. Blumenstein, K. Schwartzkopf and J. Metzger, *Angew. Chem.*, 1997, **109**, 245 (*Angew. Chem., Int. Ed. Engl.*, 1997, **36**, 235); (b) K. Schwartzkopf, M. Blumenstein, A. Hayen and J. Metzger, *Eur. J. Org. Chem.*, 1998, 177.

23. D. Dakternieks, K. Dunn, V. T. Perchyonok and C. H. Schiesser, *J. Chem. Soc., Chem. Commun.*, 1999, 1665.

24. (a) J. H. Wu, R. Radinov and N. A. Porter, *J. Am. Chem. Soc.*, 1995, **117**, 11029; (b) J. H. Wu, G. Zhang and N. A. Porter, *Tetrahedron Lett.*, 1997, **38**, 2067; (c) N. A. Porter, J. H. Wu, G. Zhang and A. D. Reed, *J. Org. Chem.*, 1997, **62**, 6702; (d) M. P. Sibi, J. Ji, J. H. Wu, S. Gürtler and N. A.

Porter, *J. Am. Chem. Soc.,* 1996, **118**, 9200; (e) M. P. Sibi and J. Ji, *J. Org. Chem.,* 1997, **62**, 3800.

25. (a) G. M. Robertson, in *Comprehensive Organic Synthesis*, ed. B. M. Trost, I. Fleming and G. Pattenden, Pergamon Press, Oxford, 1991, vol. 3, p. 563; (b) T. Wirth, *Angew. Chem.*, 1996, **108**, 65 (*Angew. Chem., Int. Ed. Engl.*, 1996, **35**, 61).

26. (a) G. A. Molander and J. A. McKie, *J. Org. Chem.*, 1991, **56**, 4112; (b) G. A. Molander and C. R. Harris, *J. Org. Chem.*, 1997, **62**, 2944.

27. (a) M. Kameyama, N. Kamigata and M. Kobayashi, *J. Org. Chem.*, 1987, **52**, 3312; (b) M. Kameyama and N. Kamigata, *Bull. Chem. Soc. Jpn.*, 1989, **62**, 648.

28. R. Fittig, *Liebigs Ann.*, 1859, **110**, 23.

29. (a) J. E. McMurry, J. G. Rico and Y. N. Shih, *Tetrahedron Lett.*, 1989, **30**, 1173; (b) J. E. McMurry and R. G. Dushin, *J. Am. Chem. Soc.*, 1990, **112**, 6942; (c) K. C. Nicolaou, Z. Yang, E. J. Sorensen and M. Nakada, *J. Chem. Soc., Chem. Commun.*, 1993, 1024; (d) K. C. Nicolaou, Z. Yang, J. J. Liu, H. Ueno, P. G. Nantermet, R. K. Guy, C. F. Claiborne, J. Renaud, E. A. Couladouros, K. Paulvannan and E. J. Sorensen, *Nature,* 1994, **367**, 630.

30. For some early work and a recent review on the chemistry of low-valent titanium species, see: (a) T. Mukaiyama, T. Sato and J. Hanna, *Chem. Lett.*, 1973, 1041; (b) J. E. McMurry and M. P. Fleming, *J. Am. Chem. Soc.*, 1974, **96**, 4708. (c) A. Fürstner and B. Bogdanovic, *Angew. Chem.*, 1996, **108**, 2582 (*Angew. Chem., Int. Ed. Engl.*, 1996, **35**, 2442).

31. H. G. Raubenheimer and D. Seebach, *Chimia*, 1986, **40**, 12.

32. A. Clerici, L. Clerici and O. Porta, *Tetrahedron Lett.*, 1996, **37**, 3035.

33. Y. Handa and J. Inanaga, *Tetrahedron Lett.,* 1987, **28**, 5717.

34. (a) M. L. H. Green and C. R. Lucas, *J. Chem. Soc., Dalton Trans.*, 1972, 1000; (b) R. S. P. Coutts, P. C. Wailes and R. L. Martin, *J. Organomet. Chem.*, 1973, **47**, 375; (c) D. Sekutowski, R. Jungst and G. D. Stucky, *Inorg. Chem.*, 1978, **17**, 1848; (d) D. W. Stephan, *Organometallics,* 1992, **11**, 996.

35. M. C. Barden and J. Schwartz, *J. Am. Chem. Soc.,* 1996, **118**, 5484.

36. T. Mukaiyama, A. Kagayama and I. Shiina, *Chem. Lett.,* 1998, 1107.

37. S. Matsubara, Y. Hashimoto, T. Okano and K. Utimoto, *Synlett,* 1999, 1411.

38. (a) H. L. Pedersen, T. B. Christensen, R. J. Enemærke, K. Daasbjerg and T. Skrydstrup, *Eur. J. Org. Chem.*, 1999, 565; (b) T. B. Christensen, D. Riber, K. Daasbjerg and T. Skrydstrup, *J. Chem. Soc., Chem. Commun.*, 1999, 2051.

39. (a) J. L. Chiara, W. Cabri and S. Hanessian, *Tetrahedron Lett.*, 1991, **32**, 1125; (b) J. Uenishi, S. Masuda and S. Wakabayashi, *Tetrahedron Lett.*, 1991, **32**, 5097; (c) T. Kan, S. Hosokawa, S. Nara, M. Oikawa, S. Ito, F. Matsuda and H. Shirahama, *J. Org. Chem.*, 1994, **59**, 5532.

40. F. A. Cotton, S. A. Duraj and W. J. Roth, *Inorg. Chem.*, 1985, **24**, 913.

41. (a) J. H. Freudenberger, A. W. Konradi and S. F. Pedersen, *J. Am. Chem. Soc.*, 1989, **111**, 8014; (b) A. W. Konradi, S. J. Kemp and S. F. Pedersen, *J. Am. Chem. Soc.*, 1994, **116**, 1316.
42. (a) A. W. Konradi and S. F. Pedersen, *J. Org. Chem.*, 1992, **57**, 28; (b) M. T. Reetz and N. Griebenow, *Liebigs Ann.*, 1996, 335; (c) B. Kammermeier, G. Beck, W. Holla, D. Jacobi, B. Napierski and H. Jendralla, *Chem.–Eur. J.*, 1996, **2**, 307.
43. M. T. Reetz, *Angew. Chem.*, 1984, **96**, 542 (*Angew. Chem., Int. Ed. Engl.*, 1984, **23**, 556).
44. L. E. Manzer, *Inorg. Chem.*, 1977, **16**, 525.
45. (a) E. J. Roskamp and S. F. Pederson, *J. Am. Chem. Soc.*, 1987, **109**, 3152; (b) E. J. Roskamp and S. F. Pedersen, *J. Am. Chem. Soc.*, 1987, **109**, 6551.
46. A. Fürstner and A. Hupperts, *J. Am. Chem. Soc.*, 1995, **117**, 4468.
47. (a) T. Hirao, T. Hasegawa, Y. Muguruma and I. Ikeda, *J. Org. Chem.*, 1996, **61**, 366; (b) T. Hirao, *Synlett,* 1999, 175.
48. A. Fürstner, *Chem.–Eur. J.*, 1998, **4**, 567.
49. (a) A. Fürstner and N. Shi, *J. Am. Chem. Soc.,* 1996, **118**, 2533; (b) A. Fürstner and N. Shi, *J. Am. Chem. Soc.,* 1996, **118**, 12349; (c) A. Fürstner, *Chem. Rev.*, 1999, **99**, 991.
50. M. Bandini, P. G. Cozzi, P. Melchiorre and A. Umani-Ronchi, *Angew. Chem.*, 1999, **111**, 3558 (*Angew. Chem., Int. Ed.*, 1999, **38**, 3357).
51. (a) O. Maury, C. Villiers and M. Ephritikhine, *New J. Chem.*, 1997, **21**, 137; (b) C. Villiers and M. Ephritikhine, *Angew. Chem.*, 1997, **109**, 2477 (*Angew. Chem., Int. Ed. Engl.,* 1997, **36**, 2380).
52. M. Ephritikhine, *J. Chem. Soc., Chem. Commun.*, 1998, 2549.
53. R. Nomura, T. Matsuno and T. Endo, *J. Am. Chem. Soc.*, 1996, **118**, 11666.
54. (a) A. Gansäuer, *J. Chem. Soc., Chem. Commun.*, 1997, 457; (b) A. Gansäuer, M. Moschioni and D. Bauer, *Eur. J. Org. Chem.*, 1998, 1923.
55. J.-H. So, M. K. Park and P. Boudjok, *J. Org. Chem.*, 1988, **53**, 5871.
56. M. S. Dunlap and K. M. Nicholas, *Synth. Commun.*, 1999, **27**, 1097.
57. (a) F. R. W. P. Wild, L. Zsolnai, G. Huttner and H. H. Brintzinger, *J. Organomet. Chem.,* 1982, **232**, 233; (b) S. Collins, B. A. Kuntz, N. J. Taylor and D. G. Ward, *J. Organomet. Chem.,* 1988, **342**, 21; (c) J. B. Jaquith, J. Guan, S. Wang and S. Collins, *Organometallics,* 1995, **14**, 1079.
58. A. Gansäuer, *Synlett,* 1997, 363.
59. A. Gansäuer and M. Moschioni, unpublished results.
60. T. Hirao, B. Hatano, M. Asahara, Y. Mugurama and A. Ogawa, *Tetrahedron Lett.*, 1998, **39**, 5247.
61. A. Svatos and W. Boland, *Synlett,* 1998, 549.
62. T. A. Lipski, M. A. Hilfiker and S. G. Nelson, *J. Org. Chem.*, 1997, **62**, 4566.

63. M. Bandini, P. G. Cozzi, S. Morganti and A. Umani-Ronchi, *Tetrahedron Lett.*, 1999, **40**, 1997.

64. Y. Yamamoto, R. Hattori and K. Itoh, *J. Chem. Soc., Chem. Commun.*, 1999, 825.

65. (a) T. Hirao, M. Asahara, Y. Muguruma and A. Ogawa, *J. Org. Chem.*, 1998, **63**, 2812; (b) T. Hirao, B. Hatano, Y. Imamoto and A. Ogawa, *J. Org. Chem.*, 1999, **64**, 7665.

66. B. Hatano, A. Ogawa and T. Hirao, *J. Org. Chem.*, 1998, **63**, 9421.

67. For some recent references, see: (a) C.-J. Li, *Tetrahedron,* 1996, **52**, 5643; (b) C.-J. Li and T. J. Chan, *Organic Reactions in Aqueous Media*, John Wiley & Sons, New York, 1997; (c) C.-J. Li and W.-C. Zhang, *J. Am. Chem. Soc.*, 1998, **120**, 9102; (d) W.-C. Zhang and C.-J. Li, *J. Org. Chem.*, 1999, **64**, 3230.

68. *Handbook of Chemistry and Physics*, ed. D. R. Lide, CRC Press, Boca Raton, FL, 78th edn, 1997, ch. 8, pp. 45–55.

69. (a) A. Gansäuer and D. Bauer, *J. Org. Chem.*, 1998, **63**, 2070; (b) A. Gansäuer and D. Bauer, *Eur. J. Org. Chem.*, 1998, 2673.

70. (a) P. Girard, J.-L. Namy and H. B. Kagan, *J. Am. Chem. Soc.*, 1980, **102**, 2693; (b) For the first preparation of SmI_2 in THF see: J.-L. Namy, P. Girard and H. B. Kagan, *New J. Chem.*, 1977, **1**, 5.

71. For some reviews on SmI_2, see: (a) H. B. Kagan and J.-L. Namy, *Tetrahedron,* 1986, **42**, 6573; (b) H. B. Kagan, *New J. Chem.*, 1990, **14**, 453; (c) J. A. Soderquist, *Aldrichimica Acta,* 1991, **24**, 15; (d) T. Skrydstrup, *Angew. Chem.*, 1997, **109**, 355 (*Angew. Chem., Int. Ed.*, 1997, **36**, 345; (e) A. Krief and A.-M. Laval, *Chem Rev.*, 1999, **99**, 745.

72. S.-I. Fukuzawa, A. Nakanishi, T. Fujinama and S. Sakai, *J. Chem. Soc., Chem. Commun.*, 1986, 624; (b) S.-I. Fukuzawa, A. Nakanishi, T. Fujinami and S. Sakai, *J. Chem. Soc., Perkin Trans. 1,* 1988, 1669; (c) K. Otsubo, J. Inanaga and M. Yamaguchi, *Tetrahedron Lett.,* 1986, **27**, 5763; (d) J. Inanaga, O. Ujikawa, Y. Handa, K. Otsubo and M. Yamaguchi, *J. Alloys Compd.*, 1993, **192**, 197.

73. (a) S.-I. Fukuzawa, M. Iida, A. Nakanishi, T. Fujinami and S. J. Sakai, *J. Chem. Soc., Chem. Commun.,* 1987, 920; (b) H. B. Kagan and J.-L. Namy, in *Lanthanides: Chemistry and Use in Organic Synthesis*, ed. S. Kobayashi, Springer, Berlin, 1999.

74. K. Otsubo, K. Kawamura, J. Inanaga and M. Yamaguchi, *Chem. Lett.*, 1987, 1487.

75. R. V. Lloyd and J. G. Causey, *J. Chem. Soc., Perkin Trans. 2*, 1981, 1143; (b) Y.-D. Wu and K. N. Houk, *J. Am. Chem. Soc.*, 1992, **114**, 1656.

76. (a) M. Kawatsura, F. Matsuda and H. Shirahama, *J. Org. Chem.*, 1994, **59**, 6900; (b) F. Matsuda, M. Kawatsura, K. Hosaka and H. Shirahama, *Chem.–Eur. J.*, 1999, **5**, 3252.

77. E. J. Enholm and A. Trivellas, *Tetrahedron Lett.*, 1989, **30**, 1063.

78. G. A. Molander and C. Kenny, *J. Am. Chem. Soc.*, 1989, **111**, 8236.

79. M. Kito, T. Sakai, K. Yamada, F. Matsuda and H. Shirahama, *Synlett,* 1993, 158.

80. K. Weinges, S. B. Schmidbauer and H. Schick, *Chem. Ber.*, 1994, **127**, 1305.

81. (a) Y.-S. Hon, L. Lu and K.-P. Chu, *Synth. Commun.*, 1991, **21**, 1981; (b) G. A. Molander and J. A. McKie, *J. Org. Chem.*, 1994, **59**, 3186.

82. S.-I. Fukuzawa, K. Seki, M. Tasuzawa and K. Mutoh, *J. Am. Chem. Soc.*, 1997, **119**, 1482.

83. G. A. Molander, J. C. McWilliams and B. C. Noll, *J. Am. Chem. Soc.*, 1997, **119**, 1265.

84. K. Mikami and M. Yamaoka, *Tetrahedron Lett.*, 1998, **39**, 4501.

85. E. J. Corey and G. Z. Zheng, *Tetrahedron Lett.*, 1997, **38**, 2045.

86. G. A. Molander and C. N. Chad, *J. Org. Chem.*, 1998, **63**, 9031.

87. (a) C. F. Sturino and A. G. Fallis, *J. Org. Chem.*, 1994, **59**, 6514; (b) C. F. Sturino and A. G. Fallis, *J. Am. Chem. Soc.*, 1994, **116**, 7447.

88. (a) J. Marco-Contelles, P. Gallego, M. Rodríguez-Fernández, N. Khiar, C. Destabel, M. Bernabé, A. Martínez-Grau and J. L. Chiara, *J. Org. Chem.*, 1997, **62**, 7397; (b) H. Miyabe, M. Torieda, K. Inoue, K. Tajiri, T. Kigushi and T. Naito, *J. Org. Chem.*, 1998, **63**, 4397.

89. (a) Y. Yamamoto, D. Matsumi and K. Itoh, *J. Chem. Soc., Chem. Commun.*, 1998, 875; (b) Y. Yamamoto, D. Matsumi, R. Hattori and K. Itoh, *J. Org. Chem.*, 1999, **64**, 3224.

90. (a) A. J. Birch, *J. Proc. R. Soc. N. S. W.*, 1950, **83**, 245; (b) A. S. Hallsworth and H. B. Henbest, *J. Chem. Soc.*, 1957, 4604; (c) A. S. Hallsworth and H. B. Henbest, *J. Chem. Soc.*, 1960, 3571; (d) H. C. Brown, S. Ikegami and J. H. Kawakami, *J. Org. Chem.*, 1970, **35**, 3243; (e) R. A. Benkeser, A. Rappa and L. A. Wolsieffer, *J. Org. Chem.*, 1986, **51**, 3391.

91. J. K. Kochi, D. M. Singleton and L. J. Andrews, *Tetrahedron,* 1968, **24**, 3503.

92. M. Matsukawa, T. Tabuchi, J. Inanaga and M. Yamaguchi, *Chem. Lett.*, 1987, 2101.

93. E. Bartmann, *Angew. Chem.*, 1986, **98**, 629 (*Angew. Chem., Int. Ed. Engl.*, 1986, **25**, 855).

94. T. Cohen, I.-H. Jeong, B. Mudryk, M. Bhupathy and M. M. A. Awad, *J. Org. Chem.*, 1990, **55**, 1528.

95. A. Bachki, F. Foubelo and M. Yus, *Tetrahedron: Asymmetry,* 1995, **6**, 1907; (b) A. Bachki, F. Foubelo and M. Yus, *Tetrahedron: Asymmetry,* 1996, **7**, 2997.

96. T. Kasuki, in *Comprehensive Asymmetric Catalysis*, ed. E. N. Jacobsen, A. Pfaltz and H. Yamamoto, Springer, Berlin, 1999, vol. 2, p. 621.

97. A. E. Dorigo, K. N. Houk and T. Cohen, *J. Am. Chem. Soc.,* 1989, **111**, 8976.

98. (a) W. A. Nugent and T. V. RajanBabu, *J. Am. Chem. Soc.*, 1988, **110**, 8561; (b) T. V. RajanBabu and W. A. Nugent, *J. Am. Chem. Soc.,* 1989,

111, 4525; (c) T. V. RajanBabu, W. A. Nugent and M. S. Beattie, *J. Am. Chem. Soc.,* 1990, **112**, 6408; (d) T. V. RajanBabu and W. A. Nugent, *J. Am. Chem. Soc.*, 1994, **116**, 986.

 99. J. I. Seeman, *Chem. Rev.,* 1983, **83**, 83.

100. R. Schobert, *Angew. Chem.*, 1988, **100**, 869 (*Angew. Chem., Int. Ed. Engl.*, 1988, **27**, 855).

101. J. S. Yadav, T. Shekharam and V. R. Gadgil, *J. Chem. Soc., Chem. Commun.*, 1990, 843.

102. B. Giese, B. Kopping, T. Göbel, G. Thoma, J. Dickhaut, K. J. Kulicke and F. Trach, in *Organic Reactions*, ed. L. A. Paquette, Wiley, New York, 1996, vol. 48, p. 301.

103. J. E. Baldwin, *J. Chem. Soc., Chem. Commun.*, 1976, 734.

104. (a) A. L. J. Beckwith, C. J. Easton and A. K. Serelis, *J. Chem. Soc., Chem. Commun.*, 1980, 482; (b) A. L. J. Beckwith, T. Lawrence and A. K. Serelis, *J. Chem. Soc., Chem. Commun.,* 1980, 484; (c) A. L. J. Beckwith, C. J. Easton, T. Lawrence and A. K. Serelis, *Aust. J. Chem.,* 1983, **36**, 545; (d) D. C. Spellmeyer and K. N. Houk, *J. Org. Chem.,* 1987, **52**, 959; (e) T. V. RajanBabu and T. Fukunaga, *J. Am. Chem. Soc.,* 1989, **111**, 296; (f) T. V. RajanBabu, *Acc. Chem. Res.,* 1991, **24**, 139.

105. B. Giese, J. Dupuis, K. Groninger, T. Hasskerl, M. Nix and T. Witzel, in *Substituent Effects in Radical Chemistry*, ed. H. G. Viehe, D. Reidel Publishing Co., Dordrecht, 1986, p. 283.

106. A. Fernández-Mateos, E. Martin de la Nava, G. Pascual Coca, A. Ramos Silvo and R. Rubio González, *Org. Lett.*, 1999, **1**, 607.

107. (a) R. L. Halterman, *Chem. Rev.,* 1992, **92**, 965; (b) R. L. Halterman, in *Metallocenes*, ed. A. Togni and R. L. Halterman, Wiley-VCH, Weinheim, 1998, vol. 1, p. 455.

108. A. Gansäuer, M. Pierobon and H. Bluhm, *Angew. Chem.*, 1998, **110**, 107 (*Angew. Chem., Int. Ed.*, 1998, **37**, 101).

109. M. A. Loreto, L. Pellacani, P. A. Tardella, *Synth. Commun.*, 1981, **11**, 287.

110. A. Gansäuer and H. Bluhm, *J. Chem. Soc., Chem. Commun.*, 1998, 2143.

111. A. Gansäuer, H. Bluhm and M. Pierobon, *J. Am. Chem. Soc.*, 1998, **120**, 12849.

112. (a) W. A. Nugent, *J. Am. Chem. Soc.*, 1992, **114**, 2768; (b) I. Paterson and D. J. Berrisford, *Angew. Chem.*, 1992, **104**, 1204 (*Angew. Chem., Int. Ed. Engl.*, 1992, **31**, 1179); (c) L. E. Martínez, J. L. Leighton, D. H. Carsten and E. N. Jacobsen, *J. Am. Chem. Soc.*, 1995, **117**, 5897; (d) J. L. Leighton and E. N. Jacobsen, *J. Org. Chem.*, 1996, **61**, 389; (e) J. F. Larrow, S. E. Schaus and E. N. Jacobsen, *J. Am. Chem. Soc.*, 1996, **118**, 7420; (f) K. B. Hansen, J. L. Leighton and E. N. Jacobsen, *J. Am. Chem. Soc.*, 1996, **118**, 10924; (g) D. M. Hodgson, A. R. Gibbs and G. P. Lee, *Tetrahedron*, 1996, **46**, 14361; (h) T. Iida, N. Yamamoto, H. Sasai and M. Shibasaki, *J. Am. Chem. Soc.*, 1997, **119**, 4783; (i) K. D. Shimizu, B. M. Cole, C. A. Krueger, K. W. Kuntz, M. L. Snapper and A. H. Hoveyda,

Angew. Chem., 1997, **109**, 1782 (*Angew. Chem., Int. Ed. Engl.,* 1997, **36**, 1704); (j) W. A. Nugent, *J. Am. Chem. Soc.,* 1998, **120**, 7139); (k) S. Sagawa, H. Abe, Y. Hase and T. Inaba, *J. Org. Chem.,* 1999, **64**, 4962; (l) M. H. Wu, K. B. Hansen and E. N. Jacobsen, *Angew. Chem.,* 1999, **111**, 2167 (*Angew. Chem., Int. Ed.,* 1999, **38**, 2012).

113. A. Gansäuer, T. Lauterbach, H. Bluhm and M. Noltemeyer, *Angew. Chem.,* 1999, **111**, 3112 (*Angew. Chem., Int. Ed.,* 1999, **38**, 2909).

114. E. Cesarotti, H. B. Kagan, R. Goddard and C. Kruger, *J. Organomet. Chem.,* 1978, **162**, 297.

115. (a) A. H. Hoveyda and J. P. Morken, *Angew. Chem.,* 1996, **108**, 1378 (*Angew. Chem., Int. Ed. Engl.,* 1996, **35**, 1262); (b) A. H. Hoveyda and J. P. Morken, in *Metallocenes,* ed. A. Togni and R. L. Halterman, Wiley-VCH, Weinheim, 1998, vol. 2, p. 625.

116. R. L. Halterman and K. P. C. Vollhardt, *Organometallics,* 1988, **7**, 883.

117. E. J. Corey and H. E. Ensley, *J. Am. Chem. Soc.,* 1975, **97**, 6908.

118. J. K. Whitesell, *Chem. Rev.,* 1992, **92**, 953.

119. (a) B. Giese, K. Heuck, H. Lenhardt and U. Luning, *Chem. Ber.,* 1984, **117**, 2132; (b) B. Giese, *Angew. Chem.,* 1989, **101**, 993 (*Angew. Chem., Int. Ed. Engl.,* 1989, **28**, 969).

CHAPTER 13

Enantioselective Radical reactions and Organocatalysis[*]

13.1 INTRODUCTION

For many years, radicals—molecules that contain a single unpaired electron—were considered too reactive to be used productively in organic synthesis. This myth has been dispelled and, somewhat ironically, it is now clear that radicals frequently offer higher levels of selectivity and predictability than analogous ionic reactions.[2] Even with increased understanding, dogma dictated that radicals could not participate in highly stereoselective reactions despite being simple organic species, subject to the same steric and electronic interactions as all other molecules. This too has proved incorrect as the last decade has seen tremendous progress in enantioselective radical reactions.[3] The majority of naturally occurring compounds are chiral and not superimposable on their mirror images. One of the major challenges for organic chemists is to develop enantioselective reactions, *i.e.* reactions that can discriminate between mirror image enantiomers. The domination of enantioselective transformations by metal-based reagents is coming to an end as it becomes clear that small, metal-free molecules, or organocatalysts,[4] can achieve complementary reactions without recourse to potentially toxic or expensive metals.

The recent introduction of radical intermediates into organocatalysis by MacMillan *et al.*[5] and Sibi and Hasegawa[6] has attracted considerable attention and there is no doubt that the principles underpinning this methodology will have a major impact on organic synthesis. It is often overlooked that radical chemistry has always been conducive to organocatalysis, with many of the general characteristics of radicals being ideally suited for a synergistic relationship with organic-based catalysts. Radicals are largely impervious to the effects of water, display greater functional group tolerance than ionic reagents

* Chapter written by V. Tamara Perchyonok.
Streamlining Free Radical Green Chemistry
V. Tamara Perchyonok, Ioannis Lykakis and Al Postigo
© V. Tamara Perchyonok, Ioannis Lykakis and Al Postigo 2012
Published by the Royal Society of Chemistry, www.rsc.org

and operate over a wider pH range. It is the aim of this chapter to briefly outline the shared history of radicals and organocatalysis and speculate on the future directions of this profitable partnership.

13.2 ORGANIC REAGENTS AND ORGANOCATALYSTS IN STEREOSELECTIVE RADICAL CHEMISTRY

When the spatial alignment of two interacting reagents is controlled, it is possible to achieve a stereoselective reaction. For this to occur, at least one of the reactants must have a well-defined shape; this can arise if either the molecule has a specific configuration or if it forms a temporary bond to a second molecule with a well-defined shape. This chapter will cover three strategies for achieving stereoselective radical reactions: the interaction of a substrate with a Lewis acid, the interaction of the substrate with a chiral reagent, and the temporary incorporation of chirality into the substrate.

13.2.1 Chiral Lewis Acid Activation

Lewis acids activate a substrate by accepting electrons and lowering the energy of the lowest unoccupied molecular orbital (LUMO) of the molecule, thus encouraging nucleophilic attack. The smallest Lewis acid is the proton and its use in catalysis is often termed Brønsted acid catalysis. The last five years have seen a renaissance in the use of Brønsted acids in asymmetric catalysis[7,8] but it is clear that the ground work was laid over a decade ago. Brønsted acids can be classified into two categories: neutral acid catalysts such as ureas, which are often termed hydrogen-bond catalysts, and stronger acid catalysts that are proton donors, such as phosphoric acids. Both classes have been employed in radical chemistry.

Since Sigman and Jacobsen's seminal work in 1998 and 1999 (thio)ureas have rapidly become one of the privileged motifs for asymmetric organocatalysis,[9] yet the first example of a urea being used as a hydrogen-bond catalyst was four years earlier, in Curran and Kuo's diastereoselective allylation of cyclic sulfinyl radicals formed from **1** (Scheme 13.1).[10] Urea **3** was shown to

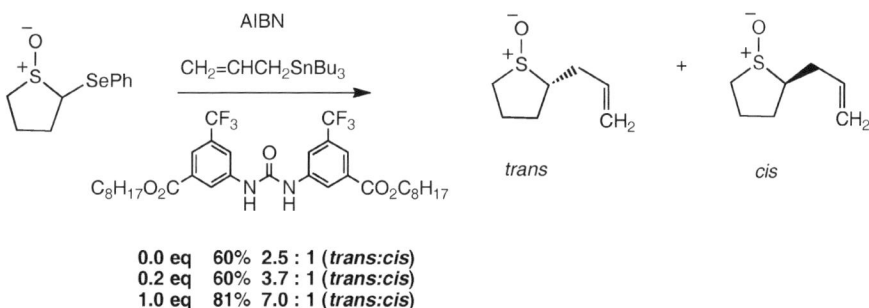

0.0 eq	60%	2.5 : 1 (*trans:cis*)
0.2 eq	60%	3.7 : 1 (*trans:cis*)
1.0 eq	81%	7.0 : 1 (*trans:cis*)

Scheme 13.1 Diastereoselective allylation of cyclic sulfinyl radicals as example or organocatalysis.

increase both the yield and the *trans/cis* ratio of **2**, with as little as 0.2 equivalents of **3** increasing the ratio from 2.5/1 with no catalyst to 3.7/1. Urea **3** is thought to clamp the sulfinyl oxygen and increase the steric bulk of one face of the molecule whilst activating the incipient radical to allylation. Considering this promising result it is shocking that there have been no other reports of the use of (thio)ureas in radical chemistry.

Chiral lactam has been employed as a hydrogen-bond catalyst for a host of enantioselective reactions including the cyclization of a corresponding iodide to afford a corresponding cyclized product in excellent yield and selectivity (Scheme 13.2).[11]

Lactam binds to the starting amide by complementary hydrogen-bonding between a carbonyl and a N–H moiety and thus induces high enantioselectivity. Whilst super-stoichiometric quantities of lactam give the best results (2.5 equiv. gives 99% ee), use of sub-stoichiometric quantities still results in chiral amplification with just 0.1 equivalents affording the final cyclized product in 55% ee. The correct choice of solvent is pivotal for high enantioselectivities in the catalytic variant; the reaction mixture must be heterogenous throughout the reaction with the substrate dissolving only on its complexation to **6** thus forcing cyclization to occur in a chiral environment. The same complexing reagent has been successfully used in the enantioselective radical cyclization of piperidines[12,13] and the synthesis of fused tetracycle cyclobutanes *via* either inter- or intramolecular [2+2]-photocycloaddition reactions.[13]

An exciting variation on this system has permitted a highly enantioselective, catalytic, photoinduced electron-transfer reaction that furnishes a tetracycle (Scheme 13.3).[14]

Key to the success of this reaction is catalyst that acts as both the chiral template and an antenna for harvesting the light required to activate the substrate. Excitation of the catalyst initiates electron transfer from the starting amine to a catalyst and permits formation of the corresponding α-aminoalkyl radical. Cyclization of the complexed α-aminoalkyl radical occurs from the top face as the catalyst blocks the bottom face of the alkene. Just 0.3 equivalents of catalyst are required for the reaction to occur in high yield and enantioselectivity.

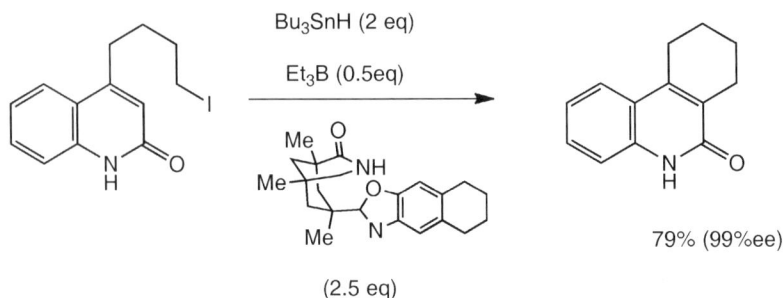

Scheme 13.2   Hydrogen bond-mediated enantioselective transformation in action.

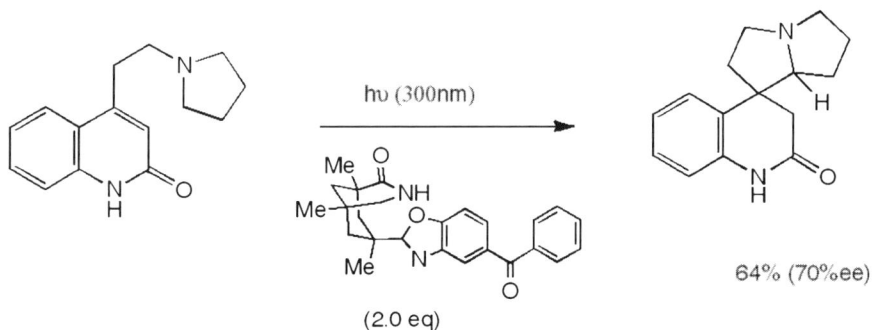

Scheme 13.3 Catalytic enantioselective radical reactions driven by photoinduced electron transfer.

Recently Bach extended the methodology to catalytic enantioselective radical reactions driven by photoinduced electron transfer to give a tetracyclic product in 55% yield and 91% ee starting from the quinolone substrate in the presence of the chiral xanthone catalyst (Scheme 13.4).[11,12,13] Although the efficiencies in these photochemical reactions remain to be improved, these methods represent a breakthrough in enantioselective photochemistry that will further stimulate research in this rapidly growing area of science. The simplicity of this methodology and its use of just two reagents coupled with the rapid increase in molecular complexity ensure that this methodology has a bright future.

Addition reactions to the double bond of α,β-unsaturated carbonyl compounds have significantly influenced the development of modern radical chemistry. The fact that nucleophilic carbon radicals exhibit high regio- and chemoselectivity in this reaction attracted the attention of organic chemists and alerted the synthetic community to the advantages of radical reactions.[1,2] If the double bond, to which a radical addition occurs, bears two different β-substituents, a stereogenic centre is formed. In acrylic acid derivatives this process can be rendered diastereoselective if the acryloyl group is attached to

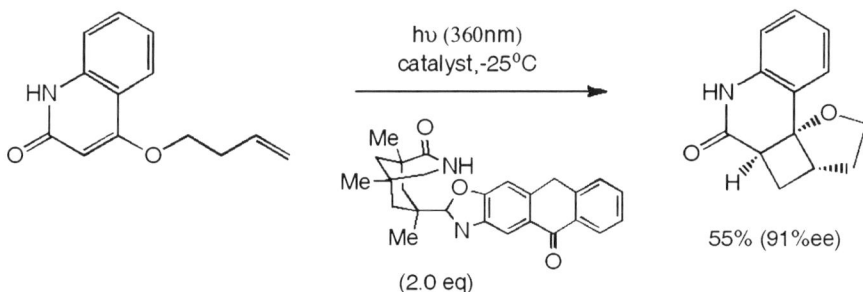

Scheme 13.4 Catalytic enantioselective radical reactions driven by photoinduced electron transfer.

a chiral auxiliary.[3] Frequently used auxiliaries include amines, amides or alcohols linked by a C–N or C–O bond. Seminal studies in the area of auxiliary-induced diastereoselective radical addition reactions were performed in the late 1980s by the groups of Giese, Porter and Curran.[3,4] In recent years, the area of radical addition chemistry was dominated by the search for methods to achieve a direct enantioselective C–C bond formation using chiral templates or chiral catalysts.[5,6]

Lewis acids have turned out to be particularly well suited for combining an efficient stereocontrol with—in some cases remarkable—rate enhancements enabling catalytic enantioselective radical reactions.[5,7] Other templates relating on different but Lewis acid/base interactions have received less attention.[8] Bach utilized the use of hydrogen bonds to bind a potential radical reaction precursor in a chiral environment. In a U-shaped template, as depicted in Scheme 13.5, binding to a lactam by two hydrogen bonds could occur and simultaneously a bulky substituent (steric shield) would prevent an approach of a given reagent from the upper face. For a conjugate radical addition the corresponding alkyl radical would approach the double bond from the lower face.

Bach extended his elegant work to reductive radical cyclization conditions employing BEt_3/O_2 as the initiator and either Bu_3SnH or TMS_3SiH as a hydride source. 4-(4-Iodobutyl)-quinolone and 4-(3-iodopropylthio)-quinolone

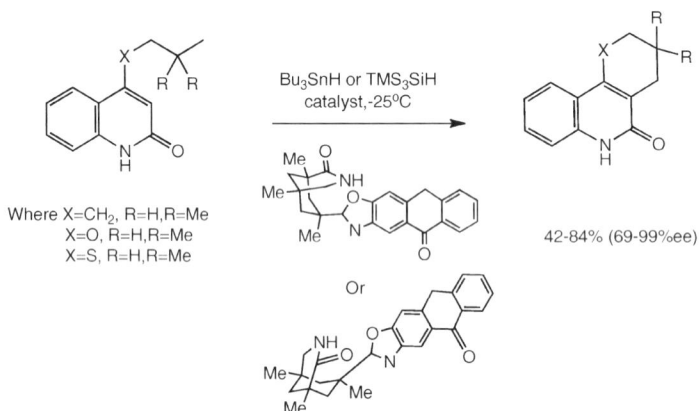

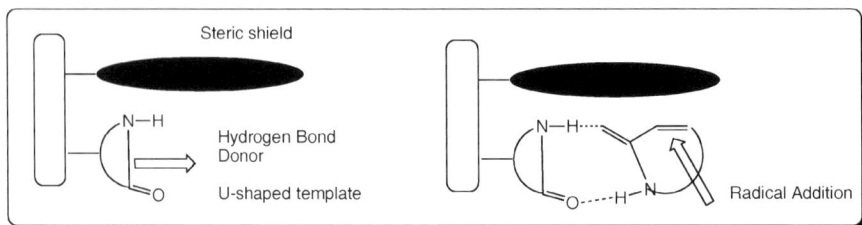

Scheme 13.5   U-shape template and enatioselective radical transformation.

gave the respective 6-*endo*-cyclisation products in good yields. 4-(3,3-Dimethyl-4-iodobutyl)-quinolone cyclized in a 5-*exo*-fashion, while the other substrates delivered only reduction products. The cyclisation reactions could be conducted in the presence of a chiral template with high enantiomeric excess (94–99% ee).

A stronger Brønsted acid catalyst is the chiral quaternary ammonium salt **12** that has been used in the synthesis of amino acid derivatives under mild, aqueous reaction conditions from 2-((benzyloxy)imino)acetic acid (Scheme 13.6).[15]

A chiral catalyst/hydrogen donor, prepared from a *Cinchona* alkaloid and hypophosphorous acid, plays a multitude of different roles: it acts as the radical chain carrier, as a surfactant to increase the water solubility of the organic components, and as a chiral additive capable of inducing high stereoselectivity in the addition (up to 98% ee). The enantioselectivity is thought to arise from hydrogen bonding between radical precursor and chiral catalyst/hydrogen donor coupled with the π-stacking of the aryl groups. Whilst a catalytic variant has not been developed yet, the reaction still has many advantages over conventional nucleophilic additions: no metal reagents are utilized, all reagents are cheap and readily available, aqueous solvent mixtures are preferred to organic solvents, and the chiral amine can be recycled readily. With these benefits in mind, further applications of this and analogous systems can be anticipated in the future along with efforts to develop catalytic variants.

13.3 ENANTIOSELECTIVE HYDROGEN ATOM TRANSFER

It is possible to transfer a hydrogen atom from a radical chain carrier to a substrate radical in a stereoselective manner. There are two distinct methods that achieve this selectivity: the first involves an achiral hydrogen source and a chiral Lewis acid, whilst the second involves the use of a chiral hydrogen donor. Curiously, whilst the former method is the more common with metal-based systems it has not been achieved under metal-free conditions. On the

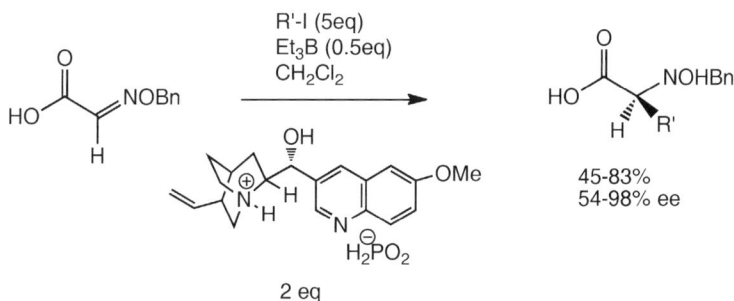

Scheme 13.6 Chiral carbon-carbon bond formation mediated by cinchona alkaloid hypophosphorous acid.

other hand, the second strategy permitted some of the earliest examples of asymmetric organocatalysis.

Nearly thirty years ago, Ohno *et al.* synthesized nictotinamide as a chiral model of the coenzyme NAD(P)H and found that it reduced certain carbonyl compounds with high selectivity (~70% ee).[16] Later, Tanner and Kharrat showed that the reaction proceeded *via* a radical pathway and was in fact the first radical chain reaction whose propagation steps contained an enantioselective hydrogen atom transfer (Scheme 13.7).[17]

Radical initiation with azobisisobutyronitrile (AIBN) abstracts a hydrogen from (*R*)-4-methyl-*N*-((*R*)-1-phenylethyl)-1-propyl-1,4-dihydropyridine-3-carboxamide to give the doubly allylic stabilized radical. Electron transfer from the newly formed allylic radical to 2,2,2-trifluoro-1-phenylethanone is driven by aromatisation that provides a (*R*)-4-methyl-3-((1-phenylethyl)carbamoyl)-1-propylpyridin-1-ium cation as an intermediate and generates the ketyl radical anion from 2,2,2-trifluoro-1-phenylethanone. Interaction of this last intermediate with chiral hydrogen source (*R*)-4-methyl-*N*-((*R*)-1-phenylethyl)-1-propyl-1,4-dihydropyridine-3-carboxamide furnishes the desired chiral alcohol and propagates the chain by reforming the methyl-3-((1-phenylethyl)carbamoyl)-1-propylpyridin-1-ium cation. Once again, this reaction appears to have been consigned to history with little study outside of biochemistry reported, even though it offers an intriguing route to enantiopure alcohols.

It is often forgotten that radical chain reactions rely on polarity effects for efficient propagation; smooth hydrogen atom transfer only occurs if an electrophilic radical interacts with a nucleophilic source of hydrogen or *vice versa*. If the polarities are mis-matched then the reaction will be a non-chain process and will be sluggish at best.

Polarity-reversal catalysts alleviate this problem, facilitating hydrogen atom transfer *via* the addition of an extra propagation step.[18,19] In the radical hydrosilylation of electron-rich alkenes such as **21**, the hydrogen atom transfer step from silane **23** to prochiral radical **22** is slow (Cycle A, Scheme 13.8) as both donor **23** and acceptor **22** are nucleophilic. Thiol **25** acts as a polarity-reversal catalyst and overcomes this problem; the slow propagation step of the previous reaction is replaced now by two fast propagation steps (Cycle B) as the thiol provides an electron deficient (or electrophilic) hydrogen atom. By making the polarity-reversal catalyst a chiral carbohydrate derivative (**25**), it is possible to enantioselectively transfer hydrogen to radical **22**. Furthermore, since **25** is regenerated in the second propagation step of Cycle *B* a sub-

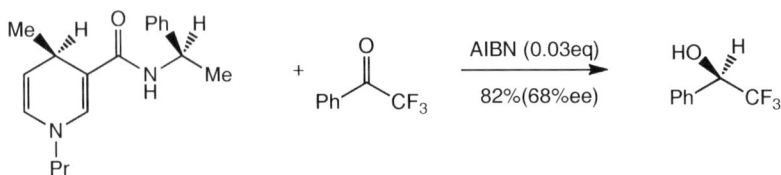

Scheme 13.7   Early example of enantioselective hydrogen atom transfer reactions.

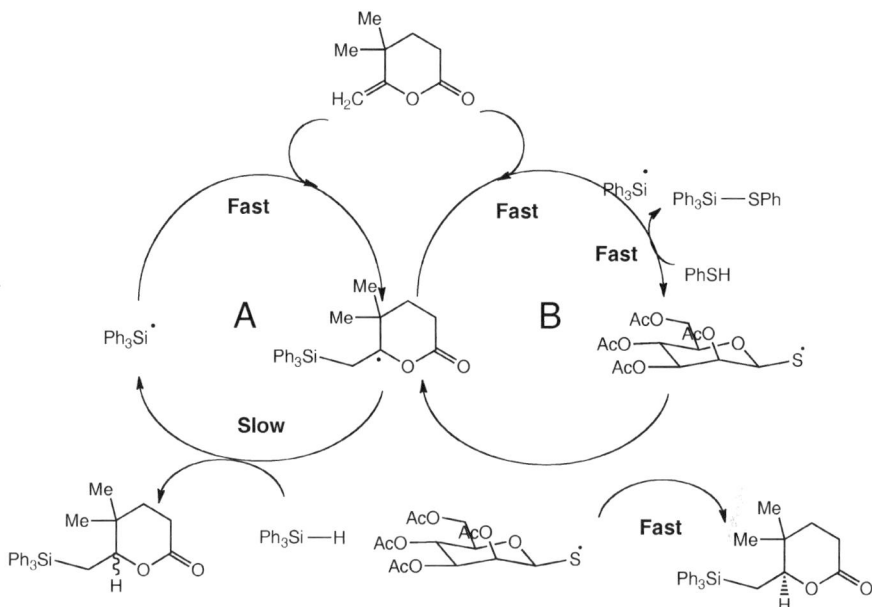

Scheme 13.8   Enantioselective catalytic cycle in action.

stoichiometric quantity of chiral reagent can be employed. In this example just 0.05 equivalents of **25** in the presence of a radical initiator gives lactone (*R*)-**24** in 90% yield and 95% ee.[18] It should be noted that this is a general principle and should allow enantioselective organocatalytic radical reactions for a range of transformations. It is somewhat surprising that, with the exception of the work of Roberts's group, very few examples of chiral polarity-reversal catalysts have been reported; therefore this appears to be an area ripe for exploitation.

13.4   AMINOCATALYSIS/ENAMINE ACTIVATION

The new radical activation strategy reported by MacMillan *et al.*[5] and Sibi and Hasegawa[6] is based on aminocatalysis or enamine catalysis popularized by List and MacMillan,[20] but has its roots in older radical methodology (Scheme 13.9).

In 1992 Narasaka *et al.* reported the cerium(IV) ammonium nitrate (CAN) mediated oxidation of an enamine (4-(3,3-dimethylbut-1-en-2-yl)morpholine) to a radical cation and the addition of this radical to electron rich alkenes such as *tert*-butyldimethyl((1-phenylvinyl)oxy)silane to give 5,5-dimethyl-1-phenylhexane-1,4-dione as a final product (Scheme 13.10).[21]

Whilst this undoubtedly laid the foundations for the current methodology, it was limited by the need to pre-form the enamine and because it was non-stereoselective. Arguably, Cossy *et al.* reported[22] the first solution to the latter

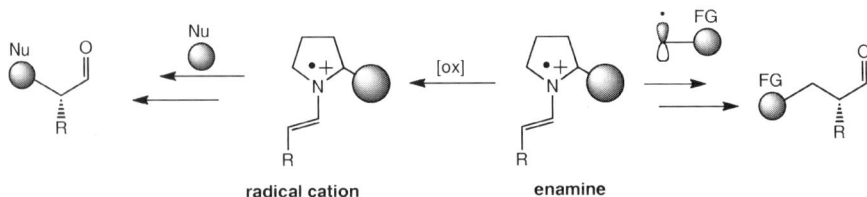

Scheme 13.9   General mode of activation in organocatalysis.

shortcoming with the manganese(III) acetate/copper(II) acetate-mediated oxidation of a chiral β-carboxamido enamine to a radical cation (Scheme 13.11).

The radical then underwent cyclization to give the spirocycle in moderate
diastereoselectivity (45% de). The enamine was formed from a primary amine
and it is entirely possible that use of a secondary amine would have led to an
iminium cation with less rotational freedom and thus would have resulted in
better stereochemical induction. Again the reaction was limited by the need to
pre-form the enamine precursor, but it clearly revealed the plausibility of this
strategy for asymmetric radical chemistry.

By refining these early examples both MacMillan *et al.*[5] and Sibi and
Hasegawa[6] have developed a truly exciting method for conducting enantioselective radical reactions that shows great potential to encompass many
different transformations. Both methodologies combine radical chemistry with
enamine-based organocatalysis to functionalize the α-position of aldehydes;
condensation of an aldehyde with a sub-stoichiometric amount of a chiral
secondary amine gives the corresponding enamine as a final product (Scheme
13.12).

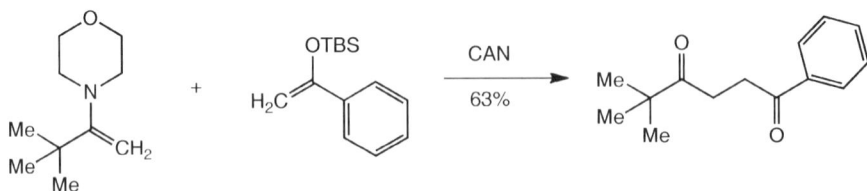

Scheme 13.10   CAN mediated oxidation of enamines.

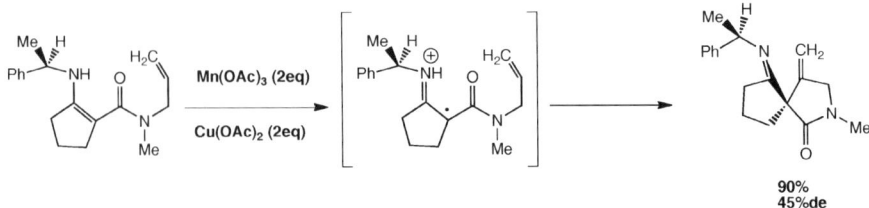

Scheme 13.11   Mn(III)/Cu(II) acetate-mediated oxidation of chiral b-carboxamido
enamine.

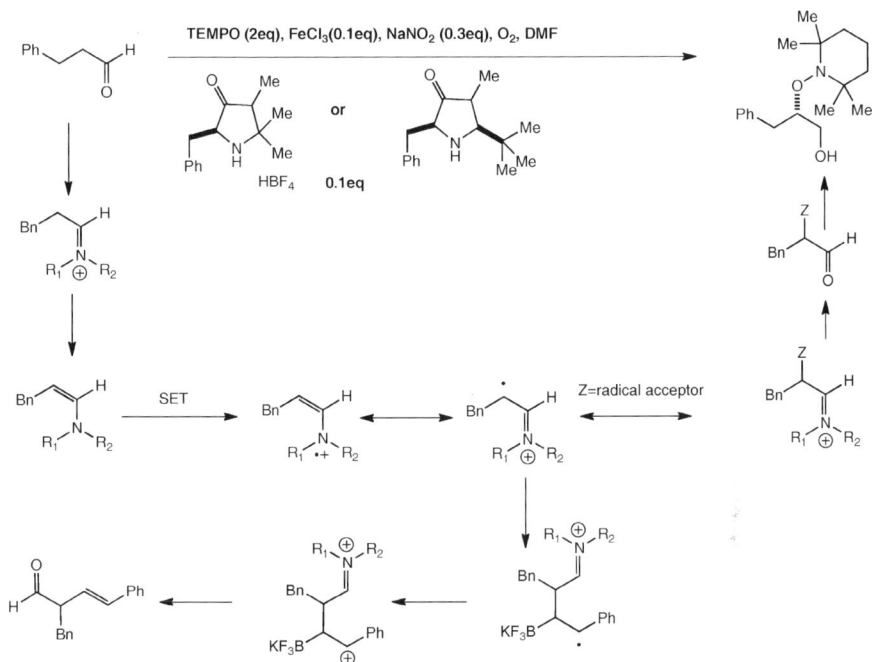

Scheme 13.12 Tempo-Mediated organocatalysis.

Oxidation of the enamine by single electron transfer then leads to formation iminium radical cation that reacts with the appropriate radical acceptor to give the corresponding cation as an intermediate, which finally undergoes hydrolysis to product and regenerates the chiral catalyst, ready to repeat the reaction. The fate of the alkyl radical differentiates the two methodologies. In Sibi and Hasegawa's methodology[6] the radical is trapped with the persistent O-centred 2,2,6,6-tetramethylpiperidine-1-oxyl radical (TEMPO), resulting in the α-oxyamination of the initial aldehyde (Scheme 13.12). The chiral imidazolidinone gives moderate to excellent en-antioselectivities for a range of aldehydes, and although a variety of aryl-substituted aldehydes were tolerated, simple alkyl aldehydes containing no aromatic rings or double bonds gave no selectivity suggesting that π inter-actions are important.[5] The benefit of this system is that oxidation is achieved with a catalytic quantity of iron(III) chloride in conjunction with a stoichiometric amount of a co-oxidant comprised of sodium nitrite and oxygen. The disadvantage is that products are accessible by more conventional chemistry.

MacMillan's group's methodology[5] appears to be more versatile and permits the reaction of the radical cation with a host of electron-rich acceptors including allylsilanes, silylenol ethers, heteroaromatics and alkenyl potassium trifluoroborate salts. Thus, reaction of radical cation with the alkenyl potassium trifluoroborate gives final radical cation in good yield (Scheme 13.12). In MacMillan's reactions two distinct oxidation steps occur: the first gives a radical cation whilst the second is required to oxidize the benzyl radical to a cation. As a result, the methodology currently needs an excess of the metal-based oxidant CAN. The reaction appears to be quite general as a range of aldehydes can be employed whilst the alkenyl component can be alkyl- or aryl-substituted with little variation in the yield or selectivity. Not only does this methodology permit the facile synthesis of enantiomerically pure homo-allylic aldehydes that would be difficult to form by conventional means, but also it is undoubtedly just the tip of the iceberg: it is easy to imagine this general strategy, the formation of a chiral radical cation from enamines, being employed in a wide range of novel transformations and it will be fascinating to see how this work progresses.

Currently, neither methodology is ideal. Sibi and Hasegawa's system[6] involves a single oxidation and so appears to be limited to the addition of persistent radicals. MacMillan's protocol[5] is far more impressive in scope but uses an excess of oxidant (Scheme 13.13).

In 2008, a solution to this problem was described by MacMillan[5] by incorporating photoredox catalysis[11] with organocatalysis. As illustrated in Scheme 13.14, under irradiation with fluorescent light an electron-deficient alkyl radical was generated from the starting halide through a reductive process mediated by the photoredox catalyst $Ru(bpy)_3^+$. Then the alkyl radical added efficiently to the electron-rich enamine (generated from the condensation of starting aldehyde and amine catalyst) to give an α-amino radical intermediate. The radical was readily oxidized by the excited catalyst, $*Ru(bpy)_3^{2+}$, to the iminium ion, which upon hydrolysis gave an α-alkylated aldehyde and liberated amine catalyst. The oxidation step also regenerated the

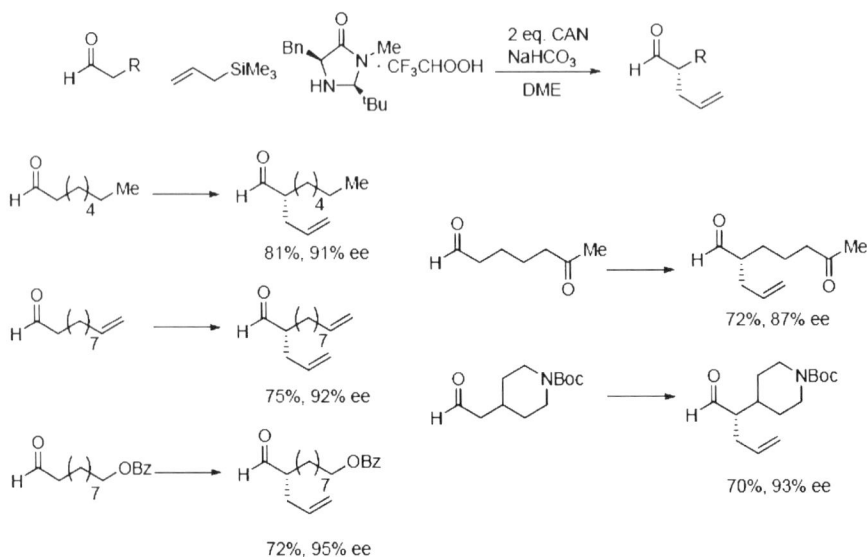

Scheme 13.13 Catalytic enantioselective organocatalysis using SOMO activation.

Ru(bpy)$_3^+$ for the photoredox catalytic cycle. The data obtained from a radical clock experiment[7] and control experiments are consistent with the proposed mechanism in Scheme 13.14.

The cooperation of photoredox catalysis and organocatalysis offers many possibilities for the discovery of novel asymmetric transformations. A catalytic, enantioselective α-trifluoromethylation of aldehydes has recently been developed using this strategy. As shown in Scheme 13.15, trifluoromethyl iodide was selected as the requisite radical precursor and high levels of efficiency and enantiocontrol were achieved at sub-ambient temperatures (−20 °C) with a photoredox catalyst and an amine catalyst.

MacMillan next performed a series of experiments to determine the scope of the aldehydic component in this asymmetric trifluoromethylation protocol.[21] In the course of the investigation MacMillan showed that mild redox conditions are compatible with a wide range of functional groups including ethers, esters, amines, carbamates, and aromatic rings. Moreover, a significant

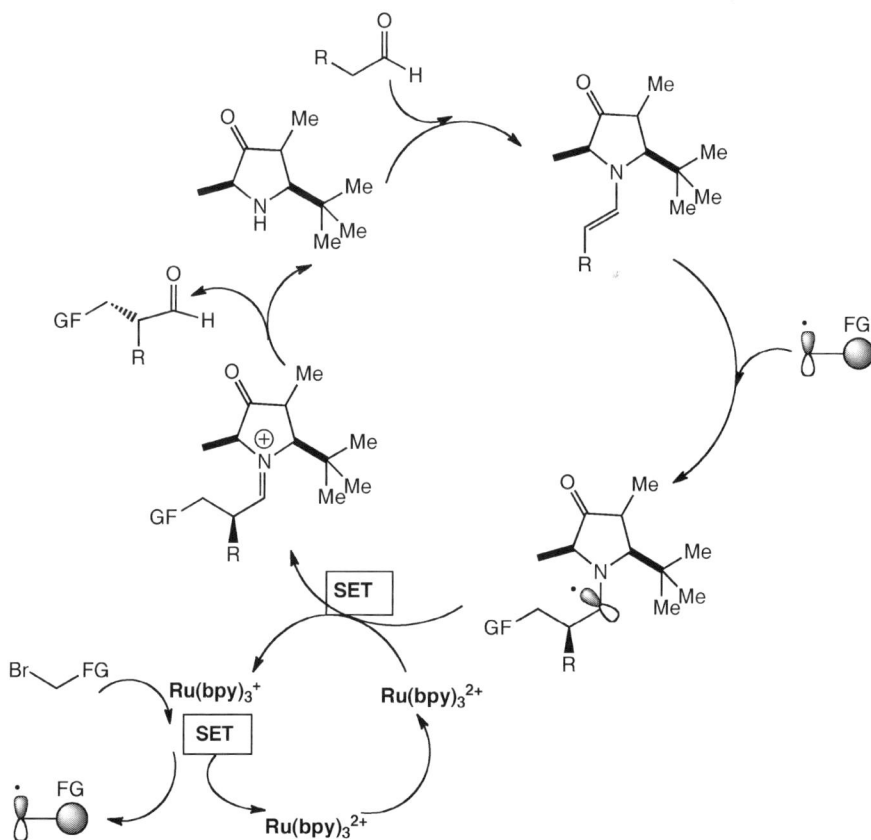

Scheme 13.14   Catalytic enantioselective intermolecular α-alkylation of aldehydes with photoredox organocatalysis.

Photoredox catalyst
15W fluorescent light

Amine catalyst
63-98%
88-99%ee

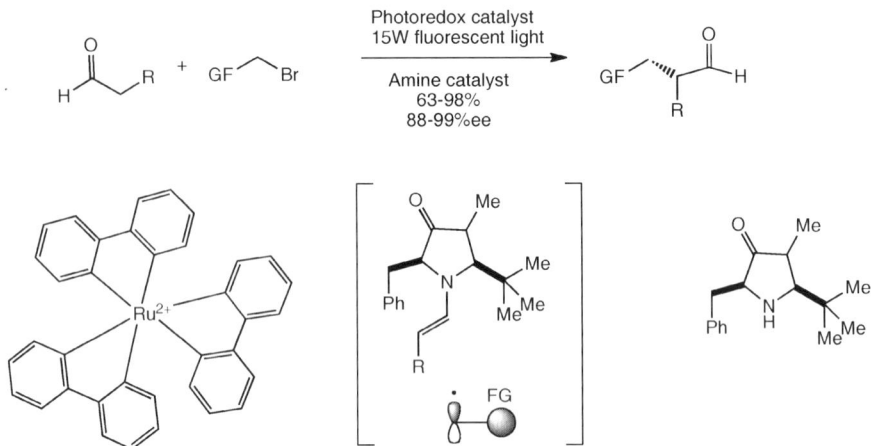

Proposed Mechanism

variation in the steric demand of the aldehyde substituent can be accommodated without loss in enantiocontrol. Notably, this protocol enables the formation of a benzylic-CF_3 R-formyl stereocenter without significant erosion in enantiopurity.[21] These transformations clearly demonstrate the synthetic advantages of catalyst-enforced induction *versus* substrate-directed stereocontrol and will undoubtedly lead to the flourishing field of asymmetric organocatalysis and wider application in synthetic chemistry and strategy.

MacMillan has found that a broad range of perfluoroalkyl iodides and bromides also participate in this enantioselective alkylation reaction. For example, *n*-perfluoroalkyl substrates of varying chain length undergo reductive radical formation and enamine addition without loss in enantiocontrol or reaction efficiency. MacMillan has also found that the aldehyde R-functionalization step can be performed with sterically demanding coupling partners such as perfluoro-isopropyl iodides. Moreover, benzylic, R-ester, and R-ether difluoromethylene carbons are readily incorporated as part of this new enantioselective catalytic R-carbonyl alkylation.

MacMillan predicted that the R-trifluoromethyl aldehyde products generated in this study will be of value for the production of a variety of organofluorine synthons. As shown in Scheme 13.16, reduction or oxidation of the formyl group is possible to generate α-hydroxy and R-trifluoromethyl acids (the latter we expect will be a key building block for the formation of heterocycles that incorporate CF_3 at the benzylic position). Moreover, these aldehyde products can be employed in a reductive amination sequence without significant loss in enantioselectivity to produce α-trifluoromethyl amines.

MacMillan has also extended the methodology to aldehyde oxidation followed by a Curtius rearrangement allows enantioselective formation of R-trifluoromethyl amine-containing stereocenters, a commonly employed amide isostere in medicinal chemistry (Scheme 13.16). In this case, careful selection of

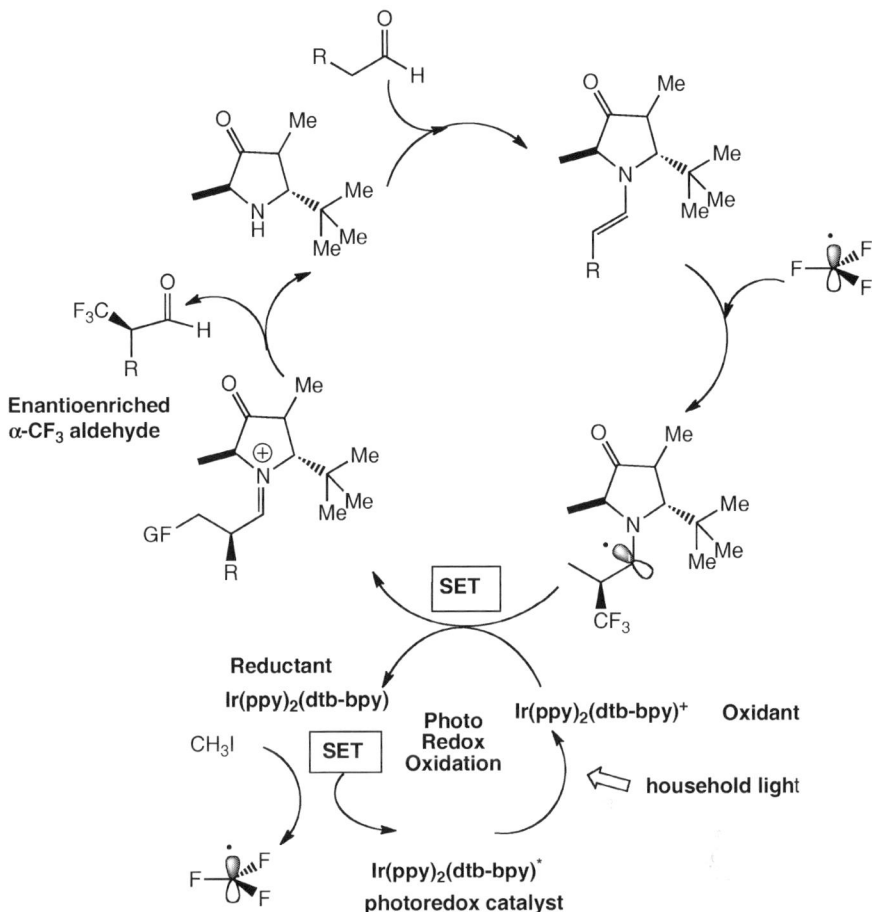

Scheme 13.15 Catalytic enantioselective α-trifluoromethylation using photoredox organocatalysis.

base and reaction temperature is essential to maintain the enantiopurity obtained in the initial alkylation step.

A combination of the two would have a major impact on both radical chemistry and organocatalysis, and is undoubtedly being investigated and explored.

13.5 FUTURE DIRECTIONS FOR ORGANOCATALYSIS IN RADICAL CHEMISTRY

Hopefully, the discussion above has shown that most forms of organocatalysis can be applied to enantioselective radical reactions. No doubt the area will continue to develop rapidly in order to take advantage of all the benefits proffered by both radical processes and organocatalysis in terms of both clean

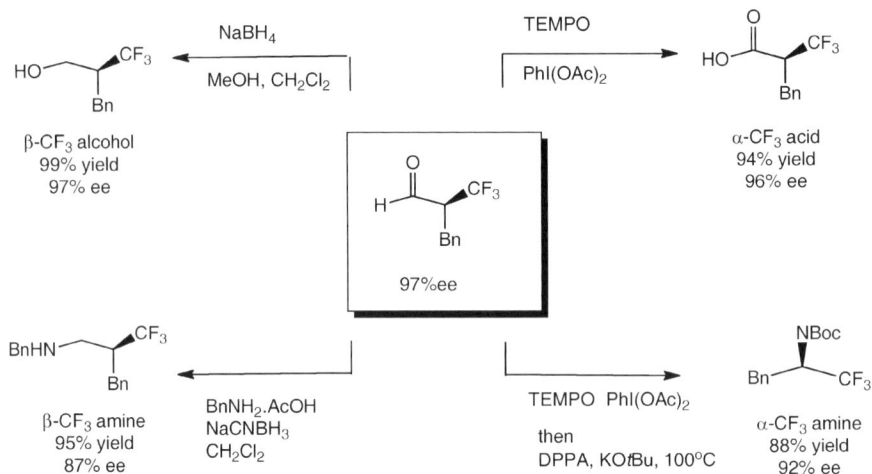

Scheme 13.16   Access to enantio-enriched organofluorine synthons.

reaction conditions and the range of transformations possible. It is obvious that the radical enamine-activation strategy will have a major impact and many new applications can be expected in the future. Improvements to the oxidation protocol that allow for use of sub-stoichiometric amounts of metal-based oxidants and more environmentally benign terminal oxidants are major goals. It will be interesting to see if the antithetical strategy, the reductive formation of radical intermediates, will be applicable to enantioselective organocatalysis. Already organic electron donors have been developed for the reduction of halides to radicals.[23] Tetraazaalkene is a neutral ground-state organic molecule capable of donating an electron to a suitable acceptor due, in part, to the considerable aromatisation energy residing in its derived radical cation (Scheme 13.17), and the stability imparted to the carbocationic centre by the adjacent nitrogen atoms. This allows tetraazaalkene to reduce un-activated aryl and alkyl iodides to C-centred radicals that can then undergo cyclization to yield a final product. The major drawback here is the high reactivity of tetraazaalkene towards air, which requires that it normally be prepared by deprotonation of the stable salt immediately prior to use. If this limitation can be overcome related reagents could be of considerable value.

The use of hydrogen-bond donor catalysts (or Brønsted acid catalysts) in radical chemistry is ripe for exploitation; the groundwork has been laid, examples of enantioselective templated reactions[11–14] and chiral proton donors[15] are known. Now it is necessary to take these precedents and pursue more general and valuable examples. It is surprising that chiral (thio)ureas have not been employed in radical additions or cyclizations. Likewise, the use of chiral phosphoric acids appears to be an obvious progression; logically, the use of chiral acids to activate various C=N moieties to radical attack would be the first step before tackling more general activation/addition methodology. It should be remembered that the last decade has seen considerable progress in

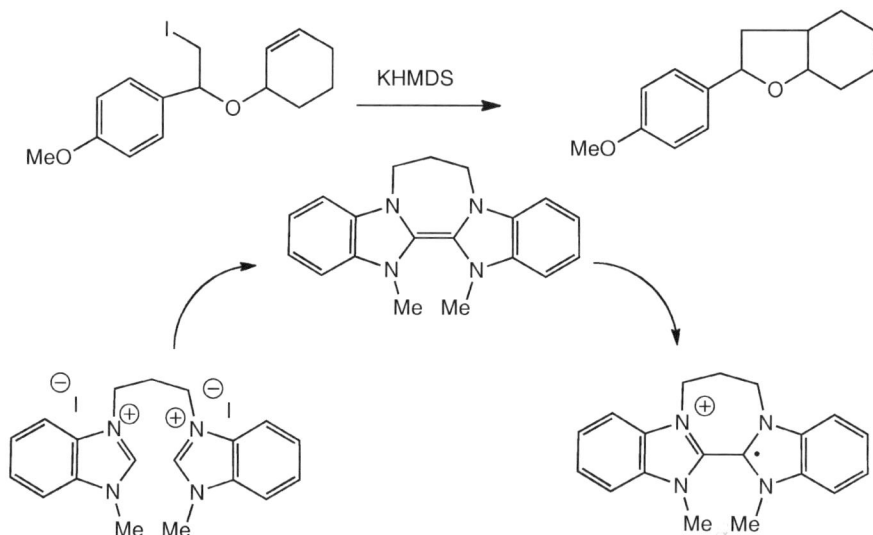

Scheme 13.17 Tetrazaalkene as an organic donor for the reduction of halides.

the use of metal-based Lewis acids for enantioselective radical reactions whilst at the same time, Brønsted acids have begun to replace Lewis acids in many ionic transformations.[7]

The development of new chiral hydrogen atom sources as simple chain carriers or polarity-reversal catalysts is likely to continue. Not only do phosphorus hydrides show great promise as radical reagents due to their stability, low toxicity and water compatibility,[24] but also they are readily incorporated into chiral frameworks offering considerable scope for optimization. Currently, the use of chiral phosphorus hydrides has met with little success[25] perhaps because of a poor choice of chiral manifold—more sterically demanding structures may result in better diastereo- and enantioselectivities.

An interesting possibility for the development of an enantioselective transformation is the thiol-catalyzed radical transfer-hydroamination of alkenes with the N-aminated Hantzsch ester (Scheme 13.18).[26] This reaction involves as both a source of an aminyl radical and a hydrogen atom donor. The latter requires hydrogen atom transfer from a carbon atom to a carbon radical, a generally inefficient transformation that can be facilitated by the use of a polarity reversal catalyst.

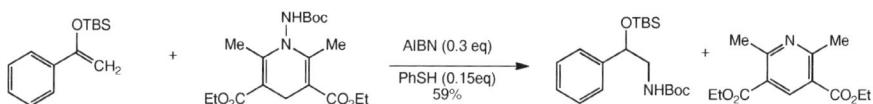

Scheme 13.18 Thiol-catalyzed radical transfer-hydroamination of alkenes with the N-aminated Hantzch Ester.

The reaction proceeds in moderate yield for a range of alkenes but, more importantly, it can proceed with good diastereoselectivity (up to 90% de) suggesting that possibly it could be developed into an enantioselective process either by the correct choice of polarity-reversal catalyst or hydrogen-bond catalyst. Such transformations hold plenty of potential especially as they are complementary to metal-based systems, which generally give the product of Markovnikov addition.

13.6 CONCLUSION

It is clear that organocatalysis is going to play an important role in the future of enantioselective radical reactions. The recent reports of Sibi and MacMillan's groups have highlighted the potential which this combination displays and it is hoped that this chapter shows that radicals, by their very nature, have always been good partners for organocatalysis, and have in fact been employed in some of the earliest examples of this now ubiquitous field. It is also hoped that the article offers an insight into the potential of this powerful marriage of chemistries and inspires others to forget their fear of radicals and enter this fascinating field of chemistry.

REFERENCES

1. C. Chatgilialoglu, *Acc. Chem. Res.,* 1992, **25**, 188.
2. G. J. Rowlands, Annu. Rep. Prog. Chem., Sect. B, 2003, **99**, 3–20; G. J. Rowlands, Annu. Rep. Prog. Chem., Sect. B, 2004, **100**, 33–49; G. J. Rowlands, Annu. Rep. Prog. Chem., Sect. B, 2005, **101**, 17–32; G. J. Rowlands, Annu. Rep. Prog. Chem., Sect. B, 2006, **102**, 17–33; G. J. Rowlands, Annu. Rep. Prog. Chem., Sect. B, 2007, **103**, 18–34; S. Z. Zard, Radical Reactions in Organic Synthesis, OUP, Oxford, 2003.
3. G. Bar and A. F. Parsons, Chem. Soc. Rev., 2003, **32**, 251–263; Stereochemistry of Radical Reactions: Concepts, Guidelines, and Synthetic Applications, ed. D. P. Curran, N. A. Porter and B. Giese, VCH, Weinheim, 1995; M. P. Sibi, S. Manyem and J. Zimmerman, Chem. Rev., 2003, **103**, 3263–3295; J. Zimmerman and M. P. Sibi, Top. Curr. Chem., 2006, **263**, 107–162.
4. A. Dondoni and A. Massi, Angew. Chem., Int. Ed., 2008, **47**, 4638–4660; H. Pellissier, Tetrahedron, 2007, **63**, 9267–9331.
5. H. Kim and D. W. C. MacMillan, J. Am. Chem. Soc., 2008, **130**, 398–399; T. D. Beeson, A. Mastracchio, J. B. Hong, K. Ashton and D. W. C. MacMillan, Science, 2007, **316**, 582–585; H. Y. Jang, J. B. Hong and D. W. C. MacMillan, J. Am. Chem. Soc., 2007, **129**, 7004–7005.
6. M. P. Sibi and M. Hasegawa, J. Am. Chem. Soc., 2007, **129**, 4124–4125.
7. T. Akiyama, Chem. Rev., 2007, **107**, 5744–5758.
8. A. G. Doyle and E. N. Jacobsen, Chem. Rev., 2007, **107**, 5713–5743.

9. M. S. Sigman and E. N. Jacobsen, J. Am. Chem. Soc., 1998, **120**, 4901–4902.
10. D. P. Curran and L. H. Kuo, J. Org. Chem., 1994, **59**, 3259–3261.
11. M. Dressel and T. Bach, Org. Lett., 2006, **8**, 3145–3147.
12. T. Aechtner, M. Dressel and T. Bach, Angew. Chem., Int. Ed., 2004, **43**, 5849–5851; M. Dressel, T. Aechtner and T. Bach, Synthesis, 2006, 2206–2214; A. Bakowski, M. Dressel, A. Bauer and T. Bach, Org. Biomol. Chem., 2011, **9**, 3516.
13. P. Selig and T. Bach, J. Org. Chem., 2006, **71**, 5662–5673.
14. A. Bauer, F. Westkamper, S. Grimme and T. Bach, Nature, 2005, **436**, 1139–1140.
15. D. H. Cho and D. O. Jang, Chem. Commun., 2006, 5045–5047.
16. A. Ohno, M. Ikeguchi, T. Kimura and S. Oka, J. Am. Chem. Soc., 1979, **101**, 7036–7040.
17. D. D. Tanner and A. Kharrat, J. Am. Chem. Soc., 1988, **110**, 2968–2970.
18. Y. D. Cai, B. P. Roberts and D. A. Tocher, J. Chem. Soc., Perkin Trans. 1, 2002, 1376–1386.
19. M. B. Haque, B. P. Roberts and D. A. Tocher, J. Chem. Soc., Perkin Trans. 1, 1998, 2881–2889; M. B. Haque and B. P. Roberts, Tetrahe-dron Lett., 1996, **37**, 9123–9126.
20. A. Erkkila, I. Majander and P. M. Pihko, Chem. Rev., 2007, **107**, 5416–5470; S. Mukherjee, J. W. Yang, S. Hoffmann and B. List, Chem. Rev., 2007, **107**, 5471–5569.
21. K. Narasaka, T. Okauchi, K. Tanaka and M. Murakami, Chem. Lett., 1992, 2099–2102.
22. J. Cossy, A. Bouzide and C. Leblanc, J. Org. Chem., 2000, **65**, 7257–7265.
23. J. A. Murphy, T. A. Khan, S.-z. Zhou, D. W. Thomson and M. Mahesh, Angew. Chem., Int. Ed., 2005, **44**, 1356–1360.
24. D. Leca, L. Fensterbank, E. Lacôte and M. Malacria, Chem. Soc. Rev., 2005, **34**, 858–865.
25. C. M. Jessop, A. F. Parsons, A. Routledge and D. J. Irvine, Tetrahe-dron: Asymmetry, 2003, **14**, 2849–2851.
26. J. Guin, R. Fröhlich and A. Studer, Angew. Chem., Int. Ed., 2008, **47**, 779–782.

CHAPTER 14

The Sunny Side of Chemistry: Green Synthesis by Solar Light[*]

14.1 INTRODUCTION

"Che l'energia solare, luce e calore, fosse necessaria alla vita, è cognizione che si disperde nella più remota antichità; negli scrittori e nei poeti di ogni epoca, che ne trovano indicazioni sicure e molteplici. Scopo nostro non è quello di occuparci dell'azione della luce nell'evoluzione organica e nella vita, ma di esaminare soltanto l'azione della luce sui fenomeni che vengono attribuiti all'affinità chimica, combinazioni e decomposizioni."[1]

"It is well known from the most remote ancient times that life needs solar energy, light and heat; we can find obvious and numerous signs in the work of writers and poets in every epoch. However, our aim is not to be interested in the action of light in the organic evolution and in life, but we want to look into the action of light on the phenomena attributable to the chemical affinity, combinations and decompositions."[1]

In this statement by Paternò, published in 1909, the author recognized the importance of solar radiation for human beings and at the same time defined the role of photochemistry in investigating all of the chemical phenomena that the matter underwent upon (solar) light absorption. Solar energy is the prerequisite for maintaining life on Earth, as recognized many thousands years ago when the sun had been an object of worship in many ancient civilizations. Sunlight-induced transformations on Earth are significantly older than life itself. It has been demonstrated that the sun produces light with a distribution similar to what would be expected from a 5525 K (5250 °C) black body (Figure 14.1). Roughly, we can say that half lies in the visible part of the

* Chapter written by V. Tamara Perchyonok.
Streamlining Free Radical Green Chemistry
V. Tamara Perchyonok, Ioannis Lykakis and Al Postigo
© V. Tamara Perchyonok, Ioannis Lykakis and Al Postigo 2012
Published by the Royal Society of Chemistry, www.rsc.org

electromagnetic spectrum and the other half in the near-infrared part. Recently, the total solar irradiance has been measured, confirming a solar constant value of 1366.1 W m^{-2} as the average intensity hitting the Earth's atmosphere.[2] As a matter of fact, the energy from sunlight reaching our planet in 1 h (4.3×10^{20} J) is more than that consumed globally in 1 year (4.1×10^{20} J, 2001 data).[3]

From the above, it can be concluded that solar radiation is the best energy source since it is cheap, clean and available throughout the entire world.[4] Unfortunately, the light flux reaching the Earth's surface is discontinuous depending on the weather conditions and not available at night. Nevertheless, such conditions were sufficient for nature to perform the photochemical reaction *par excellence, viz.* chlorophyllian photosynthesis used by organisms and plants to fix atmospheric carbon dioxide and produce biomass (especially sugars).[5] The use of solar light for the synthesis of valuable compounds is a challenge since the solar spectrum must necessarily match that absorbed by starting materials. This is not trivial since many compounds (*e.g.* organics) do not absorb in the visible or in the limited UV portion region of the solar spectrum (Figure 14.1).

On the other hand, the sun was the only accessible light source in the earliest experiments of organic photochemistry since artificial sources such as UV lamps were not yet available. Moreover, the use of photochemical reactions mediated by sunlight is obviously an appealing green way for storing the solar energy into the newly formed chemical compounds.[6] As for the use in synthesis, solar radiation is clean and renewable, and further leaves no residues in the reaction mixture, in accord with the postulates of green or eco sustainable chemistry.[7]

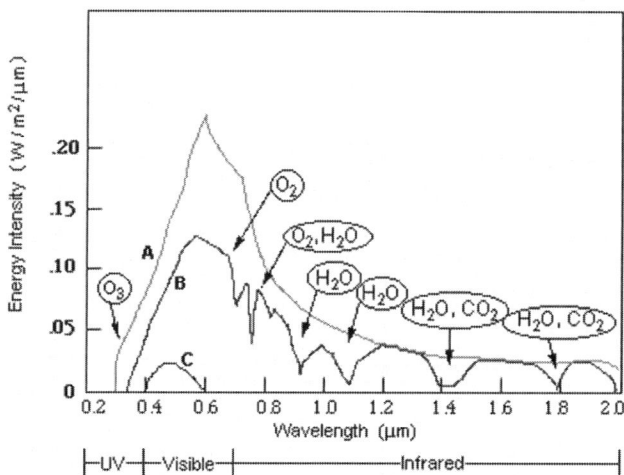

Figure 14.1 Solar radiation spectrum. Image created by Robert A. Rohde/Global Warming Art. See: http://www.globalwarmingart.com/wiki/Image: Solar_Spectrum_png.

Despite its long tradition, however, solar light is rarely applied in chemistry and the available reviews are that of Scharf and coworkers at the Plataforma Solar de Almería (PSA, Spain), dating back 15 years,[8] and the reports of Oelgemöller *et al.* on the solar chemistry performed in Cologne (Germany) and Dublin (Ireland).[9] As a matter of fact, no systematic review is available, while a large number of examples has been accumulating in recent years. Thus, it deemed worthwhile to highlight the new trends on solar-mediated synthesis of valuable compounds by reviewing recent examples of the preparation of organic and inorganic materials (photopolymerizations not included), and emphasizing the experimental advantage. Thus, after a short historical back-ground, chemical processes occurring by exposure to non-concentrated solar radiation, with no need of an elaborated apparatus, will be discussed. When appropriate, examples from the older literature will be quoted. As will be apparent in the following, many reactions (some of them important key steps in multistep syntheses)[10] are accessible under these conditions. In the last part, the scaling up and optimization and the use of solar collectors will be discussed.

14.2   HISTORICAL BACKGROUND

The use of photochemical reactions can be dated back to about 4000 years ago when ancient Egyptians and Babylonians used sunlight for the mummification and for waterproofing papyrus boats *via* photopolymerization.[11] The attribu-tion of these processes to the absorption of visible and UV light by matter and the ensuing chemical transformations observed is quite recent.[12] The beginning of organic photochemistry up to the early 20th century has been reviewed by Roth.[13] Many classes of photochemical reactions were discovered at the end of that period. An example is the sunlight-induced transformation of santonin in 80% acetic acid into photosantonic acid through three consecutive photo-chemical steps (Scheme 14.1; notice that the structure of compound was recognized only in 1958).

The [2+2] dimerization of anthracene and thymoquinone in the solid state,[14] the *Z* to *E* isomerizations of cinnamic and related unsaturated acids[15] and the photoreduction of carbonyls were likewise investigated in that time period. A possible industrial application of photochemistry emerged after the discovery

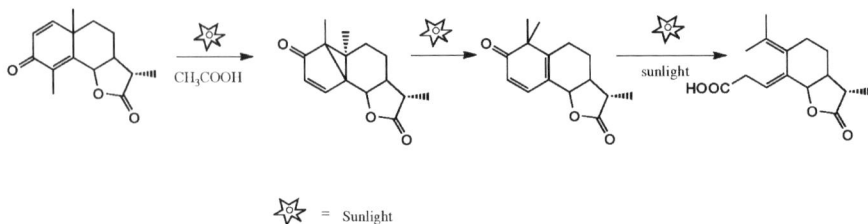

Scheme 14.1   Sunlight induced transformation of santonin into photosantonic acid.

of the easy bromination of alkyl benzenes[16] and of the rearrangement of naphthoquinone diazides. In the last case, a Wolff rearrangement with loss of nitrogen and ring contraction took place with concomitant formation of a carboxylic acid. This was one of the early examples of a photoacid generator that still has a wide application for positive photoresist materials.[17] Giacomo Ciamician is considered the pioneer of modern photochemistry for his systematic study mainly devoted to the photoreactivity of the carbonyl group, including a and b cleavage, pinacolization, reduction and intramolecular cycloadditions.[18–20] He demonstrated that photochemistry represents a major branch in organic chemistry. The image of the balconies of the University of Bologna *"where hundreds of bottles and glass pipes containing various substances and mixtures were exposed to the sun rays"* is familiar.[19,21] Inspired by the ability of plants to behave as light-driven chemical factories, he published 33 papers in cooperation with Paul Silber on "the chemical action of light". Ciamician's presentation of his photochemical work to the French Chemical Society in 1908 is a milestone in the history of photochemistry.[20] Also important is the series of papers on synthesis in organic chemistry by means of light published by Paternò in the same years[1] which, apart from the cycloaddition of ketones to alkenes (now known as the Paternò–Buchi reaction), includes the oxidative dimerization of alkylbenzenes and the cross-coupling reaction between aromatic ketones with alkylbenzenes or amines.[1]

As one would expect, during the period of both World Wars the development of photochemistry slowed down. Nevertheless, in 1937, Schönberg was forced to move to Cairo,[22] where the favourable solar conditions allowed the development of preparative photochemistry.[8] The photoreaction of diketones in the presence of olefins to give 1,4-dioxins, among others, was one of the main topics investigated.[23] Another milestone was the preparation of the anthelmintic compound ascaridol by Schenck and Ziegler in 1943 (Scheme 14.2) by using natural chlorophyll (extracted from the leaves of stinging nettles) as the sensitizer and not hazardous air as the oxidant. A pioneering solar chemical pilot plant was built in Heidelberg (Germany) for this purpose.[8a,24]

As evidenced above, many photochemical reactions have been discovered by using sunlight. Despite this unquestionable fact, there is a common prejudice that solar light is unsuitable for a photochemical reaction since a limited number of organic compounds absorb solar radiation. Accordingly, the development of solar photochemistry gradually slowed down, especially when

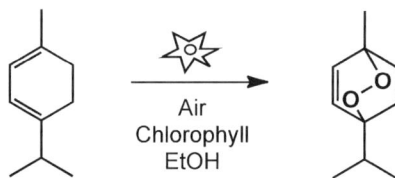

Scheme 14.2 Preparation of ascaridol by using mnatural chlorophyll as sensitizer.

artificial UV lamps became available at a reasonable price. The potential applications of solar chemistry, however, were again recognized in the 1980s and 1990s thanks to the work performed in Almería and Cologne by Scharf and co-workers (see also below).[8]

14.3   SYNTHESIS USING NON-CONCENTRATED SUNLIGHT

In this section, we present some examples in which the reaction occurs by simply exposing the reaction vessel to solar light (*e.g.* on a window ledge) with no desire of concentrating the light flux. These examples give an idea of how many photochemical processes are actually accessible in this way. At least in principle, all of the reactions mentioned could be improved and scaled up for large scale application.

14.4   PHOTOCATALYTIC/PHOTOMEDIATED PROCESSES

Photocatalysis is a promising approach for solar-mediated synthesis since light is absorbed by the photocatalyst and not by organic substrates, inducing reaction also in molecules transparent to solar light (or generally UVA radiation).[25,26] Many examples are known regarding photoamidation reactions. In fact, one of the earliest examples of the photocatalyzed formation of C–C bond involved the photoamidation of olefins developed by Elad and coworkers.[27] Acetone acted in this case both as the photomediator and as the (co)solvent. Formamide was generally used as the hydrogen donor and added both to terminal (*e.g.* 1-octene[27a]) and non-terminal (*e.g.* cyclohexene[27c]) olefins to give mainly the 1:1 adducts resulting from anti-Markovnikov addition. The amidation yields in some cases were higher when using sunlight rather than a UV lamp. The reaction proceeded *via* the carbamoyl radical, $^{\cdot}CONH_2$, generated *via* hydrogen abstraction by the acetone excited state. This added to norbornene forming norbornane-2-*exo*-carboxyamide in a high yield (87%) and in a stereospecific fashion (Scheme 14.3).[27b]

A cleaner reaction has been reported in the benzophenone photocatalyzed addition of formamide to α,β-unsaturated esters. The TiO_2 photocatalyzed amidation of various heterocyclic bases can likewise be induced by sunlight.[28,29] The reactions were successful despite the fact that TiO_2 is able to absorb only <5% of the solar radiation. No reactions occurred in the dark either at room temperature or by refluxing the reaction mixture. The reaction

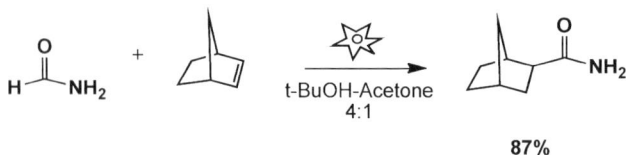

Scheme 14.3   Photoamidation reacton of norbornene into norbornane-2-exo-carboxyamide.

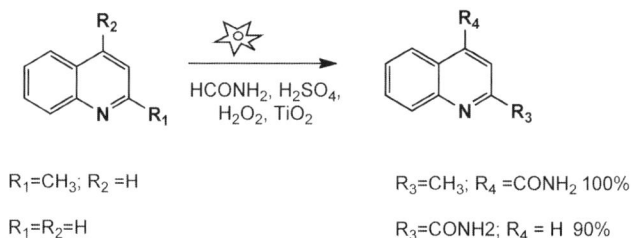

R$_1$=CH$_3$; R$_2$ =H

R$_1$=R$_2$=H

R$_3$=CH$_3$; R$_4$ =CONH$_2$ 100%

R$_3$=CONH2; R$_4$ = H 90%

Scheme 14.4   TiO$_2$ photocatalytic amidation of heterocycles.

required the presence of hydrogen peroxide to occur and the authors claimed that OH radicals could be generated *via* the reaction of H$_2$O$_2$ with excited TiO$_2$. These radicals were able to abstract hydrogen from the amide to form carbon-centered radicals, thus initiating the reaction.

Some examples are depicted in Scheme 14.4. Quinoline and 2-methylquinoline were photoamidated in a regioselective fashion in more than 90% yield upon illumination of a solution of the heteroaromatic base containing H$_2$O$_2$ and H$_2$SO$_4$ in the presence of TiO$_2$.

Photocatalyzed reactions induced by sunlight between various heterocyclic bases and ethers have been reported, for example for the functionalization of quinoline. Particularly interesting were the derivatives obtained with trioxane that gave, after acidic hydrolysis, excellent yields (88–95%) of the corresponding heterocyclic aldehydes. For instance, attack of the trioxane radical to position 4 in quinaldine gave 2-methyl-4-[1,3,5]trioxan-2-yl-quinoline in 44% yield after 35 h of sunlight exposure.[29a] Aliphatic and aromatic aldehydes can be photoadded to the same heterocyclic bases by using the same protocol, although a mixture of acylated and alkylated compounds was usually formed.[29b] A visible light absorbing photocatalyst such as Ru(bipy)$_3$$^{2+}$ was for [2+2] enone cycloaddition reactions. The photoexcited state (Ru(bipy)$_3$ $^{2+*}$) was able to reduce an α,β-unsaturated ketone moiety in bis(enone) was subsequently formed on a gram scale by a cycloaddition reaction under ambient sunlight in a laboratory window as the sole illumination (Scheme 14.5). Although the reaction was slightly slower than that performed under high-pressure UV-lamp irradiation, it was completed in 1 h and gave Z-dione 7 in excellent yield (94%) and diastereoselectivity (10 : 1).[30]

Really green conditions mimicking biomimetic redox processes were recently applied in the photocatalytic oxidation of alkenes in water with sunlight and oxygen as clean reagents.[31,32] Cyclooctene was smoothly oxidized to the corresponding epoxide, with a selectivity higher than 90% by using an iron(III)-based catalyst (*meso*-tetrakis(2,6-dichlorophenyl)-porphyrin [Fe(III)(TDCPP)]) hosted in a properly designed surfactant such as *N,N*-dimethyltetradecylamine *N*-oxide (DTAO) to allow the dissolution of hydrophobic alkenes in the aqueous medium.[31] Immobilization (and heterogenization) of the iron catalyst into Nafion membranes was another approach that markedly increased both the photocatalytic efficiency (by about ten times) and the stability of the

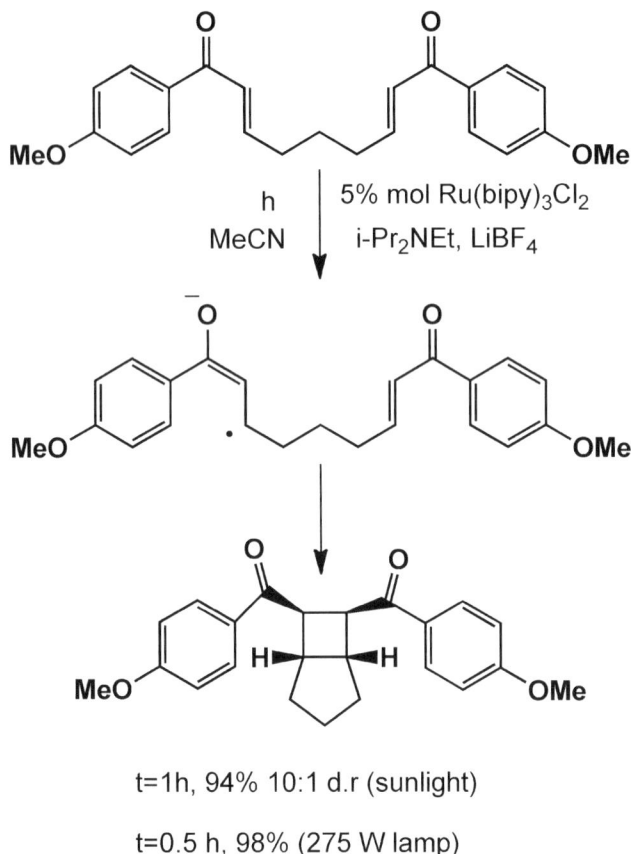

t=1h, 94% 10:1 d.r (sunlight)

t=0.5 h, 98% (275 W lamp)

Scheme 14.5 Photocatalytic oxidation of alkenes in water with sunlight and water as clean reagents.

catalyst (turnover value > 1000). In such a way, the functionalization of hydrocarbons (mainly cycloalkenes) *via* oxidation by sunlight and O_2 at room temperature has been obtained.[32] The direct oxidation of alkanes represents an important problem in the chemical industry. Takami *et al.* found that the $CuCl_2$ (or $FeCl_3$) photocatalyzed oxidation of cyclododecane (to give the corresponding alcohol and ketone as the main products) under sunlight took place more efficiently than using halogen lamps as the light source.[33]

14.5 PHOTODIMERIZATION

Sunlight-mediated dimerization reactions have been known since 1867 (this may be the first well-characterized photochemical reaction), when Fritzsche obtained the photodimer of anthracene as an insoluble compound upon illumination of a benzene solution of anthracene.[14] Several early examples belonging to this class were collected by Mustafa about 50 years ago.[14] As

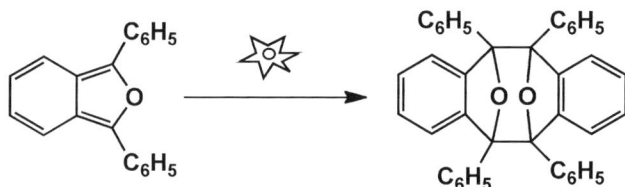

Scheme 14.6 Dimerization of 1,3-diphenylisobenzofuran under sunlight conditions.

an example, the yellow-orange 1,3-diphenylisobenzofuran (8) was smoothly converted into the dimer by the action of sunlight (Scheme 14.6). It is noteworthy that the same reaction took place thermally under drastic conditions, *i.e.* when was heated at 270 °C in a sealed tube.

Strained polycyclic molecules have attracted considerable attention in recent years as possible models for the reversible storage of light energy. An example is illustrated in Scheme 14.7 where a trishomocubane system (11) was accessed by exposure of 10 to sunlight for a few hours. This led to intramolecular [2+2] cage cyclization. A catalytic acid-induced cycloreversion was able to restore the starting bis(enone).[34]

Photoexcitation of anthracene derivatives in solution leads to dimers, usually in a head-to-tail fashion. In some cases the starting material is coloured, thus allowing the use of sunlight. As an example, yellow *syn*-[2.2](1,4)-anthracenophane 12 gave the corresponding colourless cage dimer in quantitative yield (Scheme 14.8).[35] Interestingly, the reaction was completely reversible either thermally (at 200 °C) or photochemically under irradiation with a shorter wavelength.

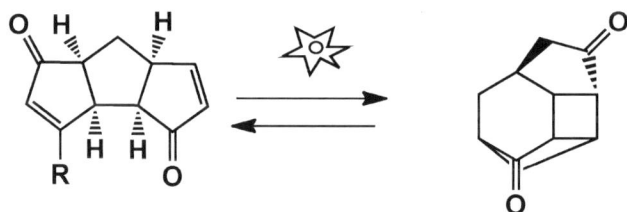

Scheme 14.7 Synthesis of strained polycyclic molecules for reversible storage of light energy.

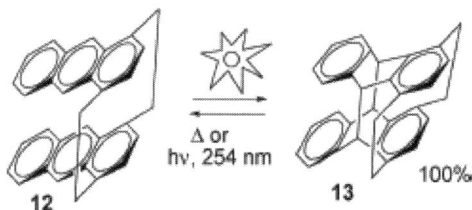

Scheme 14.8 Photoexcitation of anthracene derivatives.

Head-to-tail [4p+4p] photodimerization ($\phi = 0.023$) and photodissociation of 9-anthroylacetone can also take place in the solid state.[36] Linear oligo-acenes have been recently studied as semiconductors in thin-film transistors. A film containing a highly soluble derivative of pentacene, namely 6,13-bis((triisopropylsilyl)ethynyl)pentacene was exposed to sunlight in Manchester (UK) and underwent a clean dimerization where oxygen or moisture seemed to have a limited role in the reaction.[37]

A regio- and stereoselective [2+2] photodimerization reaction has been reported by sunlight exposure (14 days) of a suspension of various diethyl 1,2 benzoxaphosphorine-3-carboxylates in water. In all cases mixtures of the corresponding *anti* head-to-tail dimers were obtained.[38]

14.6 CYCLOADDITIONS

Although cycloadditions are one of the most widely used photochemical processes, solar light has seldom been exploited in this case. As an example, the photocycloaddition of enol ethers to 2-substituted anisoles (*e.g.* 2-cyanoanisole) smoothly occurred in a regioselective fashion upon sunlight exposure.[39] The thus-formed bicyclic compound 14 was thermally transformed to cyclooctatriene 15 that underwent a second photochemical step to bicyclo[4.2.0]octa-2,7-diene 16. This one-pot procedure led to a 20:1 mixture 16:15 in 92% overall yield (Scheme 14.9).

[2+2] cycloadditions likewise took place under photocatalytic conditions[30] and [4+2] cycloadditions were also involved in the synthesis of endoperoxides. Furthermore, sunlight may be employed for the generation of ground-state reagents such as nitrosocarbonyl intermediates that react smoothly both in hetero Diels–Alder cycloadditions and in ene reactions. These 1,3-dipoles could be thermally generated from 1,2,4-oxadiazole-4-oxides at 100–150 °C, but these conditions did not allow the use of low boiling alkenes as the trap.[40]

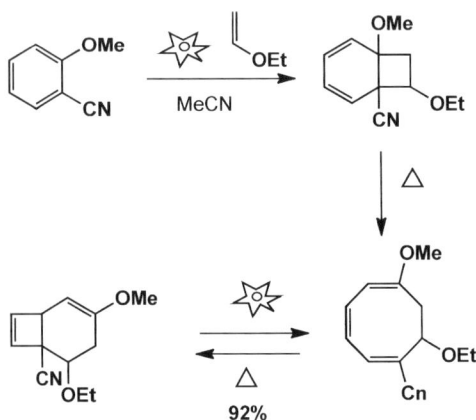

Scheme 14.9   Photocycloaddition of enol ethers to 2-substituted anizoles.

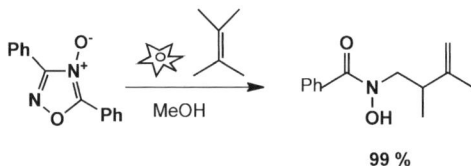

Scheme 14.10   Sunlight mediated synthesis of substituted benzoates.

Exposure of a methanolic solution of 3,5-diaryl-1,2,4-oxadiazole (17) to sunlight in the presence of tetramethylethylene, however, led to the ene adduct 18 in an almost quantitative yield (Scheme 14.10).[41]

Interestingly the same procedure was also performed in the solid phase, although compound 18 was formed in a markedly lower yield (20%) in that case.[42]

14.7   CYCLIZATIONS

An example of the Norrish–Yang photocyclization, a useful tool for the formation of C–C bonds under regio- and stereocontrolled conditions, has been reported under solar light illumination. When a solution of α-(o-tolyl)acetophenone was photolyzed at 313 nm in benzene, γ-hydrogen abstraction and formation of 2-phenyl-2-indanol (19) took place.[43] The reaction was repeated under sunlight in polymer beads obtained by cross-linking of a commercially available ethylene–vinyl acetate copolymer (Elvax 150 N, 20% vinyl acetate content) meant to transfer energy to the selected substrates. A quantum yield close to unity was claimed, even in the solid polymer matrix. The reaction carried out in a 1 hectare (10^8 cm^2) illuminated area was estimated to produce 525 kg of product per day (Scheme 14.11).[44]

Recently, compounds containing the 1,2-diketone moiety have been employed, as for the case of the 1-glycosyl-2,3-butanedione derivative 20, where a new spirocyclic monosaccharide 21 was formed in a diastereoselective manner and in a quantitative yield.[45] Interestingly, compound 21 was also formed under solvent-free conditions, although in a lower yield (42%, Scheme 14.12).

The remarkable regioselectivity of the photochemical intramolecular 1,5-hydrogen atom transfer in 1,2-diketones was also exploited for the diastereoselective synthesis of functionalized cyclopentitols. Outdoors illumination of a nono-2,3-diulose (*e.g.* 22, Scheme 14.13) in a NMR tube led to a 1,4-biradical

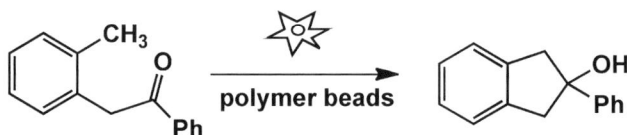

Scheme 14.11   Sunlight mediated large scale synthesis of 2-phenyl-2-indanol.

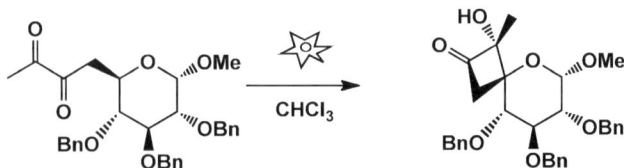

Scheme 14.12   Sunlight mediated synthesis of novel spirocyclic monosaccharides.

intermediate 23 that underwent Norrish type II photoelimination to give enol 24. An intramolecular enolexo aldol reaction followed by intramolecular acetalization gave bicyclic compound 25. The acetalization step was crucial since it avoided absorption of visible light by the 1,2-diketone intermediate.[46]

14.8   PHOTOPINACOLIZATION (PHOTOREDUCTION)

Photopinacolization, *viz.* the bimolecular reduction of an aromatic ketone or aldehyde by hydrogen abstraction followed by radical coupling, was reported more than 100 years ago by Ciamician and Silber.[47] They found that an ethanolic solution of benzophenone was efficiently reduced to benzopinacol by exposure to sunlight for a few months. The reaction in alcohols (MeOH, EtOH *i*-PrOH, *etc.*) has been explored in detail by Cohen, who extended the reaction to a variety of substituted benzophenones.[48]

Hydrocarbon solvents such as toluenewere likewise found suitable for the reaction.[49,50] In addition, fine tuning of the reduction pathway could be achieved by varying the reaction conditions, as reported by Bachmann who found that prolonged illumination in the presence of a small amount of sodium alcoholate led selectively to benzhydrols rather than benzopinacols (Scheme 14.14a).[51]

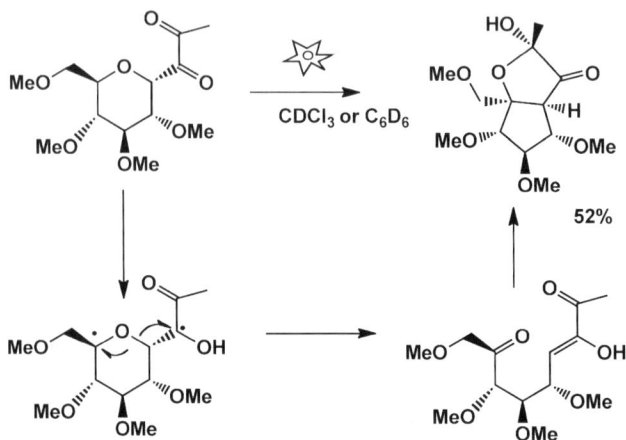

Scheme 14.13   Outdoor illumination of nono-2,3-diulose.

a.

93%

97%

b.

55-95% yield

$Ar=C_6H_5$; $4-ClC_6H_4$, $3-ClC_6H_4$, $2,4-Cl_2C_6H_3$
$R=H$, CH_3, C_6H_5, CH_2Cl

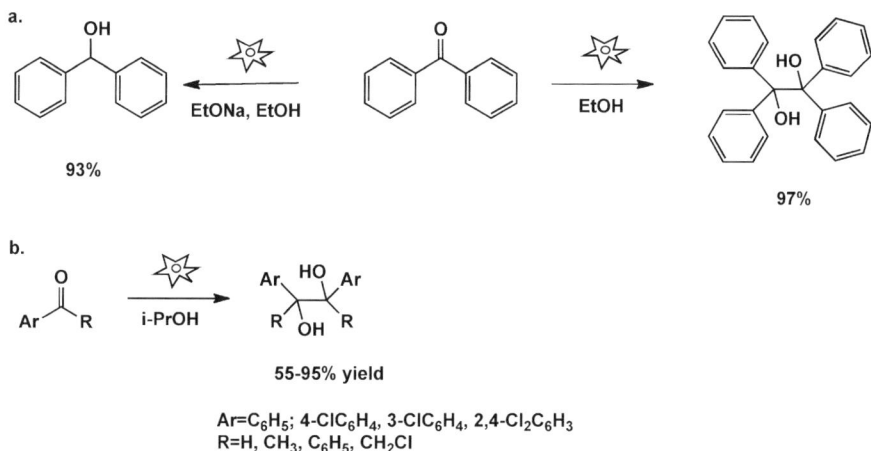

Scheme 14.14 Benzophenones to benzopinacolones with sunlight.

Very recently a general procedure for the reductive coupling of aromatic aldehydes and ketones has been reported (Scheme 14.14b). The formation of 1,2-diols by this protocol represents a valid green alternative to thermal methods, especially for the case of benzaldehyde where the solvent (isopropanol and acetone) was recycled three times.[52]

14.9 SYNTHESIS *VIA* ELIMINATION OF A GROUP

The sunlight-induced cleavage of alkynyl(phenyl)iodonium salts has been employed for the synthesis of alkynylphosphonium derivatives, excellent Michael acceptors towards a large variety of nucleophiles.[53] A few years ago, a mild, one-pot protocol for the preparation of a-substituted nitrogen heterocycles from α-amino acids has been reported and was applied to the synthesis of different alkaloid analogues in good to excellent yields. The process was based on an initial sunlight-induced radical decarboxylation, which generated a carbon radical, which was in turn readily oxidized *in situ* to an acyliminium ion. Trapping of these ions by different nucleophiles offered an easy access to a-substituted heterocycles.[54,55] Spirocyclic compound 27 was then obtained by a new tandem decarboxylation/hetero-Diels–Alder reaction (Scheme 14.15) by treating substrate 26 with DIB/iodine followed by addition of boron trifluoride and 2,3-dimethylbutadiene.

Potentially bioactive (α-amino) phosphonates have been likewise synthesized by a radical fragmentation-phosphorylation reaction starting from the corresponding a-amino acids.[54] Some of the reactions from this group have been carried out in the crystals, where, as sometimes happens, a clear-cut reaction was obtained, in contrast to the formation of a mixture of products in solution. In some cases, a clean crystal-to-crystal reaction took place. This was the case of a suspension of carboxylate microcrystals (28, 1.5 g) that upon

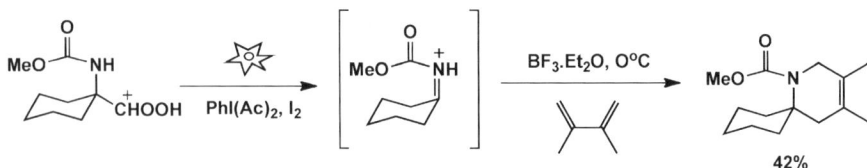

Scheme 14.15 Synthesis of alpha-substituted heterocycles via novel cascade reaction.

sunlight exposure in 100 mL of *n*-hexane with a medium-pressure Hg lamp (Pyrex-filtered light) lost CO (>97% conversion in 12 h) forming the pure photoproduct 29 in a completely stereoselective fashion (Scheme 14.16).[56] This protocol was ideal for large-scale reactions since compound 29 was simply collected by filtration. The feasibility of the reaction under solar light was assessed by exposing a small sample to sunlight and yielded a quantitative reaction within 2 h.

Another example of selective carbon–carbon bond cleavage in the solid state was observed upon exposure of crystals of a (diphosphine)Pt^0 complex, namely [(1,2-bis(diphenylphosphino)-benzene) (η-2,2,2′-dibromotolane)Pt^0] to sunlight. This alkyne complex underwent a highly selective insertion of a (diphosphine)Pt^0 complex fragment into an aryl–alkynyl C–C bond.[57]

The photoinduced loss of a Ph$_2$S group in N-conjugated sulfilimines was adopted for the synthesis of various pyrrolo[2,3-d]pyrimidine-2,4-diones as depicted in Scheme 14.17. Interestingly, an almost quantitative yield of 31 was obtained by exposing to sunlight a stirred, aerated solution of 30 in ethanol for only 40 min, whereas refluxing compound 30 in toluene for 30 h caused no reaction.[58]

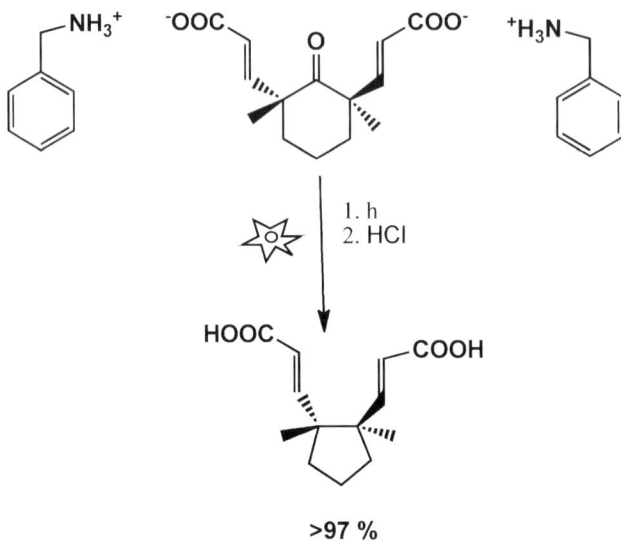

Scheme 14.16 Clean, sunlight mediated crystal to crystal reaction.

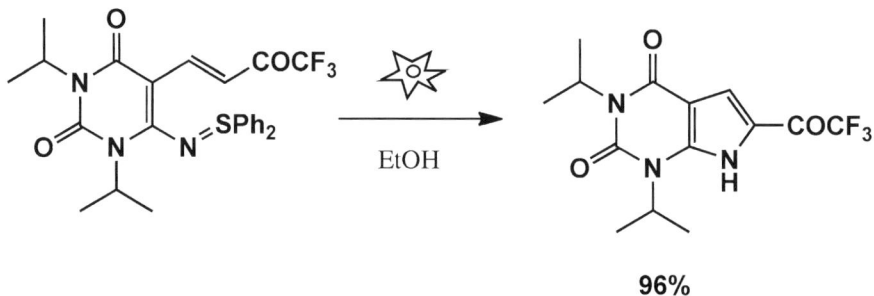

96%

Scheme 14.17   Synthesis of various pyrrolo[2,3[pyrimidine-2,4-diones using sunlight.

Sunlight-induced hydrolysis of a Schiff base was also used to liberate an aromatic amine (or a sulfadiazine) from its benzalimine. Aqueous acetone was the elective solvent of the reaction.[59]

14.10   ARYLATION REACTIONS

To our knowledge, only a few cases of sunlight-induced formation of an Ar–X bond have been reported in the literature. For example, Rajan and Muralimohan described an efficient phenylation of dimsyl anion by a photo-stimulated Ar_{SRN1} reaction in DMSO in the presence of iodo or bromobenzene.[60] The solar-induced nucleophilic substitution of an alkoxy group by butylamine has been successfully applied to the synthesis of nitrobenzacrown aminoether 32 (Scheme 14.18).[61]

14.11   ISOMERIZATIONS

The sunlight-induced interconversion of the Z and E isomers of organic molecules is a well known process.[15] This photoreaction is usually adopted when the less easily accessible Z isomer is required. As an example, the Z isomer of both 1-phenyl-1,3-butadiene and of p-benzoyl- and (4-bromobenzoyl)acrylic acids and of their methyl esters and amides have been smoothly prepared starting from the corresponding E isomers.[15] 6-(Z)-Styrylpurine was likewise prepared by photoisomerization of the E isomer.[62]

3-Aryl-2-quinolinones are a class of compounds that find use in the treatment of osteoporosis and can inhibit the migration of tumour cells. A straightforward synthesis of such compounds involved sunlight-induced isomerization

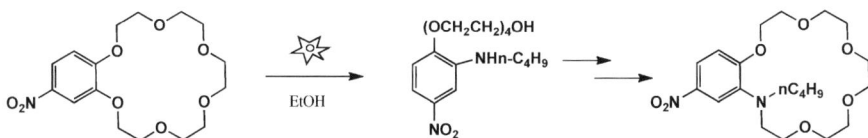

Scheme 14.18   Synthesis of nitrobenzacrown aminoethers under sunlight conditions.

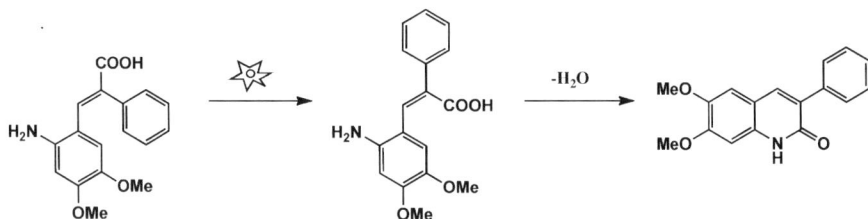

Scheme 14.19   Synthesis of 3-aryl-2-quinolinones *via* sunlight.

of (*E*)-2-amino-a-phenylcinnamic acids (*e.g.* 33) and the intramolecular amidation, as depicted in Scheme 14.19.[63]

In some cases the process can be sensitized by triplet energy transfer either from a coloured (rose bengal) or a colourless (α-acetonaphthone) sensitizer. In the last case, a 1.2 g sample of β-ionol gave the corresponding *Z* isomer in a de-oxygenated ethanol solution after 15 h of exposure to late November sunlight in Honolulu (Hawaii).[64a] A simple floating solar reactor was designed for this purpose with no need of circulating cooling water. The so-designed "kick-board reactor" has been employed for the closely quantitative one-way *E* to *Z* photosensitized isomerization of several dienes and styryl derivatives.[64b]

14.12   HALOGENATIONS

Early reports dealt with the solar induced synthesis of variable amounts of chlorinated hydrocarbons by reaction between simple hydrocarbons and chlorine gas.[16] On the other hand, activated C–H bonds are well suited for halogenation reactions. A fast and efficient α-bromination of aromatic and aliphatic carbonyl compounds has been achieved through solar exposure by using *N*-bromosuccinimide (NBS) as the halogen source.[65] The mechanism was thought to involve a radical pathway, and occurred at 30 °C, with no need of catalyst or radical initiator. Et_2O and THF appeared to be the most suitable solvents for the reaction. Cyclododecanone was completely consumed in 2 min under portionwise addition of NBS upon sunlight illumination and gave a high yield along with traces of the dibrominated derivative (Scheme 14.20).

Substitution of a benzylic hydrogen with a halogen has been likewise reported, as illustrated in Scheme 14.21. An easy method to synthesize

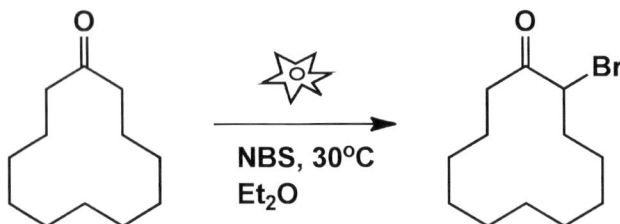

Scheme 14.20   NBS mediated bromination under sunlight.

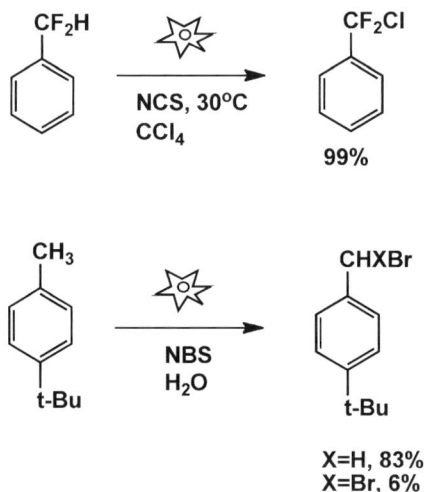

Scheme 14.21 Synthesis of halogenated substituted toluenes under sunlight conditions.

α,α,α-chlorodifluorotoluene in high yield involved a sunlight- (or sunlamp) initiated chlorination of α,α-difluorotoluene with *N*-chlorosuccinimide (Scheme 14.21a).[66]

NBS led to the monobromination of methylbenzenes by using water as an eco-friendly solvent. As an example, a mixture of water insoluble 4-*t*-butyl toluene (that formed a layer 'on water') and NBS was vigorously stirred upon sunlight exposure. The increased specific weight of the hydrophobic brominated product formed (35a) caused the organic phase to sink to the bottom of the flask, whereas succinimide formed as a by-product remained in the water phase. As a result, 35a was obtained in a high yield along with a small amount of dibrominated compound 35b and easily separated by phase separation or filtration (Scheme 14.21b).[67]

14.13 SYNTHESIS OF ENDOPEROXIDES

The importance of endoperoxides is known, in particular with regard to the significant biological properties. Natural compounds having an endoperoxide group such as artemisinin (from *Artemisia annua*) and artellene, belong to a promising new class of antimalarial drugs. Endoperoxides can be easily accessed by solar dye photosensitized [4+2] reaction of singlet oxygen with a diene moiety. As an example, the diene lactam 36 was exposed to sunlight (eosin as the sensitizer) while oxygen was bubbled through the solution forming the unstable endoperoxide 37 that was readily hydrogenated in one-pot to give the diol lactam 38. This reaction was applied to the multigram synthesis of thebaine analogue 39 in view of its potential analgesic properties (Scheme 14.22).[68]

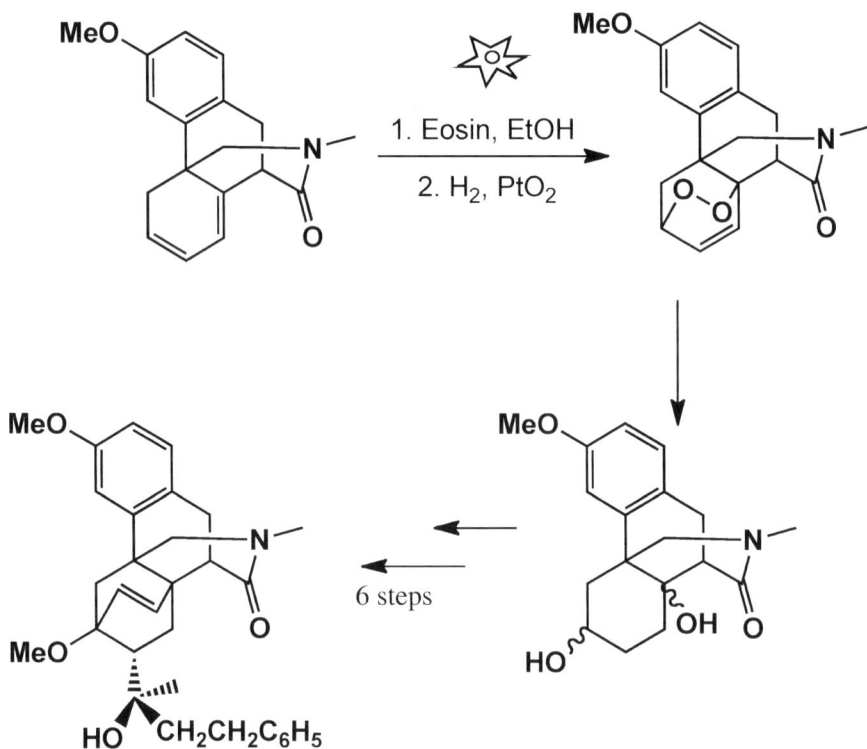

Scheme 14.22   Multigram synthesis of thebaine analogues under sunlight conditions.

Sunlight-mediated photooxygenations of furanose ringsubstituted enones have been used to prepare new spiroendoperoxide compounds. These are appealing because of their potential biological activities since polyether anti-biotics such as salinomycin and narasin contain spiroketal rings.[69] It is noteworthy that no sensitizer was required for the reaction. In fact, the direct sun illumination (4 h a day for about 20–40 days) was carried out by exposing dilute (0.1%) carbon tetrachloride solutions with occasional stirring of the flask for ensuring saturation with air. In the frame of the study of organic semiconductors, the development of an ecological film-forming technology is of great importance, as is the study of pentacene derivatives as possible components in organic field-effect transistors. Recently, the exposure of 6,13-dithienylpentacenes solutions to sunlight in air afforded the endoperoxides in good yield. Interestingly, the thin film formed with the endoperoxides underwent a deoxygenation process under UV light exposure.[70]

14.14   OXIDATIONS/OXYGENATIONS

Solar-mediated oxidations have been used for the synthesis of valuable compounds. Brasiliquinones, for example, are natural substances known to

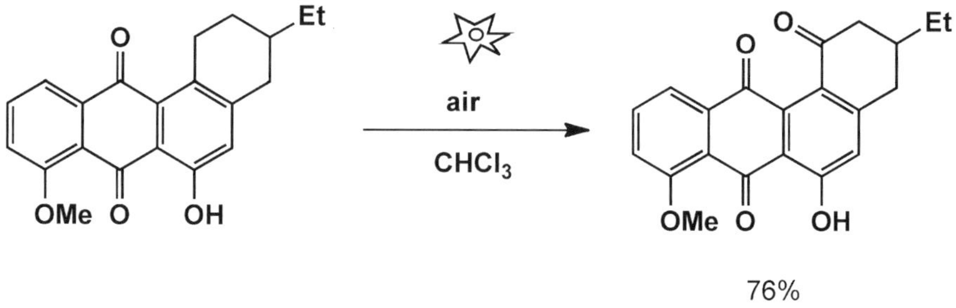

76%

Scheme 14.23   Synthesis of brasiliquinones under sunlight conditions.

possess both antibacterial and anti-tumour activities.The last step of the synthesis of brasiliquinone B (41) was accomplished by a simple exposure of a solution of antraquinone derivative 40 in aerated chloroform for 6 h to sunlight. Oxygen oxidized one benzylic position in a chemo-selective fashion (Scheme 14.23).[71]

The mild photochemical conditions allow the one-step silyl deprotection of a hydroxy group and the subsequent oxidation to a carbonyl group and aromatization. This strategy was followed in the synthesis of rubiginones, a family of cytotoxic natural products recently investigated for their use in treatment of AIDS and Alzheimer's disease. Rubiginone C2 (43) has been easily prepared from 42 by sunlight exposure under solvent-free conditions and by using air as the oxidant; rubiginone A2 (44) has been synthesized in a high yield by treatment of 43 under basic conditions (Scheme 12.24).[72]

Very recently, the dye-sensitized photooxygenation of some 5-amido-1-naphthols was performed under sunlight in July and August 2008 at Dublin City University. 5-Amido-1,4-naphthoquinones were isolated in yields form moderate to excellent. The rate of formation was higher than under artificial illumination and consumed only 0.002 kWh (or 7.2 kJ) of energy for the air pump, since no cooling was required. In contrast, the indoor procedure consumed a total of 0.45 kWh, mainly for the operation of the artificial light source.[73]

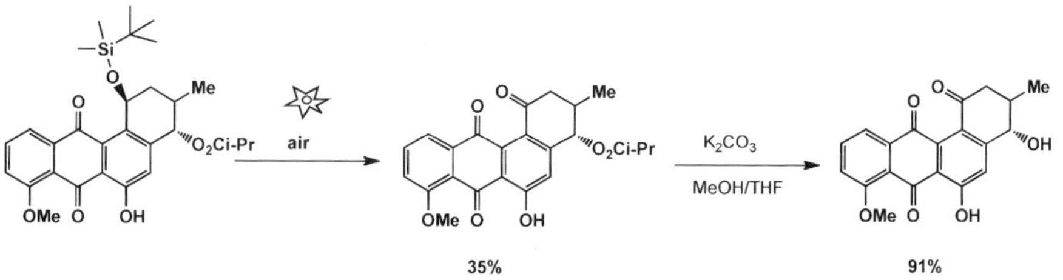

35%   91%

Scheme 14.24   Synthesis of brasiliquinones under sunlight conditions.

14.15 CONCENTRATED SUNLIGHT

14.15.1 General remarks

In early designed solar reactors (*e.g.* the Schenk's plant for the synthesis of Ascaridol),[8] the light (both direct and diffuse radiation) was absorbed without prior concentration. The simple reactor design represents what was available in the post-war period, and was not optimized with respect to the exploitation of light. The intensity of radiation was relatively low and thus carrying out reactions in reasonable times required a high optical window area. On the contrary, during recent decades the use of more elaborate reactors has enabled the concentration of sunlight[74] with a considerable increase of the space–time yield, depending on the concentration factor chosen. Pioneering experiments employing solar concentration systems have been carried out at the Plataforma Solar de Almería (PSA, Spain). Different apparatuses (characterized by different concentration factors) have been built: for gram scale reactions the SOLFIN (SOLar synthesis of FINe chemicals, Figure 14.2) facility, based on a CPC (compound parabolic collector),[75] employed low concentrated sunlight (concentration factor, CF = 2–3). The SOLFIN was a simple reactor where a coaxial cylindrical mirror angled towards the sun collected the light that was directed onto a tube placed in the focus line through which the solution (up to *ca.* 1 L) was circulated by means of a centrifugal pump (Figure 14.2).

On the other hand, multi-gram scale reactors based on the use of the Helioman Module were also tested. Here a concentration factor up to 32 suns

Figure 14.2   Image of the SOLFIN apparatus taken at Plataforma Solar de Almería.

can be reached. This very effective apparatus consists in parabolic trough-collectors having aluminium mirrors that track the position of the sun or by holographic elements as concentrating units.[76] The photolyzed solution is located in the focus line (see Figure 14.3).

This idea was first developed by G. Schneider who designed the SOLARIS reactor operating with a maximum volume of 70 L at the PSA. An improved design of the SOLARIS reactor (that was dismantled at the PSA) is the PROPHIS[77] (parabolic trough-facility for organic photochemical syntheses) reactor, then rebuilt in Cologne, Germany. In both SOLARIS and PROPHIS reactors the Helioman module was equipped with four troughs (total aperture 32 m2), though in the first case only one was used by Scharf for practical reasons.[8] At any rate, the design of this type of reactor is flexible and depends on each application. The troughs were fitted with Pyrex glass reaction tubes in the focal lines. The direct beam solar radiation was concentrated with a ratio of about 30–32. As an alternative, a parabolic dish reactor has been erected at the Institute of Solar Energy in Bornova (Turkey).[78] Using concentrating systems, various target products have been synthesized up to the kilogram scale, depending on the dimension of the parabolic collector. Contrary to CPC collectors, the latter light-concentrating reactors can only concentrate the direct fraction of global radiation missing the diffuse UV-sunlight, which makes them more dependant on natural illumination conditions. Some examples on the use of solar collectors for the synthesis of fine chemicals are reviewed in the following paragraphs.

14.15.2 Photooxidations and photooxygenations

Photooxidation is a class of photoreactions that can be easily scaled up. One of the best known examples was reported by Scharf and involves the dye-

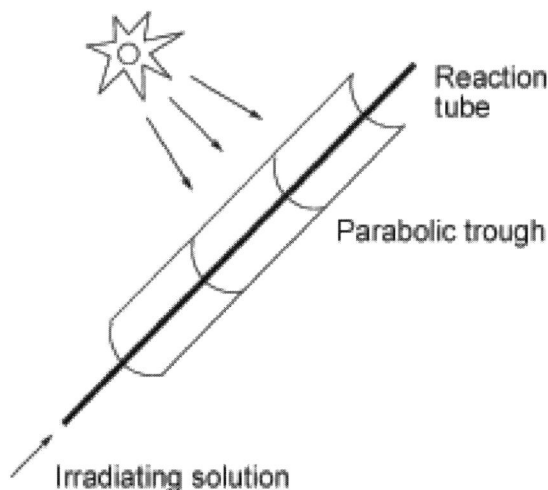

Figure 14.3 Diagram of multigram scale reactor based on Helioman module.

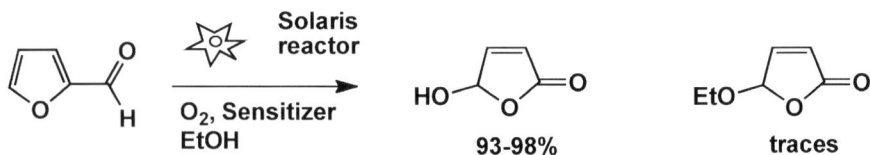

sensitizer: Rose Bengal or Methylene blue

Scheme 14.25   Dye-sensitized photooxidation of fulfural to yield 5-hydroxy furanone.

sensitized photooxidation of furfural to give 5-hydroxyfuranone (Scheme 14.32).[8]

Apart from the synthetic interest of the furanone formed (a versatile C4 building block), the importance of the reaction resides in the use of high visible light absorbing species (rose bengal or methylene blue) as singlet oxygen photo-generators.

A small scale parabolic trough reactor equipped with aluminium mirrors was used for the synthesis of versatile biologically active quinones such as the Juglone (5-hydroxy-1,4-naphthoquinone, *ca.* 79% yield) by rose bengal or methylene blue-sensitized photooxygenation of 1-naphthols under a stream of oxygen (Scheme 14.33).[79,80]

Gram-scale synthesis of final product has been achieved after just 4 h of exposure owing to the favourable absorption in the visible spectrum of the dyes employed. This powerful and environmentally friendly alternative to existing thermal processes takes advantages of the geometric concentration factor of the reactor. Although this arrangement offered a geometric concentration factor of *ca.* 15, optical losses reduced this value. The PROPHIS plant was employed in the rose bengal photosensitized oxygenation of citronellol in isopropanol under oxygen atmosphere that gave the isomeric hydroperoxides (*ca.* 1 : 1 ratio) in an almost quantitative yield. Reduction to the corresponding alcohols followed by acid-catalyzed treatment of the end mixture allowed a cyclization of one of the regioisomer to give rose oxide, an important fragrance.[81] A mixture of rose oxide (*Z* and *E* isomers, 82% yield)

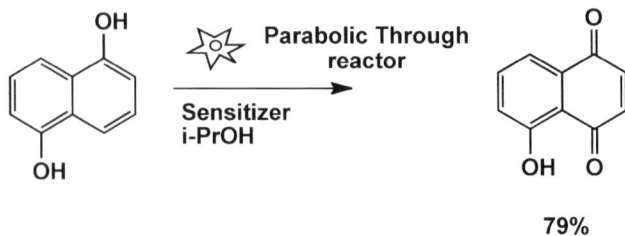

sensitizer: Rose Bengal or Methylene blue

Scheme 14.26   Methylene blue sensitized photo-oxygenation of 1-naphthols.

was obtained in a similar way by PET oxidation of citronellol with perylene diimide as the photosensitizer after only 2 h of exposure to sunlight in a dish reactor; the super oxide radical anion (generated by reduction from the diimide radical anion) mediated the photooxidation.[82]

14.15.3 Cycloadditions

Using the CPC reactor, a quantitatively cyclobutane formation was accomplished by a regio- and stereospecific (2p + 2p) photochemical addition of 2-acetoxy-1,4-naphthoquinone with styrene (or 1,1-diphenylethene). The reaction could be easily scaled-up since concentration of the quinone can be increased to 6% w/v without any adverse effects. Moreover, the water cooling of the solar-irradiated solutions could be omitted since the efficiency of the photoreactionwas not affected by temperature up to 60 °C. It is noterworther that this reaction did not require any purification procedure and the product was obtained in >99% purity directly by evaporation of the solvent *in vacuo*.[83]

Some interesting reactions have been performed by utilizing the SOLARIS apparatus.[84] As an example, a [2 + 2] cycloaddition between ethene and 5-alkoxy- [5*H*]-furan-2-onewas successful under acetone sensitization.[8] This was a remarkable result since an almost quantitative yield of the cycloadduct was formed despite the fact that the sensitizer had a threshold wavelength of 330 nm. The pheromone grandisol and its analogue demethylgrandisol were finally obtained after elaboration of the cycloadducts. The same reaction, when applied to 5-ethoxyfuranone, led to incomplete conversion due to the low solar photon yield (0.1%).[85] Large amounts of Paterno–Buchi adducts were also formed in a regiochemical and stereochemical fashion in the reaction between (*tert*-butyl)phenylglyoxylate (*ca.* 1 kg) in cyclohexane in the presence of an excess of furan (5.4 equiv., Scheme 34).[8]

14.16 PHOTOCATALYTIC REACTIONS

Photocatalysis is well suited for solar light induced reactions since a conveniently absorbing photocatalyst can be chosen that overcomes the drawbacks related to the matching between the solar radiation and the absorption spectra of photoreactive molecules.[86] Various photocatalyzed alkylation reactions *via* alkyl radicals generated both by hydrogen abstraction (from an alkane or an alcohol) or by electron transfer (from a silane donor) have been used. In the first case benzophenone or its disulfonated derivative (BPSS) were adopted. Thus, solar exposure of a solution of benzophenone and dimethylacetylendicarboxylate in neat cyclopentane gave dimethyl (*Z*)-2-cyclopentyl-2-butenedioate (along with a negligible amount of the *E* isomer) in 78% yield after 4 h of illumination.[87] Fourteen hours (in three consecutive days by using the SOLFIN facility) were required for the photocatalytic synthesis of terebic acid starting from maleic acid and BPSS in an *i*-PrOH–water mixture (Scheme 14.35).[88]

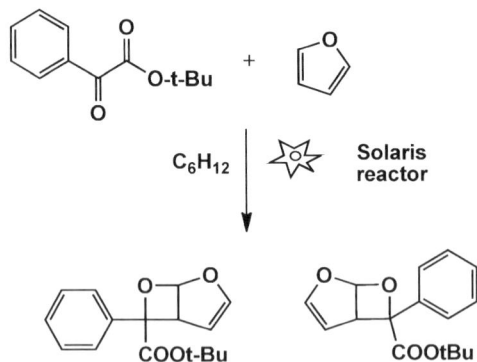

Scheme 14.27 Large scale Patterno-Buchi transformation under sunlight.

Photocatalysis can be carried out under heterogeneous conditions as well by using a semiconductor such as TiO$_2$. Accordingly, benzylated succinic acid (or anhydride) has been prepared on a gramscale by a PETreaction between 4-methoxybenzyl(trimethyl) silane and maleic acid (or anhydride). The reaction was based on the photocatalytic oxidation of the silane that caused the fragmentation of the resulting radical cation generating a benzyl radical. In turn, the last species attacked the electron-poor olefin.[112] The (C$_5$H$_5$)CoCOD photocatalyzed syntheses of 2-substituted pyridines has been performed by cyclization of ethyne and nitriles (Scheme 14.36).[113]

Solar light was the elective light source for the reaction since the catalyst was photodegraded when using harsh artificial UV light. The reaction was characterized by a high atom economy and, although the conversion of the reagents did not exceed 50%, it was found to be highly selective (a trace amount of benzene was also detected) and competitive if compared to the analogous thermal reaction. Contrary to the latter case, ambient pressure and temperature are sufficient for the occurrence of the reaction.

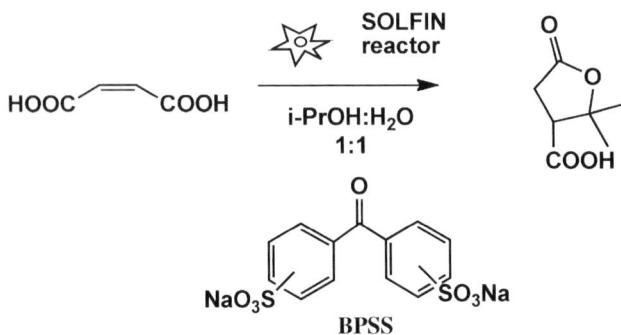

Scheme 14.28 Photocatalytic synthesis of terebic acid.

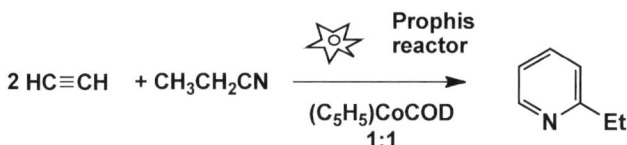

2 HC≡CH + CH₃CH₂CN → (C₅H₅)CoCOD 1:1 → [structure: 2-ethylpyridine] (Prophis reactor)

Scheme 14.29 Photocatalyzed synthesis of 2-substituted pyridines.

14.17 PHOTOACYLATIONS

Photoinduced acylations could represent an eco-sustainable alternative to classical Friedel–Craft reactions, avoiding the use of large amounts of Lewis acids and noxious acyl chlorides or anhydrides as reactants. The pioneer studies by Klinger on the direct addition under sunlight of both aliphatic and aromatic aldehydes to quinones[89] were reconsidered by Krauss,[90,91] who studied the acylation of quinones by prolonged UV irradiation (5 days). However, this procedure led to a sluggish reaction and benzene was adopted as the reaction solvent. These limitations have been recently overcome by Mattay *et al.*, who found that 500 g of naphthoquinone and 4 kg of butyraldehyde in 80 L of a *tert*-butanol–acetone mixture (3 : 1) when exposed for 3 days to sunlight formed the corresponding acylated derivativ in 90% yield and in half-kilogram scale (Scheme 14.37).[92]

This reaction type was used as a model reaction to test three different solar reactors, *viz.* the PROPHIS having a concentration factor (CF) of 15, the CPC (CF = 2–3) and a flat bed reactor (CF = 1). The CPC reactor was found to give the best results due to its robustness in terms of weather dependence.[92]

14.18 *E/Z* ISOMERIZATIONS

Solar collectors also allowed the clean *E/Z* isomerization of olefins on a kilogram scale. The synthesis of large amounts of olefins in the *Z* configuration by using thermal methods is in many cases troublesome. On the contrary, these compounds can be formed in high purity fromthe corresponding *E* isomer by photosensitized (benzil as sensitizer) isomerization, as demonstrated in the *E/Z* stilbene isomerization.[93]

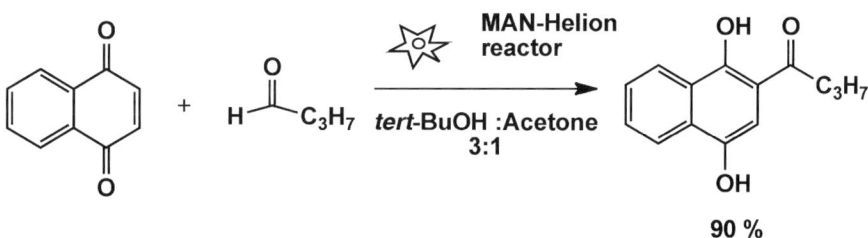

Scheme 14.30 Synthesis of 1-(1,4-dihydroxynaphthalene-2-yl) butan-1-one).

14.19   POTENTIAL INDUSTRIAL APPLICATIONS

The main handicaps of conventional industrial photochemistry are high investment costs (light sources and related installations) and high operation and maintenance (O&M) costs (electricity, cooling energy, lamp replacement).[94,95] In contrast, solar photochemistry does not need expensive lamps and has significantly lower O&M costs, assuming that the photochemical system can absorb the visible radiation of the solar spectrum. The direct use of solar radiation rather than of artificial lamps is highly attractive for the photochemical production of industrially important chemicals. The opportunities for the industries are twofold: solar chemical techniques can be employed directly at locations in industrialized countries if there is enough sunshine available, or can be exported to those industrializing countries that have good solar conditions.[96]

The best known example of an industrial photochemical process is the synthesis of cyclohexanone oxime hydrochloride (the precursor of ε-caprolactame). This procedure is based on the photochemical conversion of cyclohexane in the presence of NOCl and was carried out on a large scale by Toray in Japan by using UV-lamps (Scheme 14.38).[97]

The feasibility of the reaction under sunlight simulating conditions (radiation wavelength *ca.* 350–550 nm)[98] and an evaluation of the energetic demand for the sunlight production on industrial scale has been recently reported.[99] Thus, the use of solar radiation led to a 4-fold lower demand for electric current and an 8-fold lower demand for cooling energy if compared to the typical Toray protocol. In addition, emissions of up to 2.5 tons of CO_2 due to the conversion of fossil fuel to electricity can be avoided. As for the economic point of view,[100] although the calculated investments for a solar plants are higher (almost two times) than those required for a lamp driven plant, the reduced economic demand for electricity and cooling energy would lead to a significant reduction of the production cost. A hybrid photoreactor system by using sunlight during the day and mercury lamps at night was also envisaged.[101] One of the most recent promising industrial applications of concentrated sunlight is the synthesis of fullerenes and carbon nanotubes. The process is based on the direct vaporization of carbon and is a valid alternative to electric-arc discharge, laser ablation and catalytic chemical vapour deposition.[102–106] The latter methods suffer from several disadvantages, in particular the high cost and the high UV radiation power of the arc plasma

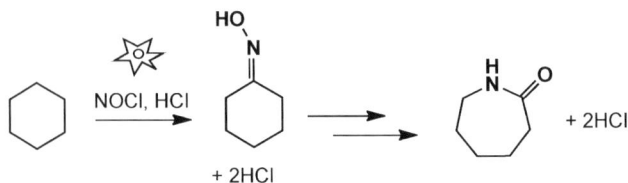

Scheme 14.31   Large scale photochemical conversion of cyclohexane in the presence of NOCl.

that induces photodestruction of the fullerenes once formed. In fact, although fullerenes absorb strongly at all wavelengths below about 700 nm, in the ultraviolet region below 350 nm the absorption is 10–100 times stronger than in the visible. As a result, a large amount of excited fullerenes reacted with other carbon species forming non-vapourizable, insoluble products. On the contrary, focused sunlight has both the role of vaporizing a carbon rod and of maintaining carbon as an atomic vapour until it moves into another light-protected compartment where the desired clustering is finally permitted to occur. This can open the way to the low cost gram-scale production of C60–C70 (or higher fullerenes) as well as carbon nanotubes providing that the costs for building and maintaining a solar furnace must be carefully evaluated. In some cases the furnace was a mosaic mirror with a parabolic shape and horizontal axis, lit by 60 flat tracking mirrors where in the focal area the focused power can reach 1000 W cm^{-2}.[106] Good quality samples of single-walled carbon nanotubes having 1.2–1.6 nm diameter have been synthesized in gram quantities (vapourization rate = 6–15 g h^{-1}) by having recourse to a 50 kW solar reactor.[107] Inorganic fullerene-like Cs_2O nanoparticles were likewise obtained in yields exceeding those from the laser-ablation method by a trivial, inexpensive, and reproducible photothermal procedure.[108]

14.20 CONCLUSION AND FUTURE DIRECTION

Solar photons can be considered *the* ideal green reagents since they are costless and leave no residue in the reaction mixture. In many cases the solar radiation could be successfully used in place of toxic or expensive chemical reagents to overcome the activation energy in organic synthesis and the future is bright for the radical chemistry under sunlight.

REFERENCES

1. E. Paternò, Synthesis in organic chemistry by means of light. I. Introduction, *Gazz. Chim. Ital.*, 1909, **39**(1), 237–250; English translation: E. Paternò, in *Synthesis in Organic Chemistry by Means of Light*, ed. M. D'Auria, Società Chimica Italiana, Rome, 2009.
2. C. A. Gueymard, The sun's total and spectral irradiance for solar energy applications and solar radiation models, *Sol. Energy*, 2004, **76**, 423–453.
3. R. M. Navarro, M. C. Sànchez-Sànchez, M. C. Alvarez-Galvan, F. del Valle and J. L. G. Fierro, Hydrogen production from renewable sources: biomass and photocatalytic opportunities, *Energy Environ. Sci.*, 2009, **2**, 35–54.
4. N. Armaroli and V. Balzani, The Future of Energy Supply: Challenges and Opportunities, *Angew. Chem., Int. Ed.*, 2007, **46**, 52–66.
5. V. Balzani, A. Credi and M. Venturi, Photochemical conversion of solar energy, *ChemSusChem*, 2008, **1**, 26–58.

6. For a basic discussion on storing solar radiation as chemical energy, including the isomerization of quadricyclane and related molecules, see: H.-D. Scharf, J. Fleischhauer, H. Leismann, I. Ressler, W.-G. Schleker and R.Weitz, Criteria for the efficiency, stability and capacity of abiotic photochemical solar energy storage systems, *Angew. Chem., Int. Ed. Engl.*, 1979, **18**, 652–662; A. Behr, W. Keim, G. Thelen, H.-D. Scharf and I. Ressler, Solar energy storage with quadricyclane systems, *J. Chem. Technol. Biotechnol.*, 1982, **32**, 627–630.

7. A. Albini and M. Fagnoni, Green chemistry and photochemistry were born at the same time, *Green Chem.*, 2004, **6**, 1–6.

8. (*a*) P. Esser, B. Pohlmann and H.-D. Scharf, The photochemical synthesis of fine chemicals with sunlight, *Angew. Chem., Int. Ed. Engl.*, 1994, **33**, 2009–2023; (*b*) B. Pohlmann, H.-D. Scharf, U. Jarolimek and P. Mauermann, Photochemical production of fine chemicals with concentrated sunlight, *Sol. Energy*, 1997, **61**, 159–168.

9. (*a*) M. Oelgemöller, C. Jung and J. Mattay, Green photochemistry: production of fine chemicals with sunlight, *Pure Appl. Chem.*, 2007, **79**, 1939–1947; (*b*) M. Oelgemöller, C. Jung, J. Ortner, J. Mattay, C. Schiel and E. Zimmermann, Green photochemistry with moderately concentrated sunlight, *The Spectrum*, 2005, **18**, 10–15.

10. N. Hoffmann, Photochemical reactions as key-steps in organic synthesis, *Chem. Rev.*, 2008, **108**, 1052–1103.

11. C. Decker and T. Bendaikha, Interpenetrating polymer networks. II. Sunlight-induced, polymerization of multifunctional acrylates, *J. Appl. Polym. Sci.*, 1998, **70**, 2269–2282.

12. V. Ramamurthy and N. Turro, Photochemistry: introduction, *Chem. Rev.*, 1993, **93**, 1–2.

13. (*a*) H. D. Roth, The Beginnings of Organic Photochemistry, *Angew. Chem., Int. Ed. Engl.*, 1989, **28**, 1193–1207; (*b*) H. D. Roth, Twentieth century developments in photochemistry. Brief historical sketches, *Pure Appl. Chem.*, 2001, **73**, 395–403.

14. C. L. Fritzsche, *I. Praict. Chem.*, 1867, **101**, 333.

15. G. M. Wyman, The *cis–trans* isomerization of conjugated compounds, *Chem. Rev.*, 1955, **55**, 625–657.

16. G. Egloff, R. E. Schaad and C. D. Lowry, The Halogenation of the Paraffin Hydrocarbons, *Chem. Rev.*, 1931, **8**, 1–80.

17. S.-Y. Moon and J.-M. Kim, Chemistry of photolithographic imaging materials based on the chemical amplification concept, *J. Photochem. Photobiol., C*, 2007, **8**, 157–173.

18. A. Albini and V. Dichiarante, The "belle époque" of photochemistry, *Photochem. Photobiol. Sci.*, 2009, **8**, 248–254.

19. M. Venturi, V. Balzani and M. T.Gandolfi, Fuels from solar energy. A dream of Giacomo Ciamician, the father of photochemistry, *Proceedings of the 2005 Solar World Congress Orlando,* American Solar Energy Society, Florida USA, 2005, pp. 1–6.

20. A. Albini and M. Fagnoni, 1908: Giacomo Ciamician and the concept of green chemistry, *ChemSusChem*, 2008, **1**, 63–66.
21. G. Bruni, *Commemorazione Solenne di Giacomo Ciamician*, Annuario della Regia Università di Bologna, 1921–22, 39.
22. (*a*) A. Schoenberg, E. Singer and W. Stephan, Extremely reactive carbon-carbon double bonds. V. 2-[Bis(methylphenylamino) methylene]-1,3-diphenylimidazolidine, a tetraaminoethylene of a new type, *Chem. Ber.*, 1987, **120**, 1589–1591; (*b*) E. Singer, Pioneers in photochemistry – Alexander Schoenberg, 1892–1985, *EPA Newsletter*, 1986, **26**, 1–11; (*c*) U. Deichmann, Chemists and biochemists during the national socialist era, *Angew. Chem., Int. Ed.*, 2002, **41**, 1310–1328.
23. A. Schönberg and A. Mustafa, Reaction of non-enolizable ketones in sunlight, *Chem. Rev.*, 1947, **40**, 181–200.
24. G. O. Schenck and K. Ziegler, The synthesis of ascaridole, *Naturwissenschaften*, 1945, **32**, 157.
25. (*a*) M. Fagnoni, D. Dondi, D. Ravelli and A. Albini, Photocatalysis for the Formation of the C–C Bond, *Chem. Rev.*, 2007, **107**, 2725–2756; (*b*) G. Palmisano, V. Augugliaro, M. Pagliaro and L. Palmisano, Photocatalysis: a promising route for 21st century organic chemistry, *Chem. Commun.*, 2007, 3425–3437.
26. D. Ravelli, D. Dondi, M. Fagnoni and A. Albini, Photocatalysis. A multi-faceted concept for green chemistry, *Chem. Soc. Rev.*, 2009, **38**, 1999–2011.
27. (*a*) D. Elad and J. Rokach, The Light-Induced Amidation of Terminal Olefins, *J. Org. Chem.*, 1964, **29**, 1855–1859; (*b*) D. Elad and J. Rokach, The Stereospecific Photoaddition of Formamide to Photoamidation. III. The, Light-Induced Amidation of Nonterminal Olefins, *J. Org. Chem.*, 1965, **30**, 3361–3364; (c) J. Rokach and D. Elad, Photoamidation. IV. The, Light-Induced Amidation of α,β- Unsaturated Esters, *J. Org. Chem.*, 1966, **31**, 4210–4215.
28. T. Caronna, C. Gambarotti, L. Palmisano, C. Punta and F. Recupero, Sunlight induced functionalisation of some heterocyclic bases in the presence of polycrystalline TiO_2, *Chem. Commun.*, 2003, 2350–2351.
29. (*a*) T. Caronna, C. Gambarotti, L. Palmisano, C. Punta and F. Recupero, Sunlight-induced reactions of some heterocyclic bases with ethers in the presence of TiO_2. A green route for the synthesis of heterocyclic aldehydes, *J. Photochem. Photobiol., A*, 2005, **171**, 237–242; (*b*) T. Caronna, C. Gambarotti, L. Palmisano, C. Punta, M. Pierini and F. Recupero, Sunlight-induced functionalisation reactions of heteroaromatic bases with aldehydes in the presence of TiO_2: A hypothesis on the mechanism, *J. Photochem. Photobiol., A*, 2007, **189**, 322–328.
30. M. A. Ischay, M. E. Anzovino, J. Du and T. P. Yoon, Efficient Visible Light Photocatalysis of [2+2] Enone Cycloadditions, *J. Am. Chem. Soc.*, 2008, **130**, 12886–12887.

31. A. Maldotti, L. Andreotti, A. Molinari, G. Varani, G. Cerichelli and M. Chiarini, Photocatalytic properties of iron porphyrins revisited in aqueous micellar environment: oxygenation of alkenes and reductive degradation of carbon tetrachloride, *Green Chem.*, 2001, **3**, 42–46.

32. A. Maldotti, A. Molinari, L. Andreotti, M. Fogagnolo and R. Amadelli, Novel reactivity of photoexcited iron porphyrins caged into a polyfluoro sulfonated membrane in catalytic hydrocarbon oxygenation, *Chem. Commun.*, 1998, 507–508.

33. K. Takami, J. Yamamoto, K. Komeyama, T. Kawabata and K. Takeira, Photocatalytic oxidation of alkanes with dioxygen by visible light and copper(II) and iron(III) chlorides: preference oxidation of alkanes over alcohols and ketones, *Bull. Chem. Soc. Jpn.*, 2004, **77**, 2251–2255.

34. G. Mehta and A. Srikrishna, Decahydro-1,3,5-methenocyclopenta-[*cd*]pentalene (Trishomocubane) Framework: Novel Photochemical Synthesis, Acid catalysed Cycloreversion, and Possible Role as a Solar Energy Storage System, *J. Chem. Soc., Chem. Commun.*, 1982, 218–219.

35. (*a*) S. Misumi, Cycloaddition reactions of cyclophanes, *Pure Appl. Chem.*, 1987, **59**, 1627–1636; (*b*) A. F. Murad, J. Kleinschroth and H. Hopf, Cyclophanes as diene components in Diels-Alder-reactions, *Angew. Chem., Int. Ed. Engl.*, 1980, **19**, 389–390.

36. F. Cicogna, G. Ingrosso, F. Lodato, F. Marchetti and M. Zandomeneghi, 9-Anthroylacetone and its photodimer, *Tetrahedron*, 2004, **60**, 11959–11968.

37. P. Coppo and S. G. Yeates, Shining Light on a Pentacene Derivative: The Role of Photoinduced Cycloadditions, *Adv. Mater.*, 2005, **17**, 3001–3005.

38. R. D. Nikolova, G. N. Vayssilov, N. Rodios and A. Bojilova, Regioand Stereoselective [2+2] Photodimerization of 3-Substituted 2- Alkoxy-2-oxo-2*H*-1,2-benzoxaphosphorines, *Molecules*, 2002, **7**, 420– 432.

39. A. Gilbert and P. Heath, Specific ortho photocycloaddition of enol ethers to 2-substituted anisoles: facile synthesis of bicyclo[4.2.0]octa- 2,7-dienes in sunlight, *Tetrahedron Lett.*, 1987, **28**, 5909–5912.

40. P. Quadrelli and P. Caramella, Synthesis and synthetic applications of 1,2,4-oxadiazole-4-oxides, *Curr. Org. Chem.*, 2007, **11**, 959–986.

41. P. Quadrelli, M. Mella and P. Caramella, A photochemical generation of nitrosocarbonyl intermediates, *Tetrahedron Lett.*, 1999, **40**, 797– 800.

42. P. Quadrelli, R. Scrocchi, A. Piccanello and P. Caramella, Photochemical generation of nitrosocarbonyl intermediates on solid phase: synthons towards Hetero Diels–Alder and Ene Adducts through photocleavage, *J. Comb. Chem.*, 2005, **7**, 887–892.

43. M. A. Meador and P. J. Wagner, 2-Indanol Formation from Photocyclization of α-Arylacetophenones, *J. Am. Chem. Soc.*, 1983, **105**, 4484–4485.

44. J. E. Guillet, W. K. MacInnis and A. E. Redpath, Prospects for solar synthesis. II. Study, of the photocyclization of a-(o-toly1-) acetophenone

in solution and in crosslinked ethylene–vinyl acetate beads, *Can. J. Chem.*, 1985, **63**, 1333–1336.

45. (*a*) A. J. Herrera, M. Rondón and E. Suárez, Stereocontrolled Photocyclization of 1,2-Diketones Applied to Carbohydrate Models: A new Entry to *C*-Ketosides, *Synlett*, 2007, 1851–1856; (*b*) A. J. Herrera, M. Rondón and E. Suárez, Stereocontrolled Photocyclization of 1,2-Diketones: Application of a 1,3-Acetyl Group Transfer Methodology to Carbohydrates, *J. Org. Chem.*, 2008, **73**, 3384–3391.

46. D. Alvarez-Dorta, E. I. León, A. R. Kennedy, C. Riesco-Fagundo and E. Suárez, Sequential Norrish Type II Photoelimination and Intramolecular Aldol Cyclization of 1,2-Diketones in Carbohydrate Systems: Stereoselective Synthesis of Cyclopentitols, *Angew. Chem., Int. Ed.*, 2008, **47**, 8917–8919.

47. G. Ciamician and P. Silber, Chemische Lichtwirkungen, *Ber. Dtsch. Chem. Ges.*, 1900, **33**, 2911–2913.

48. W. D. Cohen, La réduction des cétones aromatiques, *Rec. Trav. Chim.*, 1920, **39**, 243–279.

49. E. Paternò and G. Chieffi, Synthesis in organic chemistry using light. Note II. Compounds, of unsaturated hydrocarbons with aldehydes and ketones, *Gazz. Chim. Ital.*, 1909, **39**, 341–361.

50. (*a*) G. S. Hammond, W. P. Baker and W. M. Moore, Mechanisms of Photoreactions in Solution. II. Reduction, of Benzophenone by Toluene and Cumene, *J. Am. Chem. Soc.*, 1961, **83**, 2795–2799; (*b*) W. M. Moore, G. S. Hammond and R. P. Foss, Mechanisms of Photoreactions in Solutions. I. Reduction of Benzophenone by Benzhydrol, *J. Am. Chem. Soc.*, 1961, **83**, 2789–2794.

51. W. E. Bachmann, The Photochemical Reduction of Ketones to Hydrols, *J. Am. Chem. Soc.*, 1933, **55**, 391–395.

52. J.-T. Li, J.-H. Yang, J.-F. Han and T.-S. Li, Reductive coupling of aromatic aldehydes and ketones in sunlight, *Green Chem.*, 2003, **5**, 433–435.

53. M. Ochiai, M. Kunishima, Y. Nagao, K. Fuji and E. Fujita, sp-Carbon-Iodine Bond Cleavage of Alkynyl(phenyl)iodonium Salts, Novel Synthesis of (Alkylethynyl)triphenylphosphonium Salts, *J. Chem. Soc., Chem. Commun.*, 1987, 1708–1709.

54. A. Boto, J. A. Gallardo, R. Hernández and C. J. Saavedra, One-pot synthesis of a-amino phosphonates from a-amino acids and b-amino alcohols, *Tetrahedron Lett.*, 2005, **46**, 7807–7811.

55. A. Boto, Y. De León, J. A. Gallardo and R. Hernández, Synthesis of Alkaloid Analogues from a-Amino Acids by One-Pot Radical Decarboxylation/Alkylation, *Eur. J. Org. Chem.*, 2005, 3461–3468.

56. C. J. Mortko and M. A. Garcia-Garibay, Green Chemistry Strategies Using Crystal-to-Crystal Photoreactions: Stereoselective Synthesis and Decarbonylation of trans-a,a'-Dialkenoylcyclohexanones, *J. Am. Chem. Soc.*, 2005, **127**, 7994–7995.

57. H. Petzold, T. Weisheit, H. Görls, H. Breitzke, G. Buntkowsky, D. Escudero, L. González and W. Weigand, Selective carbon–carbon bond cleavage of 2,2′-dibromotolane *via* photolysis of its appropriate (diphosphine)Pt0 complex in the solid state, *Dalton Trans.*, 2008, 1979–1981.

58. N. Matsumoto and M. Takahashi, Synthesis of pyrrolo[2,3- d]pyrimidine-2,4-diones by sunlight photolysis of *N*-(5-vinyluracil-6- yl)sulfilimines, *Tetrahedron Lett.*, 2005, **46**, 5551–5554.

59. Z. Huang, D.Wan and J. Huang, Hydrolysis of Schiff Bases Promoted by UV Light, *Chem. Lett.*, 2001, 708–709.

60. S. Rajan and K. Muralimohan, Aromatic SRN1 reactions stimulated by sunlight-phenylation of dimsyl anion, *Tetrahedron Lett.*, 1978, **19**, 483–486.

61. A. V. Samoshin and V. V. Samoshin, Green Photochemistry: sun-induced aromatic nucleophilic substitution of alkoxy groups by alkylamines, *J. Undergraduate Chem. Res.*, 2006, **2**, 13–16.

62. S. Koyama, Z. Kumazawa, N. Kashimura and R. Nishida, Photo-chemical Synthesis and biological activity of 6-*cis*-styrylpurine, *Agric. Biol. Chem.*, 1985, **49**, 1859–1861.

63. Y. Luo, F. Tao, Y. Liu, B. Li and G. Zhang, Intramolecular amidation: an efficient synthesis of 3-aryl-2-quinolines, *Can. J. Chem.*, 2006, **84**, 1620–1625.

64. (*a*) Y.-P. Zhao, R. O. Campbell and R. S. H. Liu, Solar reactions for preparing hindered 7-*cis*-isomers of dienes and trienes in the vitamin A series, *Green Chem.*, 2008, **10**, 1038–1042; (*b*) Y.-P. Zhao, L.-Y. Yang and R. S. H. Liu, Designing systems for one-way *trans* to *cis* photo-isomerization for solar reactions, *Green Chem.*, 2009, **11**, 837–842.

65. S. S. Arbuj, S. B. Waghmode and A. V. Ramaswamy, Photochemical a-bromination of ketones using *N*-bromosuccinimide: a simple, mild and efficient method, *Tetrahedron Lett.*, 2007, **48**, 1411–1415.

66. J. He and C. U. Pittman, Jr, The photochemical synthesis of a,a,abromodifluorotoluene and a,a,a-chlorodifluorotoluene, *Synth. Commun.*, 1999, **29**, 855–862.

67. A. Podgoršek, S. Stavber, M. Zupan and J. Iskra, Visible light induced 'on water' benzylic bromination with *N*-bromosuccinimide, *Tetrahedron Lett.*, 2006, **47**, 1097–1099.

68. K. Wiesner, J. G. McCluskey, J. K. Chang and V. Šmüla, The total synthesis of a thebaine analogue, and of the corresponding "Bentley adduct", *Can. J. Chem.*, 1971, **49**, 1092–1098.

69. F. Cetin, N. Yenil and L. Yüceer, Stable spiro-endoperoxides by sunlight-mediated photooxygenation of 1,2-O-alkylidene-5(*E*)-eno-5,6,8-trideoxy-a-D-xylo-oct-1,4-furano-7-uloses, *Carbohydr. Res.*, 2005, **340**, 2583–2589.

70. (*a*) K. Ono, T. Hiei, M. Tajika, K. Taga, K. Saito, M. Tomura, J. Nishida and Y. Yamashita, Synthesis of 6,13-dithienylpentacenes by photolysis of

their endoperoxides, *Lett. Org. Chem.*, 2008, **5**, 522–526; (*b*) K. Ono, H. Totani, T. Hiei, A. Yoshino, K. Saito, K. Eguchi, M. Tomura, J. Nishida and Y. Yamashita, Photooxidation and reproduction of pentacene derivatives substituted by aromatic groups, *Tetrahedron*, 2007, **63**, 9699–9704.

71. D. Mal and H. Nath Roy, A concise total synthesis of brasiliquinones B and C and 3-deoxyrabelomycin, *J. Chem. Soc., Perkin Trans. 1*, 1999, 3167–3171.

72. M. C. Carreño, A. Somoza, M. Ribagorda and A. Urbano, Asymmetric Synthesis of Rubiginones A2 and C2 and Their 11-Methoxy Regio-isomers, *Chem.–Eur. J.*, 2007, **13**, 879–890.

73. E. Haggiage, E. E. Coyle, K. Joyce and M. Oelgemöller, Green photochemistry: solarchemical synthesis of 5-amido-1,4-naphthoquinones, *Green Chem.*, 2009, **11**, 318–321.

74. (*a*) S. Malato, J. Blanco, A. Vidal and C. Richter, Photocatalysis with solar energy at a pilot plant scale: an overview, *Appl. Catal., B*, 2002, **37**, 1–15; (*b*) K.-H. Funken and J. Ortner, Technologies for the Solar Photochemical and Photocatalytic Manufacture of Specialities and Commodities: A Review, *Z. Phys. Chem.*, 1999, **213**, 99–105.

75. J. Blanco, S. Malato, P. Fernández, A. Vidal, A. Morales, P. Trincado, J. C. Oliveira, C. Minero, M. Musci, C. Casalle, M. Brunotte, S. Tratzky, N. Dischinger, K.-H. Funken, C. Sattler, M. Vincent, M. Collares-Pereira, J. F. Mendes and C. M. Rangel, Compound parabolic concentrator technology development to commercial solar detoxification applications, *Sol. Energy*, 1999, **67**, 317–330.

76. J. Ortner, D. Faust, K.-H. Funken, T. Lindner, J. Schulat, C. G. Stojanoff and P. Fröning, New developments using holographic concentration in solar photochemical reactors, *J. Phys. IV*, 1999, **9**(pr3), 379–383.103C. Jung, K.-H. Funken and J. Ortner, PROPHIS: parabolic trough-facility for organic photochemical syntheses in sunlight, *Photochem. Photobiol. Sci.*, 2005, **4**, 409–411.

77. D. Dindar and S. Içli, Unusual photoreactivity of zinc oxide irradiated by concentrated sunlight, *J. Photochem. Photobiol., A*, 2001, **140**, 263–268.

78. M. Oelgemöller, N. Healy, L. de Oliveira, C. Jung and J. Mattay, Green photochemistry: solar chemical synthesis of Juglone with medium concentrated sunlight, *Green Chem.*, 2006, **8**, 831–834.

79. M. Oelgemöller, C. Jung, J. Ortner, J. Mattay and E. Zimmermann, Green photochemistry: solar photooxygenations with medium concentrated sunlight, *Green Chem.*, 2005, **7**, 35–38.

80. H. Dincalp and S. Içli, Photosynthesis of rose oxide by concentrated sunlight in the absence of singlet oxygen, *J. Photochem. Photobiol., A*, 2001, **141**, 147–151.

81. C. Covell, A. Gilbert and C. Richter, Sunlight-induced Regio- and Stereo-specific (2π+2π) Cycloaddition of Arylethenes to 2-Substituted-1,4-naphthoquinones, *J. Chem. Res. (S)*, 1998, 316–317.

82. A. Hülsdünker, A. Ritter and M. Demuth, Lineraly Focussing Reactor for Chemical Purpose, *EPA Newsletter*, 1992, **45**, 23–25.110R. A. Doohan and N. W. A. Geraghty, A comparative analysis of the functionalisation of unactivated cycloalkanes using alkynes and either sunlight or a photochemical reactor, *Green Chem.*, 2005, **7**, 91–96.

83. D. Dondi, S. Protti, A. Albini, S. Mañas Carpio and M. Fagnoni, Synthesis of γ-lactols, γ-lactones and 1,4-monoprotected succinaldehydes under moderately concentrated sunlight, *Green Chem.*, 2009.

84. L. Cermenati, C. Richter and A. Albini, Solar light induced carbon–carbon bond formation *via* TiO$_2$ photocatalysis, *Chem. Commun.*, 1998, 805–806.

85. G. Oehme, B. Heller and P. Wagler, A solar-driven, complex catalyzed pyridine synthesis as alternative for a thermally-driven synthesis, *Energy*, 1997, **22**, 327–336.

86. H. Klinger and W. Standke, Ueber die Einwirkung des Sonnenlichts auf organische Verbindungen, *Ber. Dtsch. Chem. Ges.*, 1891, **24**, 1340–1346; H. Klinger and W. Kolvenbach, Die Bildung von Acetohydrochinon aus Acetaldehyd und Benzochinon im Sonnenlicht, *Ber. Dtsch. Chem. Ges.*, 1898, **31**, 1214–1216.

87. G. A. Kraus and M. Kirihara, Quinone photochemistry. A general synthesis of acylhydroquinones, *J. Org. Chem.*, 1992, **57**, 3256–3257.

88. M. Oelgemöller and J. Mattay, The photochemical Friedel-Crafts acylation of quinones—From the beginnings of organic photochemisty to modern solarchemical applications, in *Handbook of Organic Photochemistry and Photobiology*, ed. W. Horspool and F. Lenci, CRC press, Boca Raton, 2nd edn, 2003, 88/1–88/45.117(*a*) C. Schiel, M. Oelgemöller and J. Mattay, Photoacylation of Electronrich Quinones. Application of the 'Photo-Friedel–Crafts Reaction', *Synthesis*, 2001, 1275–1279; (*b*) C. Schiel, M. Oelgemöller, J. Ortner and J. Mattay, Green photochemistry: the solar-chemical 'Photo–Friedel–Crafts acylation' of quinones, *Green Chem.*, 2001, **3**, 224–228.

89. (*a*) C.-L. Cian and C. G. Bochet, Clean and easy photochemistry, *Chimia*, 2007, **61**, 650–654; (*b*) H. Dürr, Industrial application of photochemistry. A promising field?, *EPA Newsletter*, 1986, **28**, 78–79.119The drawbacks related to the industrial application of a photochemical process, with particular attention to the scale-up process and to the reactor design have been discussed by Fischer for the case of the photonitrosation of cyclohexane, see: M. Fischer, Industrial applications of photochemical syntheses, *Angew. Chem., Int. Ed. Engl.*, 1978, **17**, 16–26.120(*a*) M. Pape, Industrial application of photochemistry, *Pure Appl. Chem.*, 1975, **41**, 535–558; (*b*) K. H. Pfoertner, Photochemistry in industrial synthesis, *J. Photochem. Photobiol., A*, 1990, **51**, 81–86.

90. J. Talukdar, E. H. S. Wong and V. K. Mathur, Caprolactam production by direct solar flux, *Sol. Energy*, 1991, **47**, 165–171.

91. (*a*) K.-H. Funken and M. Becker, Solar chemical engineering and solar materials research into the 21st century, *Renewable Energy*, 2001, **24**, 469–474; (*b*) K.-H. Funken, F.-J. Müller, J. Ortner, K.-J. Riffelmann and C. Sattler, Solar collectors *versus* lamps – a comparison of the energy demand of industrial photochemical processes as exemplified by the production of ε-caprolactame, *Energy*, 1999, **24**, 681–687.

92. C. Sattler, F.-J. Müller, K.-J. Riffelmann, J. Ortner and K.-H. Funken, Concept and economic evaluation of an industrial synthesis of ε-caprolactam *via* solar photooximation of cyclohexane, *J. Phys. IV*, 1999, **9**(Pr3), 723–727.124L. P. F. Chibante, Andreas Thess, J. M. Alford, M. D. Diener and R. E. Smalley, Solar Generation of the Fullerenes, *J. Phys. Chem.*, 1993, **97**, 8696–8700.

93. C. L. Fields, J. R. Pitts, M. J. Hale, C. Bingham, A. Lewandowski and D. E. King, Formation of Fullerenes in Highly Concentrated Solar Flux, *J. Phys. Chem.*, 1993, **97**, 8701–870.

94. D. Luxembourg, G. Flamant and D. Laplaze, Solar synthesis of single-walled carbon nanotubes at medium scale, *Carbon*, 2005, **43**, 2302–2310.

95. T. Guillard, L. Alvarez, E. Anglaret, J.-L. Sauvajol, P. Bernier, G. Flamant and D. Laplaze, Production of fullerenes and carbon nanotubes by the solar energy route, *J. Phys. IV*, 1999, **9**(Pr3), 399–404.128D. Laplaze, P. Bernier, G. Flamant, M. Lebrun, A. Brunelle and S. Della-Negra, Solar energy: application to the production of fullerenes, *J. Phys. B: At., Mol. Opt. Phys.*, 1996, **29**, 4943–4954.

96. A. Albu-Yaron, T. Arad, M. Levy, R. Popovitz-Biro, R. Tenne, J. M. Gordon, D. Feuermann, E. A. Katz, M. Jansen and C. Mühle, Synthesis of Fullerene-like Cs$_2$O Nanoparticles by Concentrated Sunlight, *Adv. Mater.*, 2006, **18**, 2993–2996.

97. J. E. Guillet, Prospects for solar synthesis, *Pure Appl. Chem.*, 1991, **63**, 917–924.

98. C. Herrero, D. Lassalle-Kaiser, W. Leibl, A. W. Rutherford and A. Aukauloo, Artificial systems related to light driven electron transfer processes in PSII, *Coord. Chem. Rev.*, 2008, **252**, 456–468.

99. M. Grätzel, Mesoscopic solar cells for electricity and hydrogen production from sunlight, *Chem. Lett.*, 2005, **34**, 8–13.

100. V. Balzani, M. Clemente-León, A. Credi, B. Ferrer, M. Venturi, A. H. Flood and J. F. Stoddart, Autonomous artificial nanomotor powered by sunlight, *Proc. Natl. Acad. Sci. U. S. A.*, 2006, **103**, 1178–1183.

101. Since a photon *substitutes but is not* a chemical reagent this can not be accounted for in the atom economy of the process. On the other hand, since most of the photochemical reactions have a quantum yield not exceeding 0.1, the "photon economy" of the process is in most cases unsatisfactory. Recently Liu and coworkers introduced the RESR value to determine the relative efficiency of a solar reaction.

102. (*a*) A. Albini and M. Fagnoni, Photochemistry as a green synthetic method, in *New methodologies and Techniques for a Sustainable Organic*

Chemistry, ed. A. Mordini and F. Faigl, Springer Science + Business Media B.V., 2008, 279–293; (*b*) S. Protti, S. Manzini, M. Fagnoni and A. Albini, The contribution of photochemistry to green chemistry, in *Eco-friendly synthesis of fine chemicals*, ed. R. Ballini, Royal Society of Chemistry, 2009, 80–111.136A. Albini and M. Fagnoni, The greenest reagent in organic synthesis: light, in *Green Chemical Reactions*, ed. P. Tundo and V. Esposito, Springer Science + Business Media B.V., 2008, 173–189.

103. (*a*) G. Kreisel, S. Meyer, D. Tietze, T. Fidler, R. Gorges, A. Kirsch, B. Schäfer and S. Rau, Leuchtdioden in der Chemie–Eine Hochzeit verschiedener Technologien, *Chem. Ing. Tech.*, 2007, **79**, 153–159 ; (*b*) S. Meyer, D. Tietze, S. Rau, B. Schäfer and G. Kreisel, Photosensitized oxidation of citronellol in microreactors, *J. Photochem. Photobiol., A*, 2007, **186**, 248–253.

104. A. G. Griesbeck, N. Maptue, S. Bondock and M. Oelgemöller, The excimer radiation system: a powerful tool for preparative organic photochemistry. A technical note, *Photochem. Photobiol. Sci.*, 2003, **2**, 450–45.

105. B. D. A. Hook, W. Dohle, P. R. Hirst, M. Pickworth, M. B. Perry and K. I. Booker-Milburn, A practical flow reactor for continuous organic photochemistry, *J. Org. Chem.*, 2005, **70**, 7558–7564.

106. E. E. Coyle and M. Oelgemöller, Micro-photochemistry: photochemistry in microstructured reactors: the new photochemistry of the future?, *Photochem. Photobiol. Sci.*, 2008, **7**, 1313–1322.

107. T. Guillard, G. Flamand, J. F. Robert, B. Rivoire, J. Giral and D. Laplaze, Scale up of a solar reactor for fullerene and nanotube synthesis, *J. Sol. Energy Eng.*, 2002, **124**, 22–27.

108. (*a*) E. A. Sosnin, T. Oppenländer and V. F. Tarasenko, Applications of capacitive and barrier discharge excilamps in photoscience, *J. Photochem. Photobiol., C*, 2006, **7**, 145–163; (*b*) A. G. Griesbeck, W. Kramer and M. Oelgemöller, Photoinduced decarboxylation reactions.Radical chemistry in water, *Green Chem.*, 1999, **1**, 205–207.

CHAPTER 15

Sonochemistry: Ultrasound Application in Radical Synthesis[*]

15.1 INTRODUCTION

Ultrasound can be used quite effectively for inducing radical reactions. Ultrasound-induced radicalization involves the thermal, homolytic dissociation of molecules present inside or close to the collapsing cavity. The benefits and drawbacks of ultrasonic initiation compared to alternative sources for radical formation should always be considered when performing radical chemistry. As for radical polymerizations, initiation mainly proceeds by means of thermal dissociation of an auxiliary initiator.[1–7] Initiator molecules contain a weak covalent bond and therefore, relatively little energy is required for radical formation. The rate of initiator decay and, hence, the radical concentration strongly depends on temperature. To obtain a significant radical concentration, reactions involving thermal initiation are typically performed at a temperature between 323 and 363 K. Organic peroxides and azo compounds represent examples of initiators frequently employed for thermal initiation of free-radical polymerizations. Furthermore, ultraviolet and visible light irradiation can be used for the dissociation of covalent bonds. Since hydrogen peroxide and aryl iodides are readily photolyzed upon ultraviolet radiation, these compounds are often used in photochemical reactions. In addition to light irradiation, so-called ionizing radiation can induce initiation in the presence or absence of an auxiliary initiator. Ionizing radiation refers to high-energy radiation using particles (e.g. electrons), X-rays, or γ-rays. Due to the higher cost and safety issues, this type of initiation is usually not preferred over the other methods. Redox initiation involves the formation of radicals by means of oxidation-reduction reactions. A well-known example is the Fenton

* Chapter written by V. Tamara Perchyonok.
Streamlining Free Radical Green Chemistry
V. Tamara Perchyonok, Ioannis Lykakis and Al Postigo
© V. Tamara Perchyonok, Ioannis Lykakis and Al Postigo 2012
Published by the Royal Society of Chemistry, www.rsc.org

reaction between iron(II) salts and hydrogen peroxide. Occasionally, several radical sources are used in conjunction to enhance radical production rates.[8–10]

In contrast to thermal initiation, radicalization by means of ultrasound or any of the other discussed methods does not require elevated bulk temperatures. It should be noted, however, that higher temperatures also imply an increase in reaction rate, especially important for polymerization reactions. Furthermore, ultrasound-, photo- and radiation-induced radicalization are intrinsically safe as radical production stops when the source is switched off and it allows control of the location of radical formation. Ultrasonic initiation can proceed in the absence of any auxiliary component due to the extreme conditions obtained inside the collapsing cavities, whereas thermal, photolytic and redox initiation often require the addition of a compound containing a weak covalent bond. For example, the photochemical oxidation of pollutants in water is frequently associated with the addition of hydrogen peroxide or ozone.[11] An exception is the halogenation of alkanes. The covalent bond between two halogen atoms is relatively weak and as a result, this reaction can proceed thermally and photochemically in the absence of an initiator.[12,13] The various sources are compared in Table 15.1.

Although ultrasonic initiation seems promising, improvement of the energy efficiency is considered crucial for commercial applications.[14,15] Nevertheless, only little is known about the relative efficiency of the various radical sources. Based on the hydrogen formation rates presented, rough estimates can be given for ultrasound-induced radical production rates. For example, sonication of water saturated with argon at 293 K resulted in a hydrogen production rate of approximately 0.07 μmol s^{-1} l^{-1} using an electrical input of 50 W and argon. Assuming hydrogen gas arises from the recombination of two hydrogen radicals and each hydrogen radical is accompanied by the formation of a hydroxide radical, this implies a minimal radical production of 0.3 μmol s^{-1} l^{-1} and an energy efficiency of 6 nmol l^{-1} J^{-1}. The majority of radicals produced inside the collapsing cavity recombines before entering the liquid, leading to the formation of *e.g.* hydrogen gas.[16] Accordingly, only a small fraction of the estimated radical production rate is actually available for initiation and a radical energy efficiency of well below 6 nmol l^{-1} J^{-1} is expected, which is in line with previously reported hydroxyl radical yields as obtained from dosimeter studies.[17] This radical energy efficiency is relatively high compared to values typically obtained for UV

Table 15.1 Comparison of various sources of radicals.

Radical sources	Ambient temperature	On–off production	Local radicalization	Auxiliary component	Safety
Thermal decomposition	X	X	X	X	Yes
Ultrasound irradiation	Yes	Yes	Yes	Yes	Yes
UV-Visible radiation	Yes	Yes	Yes	X	Yes
Ionization radiation	Yes	Yes	Yes	Yes	X
Redox reaction	Yes	X	X	X	Yes

irradiation. Irradiation of a hydrogen peroxide solution using a mercury lamp (256 nm) results in a maximum radical energy efficiency of 0.1 nmol l^{-1} J^{-1}.[11] The values for ultrasonic and photochemical initiation should be compared to radical energy efficiencies encountered in polymerizations induced by thermal decomposition of an initiator. The decomposition constant of potassium persulfate in water at 353 K equals 6.9 × 10^{-5} s^{-1}.[18] For an initial concentration of 10 mmol l^{-1} and an initiator efficiency of 0.5, this decomposition constant implies an average radical production rate during the first hour of 0.6 µmol s^{-1} l^{-1}. The amount of energy required for radicalization depends on the heat that has to be supplied to the liquid to raise and maintain the temperature. Supposing that the reactor is thermally isolated and the energy required only consists of heating the mixture to the desired temperature, the average radical energy efficiency during the first hour resembles 9 nmol l^{-1} J^{-1}.

The calculated radical energy efficiencies for thermal and ultrasonic initiation are roughly in the same order of magnitude. Since the energy efficiency estimated for ultrasonic initiation represents an upper limit, thermal initiation seems favored. Although this thesis has demonstrated that ultrasound-induced reaction rates can be optimized by changing experimental conditions, *e.g.* liquid temperature and saturation gas, it is improbable that the maximum radical energy efficiency for a conventional sonication system will exceed that obtained for thermal initiation. With respect to emulsion polymerizations, ultrasound irradiation provides additional benefits which have not been taken into account in the preceding analysis. The shearing action of ultrasound leads to the formation of polymeric radicals from polymer scission and to smaller monomer droplets.[19,20,21] As a consequence, ultrasound-initiated emulsion polymerizations proceed at reasonable reaction rates.

When comparing the various sources for inducing radical reactions, the following observations can be made:

- Halogenation reactions: ultrasonic initiation seems to be a promising technique for the dissociation of halogen compounds.
- Waste water treatment processes: ultrasound, UV-Visible and redox initiation have proven to be effective in the oxidation of organic pollutants; nevertheless, ultrasound and UV-Visible initiation are less energy efficient.
- Free-radical polymerizations: ultrasound irradiation provides unique advantages; however, its energy efficiency is relatively low compared to that of thermal initiation.

15.2 ENERGY EFFICIENCY

For commercial applications, the energy efficiency has to improve dramatically before ultrasound will be utilized as enabling technology or to favor this technique over conventional methods, such as thermal initiation and stirring. The transformation of electrical energy into the desired physical and chemical effects involves several steps, which are schematically depicted in Figure 15.1.[24,25]

Each of these energy conversion processes is accompanied by an energy loss. The first step comprises the conversion of electrical energy into acoustic energy. By means of a frequency generator, transducer and magnifying horn. The energy coupling between the horn and the medium (decoupling losses) also influences the efficiency of this step. Liquid energy refers to the amount of energy transferred to the liquid, which can be estimated from calorimetric measurements.

15.3 SONOCHEMICAL INITIATION OF RADICAL CHAIN REACTIONS: HYDROSTANNATION AND HYDROXYSTANNATION OF C–C MULTIPLE BONDS

"Non-homogeneous" systems are a class of homogeneous medium where the local properties are heterogeneous. Such a system has rarely been used intentionally for synthetic purposes, and hence its utility has not yet been explored. Irradiation of homogeneous liquid with ultrasound produces characteristic thermal non-equilibrium conditions by creating localized superheated sonochemical cavities, wherein a maximum temperature over 2000 K can be

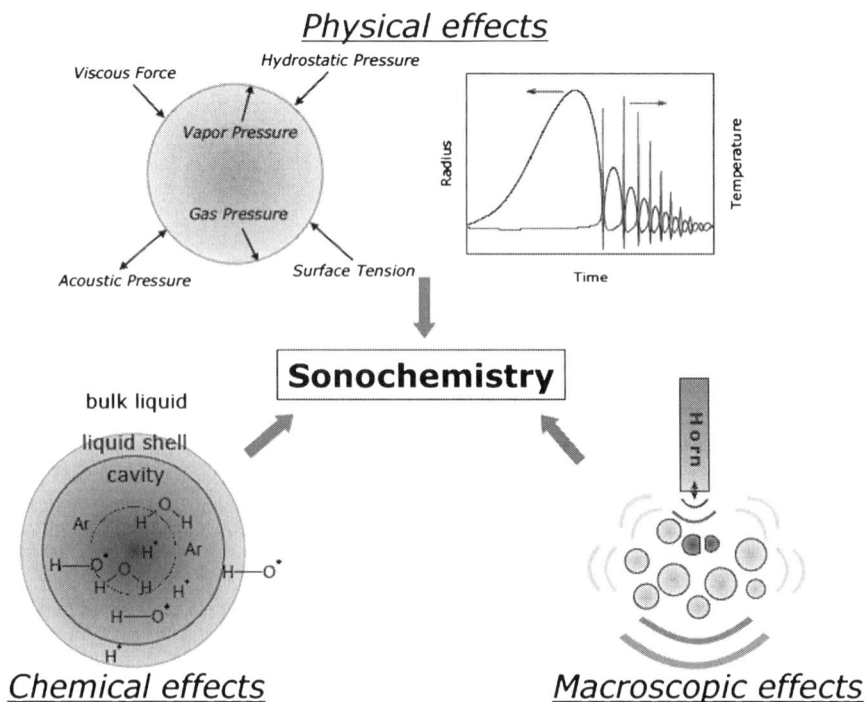

Figure 15.1   Physical, chemical and macroscopic effects in sonochemistry.

generated.[5] It is commonly accepted that the unusual observations made in the ultrasound-driven chemical reactions are largely due to this acoustic cavitation.

In contrast to heterogeneous sonochemistry, which has been widely employed in organic sonochemistry in homogeneous media (*i.e.*, under a non-homogeneous environment, *vide supra*)[9] has not been explored.[1] This may be due in part to poor efficiency of acoustic cavitation in homogeneous liquid phase and in part to the fact that the cavities themselves are too small in size (diameter of <1 pm) and too short-lived (<2 ms) to serve as a medium for product formation on a synthetically useful scale.[2]

Nakamura has shown that homogeneous sonochemistry at low temperatures allows the tin radical species[11,12] to react in a previously unknown manner. Thus, when an aerated solution of R_3SnH and an olefin is irradiated at 0 to 10 °C, hydroxystannation rather than the conventional hydrostannation of the C=C double bond takes place to produce a hydroxystannane (Scheme 15.1, X = R_3Sn).

The hydroxystannation of 1,3-dienes may proceed either in a 1,2- or 1,4-manner[13] to provide P-hydroxystannanes **2** or hydroxylated allylic stannanes **3** (Scheme 15.2), respectively.

These transformations indicate that, at low temperatures, tin radicals are compatible with molecular oxygen, with which they are intrinsically reactive. The addition of two heteroatoms across an olefinic bond (Scheme 15.1) is an important class of synthetic transformations as represented by the classical

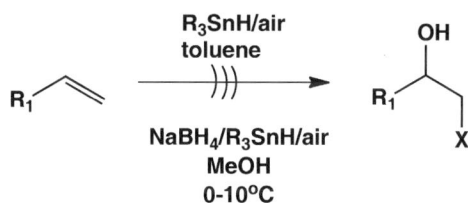

where R₁=vinyl, aryl, alkoxycarbonyl

Scheme 15.1 Homogenous sonochemistry at low temperature involving hydrostannation.

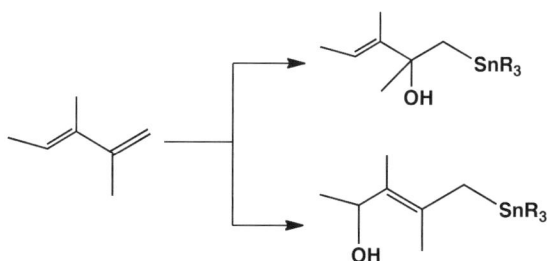

Scheme 15.2 Hydroxystannation of 1,3-dienes, proceeding *via* 1,2- and 1,4-manner.

halohydrin formation and oxymercuration reaction (X = halogen or Hg(II) for **1**).[14,15]

The "hydroxystannation" reaction provides the first access to hydroxylated organotin compounds through addition of stannyl and hydroxyl groups to an olefin. In order to put these results into perspective, Nakamura first summarized the sonochemical generation of tin radical species in the reaction of R$_3$SnH with a C–C multiple bond *under argon* and then detailed the new reaction that takes place *in the presence of air*.

15.4 HOMOGENEOUS SONOCHEMISTRY OF HYDROSTANNATION IN DETAIL

The addition of a tin hydride reagent to an acetylene is a useful synthetic reaction (Scheme 15.3),[16] representing the simplest member of radical chain reactions involving turnover of a tin radical species. A straightforward reaction mechanism and sensitivity of product stereochemistry to the reaction conditions made this reaction a useful probe to examine the effects of ultrasound irradiation on a radical chain reaction. The reaction is normally carried out by heating a neat mixture of a tin hydride and an acetylene in the presence of AIBN to obtain a vinylstannane (**6**) in high yield. Nakamura found that sonochemical irradiation dramatically accelerates this reaction.

Taking the reaction of 1-hexyne (5 equiv.) with Ph$_3$SnH under argon as a test reaction, Nakamura investigated the product yield and the *cis/trans* isomer ratio against variation of the reaction conditions (*e.g.*, on/off of ultrasound, temperature, and solvent). Under the usual thermal conditions, the reaction of a neat mixture of two reactants proceeded at 90 °C and gave a 100% yield. The reaction was extremely slow in an ice bath (1% yield at 0 °C, 3 h). However, when the reaction mixture was sonicated for 3 h under argon with a titanium immersion horn, 1-(triphenylstannyl)hex-1-ene formed in 88% isolated yield, predominantly (92%) as the *cis*-isomer.

During this experiment, the internal temperature rose slightly after a few minutes and remained constant (7 °C) throughout the reaction period. The sonochemical reaction proceeded equally well in toluene at 7 °C (entry 2, 86% yield, 91% *cis*, 2 h), whereas the reaction without irradiation proceeded only to 3% conversion. No induction time was observed in the sonochemical reaction, and the reaction took place only during ultrasound irradiation, without which the reaction almost completely stopped (Figure 15.2). The sonochemical

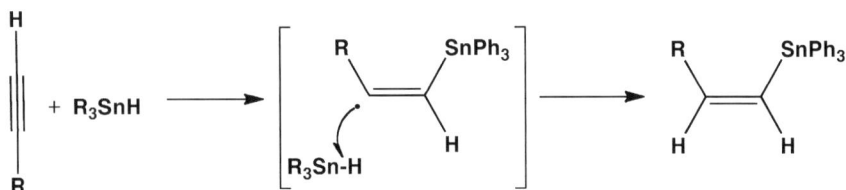

Scheme 15.3 Tin hydride mediated addition to acetylene.

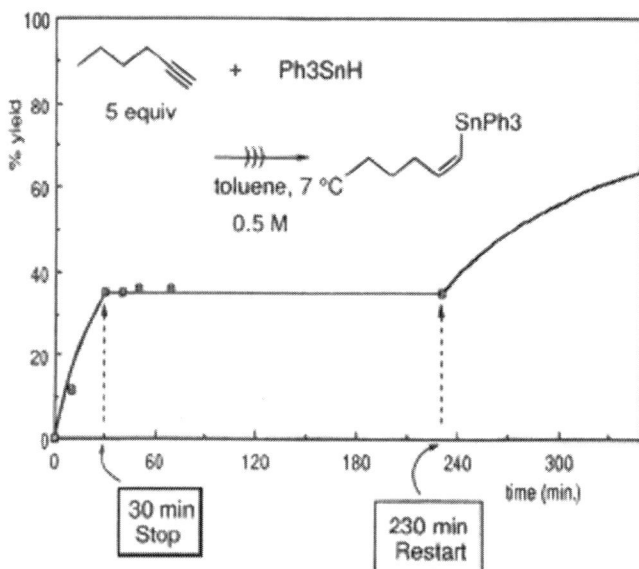

Figure 15.2 Progress of the hydrostannation reaction (Scheme 15.3) with and without ultrasound irradiation.

reaction does not necessarily need a radical initiator, but it proceeded faster in the presence of one. The sonochemical reaction was inhibited by hydroquinone.

In light of the physics of ultrasound irradiation, Nakamura can account for the foregoing observations by a hypothesis (Scheme 15.4) that an initiating

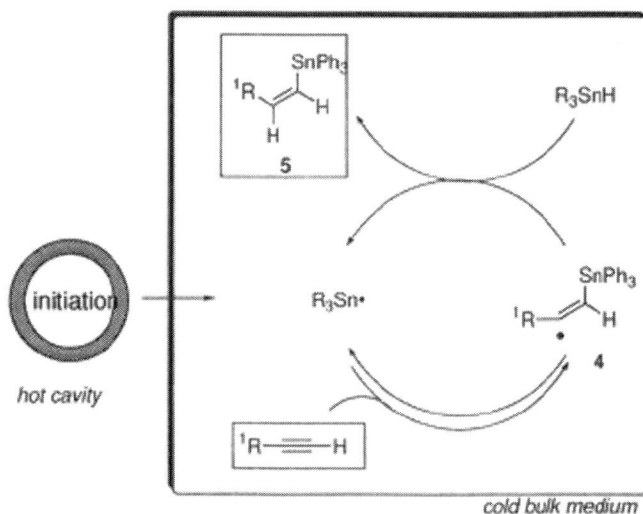

Scheme 15.4 Mechanistic aspects of the triorganotin mediated addition to triple bond.

radical species formed in the hot cavity 1* diffuses into a cold hulk medium after adiabatic collapse of the cavity, and the product-forming propagation reaction continues there.

Several lines of evidence supported this hypothesis. First, the high degree of kinetic *cis* selectivity of the hydrostannation strongly supports the notion that the vinylstannanes are formed in the low-temperature medium rather than in the hot cavity. Second, Nakamura found that the overall reaction rate is sensitive to the temperature of the bulk medium. Thus, the hydrostannation reaction of 1-hexyne with Ph$_3$SnH in toluene slowed down significantly as the temperature was lowered from 7 to -15 to -30 °C (Scheme 15.5). If the product were formed in the hot cavity, such a small temperature change would not have an effect on the reaction rate. In addition, high temperature raises the vapor pressure of a solvent and makes acoustic cavitation less effective and hence should slow down the reaction.[20]

Nakamura next compared the sonochemical radical initiation with two other common methods of radical initiation: photochemical (100 W high-pressure mercury lamp with Pyrex filter) and chemical (10 mol% Bu$_3$B) for the reaction of Ph$_3$SnH with excess 1-hexyne (5 equiv.). For this unique chemistry only stereochemistry can be controlled. In the following studies, we explored it for the development of a new reaction. In the hydrostannation reaction, the intermediate carbon radicals **4** (Scheme 15.5) and **6** (Scheme 15.6)

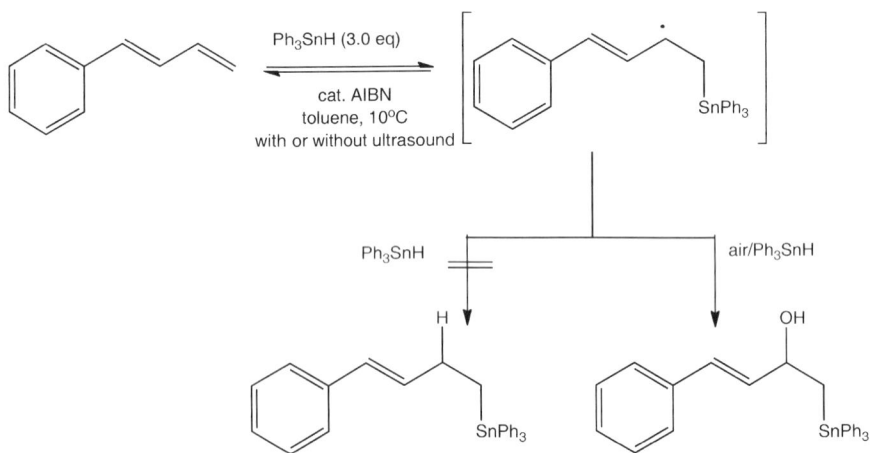

Scheme 15.5   Hydrostannation reaction and sonochemistry.

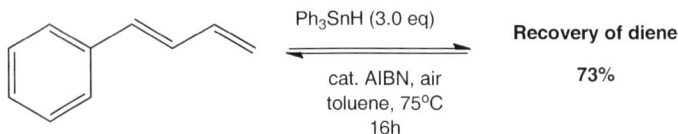

Scheme 15.6   Hydrostannation reaction under conventional conditions.

abstract hydrogen from the tin hydride. During the course of the investigations, it occurred to us that one can trap these intermediates with molecular oxygen to achieve double functionalization of C–C multiple bond (Scheme 15.1). For this "hydroxystannation" reaction, one can draw the radical chain mechanism shown in Scheme 15.2. Realization of such a reaction appeared, by no means, to be easy. The reaction of R_3Sn with molecular oxygen is an extremely fast reaction with a rate constant of $>10^9$ M^{-1} s^{-1} for Bu_3Sn,[22] while the addition of this tin radical to methyl methacrylate is 60 times slower.[23] However, there is a good chance that the latter reaction can effectively compete with the former, since the solubility of oxygen is very low in an organic solvent (8.3×10^{-3} M at 20 °C)[24] and the olefin concentration can be made as high as 0.1–1 M on the other hand. Nakamura found that it is indeed the case; namely, sonochemical hydroxystannation of an olefin can he achieved by oxygen trapping of the radical **8,** when the olefin is activated in conjugation with an electron-withdrawing group. Formation of the hydrostannation product **7** was almost completely suppressed by the predominance of the oxygen-trapping pathway. However, the vinyl radical **4** formed in the reaction of an acetylene (Scheme 15.5) reacted faster with the tin hydride reagent than with oxygen and therefore failed to give hydroxystannation products.

Details of the hydroxystannation of an olefin with a tin hydride reagent in the presence of molecular oxygen were investigated for the reaction of 1-phenylbutadiene with Ph_3SnH (3 equiv.) in toluene (Scheme 15.6). Hydrostannation of the diene is a slow reaction at low temperature. Thus, under argon or nitrogen, a mixture of the diene, Ph_3SnH (3 equiv.), and a small amount of AIBN did not react to any appreciable extent at *ca.* 10 °C (note: no reaction took place even under sonication). Aeration did not cause any appreciable consumption of starting materials, either. However, when the aerated solution was irradiated with ultrasound at 10 °C, smooth hydroxystannation took place and produced the β-hydroxystannane (Scheme 15.5) in 71% isolated yield. Simple Hydrostannation products were not produced at all. When the same reaction was carried out at a higher temperature (75 °C), oxidative loss of the tin hydride reagent took place and the starting diene was recovered (73%, Scheme 15.6). The high-temperature reaction without irradiation also resulted in the simple loss of the tin hydride, indicating that the temperature is as important as the ultrasound (cavitation and mixing). When pure oxygen gas was used instead of air, oxidative loss of the tin hydride predominated over the hydroxystannation, indicating that the low partial pressure of oxygen in air is just suitable for this subtly balanced reaction.

The above experiments can be rationalized by assuming reversible generation of an allylic radical (Scheme 15.7), which is reactive only to oxygen and not to Ph_3SnH at low temperature. This was supported by the following experiments. First, at low temperature, an allylic bromide is not reduced by a tin hydride reagent but is only oxidized by molecular oxygen.[4] Thus, irradiation of a mixture of an allylic bromide and Bu_3SnH at 10 °C under

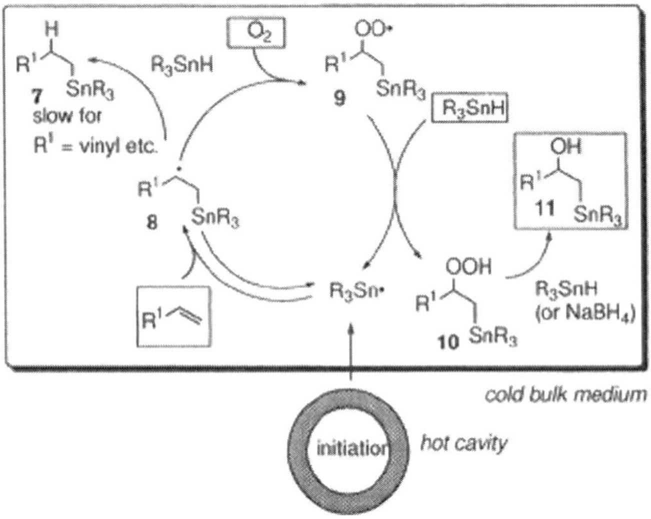

Scheme 15.7   Reversibility in generation of allylic radical under the spotlight.

argon results in the recovery of the starting materials; yet, upon aeration, the irradiated solution starts smooth conversion of the bromide to oxygenated products (Scheme 15.8).

The foregoing observations indicate that molecular oxygen drives the chain reaction (Scheme 15.7) by selectively trapping the allylic radical (R^1 = vinyl) to form the peroxy radical and eventually the alcohol. More than 2 equiv. of the tin hydride reagent is necessary for this reaction, since 1 equiv. is consumed for the conversion of the hydroperoxide to the alcohol. This is obviously inconvenient from a synthetic viewpoint. Nakamura found that this problem can be circumvented with $NaBH_4$, which allows the use of only 1 equiv. of more experimentally convenient Ph_3SnCl as the tin source (Scheme 15.9).

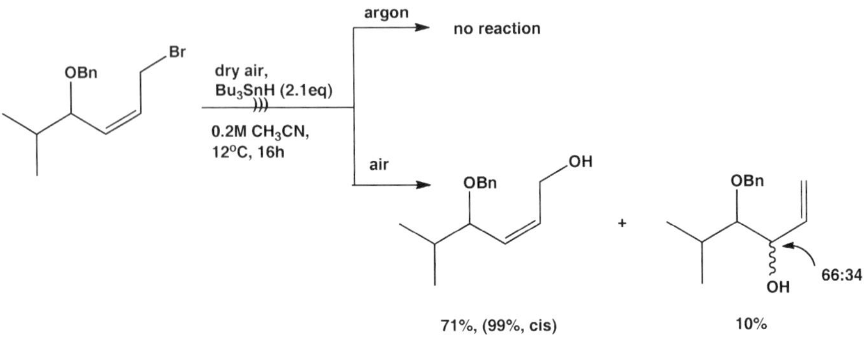

Scheme 15.8   Conversion of bromide into oxygenated final product.

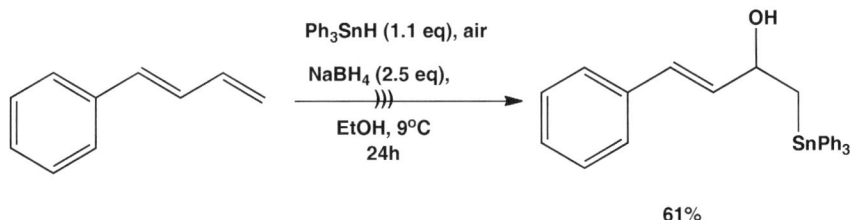

Scheme 15.9 Conversion of hydroperoxide into alcohol.

Thus, the tin chloride is reduced *in situ* with $NaBH_4$ to the tin hydride and the second equivalent of $NaBH_4$ rapidly reduces the hydroperoxide intermediate to afford the stannyl alcohol. Typically, a nearly stoichiometric mixture of a diene and Ph_3SnCl in ethanol was aerated first, sonication started, $NaBH_4$ added, and sonication continued for several hours in an ice-cold bath to obtain the product in a yield slightly lower than that by the stoichiometric procedure.

In summary, Nakamura *et al.* have shown that radical reactions initiated in the hot acoustic cavity undergo chain propagation involving a tin radical as a chain carrier. Since the bulk temperature may be varied simply by changing the temperature of the cooling bath, and the cavity temperature can be controlled by suitable choice of the medium (*i.e.*, its vapor pressure), we can control, in principle, the initiation and the propagation temperatures independently to achieve selectivity previously unavailable for the radical reactions. Among other methods for radical initiation such as thermal, photochemical, and chemical ones, the sonochemical initiation appears to be the least effective in term of the number of radicals generated in the hot cavity. On the other hand, the sonochemical method can be more selective than others.

15.5 SONICATION-INDUCED HALOGENATIVE DECARBOXYLATION OF THIOHYDROXAMIC ESTERS

The sonication of primary, secondary, and tertiary thiohydroxamic esters in CCl_4 has led to their synthetic transformation to alkyl chlorides, bromides, or iodides. The high yields were comparable to the previous thermal and photo-induced version of this same reaction. This radical reaction calls attention to the utility of ultrasound in the production of the trichloromethyl radical, which was concluded to initiate the decomposition of the thiohydroxamic esters. To investigate the sonochemistry of thiohydroxamic ester, carbon tetrachloride (CCl_4) was chosen as the solvent. The radical process by which thiohydroxamic esters decompose in CCl_4 has been studied both under light conditions and thermal conditions.[23,24] The mechanism involves initiation by decomposition of the thiohydroxamic esters, followed by propagation in chain fashion to give 1-chloropentadecane. This reaction sequence is a very useful synthetic procedure and also could be a sensitive probe of sonication, since thiohydroxamic esters are thermally labile and the process follows a chain mechanism.

15.6 AEROBIC CONVERSION OF ORGANIC HALIDES TO ALCOHOLS: AN OXYGENATIVE RADICAL CYCLIZATION

Reductive cleavage of a carbon–halogen bond by a tin hydride reagent generates a carbon radical, and the subsequent synthetic sequence generally ends with the formation of a carbon-hydrogen bond.[1,2] Nakamura *et al.* report a unique tin hydride-mediated reaction that aerobically converts a carbon–halogen bond to a synthetically valuable carbon–oxygen bond. A striking synergetic action of molecular oxygen and a tin hydride at low temperatures (0–20 °C) effects an efficient conversion of an organic halide to the corresponding alcohol under neutral conditions through oxygenation of an intermediate radical (Scheme 15.10).

The reaction tolerates a wide range of functional groups, thus complementing the classical conditions employed for this standard, yet sometimes nontrivial transformation. The radical nature of the reaction permits a new oxygenative radical cyclization of an olefinic iodide (Scheme 15.11) with incorporation of a hydroxy group at the cyclization terminus. Aerobic conversion of (*E*)-cinnamyl bromide to cinnamyl alcohol illustrates the simplicity and effectiveness of the reaction.

The aerobic conversion of halides to alcohols provides an especially powerful synthetic strategy for intramolecular radical cyclization and was

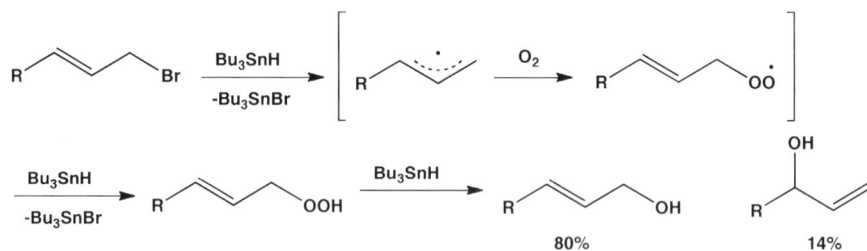

Scheme 15.10   Aerobic conversion of halides to alcohols.

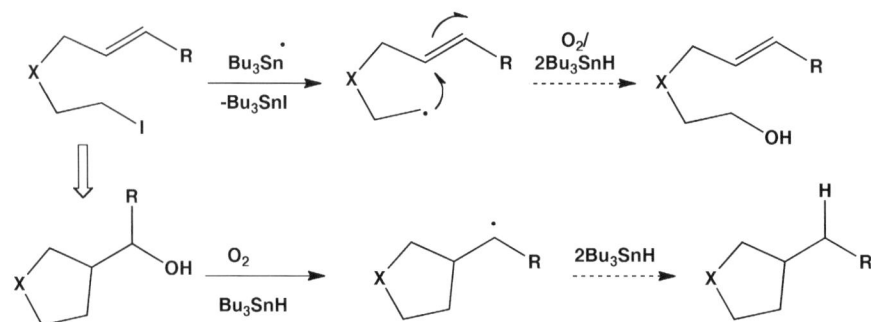

Scheme 15.11   Mechanistic implications of aerobic conversion of halide to alcohols.

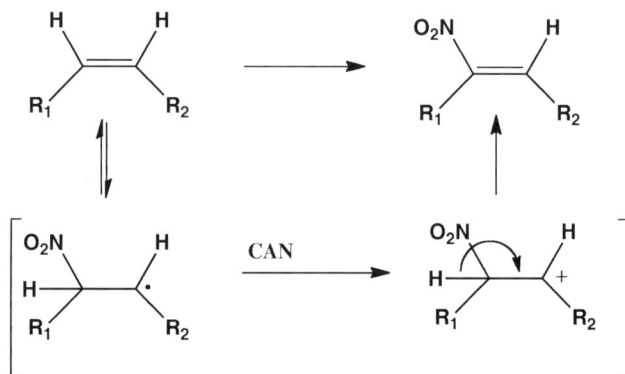

Scheme 15.12 Sonochemical nitration of alkenes to give the corresponding α-unsaturated nitroalkenes in a sealed tube by use of the following conditions: $NaNO_2$ (10 equiv.), $Ce(NH_4)_2(NO_3)_6$ (2.0 equiv.), AcOH, chloroform; 25–73 °C; 4 h.

investigated by Nakamura. Despite the presence of several competitive reaction pathways, the reaction gave the cyclization product as a single predominant product, together with small amounts (5–20%) of the uncyclized product and/or reduced product. Unlike the conventional *reductive* cyclization, which generates one ring at the expense at two functional groups (halogen and olefin), the present *oxygenative* cyclization generates a ring *and* a hydroxy group. In light of this functional group economy, good chemoselectivity and procedural simplicity, the present reaction will add to the versatility of radical-based ring formation strategies.

15.7 A NEW METHOD FOR NITRATION OF ALKENES TO A,B-UNSATURATED NITROALKENES

A sonochemical method for the synthesis of α,β-unsaturated nitroalkenes by use of sodium nitrite, CAN and acetic acid possesses the following advantages: mild reaction conditions, high regioselectivity, good to excellent yields and a short period of reaction time (Scheme 15.12).[25]

15.8 CONCLUSION AND FUTURE DIRECTION

Sonochemistry is gaining significance based on laboratory results and the availability of scale-up systems. Sonochemical applications can be envisaged in all types of systems, including homogenous reactions. Wide acceptance has been gained at a practical/empirical level; however, the theoretical understanding still lags significantly behind. The usefulness of sonochemistry continues to expand into the arena of electrochemistry, photochemistry and biotechnology.

414   *Chapter 15*

REFERENCES

1. For recent reviews, see A. G. M. Barrett, *Chem. Soc. Rev.,* 1991, **20**, 95; G. W. Kabalka and R. S. Varrna, *Org. Prep. Proced. Int.,* 1987, **19**, 283; A. G. M. Barrett and G. G. Graboski, *Chem. Rev.,* 1426; S. G. Pyne, Z. Dong, B. W. Skelton and A. H. White, *J. Chem. Soc., Chem. Commun.,* 1994, 751; R. S. Varma and G. W. Kabalka, *Heterocycles,* 1986, 24, 2645.

2. For recent, representative works, see H.-H. Tso, B. A. Gilbert and J. R. Hwu, *J. Chem. Soc., Chem. Commun.,* 1993, 669; S. E. Denmark and L. R. Marcin, *J. Org. Chem.,* 1993, **58**, 3857; R. F. W. Jackson. J. M. Kirk, N. J. Palmer, D. Waterson and M. J. Wythes, *J. Chem. Soc., Chem Commun.,* 1993, 889; M. L. Morris and M. A. Sturgess, *Tetrahedron Lett.,* 1993, **34**, 43; B. C. Ranu and R. Chakraborty, *Tetrahedron,* 1992, **48**, 5317; D. Y. Kim, K. Lee and D. Y. Oh, *J. Chem. Soc., Perkin Trans. 1,*1992, 2451; K. Fuji, K. Tanaka, H. Abe and A. Itoh, *Tetrahedron: Asymmetry,* 1991, **2**, 179; F. Felluga, P. Nitti, G. Pitacco and E. Valentin, *J. Chem. Res.,* 1992, (S)86, (M)0401; J. R. Hwu and N. Wang, *Synth. Commun.,* 1988, **18**, 21; J. R. Hwu and N. Wang, *J. Chem. Soc., Chem. Commun.,* 1987, 427.

3. For recent works on the preparation of general nitroalkenes, see: S. E. Denmark and L. R. Marcin, *J. Org. Chem.,* 1993, **58**, 3850; M. Node, A. Itoh, K. Nishide, H. Abe, T. Kawabata, Y. Masaki and K. Fuji, *Synthesis,* 1992, 1119; R. Ballini, R. Castagnani and M. Petrini, *J. Org. Chem.,* 1992, **57**, 2160; N. Ono, A. Kamimura, T. Kawai and A. Kaji, *J. Chem. Soc., Chem. Commun.,* 1987, 1550.

4. S.-S. Jew, H.-D. Kim, Y.-S. Cho and C.-H. Cook, *Chem. Lett.,* 1986, 1747; T. Hayama, S. Tomoda, Y. Takeuchi and Y. Nomura, *Chem.*

5. Compounds are available from Aldrich Chemical Co.

6. A. I. Meyers and J. C. Sircar, *J. Org. Chem.,* 1967, **32**, 4134.

7. H.-O. Kalinowski, S. Berger and S. Braun, *Carbon-13 NMR Spectroscopy,* Wiley, Chichester, 1988, p. 292.

8. H. Shechter, J. J. Gardikes, T. S. Cantrell and G. V. D. Tiers, *J. Am. Chem. Soc.,* 1967, **89**, 3005.

9. A. Tromelin, P. Demerseman and R. Royer, *Synthesis,* 1985, 1074.

10. D. E. G. Shuker, *Nitrosamines: Toxicology and Microbiology,* ed. M. J. Hill, Ellis Honvood, Chichester, 1988, pp. 49–50.

11. P. G. Ashmore and B. J. Tyler, *J. Chem. Soc.,* 1961, 1017.

12. T.-L. Ho, *Organic Syntheses by Oxidation with Metal Compounds,* ed. W. J. Mijs and C. R. H. I. de Jonge, Plenum, New York, 1986, ch. 11 and references cited therein.

13. S. V. Ley and C. M. R. Low, *Ultrasound in Synthesis,* Springer, Berlin, 1989, ch. 2.

14. R. M. Silverstein, G. C. Bassler and T. C. Morrill, *Spectrometric Identification of Organic Compounds,* Wiley, New York, 5th edn, 1991, p. 221.

15. See ref. 4 and 6; *cf.* D. Seebach and P. Knochel, *Helv. Chim. Acta,* 1984, **67**, 261.

16. A. R. Forrester, J. M. Hay and R. M. Thomson, *Organic Chemistry of Stable Free Radicals*, Academic Press, London, 1967.

17. E. G. Rozantsev, *Free Nitroxyl Radicals*, Plenum Press, New York, 1970.

18. *Spin Labeling*, ed. L. J. Berlinger, Academic Press, London, 1975.

19. (a) G. Sosnovsky and M. Konieczny, *Z. Naturforsch*, 1976, **31B**, 1376–1378; (b) E. S. Kagan, V. I. Mikhailov, E. G. Rozantsev, V. D. Scholle, *Synthesis*, 1984, 895–916.

20. (a) T. Toda, E. Mori and K. Murayama, *Bull. Chem. Soc. Jpn.*, 1972, **45**, 1904–1908; (b) J. A. Cella, J. A. Kelley and E. F. Kenehan, *J. Org. Chem.*, 1975, **40**, 1860–1862.

21. N. T. Caproin, N. Negoita and A. T. Balaban, *Tetrahedron Lett.* 1977, 1825–1826.

22. A. M. Freeman and A. K. Hoffman, *Fr. 136030*; *Chem. Abstr.*, 1964, **61**, 13289d.

23. L. A. Kalashnikova, M. B. Neyman, E. G. Rozantsev and L. A. Skripko, *Zh. Org. Khim.*, 1966, **2**, 1529–1532; *Chem. Abstr.*, 1967, **66**, 75550a.

24. V. Kaliska, S. Toma and J. Lesko, *Chem. Papers*, 1988, **42**, 243–248.

25. V. Kaliska, S. Toma and J. Lesko, *Coll. Czech. Chem. Commun.*, 1987, **52**, 2266–2273.

Black-light-initiated Free Radical Reactions for Synthetic Applications, Micro-reactors and Modified Nucleoside Synthesis*

16.1 INTRODUCTION: WHY BLACK LIGHT IS SO IMPORTANT

It is well recognized that non-ionizing radiation can react photochemically with biological chromophores, producing end products that are toxic and/or mutagenic in mammalian cells. Most studies have concentrated on the role of UV irradiation due to its high energy, photo-reactivity, wide range of biological chromophores, specific cellular responses, and association with pathologies such as skin melanoma and cataracts.[1–4] However, the role of visible light has been less extensively investigated, even though studies have demonstrated that visible light can induce cellular dysfunction and cell death both *in vitro* and *in vivo*.[1–3,5–7] The blue region (400–500 nm) of the visible spectrum is likely to be particularly important because it has a relatively high energy, can penetrate tissue(s), and is associated with the occurrence of malignant melanoma in animal models.[6,7] Surprisingly, despite its potential for damage, the use of blue light blockers in sunscreens and spectacle lenses has, until recently, received only limited attention. Studies have shown that irradiation of mammalian cells with visible light induces cellular damage primarily *via* reactive oxygen species (ROS).[8] ROS such as the hydroxyl radical, superoxide anion, and singlet oxygen can be produced when visible light excites cellular photosensitizers.[9,10] Whereas photosensitizers such as melanin and lipofuscin in pigmented cells and retinoids in photoreceptor cells have been identified, the identity and location of photosensitizers in non-pigmented cells remain largely unknown. However, a number of options exist, including flavin-containing oxidases, the cytochrome system, heme-containing proteins, and tryptophan-rich proteins.

---

* Chapter written by V. Tamara Perchyonok.
Streamlining Free Radical Green Chemistry
V. Tamara Perchyonok, Ioannis Lykakis and Al Postigo
© V. Tamara Perchyonok, Ioannis Lykakis and Al Postigo 2012
Published by the Royal Society of Chemistry, www.rsc.org

The interaction of these chromophores with light can generate ROS, which in turn can damage lipids, proteins, and DNA. This is emphasized by the study of The violet–blue light stimulated H_2O_2 production from peroxisomes and mitochondria in cultured 3T3 and CV1 mammalian cells.[11] Hydrogen peroxide production was enhanced by over-expression of flavin-containing oxidases, which proposes that violet–blue light initiates photoreduction of flavins, which activate flavin-containing oxidases in mitochondria and peroxisomes, resulting in H_2O_2 production. Furthermore, the mechanism by which photosensitization leads to cellular dysfunction is unclear but may center on DNA damage.

The role of DNA damage in aging mammals appears to be pivotal, and there is increasing evidence that oxidative damage is an important factor in producing mutations in genes, shortening telomeres, and damaging mitochondrial DNA. To support a role for visible light in DNA damage, Pflaum *et al.*[13] have previously shown that oxidative damage induced by visible light does yield DNA modifications.

16.2 C2′,3′-CYCLIC CARBONATES DERIVED FROM NUCLEOSIDES WHY THEY ARE IMPORTANT

The need for efficient and novel antiviral molecules has encouraged the synthesis of nucleoside analogues over the last decades.[1] Reactions carried out under mild experimental conditions and displaying regioselectivity are powerful tools in the modification of natural nucleosides, since these polyfunctional compounds are labile in many reaction media and bear several hydroxyl groups having similar reactivity. On the other hand, alkyl carbonates are widely used as protecting groups for diols and polyols.[2] In addition to their synthetic applications, they are employed in polymer and medicinal chemistry.[3] In this latter field, the lipophilic nature of carbonates can provide prodrugs of pharmacological active compounds, as previously reported for penciclovir.[4] However, carbonates are not extensively applied to nucleoside chemistry because they are generally introduced either through chloroformates,[5,6] which are very reactive agents, or enzymatically using oxime carbonates.[7] Moreover, the removal of alkyl carbonates usually requires strong nucleophile reagents or is carried out by taking advantage of the properties of the second alkyl substituent.[2]

16.3 C5′ GENERAL COMMENTS AND HISTORY

Hydrogen abstraction from the 2-deoxyribose moiety of DNA produces carbon-centered radicals whose fate depends upon the environment.[1] The accessibility of the C–H bond in sugar moieties determines the preferential site of attack even by the highly reactive HO• radicals. In B-DNA, the H5′ and H5″ are the most exposed ones, and therefore, the preference for abstraction of these hydrogens is higher.[2,3] Hydrogen abstraction from the 5′-position of purine nucleosides can lead to 5′,8-cyclonucleosides. They have been observed among the decomposition products of DNA, when it is exposed to ionizing radiations[4–6] or treated chemically by highly oxidizing radical species.[7] They have also been identified in mammalian cellular DNA *in vivo*, where their level is enhanced by conditions of oxidative stress.[8,9] The difficulty of repair and the amenability to mutation render these lesions biologically significant and the study of their formation necessary.[10–12] Under aerobic conditions, the fate of the C5′ radical is not well understood. All the proposed intermediates are based on rationalization of the products observed in DNA degradation by the neocarzinostatin chromophore (NCS-chromophore) in the presence of glutathione,[1,13,14] although related studies with metalloporphyrins have also been reported.[2,15] Moreover, strand breakage was observed to be base-selective, preferentially occurring (75%) on the thymidine unit.

The chemistry of carbon-centered radicals resulting from hydrogen atom abstraction from the sugar moieties has been the subject of many recent studies. Selective generation of these species is mainly obtained by photoreactive precursors using nucleosides or oligonucleotides (ODNs). Indeed, generation of a single radical species on duplex ODNs provides a powerful tool for elucidating the role of reactive intermediates in the formation of nucleic acid lesions. For example, C1′ and C4′ positions have been studied in detail by photolysis of the corresponding *tert*-butyl ketones.[16–20] Photolabile precursors of C5′ radicals are missing. It was recently found that 8-bromo-2′-deoxyadenosine captures electrons and rapidly loses a bromide ion to give the corresponding C8 radical. This intermediate abstracts intramolecularly a hydrogen atom from the C5′ position affording selectively the 2′-deoxyadenosin-5′-yl radical. This allowed us for the first time to verify that the C5′ radical attacks the double bond of the base moiety intramolecularly, and this occurs with a rate constant of $k_c = 1.6 \times 10^5$ s^{-1} to form, after oxidation, a 5′,8-cyclo-2′-deoxyadenosine as the final product.[21–23] The analogous sequence of 8-bromo-2′-deoxyguanosine does not operate because the electron adduct undergoes fast protonation at C8 ejecting Br- and affording the one-electron oxidized 2′-deoxyguanosine transient species.[24,25] Interestingly, the 8-bromo-2′-deoxyadenosine moieties in a series of DNA hairpins containing a light-dependent flavin electron injector in the loop region of the hairpin are found to capture electrons with quantitative formation of the corresponding debrominated oligonucleotides, similar to the analogous 8-bromo-2′-deoxyguanosine derivatives.[26]

16.4 BLACK-LIGHT-INDUCED RADICAL CYCLIZATION APPROACH TO CYCLONUCLEOSIDES: AN INDEPENDENT APPROACH

Cyclopurine and cyclopyrimidine lesions are observed among the decomposition products of DNA, when exposed to ionising radiation or to some anti-tumour agents.[1] Recently several examples of independent formation of 5′,6-cyclo-5,6-dihydrothymidine and 5′,8-cyclo-2′-deoxyadenosine have been reported.[2] The unique structural features of the cyclo-nucleosides incorporate an additional base-sugar linkage between the C6 position of pyrimidine or C8 position of purine and the C5′ position of the 2′-deoxyribose.[3] Conformationally fixed nucleosides analogues are a unique class of compounds that can be used as a tool for investigating steric interactions between nucleosides or nucleotides and the enzymes that utilize them (Figure 16.1).[4]

To date, various modified 2-deoxynucleosides containing specific DNA lesions and their incorporation into a defined sequence of oligonucleotides has been an outstanding approach to investigate the biological consequences. Synthetic oligonucleotides that contain the modified nucleosides[5–7] as well as similar cyclopurine[7] and cyclopyrimidine[8] moieties were also prepared. Recent studies have shown that the chemical synthesis of these lesions and their incorporation on specific sites of DNA are of considerable importance in order to investigate, in detail, the biochemical and biophysical features of the double helix damage.[6,9,10]

Several key contributions in the area have recently reported synthetically useful cascade methodologies for effective generation of the cyclonucleosides in order obtain a general procedure for the preparation of some of the diatereoisomers of the cyclonucleosides as well as try overcomes the limitations of the existing approaches, due to low yield multiple-step synthesis, problematic chromatographic and separation properties.[6–10]

Photolysis is one of the most appropriate and wide used methods of generating nucleosidyl radicals, with numerous reports in the literature reported of sources and setups for their generation; however, most of the methods require specialized experimental setups, glassware and rather expensive light sources for the desired transformations.

In order to overcome these problems, Perchyonok decided to investigate the use of a black light, which has a maximum peak wavelength at 352 nm, and performs superbly in broad range of free radical transformations as reported by Ryu *et al.*[10] and is applicable to the generation of C5′ nucleosidyl purine and pyrimidyl radicals in organic and aqueous media in the presence of derivatives of hypophosphorous acids in organic and aqueous media (Scheme 16.1).

Commercially available nucleosides such as uridine, adenosine and quanosine were reacted with freshly prepared Appel bromide in CH_2Cl_2 in order to

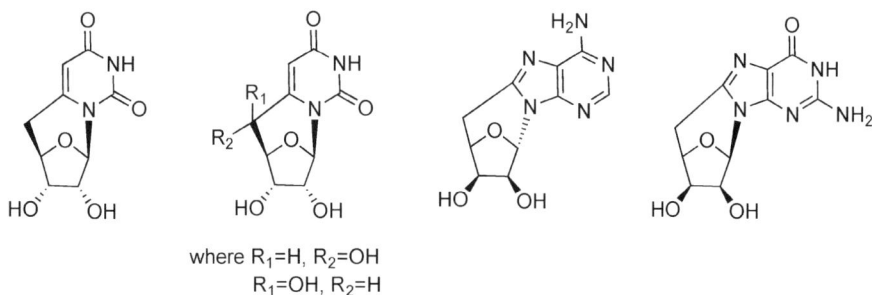

where R_1=H, R_2=OH
R_1=OH, R_2=H

Figure 16.1 Structure of C-Cyclonucleosides.

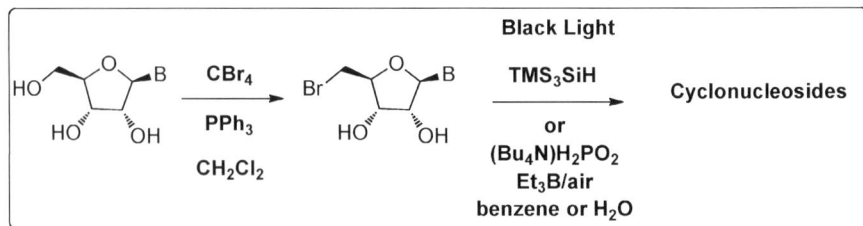

Scheme 16.1 General synthetic approaches and methodology.

prepare the corresponding 5'-bromoderivatives (Scheme 16.2). The compounds were prepared in moderate to good yields. The 5'-bromoderivatives were purified on the C18 HPLC chromatography column and identified through comparison of spectral data reported in the literature.

Following the initial plan, Perchyonok examined tandem radical 1,6-HAT-cyclization of 5'-bromouridine in the presence of (a) TMS$_3$SiH, (b) H$_3$PO$_2$/Bu$_4$N$^+$Cl$^-$, (c) Bu$_3$SnH; and in benzene and water under black light initiation, Et$_3$B/air and AIBN at 80 °C conditions and results are summarized in Scheme 16.3 and Table 16.1.

The major reaction product of all transformations summarized in the Table is indicative of the intra-molecular radical 6-*exo*-trig reaction being predominant, suggesting that intermolecular hydrogen transfer reaction is much

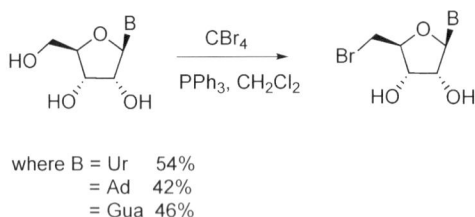

where B = Ur   54%
    = Ad   42%
    = Gua 46%

Scheme 16.2 Preparation of 5'-bromonucleosides.

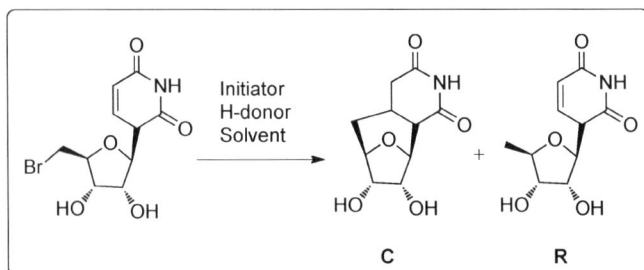

Scheme 16.3 Radical cyclization reaction of C5'-bromouridine in benzene and water under various initiation conditions and in the presence of free radical hydrogen donors.

Table 16.1 Radical cyclization reactions of C5′-bromouridine under black light, Et_3B/air (rt) and AIBN at 80 °C initiation conditions with various free radical hydrogen donors.

Entry	H-Donor	Initiation	Solvent	% yield C : R
1	TMS_3SiH	Black light	Benzene	78 : 5
2		Et_3B/air (rt)	Benzene	72 : 5
3		AIBN/80 °C	Benzene	78 : 4
4		Black light	H_2O	72 : 6
5		Et_3B/air (rt)	H_2O	70 : 4
6		AIBN/80 °C	H_2O	79 : 4
7	H_3PO_2/$Bu_4N^+Cl^-$	Black light	Benzene	73 : 5
8		Et_3B/air (rt)	Benzene	70 : 6
9		AIBN/80 °C	Benzene	67 : 3
10		Black light	H_2O	72 : 4
11		Et_3B/air (rt)	H_2O	70 : 6
12		AIBN/80 °C	H_2O	71 : 5
13	Bu_3SnH	Black light	Benzene	60 : 30
14		Et_3B/air (rt)	Benzene	58 : 25
15		AIBN/80 °C	Benzene	50 : 30
16		Black light	H_2O	50 : 30
17		Et_3B/air (rt)	H_2O	55 : 35
18		AIBN/80 °C	H_2O	54 : 30

slower process in comparison to intra-molecular radical 6-*exo*-trig cascade reaction. However, in the case of Bu_3SnH, the yield of product R (minor product) accounts for 30% in the final product mixture, which is consistent with the difference in the hydrogen donor ability of Bu_3SnH *vs.* TMS_3SiH *vs.* H_3PO_2/$Bu_4N^+Cl^-$ highlighting the potential for further development of the methodology. This observation is consistent with the previously reported by Navacchia, Chatgilialoglu and Cadet and their coworkers in their pioneering work on the understanding of the mechanistic aspects of C5′ radicals.[11,12]

The proposed mechanism for the transformation in question is represented in Scheme 16.4. Initiator decomposes to abstract a hydrogen atom from the radical mediator such as $(TMS)_3SiH$, H_3PO_2 or Bu_3SnH which generates a corresponding silyl, phosphorous centered or stannyl radical. The later abstracts the halogen atom to form a C5′ radical. In the case of $(TMS)_3SiH$, H_3PO_2 or Bu_3SnH as hydrogen donors, the C5′ radical undergoes a 6-*exo*-trig cyclization adding to the double bond of the base. The resulting C5-radical abstracts hydrogen from the hydrogen donor, yielding a *cyclo*-compound, while completing the radical chain. The minor product, which is the product of the direct reduction is also formed due to much less but still competitive H-abstraction reaction and is observed in all transformations in different amounts as pointed out in Table 16.1.

In conclusion, a short and efficient synthetic sequence, based on consecutive radical reactions in aqueous media, for the preparation of cyclonucleosides was found. The C5′ radicals, generated by tandem homolytic bond cleavage followed by carbon–carbon bond formation, are the key intermediates in these

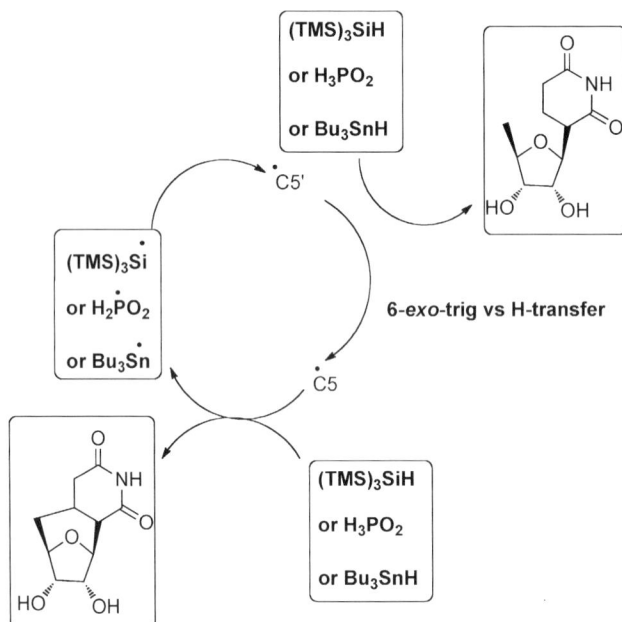

Scheme 16.4 Proposed Radical Chain Mechanism for the formation of the cyclized product (major product) and the product of direct reduction (minor product).

transformations. The chemical biology approach used for studying purine 5′,8-cyclonucleoside lesions has brought significant achievements so far.[13] The site-specific generation of sugar radicals has been the key approach for a better understanding of chemical molecular mechanisms occurring at the biological level. The chemistry of the C5′ radicals in the purine nucleosides is fairly well understood and it is now clear that the fate of the C5′ radicals is partitioned between uni-molecular processes (cyclizations) and bimolecular processes (reactions with oxygen, thiols, or oxidants).[13] Therefore, the local concentration of these components and pH are extremely important in selecting the preferred pathway. Work is currently on the way to gain further insight into these important transformations at a molecular level under aerobic and anaerobic conditions.

16.5 RADICAL CYCLIZATION "TIN-FREE" APPROACH TO C2′,C3′-CYCLIC CARBONATES DERIVED FROM NUCLEOSIDES: AN INDEPENDENT APPROACH

The search for agents that restore or enhance the ability of the human immune system to ward off infection or other invasive challenges is a key aspect of therapeutic research. The AIDS epidemic and the need for adjuvant therapy to boost the immune system of the elderly and cancer patients has brought this area of immuno-potential into sharp focus.[14,15] Many different types of compounds have been demonstrated to possess immuno-stimulatory properties. These include glycoproteins such as interferon, peptides such as tufsin, polynucleotides, and small heterocycles such as levamisole.[15,16]

Until recently, synthetic methodologies to prepare 2′- or 3′-substituted nucleosides were restricted to direct nucleophilic (S_N2) displacement, nucleophilic addition to appropriately protected 2′- or 3′-keto nucleosides, procedures involving rearrangements as well as pioneering reports involving free-radical carbon–carbon bond formation involving tributyltin hydride.[17–23]

An important area of modern research is the development of synthetic radical reactions that use non-metal hydrides. This research is partly driven by the need to move away from toxic metal hydrides, primarily tributyltin hydride. Tributyl hydride is still used extensively in small-scale synthesis, but the problems of toxicity and product purification have limited its use, particularly on a large scale in the chemical industry.[24–27]

In recent years, the use of phosphorous hydrides as replacements for tributyltin hydride in radical reductions has been investigated. Of particular interest has been the use of phosphorous hydrides to reduce organohalides and xanthates. Following Barton's seminal work on reductions using dialkyl phosphites [$(RO)_2P(O)H$], the use of other phosphorous hydrides, most notably hypophosphorous acid (and its salts), have been reported. The mechanisms of these reactions involve intermediate phosphorus–centered radicals that, for example, abstract halogen atoms from organohalides.[28,29]

Perchyonok has developed an effective radical cyclization "tin-free" approach mediated by hypophosphites and TMS_3SiH to C2′,C3′-cyclic carbonate derivatives from nucleosides in organic and aqueous media under various initiation conditions to yield novel cyclonucleosides. The reaction proceeded in good to excellent yields and mechanistic aspects and diastereospecific aspects of the transformations are also being investigated (Scheme 16.5).

The starting material (**1a**, **1b** and **1c**) was prepared from the commercially available nucleoside precursors by the optimized method of Gimisis *et al.*[30] (Scheme 16.6) in good to excellent overall yields. Compounds **1a**, **1b** and **1c** were subsequently carefully alkylated with allyl bromide in DMF in the presence of the sodium hydride at room temperature to give 2′-iodo-3′-O-allyl ethers **2a**, **2b** and **2c** in good yields.

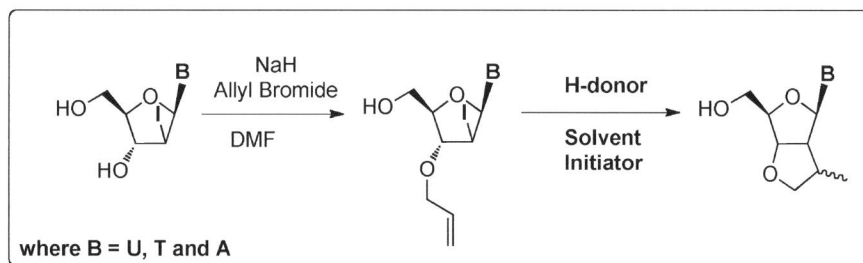

Scheme 16.5   General synthetic approaches and methodology.

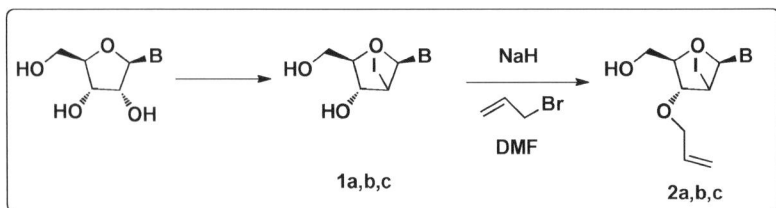

Scheme 16.6   Preparation of a radical precursor.

Next Perchyonok examined tandem radical cyclization/atom transfer reaction in the presence of (a) TMS$_3$SiH, (b) H$_3$PO$_2$/Bu$_4$N$^+$Cl$^-$, (c) H$_3$PO$_2$/Et$_4$N$^+$Cl$^-$ and (d) H$_3$PO$_2$/Bu$_2$Ph$_2$N$^+$Cl$^-$ in benzene, cyclohexane and acetonitrile (ACN):water (3:1) under black light, Et$_3$B/air and AIBN/80 °C conditions and results are summarized in Scheme 16.7 and Table 16.2.

The radical intermediates have been generated at the carbons bearing iodogroup by treatment of the precursor **2a** under a broad range of conditions involving a broad range of free radical hydrogen donors, different solvents as well as variety of common initiation methods such as Et$_3$B/air, AIBN/△ and black-light-initiation conditions. All reactions proceeded smoothly to form desired cyclized nucleosides in good to very good yield as a major product of the transformation. The reaction products were fully characterized and identified by ^{1}H NMR spectroscopy and ESI/MS analysis. Its is interesting to note that as the radical generated at the 2′ position was trapped by the double bond of the allyl ether at the a-face, it produced an inseparable diastereomeric mixture of products containing the *cis*-fused five-membered rings. The observation is interesting and consistent with observations previously on tributyltin mediated radical cyclizations of the similar nucleoside systems.

In order to expand the scope of the methodology precursors **2b** (B = A) and **2c** (B = T) have been subjected to the previously described conditions and results of the transformations is summarized in Scheme 16.8. As expected the cyclization proceeded smoothly in excellent yields with final products being isolated as a unseparable diastereomeric mixture, supporting the original observation that the radical generated at 2′ position was trapped by the double

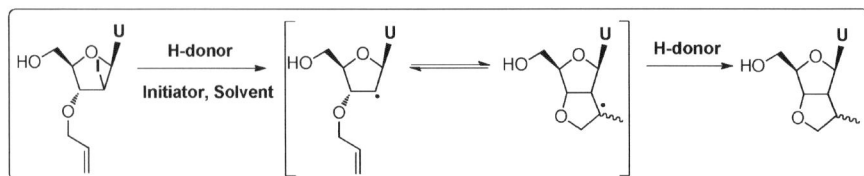

Scheme 16.7   Radical cyclization reaction of corresponding C2' radical precursor under alternative free radical reductions conditions.

Table 16.2   Radical cyclization reactions of 2-iodo-3′-allylether **2a** (B = U), Et$_3$B/air (rt) and AIBN/80 °C initiation conditions with various free radical hydrogen donors.

Entry	H-Donor	Initiation	Solvent	% yield
1	TMS$_3$SiH	Black light	C$_6$H$_{12}$	75
2		Et$_3$B/air (rt)		
3		AIBN/80 °C	C$_6$H$_6$	70
4		Et$_3$B/air (rt)	C$_6$H$_6$	75
5		AIBN/80 °C	C$_6$H$_{12}$	63
6		Et$_3$B/air (rt)	C$_6$H$_{12}$	67
		AIBN/80 °C	ACN : H$_2$O	60
			ACN : H$_2$O	55
7	H$_3$PO$_2$/Bu$_4$N$^+$Cl$^-$	Et$_3$B/air (rt)	C$_6$H$_6$	56
8		AIBN/80 °C	C$_6$H$_6$	60
9		Et$_3$B/air (rt)	C$_6$H$_{12}$	67
10		AIBN/80 °C	C$_6$H$_{12}$	68
11		Et$_3$B/air (rt)	ACN : H$_2$O	63
12		AIBN/80 °C	ACN : H$_2$O	68
13	H$_3$PO$_2$/Et$_4$N$^+$Cl$^-$	Et$_3$B/air (rt)	C$_6$H$_6$	67
14		AIBN/80 °C	C$_6$H$_6$	65
15		Et$_3$B/air (rt)	C$_6$H$_{12}$	53
16		AIBN/80 °C	C$_6$H$_{12}$	68
17		Et$_3$B/air (rt)	ACN : H$_2$O	56
18		AIBN/80 °C	ACN : H$_2$O	59
19	H$_3$PO$_2$/Bu$_2$Ph$_2$N$^+$Cl$^-$	Et$_3$B/air (rt)	C$_6$H$_6$	63
20		AIBN/80 °C	C$_6$H$_6$	65
21		Et$_3$B/air (rt)	C$_6$H$_{12}$	66
22		AIBN/80 °C	C$_6$H$_{12}$	68
23		Et$_3$B/air (rt)	ACN : H$_2$O	69
24		AIBN/80 °C	ACN : H$_2$O	67

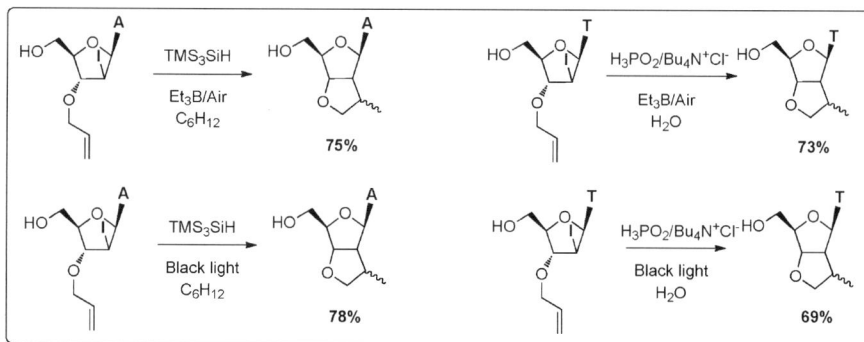

Scheme 16.8   Radical cyclization reactions of precursors **2b** and **2c**.

bond of the allyl ether at the a-face, and produced an inseparable diastereo-meric mixture of products containing the *cis*-fused five-membered rings.

In conclusion, Perchyonok has demonstrated the ease of preparation of broad range of cyclic nucleosides in good to excellent yields utilizing conventional free radical chemistry in organic and aqueous media using hypophosphites and TMS$_3$SiH as environmentally friendly alternatives to tributyltin hydride under a variety of free radical initiation techniques.

16.6   BLACK-LIGHT-INDUCED DIRECT GENERATION OF C2′-NUCLEOSIDYL RADICALS IN ADENOSINE, THYMIDINE AND URIDINE IN ORGANIC AND AQUEOUS MEDIA

Hydrogen abstraction from the 2-deoxyribose moiety of DNA produces carbon-centered radicals whose fate depends upon the environment.[31] The accessibility of the C–H bond in the sugar moiety determines the preferential site of attack even by the highly reactive hydroxyl radicals. The 2′-position of deoxyribose is less prone to H-abstraction, either because of the low accessibility or the low reactivity of these hydrogens in the dsDNA.[32] Nevertheless, the abstraction of C2′-H may be facilitated in RNA due to the presence of the additional OH-group or eventually by the presence of adjacent halogenated bases.[31,32] The chemistry of carbon-centered radicals resulting from hydrogen atom abstraction from the sugar moieties has been the subject of many recent studies.[33] Selective generation of these species is mainly obtained by photo-reactive precursor using nucleosides or oligodeoxynucleo-tides (ODNs).[33] Indeed, the generation of a single radical species on duplex ODNs provided a powerful tool for elucidating the role of reactive intermediates in the formation of nucleic acid lesions. For example C1′, C4′ and C5′ are known and have been studied in some detail by photolysis of the corresponding radical precursors.[31,32,33] In depth investigation of C2′ radicals is an emerging and unexplored area with particular focus on developing methodologies compatible with aqueous media.

2′-Deoxynucleosides are currently produced in limited amounts from salmon milt, which makes it difficult to satisfy the growing need for these compounds in the production of new diagnostic reagents such as a DNA chip and pharmaceuticals such as antisense oligonucleotides.[34] Although chemi-cal transformations from the corresponding ribose to 2′-deoxynucleoside by

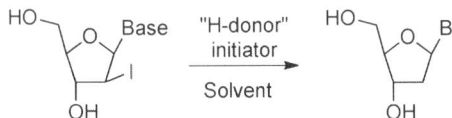

where Base = Uridine, Thimidine and Adenine

Scheme 16.9   Radical dehalogenation of nucleoside derivatives.

HO—O—Base
HO OH
where Base = Uridine,
Thimidine
Adenine

TBDMSCI
AgNO₃/DABCO
THF, rt., 48h

RO—O—Base
RO OH

Base = Uridine, **88%**
Thimidine **83%**
Adenine **85%**

I₂,PPh₃,imidazole
Toluene, reflux

RO—O—Base
—I
OR

Base = Uridine **60%**
Thimidine **55%**
Adenine **68%**

1M TBAF (2eq)
THF, -20°C, 2h

HO—O—Base
—I
OH

Base = Uridine **quant**
Thimidine **quant**
Adenine **quant**

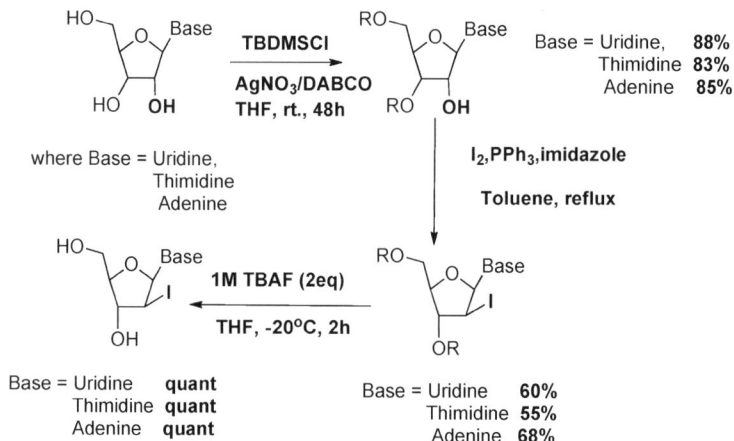

Scheme 16.10 Preparation of the C2′ radical precursors.

radical means is known in some instances, the scope of reagents and experimental setups is limited to organostannanes (Bu_3SnH) which are toxic, TMS_3SiH in organic and aqueous media under thermally initiated conditions, and reported radical deoxygenation reactions using commercial H_3PO_2 under thermally initiated conditions and aqueous buffered media.[34,35]

Hydrogen transfer reactions were performed using $[Et_4N]^+H_2PO_2^-$, $[Bu_4N]^+H_2PO_2^-$ and TMS_3SiH under black-light-induced generation of C2′ radicals in organic and aqueous media and the results are summarized in Table 16.3.

Table 16.3 lists data obtained for the hydrogen atom transfer of substrates **1**, **2** and **3** as a model compound in the presence of various hydrogen donors in water and acetonitrile as representatives of aqueous and organic media. As expected, reaction of **1**, **2** or **3** in the absence of $[NBu_4]^+H_2PO_2^-$ yielded complete recovery of unreacted starting material. Reactions performed either in the absence of a free radical initiator (Et_3B/air) or in the presence of phenol as a free radical inhibitor failed to form the reduced product, confirming the free radical nature of the transformation in question. All reactions proceeded smoothly in good to very good yields, with several transformations being performed on the preparative scale and reaction products being isolated and identified using ¹H NMR and HPLC/MS analysis. In parallel, Et_3B/air was used as alternative free radical initiation method and obtained results similar to that observed for the transformations performed under black-light-initiation conditions. Work is currently on the way in our laboratory to adapt the methodology to an efficient continuous flow-through process.

In summary, Perchyonok demonstrated a quick and efficient method of generating C2′ nucleosidyl radicals under low power black light as light source/initiator in the presence of alternative free radical hydrogen donors in aqueous and organic media at room temperature in good to excellent yield.

Table 16.3 Hydrogen atom transfer reactions of 2′-iodonucleosides (**1**(U), **2**(Th) and **3**(Ad)) using $[Et_4N]^+H_2PO_2^-$, $[Bu_4N]^+H_2PO_2^-$ and TMS_3SiH in aqueous media using black light in organic and aqueous media at 25 °C.

Entry	Substrate	M–H 2.0 equiv.	Initiator	Solvent	Time	Yield (%)[a]
1	**1**	$[Et_4N]^+H_2PO_2^-$	Black light	H_2O	3 h	78[b]
				ACN	3 h	75
2	**1**	$[Bu_4N]^+H_2PO_2^-$	Black light	H_2O	3 h	73
				ACN	3 h	74[b]
3	**1**	TMS_3SiH	Black light	H_2O	3 h	72
				ACN	3 h	70
4	**1**	$[Bu_4N]^+H_2PO_2^-$	Et_3B/air	H_2O	3 h	65[b]
				ACN	3 h	61
5	**1**	TMS_3SiH	Et_3B/air	H_2O	3 h	66
				ACN	3 h	63[b]
6	**2**	$[Et_4N]^+H_2PO_2^-$	Black light	H_2O	3 h	70
				ACN	3 h	71
7	**2**	$[Bu_4N]^+H_2PO_2^-$	Black light	H_2O	3 h	64
				ACN	3 h	62
8	**2**	TMS_3SiH	Black light	H_2O	3 h	65
				ACN	3 h	67
9	**2**	$[Bu_4N]^+H_2PO_2^-$	Et_3B/air	H_2O	3 h	62[b]
				ACN	3 h	64
10	**2**	TMS_3SiH	Et_3B/air	H_2O	3 h	65[b]
				ACN	3 h	66
11	**3**	$[Et_4N]^+H_2PO_2^-$	Black light	H_2O	3 h	62
				ACN	3 h	60
12	**3**	$[Bu_4N]^+H_2PO_2^-$	Black light	H_2O	3 h	62
				ACN	3 h	62
13	**3**	TMS_3SiH	Black light	H_2O	3 h	70
				ACN	3 h	66
14	**3**	$[Bu_4N]^+H_2PO_2^-$	Et_3B/air	H_2O	3 h	72
				ACN	3 h	68
15	**3**	TMS_3SiH	Et_3B/air	H_2O	3 h	74[b]
				ACN	3 h	67

[a]The yield is calculated from the results of HPLC analysis. [b] The yield is reported for the isolated compound.

16.7 BLACK-LIGHT-INDUCED RADICAL/IONIC HYDROXYMETHYLATION OF ALKYL IODIDES WITH ATMOSPHERIC CO IN THE PRESENCE OF TETRABUTYLAMMONIUM BOROHYDRIDE

CO is among the most prominent sources of C-1 in organic synthesis. A number of valuable transformations utilizing CO have been investigated thus far, irrespective of the reactive species.[39,40] One-carbon homologation generating alcohol derivatives is an important class of synthetic transformations,[39] and the use of CO for such processes is highly desirable. The related hydroxymethylation reaction process using organohalides was achieved by methodologies based on radical carbonylation and *in situ* reduction. In these processes, a combination of group 14 metal hydrides and cyanoborohydride reagents was used (Scheme 16.11, eqn (1)).[41]

Recently, the Giese reaction and related radical carbonylation reactions have been reported.[42]

The reaction is thought to involve iodine atom transfer from alkyl iodides to Giese adduct radicals followed by hydride reduction of the resulting carbon–iodine bond by cyanoborohydride. These results led Ryu to consider that hydroxymethylation would also be possible without the aid of organotin reagents (Scheme 16.11, eqn (2)). Ryu reports that tertiary and secondary alkyl iodides can efficiently undergo a tin-free radical/ionic hydroxymethylation in the presence of tetrabutylammonium borohydride (n-Bu$_4$NBH$_4$) as a hydrogen donor. It is important to note that in many cases *carbonylation proceeded under atmospheric pressure of CO*. This method is simple and less toxic compared to the traditional conditions. n-Bu$_4$NBH$_3$CN proved to be optimal in the tin-free Giese reactions (Scheme 16.12);[43] therefore, Ryu initially examined the reaction of 1-iodoadamantane (**1a**) and CO (82 atm) in the presence of n-Bu$_4$NBH$_3$CN (3 equiv.) under thermal initiation conditions using AIBN as the radical initiator. The expected hydroxymethylation product **2a** was obtained in 48% yield along with 1-adamantane carboxaldehyde (5%) and, unexpectedly, its cyanohydrin (25%). It is noteworthy that the direct reduction product, adamantane (**3a**), was obtained in less than 1% yield and that the expected course of the reaction progressed even with a high concentration of the substrate (0.5 M). Another advantage is that this reaction does not require a very efficient hydrogen source, such as tin hydride. Hydrogen abstraction by alkyl radicals commonly competes with CO trapping, and high dilution conditions are typically required to make the relative concentrations of CO to tin hydride high.[44,45] It is reasonable to assume that the radical forms by reduction of the initially formed 1-adamantane carboxaldehyde. On the other hand, the unexpected cyanohydrin was thought to form as a result of transfer of the cyano group from the borohydride reagent.[46] However, a notable

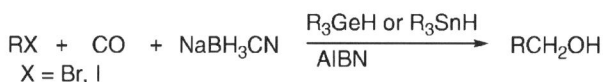

$$\begin{array}{c} \text{RX} + \text{CO} + \text{NaBH}_3\text{CN} \xrightarrow[\text{AIBN}]{\text{R}_3\text{GeH or R}_3\text{SnH}} \text{RCH}_2\text{OH} \\ \text{X = Br, I} \end{array}$$

This work

$$\text{RI} + \text{CO} + n\text{-Bu}_4\text{NBH}_4 \xrightarrow[\text{AIBN or light}]{\text{"none"}} \text{RCH}_2\text{OH}$$

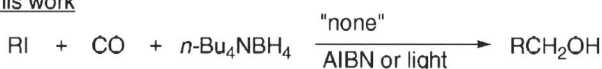

Scheme 16.11   Hydroxymethylation of RX.

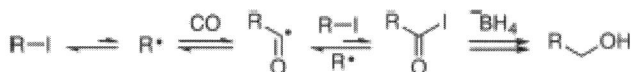

Scheme 16.12   Atom-transfer-based radical chain mechanism.

improvement was achieved by using photoirradiation conditions in place of thermal radical initiation. A similar result was obtained when a more energy-saving 15 W black light (peak wavelength 352 nm) was used.

Thus, Ryu *et al.* have succeeded in efficient hydroxymethylation of tertiary and secondary iodides without the use of group 14 radical mediators. It seems unusual that a significant amount of reduction product was formed in the reaction of a bulky tertiary iodide, in the absence of a powerful hydrogen donor such as *n*-Bu$_3$SnH. Ryu rationalized this observation as 1-iodoadamantane cannot undergo direct reduction through an ionic S$_N$2 process; therefore, the adamantyl radical arising from 1-iodoadamantane directly abstracted a hydrogen from borohydride and certain electron-transfer processes were thought to propagate the radical chain.

The second mechanism shown in Scheme 16.13 involves an electron transfer mediated by borohydride reagent[47] (S$_{RN}$1 mechanism). Thermal initiation or photoirradiation of alkyl iodides generates the initiating alkyl radicals (R$^\bullet$), which react with CO to form acyl radicals (RCO$^\bullet$). The acyl radicals abstract hydrogens from borohydride (BH$_4^-$) to form aldehydes and generates the borane radical anions (BH$_3^{\bullet-}$). While aldehydes undergo hydride reduction by borohydride anions to give alcohols, the generated borane radical anions (BH$_3^{\bullet-}$) react with alkyl iodides (R–I) through electron transfer to give radical anions ([R–I]$^{\bullet-}$) that fragment to alkyl radicals (R$^\bullet$) and iodide ions (I$^-$), thus completing a radical chain. The reduction product formed at low CO pressure gives strong indication that this mechanism is feasible.

In conclusion, Ryu has developed a novel hydroxymethylation reaction using CO and borohydride reagents without the use of toxic radical mediators such as trialkyltin hydrides or its precursors.[48] The reaction can be applied to tertiary and secondary iodides. Furthermore, a combined system involving atmospheric pressure of CO and black light irradiation was successfully employed. Ryu *et al.* have proposed a mechanism in which the borohydride reagents work both as a hydrogen source and a hydride source, and therefore act as a radical mediator.[49] The reaction conditions are simple and mild.

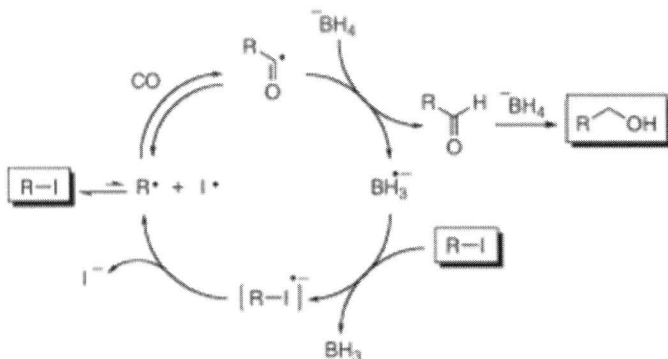

Scheme 16.13   Alternative radical chain mechanism involving electron transfer.

Therefore, this reaction represents a useful method for introducing the hydroxymethyl unit into organic molecules.

16.8   TOWARDS THE SYNTHESIS OF ALKYL ALKYNYL KETONES BY PD/LIGHT-INDUCED THREE-COMPONENT COUPLING REACTIONS OF IODOALKANES, CO, AND 1-ALKYNES

Alkynyl ketones are contained in several biologically active molecules[50] and play a crucial role as key intermediates in the synthesis of natural compounds[51] and in heterocyclic systems.[52] Although the most common route for the synthesis of alkynyl ketones is the acylation reaction of alkynyl organometallic reagents with acid chlorides,[49] a three-component approach comprising organo halides, carbon monoxide, and terminal alkynes based on transition-metal-catalyzed carbonylation represents a more straightforward approach in terms of simplicity, which makes it possible to avoid the preparation of somewhat unstable acid chlorides.[53,54]

For the past decade, significant efforts have been made to improve carbonylative approaches, and several mild reaction conditions have been found thus far. However, the system remains largely restricted to the use of aryl or vinyl halides; aliphatic halides are not necessarily applicable due to the low reactivity of sp^3 carbon–halogen bonds toward oxidative addition with palladium.[55]

Ryu *et al.* recently reported that radical/metal hybrid reactions under photo-irradiation conditions[56] accelerated the atom-transfer carbonylation of alkyl halides leading to carboxylic acid esters, amides, and related heterocycles.[57,58] In these reactions, it is assumed that the intervention of a free-radical mechanism will generate acylpalladium intermediates, thus representing a radical/metal collaboration system. Herein, Ryu *et al.* have utilized a radical/metal collaboration strategy to a carbonylative Sonogashira reaction using alkyl iodides, CO, and terminal alkynes, which gave good yields of alkyl alkynyl ketones (Scheme 16.14).

Ryu also examined radical cascade sequences accompanying ring-opening or ring-closing processes (Scheme 16.15).

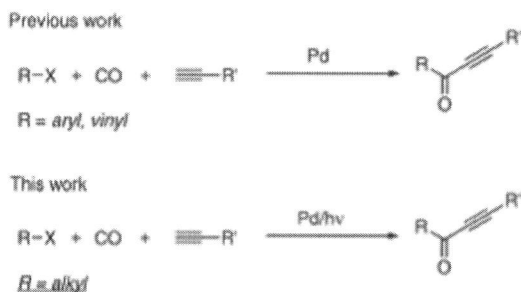

Scheme 16.14   Carbonylative Sonogashira coupling reaction leading to alkynyl ketones.

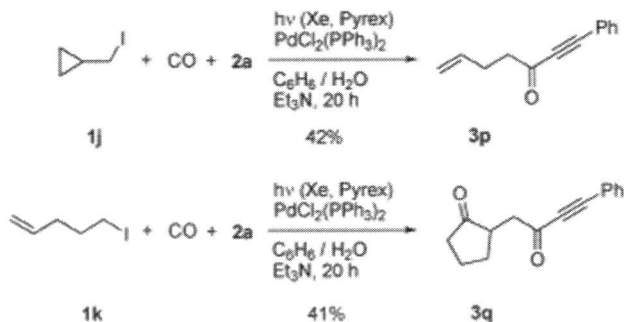

Scheme 16.15   Cascade radical carbonylation reactions.

Ryu *et al.* have proposed a mechanism for Pd/light-assisted carbonylation as shown in Scheme 16.16, using 5-iodoalkene to yield the corresponding alkyne as an example. In the first step, alkyl iodide reacts with Pd(0) under irradiation to afford alkyl radical **A** and Pd(ı) *via* a one-electron transfer.[59] The resulting alkyl radical **A** traps CO to form acyl radical **B** and then affords alkyl radical **C** through 5-*exo* cyclization of acyl radical **B**. The acyl radical species **D** generated from CO and alkyl radical **C** couples with Pd(ı) to lead to acyl-

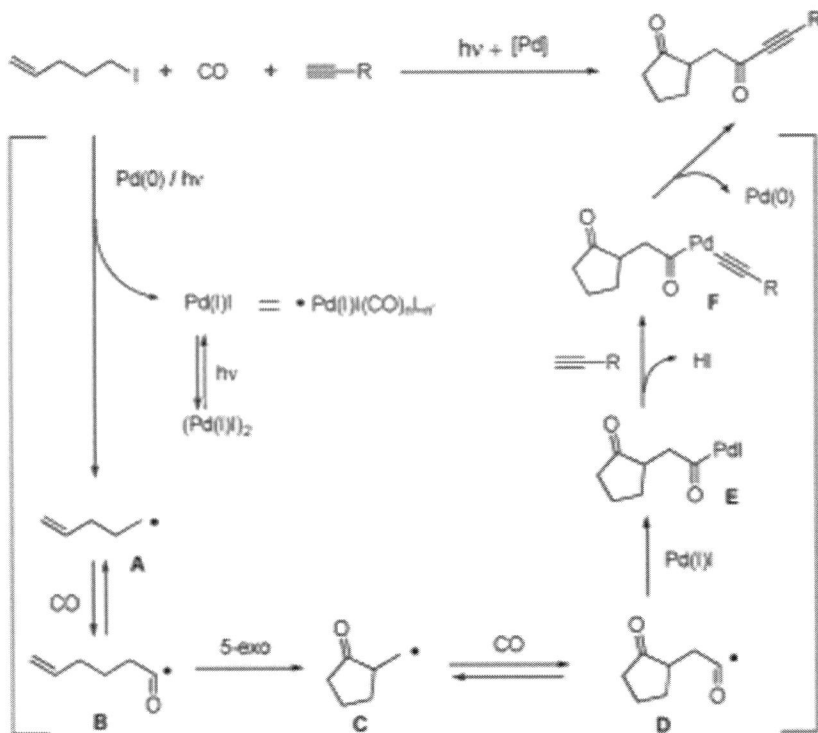

Scheme 16.16   Possible reaction mechanism.

palladium intermediate **E** and gives the alkynyl ketone through an intermediate **F**. In this reaction mechanism, a rather unfamiliar Pd(I)-type species is postulated. This may exist in an equilibrium with a Pd^I dimer under photoirradiation conditions.[60]

In summary, Ryu *et al.* have demonstrated a three-component coupling reaction of alkyl iodides, CO, and alkynes leading to alkynyl ketones that was affected by a Pd/light combined system. This protocol employs readily available alkyl iodides and terminal alkynes and could become a standard method for alkynyl ketone synthesis.

16.9 VICINAL C-FUNCTIONALIZATION OF ALKENES: PD/LIGHT-INDUCED MULTICOMPONENT COUPLING REACTIONS LEADING TO FUNCTIONALIZED ESTERS AND LACTONES

In terms of the high throughput and efficiency required to construct organic compounds with structural diversity in one-pot, multicomponent reactions (MCRs) have attracted more and more interest in recent years.[61,62] MCRs involving CO as one of the components allow for the direct incorporation of CO as a carbonyl function into carbonyl-containing products, and Ryu is particularly interested in strategies involving radical reactions[63,64] and previously found that metal/hv-induced systems[5] caused acceleration of the atom-transfer carbonylation of alkyl iodides leading to carboxylic acid esters, amides, and alkynyl ketones.[65]

Vicinal carbon functionalization of alkenes is an important challenge in multicomponent reactions. Whereas typical processes with incorporation of CO involve the double alkoxycarbonylation of alkenes,[63] reactions attaining the introduction of an alkyl unit and CO into vicinal carbons of alkenes are scarce.[66] In a related study, Ryu previously reported on the photoinduced addition of R-phenylseleno-substituted esters to alkenes and CO to give 4-keto acyl selenides as products.[65] However, a simple ester synthesis has yet to be attained. Ryu observed that Pd/light-induced radical carbonylation of various R-substituted iodoalkanes allows for vicinal carbon-functionalization of alkenes leading to esters (Scheme 16.17).

When a benzene solution of ethyl iodoacetate, 1-octene, and ethanol was exposed to photoirradiation conditions (irradiation with a 500 W xenon lamp through Pyrex) under 45 atm of CO pressure in the presence of PdCl$_2$(PPh$_3$)$_2$ (5 mol%), a base (1.1 equiv. of NEt$_3$ and 10 mol % of DMAP), and a small amount of water (*ca.* 1 equiv.), the desired diester was obtained in 72% yield after chromatographic purification (Scheme 16.18).

Ryu next examined cyclization three-component coupling reactions, using alkenyl alcohols, which also worked well to give the desired ester-functionalized lactones (Scheme 16.19).

A possible reaction mechanism for the present multicomponent coupling reaction is shown in Scheme 16.20. Alkyl radicals are formed *via* cleavage of the I–C bond of **1a**, which may be triggered by single electron transfer from the photo-irradiated Pd(0) complex (Scheme 16.20).[67,68]

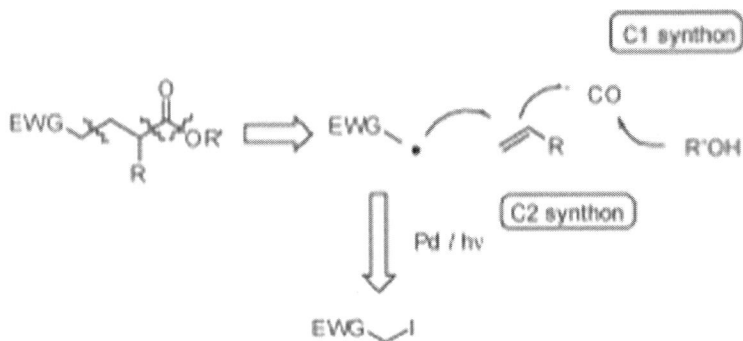

Scheme 16.17   Strategies for four-component coupling reactions of R-iodoalkanes, alkenes, CO and alcohols leading to esters.

Addition of the radicals to the alkene then takes place to give alkyl radicals. The subsequent iodine atom transfer from **1a** affords iodoalkane **5**. The reaction between alkyl radical and Pd(I) to give alkyl palladium might take place but scarcely contributes in this reaction mechanism, judging from the fact that β-hydrogen elimination product was not observed. Probably pressurized CO drives these equilibriums to afford the acyl radical intermediate, which would be trapped by Pd(I) to form acyl palladium species, precursors for the diester **4a**. The persistent radical character of Pd(I) species may be supported by the dimerization behavior.[69,70]

16.10   CLOSING THE GAP: FROM SINGLE MOLECULE SYNTHESIS THE CONVENTIONAL WAY TO MICROREACTORS—THE POWER OF BLACK LIGHT

Research chemists typically perform chemical transformations in traditional glass round-bottomed flasks. Since the optimization of chemical transforma-

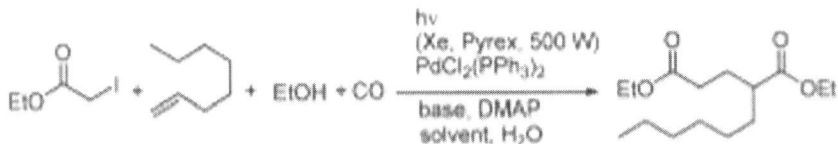

Scheme 16.18   Pd/Light-induced four-component coupling reaction of ethyl iodoaceate with 1-octene, ethanol and carbon monoxide.

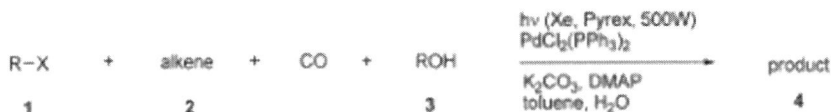

Scheme 16.19   Four-component coupling reactions leading to esters.

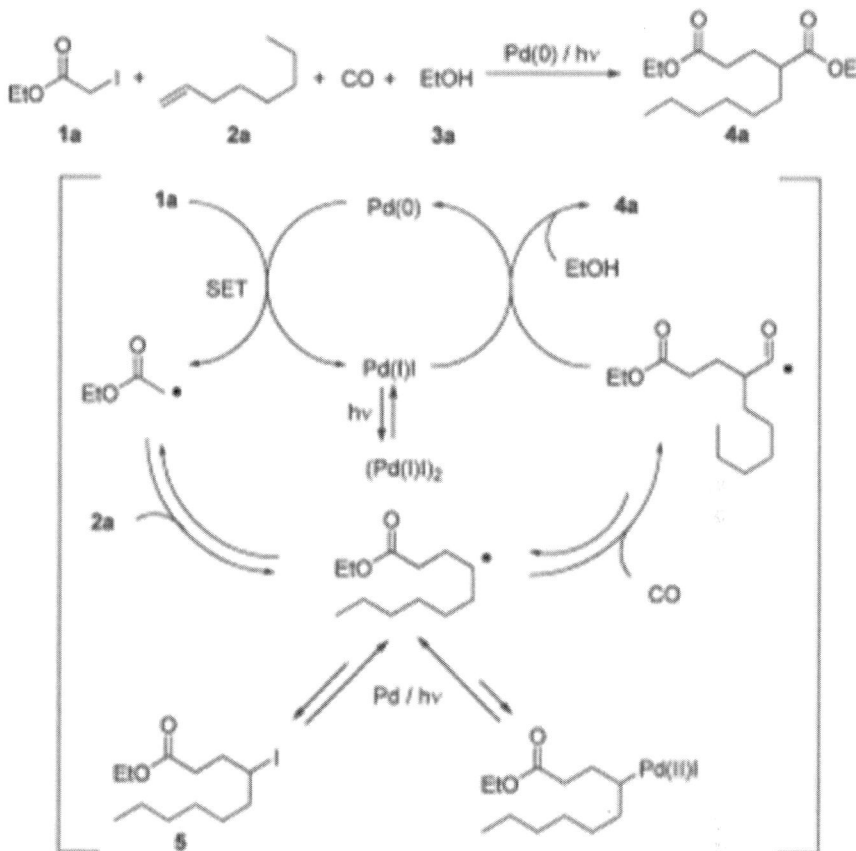

Scheme 16.20   Possible reaction mechanism.

tions in batch reactors often consumes substantial amounts of starting materials, a lot of precious building blocks as well as effort and time are required in order to identify the ideal reaction conditions for a particular reaction. Having found the optimal conditions to achieve a certain reaction on a small scale, process scale-up often poses additional challenges and requires further adjustment of the reaction parameters. To overcome these hurdles in synthetic chemistry, microstructured continuous-flow reactors and chip-based microreactors are becoming increasingly popular.[71]

Micro-structured continuous-flow devices consist of a miniaturized channel system etched into or established on materials such as metals, silicon, glass, ceramics or polymers, which allows the chemical reaction to take place in a relatively narrow pore. These reaction systems have a high interfacial area per volume (only depending on the radius of the channel) and chemical reactions therefore profit from rapid heat and mass transfer. Due to the high heat and mass transfer rates, the reaction time, yield and selectivity are strongly influenced, often rendering processes to be more efficient and selective and thereby avoiding the generally required purification processes. Besides the facile control of the physical parameters, microreactors allow for low operational volumes to minimize reagent consumption, the integration of on-line detection modules and an excellent process safety profile in case highly exothermic reactions are undertaken, enhanced by the fact that only small amounts of hazardous/explosive intermediates may be formed at any given time. To obtain synthetically useful amounts of product, the reactors are simply run longer ("scale-out" principle[71e]) or several reactors are placed in parallel ("numbering up"), assuring identical conditions for the analytical and preparative modes. Many different applications of microstructured devices in synthetic chemistry have been reported and reviewed;[72,73] here, some concepts and recent developments will be presented.

16.11   SYNTHESIS IN MICROCHEMICAL SYSTEMS

Due to the small channel dimensions and the increased surface to volume ratio of microreactors, mass and heat transport are significantly more efficient than in the classic round-bottomed flask. The mixing of reagents by diffusion occurs very quickly, and heat exchange between the reaction medium and reaction vessel is highly efficient. As a result, the reaction conditions in a continuous-flow microchannel are homogenous, and can be controlled precisely. Therefore, highly exothermic and even explosive reactions can be readily harnessed in a microreactor. The careful control of reaction temperature and residence time has a beneficial effect on the outcome of a reaction with respect to yield, purity and selectivity. Below, selected examples described to illustrate the potential application of microreactors to organic synthesis. The application of microreactors to small-scale medicinal and academic total synthesis is just beginning.

16.12   MICROFLOW PHOTO-RADICAL CHLORINATION
OF CYCLOALKANES

The recent rapid progress of microflow reaction technology led synthetic organic chemists to examine a variety of organic reactions using micro-reactors.[74,75] Since microflow reactors provide high heat and mass transfer rate for organic reactions, those reactions which are difficult to control using conventional batch equipments can be a potential target for the improvement. Work in the Ryu laboratories has been directed toward the design and application of microflow reaction technologies to conventional organic synthesis[76] and, to this end, we previously reported on effective continuous microflow reactions, which include Pd-catalyzed coupling reactions in ionic liquids,[77] thermally induced radical reactions[78] and photo-irradiation reactions.[79] Photoreactions using microreactors have some promise in terms of using light energy at a minimum because the light penetration is ideal, inherent to the thinness of the microreactor, and the resultant short residence time even allows the avoidance of undesirable second reactions.[80,81] As such an application of photo-irradiation reactions, Ryu recently reported on a successful multi-gram-scale production of a steroidal compound by the Barton reaction using a micro-flow reactor in combination with an energy saving light source such as black light.[82] Ryu also achieved diastereoselective [2+2] photoaddition reactions using a microflow system.[83]

As part of Ryu's group's ongoing interest in organic synthesis by microflow photo irradiation technologies, they began to investigate the radical chlorination of alkanes using a microreactor under photo-irradiation conditions. The radical chlorination of alkanes is one of the most basic organic reactions, and studies on this reaction have continued for more than half a century.[84,85] While various reagents including molecular chlorine[86] and sulfuryl chloride[87] are used for radical chlorination of alkanes with variable selectivity,[88] in order to achieve reproducible selectivity for single chlorination over competing polychlorination, strict control of molar ratios, light power, and reaction time are required for batch systems. Ryu thought that the radical chlorination in a microflow system would be useful even for alkane chlorination in terms of controlled selectivity and efficiency with energy-saving light sources. Although radical chlorination using a microflow system was previously reported for chlorination of a benzylic position,[89–90] no work on chlorination of simple alkanes using a microreactor has been reported; Ryu discovered that the radical chlorination of cyclic alkanes using sulfuryl chloride and molecular chlorine can be effectively carried out using a continuous microflow system and visible lights.

16.13 CONTINUOUS MICROFLOW CHLORINATION OF CYCLOHEXANE WITH MOLECULAR CHLORINE IN DETAIL

The continuous microflow system was comprised of a T-shaped micromixer and the DNS reactor, two gas/liquid feed stainless tube and an outlet. A 10 mL gas tight syringe containing cyclohexane (5 mL, 46 mmol) was attached to an inlet of the micromixer and a 25 mL gas tight syringe containing molecular chlorine (25 mL, 1.1 mmol) was connected to the other inlet of the micromixer through a check valve. The syringes were pumped using syringe pumps at flow rates of 1.32 mL h^{-1} and 7.68 mL h^{-1} for cyclohexane and molecular chlorine, respectively (residence time: 19 min). The mixture of product eluted from the outlet was collected in a flask containing 10% sodium sulfite. The yield of chlorocyclohexane was determined by gas chromatography (GC) analysis.

Ryu started a chlorination study using a microflow system with cyclohexane as the substrate and chose a DNS microreactor F005D03-HB composed of photo-etched stainless steel microchannels with covered Pyrex glass (Scheme 16.21, 1000 m width, 300 m depth and 2.35 m length, and total hold-up volume: 0.7 mL) for this reaction. The total figure of the flow reaction system is shown in Scheme 16.21. Thus, the DNS microreactor was connected with a T-shaped micromixer (i.d. = 500 m), and a check valve (6.9 MPa) was fitted to

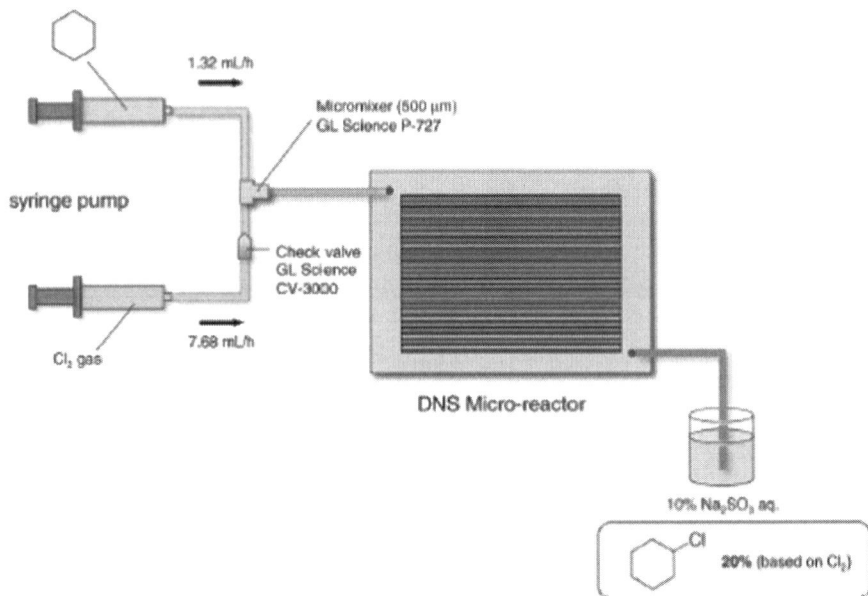

Scheme 16.21   Microflow chlorination with molecular chlorine.

one of the inlets of the micromixer for avoiding backward flow of chlorine. As molecular chlorine is known to react with alkanes violently under UV irradiation, Ryu used natural room light as the light source. Thus, Ryu *et al.* carried out the chlorination of cyclohexane with molecular chlorine at room temperature in natural light with residence time of 19 min (measured value). As Ryu envisioned, the reaction gave chlorocyclohexane with nearly complete selectivity for the single chlorination product (>95%) with a yield of 20%.

16.14   MICROFLOW CHLORINATION WITH SULFURYL CHLORIDE AND BLACK LIGHT

Ryu then examined chlorination using sulfuryl chloride as a chlorination reagent. As sulfuryl chloride is known to react more gently than molecular chlorine, a 15 W black light (peak wavelength: 352 nm) was employed for this reaction, in which a Mikroglas Dwell Device (total hold-up volume: 0.95 mL) was used as a microreactor. The total system for this reaction is shown in Scheme 16.22. When a solution containing sulfuryl chloride and 40 molar equivalents of cyclohexane was introduced to a microreactor with a residence time of 19 min (measured value, flow rate: 3.0 mL h^{-1}) at room temperature, the reaction gave chlorocyclohexane selectively in 22% yield. While increasing molar ratio of cyclohexane to 80 equiv. did not affect the yield of the product (20%), extending residence time (57 min, flow rate: 1.0 mL h^{-1}) raised product yield to 35%.

As shown in Scheme 16.22, plug flow was observed due to gases of SO$_2$ and HCl generated during the reaction. To avoid occasional disturbance by the

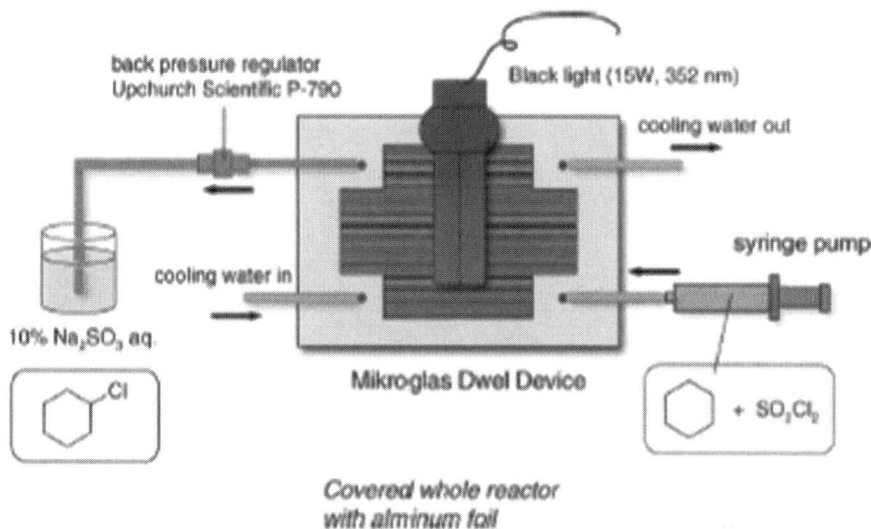

Scheme 16.22   Microflow chlorination with sulfuryl chloride.

plug flow in the microreactor, a back-pressure regulator (BPR: Upchurch Scientific, P-790, 0.3 bar) was fitted, resulting in the effective suppression of the plug flow. The reaction using two black lights resulted in slightly better yield of the product. Under the employed conditions all chlorination reactions proceeded well to give mono-chlorocycloalkanes with high selectivity (>95%).[91]

16.15 THE BARTON REACTION USING A MICROREACTOR AND BLACK LIGHT: CONTINUOUS-FLOW SYNTHESIS OF A KEY STEROID INTERMEDIATE FOR AN ENDOTHELIN RECEPTOR ANTAGONIST

The Barton reaction (nitrite photolysis), which represents the remote functionalization of saturated alcohols, uses photo-irradiation conditions for nitrite esters, prepared from the corresponding alcohols with nitrosyl chloride.[92] Having the unique potential of site-selective C–H bond cleavage at the d position *via* a 1,5-radical translocation from O to C,[93] the Barton reaction has found widespread applications in synthesis, including steroid functionalization.[94] The recent rapid progress in the area of microreaction technology[95] prompted us to examine such synthetic reactions using a compact continuous microflow system.[96] Ryu *et al.* report that the Barton reaction of a key steroidal substrate **1**, to give **2**, a key intermediate in the synthesis of an endothelin receptor antagonist (Scheme 16.23),[97] can be successfully carried out by using a glass-covered stainless-steel microreactor (Dainippon Screen Mfg.), coupled with the use of an energy-saving compact light source.[98]

Photo-microreactors have advantages over conventional batch reactors from several viewpoints:[99] (1) the efficiency of the photoenergy is improved

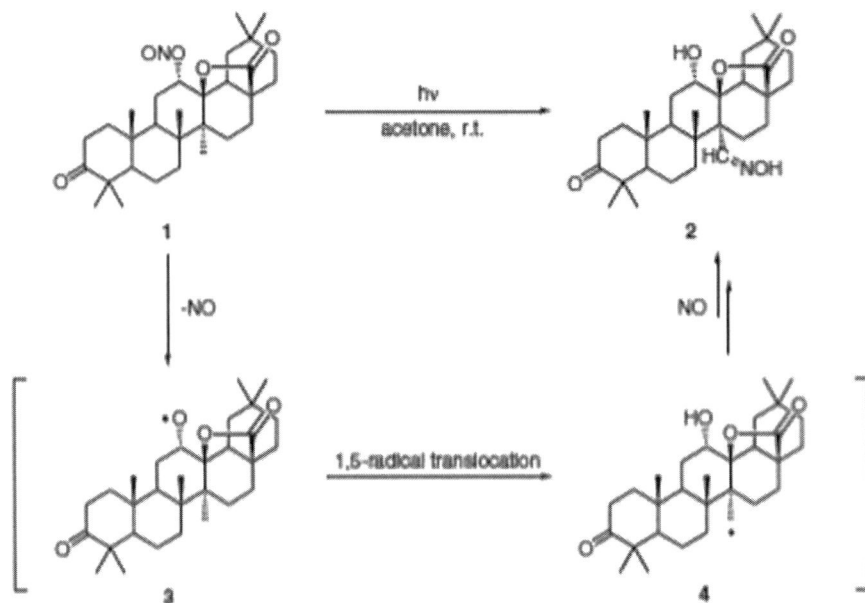

Scheme 16.23   Barton nitrite photolysis of a steroidal compound **1** leading to an
                    oxime **2**.

because of the thinness of the reaction mixture in the micro space, (2) the low
residence time avoids undesirable side reactions, (3) a continuous-flow system
can be created which allows for the use of the same micro-devices for large
quantity production, and (4) an energy-saving compact light irradiation system
can be accommodated by the reaction system. Thus, we hypothesized that the
Barton reaction could be carried out using a downsized reactor and an
inexpensive light source with good energy efficiency.

The Barton reaction typically uses a high-pressure mercury-vapor lamp as
the light source. Thus, we began with the use of a 300 W high-pressure mercury
lamp in combination with a stainless-steel microreactor (Type A) having a
serpentine single lane microchannel with hold-up volume 0.2 mL, the top of
which was covered by a glass plate. Whereas the use of a quartz cover glass
resulted in a complex mixture of products due to the low wavelength (high
energy) of the mercury light source, the use of soda lime glass as a top cover
gave good yields of the rearranged product **2**. Probably the use of Pyrex glass
has the advantage of better transparency at the wavelength used over soda lime
glass, since the shorter wavelengths produced by a black light are weak. Since
the power of the black light (15 W) is considerably weaker than that of
mercury lamp (300 W), we adjusted the residence time so as to compensate for
this deficiency.

Ryu performed the Barton reaction (nitrite photolysis) of a steroidal
substrate **1**, to give **2**, a key intermediate for the synthesis of an endothelin

receptor antagonist; it was successfully carried out in a continuous microflow system using a pyrex glass-covered stainless-steel microreactor having a microchannel. The authors found that a 15 W black light (peak wavelength: 352 nm) as the light source, suffices for the Barton reaction, creating a compact photo-micro reaction system. Multi-gram scale production was attained using two serially connected, multi-lane microreactors (Type B) (Scheme 16.24).

In summary, Ryu demonstrated that the Barton reaction (nitrite photolysis) of a steroidal substrate **1** can be successfully carried out using a stainless-steel microreactor covered by Pyrex glass, a low power black light as the light source, and DMF as the solvent. Thus, a gram-scale synthesis of oxime product **2** was attained in a continuous-flow reaction. Among a variety of organic reactions, photo-chemical transformation using a microreactor has promise regarding the efficient use of light energy.[96,97,98,99]

Photo-microreactors have advantages over conventional batch reactors from several viewpoints:

- the efficiency of light penetration is improved because of the thinness of the reaction mixture in the microspace;
- the short residence time allows the avoidance of undesirable side reactions;

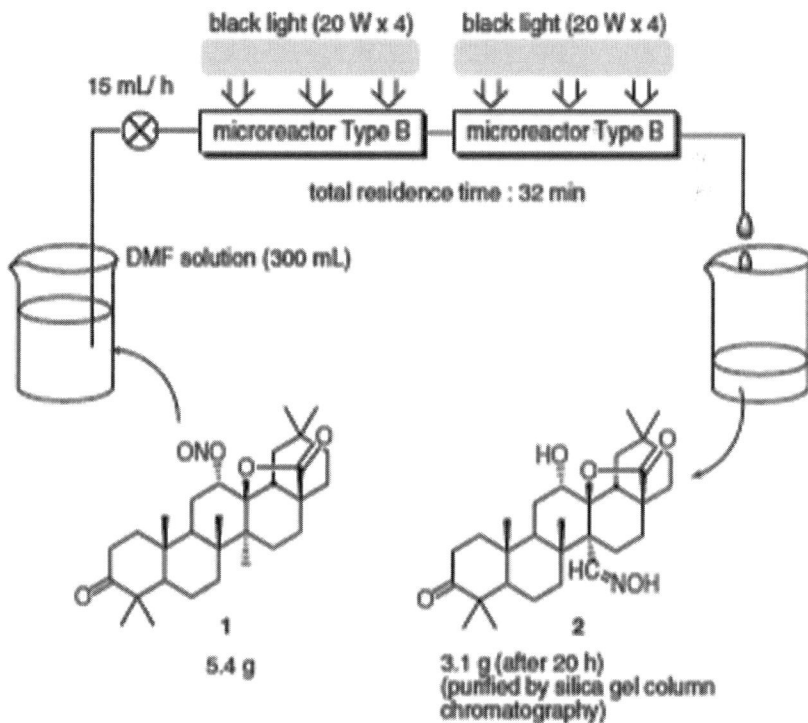

Scheme 16.24 Gram-scale production of **2**.

- a continuous-flow system can be created and allows for the use of the same micro-devices for large quantity production; and
- an energy-saving compact light irradiation system can be accommodated by the reaction system.

16.16 CONCLUSION

Light can be considered an ideal reagent for environmentally friendly, 'green' chemical synthesis; unlike many conventional reagents, light is non-toxic, generates no waste, and can be obtained from renewable sources. Nevertheless, the need for high-energy ultraviolet radiation in most organic photochemical processes has limited both the practicality and environmental benefits of photochemical synthesis on industrially relevant scales. Given the remarkable photophysical properties of synthetically useful precursors in free radical chemistry, these classical transformations involving black-light-mediated photosynthesis represent a promising strategy towards the development of practical, scalable industrial processes with great environmental benefits.

REFERENCES

1. (a) G. Pratviel, J. Bernadou and B. Meunier, Carbon–Hydrogen Bonds of DNA Sugar Units as Targets for Chemical Nucleases and Drugs, *Angew. Chem., Int. Ed.*, 1995, **34**, 746–769; (b) J. Cadet, T. Douki, D. Gasparutto and J.-L. Ravanat, Radiation-induced damage to cellular DNA: measurement and biological role, *Radiat. Phys. Chem.*, 2005, **72**, 293–299; (c) J. Cadet, T. Douki and J.-L. Ravanat, in *Redox-Genome Interactions in Health and Disease*, ed. J. Fuchs, M. Podda and L. Packer, Dekker, New York, 2003, pp. 143–189; (d) C. Chatgilialoglu and P. O'Neill, Free radicals associated with DNA damage, *Exp. Gerontol.*, 2001, **36**, 1459–1471; (e) M. Dizdaroglu, P. Jaruga, M. Birincioglu and H. Rodriguez, H., Free radical-induced damage to DNA: mechanisms and measurement, *Free Rad. Biol. Med.*, 2002, **32**, 1102–1115.
2. C. Chatgilialoglu, M. Guerra and Q. G. Mulazzani, Model Studies of DNA C5′ Redicals. Selective Generation and Reactivity of 2′-Deoxyadenosin-5′yl Radical, *J. Am. Chem. Soc.*, 2003, **125**, 3839–3848.
3. (a) R. Flyunt, R. Bazzanini, C. Chatgilialoglu and Q. G. Mulazzani, Fate of the 2′-Deoxyadenosin-5′-yl Radical under Anaerobic Conditions, *J. Am. Chem. Soc.*, 2000, **122**, 4225–4226; (b) L. B. Jimenez, S. Encinas, M. A. Miranda, M. L. Navacchia and C. Chatgilialoglu, The photochemistry of 8-bromo-2-deoxyadenosine. A direct entry to cyclopurine lesions, *Photochem. Photobiol. Sci.*, 2004, **3**, 1042–1046.
4. M. M. Greenberg, Elucidating DNA damage and repair processes by independently generating reactive and metastable intermediates, *Org. Biomol. Chem.*, 2007, **5**, 18–30.

5. A. Manetto, D. Georganakis, T. Gimisis, L. Leondiadis, T. Carell and C. Chatgilialoglu, Independent Generation of C5′-Nucleosidyl Radicals in Thymidine and 2′-Deoxyguanosine, *J. Org. Chem.*, 2007, **72**, 3659–3666.
6. (a) A. Manetto, S. Breeger, C. Chatgilialoglu and T. Carell, Complex Sequence Dependence by Excess-Electron Transfer through DNA with Different Strength Electron Acceptors, *Angew. Chem., Int. Ed.*, 2006, **45**, 318–321; (b) M. L. Navacchia, A. Manetto, P. C. Montevecchi and C. Chatgilialoglu, Radical Cyclization Approach to Cyclonucleosides, *Eur. J. Org. Chem.*, 2005, 4640–4648.
7. P. C. Montevecchi, A. Manetto, M. L. Navacchia and C. Chatgilialoglu, Thermal decomposition of the tert-butyl perester of thymidine-5′-carboxylic acid. Formation and fate of the pseudo-C4′ radical, *Tetrahedron*, 2004, **60**, 4303–4308.
8. K. Randerath, G.-D. Zhou, R. L. Somers, J. H. Robbins and P. J. Brooks, A ^{32}P-Postlabeling Assay for the Oxidative DNA Lesion 8,5′-Cyclo-2′-deoxyadenosine in Mammalian Tissues. Evidence that Four Type II I-compounds are Dinucleotides Containing the Lesion in the 3′ Nucleotide, *J. Biol. Chem.*, 2001, **276**, 36051–36057.
9. (a) A. Romieu, D. Gasparutto and J. Cadet, Synthesis and Characterization of Oligonucleotides Containing 5′,8-Cyclopurine 2′-Deoxyribonucleosides: (5′R)-5′,8-Cyclo-2′-deoxyadenosine, (5′S)-5′,8-Cyclo-2′-deoxyguanosine, and (5′R)-5′,8-Cyclo-2′-deoxyguanosine, *Chem. Res. Toxicol.*, 1999, **12**, 412–421; (b) A. Romieu, D. Gasparutto, D. Molko and J. Cadet, Site-Specific Introduction of (5′S)-5′,8-Cyclo-2′-deoxyadenosine into oligodeoxyribonucleotides, *J. Org. Chem.*, 1998, **63**, 5245–5249; (c) E. Muller, D. Gasparutto and J. Cadet, Chemical Synthesis and Biochemical Properties of Oligonucleotides that Contain the (5′S,5S,6S)-5′,6-Cyclo-5-hydroxy-5,6-dihydro-2′-deoxyuridine DNA Lesion, *ChemBioChem*, 2002, **3**, 534–542; (d) E. Muller, D. Gasparutto, M. Jaquinod, A. Romieu and J. Cadet, Chemical and Biochemical Properties of Oligonucleotides that Contain (5′S,6S)-Cyclo-5,6-dihydro-2′-deoxyuridine and (5′S,6S)-Cyclo-5,6-dihydrothymidine, Two Main Radiation-Induced Degradation Products of Pyrimidine 2′-Deoxyribonucleosides, *Tetrahedron*, 2000, **56**, 8689–8701.
10. (a) T. Fukuyama, M. Kobayashi, Md. Taifur Rahman, N. Kamata and I. Ryu, Spurring radical reactions of organic halides with tin hydride and ttmss using microreactors, *Org. Lett.*, 2008, **10**, 533–536; (b) A. Sugimoto, M. Takagi, Y. Sumito, T. Fukuyama and I. Ryu, The Barton reaction using a microreactor and black light. Continuous flow synthesis of a key steroid intermediate for an endothelin receptor antagonist, *Tetrahedron Lett.*, 2006, **47**, 6197–6200.
11. M. L. Navacchia, P. C. Montevecchi and C. Chatgilialoglu, C5′-Adenosinyl Radical Cyclization. A Stereochemical Investigation, *J. Org. Chem.*, 2006, **71**, 4445–4452.

12. (a) J. Cadet and M. Berger, Radiation-induced decomposition of the purine bases within DNA and related model compounds, *Int. J. Radiat. Biol.*, 1985, **47**, 127–143; (b) N. Mariaggi, J. Cadet and R. Teoule, Cyclisation radicalaire de la desoxy-2'-adenosine en solution aqueuse, sous l'effet du rayonnement gamma, *Tetrahedron,* 1976, **32**, 2385–2387.

13. C. Chatgilialoglu, C. Ferreri and M. A. Tetzidis, Purine 5',8-cyclonucleoside lesions: chemistry and biology, *Chem. Soc. Rev.*, 2011, **40**, 1368–1382 and references cited therein.

14. (a) H. Mitsuya, K. J. Weinhold, P. A. Furman, M. H. St Clair, S. Nusinoff-Lehrman, R. C. Gallo, D. P. Bolognesi, D. W. Barry and S. Broder, 3'-Azido-3'-deoxythymidine (BW A509U): An Antiviral Agent That Inhibits the Infectivity and Cytopathic Effect of Human T-Lymphotropic Virus Type III/Lymphadenopathy- Associated Virus in vitro, *Proc. Natl. Acad. Sci. U. S. A.,* 1985, **82**, 7096–7100; (b) R. Yarchoan and S. Broder, Anti-retroviral therapy of AIDS and related disorders: General principles and specific development of dideoxynucleosides, *Pharmacol. Ther.*, 1989, **40**, 329–348.

15. E. DeClerq, Toward Improved Anti-HIV Chemotherapy: Therapeutic Strategies for Intervention with HIV Infections, *J. Med. Chem.*, 1995, **38**, 2491–2517 and references cited therein.

16. (a) X. Tan, C. K. Chu and F. D. Boudinot, Development and optimization of anti-HIV nucleoside analogs and prodrugs: A review of their cellular pharmacology, structure-activity relationships and pharmacokinetics, *Adv. Drug Delivery Rev.*, 1999, **39**, 117–151 and references cited therein; (b) M. J. Bamford, P. L. Coe and R. T. Walker, Synthesis and antiviral activity of some 3'-C-difluoromethyl- and 3'-deoxy-3'-C-fluoromethyl nucleosides, *J. Med. Chem.*, 1990, **33**, 2488–2494; (c) M. K. Gurjar and K. Maheshwar, Stereoselective Synthesis of a Novel Carbocyclic Nucleoside, *J. Org. Chem.*, 2001, **66**, 7552–7554; (d) M. A. Crimmins, New developments in the enantioselective synthesis of cyclopentyl carbocyclic nucleosides, *Tetrahedron,* 1998, **54**, 9229–9272; (e) T. S. Lin, M. Z. Luo, M. C. Liu, S. B. Pai, G. E. Dutschman and Y. C. Cheng, Synthesis and Biological Evaluation of 2',3'-Dideoxy-L-pyrimidine Nucleosides as Potential Antiviral Agents against Human Immunodeficiency Virus (HIV) and Hepatitis B Virus (HBV), *J. Med. Chem.*, 1994, **37**, 798–803

17. M. Ikehara and M. Hiioto, Studies of Nucleosides and Nucleotides. LXXXII. Cyclonucleosides. Synthesis and Properties of 2'-Halogeno-2'-deoxyadenosines, *Chem. Pharm. Bull.,* 1978, **26**, 2449.

18. R. Ranganathan and D. Larwood, Facile conversion of adenosine into new 2'-substituted-2'-deoxy-arabinofuranosyladenine derivatives: Stereospecific syntheses of 2'-azido-2'-deoxy-,2'-amino-2'- deoxy-, and 2'-mercapto-2'-deoxy-β-d-arabinofuranosyladenines, *Tetrahedron Lett.*, 1978, 4341.

19. A. Mieczkowski, V. Roy and A. Agrofoglio, Preparation of Cyclonucleosides, *Chem. Rev.*, 2010, **110**, 1828–1856 and references cited therein.
20. X.-F. Zhu, The Latest Progress in the Synthesis of Carbocyclic Nucleosides, *Nucleosides, Nucleotides Nucleic Acids*, 2000, **19**(3), 651–690.
21. L. Colla, P. Herdewijn, E. de Clercq, J. Balzarini and H. Vanderhaeghe, *Eur. J. Med. Chem.*, 1985, **20**, 295.
22. M. Hirata, Studies on Nucleosides and Nucleotides. IX. Nucleophilic Substitution of Secondary Sulfonyloxy Groups of Pyrimidine Nucleosides. II. Reaction of 2,2'-Anhydro-1-(3'-O-tosyl-β-D-arabinofuranosyl) uracil with Sodium Bromide, Sodium Ethanethiol, and Sodium Azide, *Chem. Pharm. Bull.*, 1968, **16**, 291.
23. T. Naito, M. Hirata, Y. Nakai, T. Kobayashi and M. Kanao, Studies on Nucleosides and Nucleotides. VIII. Nucleophilic Substitution of Secondary Sulfonyloxy Groups of Pyrimidine Nucleosides, *Chem. Pharm. Bull.*, 1968, **16**, 285.
24. J.-C. Wu, Z. Xi, C. Gioeli and J. Chattopadhyaya, Intramolecular cyclization-trapping of carbon radicals by olefins as means to functionalize 2'- and 3'-carbon in b-D-nucleosides, *Tetrahedron*, 1991, **47**, 2237–2254.
25. (a) M. M. Greenberg, Elucidating DNA damage and repair processes by independently generating reactive and metastable intermediates, *Org. Biomol. Chem.*, 2007, **5**, 18–30; (b) A. Adhikary, A. Y. S. Malkhasian, S. Collins, J. Koppen, D. Becker and M. D. Sevilla, UV-visible photoexcitation of guanine radical cations produces sugar radicals in DNA and model structures, *Nucleic Acids Res.*, 2005, **33**, 5553–5564; (c) A. Manetto, D. Georganakis, T. Gimisis, L. Leondiadis, T. Carell and C. Chatgilialoglu, Independent Generation of C5'-Nucleosidyl Radicals in Thymidine and 2'-Deoxyguanosine, *J. Org. Chem.*, 2007, **72**, 3659–3666; (d) M. L. Navacchia, A. Manetto, P. C. Montevecchi and C. Chatgilialoglu, Radical Cyclization Approach to Cyclonucleosides, *Eur. J. Org. Chem.*, 2005, 4640–4648.
26. S. Takamatsu, S. Katayama, N. Hirose, M. Naito and I. Kunisuke, Radical deoxygenation of nucleoside derivatives with hypophosphorous acid and dialkyl phosphites, *Tetrahedron Lett.*, 2001, **42**, 7605–7608.
27. (a) *Radicals in Organic Synthesis*, ed. P. Renaud and M. P. Sibi, Wiley-VCH, Weinheim, 2001, vol. 1; (b) A. Postigo, C. Ferreri, M. L. Navacchia and C. Chatgilialoglu, The radical-based reduction with $(TMS)_3SiH$ on water, *Synlett*, 2005, 2854–2856; (c) A. Postigo, C. Ferreri, M. L. Navacchia and C. Chatgilialoglu, The radical-based reduction with $(TMS)_3SiH$ on water, *Synlett*, 2005, 2854–2856; (d) C. Chatgilialoglu, C. Ferreri, Q. C. Mulazzani, M. Ballestri and L. Landi, *Cis-trans* isomerization of monounsaturated fatty acid residues in phospholipids by thiyl radicals, *J. Am. Chem. Soc.*, 2000, **122**, 4593–4601; (e) T. Yorimitsu, H. Nakamura, K. Shinokubo, K. Oshima and H. Fujimoto, Powerful solvent effect of water in radical reaction: triethylborane-induced atom-

transfer radical cyclization in water, *J. Am. Chem. Soc.*, 2000, **122**, 11041–11047.

28. (a) T. Fukuyama, M. Kobayashi, Md. Taifur Rahman, N. Kamata and I. Ryu, Spurring radical reactions of organic halides with tin hydride and ttmss using microreactors, *Org. Lett.*, 2008, **10**, 533–536; (b) A. Sugimoto, M. Takagi, Y. Sumito, T. Fukuyama and I. Ryu, The Barton reaction using a microreactor and black light. Continuous flow synthesis of a key steroid intermediate for an endothelin receptor antagonist, *Tetrahedron Lett.*, 2006, **47**, 6197–6200.

29. (a) V. T. Perchyonok and I. N. Lykakis, Radical reactions in aqueous media: origins, reason and applications, *Curr. Org. Chem.*, 2009, **13**, 573; (b) V. T. Perchyonok, I. N. Lykakis and K. L. Tuck, Recent advances in C−H bond formation in aqueous media: a mechanistic perspective, *Green Chem.*, 2008, **10**, 153; (c) V. T. Perchyonok, K. L. Tuck, S. J. Langford and M. W. Hearn, On the scope of radical reactions in aqueous media utilizing quaternary ammonium salts of phosphinic acids as chiral and achiral hydrogen donors, *Tetrahedron Lett.*, 2008, **49**, 4777; (d) C.-J. Li and L. Chen, Organic chemistry in water, *Chem. Soc. Rev.*, 2006, **35**, 68 and references therein; (e) H. Yorimitsu, T. Nakamura, H. Shinokubo, K. Oshima, K. Omoto and H. Fujimoto, Powerful solvent effect of water in radical reaction: triethylborane-induced atom-transfer radical cyclization in water, *J. Am. Chem. Soc.*, 2000, **122**, 11041. For an interesting example of the use of supercritical CO_2, see: (f) S. Hadida, M. S. Super, E. J. Beckman and D. P. Curran, Radical reactions with alkyl and fluoroalkyl (fluorous) tin hydride reagents in supercritical CO_2, *J. Am. Chem. Soc.*, 1997, **119**, 7406; see also: (g) J. M. Tanko, Free-Radical Chemistry in Supercritical Carbon Dioxide, in *Green Chemistry using Liquid and Supercritical Carbon Dioxide*, ed. J. M. DeSimone and W. Tumas, Oxford University Press, New York, 2003, ch. 4, p. 64; (h) I. N. Lykakis and V. T. Perchyonok, Thiols as an efficient hydrogen atom donor in free radical transformations in aqueous media, *Curr. Org. Chem.*, 2010, in print; (i) A. E. Johnson and V. T. Perchyonok, Recent Advances in Free Radical Chemistry in unconventional medium: ionic liquids, microwaves and solid state to the rescue, review article, *Curr. Org. Chem.*, 2009, **13**(17), in print; (j) A. Postigo, S. Kopsov, S. S. Zlotsky, C. Ferreri and C. Chatgilialoglu, Hydrosilylation of C-C multiple bonds using $(Me_3Si)_3SiH$ in water. Comparative study of the radical initiation step, *Organometallics,* 2009, **28**, 3282.

30. T. Gimisis, G. Ialongo and C. Chatgilialoglu, C., Generation of C-1′ Radicals through a b-(Acyloxy)alkyl Rearrangements in Modified Purine and Pyrimidine Nucleosides, *Tetrahedron*, 1998, **54**, 573–592.

31. (a) Y. Xu and H. Sugiyama, Photochemical Approach to Probing Different DNA Structures, *Angew. Chem., Int. Ed.*, 2006, **45**, 1354–1362; (b) B. Halliwell and O. I. Aruoma, *DNA and Free Radicals*, Ellis Horwood, Chichester, 1993; (c) P. O'Neill and E. M. Fielden, Primary free

radical processes in DNA, *Adv. Radiat. Biol.*, 1993, **17**, 53–120; (d) G. Pratviel, J. Bernadou and B. Meunier, Carbon–Hydrogen Bonds of DNA Sugar Units as Targets for Chemical Nucleases and Drugs, *Angew. Chem., Int. Ed. Engl.*, 1995, **34**, 746–769; (e) J. Cadet, T. Douki, D. Gasparutto and J.-L. Ravanat, Radiation-induced damage to cellular DNA: measurement and biological role, *Radiat. Phys. Chem.*, 2005, **72**, 293–299; (f) J. Cadet, T. Douki and J.-L. Ravanat, in *Redox-Genome Interactions in Health and Disease,* ed. J. Fuchs, M. Podda and L. Packer, Dekker, New York, 2003, pp. 143–189; (g) C. Chatgilialoglu and P. O'Neill, Free radicals associated with DNA damage, *Exp. Gerontol.*, 2001, **36**, 1459–1471.

32. (a) L. Pardo, J. T. Banfelder and R. Osman, Theoretical studies of the kinetics, thermochemistry, and mechanism of hydrogen-abstraction from methanol and ethanol, *J. Am. Chem. Soc.*, 1992, **114**, 2382–2390; (b) A.-O. Colson and M. D. Sevilla, Structure and Relative Stability of Deoxyribose Radicals in a Model DNA Backbone: Ab Initio Molecular Orbital Calculations, *J. Phys. Chem.*, 1995, **99**, 3867–3874; (c) K. Miaskiewicz and R. Osman, Theoretical study on the deoxyribose radicals formed by hydrogen abstraction, *J. Am. Chem. Soc.*, 1994, **116**, 232–238.

33. (a) M. M. Greenberg, Elucidating DNA damage and repair processes by independently generating reactive and metastable intermediates, *Org. Biomol. Chem.*, 2007, **5**, 18–30; (b) A. Adhikary, A. Y. S. Malkhasian, S. Collins, J. Koppen, D. Becker and M. D. Sevilla, UV-visible photo-excitation of guanine radical cations produces sugar radicals in DNA and model structures, *Nucleic Acids Res.*, 2005, **33**, 5553–5564; (c) A. Manetto, D. Georganakis, T. Gimisis, L. Leondiadis, T. Carell and C. Chatgilialoglu, Independent Generation of C5'-Nucleosidyl Radicals in Thymidine and 2'-Deoxyguanosine, *J. Org. Chem.*, 2007, **72**, 3659–3666 (d) M. L. Navacchia, A. Manetto, P. C. Montevecchi and C. Chatgilialoglu, Radical Cyclization Approach to Cyclonucleosides, *Eur. J. Org. Chem.*, 2005, 4640–4648.

34. S. Takamatsu, S. Katayama, N. Hirose, M. Naito and I. Kunisuke, Radical deoxygenation of nucleoside derivatives with hypophosphorous acid and dialkyl phosphites, *Tetrahedron Lett.*, 2001, **42**, 7605–7608.

35. (a) *Radicals in Organic Synthesis*, ed. P. Renaud and M. P. Sibi, Wiley-VCH, Weinheim, 2001, vol. 1; (b) A. Postigo, C. Ferreri, M. L. Navacchia and C. Chatgilialoglu, The radical-based reduction with $(TMS)_3SiH$ on water, *Synlett*, 2005, 2854–2856; (c) A. Postigo, C. Ferreri, M. L. Navacchia and C. Chatgilialoglu, The radical-based reduction with $(TMS)_3SiH$ on water, *Synlett*, 2005, 2854–2856; (d) C. Chatgilialoglu, C. Ferreri, Q. C. Mulazzani, M. Ballestri and L. Landi, *Cis-trans* isomeri-zation of monounsaturated fatty acid residues in phospholipids by thiyl radicals, *J. Am. Chem. Soc.,* 2000, **122**, 4593–4601; (e) T. Yorimitsu, H. Nakamura, K. Shinokubo, K. Oshima and H. Fujimoto, Powerful solvent effect of water in radical reaction: triethylborane-induced atom-

transfer radical cyclization in water, *J. Am. Chem. Soc.*, 2000, **122**, 11041–11047.

36. T. Fukuyama, M. Kobayashi, Md. Taifur Rahman, N. Kamata and I. Ryu, Spurring radical reactions of organic halides with tin hydride and ttmss using microreactors, *Org. Lett.*, 2008, **10**, 533–536; (b) A. Sugimoto, M. Takagi, Y. Sumito, T. Fukuyama and I. Ryu, The Barton reaction using a microreactor and black light. Continuous flow synthesis of a key steroid intermediate for an endothelin receptor antagonist, *Tetrahedron Lett.*, 2006, **47**, 6197–6200.

37. (a) V. T. Perchyonok and I. N. Lykakis, Radical reactions in aqueous media: origins, reason and applications, *Curr. Org. Chem.*, 2009, **13**, 573; (b) V. T. Perchyonok, I. N. Lykakis and K. L. Tuck, Recent advances in C−H bond formation in aqueous media: a mechanistic perspective, *Green Chem.*, 2008, **10**, 153; (c) V. T. Perchyonok, K. L. Tuck, S. J. Langford and M. W. Hearn, On the scope of radical reactions in aqueous media utilizing quaternary ammonium salts of phosphinic acids as chiral and achiral hydrogen donors, *Tetrahedron Lett.*, 2008, **49**, 4777; (d) C.-J. Li and L. Chen, Organic chemistry in water, *Chem. Soc. Rev.*, 2006, **35**, 68 and references therein; (e) H. Yorimitsu, T. Nakamura, H. Shinokubo, K. Oshima, K. Omoto and H. Fujimoto, Powerful solvent effect of water in radical reaction: triethylborane-induced atom-transfer radical cyclization in water, *J. Am. Chem. Soc.*, 2000, **122**, 11041. For an interesting example of the use of supercritical CO_2, see: (f) S. Hadida, M. S. Super, E. J. Beckman and D. P. Curran, Radical reactions with alkyl and fluoroalkyl (fluorous) tin hydride reagents in supercritical CO_2, *J. Am. Chem. Soc.*, 1997, **119**, 7406; see also: (g) J. M. Tanko, Free-Radical Chemistry in Supercritical Carbon Dioxide, in *Green Chemistry using Liquid and Supercritical Carbon Dioxide*, ed. J. M. DeSimone and W. Tumas, Oxford University Press, New York, 2003, ch. 4, p. 64; (h) I. N. Lykakis and V. T. Perchyonok, Thiols as an efficient hydrogen atom donor in free radical transformations in aqueous media, *Curr. Org. Chem.*, 2010, in print; (i) A. E. Johnson and V. T. Perchyonok, Recent Advances in Free Radical Chemistry in unconventional medium: ionic liquids, microwaves and solid state to the rescue, review article, *Curr. Org. Chem.*, 2009, **13**(17), in print; (j) A. Postigo, S. Kopsov, S. S. Zlotsky, C. Ferreri and C. Chatgilialoglu, Hydrosilylation of C-C multiple bonds using $(Me_3Si)_3SiH$ in water. Comparative study of the radical initiation step, *Organometallics,* 2009, **28**, 3282.

38. C. Chatgilialoglu and T. Gimisis, Fate of the C-1 peroxyl radical in the 2-deoxyuridine system, *Chem. Commun.*, 1998, 1249–1250.

39. (a) C.-H. Fawcett, R. D. Firu and D. M. Spencer, *Physiol. Plant Pathol.,* 1971, **1**, 163; (b) K. Imai, *J. Pharm. Soc. Jpn.,* 1956, **76**, 405; (c) C. Chowdhury and N. G. Kundu, *Tetrahedron,* 1999, **55**, 7011; (d) C. A. Quesnelle, P. Gill, M. Dodier, D. St Laurent, M. Serrano-Wu, A. Marinier, A. Martel, C. E. Mazzucco, T. M. Stickle, J. F. Barrett,

D. M. Vyas and B. N. Balasubramanian, *Bioorg. Med. Chem. Lett.,* 2003, **13**, 519.

40. (a) V. N. Kalinin, M. V. Shostakovsky and A. B. Ponamaryov, *Tetrahedron Lett.,* 1990, **31**, 4073; (b) P. G. Ciattini, E. Morera, G. Ortar and S. S. Rossi, *Tetrahedron,* 1991, **47**, 6449; (c) S. Torri, H. Okumoto, L.-H. Xu, M. Sadakane, M. V. Shostakovsky, A. B. Ponomaryov and V. N. Kalinin, *Tetrahedron,* 1993, **49**, 6773; (d) D. Bernard, C. Daniel and L. Robert, *Tetrahedron Lett.,* 1996, **37**, 1019; (e) J. Marco-Contelles and E. de Opazo, *J. Org. Chem.,* 2002, **67**, 3705; (f) C. J. Forsyth, J. Xu, S. T. Nguyen, I. A. Samdai, L. R. Briggs, T. Rundberget, M. Sandvik and C. O. Miles, *J. Am. Chem. Soc.,* 2006, **128**, 15114; (g) L. F. Tietze, R. R. Singidi, K. M. Gericke, H. Bockemeier and H. Laatsch, *Eur. J. Org. Chem.,* 2007, 5875.

41. (a) H. Sheng, S. Lin and Y. Huang, *Tetrahedron Lett.,* 1986, **27**, 4893; (b) S. Torii, H. Okumoto, L. H. Xu, M. Sadakane, M. V. Shostakovsky, A. B. Ponomaryov and V. N. Kalinin, *Tetrahedron*, 1993, **49**, 6773; (c) S. Mohamed, A. Mohamed, K. Kobayashi and A. Mori, *Org. Lett.,* 2005, **7**, 4487; (d) A. Aradi, M. Aschi, F. Marinelli and M. Verdecchia, *Tetrahedron,* 2008, **64**, 5354; (e) P. Bannwarth, A. Valleix, D. Gree and R. Gree, *J. Org. Chem.,* 2009, **74**, 4646.

42. (a) R. B. Davis and D. H. Scheiber, *J. Am. Chem. Soc.,* 1956, **78**, 1675; (b) J. F. Normant, *Synthesis,* 1972, 63; (c) M. W. Logue and G. L. Moore, *J. Org. Chem.,* 1975, **40**, 131; (d) U. Schmidt and M. Schwochau, *Chem. Ber.,* 1964, **97**, 1649; (e) L. Birkofer, A. Ritter and H. Uhlenbrauck, *Chem. Ber.,* 1963, **96**, 3280; (f) D. R. M. Walton and F. Waugh, *J. Organomet. Chem.,* 1972, **37**, 45; (g) M. W. Logue and K. Teng, *J. Org. Chem.,* 1982, **47**, 2549.

43. (a) T. Kobayashi and M. Tanaka, *J. Chem. Soc., Chem. Commun.,* 1981, 333; (b) L. Delaude, A. M. Masdeu and H. Alper, *Synthesis,* 1994, 1149; (c) P. G. Ciattini, E. Morera and G. Ortar, *Tetrahedron Lett.,* 1991, **32**, 6449; (d) M. S. Mohamed Ahmed and A. Mori, *Org. Lett.,* 2003, **5**, 3057; (e) B. Liang, M. Huang, Z. You, Z. Xiong, K. Lu, R. Fathi, J. Chen and Z. Yang, *J. Org. Chem.,* 2005, **70**, 6097; (f) M. Iizuka and Y. Kondo, *Eur. J. Org. Chem.,* 2007, 5180; (g) J. Liu, J. Chen and C. Xia, *J. Catal.,* 2008, **253**, 50.

44. For recyclable reaction media and catalysts, see: (a) T. Fukuyama, R. Yamaura and I. Ryu, *Can. J. Chem.,* 2005, **83**, 711; (b) M. T. Rahman, T. Fukuyama, N. Kamata, M. Sato and I. Ryu, *Chem. Commun.,* 2006, 2236.

45. (a) M. Eckhardt and G. C. Fu, *J. Am. Chem. Soc.,* 2003, **125**, 13642. Also see the following reviews: (b) A. C. Frisch and M. Beller, *Angew. Chem., Int. Ed.,* 2005, **44**, 674; (c) M. R. Netherton and G. C. Fu, *Adv. Synth. Catal.,* 2004, **346**, 1525; (d) A. Rudolph and L. Lautens, *Angew. Chem., Int. Ed.,* 2009, **48**, 2656.

46. For previous efforts on metal-catalyzed carbonylation under photoirradiation conditions, see: (a) T. Kondo, Y. Sone, Y. Tsuji and Y. Watanabe,

J. Organomet. Chem., 1994, **473**, 163; (b) T. Ishiyama, M. Murata, A. Suzuki and N. Miyaura, *J. Chem. Soc., Chem. Commun.,* 1995, 295.

47. For atom transfer carbonylation reactions, see: (a) I. Ryu, K. Nagahara, N. Kambe, N. Sonoda, S. Kreimerman and M. Komatsu, *Chem. Commun.,* 1998, 1953; (b) K. Nagahara, I. Ryu, M. Komatsu and N. Sonoda, *J. Am. Chem. Soc.,* 1997, **119**, 5465; (c) S. Kreimerman, I. Ryu, S. Minakata and M. Komatsu, *Org. Lett.,* 2000, **2**, 389. Also see a review: (d) I. Ryu, *Chem. Soc. Rev.,* 2001, **30**, 16.

48. For radical/metal hybrid reactions, see: (a) I. Ryu, S. Kreimerman, F. Araki, S. Nishitani, S. Oderaotoshi, S. Minakata and M. Komatsu, *J. Am. Chem. Soc.,* 2002, **124**, 3812; (b) T. Fukuyama, S. Nishitani, T. Inouye, K. Morimoto and I. Ryu, *Org. Lett.,* 2006, **8**, 1383; (c) T. Fukuyama, T. Inouye and I. Ryu, *J. Organomet. Chem.,* 2007, **692**, 685. Also see a review: (d) I. Ryu, *Chem. Rec.,* 2002, **2**, 249.

49. Acceleration by water may be concerned with the fact that in the absence of water the reaction mixture became a suspension due to the gradually formed amine–HI salt, which is supposed to prevent effective photo-irradiation. In the presence of water, we observed that a clear solution was maintained during the photoirradiation reaction.

50. (a) V. W. Bowry and K. U. Ingold, *J. Am. Chem. Soc.,* 1991, **113**, 5699; (b) M. Newcomb, *Tetrahedron,* 1993, **49**, 1151.

51. For precedents of electron transfer from low-valent palladium and platinum complexes, see: (a) A. V. Kramer and J. A. Osborn, *J. Am. Chem. Soc.,* 1974, **96**, 7832; (b) A. V. Kramer, J. A. Labinger, J. S. Bradley and J. A. Osborn, *J. Am. Chem. Soc.,* 1974, **96**, 7145. Miyaura and co-workers also proposed an electron transfer from Pd(0) to alkyl iodides to afford a radical pair (Pd(I) + alkylradical); see ref. 8*b*.

52. For the photogeneration of a Pd radical from a Pd dimer complex, see: F. R. Lemke and C. P. Kubiak, *J. Organomet. Chem.,* 1989, **373**, 391.

53. It would be also possible to assume that ˙C(O)Pd(II)I might exist as a persistent radical. For the persistent radical effect, see the following reviews: (a) H. Fischer, *Chem. Rev.,* 2001, **101**, 3581; (b) A. Studer, *Chem.– Eur. J.,* 2001, **7**, 1159; (c) A. Studer, *Chem. Soc. Rev.,* 2004, **33**, 267; (d) A. Studer and T. Schulte, *Chem. Rec.,* 2005, **5**, 27. Also see: (e) K. S. Focsaneanu, C. Aliaga and J. C. Scaiano, *Org. Lett.,* 2005, **7**, 4979.

54. (a) H. Bienayme, C. Hulme, G. Oddon and P. Schmitt, *Chem.–Eur. J.,* 2000, **6**, 3321; (b) *Multicomponent Reactions,* ed. J. Zhu and H. Bienayme, Wiley-VCH, Weinheim, 2005.

55. For radical multicomponent reactions, see: (a) *Multicomponent Reactions,* ed. J. Zhu and H. Bienayme, Wiley-VCH, Weinheim, 2005, pp. 169–198. Also see reviews: (b) M. Malacria, *Chem. Rev.,* 1996, **96**, 289; (c) E. Godineau and Y. Landais, *Chem.–Eur. J.,* 2009, **15**, 3044.

56. For reviews on radical carbonylation reactions using carbon monoxide, see: (a) I. Ryu and N. Sonoda, *Angew. Chem., Int. Ed.,* 1996, **35**, 1050; (b) I. Ryu, N. Sonoda and D. P. Curran, *Chem. Rev.,* 1996, **96**, 177. For a

review on acyl radial chemistry, see: (c) C. Chatgilialoglu, D. Crich, M. Komatsu and I. Ryu, *Chem. Rev.,* 1999, **99**, 1991.

57. For atom-transfer carbonylations, see: (a) K. Nagahara, I. Ryu, M. Komatsu and N. Sonoda, *J. Am. Chem. Soc.,* 1997, **119**, 5465; (b) I. Ryu, K. Nagahara, N. Kambe, N. Sonoda, S. Kreimerman and M. Komatsu, *Chem. Commun.,* 1998, 1953; (c) S. Kreimerman, I. Ryu, S. Minakata and M. Komatsu, *Org. Lett.,* 2000, **2**, 389; (d) I. Ryu, T. Niguma, S. Minakata and M. Komatsu, *Tetrahedron Lett.,* 1997, **38**, 7883; (e) S. Kobayashi, T. Kawamoto, S. Uehara, T. Fukuyama and I. Ryu, *Org. Lett.,* 2010, **12**, 1548. Also see a review: (f) I. Ryu, *Chem. Soc. Rev.,* 2001, **30**, 16.

58. For previous efforts on metal-catalyzed carbonylation under photoirradiation conditions, see: (a) T. Kondo, Y. Sone, Y. Tsuji and Y. Watanabe, *J. Organomet. Chem.,* 1994, **473**, 163; (b) T. Ishiyama, M. Murata, A. Suzuki and N. Miyaura, J. Chem. Soc., *Chem. Commun.,* 1995, 295.

59. (a) I. Ryu, S. Kreimerman, F. Araki, S. Nishitani, S. Oderaotoshi, S. Minakata and M. Komatsu, *J. Am. Chem. Soc.,* 2002, **124**, 3812; (b) T. Fukuyama, S. Nishitani, T. Inouye, K. Morimoto and I. Ryu, *Org. Lett.,* 2006, **8**, 1383; (c) T. Fukuyama, T. Inouye and I. Ryu, *J. Organomet. Chem.,* 2007, **692**, 685; (d) A. Fusano, T. Fukuyama, T. Nishitani, T. Inouye and I. Ryu, *Org. Lett.,* 2010, **12**, 2410; (e) I. Ryu, *Chem. Rec.*, 2002, **2**, 249.

60. (a) Bo. Liang, J. Liu, Y. X. Gao, K. Wongkhan, D. X. Shu, Y. Lan, A. Li, A. S. Batsanov, J. A. H. Howard, T. B. Marder, J. H. Chen and Z. Yang, *Organometallics,* 2007, **26**, 4756; (b) M. Dai, C. Wang, G. Dong, J. Xiang, T. Luo, B. Liang, J. Chen and Z. Yang, *Eur. J. Org. Chem.,* 2003, 4346; (c) Y. Yamamoto, H. Maekawa, S. Goda and I. Nishiguchi, *Org. Lett.,* 2003, **5**, 2755; (d) T. Yokota, S. Sakaguchi and Y. Ishii, *J. Org. Chem.,* 2002, **67**, 5005; (e) P. Brechot, Y. Chauvin, D. Commereuc and L. Saussine, *Organometallics,* 1990, **9**, 26; (f) Y. Tamaru, M. Hojo, H. Higashimura and Z. Yoshida, *J. Am. Chem. Soc.,* 1988, **110**, 3994.

61. (a) H. Urata, Y. Kinoshita, T. Asanuma, O. Kosukegawa and T. Fuchikami, *J. Org. Chem.,* 1991, **56**, 4996; (b) J. Tsuji, K. Sato and H. Nagashima, *Tetrahedron Lett.,* 1982, **23**, 893; (c) J. Tsuji, K. Sato and H. Nagashima, *Tetrahedron,* 1985, **41**, 5003.

62. I. Ryu, H. Muraoka, N. Kambe, M. Komatsu and N. Sonoda, *J. Org. Chem.,* 1996, **61**, 6396.

63. When the reaction was performed without DMAP, the yield of diester **4a** dropped to 53%. In this case, the iodoalkane **5** arising from the addition of **1a** to **2a** was also obtained in 24% yield.

64. For examples of an electron transfer from low-valent palladium or platinum complexes to iodoalkanes, see: (a) A. V. Kramer and J. A. Osborn, *J. Am. Chem. Soc.,* 1974, **96**, 7832; (b) A. V. Kramer, J. A. Labinger, J. S. Bradley and J. A. Osborn, *J. Am. Chem. Soc.,* 1974, **96**,

7145; (c) P. Knochel and G. Manolikakes, *Angew. Chem., Int. Ed.,* 2009, **48**, 205. Also see ref. 6*b*.

65. For SET-induced radical reactions of perfluoroalkyl iodides, see: (a) Q. Y. Chen, Z. Y. Yang, C. X. Zhao and Z. M. Qiu, *J. Chem. Soc., Perkin Trans. 1,* 1988, 563; (b) Z. M. Qiu and D. J. Burton, *J. Org. Chem.,* 1995, **60**, 5570; (c) M. Yoshida and M. Iizuka, *J. Fluorine Chem.,* 2009, **130**, 926.

66. For reviews on the persistent radical effect, see: (a) H. Fischer, *Chem. Rev.,* 2001, **101**, 3581; (b) A. Studer, *Chem.–Eur. J.,* 2001, **7**, 1159; (c) A. Studer, *Chem. Soc. Rev.,* 2004, **33**, 267; (d) A. Studer and T. Schulte, *Chem. Rec.,* 2005, **5**, 27. Also see: (e) K. S. Focsaneanu, C. Aliaga and J. C. Scaiano, *Org. Lett.,* 2005, **7**, 4979.

67. For photogeneration of a Pd radical from a Pd dimer complex, see: F. R. Lemke and C. P. Kubiak, *J. Organomet. Chem.,* 1989, **373**, 391.

68. For recent reviews, see; (a) K. Geyer, T. Gustafsson and P. H. Seeberger, *Synlett,* 2009, 2382–2391; (b) J. Yoshida, A. Nagaki and T. Yamada, *Chem.–Eur. J.,* 2008, **14**, 7450–7459; (c) B. K. Singh, N. Kaval, S. Tomar, E. Van der Eycken and V. S. Parmar, *Org. Process Res. Dev.,* 2008, **12**, 468–474; (d) P. Watts and C. Wiles, *Org. Biomol. Chem.,* 2007, **5**, 727–732; (e) B. Ahmed-Omer, J. C. Brandt and T. Wirth, *Org. Biomol. Chem.,* 2007, **5**, 733–740; (f) K. Geyer, J. D. C. Codée and P. H. Seeberger, *Chem.–Eur. J.,* 2006, **12**, 8434–8442; (g) *Microreactors in Organic Synthesis and Catalysis*, ed. T. Wirth, Wiley-VCH, Weinheim, 2008.

69. For recent work on organic synthesis using microreactors, see; (a) M. W. Bedore, N. Zaborenko, K. F. Jensen and T. F. Jamison, *Org. Process Res. Dev.,* 2010, **14**, 432–440; (b) A. Herath, R. Dahl and N. D. P. Cosford, *Org. Lett.,* 2010, **12**, 412–415; (c) M. Damm, T. N. Glasnov and C. O. Kappe, *Org. Process Res. Dev.,* 2010, **14**, 215–224; (d) A. R. Bogdan, S. L. Poe, D. C. Kubis, S. J. Broadwater and D. T. McQuade, *Angew. Chem., Int. Ed.,* 2009, **48**, 8547–8550; (e) A. Nagaki, H. Kim and J. Yoshida, *Angew. Chem., Int. Ed.,* 2009, **48**, 8063–8065; (f) J. Jin, M. Cai and J. Li, *Synlett,* 2009, 2534–2538; (g) L. Kong, Q. Lin, X. Lv, Y. Yang, Y. Jia and Y. Zhou, Green Chem., 2009, **11**, 1108–1111; (h) A. Nagaki, E. Takizawa and J. Yoshida, *Chem. Lett.,* 2009, **38**, 486–487; (i) B. Ahmed-Omer, D.A. Barrow and T. Wirth, *Tetrahedron Lett.,* 2009, **50**, 3352–3355; (j) Y. Lan, M. Zhang, W. Zhang and L. Yang, *Chem.–Eur. J.,* 2009, **15**, 3670–3673; (k) A. Odedra and P. H. Seeberger, *Angew. Chem., Int. Ed.,* 2009, **48**, 2699–2702; (l) C. B. McPake, C. B. Murray and G. Sandford, *Tetrahedron Lett.,* 2009, **50**, 1674–1676; (m) M. Baumann, I. R. Baxendale and S. V. Ley, *Synlett,* 2008, 2111–2114.

70. For a review of our work, see T. Fukuyama, M. T. Rahman, M. Sato and I. Ryu, *Synlett,* 2008, 151–163.

71. For our recent work on Pd-catalyzed coupling reactions using a micro-reactor, see: (a) T. Fukuyama, M. Shinmen, S. Nishitani, M. Sato and I. Ryu, *Org. Lett.,* 2002, **4**, 1691–1694; (b) S. Liu, T. Fukuyama, M. Sato and I. Ryu, *Org. Process Res. Dev.,* 2004, **8**, 477–481; (c) M. T. Rahman,

T. Fukuyama, N. Kamata, M. Sato and I. Ryu, *Chem. Commun.,* 2006, 2236–2238.

72. T. Fukuyama, M. Kobayashi, M.T. Rahman, N. Kamata and I. Ryu, *Org. Lett.,* 2008, **10**, 533–536.

73. T. Fukuyama, Y. Hino, N. Kamata and I. Ryu, *Chem. Lett.,* 2004, **33**, 1430–1431.

74. For reviews on photo-reactions using a microreactor, see; (a) Y. Matsushita, T. Ichimura, N. Ohba, S. Kumada, K. Sakeda, T. Suzuki, H. Tanibata and T. Murata, *Pure Appl. Chem.,* 2007, **79**, 1959–1968; (b) E. E. Coyle and M. Oelgemöller, *Photochem. Photobiol. Sci.,* 2008, **7**, 1313–1322.

75. For recent work on photoreactions using microreactors, see; (a) Y. Matsushita, M. Iwasawa, T. Suzuki and T. Ichimura, *Chem. Lett.,* 2009, **38**, 846–847; (b) A. Ouchi, H. Sakai, T. Oishi, M. Kaneda, T. Suzuki, A. Saruwatari and T. Obata, *J. Photochem. Photobiol. A,* 2008, **199**, 261–266; (c) Y. Matsushita, N. Ohba, T. Suzuki and T. Ichimura, *Catal. Today,* 2008, **132**, 153–158; (d) H. Mukae, H. Maeda, S. Nashihara and K. Mizuno, *Bull. Chem. Soc. Jpn.,* 2007, **80**, 1157–1161; (e) H. Maeda, H. Mukae and K. Mizuno, *Chem. Lett.,* 2005, **34**, 66–67.

76. (a) A. Sugimoto, Y. Sumino, M. Takagi, T. Fukuyama and I. Ryu, *Tetrahedron Lett.,* 2006, **47**, 6197–6200; (b) A. Sugimoto, T. Fukuyama, Y. Sumino, M. Takagi and I. Ryu, *Tetrahedron,* 2009, **65**, 1593–1598.

77. K. Tsutsumi, K. Terao, H. Yamaguchi, S. Yoshimura, T. Morimoto, K. Kakiuchi, T. Fukuyama and I. Ryu, *Chem. Lett.,* 2010, **39**, 828–829.

78. For chlorination of alkanes with molecular chlorine; (a) K. U. Ingold, J. Lusztyk and K. D. Raner, *Acc. Chem. Res.,* 1990, **23**, 219–225, and references cited therein; (b) B. Fletcher, N. K. Suleman and J. M. Tanko, *J. Am. Chem. Soc.,* 1998, **120**, 11839–11844; (c) N. Sun and K. J. Klabunde, *J. Am. Chem. Soc.,* 1999, **121**, 5587–5588.

79. For radical chlorination of alkanes with sulfuryl chloride; (a) M. S. Kharasch, H. C. Brown and A. B. Ash, *J. Am. Chem. Soc.,* 1939, **61**, 2142–2150; (b) H. C. Brown and A. B. Asii, *J. Am. Chem. Soc.,* 1955, **77**, 4019–4024; (c) G. A. Russell and H. C. Brown, *J. Am. Chem. Soc.,* 1955, **77**, 4031–4035; (d) I. Tabushi, Z. Yoshida and Y. Tamaru, *Tetrahedron,* 1973, **29**, 81–84; (e) A. F. Andrews, C. Glidewell and J. C. Walton, *J. Chem. Res. Synop.,* 1978, 294; (f) A. S. Dneprovskii, E. V. Eliseenkov and S. A. Mil'tsov, *Z. Org. Khim.,* 1980, **16**, 1086–1087; (g) V. Khanna, P. Tamilselvan, S. J. S. Kalra and J. Iqbal, *Tetrahedron Lett.,* 1994, **35**, 5935–5938.

80. For radical chlorination of alkanes using other than sulfuryl chloride and molecular chlorine, see; (a) J. L. Brokenshire, A. Nechvatal and J. M. Tedder, *Trans. Faraday Soc.,* 1970, **66**, 2029–2037; (b) J. M. Krasniewski Jr and M. W. Mosher, *J. Org. Chem.,* 1974, **39**, 1303–1306; (c) G. A. Olah, P. Schilling, R. Renner and I. Kerekes, *J. Org. Chem.,* 1974, **39**, 3472–3478; (d) R. A. Johnson and F. D. Greene, *J. Org. Chem.,* 1975, **40**, 2192–

2196; (e) M. W. Mosher and G. W. Estes, *J. Am. Chem. Soc.*, 1977, **99**, 6928–6932; (f) N. C. Deno, E. J. Gladfelter and D. G. Pohl, *J. Org. Chem.*, 1979, **44**, 3728–3729; (g) A. Arase, M. Hoshi and Y. Masuda, *Chem. Lett.*, 1979, 961–964; (h) M. W. Mosher and G. W. Estes, *J. Org. Chem.*, 1982, **47**, 1875–1879; (i) R. Nouguier, J. M. Surzur, A. Virgili, *Anal. Quim. Ser. C: Quim. Org. Bioquim.*, 1982, **78**, 261–262; (j) M. Hoshi, Y. Masuda and A. Arase, *Chem. Lett.*, 1984, 195–198; (k) R. Davis, J. L. A. Durrant and C. C. Rowland, *J. Organomet. Chem.*, 1986, **316**, 147–162.

81. M. B. Smith and J. March, *March's Advanced Organic Chemistry: Reactions, Mechanisms and Structure*, Wiley, New Jersey, 6th edn, 2007, pp. 954–955.

82. H. Ehrich, D. Linke, K. Morgenschweis, M. Baerns and K. Jahnisch, *Chimia*, 2002, **56**, 647–653.

83. For a review of halogenations using a microreactor, see P. Löb, H. Löwe and V. Hessel, *J. Fluorine Chem.*, 2004, **125**, 1677–1694.

84. For recent papers on fluorination using a microreactor, see; (a) G. Sandford, *J. Fluorine Chem.*, 2007, **128**, 90–104; (b) M. Baumann, I. R. Baxendale and S. V. Ley, *Synlett*, 2008, 2111–2114; (c) K. Geyer, T. Gustafsson and P. H. Seeberger, *Synlett*, 2009, 2382–2391; (d) N. de Mas, A. Günther, M. A. Schmidt and K. F. Jensen, *Ind. Eng. Chem. Res.*, 2009, **48**, 1428–1434; (e) M. Baumann, I. R. Baxendale, L. J. Martin and S. V. Ley, *Tetrahedron*, 2009, **65**, 6611–6625; (f) T. Gustafsson, R. Gilmour and P. H. Seeberger, *Chem. Commun.*, 2008, 3022–3024.

85. Taking the sensitivity limit of detection by GC, we used a rough number >95%. Only a small amount of dichlorides was detected by GC and GC-MS.

86. (a) D. H. R. Barton, *Pure Appl. Chem.*, 1968, **16**, 1; (b) A. Studer, *Chem.–Eur. J.*, 2001, **7**, 1159; (c) L. Grossi, *Chem.–Eur. J.*, 2005, **11**, 5419.

87. G. Majetich and K. Wheless, *Tetrahedron*, 1995, **51**, 7095.

88. H. Suginome, in *CRC Handbook of Organic Photochemistry and Photobiology*, ed. W. M. Horspool and F. Lenci, CRC Press, Boca Raton, 2nd edn, 2004, p. 102.

89. (a) W. Ehrfeld, V. Hessel and H. Loanwe, *Microreactors: New Technology for Modern Chemistry*, Wiley-VCH, Weinheim, 2000; (b) K. Jaanhnisch, V. Hessel, H. Loanwe and M. Baerns, *Angew. Chem., Int. Ed.*, 2004, **43**, 406; (c) H. Pennemann, P. Watts, S. J. Haswell, V. Hessel and H. Loanwe, *Org. Process Res. Dev.*, 2004, **8**, 422; also see a review on continuous flow reactions: (d) G. Jas and A. Kirshning, *Chem.–Eur. J.*, 2003, **9**, 5708.

90. For Ryu's recent work on catalytic reactions using microreactors, see: (a) T. Fukuyama, M. Shinmen, S. Nishitani, M. Sato and I. Ryu, *Org. Lett.*, 2002, **4**, 1691; (b) S. Liu, T. Fukuyama, M. Sato and I. Ryu, *Org. Process Res. Dev.*, 2004, **8**, 477; (c) M. T. Rahman, T. Fukuyama, N. Kamata, M. Sato and I. Ryu, *Chem. Commun.*, 2006, 2236; (d) T. Konoike, K. Takahashi, Y. Araki and I. Horibe, *J. Org. Chem.*, 1997, **62**, 960.

91. For [2+2] cycloaddition using a photo-microreactor, see: T. Fukuyama, Y. Hino, N. Kamata and I. Ryu, *Chem. Lett.,* 2004, **33**, 1430.

92. For photoreactions using microreactors, see: (a) H. Lu, M. A. Schmidt and K. F. Jensen, *Lab Chip,* 2001, **1**, 22; (b) K. Ueno, F. Kitagawa and N. Kitamura, *Lab Chip,* 2002, **2**, 231; (c) H. Ehrich, D. Linke, K. Morgenschweis, M. Baerns and L. Jaanhnisch, *Chimia,* 2002, **56**, 647; (d) R. C. R. Wootton, R. Fortt and A. J. de Mello, *Org. Process Res. Dev.,* 2002, **6**, 187; (e) H. Maeda, H. Mukae and K. Mizuno, *Chem. Lett.,* 2005, **34**, 66; (f) Y. Matsushita, S. Kumada, K. Wakabayashi, K. Sakeda and T. Ichimura, *Chem. Lett.,* 2006, **35**, 410.

93. The continuous-flow reaction was performed by irradiating a solution of nitrite 1 (5.4 g, 10.8 mmol) in DMF (300 mL) containing a small amount of pyridine (0.2 mol equiv. of 1) with two microreactors (Type B) and eight 20 W black lights through a Pyrex glass cover (flow rate: 15 mL h^{-1}, residence time: 32 min, reaction time: 20 h). Water (600 mL) was added to the photoreaction mixture and the resulting slurry was collected by filtration and washed with water (100 mL) to give a white solid. The solid was purified by silica gel column chromatography to give oxime 2 in 60% isolated yield (3.1 g).

94. Photo-induced intermolecular [2+2] cycloaddition in a microflow system: T. Fukuyama, Y. Hino, N. Kamata and I. Ryu, *Chem. Lett.,* 2004, **33**, 1430.

95. Preliminary report on the Barton reaction using a microreactor, see: A. Sugimoto, M. Takagi, Y. Sumino, T. Fukuyama and I. Ryu, *Tetrahedron Lett.,* 2006, **47**, 6197.

96. Seeberger *et al.* also reported radical reaction in a microflow system, see: Odedra, K. Geyer, T. Gustafsson, R. Gilmour and P. H. Seeberger, *Chem. Commun.,* 2008, 3025.

97. A review on photo-reactions using a microreactor, see: (a) Y. Matsushita, T. Ichimura, N. Ohba, S. Kumada, K. Sakeda, T. Suzuki, H. Tanibata and T. Murata, *Pure Appl. Chem.,* 2007, **79**, 166.

98. Examples of photo-reactions using a microreactor, see: pinacol coupling of benzophenone: (a) H. Lu, M. A. Schmidt and K. F. Jensen, *Lab Chip,* 2001, **1**, 22; cyanation of pyrene: (b) K. Ueno, F. Kitagawa and N. Kitamura, *Lab Chip,* 2002, **2**, 231; photo-chlorination of 2,4-diisocyanatotoluene: (c) H. Enrich, D. Linke, K. Morgenschweis, M. Baerns and L. Jähnisch, *Chimia,* 2002, **56**, 647; generation of singlet oxygen and [4+2] cycloaddition with dienes: (d) R. C. R. Wootton, R. Fortt and A. J. de Mello, *Org. Process Res. Dev.,* 2002, **6**, 187. Intramolecular [2:2] cycloaddition: (e) H. Maeda, H. Mukae and K. Mizuno, *Chem. Lett.,* 2005, **34**, 66; (f) H. Maeda, H. Mukae and K. Mizuno, *Angew. Chem., Int. Ed.,* 2006, **45**, 6558; (g) H. Mukae, H. Maeda, C. Nashihara and K. Mizuno, *Bull. Chem. Soc. Jpn.,* 2007, **80**, 1157; photo-catalytic redox-combined synthesis of L-pipecolinic acid: (h) G. Takei, T. Kitamori and H.-B. Kim, *Catal. Commun.,* 2005, **6**, 357;

photo-catalytic reduction: (i) Y. Matsushita, S. Kumada, K. Wakabayashi, K. Sakeda and T. Ichimura, *Chem. Lett.,* 2006, **35**, 410; N-alkylation of amines: (j) Y. Matsushita, N. Ohba, S. Kumada, T. Suzuki and T. Ichimura, *Catal. Commun.,* 2007, **8**, 2194; (k) Y. Matsushita, N. Ohba, T. Suzuki, and T. Ichimura, *Catal. Today,* 2008, **132**, 153.

99. Barton reaction under oxygen leading to nitrates, see: (a) J. Allen, R. B. Boar, J. F. McGhie and D. H. R. Barton, *J. Chem. Soc., Perkin Trans. 1,* 1973, 2402; (b) D. H. R. Barton, M. J. Day, R. H. Hesse and M. M. Pechet, *J. Chem. Soc., Perkin Trans. 1,* 1975, 2252.

CHAPTER 17

Photo-catalysis and Metal–Oxygen-anion Cluster Decatungstate in Organic Chemistry: a Manifold Concept for Green Chemistry*

17.1 INTRODUCTION

In this chapter we have outlined some of the progress made in the emerging field of decatungstate catalysis, with emphasis on synthetic implications in organic chemistry. Thus, we have traced the developments of decatungstate from its employment as a homogeneous catalytic system under either aerobic or anaerobic conditions to its heterogenization for the functionalization/oxidation of alkanes, alkenes, alcohols, aldehydes, and sulfides, and its applications in water decontamination technology. Some of the newest developments outlined above as well as the ongoing interest on this subject convey that there are more yet unexplored applications of decatungstate catalysis that will be likely to emerge in the future. In essence, this review provides an important impulse for further studies in the field of decatungstate catalysis, so that the synthesis of new decatungstate-based materials for further synthetic applications can be envisaged.

Polyoxometalates (POMs, transition metal–oxygen-anion clusters) are a large and rapidly growing class of inorganic compounds with significant applications in a range of areas. Such materials have been studied in detail over the past decades with some of them possessing interesting applications in catalysis.[1] Among them, the decatungstate anion ($W_{10}O_{32}^{4-}$) is one of the most promising examples, thanks to its high photocatalytic activity (Figure 17.1).[2] Although the mechanistic aspects of decatungstate photoexcitation and subsequent activity have been intensively studied for almost two decades, its appli-

* Chapter written by Ioannis N. Lykakis.
Streamlining Free Radical Green Chemistry
V. Tamara Perchyonok, Ioannis Lykakis and Al Postigo
© V. Tamara Perchyonok, Ioannis Lykakis and Al Postigo 2012
Published by the Royal Society of Chemistry, www.rsc.org

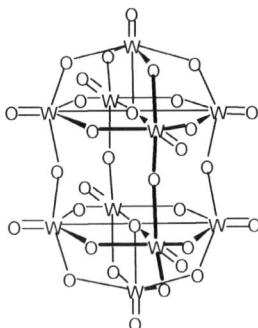

Figure 17.1   Structure of the decatungstate anion ($W_{10}O_{32}^{4-}$).

cation in organic synthesis is a subject of considerable ongoing interest; there is a plethora of examples of decatungstate-catalyzed reactions and presumably many more remain to be explored.

The oxidation of organic substrates, such as alkanes, alkenes and alcohols, is one of the first and most widely investigated decatungstate-catalyzed reactions with a high synthetic value. More recent developments in this field have been centered on the immobilization of decatungstate on a solid support (*e.g.*, silica, alumina or polymeric membranes) thus providing an efficient and recyclable photocatalytic system, especially for oxidation reactions. On the other hand, anaerobic conditions have been also adopted to fully exploit the catalytic versatility of decatungstate. Early examples include the functionalization of unactivated C–H bonds of organic compounds (*i.e.*, alkanes), while more recent examples such as the functionalization of aldehydes and amides, as well as the employment of decatungstate in nanocarbon functionalization chemistry, reflect the ongoing synthetic interest on this subject. As a final point, decatungstate has been also applied in the field of water decontamination, as an effective "green" catalyst for the photochemical degradation of pollutants (*e.g.*, chlorophenols), a subject that is now receiving much attention.

An overview of these diverse applications of decatungstate anion as an efficient catalyst in organic chemistry is given in this chapter, giving special attention to the most recent and outstanding contributions in the area.

17.2   C–C BOND FORMATION *VIA* C–H BOND FRAGMENTATION UNDER ANAEROBIC CONDITIONS

17.2.1   Functionalization of Alkanes by Homolytic C–H Bond Cleavage

Many research efforts have been directed toward the exploitation of decatungstate catalyst in organic synthesis, under anaerobic conditions. In this context, the activation of unactivated C–H bonds by decatungstate was one of the first examples reported in literature. Thus, in their pioneer studies

Hill and co-workers showed that the catalytic photochemical functionalization of alkanes by decatungstate under anaerobic conditions facilitates high yield, high selectivity routes to several useful alkyl derivatives (Scheme 17.1).[3,4] An anaerobic atmosphere is a prerequisite to avoid the extremely facile trapping of alkyl radicals by molecular oxygen (O_2).

Initially, almost two decades ago, Hill and others reported the decatungstate-mediated acylation of an unactivated carbon–hydrogen bond in a caged hydrocarbon (heptacyclotetradecane).[3] The same year, it was shown that the anaerobic decatungstate-catalyzed dehydrogenation of alkanes affords in each case the corresponding alkene isomer having the less-substituted double bond, in reasonably high selectivity (Scheme 17.1).[3] Next, in the early 1990s, decatungstate was effectively applied, by the same research group, in the catalytic radical addition reactions of unactivated C–H bonds to both acetylene (vinylation) and ethylene (ethylation).[3] Although this latter reaction, namely the alkylation of simple alkenes, proceeded with a low yield, the use of electrophilic alkenes improved the efficiency of this process greatly (*vide infra*) (Scheme 17.1). Later, in the mid-1990s, Hill reported the selective radical carbonylation of alkanes with CO.[3] In their investigation they first generated alkyl radicals from alkanes (*i.e.*, cyclohexane, cyclooctane, and *n*-hexane) *via*

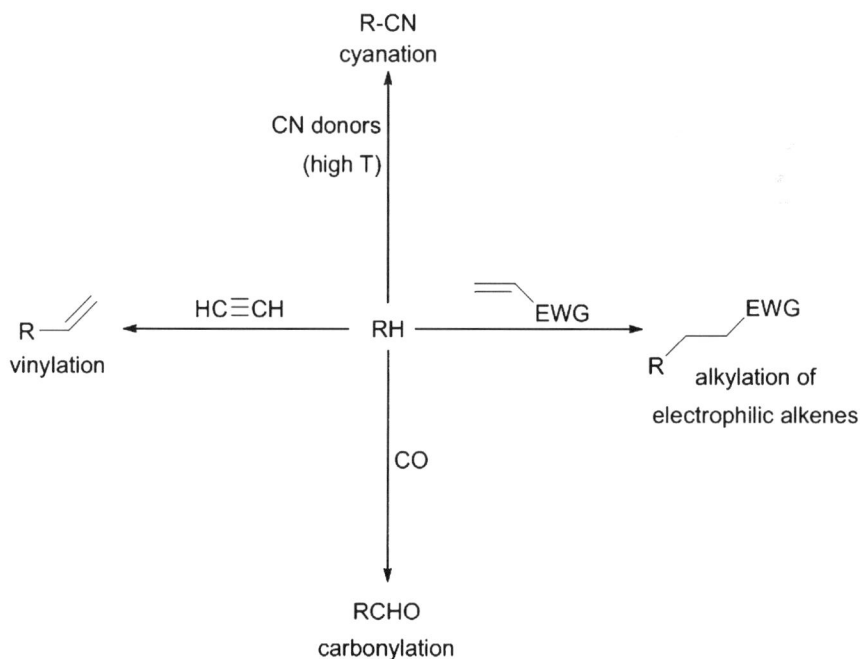

Scheme 17.1 Functionalization of unactivated C–H bonds in alkanes photocatalyzed by decatungstate under anaerobic conditions (EWG = electron-withdrawing group).

decatungstate photocatalysis, followed by carbonylation under CO atmosphere. This reaction exhibited high turnover numbers and selectivities for the production of the corresponding aldehydes, although the facile decarbonylation of these products, under the same conditions, was found to be a yield limiting factor (Scheme 17.1). Finally, in the late 1990s, Hill's group developed a convenient method for the preparation of nitriles and α-iminoesters starting from alkanes.[4] This method consisted of the decatungstate photo-mediated cleavage of unactivated C–H bonds of a variety of alkanes and subsequent coupling with methyl cyanoformate affording the corresponding nitriles and α-iminoesters at high and lower temperatures, respectively, in good yields. For example, the iminoesters derived from 2,3-dimethylbutane and *cis*-1,2-dimethylcyclohexane were isolated in 59 and 67% yield, respectively (at 28 and 61% conversion of alkane). Nitriles and α-ketoesters (readily derived from hydrolysis of α-iminoesters) are important groups of compounds with wide applications in organic synthesis.[4]

Following-on from these early studies, Albini and co-workers investigated the alkylation of electrophilic alkenes by alkanes, upon decatungstate photocatalysis (Scheme 17.1).[5,6] Although the yields in most cases were moderate (30–66%), the synthetic significance of this method was indicated by the mild activation and subsequent functionalization of aliphatic C–H bonds. Thus, alkyl radicals obtained by irradiation of tetrabutylammonium decatungstate in acetonitrile in the presence of alkanes (*i.e.*, cycloalkanes such as C_5H_{10}, C_6H_{12}, and C_7H_{14}) were efficiently trapped by electrophilic alkenes including α,β-unsaturated nitriles, esters, and ketones, to afford the corresponding alkylated aliphatic nitriles, esters and ketones, respectively (Scheme 17.1). Other oxygen-substituted radicals ($-O-C^{\bullet}$) such as those from propan-2-ol, terahydrofuran, and dioxolane were likewise efficiently trapped by α,β-unsaturated nitriles in good yields (50–78% yield, with up to complete conversion of nitrile).[6]

The basic photochemical reactions of decatungstate catalyst have been fully clarified by extensive mechanistic investigations carried out by several research groups.[2,3,7–10] Thus, eqn (17.1) and (17.2) are a vastly oversimplified representation of the decatungstate excitation and subsequent activation of C–H bonds in alkanes. The overall sequence of chemical transformations depicted in eqn (17.1)–(17.4) supports a radical mechanism pertinent to the observed C–C bond formation in the aforementioned alkylation reactions.[3,5] As described in eqn (17.1) and (17.2), the photoexcited decatungstate catalyst ($W_{10}O_{32}^{4-*}$) abstracts a hydrogen atom from the alkane (RH), producing an alkyl radical ($R^{\bullet}$) and the protonated reduced catalyst $W_{10}O_{32}^{5-} H^+$ (see also eqn (17.7) below and the accompanying text). The resulting alkyl radical is subsequently trapped by an unsaturated substrate (C=C), yielding a new radical intermediate (eqn (17.3)). This radical intermediate is then converted to the final product by reductive $H^{\bullet}$ donation from $W_{10}O_{32}^{5-} H^+$ (eqn (17.4)), in which the reduced catalyst is reoxidized to complete the catalytic cycle. It should be also noted that reverse hydrogen atom transfer (eqn (17.2)) as well as coupling of radicals ($R^{\bullet}$) or disproportionation to alkenes competes with the radical addition process (eqn (17.3)). A similar mechanism has been also proposed for the alkyl radical addition to carbon monoxide[3] and nitriles or α-iminoesters.[4]

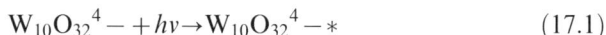

$$W_{10}O_{32}{}^{4-} + h\nu \rightarrow W_{10}O_{32}{}^{4-} - * \tag{17.1}$$

$$RH + W_{10}O_{32}{}^{4-} - {*}W_{10}O_{32}{}^{5-} - H^+ + R \tag{17.2}$$

$$R^{\bullet} + C = C \rightarrow RCC^{\bullet} \tag{17.3}$$

$$RCC^{\bullet} + W_{10}O_{32}{}^{5-} - H^+ \rightarrow RCCH + W_{10}O_{32}{}^{4-} - \tag{17.4}$$

17.2.2 Functionalization of Aldehydes by Homolytic C–H Bond Cleavage

The inter- and intramolecular addition of nucleophilic acyl radicals to olefins or alkynes is one of the most useful approaches for the synthesis of cyclic and acyclic ketones. Thus, numerous methods of generating acyl radicals have been developed and applied in organic synthesis. Among them, decatungstate catalysis has been recently applied for the acylation of electrophilic alkenes[11] and [60]fullerene.[12] In this process, the photocatalytic role of decatungstate is to produce acyl radicals by activating the C(O)–H bond of the corresponding aldehydes. Trapping of the so-obtained acyl radicals by electrophilic alkenes has led to the synthesis of unsymmetrical ketones in moderate to good yields (Scheme 17.2).[11] The competing decarbonylation of acyl radicals prior to acylation, when secondary or tertiary aldehydes were used at room temperature, initially proved to be a limiting factor for the scope of this reaction. However, at low reaction temperatures (*i.e.*, at −20 °C) the decarbonylation pathway of secondary aldehydes decreases dramatically with simultaneous increase of the acylation products.[11,12]

More recently, a direct acylation of [60]fullerene with acyl radicals was achieved *via* decatungstate catalysis (Scheme 17.3).[12] This free-radical acylation method was highly effective when using a wide range of aldehydes (*e.g.*, aromatic, alkyl, and cyclopropyl substituted aldehydes). Additionally, a simple low-temperature approach has been successfully applied for overcoming decarbonylations encountered in certain cases where the acyl radical intermediates are prone to undergo rapid decarbonylation (*i.e.*, pivaloyl or phenylacetyl radical). Mechanistic investigations, based mainly on intra- and

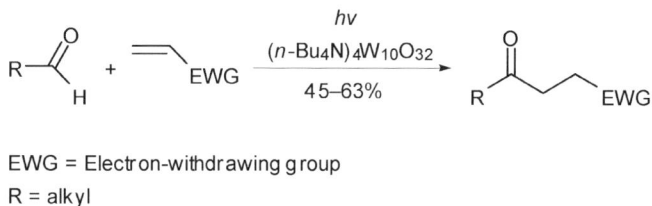

EWG = Electron-withdrawing group
R = alkyl

Scheme 17.2 Photocatalyzed acylation of electron-poor olefins.

R = alkyl, aryl

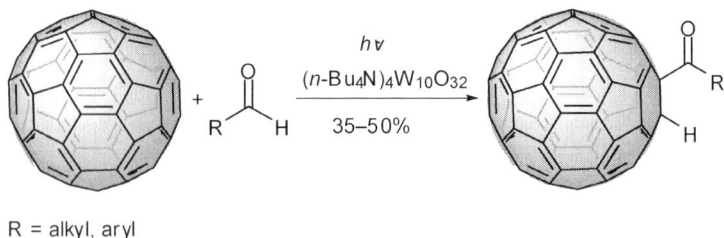

Scheme 17.3   Decatungstate-mediated acylation of [60]fullerene.

intermolecular deuterium isotope effect studies, indicated a C(O)–H bond breaking in the rate-determining step of this reaction.

17.2.3   Functionalization of Amides by Homolytic C–H Bond Cleavage

Recently, Fagnoni and co-workers reported that tetrabutylammonium decatungstate can be utilized for the C–H bond functionalization of amides.[13] In their study, different carbon-centered radicals, depending on the amide structure, were generated *via* decatungstate catalysis, and subsequently trapped by electrophilic alkenes. In particular, it was found that in a tertiary amide (*i.e.*, DMF or *N,N*-dimethylacetamide) the N-methyl C–H bond was chemoselectively cleaved, with no competition by the C–H bond of the formyl group or the α-hydrogen of the carboxamide group (Scheme 17.4, eqn (17.5)). On the contrary, a complete change in the chemoselectivity was observed when passing to a secondary amide (*i.e.*, N-methylformamide). In this case, only the C–H bond of the formyl group was homolytically cleaved (Scheme 17.4, eqn (17.6)). The same method was extended to carbamates, affording the corresponding alkylation products (26–59% yields).

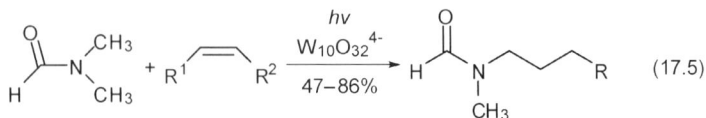

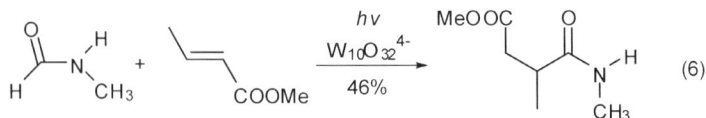

$$R^1 = R^2 = COOMe$$
$$R^1 = H, R^2 = COOEt, CN, COMe$$

Scheme 17.4   Decatungstate-mediated amidation of electron-poor olefins.

17.2.4 Functionalization of Toluenes, Anisoles and Thioanisole by Homolytic C–H Bond Cleavage

The recent introduction of decatungstate catalysis in nanocarbon functionalization chemistry has established a convenient and highly efficient route to C–C bond formation in fullerenes.[14] One of the major outcomes of this investigation was that decatungstate promotes the generation of carbon-centered radicals by activating the methyl C–H bond in substituted toluenes, anisoles and thioanisole. The so-obtained radical intermediates were successfully trapped by C_{60} to afford selectively the corresponding monofunctionalized fullerenes in moderate to good yields (Scheme 17.5). The observed single addition of radicals to C_{60} is of particular interest, given the well-established high affinity of C_{60} with free-radicals (C_{60} has been described as a "radical sponge"). In particular, [60]fullerene has 30 carbon–carbon double bonds which means that multiple additions of a variety of radicals can take place very readily, thus affording a complex mixture of multi-addition products. Apart from the synthetic applications of this reaction, a mechanistic approach has been also provided in which a hydrogen atom abstraction from the CH_3 or the CH_3O group occurs in the rate-determining step of the reaction.[14]

17.3 HOMOGENEOUS OXIDATION OF ORGANIC COMPOUNDS BY DECATUNGSTATE

Catalytic oxo-functionalization plays an important role in organic chemistry and has received considerable attention over the last years. An attractive

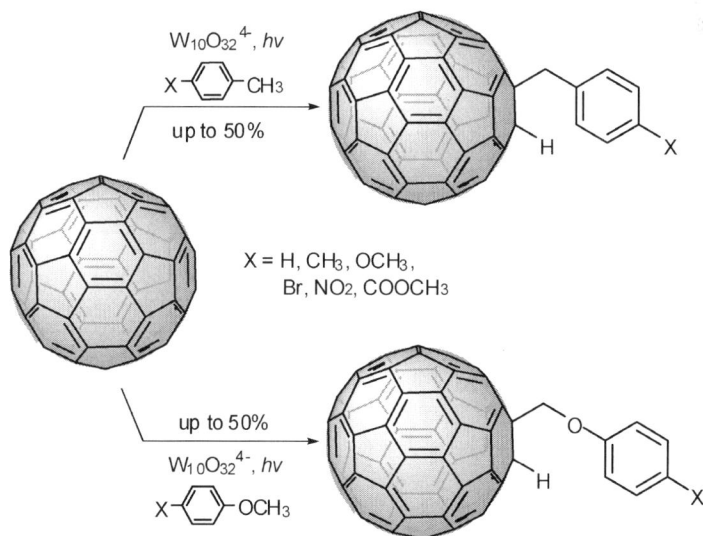

Scheme 17.5 Decatungstate-mediated radical reactions of C_{60} with substituted toluenes and anisoles.

approach for these catalytic oxidations is the use of an effective catalyst and an environmentally friendly oxidant, such as molecular oxygen. Polyoxometalates have attracted much attention as efficient photocatalysts in oxidation reactions due to the wide range of their redox potentials, as well as the reversibility in their multi-electron reductions. Among them, the decatungstate anion $W_{10}O_{32}^{4-}$ is of particular interest by virtue of its high photocatalytic activity. Thus, decatungstate ($W_{10}O_{32}^{4-}$) has been successfully applied in the homogeneous photocatalytic oxidation of various organic substrates, such as alkanes, alkenes, alcohols, and amines.

It is generally accepted that illumination of $W_{10}O_{32}^{4-}$ leads to the formation of a charge transfer excited state ($W_{10}O_{32}^{4-*}$) that decay in *ca.* 30 ps to a longer-lived transient designated as wO.[15] This transient species does not react with O_2, but exclusively with organic substrates (XH) to give the radical $X^{\bullet}$ and the one-electron reduced form of decatungstate (eqn (17.7)).

$$W_{10}O_{32}^{4-} \xrightarrow{h\nu} W_{10}O_{32}^{4-*} \rightarrow wO \xrightarrow{XH} W_{10}O_{32}^{5-} H^+ + X^- \quad (17.7)$$

Mechanistic studies have shown that there are two distinct mechanisms that may operate in the photosensitized oxidations of organic substrates with decatungstate, in the presence of O_2, namely an electron transfer (ET) or a hydrogen atom transfer (HAT) mechanism.[7,8,10,15] In particular, aromatic hydrocarbons (*e.g.*, naphthalene, anthracene) and aromatic amines react with decatungstate *via* an ET mechanism, due to their low redox potentials and/or the absence of a weak Z–H bond (Z = C, N, O).[8,10] On the other hand, a HAT mechanism was found to occur with weaker Z–H bonds in the case of alkanes and aliphatic alcohols.[2,10,15] The reaction of some substrates by both ET and HAT mechanisms, however, could not be excluded.[7,8] In any case, both mechanisms give rise to the same one-electron reduced form of decatungstate ($W_{10}O_{32}^{5-} H^+$), and to the corresponding substrate-derived radical. These latter species may react quantitatively with oxygen to form hydrogen peroxide and/or organic hydroperoxides as final products.

17.3.1 Oxidation of Aliphatic Alcohols and Alkanes

Aliphatic alcohols and alkanes are among the first organic compounds that have been oxidized *via* decatungstate catalysis in the presence of molecular oxygen. Initially, the mechanism for the oxidation of aliphatic secondary alcohols and alkanes to the corresponding ketones and hydroperoxides, respectively, was studied by Tanielian and co-workers by using a laser flash photolysis technique.[15] In this study, it was found that the reduced form of decatungstate ($W_{10}O_{32}^{5-}$) can be reoxidized to $W_{10}O_{32}^{4-}$ in the presence of O_2 with parallel formation of peroxy compounds. For example, propan-2-ol was converted into acetone and hydrogen peroxide upon decatungstate photocatalysis, whereas oxidation of adamantane afforded the corresponding hydroperoxides (Scheme 17.6, eqn (17.8) and (17.9)).[15] The

$$\text{(17.8)}$$

$$\text{(17.9)}$$

$$\text{(17.10)}$$

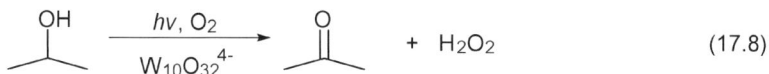

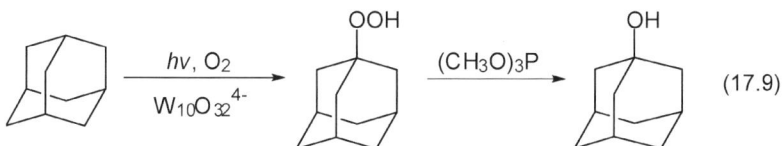

Scheme 17.6   Decatungstate-catalyzed oxidation of aliphatic alkanes and alcohols.

decatungstate-catalyzed oxidation of aliphatic alkanes was also investigated by Giannotti and Richter.[16] For instance, it was shown that the oxidation of adamantane afforded selectively 1- and 2-adamantanol (after reduction of the corresponding hydroperoxides with trimethylphosphite) as the only products, albeit only in low to medium conversions. In this case, the tertiary alcohol (adamantanol-1) was formed in almost 80% relative yield (Scheme 17.6, eqn (17.9)). On the other hand, at conversion higher than 70% (based on the starting material), the accumulation of poly-oxygenated products lowered the efficiency of the oxidation. The efficiency of this reaction was also found to be depended on both the decatungstate counter anion, as well as the solvent.[16] In particular, it was found that nitrile solvents (*i.e.*, CH_3CN, CH_3CH_2CN) and organic counter-ions of the catalyst behave like substrates and compete in this photooxidation with alkane substrate. This was evidenced by the decrease on the overall hydroperoxide formation, when a counter-ion that bears no oxidizable C–H bonds (*i.e.*, Na^+) or a solvent with a high C–H bond strength (*i.e.*, CH_3CN) was tested in separate experiments. Similarly, Maldotti and co-workers studied the homogeneous oxidation of cyclohexane (Scheme 17.6, eqn (17.10)). The only products obtained in this reaction were the corresponding cyclic alcohols and ketones. Moreover, it was found that the O_2 pressure had a severe effect on the cyclohexanone/cyclohexanol ratio. In particular, the high O_2 pressure was found to favour the formation of cyclohexanone, whereas the low O_2 pressure led selectively to the formation of cyclohexanol.[17]

17.3.2   Oxidation of Aromatic Alcohols and Alkanes

Decatungstate has been successfully used as an efficient catalyst for the selective oxidation of aromatic hydrocarbons (*i.e.*, cumene, ethyl benzene, and methyl fluorene). The major products obtained from the oxidation of these aromatic hydrocarbons were the corresponding alcohols (Scheme 17.7, eqn (17.11) and eqn (17.12)). In particular, the oxidation of *para*-substituted cumenes afforded the corresponding tertiary alcohols (after reduction of the initially formed hydroperoxides with PPh$_3$) as the major products, along with a small amount of the corresponding aryl ketones (derived from further photo-degradation). Kinetic isotope effect studies supported a hydrogen atom abstraction (*H*AA) from the aromatic substrate in the rate-determining step of this oxidative transformation.[18] In addition, the presence of a hydrogen donor, such as Et$_3$SiH, was found to increase substantially the proportion of the hydroperoxide product. For example, the catalytic oxidation of ethyl benzene in the presence of Et$_3$SiH afforded selectively, at low conversions, the corresponding secondary hydroperoxide adduct. At higher conversions the corresponding ketone was obtained as the major product.

Apart from the decatungstate-catalyzed oxidation of aromatic alkanes, the oxidation of benzyl alcohols under similar conditions has been studied.[19] On the basis of several investigations including time-resolved techniques, kinetic isotope effects, Hammett kinetics, and product analysis, a HAT mechanism was proposed for this reaction. Typically, the corresponding aromatic ketones were selectively formed as the only products in good yields (Scheme 17.7, eqn (17.13)).

17.3.3 Oxidation of Aliphatic and Aromatic Alkenes

Another class of organic compounds that have been studied under the same oxidative conditions *via* decatungstate catalysis, in addition to alcohols and alkanes, is olefins. Earlier studies by Maldotti and co-workers[20] showed that

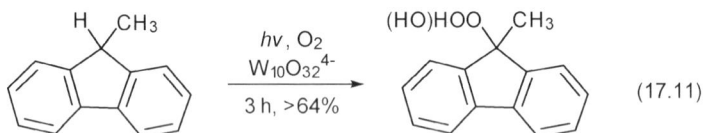

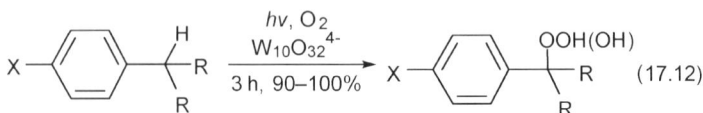

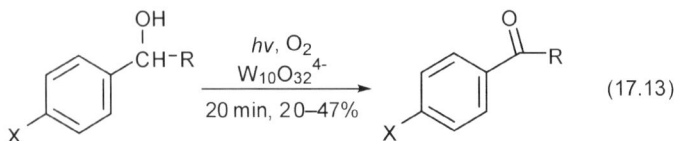

X = NO$_2$, CF$_3$, Br, H
R = H, Me

Scheme 17.7   Decatungstate-catalyzed oxidation of aromatic alkanes and alcohols.

the oxidation of cyclohexene and cyclooctene in the presence of $W_{10}O_{32}^{4-}$ afforded the corresponding secondary hydroperoxides and α,β-unsaturated cycloketones, in moderate to high yields. Moreover, the presence of Fe^{III}-[meso-tetrakis(2,6-dichlorophenyl)porphyrin] chloride as a co-catalyst in the oxidation of these cycloalkenes was found to affect both the efficiency and chemoselectivity of the reaction. The decomposition of the allylic hydro-peroxides to the corresponding allylic alcohols was also catalyzed by this co-catalyst.[20]

The selectivity in the decatungstate-catalyzed oxidation of various alkyl and phenyl substituted cycloalkenes (*e.g.*, 1-alkyl- or 1-aryl-substituted cycloalk-enes) in the presence of O_2, was also reported.[21] In particular, it was shown that a hydrogen atom on the less hindered side of the double bond was preferentially abstracted by decatungstate, affording the corresponding allylic hydroperoxides and enones as the major products (Scheme 17.8). The reason-ably high yields observed, makes this catalytic reaction synthetically useful.[21]

17.4 HETEROGENEOUS OXIDATION OF ORGANIC COMPOUNDS BY DECATUNGSTATE

The main shortcoming of homogeneous catalytic processes comes from the requirement of catalyst recovery or separation from the reaction mixture. Moreover, there is a growing need for the development of environmentally benign, recoverable, catalysts that could replace the current homogeneous chemical procedures. One way to attain this goal is to immobilize one or more components of the catalytic systems onto a large surface area of a solid carrier.

In this challenging field of research, a number of groups are investigating decatungstate as a catalyst with significant activity in the heterogeneous photooxidation of various organic compounds including alkanes, alkenes, alcohols, and sulfides. Research published since *ca.* 2000 has demonstrated

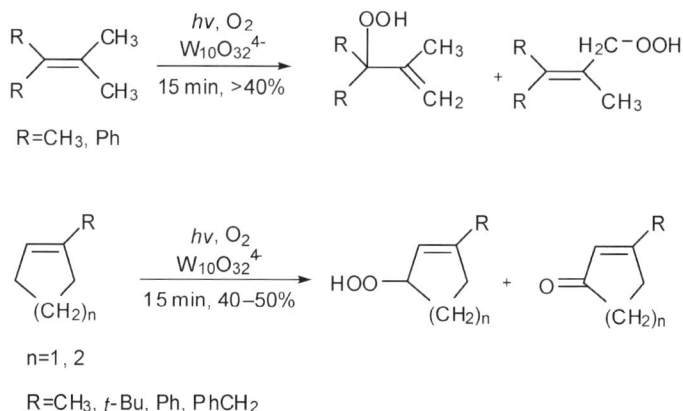

Scheme 17.8 Decatungstate-catalyzed oxidation of alkenes.

that the use of decatungstate in heterogeneous catalysis offers several distinct advantages. Some of these advantages are:

(i)   there is more freedom in the choice of the dispersing medium;
(ii)   the decatungstate as heterogeneous catalyst is able to simulate homogeneous reaction conditions with comparable efficiency, and at the same time has the advantage of easy separation and recovery;
(iii) the photocatalyst can be reused several times without any significant loss of photochemical efficiency; and
(iv)   it, sometimes, controls the selectivity of the processes.

The ongoing interest on the decatungstate heterogeneous catalysis has in turn triggered the development of a variety of techniques for the synthesis of decatungstate-based catalytic systems. Thus, different procedures have been developed, such as (a) impregnation on a solid support,[22–24] (b) immobilization inside the silica network,[25–27] (c) on silica previously functionalized with different ammonium cations,[28,29] (d) onto organic ion-exchange resins,[30,31] and (e) in polymeric membranes.[32–34]

17.4.1   Immobilization on a Solid Support

A decade ago, Maldotti and co-workers reported the first application of heterogeneous $(n\text{-}Bu_4N)_4W_{10}O_{32}$ photo-catalysis.[35] To prepare this catalyst an 'impregnation' procedure was followed and resulted in the adsorption of $(n\text{-}Bu_4N)_4W_{10}O_{32}$ on silica through electrostatic interactions; the tetrabutylammonium cations presumably act as a bridge between the negative surface and the decatungstate anion.[35] The catalytic performance of this catalytic system was evaluated in the oxidation of cyclohexane to cyclohexanol and cyclohexanone. This transformation is of considerable industrial interest since the resultant products are precursors in the synthesis of adipic acid, which is in turn an intermediate in the production of nylon. Later, in 2002, the same authors reported the immobilization of $(n\text{-}Bu_4N)_4W_{10}O_{32}$ on amorphous and mesoporous MCM-41 (pore size 20–100 Å, surface area $\sim 1000 \text{ m}^2 \text{ g}^{-1}$) silicas following the same impregnation procedure.[23] In their study, the different factors affecting the efficiency and selectivity in the photooxidation of cycloalkanes, related to the morphology of the support, decatungstate loading, and heterogenization were investigated. In particular, it was observed that the ketone/alcohol ratio increased significantly when decatungstate was loaded on both amorphous and MCM-41 silicas. Additionally, the model photooxidation of cyclohexane and cyclododecane showed that the decatungstate supported on the more ordered MCM-41 silica promotes the process with higher chemoselectivity than when $(n\text{-}Bu_4N)_4W_{10}O_{32}$ was supported on amorphous silica (Scheme 17.9). More importantly, the photocatalytic efficiency of $(n\text{-}Bu_4N)_4W_{10}O_{32}$ was not reduced after heterogenization. In fact, decatungstate immobilized on amorphous silica was found capable of oxidizing cyclododecane with a markedly higher efficiency than in the homogeneous phase while the supported photocatalysts could be used at least three times without any loss of activity. Finally, $(n\text{-}Bu_4N)_4W_{10}O_{32}/SiO_2$ was not found to cause any mineralization of the substrate. This is, typically, a major advantage since the total photodegradation of the substrate is often a competing reaction that limits the reaction yield.

The effect of the nature of the cations in the tetralkylammonium and sodium decatungstate supported on silica on the photocatalytic oxidation of organic substrates (*i.e.*, cyclohexane) has also been reported.[24] Thus, immobilization of $W_{10}O_{32}{}^{4-}$ decatungstate on amorphous silica afforded stable photocatalytic systems, with different and tunable activity depending on the nature of the counterion (*i.e.*, organic or inorganic, size, hydrophobicity). For example, the efficiency of cyclohexane photooxidation was substantially enhanced when tetralkylammonium cations were used instead of their ammonium or sodium analogues. This was explained in terms of surface polarity induced by the cation. Accordingly, the cation makes the environment around the photoactive anion centre $(W_{10}O_{32}{}^{4-})$ more or less hydrophilic so influencing the possibility of the substrate to reach the surface, and hence the photocatalytic efficiency of the process.[24] In these latter studies, the structural characterization of the $W_{10}O_{32}{}^{4-}/SiO_2$ systems has been also carried out by spectroscopic and N_2 adsorption–desorption techniques.[23,24] One of the main conclusions was that the structures of both anchored species, *viz.* $W_{10}O_{32}{}^{4-}$ anion and its counterca-tions, had been preserved after heterogenization. It was also suggested that $n\text{-}Bu_4N^+$ cations may act as a bridge between the negative surface of silica and the $W_{10}O_{32}{}^{4-}$ anions, according to an ion-mediated adsorption pathway of the type $SiO^-/\text{counter-ion}^+/W_{10}O_{32}{}^{4-}$.

In 2000, Maldotti and co-workers reported the heterogeneous oxidation of cyclohexene and cyclooctene with $(n\text{-}Bu_4N)_4W_{10}O_{32}$ supported on silica. In addition, the photocatalytic properties of the $(n\text{-}Bu_4N)_4W_{10}O_{32}/SiO_2$ system were investigated in the presence of $Fe^{III}[\textit{meso}\text{-tetrakis}(2,6\text{-dichlorophenyl})\text{-}$ porphyrin] chloride [Fe(TDCPP)Cl] as cocatalyst.[20] In particular, the decatungstate photocatalyzed oxidation of cyclohexene and cyclooctene, under heterogeneous conditions, afforded the corresponding allylic hydro-peroxide and cyclooctene epoxide, respectively, as the major reaction product. In both oxidation reactions, the presence of Fe(TDCPP)Cl was found to increase the efficiency of the photocatalytic process. Moreover, in the case of

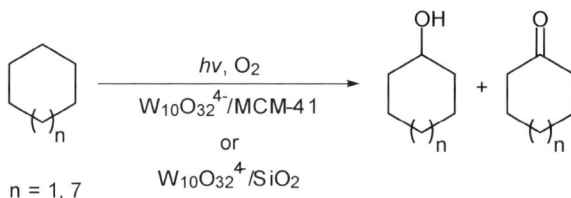

Scheme 17.9 Photocatalytic oxidation of cycloalkanes by decatungstate supported on mesoporous MCM-41 and amorphous silicas.

cyclooctene, contrary to what was observed in the case of cyclohexene, the heterogenization of decatungstate affected the chemoselectivity of the photocatalytic process as evidenced by the increased epoxide/hydroperoxide concentration ratio (Scheme 17.10).[20]

Later on, the same authors reported again the oxidation of the previously mentioned cycloalkenes (cyclohexene and cyclooctene) by $(n\text{-Bu}_4\text{N})_4\text{W}_{10}\text{O}_{32}$ supported on silica, but this time in the presence of CH_2Cl_2 as co-substrate.[36] In this latter study, it was found that the presence of CH_2Cl_2, instead of CH_3CN, significantly affects the photocatalytic activity of $W_{10}O_{32}{}^{4-}/SiO_2$, both in terms of product distribution and overall conversion yield. Thus, the reverse trend in the formation of products was obtained, that is the major products derived from the photooxidation of both cyclohexene and cyclooctene, under these conditions, were the corresponding epoxides. When these experiments had been carried out under the same conditions in CH_3CN, the corresponding allylic hydroperoxides were predominantly formed. Cyclooctene epoxide is stable enough to be accumulated in the irradiated solution, in contrast to labile cyclohexene epoxide which undergoes ring opening to form 2-chlorocyclohexanol. However, it is known that halohydrines are readily converted to the corresponding epoxides under alkaline conditions. Accordingly, transformation of 2-chlorocyclohexanol into cyclohexene epoxide could be easily performed by addition of an excess of NaOH to the irradiated sample (Scheme 17.11). Moreover, in the case of the CH_2Cl_2-assisted oxidation of cyclohexene, the overall yield was increased by ca. three times.

In a recent report, decatungstate was deposited for the first time on the surface of γ-alumina and silica by wet impregnation at various pH values (below and above of the point of zero charge of the support).[22] In this study it was shown that the impregnation pH as well as the nature of the support had a severe effect on both the structure and state of dispersion of the decatungstate phase. Notably, the photocatalytic activity of the

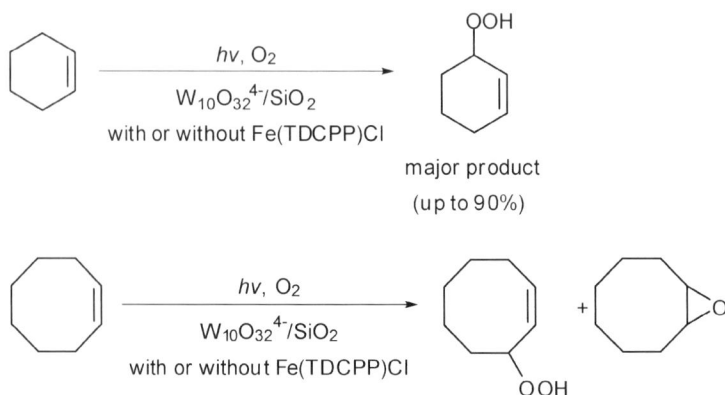

Scheme 17.10 Heterogeneous photocatalyzed oxidation of cyclohexene and cyclooctene by $W_{10}O_{32}{}^{4-}/SiO_2$.

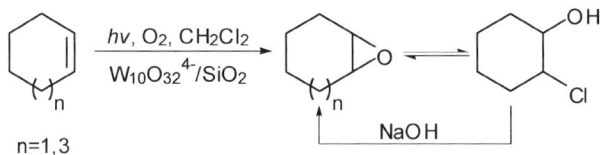

Scheme 17.11 CH$_2$Cl$_2$-assisted functionalization of cycloalkenes by photoexcited (*n*-Bu$_4$N)$_4$W$_{10}$O$_{32}$ heterogenized on SiO$_2$.

silica-supported catalysts was found to be as high as that of the homogeneous precursor. The Al$_2$O$_3$-supported catalyst exhibited lower activity compared with the homogeneous or silica-supported catalysts. However, both the Al$_2$O$_3$- and SiO$_2$-based catalysts were very effective in the photooxidation of a series of secondary and primary benzyl alcohols, in which aryl ketones and *p*-substituted benzoic acids were selectively formed as the only or major products, respectively, in quantitative yields (Scheme 17.12). Among the alcohols tested, a series of *p*-alkyl-substituted benzyl alcohols were also included to evaluate the chemoselectivity of these catalytic systems. In particular, these substrates bear two distinguishable benzylic hydrogen atoms, one on the alcohol carbon and one on the *p*-alkyl substituent, that both can be potentially cleaved under the photooxidation conditions. Interestingly, the photooxidation of these benzyl alcohols showed a strong preference for oxidation of the carbon that bears the hydroxyl group, thus a highly selective oxidation to the corresponding aryl ketones was observed; on the other hand, the formation of dioxygenated products derived from hydrogen abstraction from the benzylic

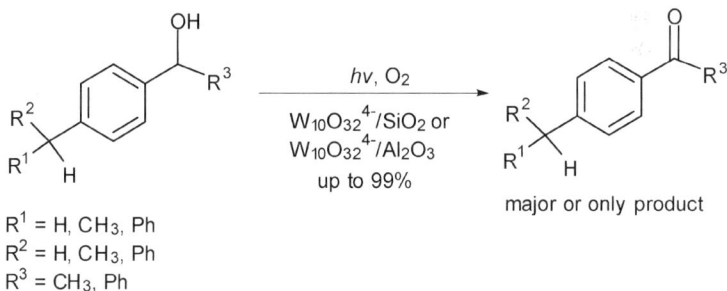

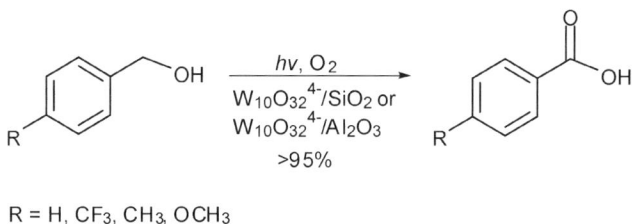

R^1 = H, CH$_3$, Ph
R^2 = H, CH$_3$, Ph
R^3 = CH$_3$, Ph

R = H, CF$_3$, CH$_3$, OCH$_3$

Scheme 17.12 Heterogeneous decatungstate photocatalyzed oxidation of secondary and primary benzyl alcohols.

carbon of the *p*-substituent, was hardly observed (Scheme 17.12). Inasmuch as these catalytic systems exhibited high activity and selectivity in the oxidations of a great variety of benzyl alcohols while retaining their activity in subsequent catalytic cycles, they may serve as a valuable tool in synthetic organic chemistry.

17.4.2 Immobilization inside the Silica or Zirconia Network

Microporous polyoxometalates (POMs) constitute another important subclass of heterogenized catalysts. Hu and co-workers were the first to report the synthesis of microporous decatungstate as a new type of heterogeneous photocatalytic materials.[25,26] The process included hydrolysis of tetraethyl orthosilicate (TEOS) in the presence of $W_{10}O_{32}^{4-}$ ion at pH 2.0 $\pm$ 0.2.[25] Formation of microporous decatungstates is due to the chemical interaction between the $M_4W_{10}O_{32}$ molecule and the silica network (M = Na^+ or *n*-Bu_4N^+). The chemically active silanol group (Si-OH) was protonated under acidic conditions to form the $Si-OH_2^+$ group (eqn (17.14)):

$$\text{>Si-OH} \quad + \quad H^+ \quad \longrightarrow \quad \text{>Si-OH}_2^+ \qquad (17.14)$$

The $Si-OH_2^+$ group should act as a counterion for the $W_{10}O_{32}^{4-}$ ion and yielded $(Si-OH_2^+)(M_3W_{10}O_{32}^-)$ by acid–base reaction. Spectroscopic analyses indicated that the structural integrity of $W_{10}O_{32}^{4-}$ cluster had been preserved after incorporation into the silica network. The activity of the resulting $W_{10}O_{32}^{4-}/SiO_2$ composites was tested for the heterogeneous photodegradation of organic pollutants (*i.e.*, pesticides) and will be discussed further in the last section of this chapter.

This sol–gel technique, reported by Hu and co-workers, has also been applied to the synthesis of zirconia-supported sodium decatungstate, $Na_4W_{10}O_{32}/ZrO_2$, as demonstrated by Farhadi and Momeni.[37] The catalytic activity of the $Na_4W_{10}O_{32}/ZrO_2$ composite was tested in the oxidation of a series of primary and secondary benzyl alcohols with molecular oxygen. It was found that these substrates were efficiently oxidized to the corresponding carbonyl compounds in moderate to high yields (60–96%) without over-oxidation of the benzaldehydes to carboxylic acids.[37] More importantly, it was found that the heterogeneous photocatalyst was more active than the unsupported $Na_4W_{10}O_{32}$, while leaching of $W_{10}O_{32}^{4-}$ cluster from the zirconia support into the reaction system was hardly observed.

17.4.3 Immobilization on Silica containing Ammonium Cations

As an alternative to the conventional impregnation or sol-gel technique for the immobilization of decatungstate on silica, it has been recently reported that decatungstate $W_{10}O_{32}^{4-}$ can be supported on silica previously functionalized with different ammonium cations covalently bound on the surface of the solid support (Figure 17.2).[28] The polyoxoanion is immobilized on the solid surface through an exchange reaction by mixing the selected surface bound alkyl-ammonium salt and an aqueous solution of $Na_4W_{10}O_{32}$. The net result is the linking of decatungstate salt to the solid support by chemical bonds rather than by simple electrostatic interactions. This catalyst proved sufficiently robust since it could be reused at least five times without leaching of the polyoxotungstate anion.

The so-obtained systems exhibited high efficiency and selectivity in the catalytic oxidation of various sulfides, including allyl phenyl sulfide and dibenzyl sulfide (both of them known to be hardly oxidized to sulfoxides, selectively), by hydrogen peroxide (Scheme 17.13); primary propylammonium decatungstate was the most active catalyst. Typically, the corresponding sulfoxides were selectively obtained in high yields, whereas overoxidation led to the corresponding sulfones.[28] The use of a low amount of heterogeneous catalyst (0.1 mol%) with a slight excess of 30% H_2O_2 (1.15 equiv.) without any other additive in a non-chlorinated solvent makes this oxidation reaction an environmentally friendly chemical process.

The same catalytic systems were later investigated as photocatalysts for the regioselective oxidation of diols.[29] For example, proper reaction conditions were found for obtaining more than 90% of 4-hydroxy-2-butanone from 1,3-butanediol and the selective oxidation of 1,4-pentanediol to 4-hydroxypentanal (Scheme 17.14). The preferential oxidation of one of the two hydroxyl groups in diols is, typically, an important transformation in synthetic organic chemistry.

17.4.4 Immobilization onto Organic Ion-exchange Resins

Organic polymeric materials have been also studied as solid supports for the heterogenization of decatungstate. For example, Fornal and Giannotti studied the possibility of decatungstate immobilization by ion-exchange onto organic ion-exchange resins (*i.e.*, poly(4-vinylpyridine)) as well as on a carbon material (Ambersorb®).[30] In contrast to the leak of decatungstate anions observed

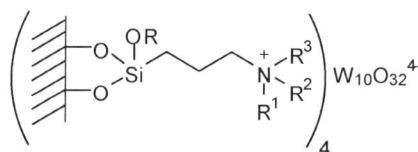

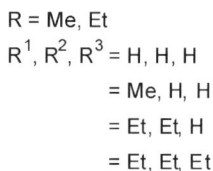

R = Me, Et
R^1, R^2, R^3 = H, H, H
= Me, H, H
= Et, Et, H
= Et, Et, Et

Figure 17.2 Silica-bound decatungstate catalyst.

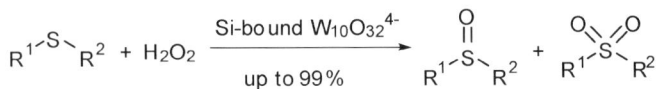

$$R^1\text{-}S\text{-}R^2 + H_2O_2 \xrightarrow[\text{up to 99\%}]{\text{Si-bound } W_{10}O_{32}^{4-}} R^1\text{-}S(=O)\text{-}R^2 + R^1\text{-}S(=O)_2\text{-}R^2$$

R^1 = aryl
R^2 = aryl, alkyl

Scheme 17.13   Heterogeneous oxidation of sulfides mediated by silica-bound decatungstate *via* H_2O_2 activation.

when the carbon material was used as a support, the immobilization was strong enough to prevent decatungstate from releasing into solution when decatungstate was supported on organic resins.[30] The photocatalytic activity of this resin–decatungstate composite was evaluated in the oxidation of cyclohexane with molecular oxygen. It was shown that the selectivity of this oxidation reaction (*i.e.*, the formation of cyclohexanone, cyclohexanol or cyclohexyl hydroperoxide) depends on the decatungstate loading on the support rather than the catalyst concentration in the irradiated solution. Thus, the lower loading of decatungstate promoted cyclohexanone production whereas the higher loading of decatungstate favored cyclohexyl hydroperoxide generation.

In addition, Maldotti and co-workers have studied the heterogenization of $(n\text{-Bu}_4N)_4W_{10}O_{32}$ with Amberlite IRA-900. This support is a macroreticular styrene–divinylbenzene copolymer, bearing $-N(CH_3)_3^+$ functional groups (commercially available in its chloride form).[31] The catalytic system thus obtained was shown to promote the conversion of phenol and anisole to the corresponding mono-brominated derivatives, and cycloalkenes to the corresponding bromohydrins and dibromides. Bromohydrins could be quantitatively transformed into the corresponding epoxides by simply adjusting the pH

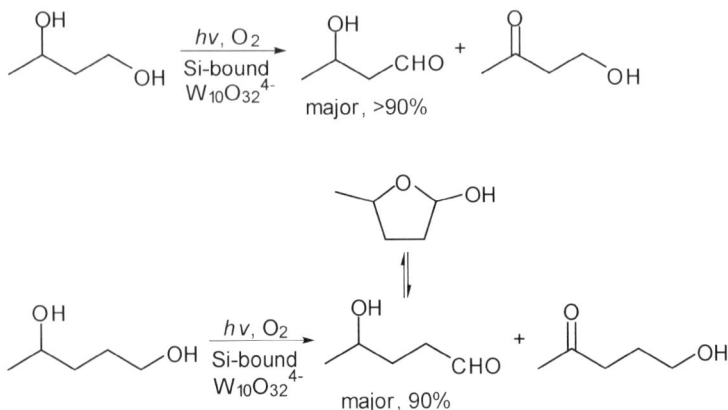

Scheme 17.14   Photooxidation of diols, in the presence of O_2, catalyzed by silica-bound decatungstate.

value (Scheme 17.15). These catalytic transformations are of considerable synthetic interest due to the otherwise difficult monobromination of activated arenes, and the diverse role of epoxides as important intermediates in organic synthesis. Finally, the solid matrix was found to play a crucial role in the selectivity of this photocatalytic process. In particular, the yields of epoxides and bromohydrins from alkenes were increased, upon heterogenization of the decatungstate, whereas the functionalization of the para position in anisole bromination was also favored.

17.4.5 Immobilization with Organic Sensitizers

In 2004, Bonchio and co-workers reported the synthesis of hybrid molecular photocatalysts consisting of an organic sensitizer and a polyoxometalate unit; these hybrid compounds were synthesized based on the charge inter-action between cationic sensitizers, namely methylene blue (MB^+) and tris(2,2'-bipyridine)ruthenium(II) ($[Ru(bpy)_3]^{2+}$), and the decatungstate anion $[W_{10}O_{32}]^{4-}$ yielding electrostatic aggregates.[38] The activity of the resulting hybrid complexes, $[Ru(bpy)_3]_2[W_{10}O_{32}]$ and $(MB)_4[W_{10}O_{32}]$, was assessed in water under heterogeneous conditions, 25 °C and O_2 (1 atm), using visible light irradiation ($\lambda > 375$ nm). To assess the potential of this method for wastewater treatment and synthetic applications, the photooxygenation of an aqueous phenol solution and the photooxygenation of L-methionine methyl ester were selected, respectively, as representative examples. The latter reaction occurred selectively to the corresponding sulfoxide in quantitative yield whilst the former reaction displayed a markedly reduced efficiency with respect to the homogeneous counterparts of the hybrid photocatalysts. Nevertheless, the ease

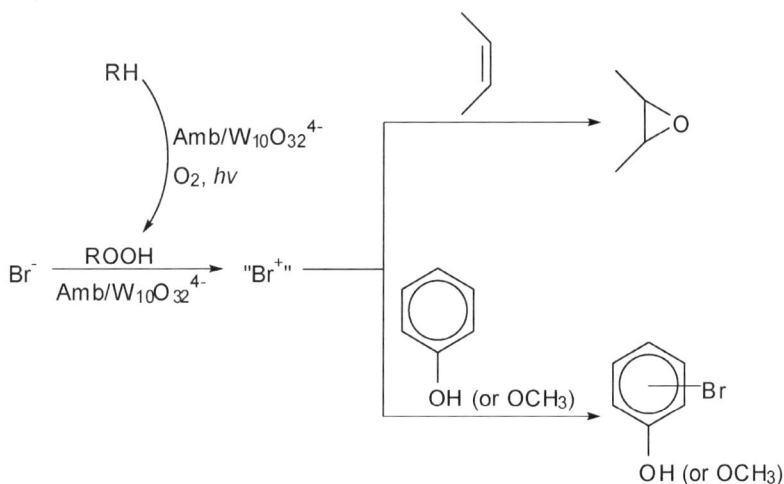

Scheme 17.15 Bromide-assisted bromination of arenes and alkenes *via* hetero-geneous decatungstate catalysis in the presence of oxygen.

of recovery, recycling and the potential flexibility of application in different media, represent a noticeable advantage outweighing this shortcoming.

17.4.6 Immobilization in Polymeric Membranes

As mentioned in the preceding section, numerous attempts have been made to disperse $W_{10}O_{32}^{4-}$ on various supporting materials. These attempts have been partially made for overcoming certain drawbacks of $W_{10}O_{32}^{4-}$ in homogeneous reactions (*i.e.*, oxidations), such as small surface area, low quantum yields, poor selectivity and limited stability at pH higher than 2.5. Among the different techniques for the immobilization of decatungstate, the entrapping of decatungstate in membranes offered new possibilities for heterogeneous catalysis in the sense that the selective transport properties of the membranes can be used to improve the yields and selectivity of the reactions.

In this context, decatungstate was initially embedded within four different polymeric membranes, thus providing new heterogeneous photocatalytic systems.[32] Among them, polyvinylidene fluoride (PVDF) and polydimethylsiloxane (PDMS) based systems were found to be the most resistant, in terms of self-induced degradation upon irradiation in water (Scheme 17.16). Therefore, these latter systems were subsequently evaluated in the oxidation of several alcohols (aliphatic or aromatic) soluble in water. Although in all cases the corresponding homogeneous oxidation was faster, both membrane-based systems exhibited good performance. Thus, the heterogeneous photooxidation of these alcohols proceeded to completion in a few hours, yielding the corresponding carbonyl derivatives with good chemoselectivity at low substrate conversion.[32]

The preparation of new hybrid photocatalysts by embedding the fluorous-tagged decatungstate $(R_fN)_4W_{10}O_{32}$ $(R_fN = [CF_3(CF_2)_7(CH_2)_3]_3CH_3N^+)$

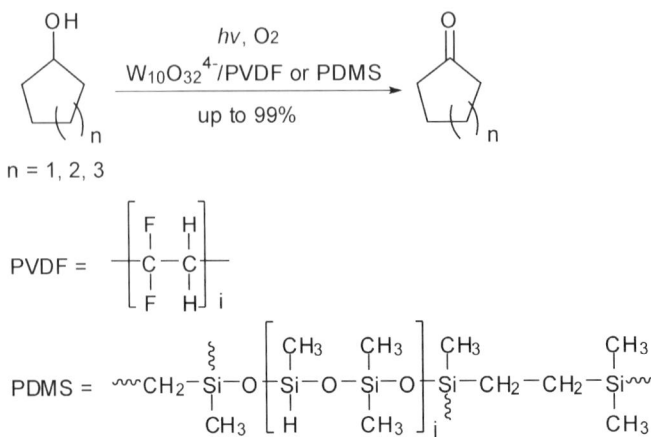

Scheme 17.16 Heterogeneous photooxidation of alcohols in water by photocatalytic membranes incorporating decatungstate.

within fluoropolymeric films, like Hyflon®, was later reported (Scheme 17.17).[33] The fluorophilic salt of decatungstate $(R_fN)_4W_{10}O_{32}$ has been isolated from the sodium salt $Na_4W_{10}O_{32}$, by cation metathesis with the $R_fNCH_3OSO_3$ salt; this fluorous-tagged decatungstate complex was then embedded in Hyflon® membranes. Perfluoropolymers, such as Hyflon®, offer several advantages when compared to other polymeric materials which, apart from their outstanding thermal and oxidative resistance, include their molecular oxygen preferential permeability. The resulting hybrid materials exhibited a remarkable improvement of both the morphology and performance, in comparison with the PVDF or Hyflon® membranes embedding the fluorous-free $(n\text{-}Bu_4N)_4W_{10}O_{32}$. The practical application of these photocatalytic membranes was demonstrated with the solvent-free oxygenation of benzylic hydrocarbons (*e.g.*, ethylbenzene, cumene, tetraline, indane) where exceptionally high turnover efficiency was observed (Scheme 17.17).

In more recent developments, plasma-treated membranes have been utilized for the heterogenization of decatungstate.[34] Typically, plasma processes have been extensively used in the past since they offer a convenient way to modify only the upper surface chemistry and properties of several polymer membranes while retaining their mechanical, physical, and chemical bulk properties. In this study, poly(vinylidene fluoride) (PVDF) membranes have been modified, by means of plasma processes, so that amino groups have been grafted at their surface, thus providing active sites for stable immobilization of decatungstate $W_{10}O_{32}{}^{4-}$.[34] In particular, the plasma grafted NH_2 groups were further

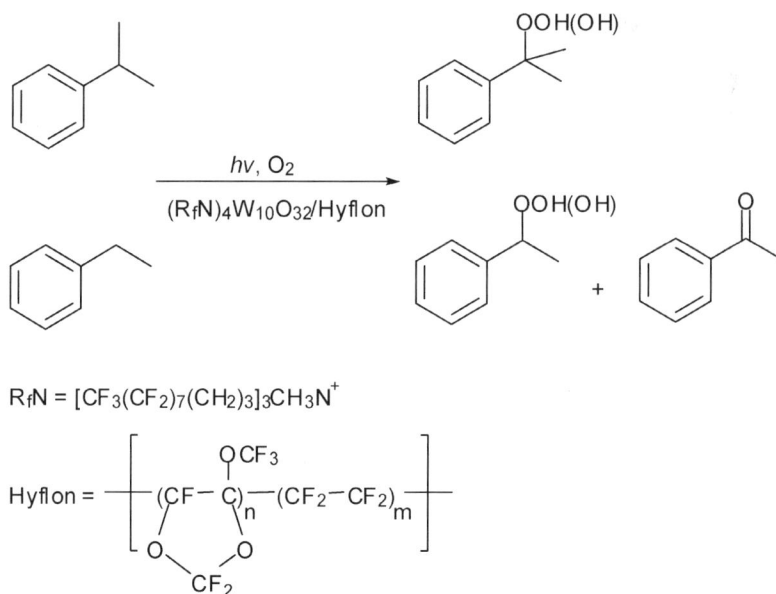

$R_fN = [CF_3(CF_2)_7(CH_2)_3]_3CH_3N^+$

$Hyflon = $

Scheme 17.17 Photooxidation of cumene and ethyl benzene by Hyflon® membranes embedding a fluorous-tagged decatungstate.

protonated to NH_3^+ groups under acidic conditions, and reacted with the tungsten anions, presumably, through electrostatic interaction. The resulting membranes were found to catalyze the complete degradation of phenol to CO_2 and H_2O very efficiently, under UV-vis irradiation. More importantly, the $W_{10}O_{32}^{4-}$ catalyst heterogenized on the surface of PVDF membranes exhibited improved catalytic performance with respect to the corresponding homogeneous system.

17.5   DEGRADATION OF ORGANIC POLLUTANTS BY DECATUNGSTATE

Pesticides are widely used in agriculture to enhance the efficiency of food production. Annually *ca.* 5 million tons of pesticides are applied to crops worldwide and eventually end up into aquatic ecosystems. Household pesticides are just as dangerous as agricultural ones. Not only do they have several toxic effects in living organisms but once groundwater is contaminated, the pesticide residues remain for long periods of time. Other organic pollutants, derived mostly from industrial wastes, including halogenated hydrocarbons, phenols, and thioethers have equally disastrous consequences for the environment.

Many of these chemicals present in aqueous media can undergo photo-chemical transformation with sunlight *via* direct or/and indirect photoreactions. Nowadays, advanced oxidation processes (AOPs) are particularly useful for waste water purification, by mineralization of biologically toxic or non-degradable materials, through the catalytic reaction with highly oxidizing reagents such as hydroxyl radicals. In the search for effective catalysts in this vital area of research, polyoxometalates (POMs) have attracted great interest, with decatungstate anion ($W_{10}O_{32}^{4-}$) possessing a rather prominent role. Additionally, the catalytic efficiency of polyoxotungstates (*i.e.*, decatungstate) is comparable with that of TiO_2, a well established catalyst with several applications in decontamination of wastewater using solar or UV irradiation; more importantly, decatungstate was proved, in certain cases, much more effective than TiO_2.

The pioneer studies of Hill and co-workers in the field of degradation of toxic pollutants, paved the way for decatungstate to be used as an effective "green" material in photocatalytic decontamination technology.[39–41] Initially, almost two decades ago, the photochemical degradation of thioether substrates catalyzed by polyoxotungstates, including $W_{10}O_{32}^{4-}$, was examined under both anaerobic and aerobic conditions.[39,40] In particular, under anaerobic conditions in solution at ambient temperature, the excited state of $W_{10}O_{32}^{4-}$ reacts with thioethers in high selectivity by α-hydrogen abstraction to form the corresponding α-carbon radicals and the one- and two-electron-reduced complexes, $W_{10}O_{32}^{5-}$ and $W_{10}O_{32}^{6-}$ (formed in a *ca.* 1:4 molar ratio, respectively). The principal reduced form of the catalyst ($W_{10}O_{32}^{6-}$) reduces thioethers to effect C–S bond cleavage. Thus, the net observed products, in anaerobic decatungstate-photocatalyzed oxidation of these substrates, are the dimers resulting from coupling of the α-carbon radicals and C–S bond cleavage products including the hydrocarbon resulting from complete desulfurization of the thioether.[39] For example, photochemical oxidation of dimethyl sulfide by $W_{10}O_{32}^{4-}$ under anaerobic conditions afforded 1,2-bis(methylthio)ethane, in high yield, *via* dimerization of the parent compound (Scheme 17.18, eqn (17.15)). In addition, photochemical oxidation of bis(chloromethyl) sulfide under the same conditions generated 1,1-bis(chloromethylthio)methane, a product implicating C–S bond cleavage (Scheme 17.18, eqn (17.16)).

Neither sulfoxides nor sulfones, the usual products of thioether oxidation by oxometal species, are produced under these anaerobic reaction conditions. On the other hand, the aerobic $W_{10}O_{32}^{4-}$-catalyzed oxidative degradation of thioethers including tetrahydrothiophene, 2-chloroethyl ethyl sulfide (half mustard), methyl phenyl sulfide, diphenyl sulfide, dibutyl sulfide, and dimethyl sulfide led, principally, to sulfoxides and sulfones by a complicated process that involves, in part, radical-chain autoxidation.[39] The oxidation of thioethers took place readily in water (even when the thioethers were immiscible), the medium most likely to be used in decontamination, as well as in acetonitrile. The reactivity of $W_{10}O_{32}^{4-}$ increased upon addition of O_2, while it was found to be dependent on the cation, with the sodium salt of the complex more reactive than the tetra-*n*-butylammonium salt.[39] Decatungstate ($W_{10}O_{32}^{4-}$) was also the most effective catalyst of all those polyoxotungstate systems evaluated under either aerobic or anaerobic conditions.[40] Furthermore, it was found that the homogeneous polyoxotungstate systems (*i.e.*, decatungstate) were quite effective in the oxidative degradation of thioethers (*e.g.*, tetrahydrothiophene) in contrast to semiconductor metal oxides (*e.g.*, TiO$_2$, SnO$_2$), which were completely ineffective.[40]

Additionally, Hill and co-workers have also reported an extensive study regarding the decatungstate-mediated cleavage of carbon–halogen bonds (C–X, X = Cl or Br).[41] The practical interest in processes that cleave carbon–halogen bonds, particularly C–Br and C–Cl bonds, stems from the generic toxicity and carcinogenicity of this ubiquitous class of organic materials (*i.e.*, pesticides, polychlorinated biphenyls (PCBs), and halogenated waste solvents). Typically, the catalytic photochemical degradation of such organic materials (*i.e.*, halocarbons and thioethers) by decatungstate is of great commercial importance. Moreover, the low toxicity of decatungstate and its

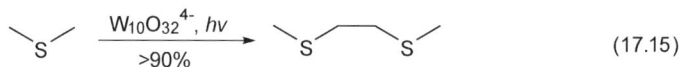

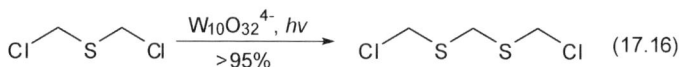

$$\diagdown_S\diagup \quad \xrightarrow[>90\%]{W_{10}O_{32}^{4-},\ h\nu} \quad \diagdown_S\diagup\diagdown_S\diagup \qquad (17.15)$$

$$Cl\diagdown_S\diagdown Cl \quad \xrightarrow[>95\%]{W_{10}O_{32}^{4-},\ h\nu} \quad Cl\diagdown_S\diagdown_S\diagdown Cl \qquad (17.16)$$

Scheme 17.18   Photochemical oxidation of sulfides under anaerobic conditions.

accessibility in quantity from inexpensive starting materials, initiated possible uses of this compound in the photocatalytic decontamination technology.

These interesting applications of polyoxotungstates in the field of water decontamination have been systematically explored by Papaconstantinou and co-workers for the photochemical degradation and also for efficient minerali- zation of a variety of organic pollutants, including chlorophenols and chloroacetic acids.[42] Chlorophenols are among the top priority pollutants. Most chlorophenolic compounds present in waste waters mainly arise from chemical intermediates or by-products in petrochemical, paper making, plastic, pesticidal, and insecticidal industries and also the conventional drinking water disinfection. In their studies, Papaconstantinou and co-workers found that decatungstate,[42] among other polyoxotungstates, was an effective photocata- lyst for a variety of organic pollutants, leading to their decomposition upon photolysis with near visible and UV light, to CO_2 and H_2O, and Cl^- in the case of chlorinated hydrocarbons (Scheme 17.19). As representative pollutants, several organic compounds including chlorinated acetic acids (*i.e.*, mono- chloro- and trichloroacetic acid), aromatic (*i.e.*, *o*-, *m*- and *p*-chlorophenol, phenol, *p*-cresol, 4-nitrophenol, and 2,4-dimethylphenol), and aliphatic (*i.e.*, 1,1,2-trichloroethane) compounds, were studied. In all these decomposition reactions, the main oxidant was shown to be hydroxyl radicals ˙OH formed by the reaction of the excited polyoxotungstate with H_2O. These ˙OH radicals may react by hydrogen abstraction (*i.e.*, by the well-known α-carbon hydrogen abstraction in carboxylic acids or alcohols) and/or addition with organic substrates leading to their photodegradation. This was further supported by product analysis (*i.e.*, hydroxylated products), by ESR-detection of ˙OH radicals upon photolysis of decatungstate in aqueous solution, as well as by the order of the chlorinated acetic acids decomposition: $Cl_3CCOOH \ll ClCH_2COOH \sim CH_3COOH$; no α-hydrogen exists in Cl_3CCOOH and as such no H-abstraction by •OH radicals is possible.

Later, in 1999, Giannotti and co-workers reported a comparative study on the photocatalytic efficiency of TiO_2 and $Na_4W_{10}O_{32}$ by means of solar photodegradation of phenols and pesticides such as bromoxynil, atrazine, imidachloprid and oxamyl pesticides in aqueous solution.[43] TiO_2 was found to be the most effective photocatalyst in terms of the degradation rate and of the mineralization of the compounds. However, the decatungstate anion appeared to be more efficient in the case of formulated pesticides, such as in the presence of surfactants.[43] The major drawback to the practical applications of this

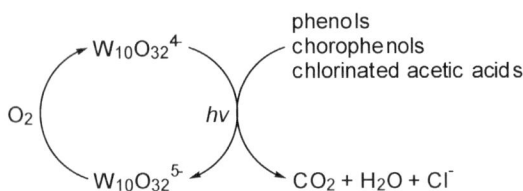

Scheme 17.19   Mineralization of organic pollutants by decatungstate in water.

homogeneous photocatalytic system stems from the high water solubility of decatungstate, which impedes its separation from the treated aqueous wastes. However, this problem could be overcome by immobilization of decatungstate on a solid support as was later pointed out by Hu and co-workers [25,26]

Accordingly, in 2001, $Na_4W_{10}O_{32}$ and $(n\text{-}Bu_4N)_4W_{10}O_{32}$ were immobilized inside the silica network *via* a sol–gel technique, resulting in the $Na_4W_{10}O_{32}/SiO_2$ and $(n\text{-}Bu_4N)_4W_{10}O_{32}/SiO_2$ composites.[25] Their photocatalytic activity was tested by degradation and mineralization of organophosphorus pesticide, trichorofon (TCF, 1-hydroxyl-2,2,2-trichloroethylphosphonate-*o,o*-dimethyl ester), under near-UV irradiation in aqueous media. In addition, their photo-chemical behavior was compared with two silica-supported decatungstates, $Na_4W_{10}O_{32}\text{-}SiO_2$ and $(n\text{-}Bu_4N)_4W_{10}O_{32}\text{-}SiO_2$. The highest photocatalytic activity was achieved for microporous $Na_4W_{10}O_{32}/SiO_2$; more importantly, this latter system appears to be sufficiently robust to prevent leaching of the immobilized decatungstate anion in the solution during the photocatalytic process. The above experiments were performed under atmospheric pressure. When the suspension was saturated with oxygen, mineralization of TCF into Cl^- ion over the microporous $Na_4W_{10}O_{32}/SiO_2$ decreased greatly. This was explained as follows: in the presence of pure oxygen, reoxidation of $W_{10}O_{32}^{5-}$ to $W_{10}O_{32}^{4-}$ was so rapid that the photooxidation of TCF with $Na_4W_{10}O_{32}/SiO_2$ was retarded severely. Therefore, TCF was only photodegraded into organic acids but not mineralized into CO_2 and H_2O continuously. Total mineralization of aqueous TCF (into CO_2, H_2O, and other inorganic ions such as Cl^-, $H_2PO_4^-$, and H^+) over the $Na_4W_{10}O_{32}/SiO_2$ composite was achieved only under atmospheric pressure. Studies on the reaction mechanism indicate that ˙OH radical attack is most likely responsible for the degradation and final mineralization of TCF. Besides, adsorption might play also a key role in heterogeneous photocatalyst chemistry, and rapid diffusion of the reactants into the micropores of the catalyst may achieve high photoactivities, so that the mineralization may be performed by direct oxidation *via* excited $W_{10}O_{32}^{4-}$.

Another example of microporous decatungstate application was later reported for the degradation of a model molecule, hydroxy-butanedioic acid (malic acid) (Scheme 17.20).[26] Among the three polyoxotungstate/SiO_2 composites studied, $Na_4W_{10}O_{32}/SiO_2$ was found to be the most photocataly-tically active material allowing for total degradation of an aqueous malic acid (100 mg l^{-1}) into several intermediates (*i.e.*, oxalic, glyceric, tartaric, butenedioic, acetic, and formic acid) in only 90 min of irradiation. After subsequent 60 min irradiation under the same conditions, these intermediates were totally mineralized into CO_2 and H_2O. The photocatalytic activity of the polyoxotungstate/SiO_2 composites was ascribed to their unique structure and microporosity, so that the photodegradation reaction could be performed effectively on the surface and in the micropores of the catalysts. Further studies showed that both ˙OH radical attack and direct oxidation of malic acid contribute in the overall mechanism.

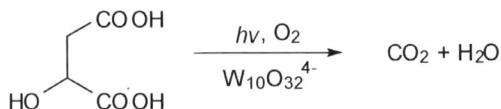

Scheme 17.20   Decatungstate-catalyzed mineralization of hydroxy-butanedioic acid.

The degradation of 4-chlorophenol by microporous polyoxotungstates, including decatungstate, was also reported in the same year.[27] The photo-catalytic degradation of this compound was tested as a probe reaction since chlorophenols represent, as mentioned above, an important class of environmental, top priority, water pollutants. The silica-immobilized polyoxotungstates were prepared by means of the sol–gel hydrothermal technique, through the hydrolysis of tetraethoxysilane in aqueous solution of the corresponding polyoxotungstate. The photocatalyzed degradation of 4-chlorophenol was performed by irradiation with near-UV light under aerobic conditions in aqueous media. During irradiation, 4-chlorophenol first dechlorinated to form hydroquinone and p-benzoquinone, and then these intermediates further mineralized to CO_2 and H_2O (Scheme 17.21). Among those polyoxometalates studied, the decatungstate-based composite, $W_{10}O_{32}^{4-}/SiO_2$, exhibited the highest activity affording a complete mineralization of 4-chlorophenol after 60 min of irradiation. The homogeneous precursor $W_{10}O_{32}^{4-}$ was also found to be the most active catalyst between the polyoxotungstates studied, under the same reaction conditions, albeit in a lower yield compared with its heterogeneous analogue. ESR studies indicated that the generation of ˙OH radicals during irradiation was responsible for the initial oxidation and subsequent mineralization of 4-chorophenol to CO_2, H_2O, and HCl. This result is in complete agreement with those from Papaconstantinou and co-workers who had also found that ˙OH radicals were the main oxidant of decomposition reactions in aqueous media.[42]

Recently, the degradation of the pesticide metsulfuron methyl (a sulfonyl urea herbicide) was studied in aqueous solutions by using decatungstate anion, $W_{10}O_{32}^{4-}$, as a photocatalyst (Figure 17.3).[44]

The influence of various parameters, such as oxygen concentration and pH, on the photocatalytic degradation was also studied.[44] It was shown that oxygen played a major role in the decomposition of the pesticide by increasing the degradation rate with increasing the oxygen concentration. Additionally,

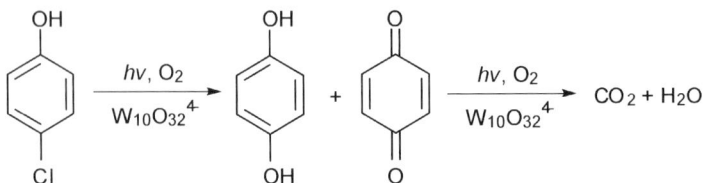

Scheme 17.21   Decatungstate-mediated degradation and subsequent mineralization of 4-chlorophenol.

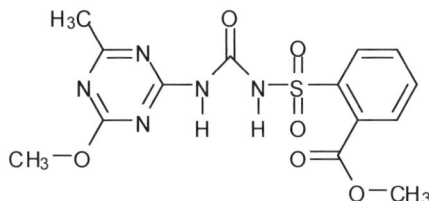

Figure 17.3 Metsulfuron methyl.

the maximum efficiency in the degradation process was observed in both acid and neutral solutions. In aerated or oxygen-saturated solutions, the system was found to operate in a photocatalytic way with the regeneration of the starting decatungstate species by molecular oxygen. A mechanistic scheme for the complete degradation of metsulfuron methyl has been proposed which, concisely, showed that the photodegradation of this pesticide occurs through three different reaction pathways involving the three main parts of the chemical structure: the aromatic ring, the sulfonyl urea bridge and (to a lesser extent) the methoxy group of the triazine moiety.

17.6 CONCLUSION AND FUTURE DIRECTION

In this chapter some of the progress made in the emerging field of decatungstate catalysis, with an emphasis on synthetic implications in organic chemistry has been outlined. Thus, we have traced the developments of decatungstate from its employment as a homogeneous catalytic system under either aerobic or anaerobic conditions to its heterogenization for the oxidation of alkanes, alkenes, alcohols, and sulfides, and its applications in water decontamination technology. Some of the newest developments outlined above, as well as the ongoing interest on this subject, show that there are more yet-unexplored applications of decatungstate catalysis that are likely to emerge in the future. In essence, this review provides an important impulse for further studies in the field of decatungstate catalysis, so that the synthesis of new decatungstate-based materials for further synthetic applications can be envisaged.

REFERENCES

1. C. L. Hill (Ed.), *J. Mol. Catal. A: Chem.*, 2007, **262**, 1 (Special Issue: Polyoxometalates in Catalysis).
2. C. Tanielian, *Coord. Chem. Rev.*, 1998, **178–180**, 1165.
3. For comprehensive reviews on the functionalization of unactivated C-H bonds in alkanes by polyoxometalates including decatungstate, see: (a) C. L. Hill, *Synlett*, 1995, 127, and references therein; (b) C. L. Hill, L. Delannoy, D. C. Duncan, I. A. Weinstock, R. F. Renneke, R. S. Reiner, R. H. Atalla, J. W. Han, D. A. Hilleshein, R. Cao, T. M. Anderson, N. M. Okun, D. G. Musaev and Y. V. Geletii, *C. R. Chimie*, 2007, **10**, 305, and

references therein; (c) D. Ravelli, D, Dondi, M. Fagnoni, A. Albini, *Chem. Soc. Rev.*, 2009, **38**, 1999, and references therein.

4. Z. Zheng and C. L. Hill, *Chem. Commun.*, 1998, 2467.
5. D. Dondi, M. Fagnoni, A. Molinari, A. Maldotti and A. Albini, *Chem.– Eur. J.*, 2004, **10**, 142.
6. D. Dondi, M. Fagnoni and A. Albini, *Chem.–Eur. J.*, 2006, **12**, 4153.
7. C. Tanielian, R. Seghrouchni and C. Schweiter, *J. Phys. Chem. A*, 2003, **107**, 1102 and references therein.
8. I. Texier, J. A. Delaire and C. Giannotti, *Phys. Chem. Chem. Phys.*, 2000, **2**, 1205 and references therein.
9. H. Duclusaud and S. A. Borshch, *Chem. Phys. Lett.*, 1998, **290**, 526.
10. D. C. Duncan and M. A. Fox, *J. Phys. Chem. A*, 1998, **102**, 4559.
11. S. Esposti, D. Dondi, M. Fagnoni and A. Albini, *Angew. Chem., Int. Ed.*, 2007, **46**, 2531.
12. M. D. Tzirakis and M. Orfanopoulos, *J. Am. Chem. Soc.*, 2009, **131**, 4063.
13. S. Angioni, D. Ravelli, D. Emma, D. Dondi, M. Fagnoni and A. Albini, *Adv. Synth. Catal.*, 2008, **350**, 2209.
14. M. D. Tzirakis and M. Orfanopoulos, *Org. Lett.*, 2008, **10**, 873.
15. C. Tanielian, K. Duffy and A. Jones, *J. Phys. Chem. B*, 1997, **101**, 4276.
16. C. Giannotti and C. Richter, *Trends Photochem. Photobiol.*, 1997, **4**, 43 and references therein.
17. A. Maldotti, R. Amadelli, V. Carassiti and A. Molinari, *Inorg. Chim. Acta*, 1997, **256**, 309 and references therein.
18. I. N. Lykakis and M. Orfanopoulos, *Tetrahedron Lett.*, 2005, **46**, 7835 and references therein.
19. I. N. Lykakis, C. Tanielian, R. Seghrouchni and M. Orfanopoulos, *J. Mol. Catal. A: Chem.*, 2007, **262**, 176 and references therein.
20. A. Molinari, R. Amadelli, V. Carassiti and A. Maldotti, *Eur. J. Inorg. Chem.*, 2000, 91.
21. I. N. Lykakis and M. Orfanopoulos, *Synlett*, 2004, 2131.
22. M. D. Tzirakis, I. N. Lykakis, G. D. Panagiotou, K. Bourikas, A. Lycourghiotis, C. Kordulis and M. Orfanopoulos, *J. Catal.*, 2007, **252**, 178.
23. A. Maldotti, A. Molinari, G. Varani, M. Lenarda, L. Storaro, F. Bigi, R. Maggi, A. Mazzacani and G. Sartori, *J. Catal.*, 2002, **209**, 210.
24. A. Molinari, R. Amadelli, A. Mazzacani, G. Sartori and A. Maldotti, *Langmuir*, 2002, **18**, 5400.
25. Y. Guo, C. Hu, X. Wang, Y. Wang and E. Wang, *Chem. Mater.*, 2001, **13**, 4058.
26. Y. Guo, C. Hu, S. Jiang, C. Guo, Y. Yang and E. Wang, *Appl. Catal., B*, 2002, **36**, 9.
27. B. Yue, Y. Zhou, J. Xu, Z. Wu, X. Zhang, Y. Zou and S. Jin, *Environ. Sci. Technol.*, 2002, **36**, 1325.
28. F. Bigi, A. Corradini, C. Quarantelli and G. Sartori, *J. Catal.*, 2007, **250**, 222.

29. A. Maldotti, A. Molinari and F. Bigi, *J. Catal.*, 2008, **253**, 312.
30. E. Fornal and C. Giannotti, *J. Photochem. Photobiol., A*, 2007, **188**, 279.
31. A. Molinari, G. Varani, E. Polo, S. Vaccari and A. Maldotti, *J. Mol. Catal. A: Chem.*, 2007, **262**, 156.
32. M. Bonchio, M. Carraro, G. Scorrano, E. Fontananova and E. Drioli, *Adv. Synth. Catal.*, 2003, **345**, 1119.
33. M. Carraro, M. Gardan, G. Scorrano, E. Drioli, E. Fontananova and M. Bonchio, *Chem. Commun.*, 2006, 4533.
34. L. C. Lopez, M. G. Buonomenna, E. Fontananova, G. Iacoviello, E. Drioli, R. d'Agostino and P. Favia, *Adv. Funct. Mater.*, 2006, **16**, 1417.
35. A. Molinari, R. Amadelli, L. Andreotti and A. Maldotti, *J. Chem. Soc., Dalton Trans.*, 1999, 1203.
36. A. Maldotti, R. Amadelli, I. Vitali, L. Borgatti and A. Molinari, *J. Mol. Catal. A: Chem.*, 2003, **204**, 703.
37. S. Farhadi and Z. Momeni, *J. Mol. Catal. A: Chem.*, 2007, **277**, 47.
38. M. Bonchio, M. Carraro, G. Scorrano and A. Bagno, *Adv. Synth. Catal.*, 2004, **346**, 648.
39. R. C. Chambers and C. L. Hill, *J. Am. Chem. Soc.*, 1990, **112**, 8427.
40. R. C. Chambers and C. L. Hill, *Inorg. Chem.*, 1991, **30**, 2776.
41. D. Sattari and C. L. Hill, *J. Am. Chem. Soc.*, 1993, **115**, 4649.
42. A. Mylonas, A. Hiskia and E. Papaconstantinou, *J. Mol. Catal. A: Chem.*, 1996, **114**, 191 and references therein.
43. I. Texier, C. Giannotti, S. Malato, C. Richter and J. Delaire, *Catal. Today*, 1999, **54**, 297.
44. S. Rafqah, P. W.-W. Chung, C. Forano and M. Sarakha, *J. Photochem. Photobiol., A*, 2008, **199**, 297.

CHAPTER 18

Radical Domino Reactions: Intermolecular Telescopic Reactions[*]

18.1 INTRODUCTION: ADVANTAGES AND LIMITS

Besides carbon–carbon-bond formations by aldol reactions, transition metal-catalyzed couplings and pericyclic reactions, transformations based on radicals have become an indispensable alternative as a non-polar method for the connection of carbon atoms. For quite some time, chemists have hesitated to apply transformations involving free radicals for the synthesis of fine chemicals, because they were afraid of the radicals' high reactivity and thus non-selectivity, as well as their unpredictability in product formation. On the other hand, radical chemistry was always very important for the synthesis of bulk chemicals. During the past few decades, radical reactions have been rehabilitated as a versatile mainstream tool for the controlled preparation of complex molecules in organic synthesis. This was made possible through intense research in physical organic chemistry, bringing light into the rather marginal knowledge of the behavior of radicals. Since radical reactions are ideal for sequencing, due to the very fundamental reason that the product of every radical reaction is a radical, they have opened the door for many efficient and elegant domino processes.

Nowadays, domino radical reactions are prized for their ability to build complex, highly substituted ring systems, and for their general tolerance of functionalities in the substrates, allowing transformations with a minimum use of protecting groups. In marked contrast to polar processes, radical transformations can proceed in most cases in the presence of free hydroxyl and amino groups, as well as keto and ester functionalities. Not surprisingly, high grades of chemo-, regio- and stereoselectivity can be obtained, a characteristic attributable to the mild reaction conditions, which are applied in radical

* Chapter written by V. Tamara Perchyonok.
Streamlining Free Radical Green Chemistry
V. Tamara Perchyonok, Ioannis Lykakis and Al Postigo
© V. Tamara Perchyonok, Ioannis Lykakis and Al Postigo 2012
Published by the Royal Society of Chemistry, www.rsc.org

chemistry. Another advantageous feature of the use of radicals is the fact that it is equally feasible to add them to either inactivated double and triple bonds or to those bearing polarizing groups. As a consequence, the development of new domino radical reactions continues at a vigorous pace, their beneficial contributions in the field of organic chemistry having been documented by a multitude of publications.

Usually, free-radical domino processes are characterized by a sequence of intramolecular steps, the overall propagation coordinate being unimolecular (excluding initiation and termination steps) (Scheme 18.1).[1] The most relevant counterpart of these unimolecular reactions is represented by reactions in which one step—in many cases the first—is an intermolecular radical addition to an appropriate functionalized acceptor (Scheme 18.2).

If the attacking radical contains an adequately placed radical acceptor functionality, the possibility of a radical cycloaddition is provided, offering a procedure to construct cyclic products from acyclic precursors. For this type of ring-forming process, in which two molecular fragments are united with the formation of two new bonds, the term "annulation" has been adopted (Scheme 18.3).

Oligomerizations and polymerizations in which many radical additions to a limited range of alkenes (or other acceptors) take place will not be discussed in this book, although they are typical domino reactions. However, they usually do not lead to single well-defined products.

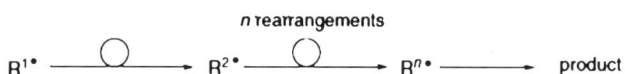

$$R^{1} \bullet \quad \longrightarrow \quad R^{2} \bullet \quad \xrightarrow{\ n \text{ rearrangements}\ } \quad R^{n} \bullet \quad \longrightarrow \quad \text{product}$$

Scheme 18.1   General scheme for an intramolecular domino radical reaction.

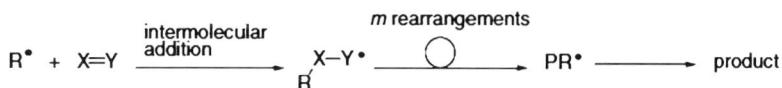

$$R^{\bullet} + X{=}Y \xrightarrow[\text{addition}]{\text{intermolecular}} \underset{R}{X{-}Y^{\bullet}} \xrightarrow{\ m \text{ rearrangements}\ } PR^{\bullet} \longrightarrow \text{product}$$

Scheme 18.2   General scheme for an intermolecular domino radical reaction.

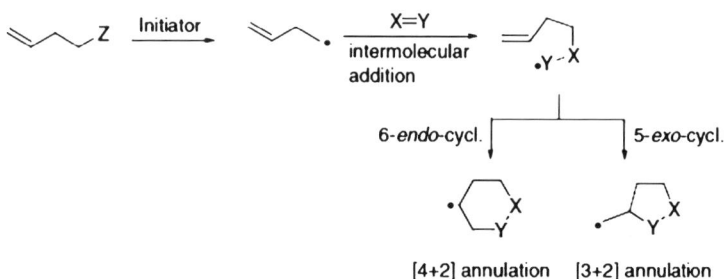

Scheme 18.3   Intermolecular radical addition leading to cyclic products.

Radical domino processes follow a general scheme involving the generation of free radicals as an initiation step. Specifically, the formation of radicals can either proceed by abstraction or substitution utilizing halides, as well as phenylthio or phenylselenium compounds as substrates and stannanes such as $n\text{Bu}_3\text{SnH}$, silanes and germanes as initiators, or by redox processes employing transition metals or lanthanides. Since the often-utilized organo tin compounds bear a toxic potential, there is an ongoing search to use alternative methodologies for the creation of radicals. These should be on the one hand non-polluting and safe, and on the other hand efficient and reliable. Nevertheless, the provided radical intermediate undergoes either an intermolecular addition to an acceptor molecule or a unimolecular rearrangement. Furthermore, the sequence can be carried on until the final radical is captured by reduction, oxidation or atom transfer, resulting in the formation of the desired product. A fundamental requirement for a radical domino reaction to proceed effectively is that the rates of each individual rearrangement must be rapid compared to the termination reactions (combination, disproportionation, redox) of the radical intermediates and additionally, in comparison with reactions with the solvent, the precursor, or initiator molecules. Another important fact is the selective reaction of the final radical product (and not the remainder of the intermediate radicals) with the designated acceptor. This elementary goal can be achieved by creating a final radical which displays a significant change in the polarity or reactivity, for example an *O*-centered or vinyl-type radical. An excellent review by Walton and McCarroll has recapitulated the various processes which can be featured during a radical cascade.[2] Moreover, these authors have elaborated a compilation of classes of unimolecular free-radical rearrangements, as illustrated in Scheme 18.4.

The first type of process is characterized by cyclization reactions, which are found in a plethora of examples and hence can be considered as the "flagship" of the different classes being discussed in this section. In spite of the fact that this reaction type distinguishes a broad scope of subsections, the 5-*exo*-trig ring closure can be regarded as the most frequent and productive one. Furthermore, 6-*endo* and 6-*exo* processes are also encountered in radical chemistry, though less often. Nevertheless, the well-established Baldwin's rules elucidate the favored and disfavored cyclization modes, providing a helpful guidance for synthetic predictions and mechanistic assumptions.[3] Another important feature that can be imputed to radical cyclization methodologies, is the ability to construct carbon–carbon bonds at centers exhibiting a high steric demand and in addition congested quaternary stereocenters—one of the most striking tasks in organic chemistry—in an efficient manner.[4] The second, albeit scarcely applied, class of radical domino reactions includes intramolecular homolytic substitutions, which can also result in the formation of the rings.[5] As a consequence, the radical is ejected with the displaced center and does not remain with the main ring; hence, this type of reaction is normally featured as terminating step.[6] The most typical complication occurring within substitutions is an undesired branching of a sequence caused by the released radical.[7]

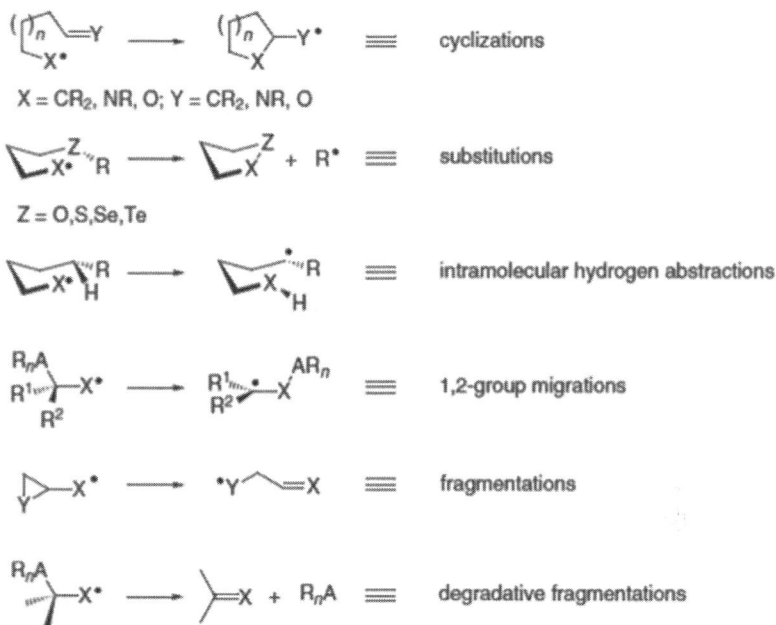

X = CR$_2$, NR, O; Y = CR$_2$, NR, O

cyclizations

Z = O,S,Se,Te

substitutions

intramolecular hydrogen abstractions

1,2-group migrations

fragmentations

degradative fragmentations

Scheme 18.4 Walton's and McCarroll's classification of unimolecular free-radical rearrangements.

Another more common group envisages intramolecular hydrogen abstraction processes. Both 1,5- and 1,6-hydrogen migrations should be mentioned at this point as the most prevalent appearing events leading to a translocation of the prior radical center.[8] Moreover, 1,2-group migration typifies a class of radical processes presented as the fourth example in Scheme 18.4. Most of them involve groups that feature some type of unsaturation such as aryl, vinyl, or carbonyl, whereas carbon centered groups, R$_3$C, with sp^3 hybridization are usually not able to afford a 1,2-migration. The final category is represented by fragmentations, whereupon ring-opening motifs, in which a single unsaturated radical is generated, is the most relevant type.

Substrates which are well-suited to this type of transformation include cyclopropylmethyl, oxiranylmethyl and cyclobutylmethyl radicals which exhibit a rapid ring opening. A special case of fragmentation is combined with a coexistent degradation such as decarboxylation.

Furthermore, Walton and McCarroll proposed a very logical and concise system for the classification of free-radical domino reactions, and this is presented in the following.[2] First, a capital letter is assigned to each of the above-described processes: cyclizations (**C**), substitution reactions (**S**), H-abstractions (**H**), 1,2-group migrations (**M**), and fragmentations (**F**). The corresponding symbols are combined with a suffix (*exo* = **x**, *endo* = **n**) clarifying the mechanistic details—for example, **C5x** codes a 5-*exo*-cyclization, and **C6n** a

6-*endo*-cyclization. By using this abbreviation system the specification of complex radical domino processes becomes much faster and convenient.

It is also worth emphasizing that the initiation and termination steps are not included in the central chain process. For instance, in metal hydride-promoted domino reactions the initial halogen abstraction (or SePh displacement, *etc.*) and the final hydrogen abstraction from R_nMH is not classified as part of the domino sequence. More precisely, only the propagation steps within the mechanism of this process will be considered as a strict integral part of the domino reaction. Nevertheless, a few hetero-radical domino processes, such as combinations with cationic, anionic, pericyclic and oxidative processes, have also been published to date, and insights into these more exotic types of procedure will be described here. As yet, however, no examples in which photochemically induced, transition metal-catalyzed, reductive or enzymatic processes were involved have been identified in the literature.

18.2 RADICAL/RADICAL DOMINO PROCESSES IN SYNTHESIS

As mentioned above, processes with two or more radical intermediates represent the majority of radical domino reactions. Of special interest is the use of this methodology for the efficient synthesis of natural products. The naturally occurring alkaloid lysergic acid has disclosed challenging pharmacological activities, and therefore has attracted the attention of both medicinal[9] and synthetic[10] chemists. Parsons and coworkers developed a radical domino approach for the facile construction of the tetracyclic ring system (Scheme 18.5).[11] After submitting the aromatic bromide to classical radical conditions (tributyltin hydride/azobisisobutyronitrile (AIBN), refluxing benzene), initially a 5-*exo*-trig cyclization took place to afford a *N*-heterocycle, which was followed by a 6-*endo*-trig cyclization. In this way, the desired product was obtained as a 3:1 mixture of two diastereoisomers, in 74% yield.

In 2002, Lee and coworkers prepared the unnatural (−)-enantiomer of lasonolide A using a domino radical cyclization procedure as the key step, and furthermore revised its structure.[12] Lasonolide A possesses a macrolactone structure with two embedded stereochemically demanding tetrahydropyran rings. The compound demonstrates antitumor activity, in particular against leukemia and the highly lethal lung carcinoma. Reaction of alkoxyacrylate

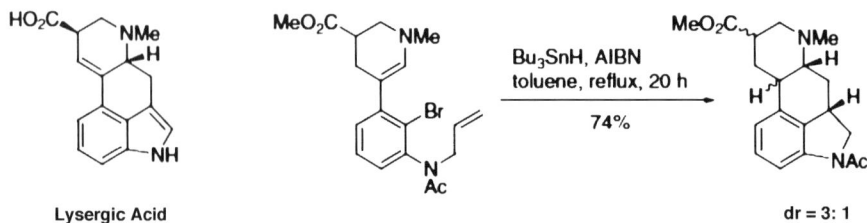

Scheme 18.5 Domino radical cyclization procedure in the synthesis of lysergic acid derivatives.

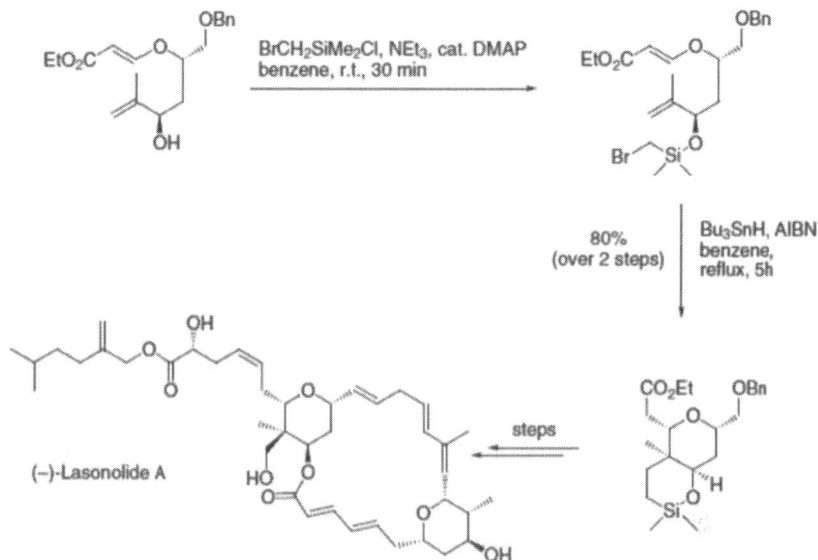

Scheme 18.6 Synthesis of (–)-lasonolide A

and (bromomethyl)-chloro-dimethylsilane provided corresponding silyl ether derivative, which was used for the following radical cyclization (Scheme 18.6).

Another effective combination of two radical cyclization steps has been demonstrated by Sha and coworkers during the course of the first total synthesis of (+)-paniculatine, a natural alkaloid belonging to the subclass of *Lycopodium* alkaloids.[13] (+)-Paniculatine has a unique tetracyclic scaffold with seven stereogenic centers.[14] Although no special features of (+)-paniculatine have so far been documented, other *Lycopodium* alkaloids are reported to be potent acetylcholinesterase inhibitors, or show promising results in the treatment of Alzheimer's disease.[15]

When standard radical conditions were applied to iodo ketone, an α-carbonyl radical is generated, which undergoes a 5-*exo*-dig cyclization forming a bicyclic vinyl radical intermediate (Scheme 18.7). This is followed by a 5-*exo*-trig ring closure onto the adjacent olefin moiety, which then provides the tricyclic core structure as a single diastereoisomer, in 82% yield.

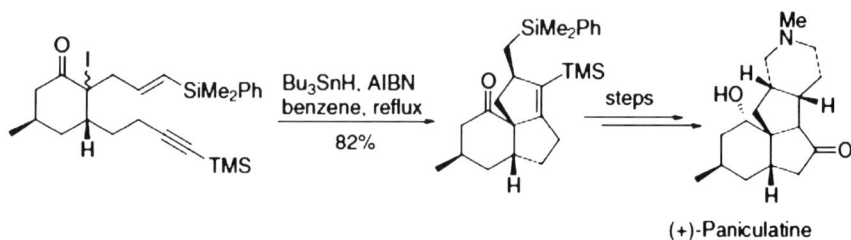

Scheme 18.7 Synthesis of (+)-paniculatine .

In 1997, a compound termed CP-263,114 was isolated from an unidentified fungal species by a group at Pfizer (Groton, USA).[16] The compound exemplified the architecture of unprecedented molecular connectivity, and possessed interesting biological activities such as cholesterol-lowering properties through the inhibition of squalene synthase.[16,17] Moreover, it was found to inhibit farnesyl transferase, an enzyme implicated in cancer.[16,18] The first total synthesis was accomplished by the Nicolaou group in 1999.[19] In another approach to CP- 263,114, the Wood group has established an intramolecular domino radical cyclization to afford the desired isotwistane core (Scheme 18.8).[20] First, the tertiary alcohol was converted into the corresponding acetal using dibromoethyl ether and N,N-dimethylaniline. Homolytic abstraction of the bromo atom furnished compounds as a mixture of diastereoisomers in 86% yield over two steps.[21] The process was suggested to proceed *via* initial formation of the primary radical, followed by a 5-*exo*-trig ring closure to give corresponding cyclized product, by reaction with the maleate moiety and subsequent addition of hydrogen from the less-hindered side.[22]

Natural products containing a spiropyrrolidinyloxindole nucleus have recently been found to exhibit interesting biological activity such as cell-cycle inhibition.[23] This observation encouraged Murphy and coworkers to design a novel domino route to horsfiline, a natural spiropyrrolidinyloxindole (Scheme 18.9).[24]

Treatment of azide with tris(trimethylsilyl)silane (TTMSS) and AIBN as radical starter led to the radical intermediate after a first ring closure, which underwent a second 5-*exo*-trig cyclization with concomitant loss of nitrogen providing a spirocycle *via* the corresponding radical intermediate. Methylation and debenzylation completed the synthesis of horsfiline. Another domino radical cyclization approach, which allows the construction of the B- and E-rings of the alkaloid aspidospermidine, has been described by the same group.[25]

Transformation of the iodoazide into the tetracycle was accomplished in 40% yield by selective attack at the carbon–iodine bond in the presence of the azide group, again using TTMSS/AIBN (Scheme 18.10).[26] The resulting aryl radical initiated a two-fold 5-*exo*-trig cyclization sequence which is accompanied by evolution of nitrogen. The skeleton of the indole alkaloid (–)-vindoline was prepared in a similar way.[27]

Since the discovery that the pyrroloquinoline alkaloid camptothecin and the related alkaloids mappicine, nothapodytine B and nothapodytine A exhibit significant anti-cancer and antiviral properties,[28] many synthetic approaches towards these compounds have been developed. Recently, the Bowman group devised a radical domino reaction protocol for the synthesis of the A–D ring system, featuring a vinyl radical cyclization onto nitriles.[29] Intramolecular cyclization of the vinyl radical, generated from vinyl iodide and hexamethylditin under photolysis, led to the iminyl radical (Scheme 18.11). This can either cyclize in a 6-*endo*-trig fashion to give the π-radical directly, or a 5-*exo*-trig

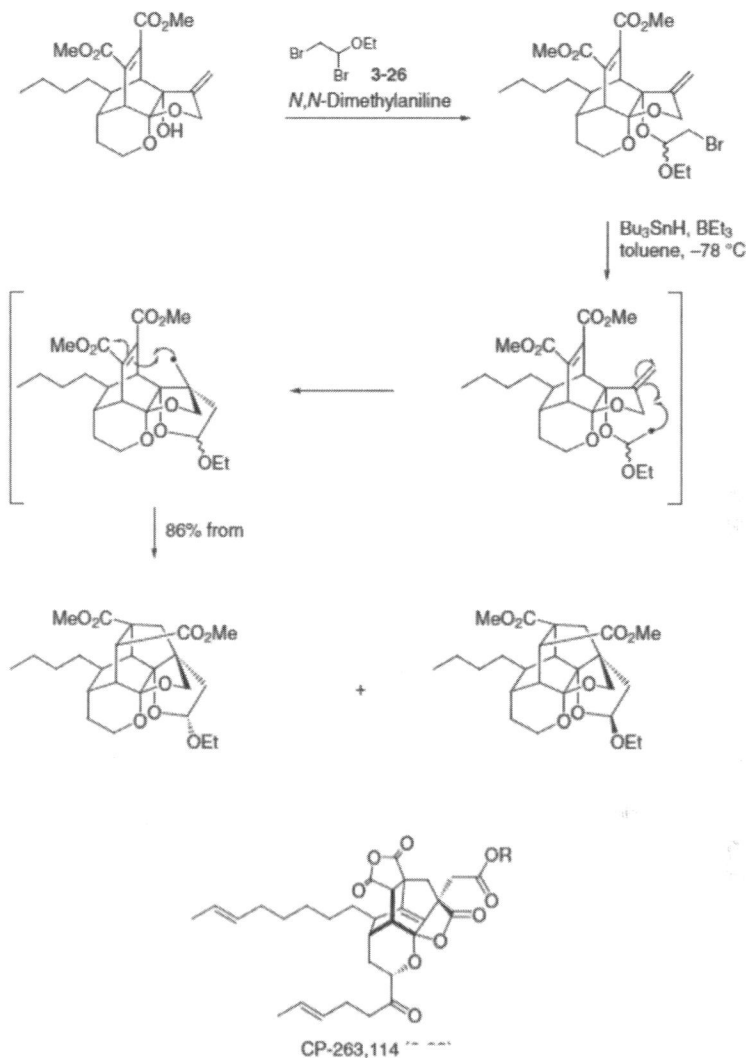

Scheme 18.8 Domino radical cyclization process towards the total synthesis of CP-263,114.

cyclization takes place leading to the spirodienyl radical, which is then converted into intermediate by a neophyl rearrangement. Finally, the loss of a hydrogen atom from the newly formed radical yields the tetracyclic product. The mechanism of the final oxidation step is still unclear; however, a H-abstraction caused by methyl radicals, formed from thermal breakdown of trimethyltin radicals, was suggested as an explanation.

Another anti-cancer agent in clinical use is podophyllotoxin; this has an aryl tetrahydronaphthalene lignan lactone skeleton, and demonstrates potent tubulin-binding, anti-mitotic properties (Scheme 18.12).[30] The Sherburn

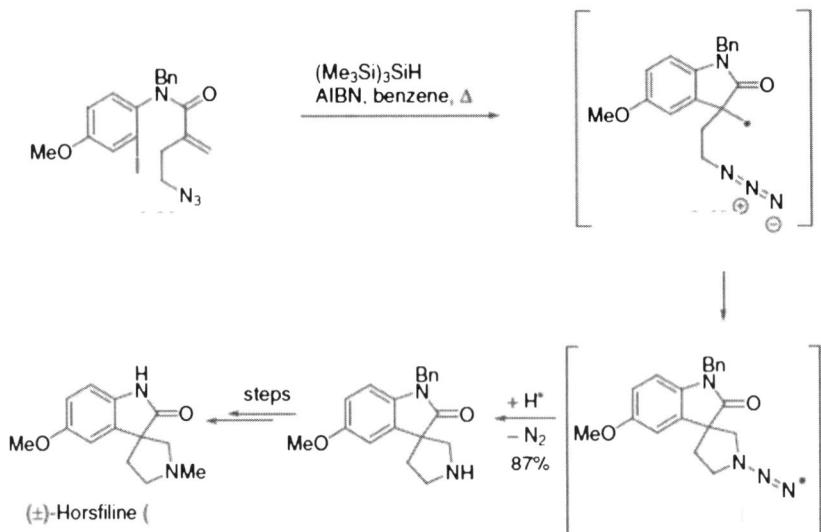

Scheme 18.9 Synthesis of horsfiline.

group[31] prepared this molecule by a tris(trimethylsilyl)silane-promoted conversion of thionocarbonate into the lactone, which proceeded with a yield of 38%. As intermediates, the corresponding radicals can be assumed.

Zard and coworkers[32] reported a simple approach to create another group of natural products, namely the lycopodium alkaloids.[15] These authors first investigated the reaction of *O*-benzoyl-*N*-allylhydroxylamide with tributyltin hydride and ACCN in refluxing toluene, which led (after formation of the corresponding *N*-radical in a 5-*exo*-trig/5-*exo*-trig cyclization) to the undesired pyrrolidine in 48% yield. Nevertheless, a small structural modification, namely the placement of a chlorine atom at the allyl moiety, induced a 5-*exo*-/6-*endo*- instead of the 5-*exo*-/5- *exo*-trig cyclization to give the wanted indolizidine in 52% yield as a single diastereomer. In this case, a second equivalent of tributyltin hydride was necessary to reductively remove the chlorine atom *in situ* after the twofold cyclization process. Corresponding cyclized product

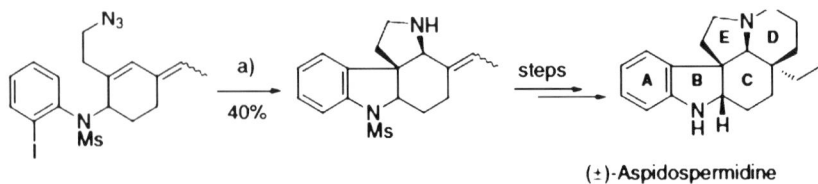

a) TTMSS, AIBN, benzene, reflux.   TTMSS = $(Me_3Si)_3SiH$

Scheme 18.10 Synthesis of ($\pm$)-aspidospermidine.

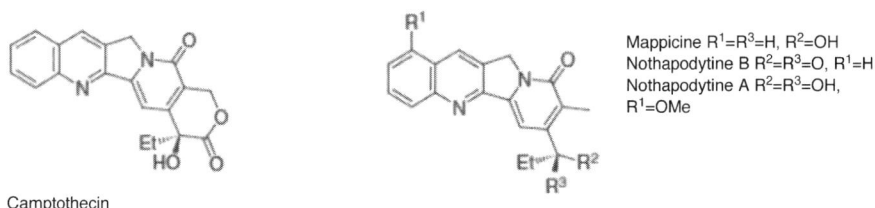

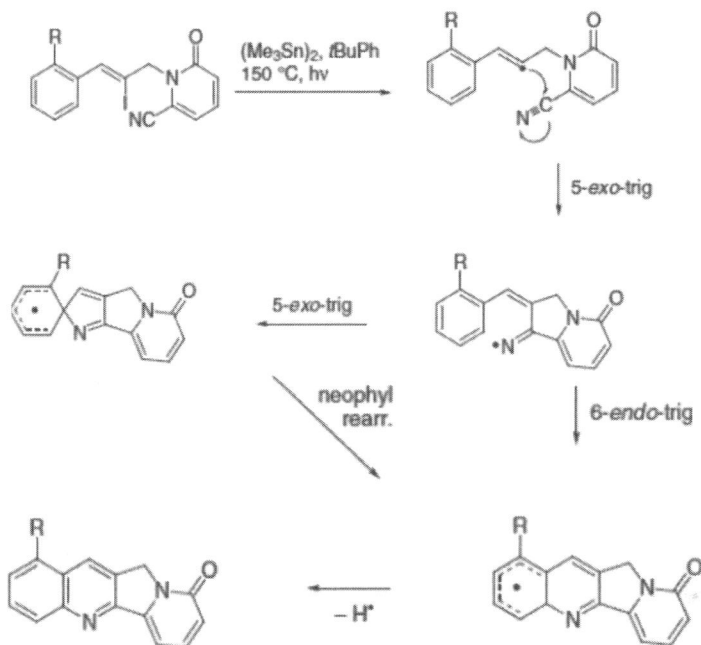

Mappicine R^1=R^3=H, R^2=OH
Nothapodytine B R^2=R^3=O, R^1=H
Nothapodytine A R^2=R^3=OH,
R^1=OMe

Camptothecin

R=H, 73%
R=OMe, 45%

Scheme 18.11 Domino radical procedure in the synthesis of camptothecin and analogs.

could be transformed into the lycopodium alkaloid 13-desoxyserratine (Scheme 18.13).

N-Aziridinylimines are valuable substrates for domino radical cyclizations since they are able to serve simultaneously as radical acceptors and donors. They allow a versatile and general construction of quaternary carbon centers from carbonyl compounds.[33] By employing this methodology, an elegant and stereoselective synthesis of modhephene, one of the rare naturally occurring [3.3.3]-propellanes, has been designed by the group of Lee and Kim (Scheme 18.14).[34] The necessary substrate was prepared by treatment of corresponding 2-oxocyclopentanecarboxylate with *N*-amino-2-phenylaziridine under acid-catalysis. In the presence of tributyltin hydride, corresponding adduct was transformed almost exclusively into the desired propellane. In a

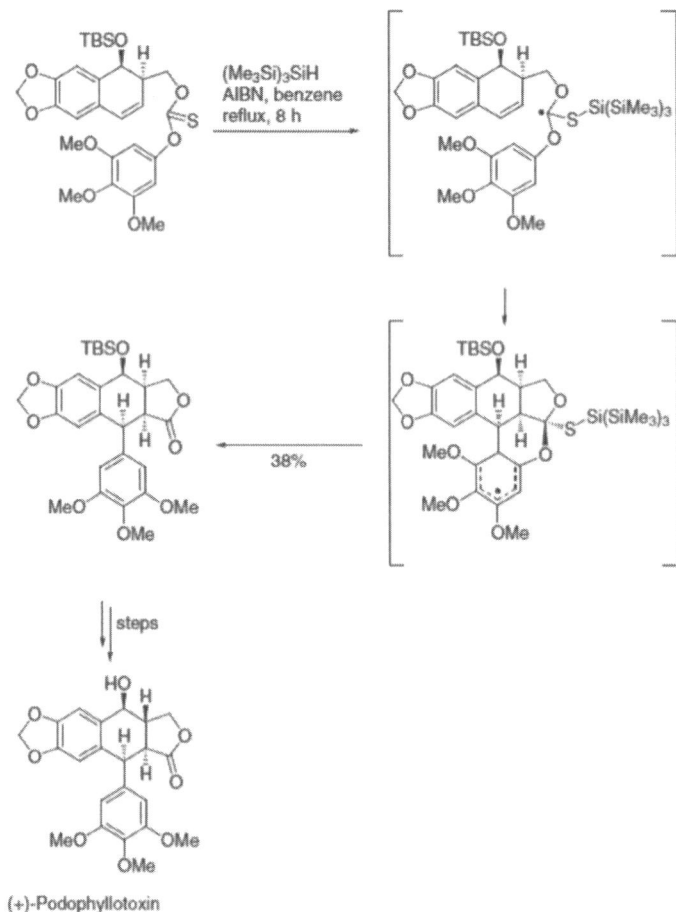

Scheme 18.12 Domino radical cyclization in the total synthesis of (+)-podophyllotoxin.

similar way, α-cedrene has been synthesized by the same group.[35] Moreover, these authors have also developed a further access to modhephene using a free-radical 6-*endo*-/5-*exo*-domino cyclization of a dieneyne.[36]

During the course of a short and efficient total synthesis of (–)-dendrobine, an alkaloid which exhibits antipyretic and hypotensive activities,[37] a new dominoradical sequence has been exploited by Cassayre and Zard which involves the cyclization of a carbamyl radical.[38] *O*-Benzoyl-*N*-hydroxyurethane reacts with tributyltin hydride and ACCN in refluxing toluene to the carbamyl radical, which directly undergoes a 5-*exo*-trig cyclization to furnish the oxazolidinone radical (Scheme 18.14). The carbon framework of then collapses by a radical fragmentation, providing the more stable radical, which is finally intercepted by tributyltin hydride. This results in formation of the annulated oxazolidinone with the desired stereochemistry, in 71% yield. In their enantioselective total synthesis of (+)-triptocallol, a naturally occurring

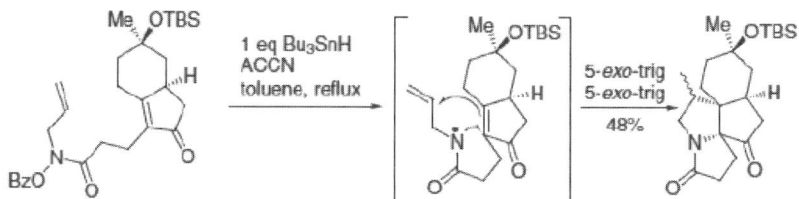

ACCN = 1,1'-azobis(cyclohexanecarbonitrile)

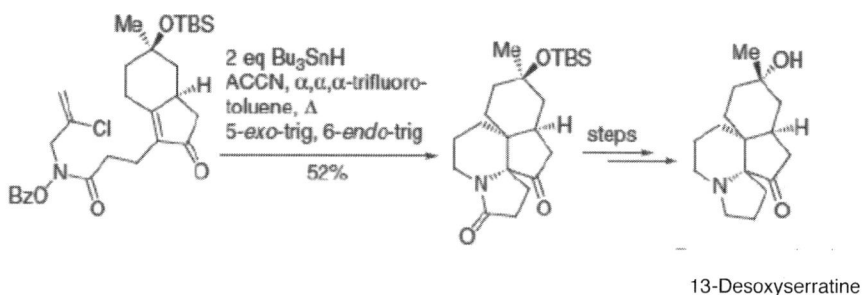

13-Desoxyserratine

Scheme 18.13 Synthesis of pyrrolidines and indolizidines as well as of the lycopodium alkaloid 13-desoxyserratine.

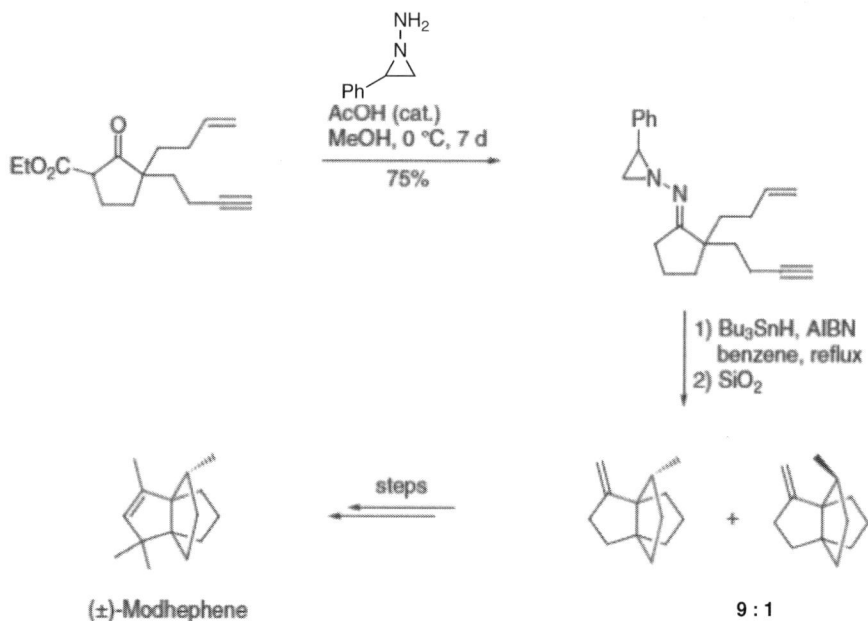

(±)-Modhephene   9 : 1

Scheme 18.14 Domino radical cyclization reaction of *N*-aziridinyl imine in the total synthesis of (±)-modhephene.

terpenoid, Yang and coworkers made use of a concise Mn(OAc)$_3$-mediated and chiral auxiliary-assisted oxidative free-radical cyclization.[39] The reaction of starting corresponding ester, bearing a (*R*)-pulegone-based chiral auxiliary, with Mn(OAc)$_3$ and Yb(OTf)$_3$ yielded a tricyclic product in a twofold ring closure in 60% yield and a diastereomeric ratio of 9.2:1 (Scheme 18.15). A further two steps led to (+)-triptocallol. For the interpretation of the stereochemical outcome, the authors proposed the hypothetical transition state, in which chelation of the keto ester moiety with Yb(OTf)$_3$ locks the two carbonyl groups in a *syn* orientation. The attack of the Mn^{III}-oxidation-generated radical onto the proximate double bond is then restricted to the more accessible (*si*)-face, as the (*re*)-face is effectively shielded by the 8-naphthyl moiety.

An exciting example of a rare radical transannular cyclization is the transformation of the iododieneyne in the presence of Bu$_3$SnH by Pattenden and coworkers which, *via* radical intermediate, led to the two diastereomeric products in a 6:1 ratio with 45–60% yield (Scheme 18.16).[40] This procedure offers a straightforward approach to the taxane system. These authors have also used an iodotrienedione in this process, but this led to the desired taxane skeleton in only 25% yield.

The [6.5.5]-ring fused tricyclic motif is found in many natural products, and has therefore become an important target in synthesis. A convenient access to this structural framework is offered by a radical domino procedure published

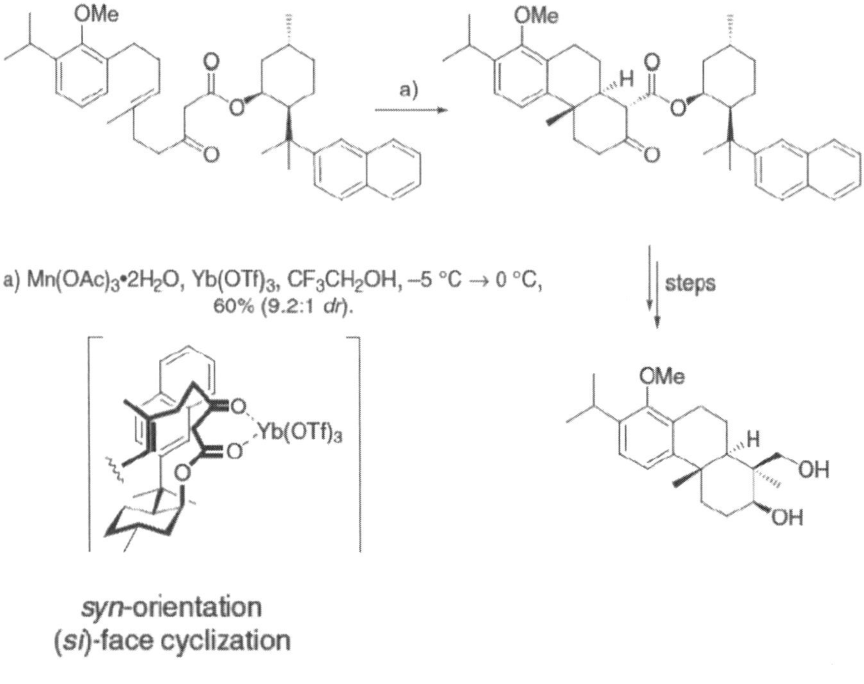

a) Mn(OAc)$_3$·2H$_2$O, Yb(OTf)$_3$, CF$_3$CH$_2$OH, −5 °C → 0 °C, 60% (9.2:1 *dr*).

syn-orientation
(*si*)-face cyclization

Scheme 18.15   Mn(OAc)$_3$-mediated chiral auxiliary-assisted enantioselective domino radical cyclization in the total synthesis of (+)-triptocallol.

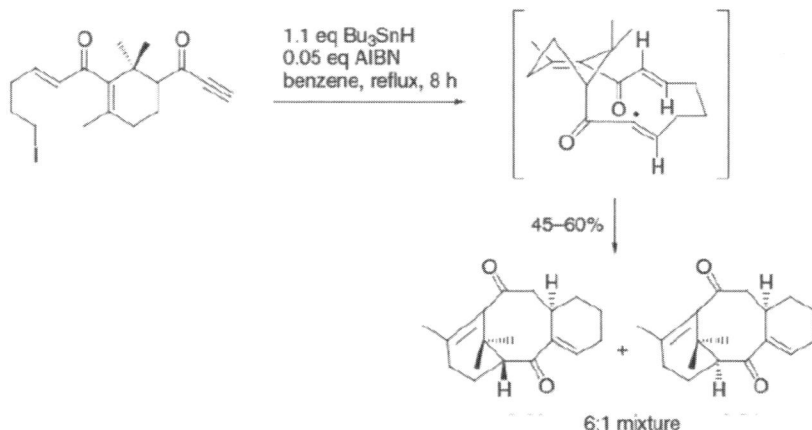

Scheme 18.16 Domino radical macro-cyclization/transannular-cyclization procedure for the synthesis of the taxane skeleton.

by the Nagano group.[41] This reaction of optical pure dibromoacetal led to the
desired tricycle *via* bicyclic intermediate as a single diastereoisomer in a very
respectable yield of 94% by applying classical radical conditions (excess
tributyltin hydride/AIBN, irradiation) (Scheme 18.17). Intermediate could be
isolated when only one equivalent of tin hydride was used.

The conversion of a glycal to enantiomerically pure [6.5.6]-dioxatricycles
containing eight stereogenic centers has been accomplished by Hoffmann and
coworkers, carrying out only three simple steps.[42] As shown in Scheme 18.18,
the stereochemically pure substrates for the cyclization (*R*)- and (*S*)- were
prepared by iodoglycosylation of glycal and the racemic silylated ene-ynols
after chromatographic separation of the formed diastereomers, followed by
desilylation.

The cyclization method of choice appeared to be the air-induced triethylborane protocol, launching the 5-*exo*-trig-/6-*endo*-dig radical cascade, which
also implies an iodine transfer. The final products, were obtained in low to
moderate yield in case of the (*R*)- isomer, and in moderate to good yields in the
case of the (*S*)-isomer. Undoubtedly, the difference in outcome of both series
is caused by steric factors. Hence, X-ray crystal structure elucidation of the

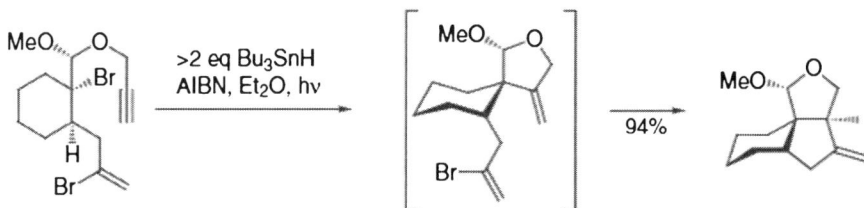

Scheme 18.17 Domino radical cyclization procedure for the synthesis of [6.5.5]-
fused tricycle

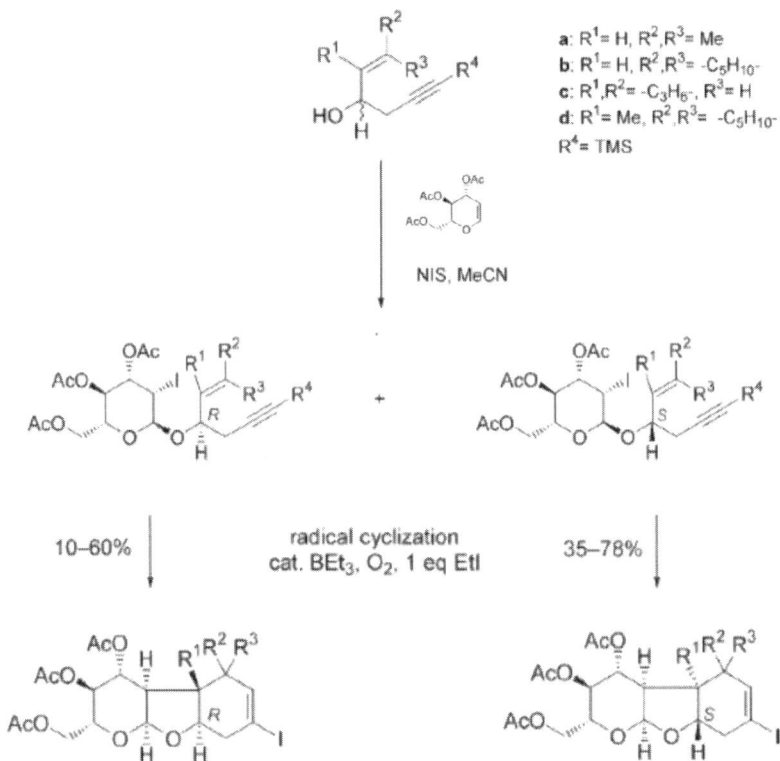

a: R^1= H, R^2,R^3= Me
b: R^1= H, R^2,R^3= -C$_5$H$_{10}$-
c: R^1,R^2= -C$_3$H$_6$-, R^3= H
d: R^1= Me, R^2,R^3= -C$_5$H$_{10}$-
R^4= TMS

NIS, MeCN

10–60%   radical cyclization   35–78%
                       cat. BEt$_3$, O$_2$, 1 eq EtI

Scheme 18.18 Radical-mediated domino reaction of glucopyranosides.

products confirmed that in the (*S*)-series the pyranoside ring is a well-developed chair, whereas in the (*R*)-series a non-chair conformation is adopted. The first examples of a consecutive radical 5-*exo*-/dig-5-*exo*-dig cyclization of 1,5-diynes have been accomplished by the same researchers.[43] These authors were able to show that their cycloisomerization procedure provides access to strained semicyclic, conjugated dienes with a functionalized dioxatriquinane framework which occurs in the aglycones of steroidal cardiac glycosides, such as isogenine[44] and C-norcardanolide (Scheme 18.19).[45] For example, exposure of the iodotetrahydrofurans to triethylborane in refluxing benzene in the presence of air and ethyl iodide triggered the formation of the diastereomeric dioxatriquinanes (Scheme 18.20).

An efficient methodology for the construction of pyrrolizidines and other polycyclic nitrogen heterocycles using a radical domino sequence has been revealed by Bowman and coworkers.[46] These authors employed sulfenamides as substrates, which easily form aminyl radicals by treatment with tributyltin hydride and AIBN. For instance, the starting amine smoothly underwent a twofold 5-*exo*-trig cyclization to give the tetracyclic pyrrolizidine product in 90% yield (Scheme 18.21). As intermediates, the radicals can be assumed.

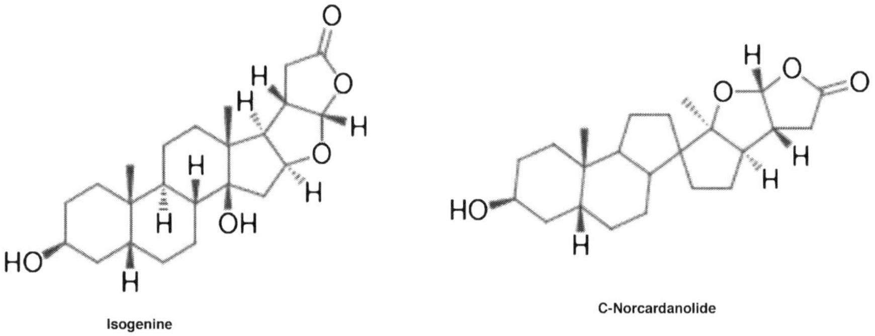

Isogenine

C-Norcardanolide

Scheme 18.19 Steroids with dioxatriquinane substructure.

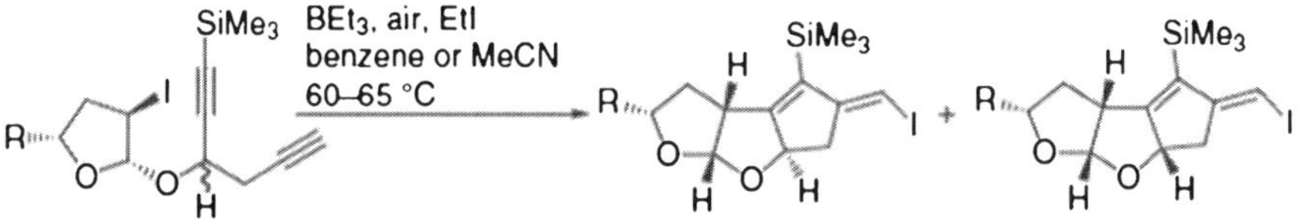

Scheme 18.20 Dioxatriquinanes by triethylborane-induced domino radical atom transfer/cyclization of 1,5-diynes.

Free-radical ring-expansion reactions have been established as attractive approaches to standard, medium-sized, and even large rings.[47] The incorporation of an additional, appropriately positioned radical acceptor offers the possibility to extend this methodology to a domino ring expansion/cyclization procedure.[48] A general reaction path is presented in Scheme 18.22, in which readily available substrates first undergo a radical ring expansion to generate corresponding radical intermediates.

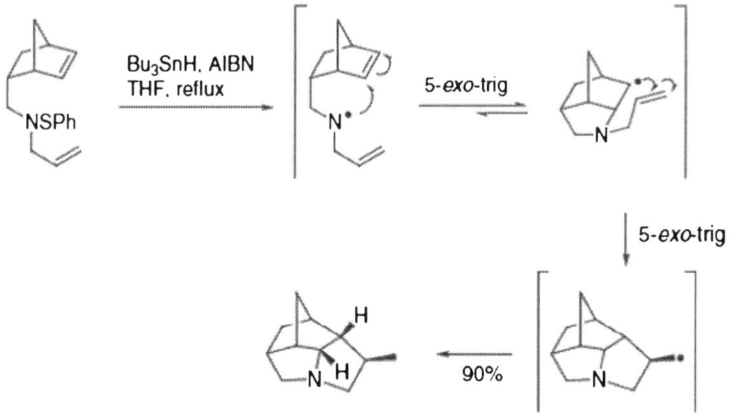

Scheme 18.21 Synthesis of pyrrolizidine by a domino radical cyclization *via* an aminyl radical.

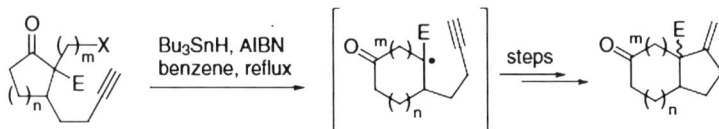

Scheme 18.22   General scheme of domino radical ring expansion/cyclization procedure.

These are then captured by the tethered alkyne moiety, furnishing the fused bicyclic product.[49] As pointed out by Curran and coworkers, the two reactions may compete with each other.Thus, initially, instead of the ring expansion a 1,5-hydrogen transfer could take place, whereas in the cyclization step with *m* = 1 a 6-*exo* cyclization might compete. However, cyclopentanone with a *cis* orientation of the iodoalkane group and the alkyne moiety was converted into the fused cyclooctanone in 82% yield (Scheme 18.23). In contrast, the corresponding *trans*-isomer only underwent 1,5-hydrogen transfer, leading to a dehalogenated starting material.

An early—but mechanistically interesting—construction of a bicyclo[3.1.0]-oxahexane by a domino radical cyclization was presented by Weng and Luh.[50] The addition of tributyl tin and AIBN to a solution of bromides in refluxing benzene gave the final product as single diastereoisomers in acceptable yields *via* the corresponding intermediates (Scheme 18.24).

Scheme 18.23   Example of a domino radical ring expansion/cyclization procedure.

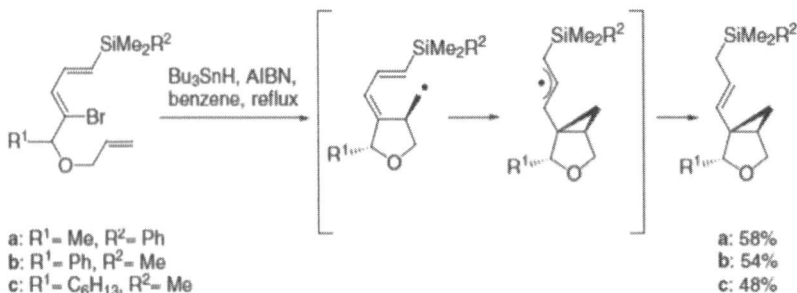

a: R^1= Me, R^2= Ph
b: R^1= Ph, R^2= Me
c: R^1= C_6H_{13}, R^2= Me

a: 58%
b: 54%
c: 48%

Scheme 18.24   Synthesis of bicyclo[3.1.0]oxahexanes.

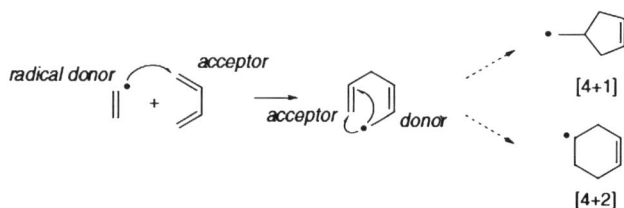

Scheme 18.25 Boomerang-type [4+1] and [4+2] radical sequences.

It is important that the cyclopropyl carbinyl radical intermediate has the correct stability and reactivity, which is achieved by the α-silyl substituent. In the last 15 years, researchers have used a new methodology based on a boomerang-type radical domino sequence (Scheme 18.25),[51] in which an iodoalkenyl can act as both radical donor and acceptor providing [4+2] or [4+1] cyclization products.[52]

Thus, the reaction of the 2-iodovinyldienoate using different reaction conditions led to the assembly of annulated product in moderate to good yields (Scheme 18.26).

The reaction proceeds *via* a 5-*exo*-trig ring closure by attack of the primarily formed vinyl radical onto the adjacent olefin moiety. Further 6-*endo*-trig cyclization involving the initial double bond furnished the desired product. Heating of 2-iodovinyldienoate in benzene did not lead to the product, which suggests that the [4+2] annulation reactions are not the result of a Diels–Alder process. Ultimately, it should be mentioned that indirect cathodic electrolysis applying Ni(cyclam)$^{2+}$ as mediator afforded only the monocyclic product instead of bicycle.

Various bicyclo[3.3.0]octanes employing a radical iodine atom transfer reaction have been successfully synthesized by Taguchi and coworkers.[53] This

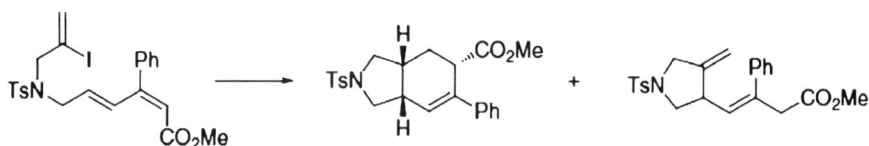

Entry	Conditions	Yield [%]	
1	Bu$_3$SnH, AIBN, benzene, reflux[a]	40	20
2	Bu$_3$SnH, AIBN, benzene, reflux[b]	63	0
3	(TMS)$_3$SiH, AIBN, benzene, reflux[b]	66	0
4	Ni(cyclam)$^{2+}$, DMF, −1.5 V, r.t.	0	85

[a] 1.2 eq. Bu$_3$SnH and 0.5 eq. AIBN were added dropwise over 1 h.

[b] 1.2 eq. hydride and 0.5 eq. AIBN were added dropwise over 2 h.

Scheme 18.26 Intramolecular [4+1] and [4+2] annulation reactions employing domino radical cyclizations.

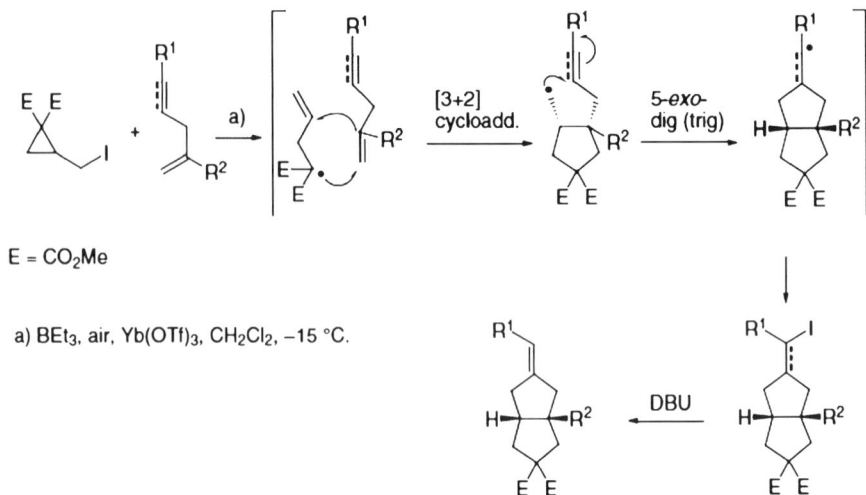

Scheme 18.27 Products and mechanism of an iodine atom transfer radical domino reaction.

procedure exemplifies the small number of radical domino processes which are initiated by an intermolecular radical addition. The process makes use of the 2-iodomethylcyclopropane derivative as precursor of a homoallylic radical, which can be captured by 1,4-dienes as well as 1,4-ene-ynes. Triethyl borane/air-mediated formation of corresponding intermediate and subsequent Lewis acid-catalyzed formal [3+2] cycloaddition involving a diene or an ene-yne led to the radical intermediate. Further 5-*exo*-dig(trig) ring closure furnished bicyclic radical, which is finally trapped by iodine to give the desired product. The products containing an iodalkyl group were converted into the corresponding olefinic compound by treatment with DBU, to allow a better separation (Scheme 18.27).

Another Lewis acid-catalyzed atom-transfer domino radical cyclization, to produce various bicyclic and tricyclic ring skeletons, has been developed by Yang and coworkers.[54] Reactions of the α-bromo-keto ester with $Yb(OTf)_3$ and Et_3B/O_2 led to the bicycle final product in 85% yield (Scheme 18.28). The reaction proceeds *via* a 6-*endo*-trig and 5-*exo*-trig cyclization after initial abstraction of the bromine atom. The reaction was also carried out in an enantioselective manner using pyboxligand **L** with 66% ee and 60% yield.

The consecutive multiple carbon–carbon bond construction with help of radical mediated intramolecular domino cyclizations is well established, and has been applied with great success. However, the extension to intermolecular procedures has been severely limited. Nonetheless, a successful example was reported by Yamago, Yoshida and coworkers using a new group-transfer coupling which avoids the problem of selective reaction of transient radicals with coupling partners in radical chain reactions.[55] The reaction was accomplished by heating a neat mixture of benzophenone, phenyl acetylene and

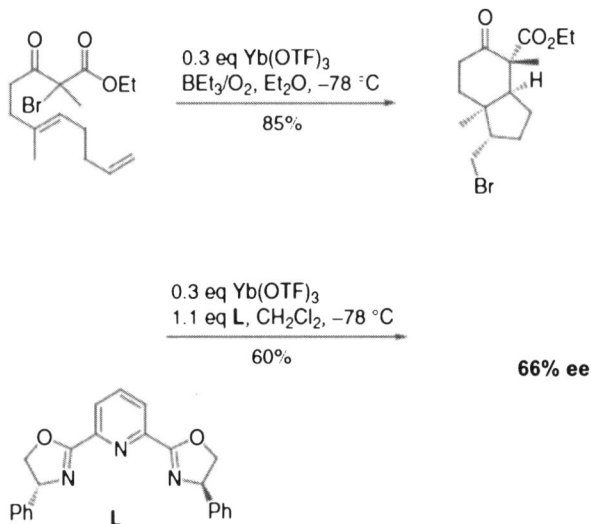

Scheme 18.28 Lewis acid-promoted free radical domino cyclization reaction and enantioselective approach.

trimethylsilyl phenyl telluride to 100 °C, providing silyl-protected allylic alcohol telluride species in excellent 93% yield and 96% (*E*)-selectivity (Scheme 18.29). The transformation is assumed to pass the radical intermediates on the way to final product. The vinylic carbon–tellurium bond in starting material can easily be cleaved by a tributyltin radical to afford vinyl radical, which can undergo further transformations as hydrogenation or C–C-bond formation, for example with dimethylfumarate in a (*Z*)-selective mode.

The addition of carbon-centered (alkyl or vinyl) radicals to alkene moieties bearing chiral auxiliaries has been extensively studied.[56] Herein, an approach is highlighted by Malacria's group, relying on the easy introduction and low cost of an enantiopure sulfoxide unit which can be regarded as a temporary chiral auxiliary.[57] As can be seen in the general approach [route (1) in

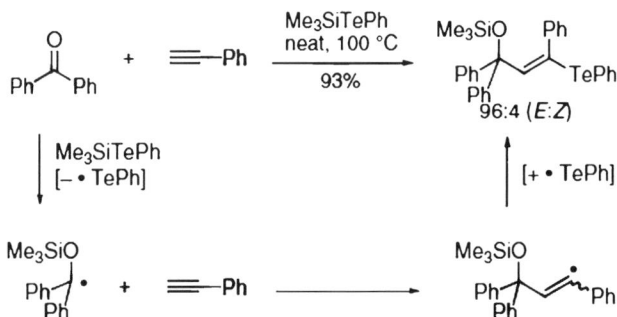

Scheme 18.29 Intermolecular domino radical addition procedure for the synthesis of silyl-protected allylic alcohols.

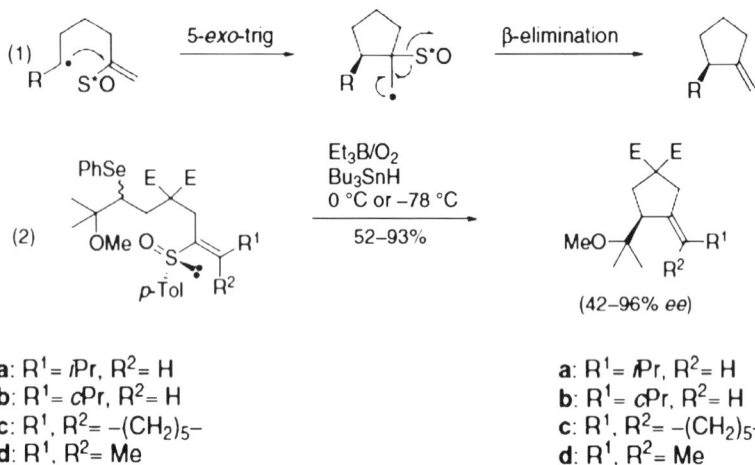

(1)

5-*exo*-trig

β-elimination

(2)

PhSe E   E

OMe O=S•
     p-Tol

R¹
R²

Et₃B/O₂
Bu₃SnH
0 °C or –78 °C

52–93%

E   E

MeO—⟨
      R¹
      R²

(42–96% *ee*)

a: R¹= *i*Pr, R²= H
b: R¹= *c*Pr, R²= H
c: R¹, R²= –(CH₂)₅–
d: R¹, R²= Me

a: R¹= *i*Pr, R²= H
b: R¹= *c*Pr, R²= H
c: R¹, R²= –(CH₂)₅–
d: R¹, R²= Me

Scheme 18.30   Domino radical cyclization/β-elimination process involving enantiopure sulfoxides.

Scheme 18.30], the sulfoxide-controlled intramolecular addition of a radical to
a proximate alkene unit furnishes a five-membered ring system in a 5-*exo*-trig
manner in an diastereoselective way. β-Elimination of the chiral auxiliary then
provides the enantiomerically enriched product. Hence, the sulfoxides were
converted under low-temperature radical conditions (Bu₃SnH, Et₃B/O₂) to
the desired products in 52–93% yield and in moderate to high enantiopurity
(42–96% ee). As early as 1986, Stork and coworkers presented an intramolecular radical cyclization/ intermolecular trapping methodology within their
synthesis of prostaglandin F2α.[58]

   The first example of an intermolecular radical addition/intermolecular
trapping domino reactions of an acyclic system in a stereo-controlled fashion
to build stereogenic centers at the α- and β-carbons was described by Sibi and
coworkers.[59] Enantioselective addition of an alkyl radical (prepared *in situ*)
to crotonate or cinnamate, facilitated by the addition of substoichiometric
amounts (30 mol%) of a Lewis acid (MgI₂, Cu(OTf)₂ or Mg(ClO₄)₂) and chiral
bisoxazoline ligand, produced a radical in the α-position which could be
trapped by a present allylic tin species in an *anti*-manner (Scheme 18.31). The
best results according to stereoselectivity were obtained when the substituents
R¹ and R² were as large as possible, whereas different Lewis acids gave almost
the same results. Thus, utilization of cinnamate (R¹ = Ph) in combination with
*t*BuI and MgI₂ as Lewis acid gave (R¹ = Ph, R² = *t*Bu) with a diastereomeric
ratio of 99:1 and 97% ee in 84% yield.

   The classical procedures for the synthesis of β-amino carbonyl compounds
are based on a Mannich reaction, which relies on iminium ions. Quite recently,
Naito and his group have designed a new type of Mannich reaction involving
free-radical chemistry.[60] Thus, submitting substrates containing two electrophilic radical acceptors to radical conditions (Et₃B, oxygen, refluxing benzene)

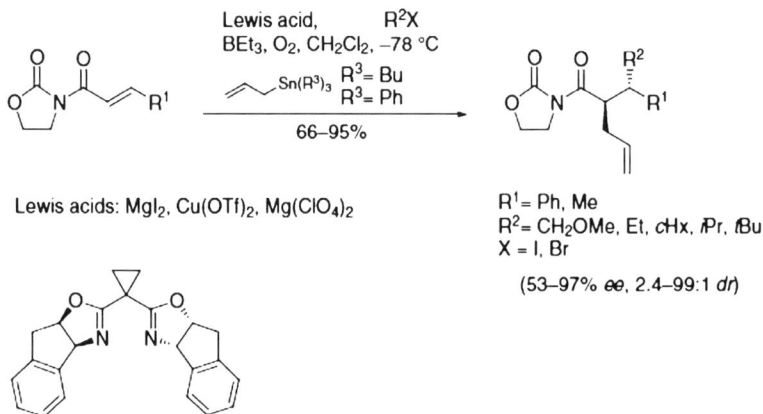

Scheme 18.31 Enantioselective intermolecular domino radical addition procedure.

in the presence of a nucleophilic alkyl radical precursor R^2I, led to a selective addition on the acrylate moiety, furnishing intermediates (Scheme 18.32). This radical stays in equilibrium with the aminyl radical. Since only the aminyl radical can be intercepted by the highly reactive triethylborane, the equilibrium is pushed to the formation of corresponding intermediate, which gives the β-aminobutyrolactone in good yields (64– 70%) and high stereoselectivity (12 : 1).

Grignon-Dubois and coworkers have shown that reduction of a quinoline using zinc and acetic acid in THF gives the dimeric compound *via* intermediate

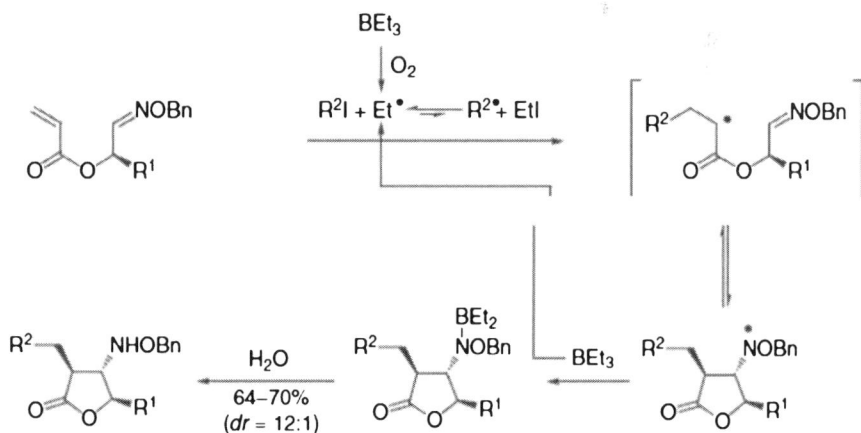

R^1= TPSOCH$_2$
R^2= Et, *i*Pr, *c*Hx, *c*Pent
reaction conditions: BEt$_3$, R^2I, benzene, reflux.

Scheme 18.32 Domino radical addition/cyclization reaction for the asymmetric synthesis of β-aminobutyrolactones.

(Scheme 18.33).[61] Usually, a mixture of the *syn-* and *anti*-products is formed; the substituent has some influence on the regioselectivity of the dimerization and cyclization step. With R = H and R = 6-Me, only the benzazepines were produced, by a head-to-head dimerization.

Finally, thermally induced isomerizations, which generate carbon-centered biradical organic molecules, have been shown to serve as alternative for conventional chemical and photochemical methods.[62] A straightforward procedure to accomplish such biradicals was described by Myers *et al.* using a thermal conversion of yne-allenes.[63] According to this scheme, Wang and coworkers[64] heated the starting precursor in 1,4-cyclohexadiene to 75 °C and obtained the final product in 22% yield *via* the biradicals as intermediates (Scheme 18.34).

The Pattenden group has recently revealed a new synthetic approach to the propellane sesquiterpen triquinane (*rac*)-modhephene, introducing an eight-membered ring system with an appropriate thioester. Thus, using standard radical conditions (Bu₃SnH, AIBN) a 5-*exo*-trig/5-*exo*-dig domino cyclization took place to generate the propellane structure in 60% yield.[65]

Another very new radical/radical domino procedure was used in the total synthesis of the alkaloid lennoxamine by Ishibasi and coworkers. Here, a 7-*endo* cyclization/ homolytic aromatic substitution reaction cascade led to the target compound in 41% yield.[66]

18.3 CONCLUSION AND FUTURE DIRECTION

Since radical reactions are ideal for sequencing, due to the very fundamental reason that the product of every radical reaction is a radical, they have opened the door for many efficient and elegant domino processes. Nowadays, domino radical reactions are prized for their capability to build complex, highly substituted ring systems, and for their general tolerance of functionalities in the substrates, allowing transformations with a minimum use of protecting groups.

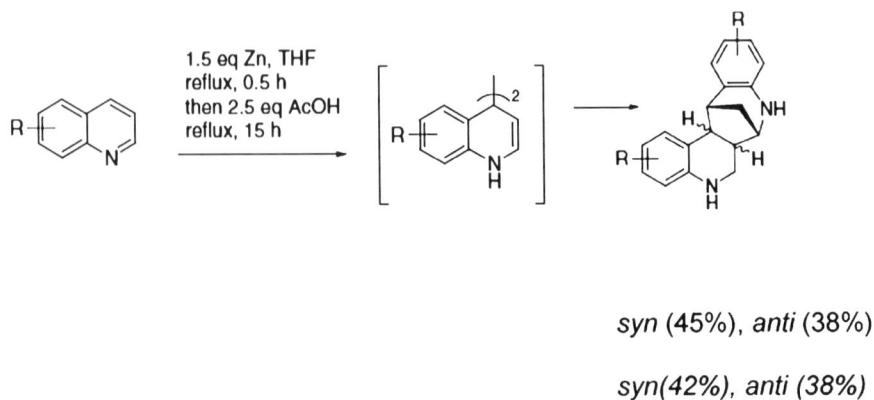

syn (45%), *anti* (38%)

syn(42%), *anti* (38%)

Scheme 18.33   Dimerization of quinoline using zinc in acetic acid.

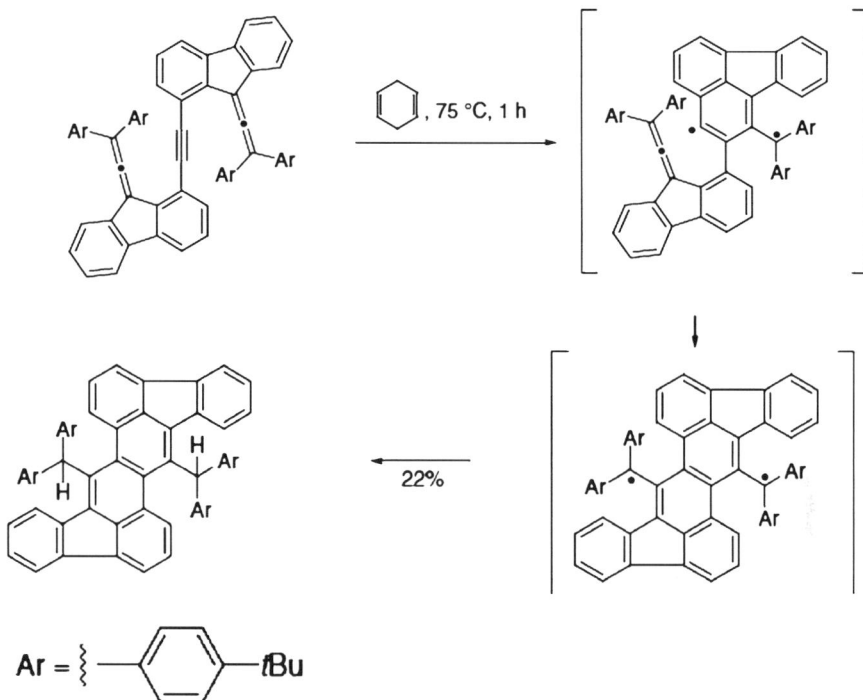

Scheme 18.34 Thermolysis of benzoenynes-allenes initiating a domino radical cyclization.

In marked contrast to polar processes, radical transformations can proceed in most cases in the presence of free hydroxyl and amino groups, as well as keto and ester functionalities. Not surprisingly, high grades of chemo-, regio- and stereoselectivity can be obtained, a characteristic attributable to the mild reaction conditions, which are applied in radical chemistry. The future of domino reactions is bright and will only be limited by the creativity and ingenuity of the researcher.

REFERENCES

1. A. J. McCarroll and J. C. Walton, *J. Chem. Soc., Perkin Trans. 1,* 2001, 3215–3229.
2. A. J. McCarroll and J. C. Walton, *Angew. Chem., Int. Ed.,* 2001, **40**, 2224–2248.
3. (a) A. L. J. Beckwith and K. U. Ingold, in Rearrangements in Ground and Excited States, ed. P. de Mayo, Academic Press, New York, 1980, vol. 1, pp. 161–310; (b) J. E. Baldwin, J. Chem. Soc., Chem. Commun., 1976, 734–736.
4. Y. Takemoto, T. Ohra, H. Koike, S.-I. Furuse and C. Iwata, J. Org. Chem., 1994, **59**, 4727–4729.

5. P. Girard, J. L. Namy and H. B. Kagan, J. Am. Chem. Soc., 1980, **102**, 2693–2698.
6. (a) G. A. Molander, Chem. Rev., 1992, **92**, 29–68; (b) G. A. Molander and C. R. Harris, Chem. Rev., 1996, **96**, 307–338.
7. A. Schwartz and C. Seger, Monatsh. Chem., 2001, **132**, 855–858.
8. S. Gross and H.-U. Reissig, Org. Lett., 2003, 4305–4307.
9. P. A. Stadler and K. A. Giger, in Natural Products and Drug Development, ed. P. K. Larsen, S. B. Christensen and H. Kofod, Munksgaard, Copenhagen, 1984, p. 463.
10. I. Ninomiya and T. Kiguchi, in The Alkaloids, ed. A. Brossi, Academic Press, New York, 1990, vol. 38, p. 1.
11. Y. Ozlu, D. E. Cladingboel and P. J. Parsons, Synlett, 1993, 357–358.
12. E. Lee. H. Y. Song, J.W. Kang, D.-S. Kim, C.-K. Jung and J. M. Joo, J. Am. Chem. Soc., 2002, **124**, 384–385.
13. (a) M. Castillo, G. Morales, L. A. Loyola, I. Singh, C. Calvo, H. L. Holland and D. B. MacLean, Can. J. Chem., 1975, **53**, 2513– 2514; (b) M. Castillo, G. Morales, L. A. Loyola, I. Singh, C. Calvo, H. L. Holland and D. B. MacLean, Can. J. Chem., 1976, **54**, 2900–2908.
14. C.-K. Sha, F.-K. Lee and C.-J. Chang, J. Am. Chem. Soc., 1999, **121**, 9875–9876.
15. (a) W. A. Ayer, Nat. Prod. Rep., 1990, **8**, 455–463; (b) W. A. Ayer, The Alkaloids, Academic Press, New York, 1994, vol. 45, pp. 233–266.
16. (a) T. T. Dabrah, T. Kaneko, W. Massefski, Jr and E. B. Whipple, J. Am. Chem. Soc., 1997, **119**, 1594–1598; (b) T. T. Dabrah, H. J. Harwood, Jr., L. H. Huang, N. D. Jankovich, T. Kaneko, J.-C. Li, S. Lindsey, P. M. Moshier, T. A. Subashi, M. Therrien and P. C. Watts, J. Antibiot., 1997, **50**, 1–7.
17. S. A. Biller, K. Neuenschwander, M. M. Ponpipom and C. D. Poulter, Curr. Pharm. Des., 1996, **2**, 1–40.
18. D. M. Leonard, J. Med. Chem., 1997, 40, 2971–2990.
19. K. C. Nicolaou, P. S. Baran, Y.-L. Zhong, H.-S. Choi, W. H. Yoon, Y. He and K. C. Fong, Angew. Chem., 1999, **111**, 1781–1784 (Angew. Chem., Int. Ed., 1999, **38**, 1669–1675).
20. J. T. Njardarson, I. M. McDonald, D. A. Spiegel, M. Inoue and J. L. Wood, Org. Lett., 2001, **3**, 2435–2438.
21. J. H. Butterworth and E. D. Morgan, Chem. Commun., 1968, 23–24.
22. K. C. Nicolaou, A. J. Roecker, H. Monenschein, P. Guntupalli and M. Follmann, Angew. Chem., Int. Ed., 2003, **42**, 3637–3642.
23. (a) A. H. Osada, C.-B. Cui, R. Onose and F. Hanaoka, Bioorg. Med. Chem., 1997, **5**, 193–203; (b) S. Edmondson, S. J. Danishefsky, L. Sepp-Lorenzino and N. Rosen, J. Am. Chem. Soc., 1999, **121**, 2147–2155.
24. D. Lizos, R. Tripoli and J. A. Murphy, Chem. Commun., 2001, 2732–2733.
25. B. Patro and J. A. Murphy, Org. Lett., 2000, **2**, 3599–3601.
26. (a) S. Kim, G. H. Joe and J. Do, J. Am. Chem. Soc., 1994, **116**, 5521–5522; (b) O. Callaghan, M. Kizil, J. A. Murphy and B. Patro, J. Org. Chem.,

1999, **64**, 7856–7862; (c) J. A. Murphy and M. Kizil, J. Chem. Soc., Chem. Commun., 1995, 1409–1410.

27. S. Zhou, S. Bommezijn and J. A. Murphy, Org. Lett., 2002, **4**, 443–445.

28. M. Potmesil and H. Pinedo, Camptothecins: New Anticancer Agents, CRC Press, Boca Raton, Florida, 1995.

29. W. R. Bowman, C. F. Bridge, P. Brookes, M. O. Cloonan and D. C. Leach, J. Chem. Soc., Perkin Trans. 1, 2002, 58–68.

30. Y. Damayanthi and J.W. Lown, Curr. Med. Chem., 1998, **5**, 205–252.

31. A. J. Reynolds, A. J. Scott, C. I. Turner and M. S. Sherburn, J. Am. Chem. Soc., 2003, **125**, 12108–12109.

32. J. Cassayre, F. Gagosz and S. Z. Zard, Angew. Chem., Int. Ed., 2002, **41**, 1783–1785.

33. S. Kim, I. S. Kee and S. Lee, J. Am. Chem. Soc., 1991, **113**, 9882–9883.

34. H.-Y. Lee, D.-I. Kim and S. Kim, Chem. Commun., 1996, 1539–1540.

35. H.-Y. Lee, S. Lee, D. Kim, B. K. Kim, J. S. Bahn and S. Kim, Tetrahedron Lett., 1998, **39**, 7713–7716.

36. H.-Y. Lee, D. K. Moon and J. S. Bahn, Tetrahedron Lett., 2005, **46**, 1455–1458.

37. L. Porter, Chem. Rev., 1967, **67**, 441–464.

38. J. Cassayre and S. Z. Zard, J. Organomet. Chem., 2001, **624**, 316–326.

39. D. Yang, M. Xu and M.-Y. Bian, Org. Lett., 2001, **3**, 111–114.

40. S. A. Hitchcock, S. J. Houldsworth, G. Pattenden, D. C. Pryde, N. M. Thomson and A. J. Blake, J. Chem. Soc., Perkin Trans. 1, 1998, 3181–3206.

41. H. Nagano, Y. Ohtani, E. Odake, J. Nakagawa, Y. Mori and T. Yajima, J. Chem. Res., Synop., 1999, 338–339.

42. H. M. R. Hoffmann, U. Herden, M. Breithor and O. Rhode, Tetrahedron, 1997, **53**, 8383–8400.

43. T. J. Woltering and H. M. R. Hoffmann, Tetrahedron, 1995, **51**, 7389–7402.

44. A. F. Krasso, M. Binder and C. Tamm, Helv. Chim. Acta, 1972, **55**, 1352–1371.

45. G. R. Pettit, T. R. Kasturi, J. C. Knight and J. Occolowitz, J. Org. Chem., 1970, **35**, 1404–1410.

46. W. R. Bowman, D. N. Clark and R. J. Marmon, Tetrahedron, 1994, **50**, 1295–1310.

47. P. Dowd and W. Zhang, Chem. Rev., 1993, **93**, 2091–2115.

48. D. P. Curran, in Comprehensive Organic Synthesis, ed. B. M. Trost and I. Fleming, Pergamon, Oxford, 1991, vol. 4, p. 779.

49. C. Wang, X. Gu, M. S. Yu and D. P. Curran, Tetrahedron, 1998, **54**, 8355–8370.

50. W.-W. Weng and T.-Y. Luh, J. Org. Chem., 1993, **58**, 5574–5575.

51. (a) A. L. J. Beckwith and D. M. O'Shea, Tetrahedron Lett., 1986, **27**, 4525–4528; (b) G. Stork and R. Mook, Jr., Tetrahedron Lett., 1986, **27**, 4529–4532; (c) D. P. Curran and S. Sun, Aust. J. Chem., 1995, **48**, 261–267; (d) B. P. Haney and D. P. Curran, J. Org. Chem., 2000, **65**, 2007–

2013; (e) O. Kitagawa, Y. Yamada, H. Fujiwara and T. Taguchi, J. Org. Chem., 2002, **67**, 922–927; (f) K. Takasu, J. Kuroyanagi, A. Katsumata and M. Ihara, Tetrahedron Lett., 1999, **40**, 6277–6280; (g) K. Takasu, S. Maiti, A. Katsumata and M. Ihara, Tetrahedron Lett., 2001, **42**, 2157–2160.

52. K. Takasu, H. Ohsato, J. Kuroyanagi and M. Ihara, J. Org. Chem., 2002, **67**, 6001–6007.

53. O. Kitagawa, Y. Yamada, A. Sugawara and T. Taguchi, Org. Lett., 2002, **4**, 1011–1013.

54. D. Yang, S. Gu, Y.-L. Yan, H.-W. Zhao and N.-Y. Zhu, Angew. Chem., Int. Ed., 2002, **41**, 3014–3017.

55. S. Yamago, M. Miyoshi, H. Miyazoe and J. Yoshida, Angew. Chem., Int. Ed., 2002, **41**, 1407–1409.

56. (a) N. A. Porter, B. Lacher, V. H.-T. Chang and D. R. Magnin, J. Am. Chem. Soc., 1989, **111**, 8309–8310; (b) N. A. Porter, D. M. Scott, B. Lacher, B. Giese, H. G. Zeitz, H. J. Lindner, J. Am. Chem. Soc., 1989, **111**, 8311–8312; (c) D. M. Scott, A. T. McPhail and N. A. Porter, Tetrahedron Lett., 1990, **31**, 1679–1682; (d) N. A. Porter, D. M. Scott, I. J. Rosenstein, B. Giese, A. Veit and H. G. Zeitz, J. Am. Chem. Soc., 1991, **113**, 1791–1799.

57. E. Lacôte, B. Delouvrié, L. Fensterbank and M. Malacria, Angew. Chem., Int. Ed., 1998, **37**, 2116–2118.

58. (a) G. Stork, P. M. Sher and H.-L. Chen, J. Am. Chem. Soc., 1986, **108**, 6384–6385; (b) G. Stork and P. M. Sher, J. Am. Chem. Soc., 1986, **108**, 303–304.

59. (a) M. P. Sibi and J. Chen, J. Am. Chem. Soc., 2001, **123**, 9472–9473. For recent applications, see: (b) M. P. Sibi and H. Hasegawa, Org. Lett., 2002, **4**, 3347–3349; (c) M. P. Sibi and H. Miyabe, Org. Lett., 2002, **4**, 3435–3438; (d) M. P. Sibi, M. Aasmul, H. Hasegawa and T. Subramanian, Org. Lett., 2003, **5**, 2883–2886.

60. H. Miyabe, K. Fujii, T. Goto and T. Naito, Org. Lett., 2000, **2**, 4071–4074.

61. J.-C. Gauffre, M. Grignon-Dubois, B. Rezzonico and J.-M. Leger, J. Org. Chem., 2002, **67**, 4696–4701.

62. W. T. Borden, Diradicals, Wiley-Interscience, New York, 1982.

63. (a) A. G. Myers, E. Y. Kuo and N. S. Finney, J. Am. Chem. Soc., 1989, **111**, 8057–8059; (b) A. G. Myers and P. S. Dragovich, J. Am. Chem. Soc., 1989, **111**, 9130–9132.

64. K. K. Wang, H.-R. Zhang and J. L. Petersen, J. Org. Chem., 1999, **64**, 1650–1656.

65. B. De Boeck, N. M. Harrington-Frost and G. Pattenden, Org. Biomol. Chem., 2005, **3**, 340–347.

66. T. Taniguchi, K. Iwasaki, M. Uchiyama, O. Tamura and H. Ishibasi, Org. Lett., 2005, **7**, 4389–4390.

CHAPTER 19

Telescopic Reactions and Free Radical Synthesis: Focus on Radical and Radical–Ionic Multicomponent Processes[*]

19.1 GENERAL INTRODUCTION: ADVANTAGES AND LIMITATIONS

In chemistry a one-pot synthesis is a strategy to improve the efficiency of a chemical reaction whereby a reactant is subjected to successive chemical reactions in just one reactor. This is much desired by chemists because avoiding a lengthy separation process and purification of the intermediate chemical compounds would save time and resources while increasing chemical yield. A sequential one-pot synthesis with reagents added to a reactor one at a time and without workup is also called a telescoping synthesis. In this chapter we will subdivide the field into radical cascade (domino reactions) and multi-component transformations in order to show the broad range, advantages and disadvantages of both important classes of free radical transformations in organic synthesis.

Multicomponent reactions (MCRs) have a long-standing history that traces back to the beginning of the last century, with the early development of processes such as the Mannich reaction.[1] Remarkable achievements such as the landmark Robinson three-component assembly of tropinone were also reported.[2] In recent years, multicomponent processes including the powerful four-component Ugi reaction have garnered a lot of attention.[3] It was shown as early as 1961 that important libraries of compounds could easily be generated with this protocol.[4] However, with the appearance of automation and parallel synthesis, MCRs underwent significant developments during the

* Chapter written by V. Tamara Perchyonok.
Streamlining Free Radical Green Chemistry
V. Tamara Perchyonok, Ioannis Lykakis and Al Postigo
© V. Tamara Perchyonok, Ioannis Lykakis and Al Postigo 2012
Published by the Royal Society of Chemistry, www.rsc.org

past two decades. Pharmaceutical and agrochemical companies have been at the onset of these developments, with MCRs allowing the everyday preparation of large collections of drug candidates, as well as high-throughput screening of thousands of lead compounds. An increasing number of transformations is now based on this strategy, providing a wide array of novel architectures.[5] A large number of MCRs rely on ionic and/or organometallic processes, but less is known about radical-mediated multicomponent processes.[6] This chapter is not intended to exhaustively list all radical MCRs, but provide an overview of the rapid manner of assembling three or more independent fragments in a one-pot process by using a combination of several radical-, radical–ionic- and radical–organometallic- mediated transformations. While cascade processes of free-radical transformations have led to some remarkable achievements, notably in the total syntheses of natural products,[7] their use is sometimes plagued by the long and tedious synthesis of the precursors. In contrast, radical MCRs are highly convergent processes leading to complex architectures, starting from structurally simple and readily available materials. Designing MCRs induces significant difficulties, among which is the potential for the formation of more than one product. Chemoselectivity during each individual intermolecular event is thus essential for success. Data gathered over the years on the kinetics of radical processes (addition, abstraction, fragmentation, *etc.*) greatly help in predicting and designing successful radical MCRs. MCRs, whether radical or not, are viewed as optimal in terms of convergency and fit perfectly with the important concept of step economy.

19.2   MNEMONIC CLASSIFICATION

MCRs are traditionally categorized according to the number of reagents involved. The description of radical and radical–ionic MCRs tends to follow this nomenclature: 3-CR, 4-CR, and 5-CR refers to three-, four-, and five-component reactions, respectively. It occurred to us that a complementary classification could be helpful in organizing the review. It is based on the electronic nature of the initial radical and that of subsequent partners. However, such a description is not always so clear-cut in radical chemistry, with some radicals having a borderline behavior.[8] The proposed classification therefore takes into account the nucleophilic (D, electron donor) or electrophilic (A, electron acceptor) character of the first radical as well as the electronic character of the two or more partners assembled intermolecularly (Scheme 19.1).

We will therefore describe MCRs as follows: (1) the number of components involved; (2) the electronic nature of the first radical (A or D); (3) the electronic nature of the second component (A or D), of the third component (A or D), and so forth (Scheme 19.2). In this classification, the reaction involves at least two elementary radical steps and the product should incorporate most of the carbon framework of each component. The initiation, atom-transfer, and reduction steps are deliberately not taken into account. As

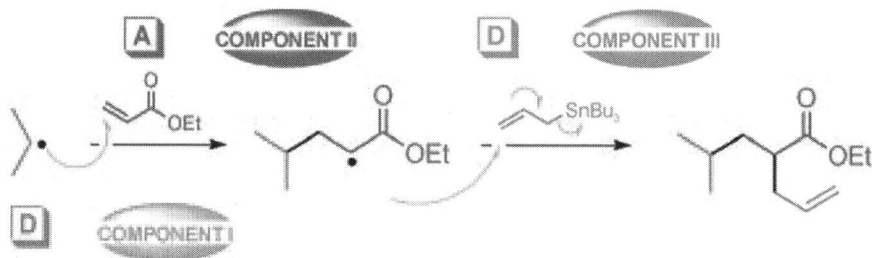

Scheme 19.1 Example of a three-component reaction between donor–acceptor–donor radicals (3-CR-DAD).

an example, the reduction of an intermediate radical by a hydride donor will not be considered and hence Bu₃SnH as one component will be excluded. The example in Scheme 19.1 illustrates this organization with the reaction between an alkyl radical (D), with nucleophilic character, and an electron-poor olefin (A). This provides an intermediate radical, which then reacts with an allyltin reagent, an electron-rich olefin (D).

This classification is not intended to be a formal nomenclature for radical MCRs due to the expected number of special cases that should arise from borderline polarity of radical species and other reagents. It should instead be regarded as a mnemonic device that (i) helps readers to find their way between the various processes described herein and (ii) provides a logical description that may be useful for designing innovative radical MCRs. Interestingly, in contrast to ionic chemistry, virtually all combinations are possible in radical processes. An electron-rich radical species may indeed react with an electron-rich olefin in a DD process, if no other option is offered. The MCRs described here start with the 3-CR, followed by the 4- and 5-CR. MCRs relying only on radical processes will be described first. A discussion relating to the combination of radical and ionic transformations will then be provided.

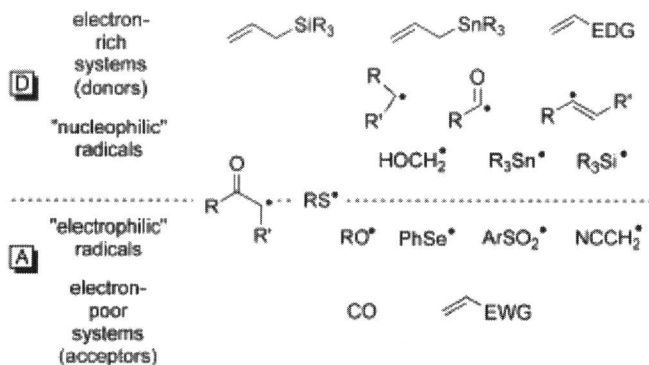

Scheme 19.2 Typical donor and acceptor radical species involved in radical MCRs.

19.3 THREE-COMPONENT RADICAL REACTIONS

19.3.1 3-CR-ADA

The majority of the developments in this area have focused on the use of carbon monoxide (CO) as a radical acceptor (A). The incorporation of CO is usually ensured by using relatively high CO pressures, thus avoiding any premature radical-chain termination. The acyl radical thus formed is nucleophilic in nature (*vide supra*) and may react further with electron-poor substrates or in group transfer processes such as SH_2.

In 1956, Lipscomb and co-workers[9a] were the first to report such a 3-CR. Upon radical addition of a thiyl radical onto propene, the resulting radical reacted with carbon monoxide (CO) as the third component (Scheme 19.3). All three steps of the transformation are reversible and produce the 3-CR aldehyde product, albeit in low yield, along with the sulfide. Better yields were later reported by Susuki and Tsuji, who used $[Co_2(CO)_8]$ as a catalyst.[9b]

The most impressive examples of MCR-ADA have been reported by Ryu, Sonoda, and co-workers.[6,10] Group transfer carbonylation constitutes an illustrative example of this class of reactions.[11] An α-seleno ester under irradiation produces an electrophilic (A) radical that adds efficiently onto olefin (D), producing a new secondary (electron-rich) radical, which then reacts with CO (A) to provide an acyl radical (Scheme 19.4). The acyl radical reacts with the starting selenide, by means of an SH2 process, and affords the final product in good yield. The use of a relatively slow group-transfer reagent (PhSe) was required here for the reaction to occur in satisfying yields.

The carboazidation of olefins recently developed by Renaud and co-workers[12] also belongs to the 3-CR-ADA series. This useful reaction involves

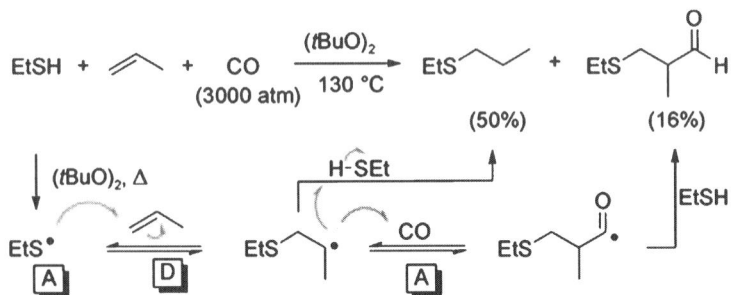

Scheme 19.3   Thiocarbonylation of olefins.

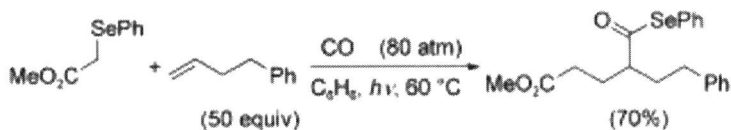

Scheme 19.4   3-CR carboformylation of olefins.

the addition of an electron-poor radical enolate to an olefin, producing a new alkyl radical (ref. 9b) possessing a nucleophilic character (Scheme 19.5). Radical then reacts efficiently with the electrophilic sulfonyl azide to produce the carboazidation product in good yield. The matched polarity of the different partners allows the reaction to be carried out simply by pre-mixing the reagents together. For instance, $PhSO_2N_3$ does not react with the initial electrophilic radical, but is only reactive towards nucleophilic alkyl radicals. Since intermediate haloester may act as a true intermediate when atom transfer is fast, a slight excess of $(Bu_3Sn)_2$ is required to ensure complete conversion of haloester back in intermediate radical form. The released phenylsulfonyl radical is unable to propagate the radical chain, requiring the intervention of tin.[12a–c,f] A tin-free procedure has however been designed with $EtSO_2N_3$.[12d,e] Upon azidation, the produced $EtSO_2$ radical is able to fragment through irreversible extrusion of SO_2 (ref. 13) furnishing an ethyl radical, which can then sustain the radical chain.

The carboazidation is not efficient with 1,2-disubstituted olefins, except with strained and reactive systems like norbornene. Landais *et al.* have shown that electron-rich chiral allylsilanes led to the corresponding β-azidosilane, not only in good yields, but also with good levels of 1,2-stereoinduction (Scheme 19.6).[14] This 3-CR thus performs well in complex settings, as testified by the total synthesis of natural products such as (+)-Hyacinthacine A1[15a] and Lepadiformine,[15b] both employing carboazidation as a key step.

A related 3-CR carbodiazenylation of olefins has recently been disclosed by Heinrich *et al.*,[16] which involves an arene diazonium salt, an α-iodoester (or a nitrile), and an olefin (Scheme 19.7). The mechanism depicted starts with the *in situ* formation of the diazonium salt, which is reduced by $TiCl_3$ into a reactive aryl radical. Iodine abstraction from the α-iodoester by the aryl radical

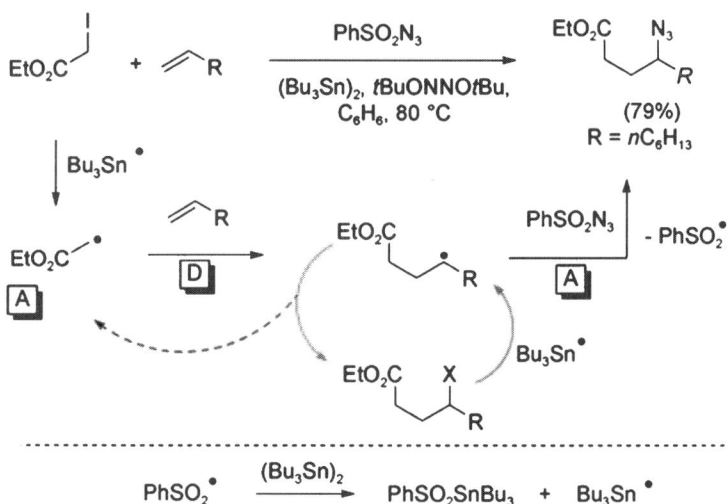

Scheme 19.5 Carboazidation of olefins.

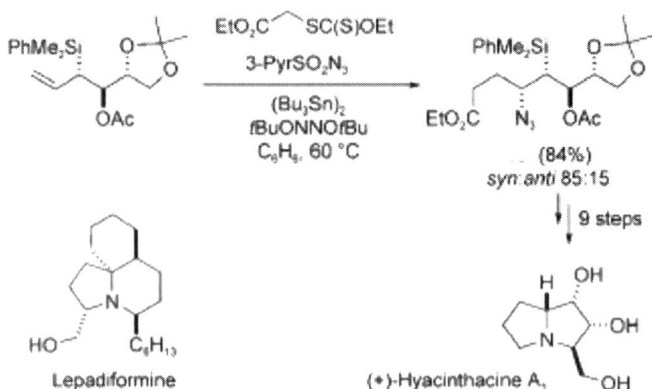

Scheme 19.6 Carboazidation of olefins—application to the total synthesis of alkaloids.

generates the corresponding radical intermediate, which adds onto olefin to provide a nucleophilic alkyl radical. Trapping of the latter radical by excess diazonium salt, followed by a second single-electron transfer (SET) from TiCl$_3$ affords final diazene product. A range of diazenes was obtained in moderate to good yield using this one-pot protocol.

Several other 3-CR-ADA have also been investigated. Heterocyclic compounds such as quinoxalines for example have been assembled using such processes based on isonitrile chemistry.[17] The addition of electron-deficient, heteroatom-centered radicals (PhS, PhSe, PhCO$_2$) to alkenes and alkynes has been described and subsequently used in intermolecular radical additions which may also be related to 3-CR processes.[18] These few examples illustrate the power of this simple and straightforward strategy to elaborate, with a high degree of convergence, functionalized and useful intermediates for organic synthesis starting from electron-deficient radical precursors.

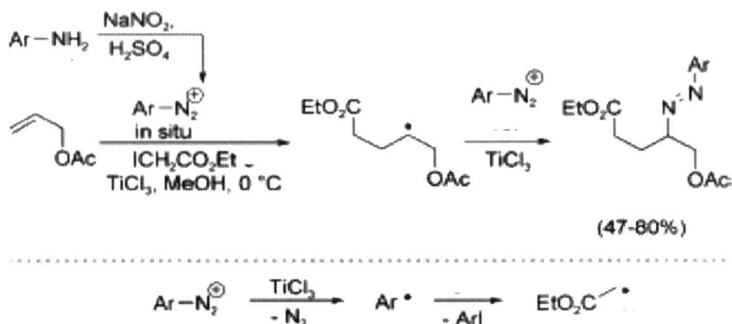

Scheme 19.7 Carboadiazenylation of olefins.

19.3.2 3-CR-DAD

The addition of electron-rich radical species onto an electron-deficient alkenes proceeds through a similar polarity matched approach (Scheme 19.1). This strategy mimics a classic in organometallic chemistry. The 1,4-addition of nucleophiles onto α,β-unsaturated carbonyl compounds produces an enolate that can be readily trapped with an external electrophile.[19] Mizuno, Otsuji, and coworkers were the first to develop a radical version of this process.[20] A nucleophilic radical generated from radical precursors such as iodides efficiently adds to electron-poor olefin in a 1,4-fashion (Scheme 19.8). This leads to a new electron-deficient radical intermediate which can be allylated with stannane to afford the final product in moderate to excellent yields, along with a tributyltin radical that sustains the radical chain.

Variations of the nature of the different partners have led to several interesting developments and further demonstrates the versatility of the Mizuno–Otsuji 3-CR (Scheme 19.9). For instance, Hosomi and co-workers have described the use of stannyl enolates as radical alkylating agents.[21] The alkyl radical originating from the starting material adds faster onto an electron-deficient alkene. The new radical enolate formed is then trapped by the electron-rich tin enolate, through a homolytic substitution (S_H2?), providing the final product together with a new useful tin propagating agent. Carbon monoxide as an acceptor has also been shown to perform well in one-pot 3-CR and provides 1,3-diketone in good yield.[22] It is worth noting that the acyl radical intermediate produced upon reaction between the alkyl radical and CO is nucleophilic and yet reacts with the electron-rich stannyl enolate, illustrating the power of radical processes as compared to their ionic counterparts. Allylsulfones have been used by Renaud and co-workers as allyltin surrogates in a multicomponent process involving three different olefins.[23] Hydroboration of cyclohexene with catecholborane produces a borane intermediate which, upon exposure to O_2, leads to the corresponding cyclohexyl radical. The cyclohexyl radical adds to succinimide, producing a radical enolate able to couple with allylsulphone, which, after fragmentation,

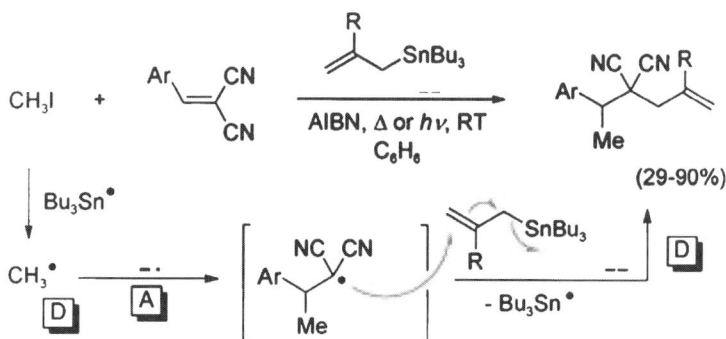

Scheme 19.8 Carboallylation of electron-deficient olefins.

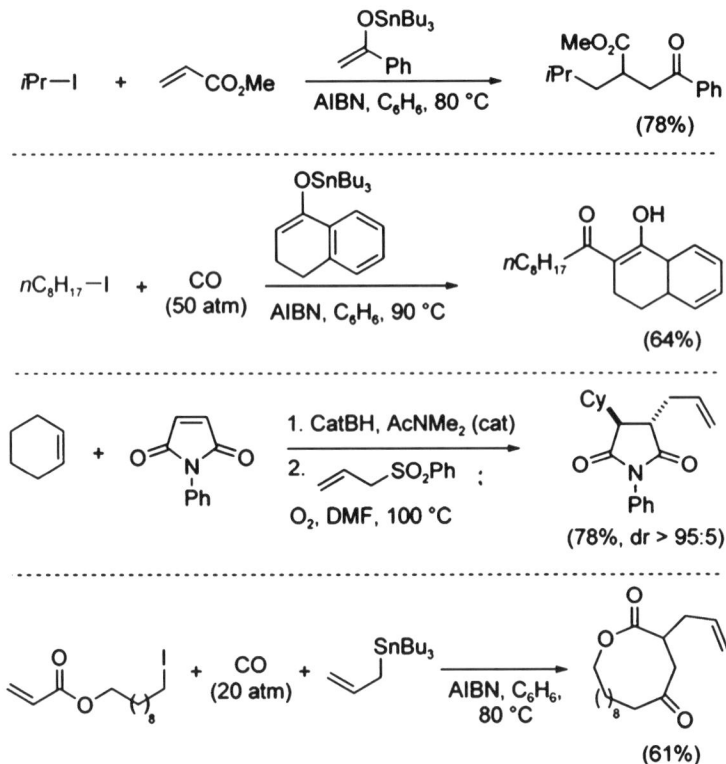

Scheme 19.9   3-CR-DAD with tin enolates, allylsulfones, and allyltin.

furnishes the final product in excellent yield and stererocontrol, along with a phenylsulfonyl radical capable of initiating the homolytic cleavage of the starting borane. Similarly, allylzirconocenes have been introduced as allyltin surrogates in such 3-CR, although excess organozirconocene is usually required.[24] The one-pot synthesis of macrolactones is also at hand by exploiting the unique reactivity of CO. The primary alkyl radical arising form corresponding radical precursor reacts under diluted conditions (0.005 mM) with CO to form an even more reactive acyl radical, which cyclizes to produce a radical enolate. Radical allylation upon exposure to 6–10 equiv. of allyltin then completes the sequence, furnishing macrocycle in 61% yield. Finally, Naito and co-workers nicely exploited the ambiphilic character of sulfide radicals by designing a three-component hydroxysulfenylation reaction involving an α,β-unsaturated oxime, a thiol, and oxygen. This led to the corresponding β-hydroxysulfides in excellent yields and good diastereoselectivities.[25]

Numerous efforts have been devoted to the development of stereoselective versions of the Mizuno–Otsuji 3-CR carboallylation process. Curran,[26a] Porter,[26a,b] and Sibi[26b,c] and colleagues have pioneered this area and demonstrated that these 3-CRs can be performed with high levels of diastereo and enantiocontrol, very much like analogous ionic processes. These approaches have been exhaustively reviewed elsewhere.[26] We will therefore focus on two representative examples, depicted in Scheme 19.10, that illustrate stereocontrolled versions of 3-CR carboallylation. Diastereoselective approaches relied on the use of chiral sultams,[27] oxazolidines,[28] and oxazolidinones[29] and auxiliaries.1,4-Addition of alkyl radicals derived from alkyl iodide onto selected radical acceptor, followed by allylation with allyltin led to a final product in high yield as a single stereoisomer. In these reactions, the Lewis acid not only promotes the reaction by lowering the LUMO of the unsaturated acceptor, but also controls the rotamer population, through bidentate coordination to both carbonyl functions. The acceptor, as well as the intermediate radical, is therefore locked in an s-*cis* conformation, in which the approach of tin reagent from the Si-face is blocked by the bulky diphenylmethyl group (see TS-I, Scheme 19.10). Allylation therefore occurs exclusively on the Re-face and rationalizes the observed selectivities.[28c] Alternatively, chiral bis-oxazoline ligands makes the enantioselective production of final product possible from a chosen radical acceptor.[29] This 3-CR was found to be both highly diastereo- and enantioselective and interestingly, copper and magnesium catalysis led to opposite enantiomers using the same chiral ligand.[29d]

19.3.3   3-CR-DAA

Most of these MCRs are again based on the use of CO as a C1 acceptor/donor synthon.[30] Addition of an alkyl radical (D) onto CO produces a nucleophilic

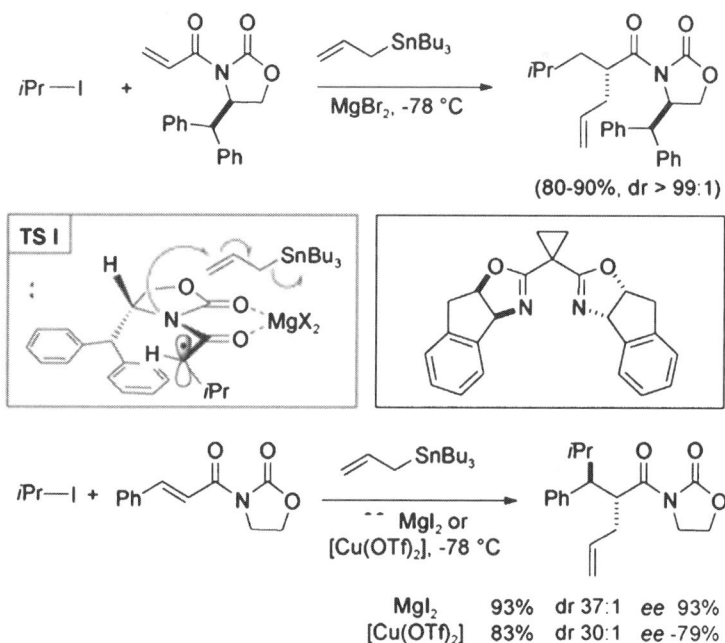

Scheme 19.10   Diastereo- and enantioselective carboallylation of olefins.

acyl radical, which can react further with an acceptor, such as acrylonitrile or methyl acrylate, to generate a further radical intermediate, which is finally reduced by Bu₃SnH (Scheme 19.11). The free-radical chain has to be carried out under diluted conditions (0.017 mM) in order to avoid the direct reaction of nuleophilic octyl radical with the electron-poor olefin. This double alkylation of CO affords unsymmetrical ketones in moderate to good yields.

This process was also applied to other acceptors such as sulfonyl cyanide (X = CN) and thioethers (X = SPh) (Scheme 19.12).[31] Tin-free cyano- and thiocarbonylation were performed starting from alkylallylsulfones. Addition of V-40 to radical precursor generated an electron-rich alkyl radical that was carbonylated. The corresponding acyl radical then reacted with the sulfonyl acceptors to provide a thioester[31a] or acylcyanide[31b] along with a sulfonyl radical able to trigger a new alkyl radical from starting material. As stated above, the direct formation of alkyl cyanide or alkyl thioethers can be decreased at high dilution (0.01 mM) and under high pressures of CO. Primary radicals reacted efficiently, while secondary and tertiary radicals produced larger amount of the corresponding direct addition products.

19.3.4 3-CR-DDA

As mentioned before, nucleophilic-type radicals may react with electron-rich substrates if no other option is available. Such is the case with nucleophilic silicon and tin radicals as illustrated with the two examples below.

Silylcarbonylation of 1,5-dienes led to cyclopentanone in excellent yield (Scheme 19.13).[32a] The initial electron-rich silyl radical adds to the less-

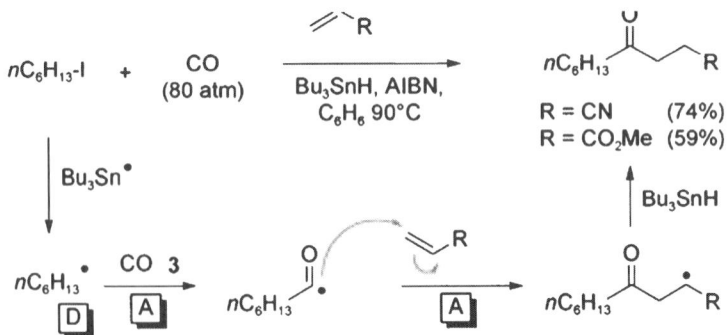

Scheme 19.11   Double alkylation of CO.

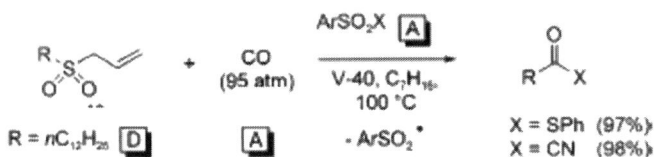

Scheme 19.12   Tin-free cyanocarbonylation and thiocarbonylation of alkylsulfones.

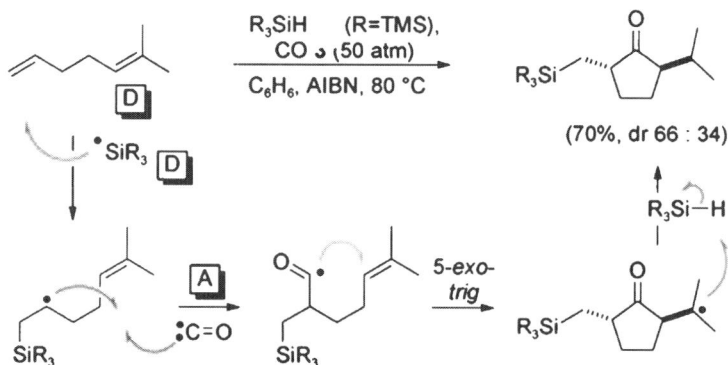

Scheme 19.13 3-CR-DDA silyl carbonylation of 1,5-dienes.

substituted end of the 1,5-diene, leading to acyl radical, after radical carbonylation. Then 5-*exo*-trig cyclization and further reduction of the tertiary alkyl radical terminate the radical chain and furnish the final chiral ketone. This silylcarbonylation, restricted to substituted dienes, has also been applied to 1,5-enynes.[32b]

As a continuation of their investigations in this area, the same laboratory described a ground-breaking stannylcarbonylation of azaenynes that led to four-, five-, six-, seven-, and eight-membered-ring lactams (Scheme 19.14).[33a-g] The unusual N-philic regioselectivity of the process was rationalized by

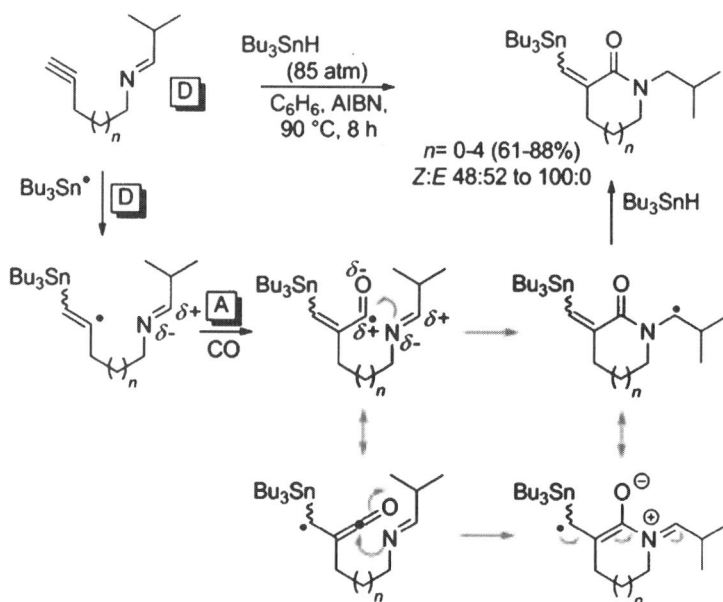

Scheme 19.14 Stannylcarbonylation of azaenynes *en route* to the synthesis of small and medium-ring lactams

invoking a polarity-matched interaction between the imino group and the ketene-like α,β-unsaturated acyl radical.[33a–f] The conversion of this apparent "mismatched" polarity into a matched acyl radical/imine combination, either by a direct radical addition or through polar intermediates, exclusively led to the corresponding carbon-centered radical and then to lactam. This innovative 3-CR synthesis of lactams can also be conducted with $(Me_3Si)_3SiH$ and PhSH, which led analogously to the expected lactams. Interestingly, these last variants proceed with complete *Z* stereochemistry in sharp contrast with the *E* selectivity observed in the tin series.[33c] The process is also efficient with alkynylamines bearing a benzylic amine moiety, in which case the benzyl group is expelled through an S_{Hi} reaction occurring at nitrogen to produce the related lactams.[33g] An intermolecular version of the sequence below was also disclosed recently,[33h,i] involving a tin radical, which upon addition onto a terminal alkyne provided the corresponding vinyl radical. This was carbonylated with CO to produce an α,β-unsaturated acyl radical. Intermolecular addition of an amine onto the ketene form led, after 1,4-H migration and tin β-elimination, to the resulting vinylamide.

19.4 FOUR- AND FIVE-COMPONENT RADICAL REACTIONS

19.4.1 4-CR-DAAD

Although not as widespread as 3-CR, several remarkable four-component processes have been designed. 4-CRs are extremely attractive and their synthetic potential is considerable (four bonds formed in an intermolecular fashion and in a single step). Once again, carbon monoxide, due to its ability to undergo transformations not possible with other reagents, holds a prominent place in this class of multicomponent reactions. Ryu *et al.* have successfully extended their 3-CR-DAA described in Scheme 19.11 and designed a versatile 4-CR-DAAD (Scheme 19.15).[34] The electrophilic radical (α to CN), generated upon assembly of the third component, can indeed serve as a starting point for the introduction of a fourth electron-rich component. Ryu *et al.* have

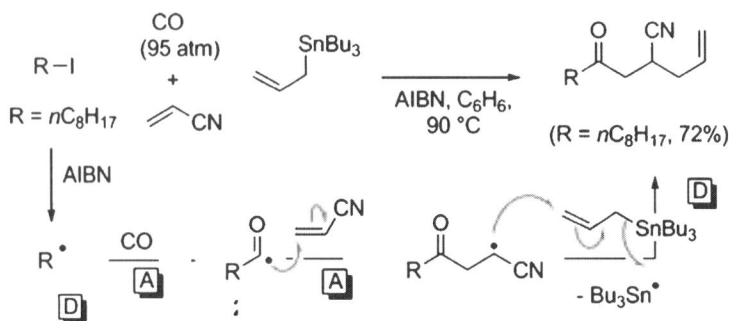

Scheme 19.15   4-CR-DAAD.

notably shown that allyltin (Scheme 19.15)[34] as well as tin enolates (*vide supra*)[22] are excellent candidates as fourth components.

On a similar basis, sulfonyloximes (Scheme 19.16), a useful class of radical acceptors introduced by Kim *et al.*,[35] can successfully take part in a 4-CR-DAAD strategy.[36] The nucleophilic acyl radical, generated as above (Scheme 19.15), efficiently adds to oxime to provide, after β-fragmentation of one of the sulfonyl groups, a monosulfonyl oxime. Additional *in situ* treatment of intermediate with ethyl iodide then affords α-ketoxime in 47% yield. Although this sequence is formally a 3-CR-DAA process, the fourth component being added in a second stage, it may be classified as a sequential or bidirectional 4-CR-DAAD (*vide infra*). Several 4-CRs have relied on the double incorporation of carbon monoxide. Kim *et al.* reported a 4-CR-DAAA-type process[31a] that provides convenient access to cyclopentanone (Scheme 19.16). The radical arising from allylsulfone leads after a first carbonylation to acyl radical, which rapidly cyclizes to the desired product. A second equivalent of CO then comes into play. With no other intramolecular options, acyl radical is readily trapped with PhSO₂SPh, yielding the final product in an impressive 66% yield.

19.4.2 4-CR-ADAA

The above synthetic plan was ingeniously modified as follows: an electrophilic malonyl-type radical derived from allylsulfone preferentially reacts with electron-rich olefins such as allylsilane. The subsequent radical cyclization of radical precursor turned out to be faster than CO incorporation at this stage and generated a primary radical to afford a thioester as a final product in excellent yield (Scheme 19.17).

19.4.3 4-CR-AADA

An example of a 4-CR-AADA has been described by Ogawa and co-workers[37] that provides efficient access, albeit with moderate stereoselectivity, to highly

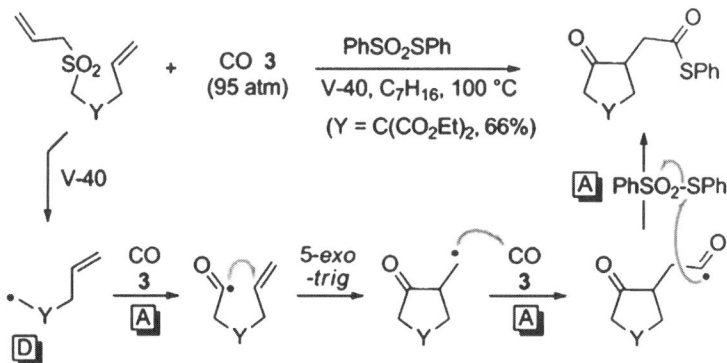

Scheme 19.16 4-CR-DAAA with incorporation of two molecules of CO.

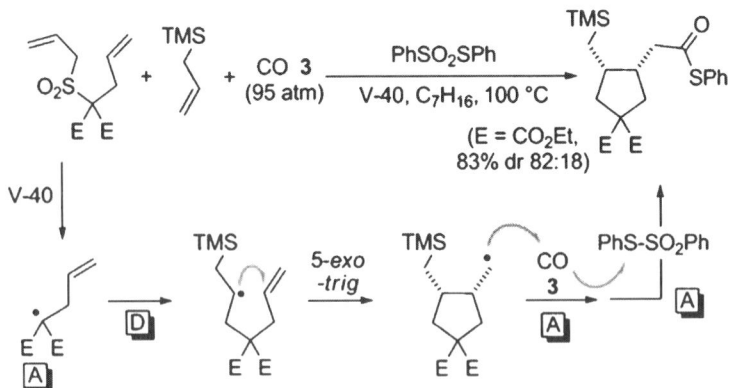

Scheme 19.17   4-CR-ADAA.

functionalized cyclopentanes using alkyne and two electronically differentiated olefins (Scheme 19.18). This process is initiated by the selective addition of PhSe radicals onto electron-poor alkyne. Such additions are known to be reversible (β-fragmentation), and although the addition of PhSe radical to a radical acceptor probably occurs at similar rates, the greater thermodynamic stability of newly formed radical increases its concentration *in situ*. The stabilized vinylic radical then reacts efficiently with the most electron-rich olefin due to more favorable SOMO–HOMO interactions. The ensuing electron-rich radical then combines with acrylonitrile to afford corresponding advanced radical intermediate. A 5-*exo*-trig cyclization followed by reaction of the α-selenoradical with diphenyldiselenide through an S_H2 process, then furnishes cyclopentane as a mixture of three stereoisomers.

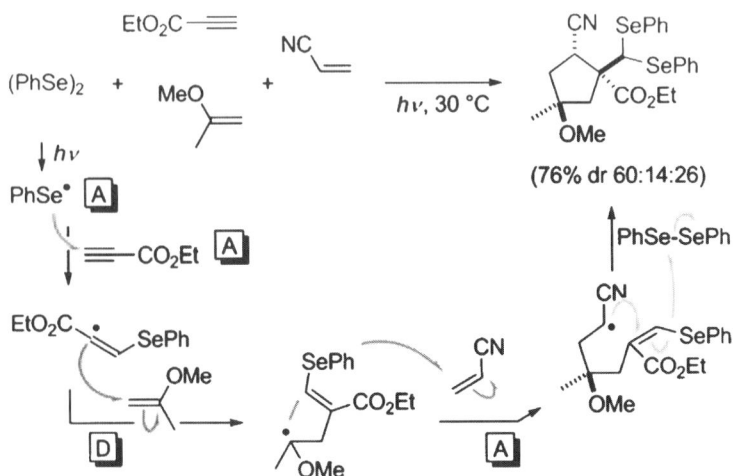

Scheme 19.18   4-CR-AADA using electronically differentiated olefins.

Finally, a rare example of a 5-CR has been reported by Kim, Ryu, and co-workers, starting from bis-sulfonyl oxime (Scheme 19.19).[36] Two molecules of CO, two molecules of octyl iodide and oxime were assembled in a bidirectional manner, with four C–C bonds formed in one pot, affording a symmetrical oxime in 70% yield.

19.5 MULTICOMPONENT RADICAL–IONIC REACTIONS

Upon the assembly of two components by means of a radical mechanism, the newly formed radical can also be reduced to generate a true nucleophilic species that can be employed in classical ionic processes. Given the relatively inert character of radicals under ionic conditions and *vice versa*, it is not surprising that multi-component reactions incorporating radical and ionic mechanisms have received considerable interest. Such processes rely on subtle relationships between the rates of the different steps involved. Successful examples of radical–ionic multicomponent reactions generally feature a metallic species able to selectively reduce or oxidize one of the radicals present.

19.5.1 Multicomponent Radical–Anionic Reactions

This class of radical–ionic MCRs has aroused very large interest and involves, after radical addition, the *in situ* generation of a nucleophilic intermediate.

19.5.1.1 3-CR-DAA

Shono *et al.* initially showed that upon addition of an alkyl radical (formed with Zn^0) to a Michael acceptor, the generated enolyl radical could be reduced with excess Zn^0 to yield an enolate that could be trapped with electrophiles.[38a] This one-pot procedure was recently extended to acylating agents by Nishiguchi and coworkers,[38b] further demonstrating the versatility of such a 3-CR. Dialkylzinc reagents act as a source of alkyl radical when exposed to traces of O_2. Under such aerobic conditions, Et_2Zn was shown to add very efficiently onto α,β-unsaturated esters and imides, further activated by the

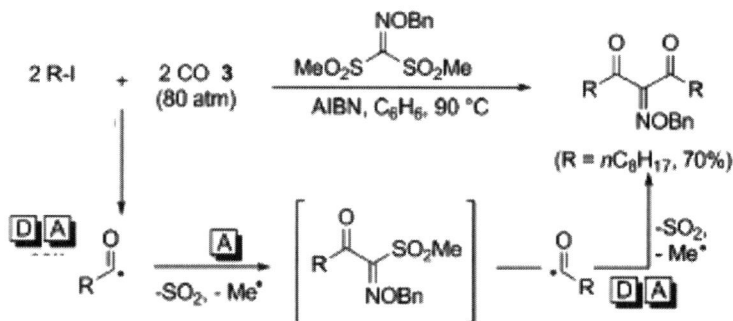

Scheme 19.19 Bidirectional 5-CR for the synthesis of ketoximes.

Lewis acidity of the zinc reagent. Homolytic cleavage of one of the ethyl groups then not only regenerates an ethyl radical, but also leaves a zinc enolate that can react further with aldehydes to afford the corresponding aldol adducts in good yields. Useful diastereoselectivities are also observed when a chiral auxiliary is embedded into the α,β-unsaturated acceptor.[39] The reduction of the radical enolate was shown by Chemla and co-workers,[40,41] to be slower than a 5-*exo*-trig cyclization, which led the authors to design the elegant multicomponent cascade reaction depicted in Scheme 19.20, featuring the following events: (1) radical addition, (2) 5-*exo*-trig radical cyclization, (3) homolytic substitution at the zinc center of $(nBu)_2Zn$, and (4) copper-mediated S_N2 displacement of an allylic bromide. Samarium iodide[42] and manganese salts[43] also have the ability to mediate such processes. In sharp contrast, radical enolates are reduced very rapidly by Mn^0, so that 5-*exo*-trig cyclization no longer has time to occur; the intermolecular ionic step thus precedes this event.[43]

Trialkylboranes, such as Et_3B are also suitable agents for radical–polar crossover reactions.[44] Upon initiation and 1,4-addition of an ethyl radical across an α,β-unsaturated carbonyl compound, the corresponding radical enolate then reacts with Et_3B to generate the boron enolate through hemolytic substitution at boron. Similarly, Barluenga and co-workers demonstrated that boron–trifluoride–molybdenum carbenes decompose through a radical pathway and trigger, by a radical addition, fast reduction of the resulting α-keto radical.[45] In all of these cases, the ensuing enolates have been successfully involved in classical ionic chemistry such as aldol reactions. Chromium(II) salts also behave in some cases as mild and highly selective reductants. Single-electron transfer (SET) in the presence of Cr(II) salts are much faster with allylic radicals than with secondary or tertiary alkyl radicals. Such a reactivity profile allowed Takai and coworkers to realize the following impressive 3-CR sequence (Scheme 19.21):[46] (1) regioselective addition of an alkyl radical across 1,3-diene, (2) chemoselective SET from Cr(II) salts to the resulting allylic radical to form allylchromium species and (3) regio- and stereoselective allylation of benzaldehyde to furnish a homoallylic alcohol in 66% yield.

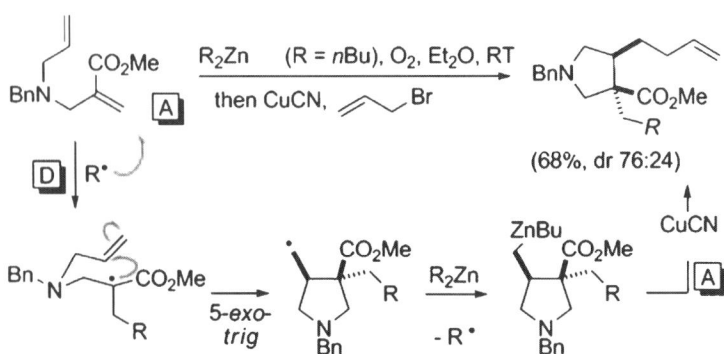

Scheme 19.20   Radical–polar crossover 3-CR.

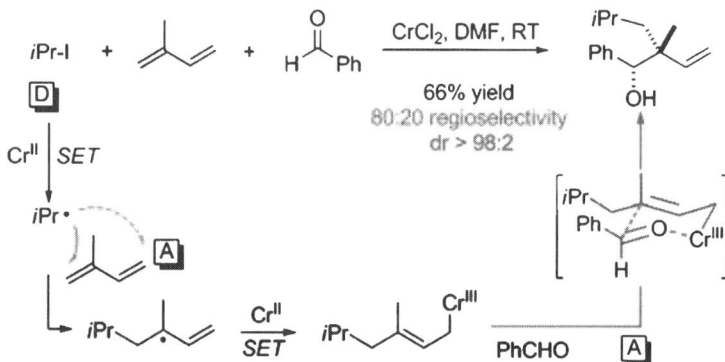

Scheme 19.21 3-CR by sequential Cr(II) radical generation.

Titanocene[47] and cobalt[48] complexes also adopt this discriminating behavior towards allylic and benzylic radicals, enabling the functionalization of dienes and styryl type olefins. Ryu, Sonoda, and co-workers have shown that the Zn(Cu) reduction system also efficiently mediates such radical/ionic hybrid 3-CRs (Scheme 19.22).[49] This alloy permits the formation of a highly nucleophilic acyl radical, arising from the radical assembly. Unlike a cyclohexyl radical, the acyl radical is readily reduced to acyl anion by the Zn(Cu)-induced system. In the presence of acrylonitrile, a Michael addition then occurs affording the corresponding adduct.

19.5.2 Multicomponent Radical–Cationic Reactions

As opposed to the radical–anionic processes discussed above, in radical–cationic MCRs, an electrophilic intermediate prone to further reactions with nucleophiles is generated *in situ* after the first radical addition.

19.5.2.1 4-CR-ADAD

Radical–cationic MCRs rely on the ability of an organometallic species to selectively oxidize one class of radicals over another. Ryu and Alper and colleagues pioneered this area and showed that SET from an acyl radical to Mn[III] proceeded at a significantly faster rate than the same process involving alkyl radicals.[50] This specificity enabled them to describe an original 4-CR that

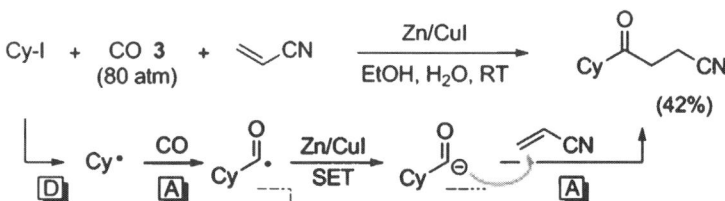

Scheme 19.22 Carbonylative Zn/Cu-mediated 3-CR.

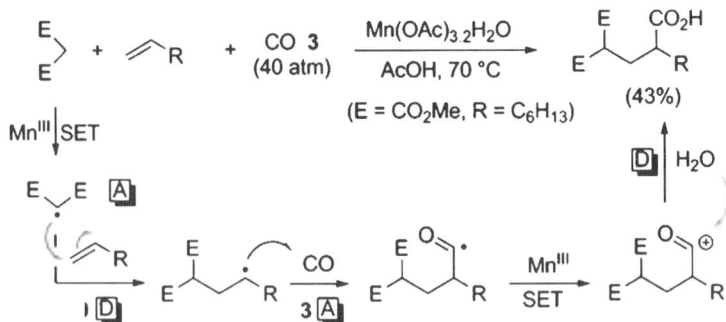

Scheme 19.23   4-CR through selective Mn(III) oxidation of acyl radicals.

judiciously exploits the multiple characteristics of Mn^{III} salts (Scheme 19.23): (1) $Mn(OAc)_3$ first generates an electrophilic electrophilic malonyl radical[51] that adds onto olefin; (2) the ensuing nucleophilic radical readily adds onto CO; and (3) $Mn(OAc)_3$ oxidizes the resulting acyl radical into an acylium cation that is immediately trapped by the nucleophilic fourth component H_2O.[52]

19.5.2.2   3-CR-DAD

Alternatively, the one-electron oxidation of the acyl radical can also occur through iodine atom transfer from an alkyl iodide (Scheme 19.24).[52,53] The resulting electrophilic acyl iodide generated *in situ* can then be trapped with alcohols or amines. When irradiated with a Xe-lamp, alkyl iodides react with CO and an alcohol or an amine to yield esters[52a] and amides,[52b] respectively. This process is further improved when a catalytic amount of $[Mn_2(CO)_{10}]$ or $[Pd(PPh_3)_4]$ is employed (Scheme 19.24).[54] The combination of Pd^0 and irradiation enables the generation of alkyl radicals by SET. Among several possibilities, the authors suggest that, after addition of the nucleophilic radical onto CO, the resulting acyl radical is transformed into an acyl-palladium species, which undergoes alcoholysis, followed by a reductive elimination to afford the desired ester with regeneration of the active Pd(0) catalyst. This 3-CR sequence has notably been applied to the concise synthesis of dihydro-capsaicin.[54a] This strategy is robust[54b] and has been nicely extended to a 4-CR protocol, following a similar design to the one described in Scheme 19.17.[54c]

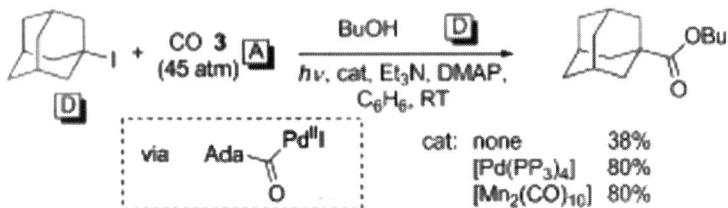

Scheme 19.24   3-CR through Pd^0-accelerated oxidation of acyl radicals.

19.5.3 Sequential Multicomponent Radical–Polar Crossover Reactions

A final category, thus far relatively unexplored, takes advantage of the orthogonality of radical and ionic processes. One-pot protocols aiming to conduct radical and ionic reactions sequentially have therefore recently begun to emerge.

19.5.3.1 4-CR-ADAD

As an example, our group recently reported a sequence in which three components were first assembled by a radical mechanism affording oxime (Scheme 19.25).[55] Upon completion of the reaction the mixture was cooled, diluted with THF, and treated with a freshly prepared solution of allylzinc bromide in THF. This resulted in a chemo- and stereoselective allylation of the oxime, in the presence of two ester groups, leading to the zinc amide, which underwent spontaneous lactamization. This one-pot protocol gave access to piperidinones, resulting from the sequential assembly of 4 components, in 59% yield as a 90:10 mixture of diastereomers. While the example above illustrates the usefulness of a strategy based on a radical cascade terminated by an anionic process, the inverse strategy relying on an ionic MCR terminated by a radical process has also been documented recently. El Kaim *et al.* described the MCR assembly of the indane skeleton based on a four-component Ugi coupling that was followed by the introduction of a fifth component through a radical pathway without isolation of the intermediate 4-CR product.[56] This constitutes a complementary approach that should also prove very useful in the near future.

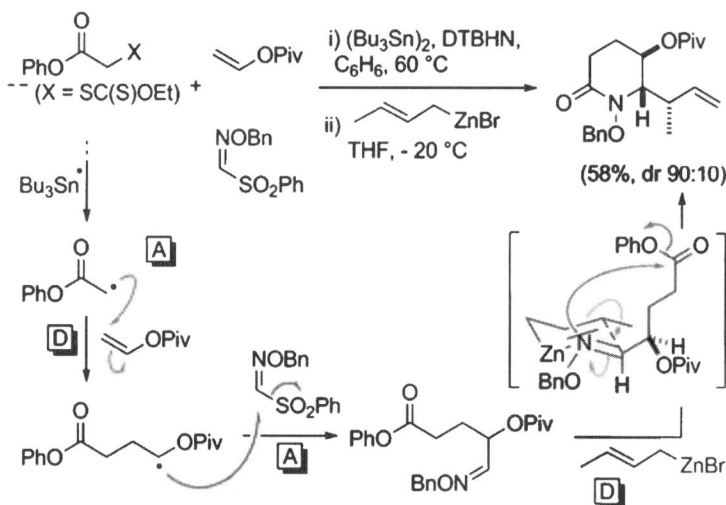

Scheme 19.25   Sequential radical–ionic 4-CR.

19.6   CONCLUSION AND FUTURE DIRECTION

Multicomponent processes have a long-standing history and are particularly useful for diversity-oriented synthesis (DOS),[57] providing access to a large variety of drug-like targets. In this context, cascade radical processes hold a special place and have yet to reveal their full potential. Radical reactions often take place under mild conditions and are generally compatible with many functional groups allowing transformations to be carried out without recourse to protecting groups.[58] In addition, the orthogonality of radical and traditional ionic processes permits the further transformations of intermediates assembled through radical MCRs, using either ionic, and/or organometallic processes. MCRs and sequential MCRs are particularly well suited to construct complex molecules in a straightforward manner and limited number of synthetic operations. They perfectly meet the requirement of the step-economy concept.[59] The few examples described above illustrate the power of this strategy and should pave the way for future discoveries in the area.

REFERENCES

1. C. Mannich and W. Kröschl, *Arch. Pharm.*, 1912, **250**, 647–667.
2. R. Robinson, *J. Chem. Soc.*, 1917, 762–768.
3. A. Dömling and I. Ugi, *Angew. Chem.*, 2000, **112**, 3300–3344 (*Angew. Chem., Int. Ed.*, 2000, **39**, 3168–3210).
4. I. Ugi and C. Steinbrückner, *Chem. Ber.*, 1961, **94**, 734–742.
5. For recent reviews on multicomponent processes, see: (a) H. Bienaymé, C. Hulme, G. Oddon and P. Schmitt, *Chem.–Eur. J.*, 2000, **6**, 3321–3329; (b) *Multicomponent Reactions,* ed. J. Zhu and H. Bienaymé, Wiley-VCH, Weinheim, 2005.
6. M. Tojino, I. Ryu, N. Sonoda and D. P. Curran, *Chem. Rev.*, 1996, **96**, 177–194; M. Tojino and I. Ryu, in *Multicomponent Reactions,* ed. J. Zhu and H. Bienaymé, Wiley-VCH, Weinheim, 2005, pp. 169–198.
7. (a) A. J. McCarroll and J. C. Walton, *Angew. Chem.*, 2001, **113**, 2282–2307 (*Angew. Chem., Int. Ed.*, 2001, **40**, 2224–2248), and references therein; (b) *Domino Reactions in Organic Synthesis*, ed. L. F. Tietze, D. G. Brasche and K. M. Gericke, Wiley-VCH, Weinheim, 2007, pp. 542–565.
8. (a) For a recent discussion on nucleophilicity *vs.* electrophilicity scale for radicals, see: F. DeVleeschouwer, V. VanSpeybroeck, M. Waroquier, P. Geerlings and F. DeProft, *Org. Lett.*, 2007, **9**, 2721–2724; (b) I. Fleming, *Frontier Orbitals and Organic Chemical Reactions*, Wiley, New York, 1976, ch. 5, pp. 182–186; (c) N. L. Arthur and P. Potzinger, *Organometallics,* 2002, **21**, 2874–2890; (d) B. Giese, *Radical Organic Synthesis: Formation of Carbon-Carbon Bonds*, Pergamon, Oxford, 1986, ch. 2, pp. 4–35.
9. (a) R. E. Foster, A.W. Larchar, R. D. Lipscomb and B. C. McKusick, *J. Am. Chem. Soc.,* 1956, **78**, 5606–5611; (b) T. Susuki and J. Tsuji, *J. Org. Chem.*, 1970, **35**, 2982–2986.

10. For exhaustive reviews on radical carbonylations, see: (a) I. Ryu and N. Sonoda, *Angew. Chem.,* 1996, **108**, 1140–1157 (*Angew. Chem., Int. Ed. Engl.,* 1996, **35**, 1050–1066); (b) I. Ryu, N. Sonoda and D. P. Curran, *Chem. Rev.,* 1996, **96**, 177–194.

11. I. Ryu, H. Muraoka, N. Kambe, M. Komatsu and N. Sonoda, *J. Org. Chem.,* 1996, **61**, 6396–6403.

12. (a) C. Ollivier and P. Renaud, *J. Am. Chem. Soc.,* 2001, **123**, 4717–4727; (b) P. Renaud, C. Ollivier and P. Panchaud, *Angew. Chem.,* 2002, **114**, 3610–3612 (*Angew. Chem., Int. Ed.,* 2002, **41**, 3460–3462); (c) P. Panchaud, C. Ollivier, P. Renaud and S. Zigmantas, *J. Org. Chem.,* 2004, **69**, 2755–2759; (d) P. Panchaud and P. Renaud, *J. Org. Chem.,* 2004, **69**, 3205–3207; (e) P. Panchaud and P. Renaud, *Chimia,* 2004, **58**, 232–233; (f) P. Panchaud, L. Chabaud, Y. Landais, C. Ollivier, P. Renaud and S. Zigmantas, *Chem.–Eur. J.,* 2004, **10**, 3606–3614.

13. F. Le Guyader, B. Quiclet-Sire, S. Seguin and S. Z. Zard, *J. Am. Chem. Soc.,* 1997, **119**, 7410–7411.

14. (a) L. Chabaud, Y. Landais and P. Renaud, *Org. Lett.,* 2002, **4**, 4257–4260; (b) L. Chabaud, Y. Landais, P. Renaud, F. Robert, F. Castet, M. Lucarini and K. Schenk, *Chem.–Eur. J.,* 2008, **14**, 2744–2756.

15. (a) L. Chabaud, Y. Landais and P. Renaud, *Org. Lett.,* 2005, **7**, 2587; (b) P. Schaer and P. Renaud, *Org. Lett.,* 2006, **8**, 1569.

16. (a) M. R. Heinrich, O. Blank and S. Wölfel, *Org. Lett.,* 2006, **8**, 3323–3325; (b) O. Blank and M. R. Heinrich, *Eur. J. Org. Chem.,* 2006, 4331–4334.

17. D. Nanni, P. Pareschi, C. Rizzoli, P. Sgarabotto and A. Tundo, *Tetrahedron,* 1995, **51**, 9045–9062.

18. (a) R. Leardini, D. Nanni and G. Zanardi, *J. Org. Chem.,* 2000, **65**, 2763–2772; (b) Y.-J. Jang, J. Wu, Y.-F. Lin and C.-F. Yao, *Tetrahedron,* 2004, **60**, 6565–6574; (c) A. Ogawa, I. Ogawa and N. Sonoda, *J. Org. Chem.,* 2000, **65**, 7682–7685.

19. R. Noyori and M. Suzuki, *Angew. Chem.,* 1984, **96**, 854–882 (*Angew. Chem., Int. Ed. Engl.,* 1984, **23**, 847–876).

20. K. Mizuno, M. Ikeda, S. Toda and Y. Otsuji, *J. Am. Chem. Soc.,* 1988, **110**, 1288–1290.

21. K. Miura, N. Fujisawa, H. Saito, D. Wang and A. Hosomi, *Org. Lett.,* 2001, **3**, 2591–2594.

22. K. Miura, M. Tojino, N. Fujisawa, A. Hosomi and I. Ryu, *Angew. Chem.,* 2004, **116**, 2477–2479 (*Angew. Chem., Int. Ed.,* 2004, **43**, 2423–2425).

23. A. P. Schaffner, K. Sarkunam and P. Renaud, *Helv. Chim. Acta,* 2006, **89**, 2450–2461.

24. K. Hirano, K. Fujita, H. Shinokubo and K. Oshima, *Org. Lett.,* 2004, **6**, 593–595.

25. M. Ueda, H. Miyabe, H. Shimizu, H. Sugino, O. Miyata and T. Naito, *Angew. Chem.,* 2008, **120**, 5682–5686 (*Angew. Chem., Int. Ed.,* 2008, **47**, 5600–5604).

26. For excellent reviews on the diastereo- and enantioselective allylations of radical species and application to 3-CR (DAD process), see: (a) N. A. Porter, B. Giese and D. P. Curran, *Acc. Chem. Res.,* 1991, **24**, 296–304; (b) M. P. Sibi and N. A. Porter, *Acc. Chem. Res.,* 1999, **32**, 163–171; (c) M. P. Sibi, S. Manyem and J. Zimmerman, *Chem. Rev.,* 2003, **103**, 3263–3296.

27. D. P. Curran, W. Shen, J. Zhang and T. A. Heffner, *J. Am. Chem. Soc.,* 1990, **112**, 6738–6740.

28. (a) N. A. Porter, J. D. Bruhnke, W. X. Wu, I. J. Rosenstein and R. A. Breyer, *J. Am. Chem. Soc.,* 1991, **113**, 7788–7790; (b) N. A. Porter, I. J. Rosenstein, R. A. Breyer, J. D. Bruhnke, W. X. Wu and A. T. McPhail, *J. Am. Chem. Soc.,* 1992, **114**, 7664–7676; (c) M. P. Sibi and J. Ji, *J. Org. Chem.,* 1996, **61**, 6090–6091.

29. (a) M. P. Sibi, C. P. Jasperse and J. Ji, *J. Am. Chem. Soc.,* 1995, **117**, 10779–10780; (b) J. H. Wu, R. Radinov and N. A. Porter, *J. Am. Chem. Soc.,* 1995, **117**, 11029–11030; (c) M. P. Sibi and J. Chen, *J. Am. Chem. Soc.,* 2001, **123**, 9472–9473; (d) M. P. Sibi, S. Manyem and R. Subramaniam, *Tetrahedron,* 2003, **59**, 10575–10580.

30. (a) I. Ryu, K. Kusano, H. Yamazaki and N. Sonoda, *J. Org. Chem.,* 1991, **56**, 5003–5005; (b) S. Tsunoi, I. Ryu, S. Yamasaki, H. Fukushima, M. Tanaka, M. Komatsu and N. Sonoda, *J. Am. Chem. Soc.,* 1996, **118**, 10670–10671.

31. (a) S. Kim, S. Kim, N. Otsuka and I. Ryu, *Angew. Chem.,* 2005, **117**, 6339–6342 (*Angew. Chem., Int. Ed.,* 2005, **44**, 6183–6186); (b) S. Kim, C. H. Cho, S. Kim, Y. Uenoyama and I. Ryu, *Synlett,* 2005, 3160–3162; (c) Y. Uenoyama, T. Fukuyama, K. Morimoto, O. Nobuta, H. Nagai and I. Ryu, *Helv. Chim. Acta,* 2006, **89**, 2483–2494.

32. (a) I. Ryu, K. Nagahara, A. Kurihara, M. Komatsu and N. Sonoda, *J. Organomet. Chem.,* 1997, **548**, 105–107; (b) T. Fukuyama, Y. Uenoyama, S. Oguri, N. Otsuka and I. Ryu, *Chem. Lett.,* 2004, **33**, 854–855.

33. (a) I. Ryu, H. Miyazato, H. Kuriyama, K. Matsu, M. Tojino, T. Fukuyama, S. Minakata and M. Komatsu, *J. Am. Chem. Soc.,* 2003, **125**, 5632–5633; (b) C. T. Falzon, I. Ryu and C. H. Schiesser, *Chem. Commun.,* 2002, 2338–2339; (c) M. Tojino, N. Otsuka, T. Fukuyama, H. Matsubara, C. H. Schiesser, H. Kuriyama, H. Miyazato, S. Minakata, M. Komatsu and I. Ryu, *Org. Biomol. Chem.,* 2003, **1**, 4262–4267; H. Kuriyama, H. Miyazato, S. Minakata, M. Komatsu and I. Ryu, *Org. Biomol. Chem.,* 2003, **1**, 4262–4267; (d) H. Matsubara, C. T. Falzon, I. Ryu and C. H. Schiesser, *Org. Biomol. Chem.,* 2006, **4**, 1920–1926; (e) C. H. Schiesser, U. Wille, H. Matsubara and I. Ryu, *Acc. Chem. Res.,* 2007, **40**, 303–313; (f) M. Tojino, N. Otsuka, T. Fukuyama, H. Matsubara and I. Ryu, *J. Am. Chem. Soc.,* 2006, **128**, 7712–7713; (g) Y. Uenoyama, T. Fukuyama and I. Ryu, *Org. Lett.,* 2007, **9**, 935–937; (h) Y. Uenoyama, T. Fukuyama, O. Nobuta, H. Matsubara and I. Ryu, *Angew. Chem.,* 2005, **117**, 1099–1102 (*Angew. Chem., Int. Ed.,* 2005, **44**, 1075–1078); (i) I. Ryu,

Y. Uenoyama and H. Matsubara, *Bull. Chem. Soc. Jpn.*, 2006, **79**, 1476–1488.

34. (a) I. Ryu, H. Yamazaki, A. Ogawa, N. Kambe and N. Sonoda, *J. Am. Chem. Soc.*, 1993, **115**, 1187–1189; (b) Y. Uenoyama, T. Fukuyama and I. Ryu, *Synlett*, 2006, 2342–2344.

35. S. Kim, I. Y. Lee, J.-Y. Yoon and D. H. Oh, *J. Am. Chem. Soc.*, 1996, **118**, 5138–5139.

36. (a) I. Ryu, H. Kuriyama, S. Minakata, M. Komatsu, J.-Y. Yoon and S. Kim, *J. Am. Chem. Soc.*, 1999, **121**, 12190–12191; (b) S. Kim, K. C. Lim, S. Kim and I. Ryu, *Adv. Synth. Catal.*, 2007, **349**, 527–530.

37. K. Tsuchii, M. Doi, I. Ogawa, Y. Einaga and A. Ogawa, *Bull. Chem. Soc. Jpn.*, 2005, **78**, 1534–1548.

38. (a) T. Shono, I. Nishiguchi and M. Sasaki, *J. Am. Chem. Soc.*, 1978, **100**, 4314–4315; (b) Y. Yamamoto, S. Nakano, H. Maekawa and I. Nishiguchi, *Org. Lett.*, 2004, **6**, 799–802.

39. (a) S. Bazin, L. Feray, D. Siri, J.-V. Naubron and M. P. Bertrand, *Chem. Commun.*, 2002, 2506–2507; (b) S. Bazin, L. Feray, N. Vanthuyne and M. P. Bertrand, *Tetrahedron*, 2005, **61**, 4261–4274; (c) S. Bazin, L. Feray, N. Vanthuyne, D. Siri and M. P. Bertrand, *Tetrahedron*, 2007, **63**, 77–85.

40. (a) F. Denes, F. Chemla and J. F. Normant, *Angew. Chem.*, 2003, **115**, 4177–4180 (*Angew. Chem., Int. Ed.*, 2003, **42**, 4043–4046); (b) F. Denes, S. Cutri, A. Perez-Luna and F. Chemla, *Chem.–Eur. J.*, 2006, **12**, 6506–6513.

41. Addition followed by 5-*exo*-trig cyclization on alkynes: (a) S. Giboulot, A. Pérez-Luna, C. Botuha, F. Ferreira and F. Chemla, *Tetrahedron Lett.*, 2008, **49**, 5322–5323; (b) A. Pérez-Luna, C. Botuha, F. Ferreira and F. Chemla, *Chem.–Eur. J.*, 2008, **14**, 8784–8788.

42. Samarium-mediated addition on isocyanides: (a) M. Murakami, T. Kawano and Y. Ito, *J. Am. Chem. Soc.*, 1990, **112**, 2437–2439; (b) D. P. Curran and M. J. Totleben, *J. Am. Chem. Soc.*, 1992, **114**, 6050–6058; (c) M. Murakami, T. Kawano, H. Ito and Y. Ito, *J. Org. Chem.*, 1993, **58**, 1458–1465; samarium-mediated addition of ketyl radicals on Michael acceptors: (d) V. Blot and H.-U. Reissig, *Synlett*, 2006, 2763–2766.

43. K. Takai, T. Ueda, N. Ikeda and T. Moriwake, *J. Org. Chem.*, 1996, **61**, 7990–7991.

44. (a) K. Nozaki, K. Oshima and K. Utimoto, *Tetrahedron Lett.*, 1988, **29**, 1041–1044; (b) K. Nozaki, K. Oshima and K. Utimoto, *Bull. Chem. Soc. Jpn.*, 1991, **64**, 403–409; (c) M. Ueda, H. Miyabe, H. Sugino, O. Miyata and T. Naito, *Angew. Chem.*, 2005, **117**, 6346–6349 (*Angew. Chem., Int. Ed.*, 2005, **44**, 6190–6193).

45. J. Barluenga, F. Rodríguez, F. J. Fañanás and E. Rubio, *Angew. Chem.*, 1999, **111**, 3272–3274 (*Angew. Chem., Int. Ed.*, 1999, **38**, 3084–3086).

46. K. Takai, N. Matsukawa, A. Takahashi and T. Fujii, *Angew. Chem.*, 1998, **110**, 160–163 (*Angew. Chem., Int. Ed.*, 1998, **37**, 152–155).

47. (a) J. Terao, N. Kambe and N. Sonoda, *Tetrahedron Lett.,* 1998, **39**, 9697–9698; (b) J. Terao, K. Saito, S. Nii, N. Kambe and N. Sonoda, *J. Am. Chem. Soc.,* 1998, **120**, 11822–11823; (c) S. Nii, J. Terao and N. Kambe, *J. Org. Chem.,* 2000, **65**, 5291–5297.

48. K. Mizutani, H. Shinokubo and K. Oshima, *Org. Lett.,* 2003, **5**, 3959–3961.

49. S. Tsunoi, I. Ryu, H. Fukushima, M. Tanaka, M. Komatsu and N. Sonoda, *Synlett,* 1995, 1249.

50. (a) I. Ryu and H. Alper, *J. Am. Chem. Soc.,* 1993, **115**, 7543–7544; (b) K. Okuro and H. Alper, *J. Org. Chem.,* 1996, **61**, 5312–5315.

51. Carbonyl-substituted radicals including malonyls are known to have ambiphilic character—see: (a) B. Giese, J. He and W. Mehl, *Chem. Ber.,* 1988, **121**, 2063–2066; (b) D. P. Curran, D. Kim and C. Ziegler, *Tetrahedron,* 1991, **47**, 6189–6196.

52. (a) K. Nagahara, I. Ryu, M. Komatsu and N. Sonoda, *J. Am. Chem. Soc.,* 1997, **119**, 5465–5466; (b) I. Ryu, K. Nagahara, N. Kambe, N. Sonoda, S. Kreimerman and M. Komatsu, *Chem. Commun.,* 1998, 1953–1954.

53. (a) I. Ryu, *Chem. Soc. Rev.,* 2001, **30**, 16–25; (b) S. Kreimerman, I. Ryu, S. Minakata and M. Komatsu, *C. R. Acad. Sci., Ser. IIc: Chim.,* 2001, 497–503.

54. a) T. Fukuyama, S. Nishitani, T. Inouye, K. Morimoto, I. Ryu, *Org. Lett.,* 2006, **8**, 1383–1386; (b) for an application in ionic liquids: T. Fukuyama, T. Inoue and I. Ryu, *J. Organomet. Chem.,* 2007, **692**, 685–690; (c) I. Ryu, S. Kreimerman, F. Araki, S. Nishitani, Y. Oderaotoshi, S. Minakata and M. Komatsu, *J. Am. Chem. Soc.,* 2002, **124**, 3812–3813.

55. E. Godineau and Y. Landais, *J. Am. Chem. Soc.,* 2007, **129**, 12662–12663.

56. L. El Kaim, L. Grimaud and E. Vieu, *Org. Lett.,* 2007, **9**, 4171–4173.

57. (a) M. D. Burke and S. L. Schreiber, *Angew. Chem.,* 2004, **116**, 48–60 (*Angew. Chem., Int. Ed.,* 2004, **43**, 46–58); (b) R. L. Strausberg and S. L. Schreiber, *Science,* 2003, **300**, 294–295.

58. (a) R. W. Hoffmann, *Synthesis,* 2006, 3531–3541; (b) P. S. Baran, T. J. Maimone and J. M. Richter, *Nature,* 2007, **446**, 404–408.

59. B. Trost, *Science,* 1991, **254**, 1471–1477; (b) R. Sheldon, *Chem. Ind.,* 1992, 903–906.

Radical–Radical–Radical Telescopic Reactions: from Rules Through Reasons to Applications[*]

20.1 INTRODUCTION

Instead of simply using two radical reactions in a domino process, the combination of three and more radical C—C or C—N bond-forming radical transformations is also possible. This makes this methodology one of the most powerful procedures in the synthesis of complex molecules starting from simple substrates.[1] During the years, several strategies have been developed, and these are depicted in Scheme 20.1. The strategies can be classified as three types:

- The so-called "zipper" strategy is characterized by starting the cyclization process in the middle of the chain and working its way back and forth across toward the ends.
- The "macrocyclization/transannular cyclization" process,[2] in contrast, is initiated at the ends of the chain and works towards the middle.
- The third, less common, "round trip radical reaction" process starts at one end of the chain and works its way back to the same end.[3]

20.2 THE "ROUND TRIP" STRATEGY IN ACTION

This latter strategy has been employed by Haney and Curran for their very short total synthesis of the natural products iso-gymnomitrene and gymnomitrene (Scheme 20.2).[4] As substrate, the easily accessible iodotriene was used; this was converted into a mixture of compounds containing 31% of iso-

* Chapter written by V. Tamara Perchyonok.
Streamlining Free Radical Green Chemistry
V. Tamara Perchyonok, Ioannis Lykakis and Al Postigo
© V. Tamara Perchyonok, Ioannis Lykakis and Al Postigo 2012
Published by the Royal Society of Chemistry, www.rsc.org

• The "zipper" strategy goes from the middle to the ends.

• The "macrocyclization/transannular cyclization" strategy goes from the ends to the middle.

• The "round trip" strategy goes from the end back to the same end.

Scheme 20.1   Strategies for Radical-Radical-Radical telescopic reactions.

gymnomitrene and 3% of gymnomitrene using a tin hydride derivative. In this process, iso-gymnomitrene is formed by a 5-*exo*-trig/6-*endo*-trig/5-*exo*-trig domino cyclization *via* corresponding advanced intermediates. Other compounds obtained are the monocycle ketone in 23%, and the corresponding bicycles were also produced in 20% and 22% yields, respectively.

A dramatic improvement in this new "round trip radical domino processes" developed by Curran's group was presented by Takasu, Ihara and coworkers. The new method relies on the introduction of a conjugated ester moiety at the terminal olefin, thereby effecting an acceleration of the domino reaction accompanied with an enhancement of the regio- and stereoselectivity.[5] Thus, the reaction of the starting iodide with Bu$_3$SnH led to a 4:3 mixture of the two diastereomeric tricycles in 83% yield. In this process, the vinyl radical is initially formed, but this smoothly cyclizes in 5-*exo*-trig manner to give intermediate radical (Scheme 20.3). Due to the high nucleophilicity of this radical, a 5-*exo*-trig addition onto the unsaturated ester moiety providing α-carboxy radical is kinetically predominant. The sequence is terminated by a 5-*exo*-trig ring closure onto the electron-rich olefin moiety.

These polycyclizations can also be performed using a radical as the initiator. Such reactions can be divided into those based on serial 6-*endo*-trig cyclizations from polyene acyl precursors,[6] radical-mediated macrocyclizations from alkyl radicals followed by radical transannulations,[7] consecutive oxy (and aminyl) radical fragmentation/transannulation/cyclization sequences,[8] and several combinations of these processes.[9] An impressive example of the

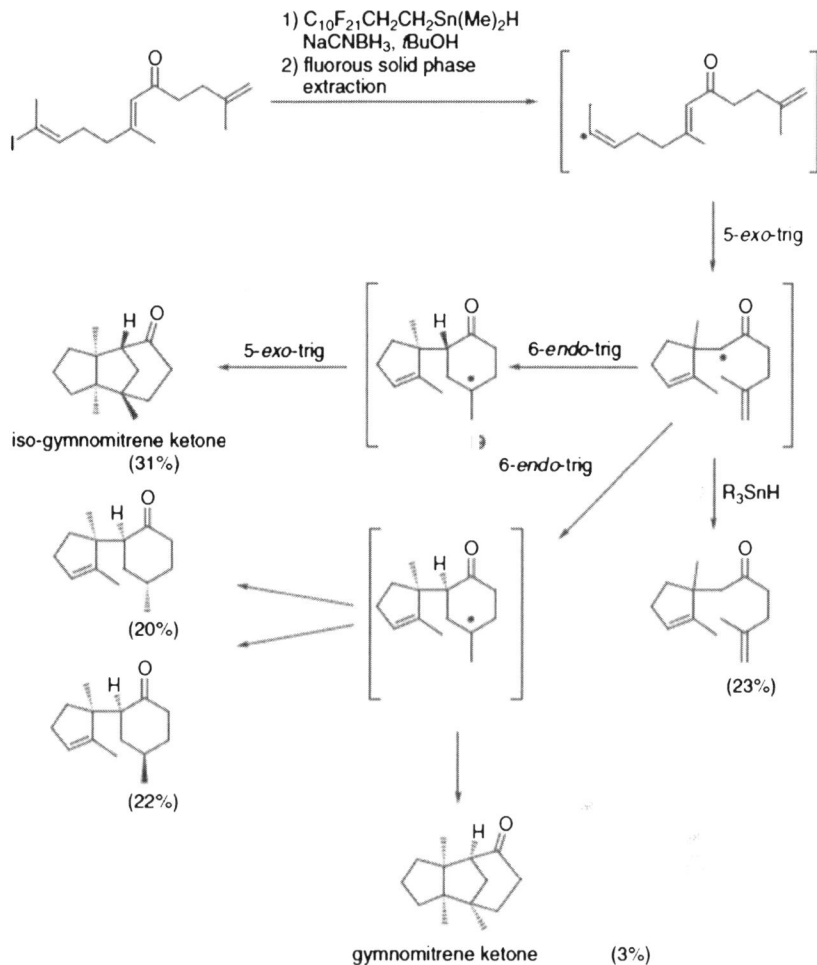

Scheme 20.2   Radical cascade in action.

power of this approach is the combination of seven 6-*endo*-trig cyclizations in one single event, as devised by Pattenden and his group (Scheme 20.4).[10] The hydroxyheptenselenoate, synthesized from all-*E*-geranylgeraniol involving two successive homologations, was treated with azobisisobutyronitrile (AIBN) and tributyltin hydride in refluxing benzene to give the all-*trans-anti* heptacyclic ketone in a yield of 17% *via* the primarily formed radical. In order to minimize the formation of reduced byproducts, the Bu₃SnH was added over 8 h using a syringe pump, a popular method in the area of radical chemistry.

It could be shown that the stereochemical outcome of such radical polycyclizations is influenced by the nature of the substituents (H, Me, CO₂R). For instance the all-(*E*)-methyl-substituted polyene gave the corresponding all-*trans-anti* polycycle in the presence of Bu₃SnH and AIBN. However, the ester-

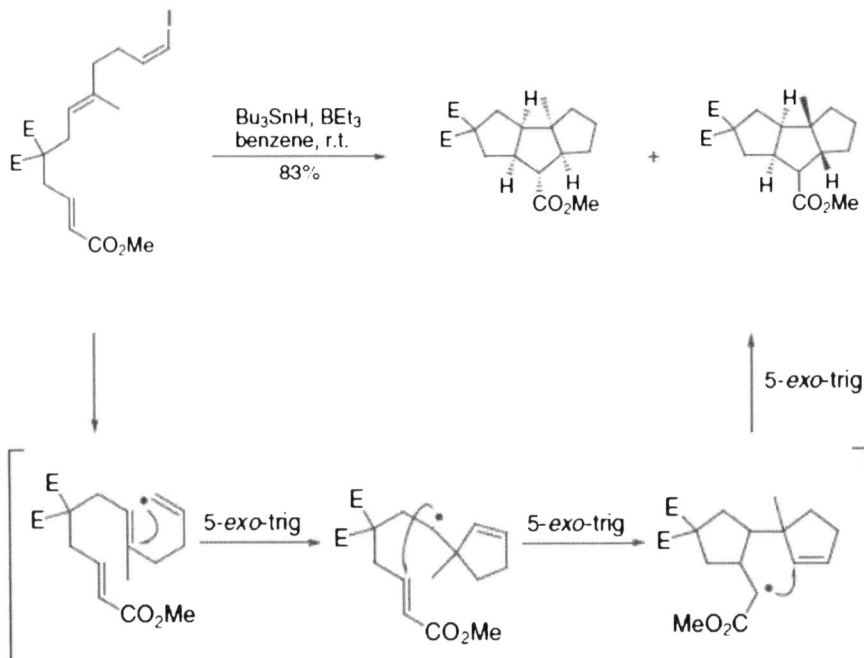

Scheme 20.3   General diagram of introducing a conjugated ester moety at terminal olefin.

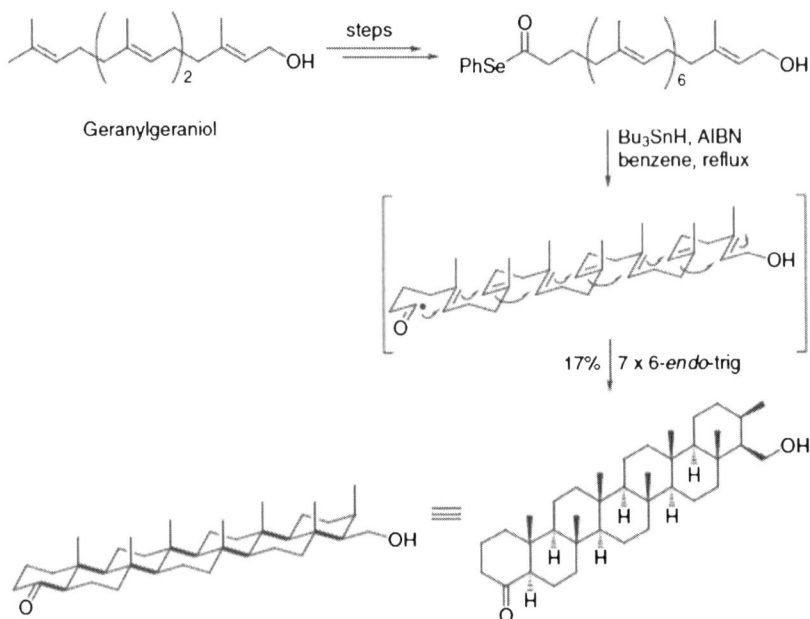

Scheme 20.4   Cascade of seven 6-endo-trig cyclizations in one single event by Pattenden *et.al.*[10]

substituted polyene led to the *cis-anti-cis-anti-cis* tetracycle under similar reaction conditions (Scheme 20.5). A certain degree of preorganization of the precursor is assumed to be the reason for this result.[11]

An exact tuning of the reactivities clearly plays an important role in the development of radical domino reactions. Thus, as shown by Pattenden and coworkers, the radical, obtained from the selenoate, does not undergo tetra-cyclization but rather forms the cyclopropyl-substituted macrocycle in about 40% yield *via* a 14-*endo*-trig-cyclization (Scheme 20.6).[12]

As mentioned above, the class of *Aspidosperma* alkaloids has attracted the attention of the chemical community, due to the compounds' challenging architecture and biological activity.[13] Here, a new approach to the ABCE-tetracycle, for example, aspidospermidine, as developed by the Jones group, is disclosed.[13] Reaction of the substrate with tributyltin hydride in refluxing 5-*tert*-butyl-*m*-xylene (*ca.* 200 °C) resulted in the formation of an aryl radical with its well-known ability to abstract a hydrogen atom *via* a six-membered ring transition state, leading to the α-amino radical. It follows an attack on the indole systems, which is facilitated by the electron-withdrawing cyano group and a 5-*exo*-trig ring closure of the intermediate to give the tetracyclic scaffold in 43% yield as an 8 : 3 : 2 : 1 mixture of diastereoisomers (Scheme 20.7).

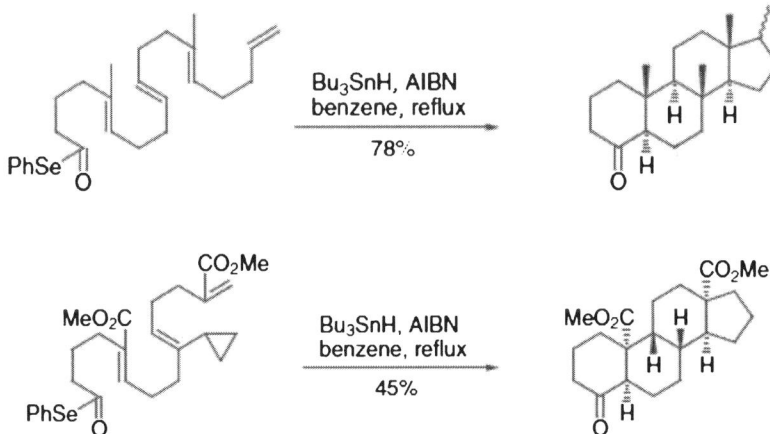

Scheme 20.5 Radical cyclization and nature of substituents.

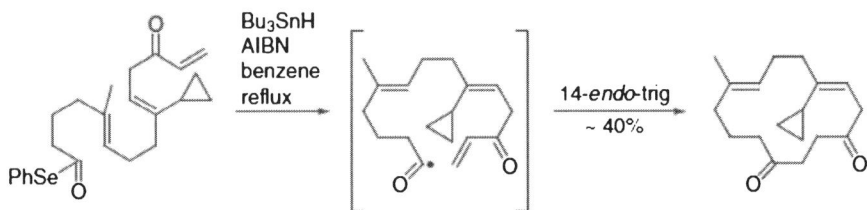

Scheme 20.6 14-endo-trig cyclization in action.

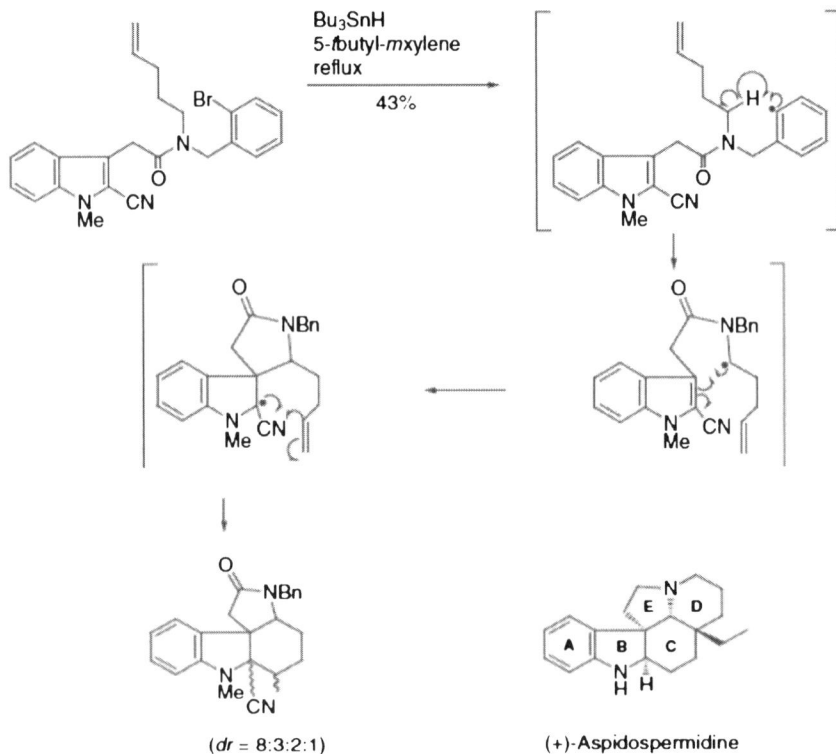

Scheme 20.7   Synthetic approach towards the class of Aspidosperma alkaloids.

The natural spironucleoside hydantocidin, which was discovered in 1991, demonstrates pronounced herbicidal and plant growth-regulatory properties. This has in turn led to a stimulation of studies related to the synthesis of anomeric spironucleosides.[15] Herein, two short and efficient synthetic approaches of the group of Chatgilialoglu,[14] both based on radical domino reactions, for the preparation of anomeric spironucleosides, are described. The first method features the conversion of protected 6-hydroxymethylribouridines into spironucleosides in 36% and 49% yields, respectively, using PhI(OAc) and I$_2$ (Scheme 20.8).

Photolysis of the primarily formed hypoiodite generates an alkoxy radical, which directly undergoes a Barton-type hydrogen migration, furnishing the anomeric C-1′ radical. This forms the corresponding oxenium salts. Finally, probably *via* an unstable anomeric iodo intermediate, nucleophilic addition of the hydroxyl group occurs to produce the desired spironucleoside. In the second example, the protected dibromovinyl deoxyuridine was transformed into a 2:1 mixture of the desired spirocompounds in 57% yield using standard radical conditions with (TMS)$_3$SiH. In addition, 25% of an *E/Z* mixture of spironucleoside was obtained (Scheme 20.9).

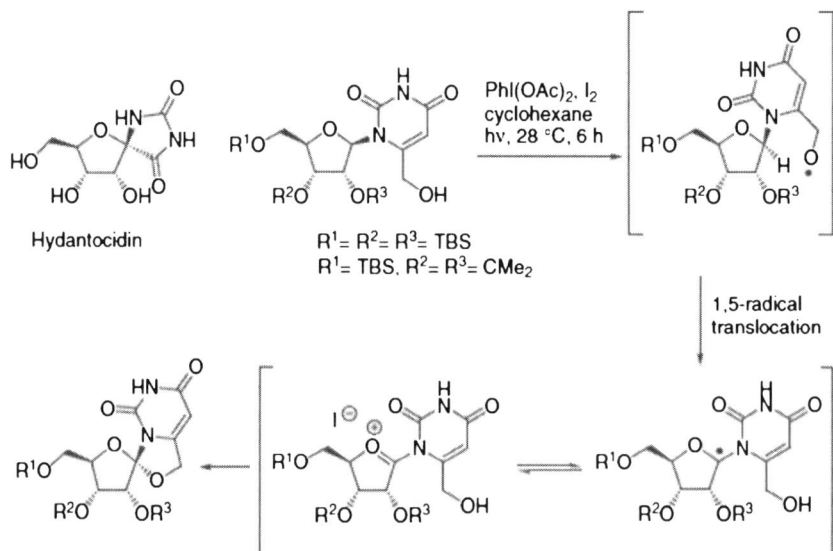

Scheme 20.8 Approaches towards spiral nucleosides.

Bromomethyldimethylsilyl (BMDMS) propargyl ethers undergo intramolecular radical 5-*exo*-dig cyclizations using tributyltin hydride; the reaction is initiated by the generation of a stabilized α-silyl radical.[17] The Malacria group

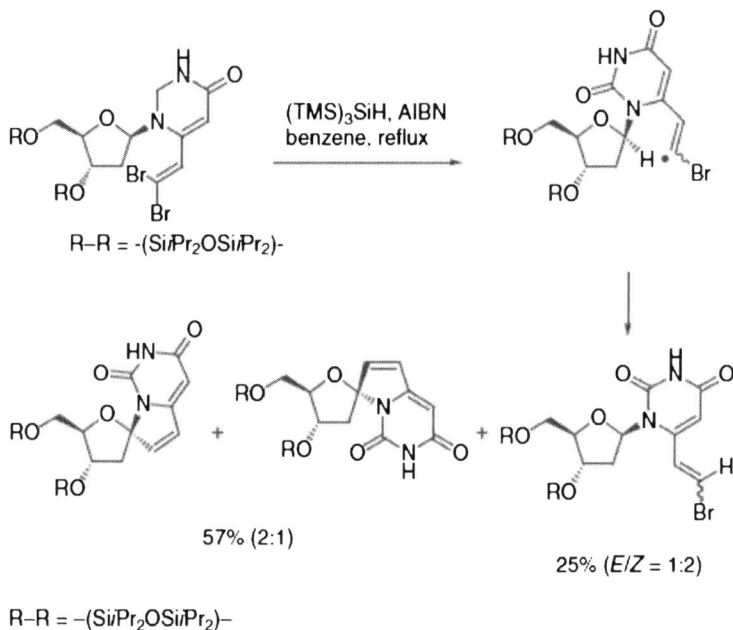

Scheme 20.9 Spiral nucleosides *via* (TMS)₃SiH cyclization.

has described the reaction of the monocyclic compound, which gave the pentacyclic product.[18] This was transformed using a Tamao oxidation and subsequent desilylation, to afford the antibiotic *epi*-illudol in 47% yield over three synthetic steps (Scheme 20.10).

It should be noted that this strategy could also be utilized to build up the linear triquinane framework in 45% yield, employing the 11-membered ring system in which the BMDMS group is simply moved from one to the other side of alkyne moiety (Scheme 20.11).[19]

In another approach, the annulated [5.6.5] ring system was obtained from suitable precursor in a radical reaction in 61% yield, (diasteromeric ration (dr) 1.5:1) (Scheme 20.12).[20] This motif is found as the central core in several important natural products.

Mascarenas and coworkers uncovered a rather exceptional procedure featuring a very effective domino radical cyclization/Beckwith⤳Dowd rearrangement sequence furnishing bicyclo[5.3.1]undecanes as products.[20] On treatment with tributyltin hydride and AIBN, the vinyl bromide underwent a smooth conversion to the desired product in 81% yield as an approximate 3:1

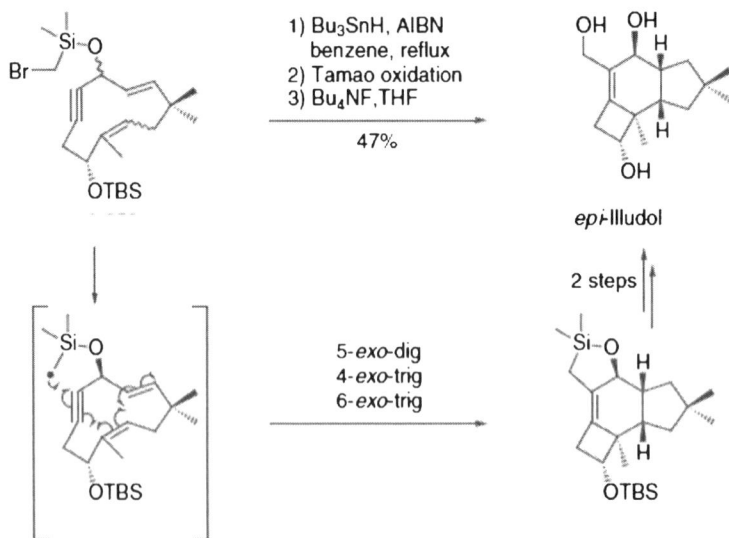

Scheme 20.10   Intramolecular radical 5-exo-dig cyclization using Bu$_3$SnH.

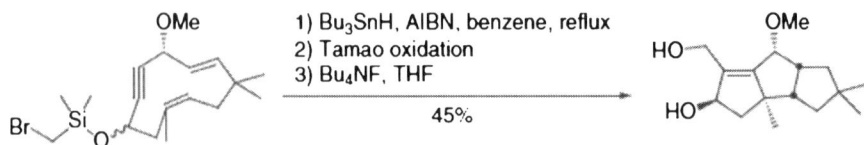

Scheme 20.11   Synthesis of 11-membered ring formation of linear triguinane framework.

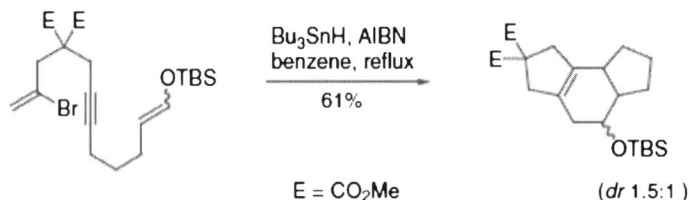

E = CO$_2$Me

Scheme 20.12 Approach toward anmelate [5,6,5] ring system.

mixture of inseparable diastereoisomers (Scheme 20.13). The authors proposed a plausible mechanism in which the primarily formed vinyl radical cyclizes in a 7-*endo*-trig mode to deliver the carbon-centered secondary radical. Afterwards, an internal 1,2-acyl transfer occurs, which in the light of the well-known Beckwith⊳Dowd ring expansion of α-halomethyl and related cyclic ketones,[22] should proceed by a 3-*exo*-trig ring closure between the secondary radical and the carbonyl group to form a strained, short-lived oxycyclopropyl radical intermediate. β-Fragmentation of the latter yielded ring-expanded radical which, after reaction with Bu$_3$SnH, provides the desired scaffold. It should be mentioned that the methodology could be extended to the preparation of the bicyclo[4.3.1]decane scaffolds, though in lower yield.

Clive and coworkers have developed a new domino radical cyclization, by making use of a silicon radical as an intermediate to prepare silicon-containing bicyclic or polycyclic compounds (Scheme 20.14).[23] After formation of the first radical from the radical precursor, a 5-*exo*-dig cyclization takes place followed by an intramolecular 1,5-transfer of hydrogen from silicon to carbon, providing a silicon-centered radical. Once formed, this has the option to undergo another cyclization to afford the radical, which can yield a stable product either by a reductive interception with the present organotin hydride

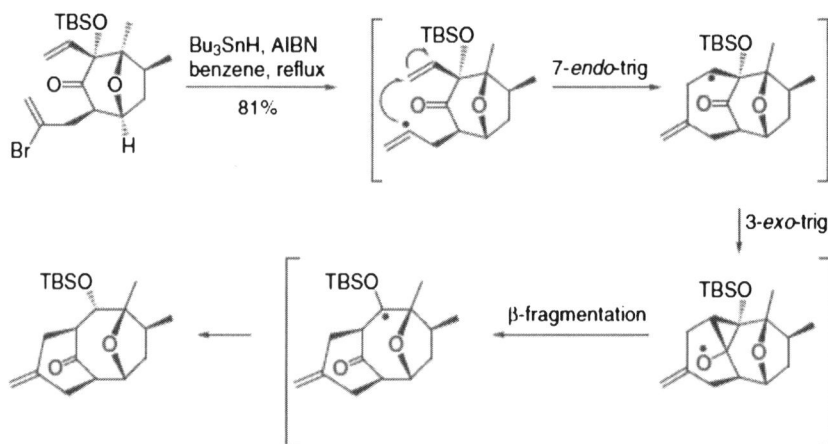

Scheme 20.13 Domino Radical cyclization/Beckwith-Dowd rearrangement to furnish bicyclo [5.3.1] undecane.

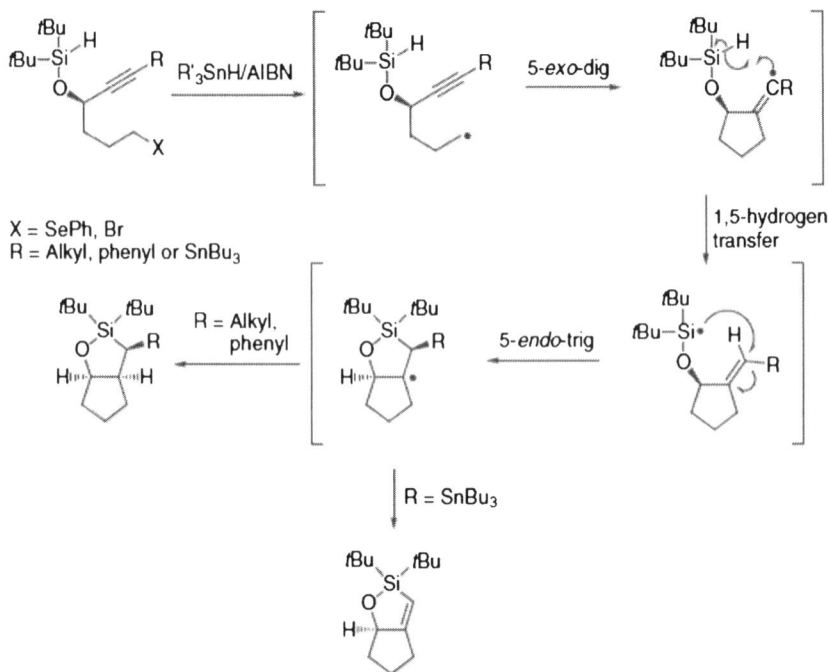

Scheme 20.14 New domino radical cyclization *via* silicon radical intermediate.

species to obtain final compounds. On the other hand, when the terminal alkyne carries a trimethylstannyl group, expulsion of a trimethylstannyl radical takes place to afford the corresponding vinyl silanes.

Scheme 20.15 illustrates three examples in which the highly efficient construction of bi- and polycyclic compounds from corresponding radical precursors, respectively, is depicted. It should be noted that the carbon–silicon bond in the obtained products can be easily cleaved[24] to achieve valuable synthons for further transformations. As discussed previously, radical ring-opening reactions of three-membered systems *via* cyclopropylmethyl and oxiranylmethyl radicals represent a fruitful method in organic synthesis.[24]

De Kimpe and coworkers have now shown that aziridines can also be used, featuring a radical one-step synthesis of pyrrolizidines from 2-(bromomethyl)-aziridines by application of standard radical conditions in 49–63% yield (Scheme 20.16).[25] The reaction sequence is assumed to be launched by the fragmentation of initially formed aziridinylmethyl radicals to give a *N*-allylaminyl radical, which undergoes a twofold 5-*exo*-trig cyclization.

The use of samarium(II) iodide in synthesis permits the assembly of complex molecules as already shown in many examples. They profit from the electron transfer ability of samarium(II) iodide; thus, if ketones are employed as substrates the furnished ketyl-radical can react in a multitude of different ways. An example, where two C–C bonds are formed and one C–C bond is broken is

(1)

(2)

(3)

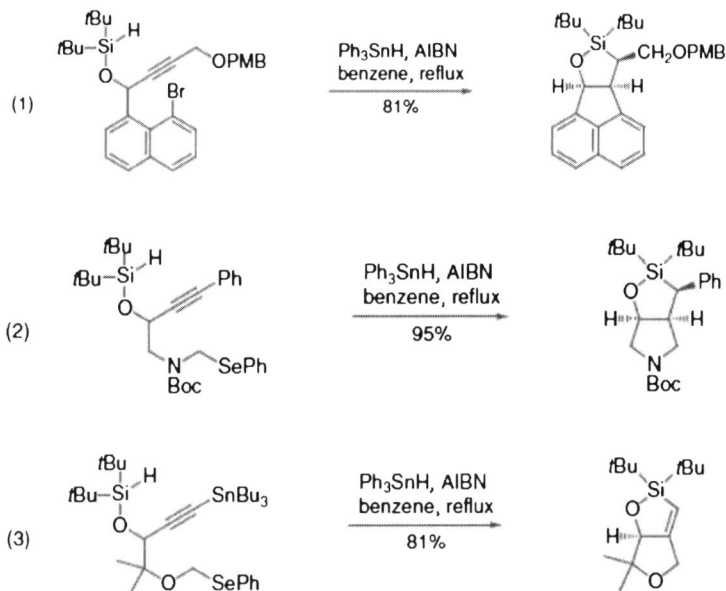

Scheme 20.15 Highly efficient constraction of bi- and polycyclic compounds *via* corresponding radical precursors.

the synthesis of the tricycle, which has some similarity with the eudesmane framework, developed by Kilburn and coworkers (Scheme 20.17).[26] Thus, exposure of the easily accessible methylenecyclopropylþcyclohexanone to samarium(II) iodide led to the generation of ketyl radical, which builds up a six-membered ring system with simultaneous opening of the cyclopropane moiety.

Subsequent capture of the formed radical by the adjacent alkyne group afforded the tricycle *via* radical intermediate as a single diastereoisomer in up to 60% yield. It should be noted that in this case the usual necessary addition of hexamethylphosphoramide (HMPA) could be omitted. The analgesic properties of paeonilactone B and its analogues have made these compounds challenging synthetic targets, and again the Kilburn group has presented a samarium(II) iodide-promoted domino radical cyclization for their synthesis.[27]

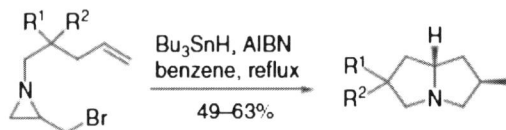

a: $R^1 = R^2 = Me$; **b**: $R^1 = R^2 = Ph$; **c**: $R^1 = Me$, $R^2 = H$

Scheme 20.16 One step synthesis of pyrolizidines from aziridines.

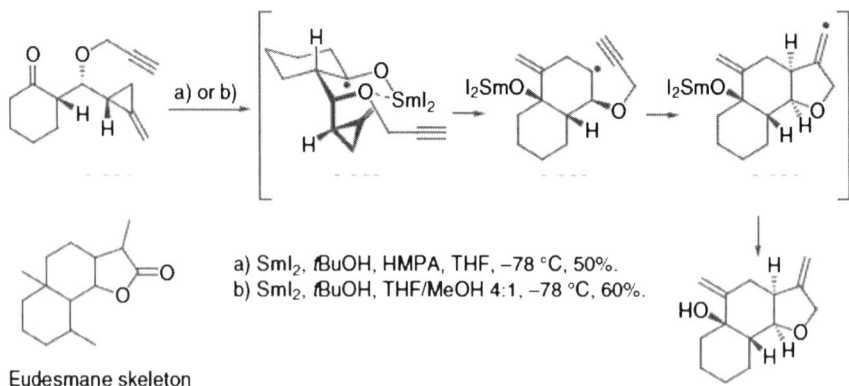

Eudesmane skeleton

a) SmI₂, ⁱBuOH, HMPA, THF, −78 °C, 50%.
b) SmI₂, ⁱBuOH, THF/MeOH 4:1, −78 °C, 60%.

Scheme 20.17 Towards synthesis of Eudesmane skeleton.

Hence, samarium(II) iodide reaction of the diastereomeric ketones led to the desired bicycles products (Scheme 20.18). The stereoselectivity depends heavily on the stereochemistry of the starting material. Thus, the (*S*)-isomeric ketone gave a mixture of the diastereomers as intermediates, with the *S*-diastereoisomer as the main product. In this case, the use of an additive such as HMPA or 1,3-dimethyl-3,4,5,6-tetrahydro-2(1H)-pyrimidinone (DMPU) is crucial for achieving satisfactory yields and high selectivities, with HMPA clearly excelling over DMPU. In the final section of this chapter, we would like to present some domino processes with a particular high number of reaction steps, and partly unusual transformations.

An unusual formation of alkyl radicals as intermediates was observed in the conversion of the silyl ether into the bicyclo[3.1.1]heptanes using Bu₃SnH, and described by the Malacria group (Scheme 20.19).[28] The obtained product

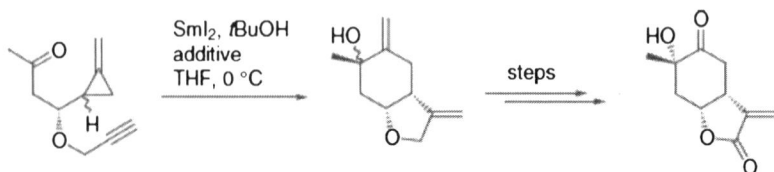

Scheme 20.18 SmI₂ mediated cyclization of the diastereomeric ketones.

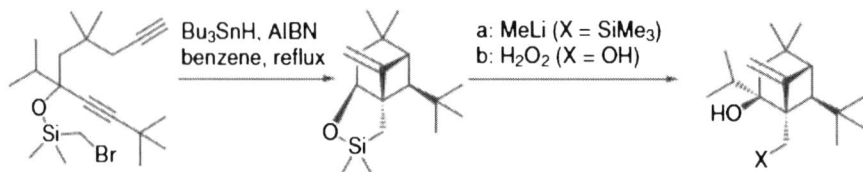

Scheme 20.19 Conversion of the silyl ether into the bicyclo[3.1.1] heptane.

was further transformed either into using methyl lithium, or by oxidative degradation. The transformation is proposed to start with the formation of radical, which undergoes a 5-*exo*-dig cyclization providing vinyl radical. This does not intercept with the alkyne moiety, but a 1,6-hydrogen transfer occurs to give radical, which by 6-*endo*-trig ring closure leads to with complete diastereoselectivity.

Xanthates serve as a reliable source of electrophilic radicals, and this was exploited by Zard and coworkers for a short synthesis of ($\pm$)-matrine, a naturally occurring alkaloid which has been claimed to have anti-ulcerogenic and anti-cancer properties.[30] Heating a mixture of xanthate and the radical

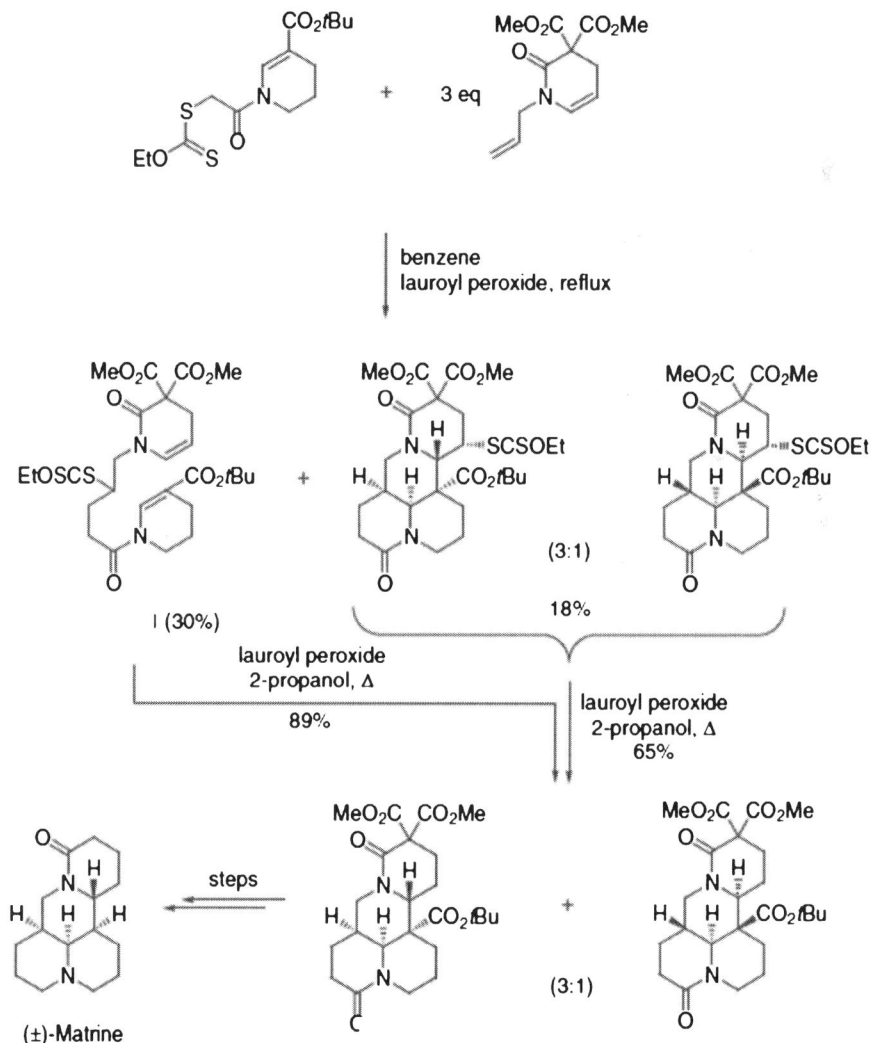

Scheme 20.20 Short synthesis of (+/−)-Matrine.

acceptor (3 equiv.) in benzene in the presence of lauroyl peroxide as initiator, gave the final product in 30% yield and a 3 : 1 mixture of the tetracylic products in 18% yield (Scheme 20.20).[30] The three compounds could be converted into the diastereomeric tetracycles utilizing lauroyl peroxide and 2-propanol as solvent. After separation, the intermediate was transformed into ($\pm$)-matrine. Although the yield was not very high, during this remarkable process four new bonds (including an intermolecular step) and five contiguous stereogenic centers were created in a single operation, with acceptable stereoselectivity.

Finally, a rather early (but, from a mechanistic viewpoint, very interesting) sequence of radical reactions has been described by Pattenden and Schulz, in which an acetylenic oxime ether was converted into the bicyclic oxime in 70% yield (Scheme 20.21).[31] Hydrolysis of bicyclic oxime led to the bicyclic enone, which in fact can also more easily be synthesized by a Robinson annulation.

20.3   CONCLUSION AND FUTURE DIRECTION

This chapter highlights an amazing progress and ingenuity demonstrated through the use of the combination of three and more radical C–C or C–N bond-forming radical transformations is also possible. This makes this

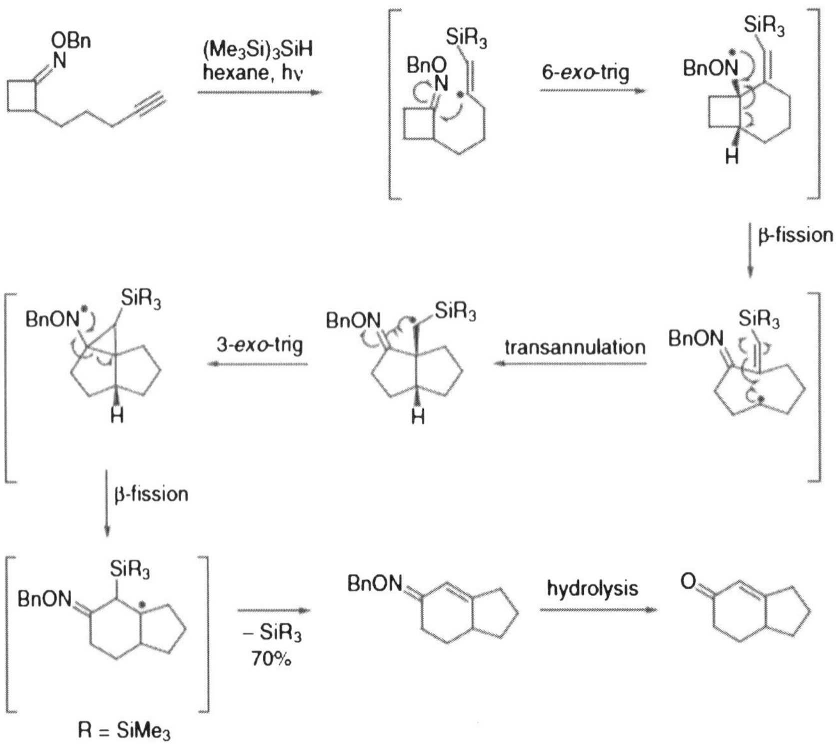

Scheme 20.21   Conversion of acetylenic oxime ether into the bicyclic oxime.

methodology one of the most powerful procedures in the synthesis of complex molecules starting from simple substrates. The progress is remarkable to date (see Chapters 26 and 27 for further examples) and will undoubtedly lead to further applications and collaborations between classical free radical chemistry and green and atom-efficient synthesis.

REFERENCES

1. (a) C. P. Jasperse, D. P. Curran and T. L. Fevig, *Chem. Rev.,* 1991, **91**, 1237–1286; (b) D. P. Curran, *Synlett,* 1991, 63–72; (c) D. P. Curran, in *Comprehensive Organic Synthesis,* ed. B. M. Trost and I. Fleming, Pergamon, Oxford, 1991, vol. 4, p. 779; (d) P. J. Parsons, C. S. Penkett and A. J. Shell, *Chem. Rev.,* 1996, **96**, 195–206; (e) M. Malacria, *Chem. Rev.,* 1996, **96**, 289–306; (f) B. B. Snider, *Chem. Rev.,* 1996, **96**, 339–363.
2. (a) N. A. Porter, V. H. T. Chang, D. R. Magnin and B. T. Wright, *J. Am. Chem. Soc.,* 1988, **110**, 3554–3560; (b) S. Handa and G. Pattenden, *J. Chem. Soc., Perkin Trans. 1,* 1999, 843–845; (c) S. A. Hitchcock, S. J. Houldsworth, G. Pattenden, D. C. Pryde, N. M. Thomson and A. J. Blake, *J. Chem. Soc., Perkin Trans. 1,* 1998, 3181–3206; (d) U. Jahn and D. P. Curran, *Tetrahedron Lett.,* 1995, **36**, 8921–8924.
3. (a) I. Ryu, N. Sonoda and D. P. Curran, *Chem. Rev.,* 1996, **96**, 177–194; (b) D. P. Curran and S. Sun, *Aust. J. Chem.,* 1995, **48**, 261–267.
4. B. P. Haney and D. P. Curran, *J. Org. Chem.,* 2000, **65**, 2007–2013.
5. K. Takasu, S. Maiti, A. Katsumata and M. Ihara, *Tetrahedron Lett.,* 2001, **42**, 2157–2160.
6. L. Chen, G. B. Gill, G. Pattenden and H. Simonian, *J. Chem. Soc., Perkin Trans. 1,* 1996, 31–43.
7. S. A. Hitchcock, S. J. Houldsworth, G. Pattenden, D. C. Pryde, N. M. Thomson and A. J. Blake, *J. Chem. Soc., Perkin Trans. 1,* 1998, 3181–3206.
8. S. Handa and G. Pattenden, *Contemp. Org. Synth.,* 1997, **4**, 196–215.
9. S. Handa and G. Pattenden, *J. Chem. Soc., Perkin Trans. 1,* 1999, 843–845.
10. (a) H. M. Boehm, S. Handa, G. Pattenden, L. Roberts, A. J. Blake and W.-S. Li, *J. Chem. Soc., Perkin Trans. 1,* 2000, 3522–3538; (b) S. Handa, P. S. Nair and G. Pattenden, *Helv. Chim. Acta,* 2000, **83**, 2629–2643.
11. S. Handa, G. Pattenden and W.-S. Li, *Chem. Commun.,* 1998, 311–312.
12. J. E. Saxton, *Nat. Prod. Rep.,* 1996, **14**, 559–590.
13. S. T. Hilton, T. C. T. Ho, G. Pljevaljcic, M. Schulte and K. Jones, *Chem. Commun.,* 2001, 209–210.
14. (a) M. Nakajima, K. Itoi, Y. Tamamatsu, T. Kinoshita, T. Okasaki, K. Kawakubo, M. Shindou, T. Honma, M. Tohjigamori and T. Haneishi, *J. Antibiot.,* 1991, **44**, 293–300; (b) H. Haruyama, T. Takayama, T. Kinoshita, M. Kondo, M. Nakajima and T. Haneishi, *J. Chem. Soc., Perkin Trans. 1,* 1991, 1637–1640; (c) H. Sano, S. Mio, M. Shindou,

T. Honma and S. Sugai, *Tetrahedron,* 1995, **46**, 12563–12572, and references therein.

15. C. Chatgilialoglu, T. Gimisis and G. P. Spada, *Chem.–Eur. J.,* 1999, **5**, 2866–2876.

16. L. Fensterbank, M. Malacria and S. M. Sieburth, *Synthesis,* 1997, 813–854.

17. C. Aissa, B. Delouvrie, A.-L. Dhimane, L. Fensterbank and M. Malacria, *Pure Appl. Chem.,* 2000, **72**, 1605–1613.

18. A.-L. Dhimane, C. Aissa and M. Malacria, *Angew. Chem., Int. Ed.,* 2002, **41**, 3284–3287.

19. L. Fensterbank, E. Mainetti, P. Devin and M. Malacria, *Synlett,* 2000, 1342–1344.

20. J. R. Rodriguez, L. Castedo and J. L. Mascarenas, *Org. Lett.,* 2001, **3**, 1181–1183.

21. P. Dowd and W. Zhang, *Chem. Rev.,* 1993, **93**, 2091–2115.

22. D. L. J. Clive, W. Yang, A. C. MacDonald, Z. Wang and M. Cantin, *J. Org. Chem.,* 2001, **66**, 1966–1983.

23. M. Sannigrahi, D. L. Mayhew and D. L. J. Clive, *J. Org. Chem.,* 1999, **64**, 2776–2788.

24. (a) D. C. Nonhebel, *Chem. Soc. Rev.,* 1993, 347–359; (b) D. J. Pasto, *J. Org. Chem.,* 1996, **61**, 252–256.

25. D. De Smaele, P. Bogaert and N. De Kimpe, *Tetrahedron Lett.,* 1998, **39**, 9797–9800.

26. F. C. Watson and J. D. Kilburn, *Tetrahedron Lett.,* 2000, **41**, 10341–10345.

27. (a) R. J. Boffey, M. Santagostino, W. G. Whittingham and J. D. Kilburn, *Chem. Commun.,* 1998, 1875–1876; (b) R. J. Boffey, W. G. Whittingham and J. D. Kilburn, *J. Chem. Soc., Perkin Trans. 1,* 2001, 487–496.

28. S. Bogen, L. Fensterbank and M. Malacria, *J. Am. Chem. Soc.,* 1997, **119**, 5037–5038.

29. K. A. Aslanov, Y. K. Kushmuradov and S. Sadykov, *Alkaloids,* 1987, **31**, 117–192.

30. L. Boiteau, J. Boivin, A. Liard, B. Quiclet-Sire and S. Z. Zard, *Angew. Chem., Int. Ed.,* 1998, **37**, 1128–1131.

31. G. Pattenden and D. J. Schulz, *Tetrahedron Lett.,* 1993, **34**, 6787–6790.

Applications of Conventional Free Radicals and Advances in Total Synthesis: from the Bench to the Future through the Vinyl Radical[*]

21.1 THE VINYL RADICAL, A PRECIOUS TOOL FOR RADICAL CASCADES IN 5-*EXO-DIG* CYCLIZATIONS

The vinyl radical resulting from the highly efficient and regioselective 5-*exo-dig* cyclization of α-silyl radical, generated from bromomethyldimethylsilyl (BMDMS) propargyl ethers has been revealed as a very versatile synthetic tool.[1] Scheme 21.1 illustrates different processes Malacria has worked out over the past years.

A simple stannane reduction on **3** provides heterocycles **4**, which upon further addition of MeLi, give trimethylenemethane (TMM) precursors.[2] Alternatively, an oxidative treatment with H_2O_2 (Tamao oxidation), or Bu_4NF (protodesilylation) yields valuable trisubstituted olefins.[3] All sorts of intramolecular trapping of the vinyl radical have been envisaged. The simplest is naturally a 5-*exo-trig* radical cyclization from **5**, giving birth to a variety of cyclopentenes **6**.[4] This process is completely diastereoselective, setting the resulting methyl group and the C–O bond *syn* to each other. When a methallyl group, instead of an allyl group, is present, a competition between the 5-*exo-trig* and 6-*endo-trig* modes of cyclization occurs. The 6-*endo-trig* cyclization becomes major when the vinyl radical is substituted by a phenyl group (cyclohexene **8**). A final diastereoselective reduction of the methine radical installs a *syn* relationship between the methyl group and the C–O bond.[4b] Vinyl radicals are also very prone to engage in hydrogen transfers. A rare 1,4-H transfer (from **9**) has been discovered.[5] However, the most useful approaches

* Chapter written by V. Tamara Perchyonok.
Streamlining Free Radical Green Chemistry
V. Tamara Perchyonok, Ioannis Lykakis and Al Postigo
© V. Tamara Perchyonok, Ioannis Lykakis and Al Postigo 2012
Published by the Royal Society of Chemistry, www.rsc.org

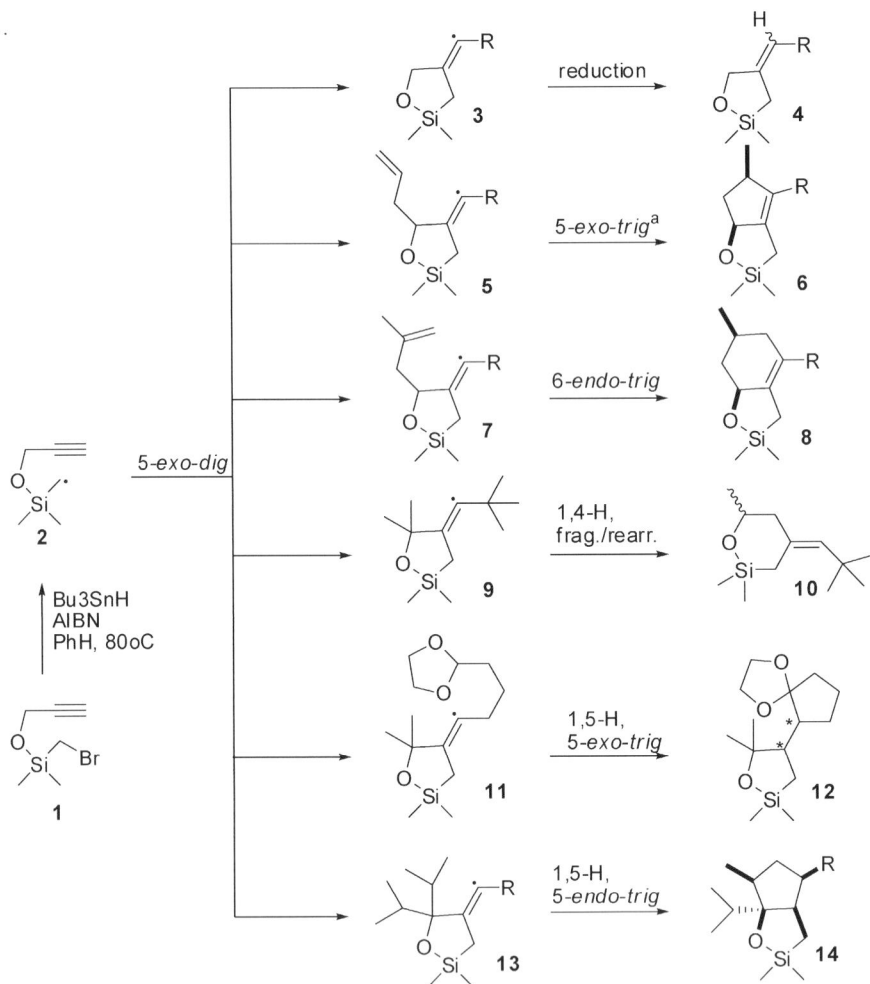

Scheme 21.1   Vinyl radicals and general possibilities.

in this domain have relied on 1,5-H transfers. An X activating group, such as a dioxolane, suitably located on the propargyl chain, promotes a 1,5-(π-*exo*)-H transfer from the vinyl radical **11**. The translocated radical then cyclizes back in a 5-*exo-trig* manner to provide α-functionalized cyclopentanone derivatives **12**. Malacria has studied the diastereoselectivity of the reduction of the final β-silyl radical,[6] and also investigated the possibility of developing an asymmetric version of this process, by introducing a homochiral dioxolane moiety. The best value of diastereomeric excess (de) obtained reached 50%.[4c] With highly sterically encumbered (gem-diisopropyl) substrates, vinyl radical **13** undergoes a diastereoselective 1,5-(π-*endo*)-H transfer with an unactivated methyl group. The resulting methylene radical then cyclizes according to the disfavored 5-*endo-trig* mode of cyclization, placing the R group *syn* to the methyl group and to the C–O bond. A diastereoselective stannane reduction of the β-silyl radical installs a *cis* ring junction and terminates the sequence.[7] This sequence has proven quite general for the synthesis of cyclopentanes, and has constituted so far the most efficient case of 5-*endo-trig* cyclization in an all-carbon system.[8] 1,6-H transfers from the vinyl radical have also been disclosed in transannular series[9] and acyclic series,[10] but their synthetic versatility has not been completely exploited.

Malacria's utilization of the propargylic BMDMS ethers has recently culminated in two cascade sequences leading to linear triquinanes and the total synthesis of *epi*-illudol. These are developed below.

21.2   LINEAR TRIQUINANES FROM ACYCLIC PRECURSORS

Previous studies in the Malacria laboratory established that the homoallyl radical, resulting from the 5-*exo-dig*, 5-*exo-trig* tandem as in the formation of cyclopentenes, could readily engage into a radical [3+2] cycloaddition, with a Michael acceptor like acrylonitrile.[11] By this methodology, a diquinane could be obtained as a single diastereomer. Malacria next focused on the possibility of constructing triquinanes.[12] This action would necessitate the addition of one more unsaturation. Malacria's initial attempts, however, were thwarted by hydrogen transfers from the initial vinyl radical.[4b] In view of the linear triquinane skeleton, we had to quaternarize the second propargylic position. Therefore, Malacria concentrated on the precursor,[13] bearing a gem-dimethyl group, whose cyclization in presence of acrylonitrile furnished the two adducts in good overall yield (Scheme 21.2).

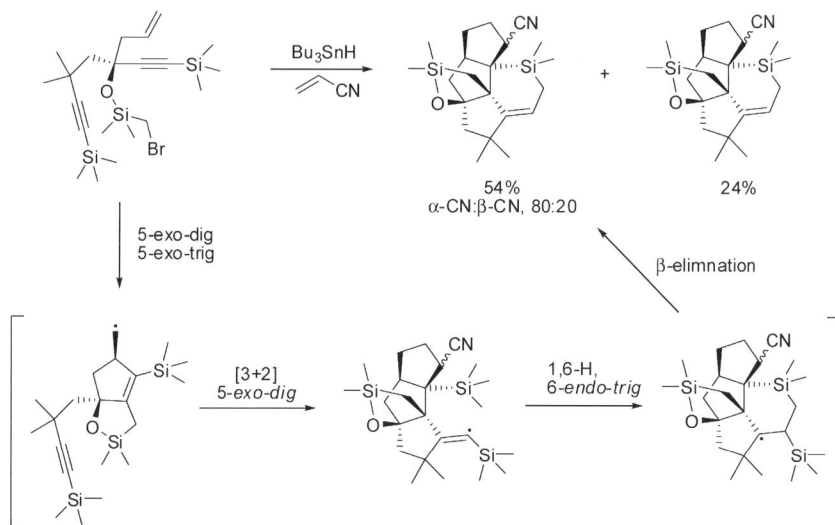

Scheme 21.2   5-exo-dig, 5-exo-trig tandem as a synthetic pathway to a broad range of triguinanes.

Both compounds originate from the initial and usual 5-*exo-dig*, 5-*exo-trig* tandem. The resulting homoallyl radical has two options. The first and major one, to provide triquinane, is a Michael condensation on acrylonitrile followed by a 5-*exo-trig* closure. The resulting β-silyl radical then cyclizes in a 5-*exo-dig* fashion to give a vinyl radical. Because of the very large surrounding steric hindrance, this radical intermediate is particularly protected from any intermolecular reaction. Instead, it engages in a 1,6-H transfer with a methylsilane group to give an α-silyl radical which then undergoes a 6-*endo-trig* cyclization. This yields a doubly neopentylic radical, whose only way out is an unprecedented β-elimination of a trimethylsilyl radical, setting the vinylsilane function of intermediate. The second option, to provide a cyclopropyl adduct, is a competing 3-*exo-trig* cyclization from the first intermediate radical, whose driving force is the subsequent 5-*exo-dig* cyclization. From this, an identical sequence as for the construction of triquinane takes place. The outcome of the sequence leading to the triquinane is remarkable: 11 elementary steps, 6 new C–C bonds, 3 contiguous quaternary centers, and 4 new stereogenic centers almost totally controlled. Only incomplete stereocontrol during the [3+2] annulation step is responsible for the formation of the minor β-CN epimer of triquinane.

Malacria next sought to obtain the linear triquinane framework without any supplementary ring.[13] To achieve this, Malacria relied on an approach ending with a favorable 1,5-H transfer from vinyl radical and followed by the β-elimination of a suitable leaving group, thus avoiding any telomerization of the final radical species (Scheme 21.3). Malacria prepared sulfoxide and sulfone because of the reported very fast β-elimination of the arylsulfinyl and -sulfonyl group. Both substrates allowed the synthesis of vinyltriquinane, with a distereoselectivity consistent with previous findings. The difference of yield between the cyclization of sulfoxide and sulfone simply reflects the poorer ability of the sulfinyl radical to propagate the radical chain.

21.3 FIRST TOTAL SYNTHESIS OF NATURAL PROTOILLUDANE, *EPI*-ILLUDOL

The unusual angular 4,6,5-tricyclic framework of protoilludanes family is built biosynthetically from humulene by a transannular carbocationic tandem cyclization (Scheme 21.4). These antibacterial natural products (*e.g.*, illudols, armillols, tsugicolines), therefore, possess a *bis*-allylic diol moiety, easily accessible from the corresponding BMDMS propargylic ether. For these reasons, Malacria has designed a biomimetic radical transannular cascade strategy from an adequately substituted cycloundecadienyne.[14]

Thus, precursor, submitted to classical Bu₃SnH-AIBN conditions, provides the expected α-silyl radical which cyclizes regioselectively[9] in a 5-*exo-dig* manner (Scheme 21.5). The resulting vinyl radical undergoes a challenging 4-(π-*exo*)-*exo-trig* transannular closure to a new radical species, which is directly involved in a second more favorable transannular 6-*exo-trig* process. This final

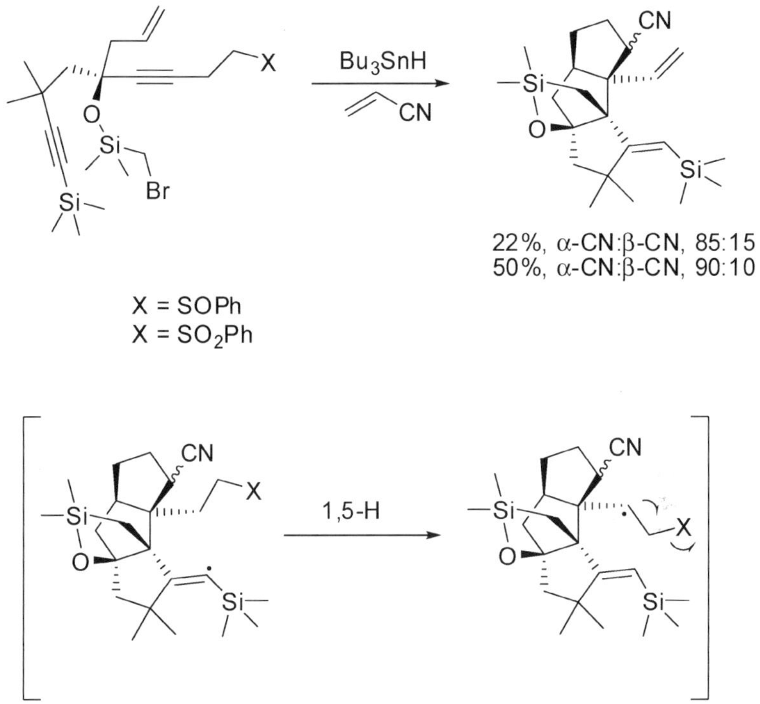

22%, α-CN:β-CN, 85:15
50%, α-CN:β-CN, 90:10

X = SOPh
X = SO$_2$Ph

Scheme 21.3 Towards linear triguinanes.

process serves as a driving force for the complete diastereoselective construction of tetracycle. A final Tamao oxidation and subsequent desilylation generate the expected *epi*-illudol.

This versatile strategy should allow a general access to the above-mentioned natural protoilludanes possessing this intriguing *exo*-methylene cyclobutane entity.[15] Malacria now focuses on how to open a new route to linear triquinane skeletons through a similar transannular strategy from an eleven-membered-ring with three judiciously positioned unsaturations (Scheme 21.6).

These new developments in radical cascades, starting from BMDMS propargylic ethers and leading to natural skeletons, indubitably show the efficiency and the great potentialities of such a trigger.[16]

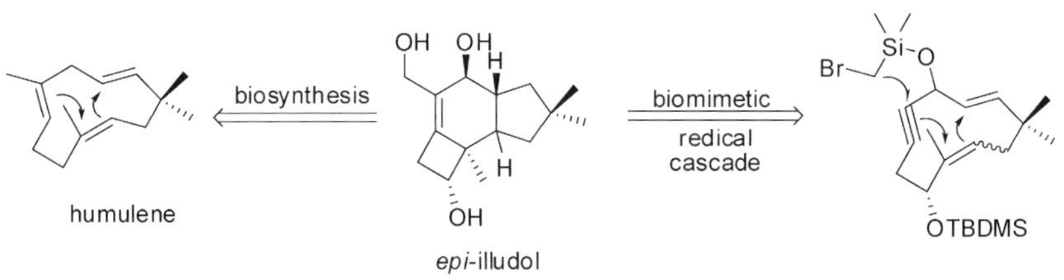

humulene

epi-illudol

OTBDMS

Scheme 21.4 New route to linear triguinane skeleton.

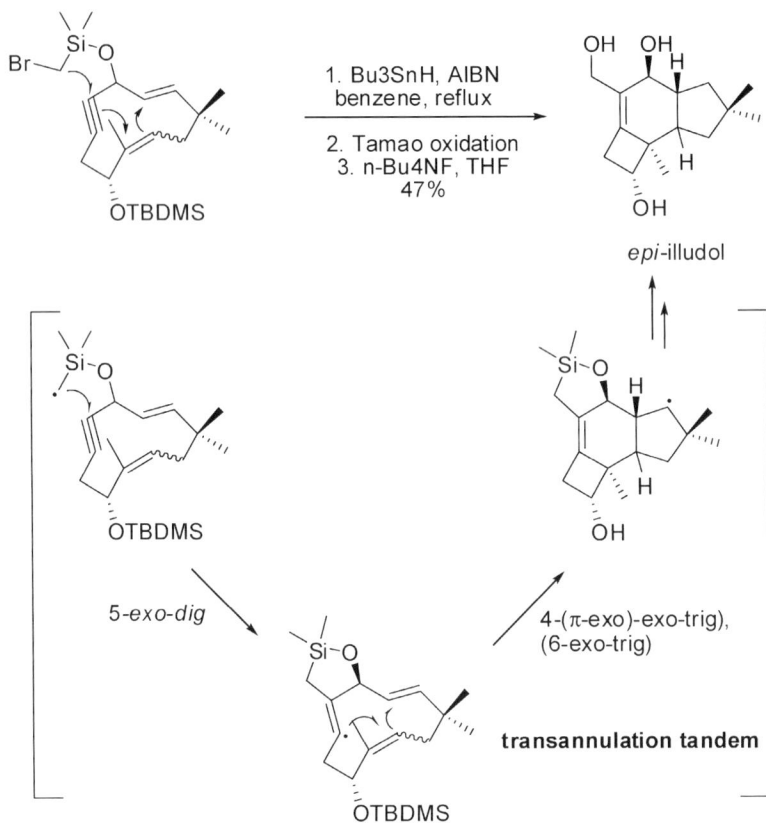

Scheme 21.5 Towards synthesis of *epi*-illudol: regioselectivity and 5exo-dig trans formation.

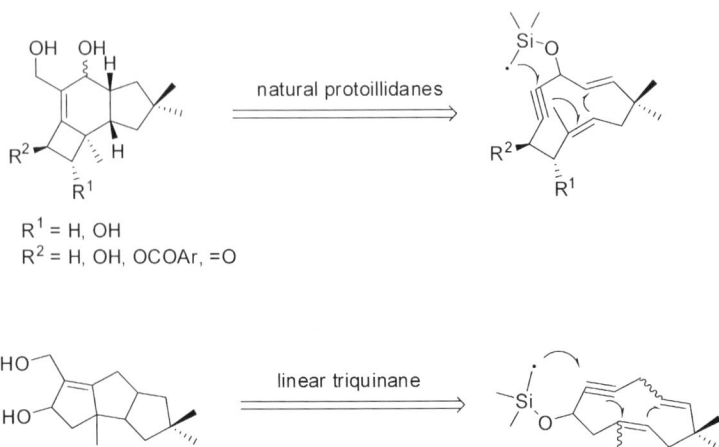

R^1 = H, OH
R^2 = H, OH, OCOAr, =O

Scheme 21.6 Transannulation approach and linear triquinanes.

21.4 ASYMMETRIC INTRAMOLECULAR RADICAL VINYLATION USING ENANTIOPURE SULFOXIDES AS TEMPORARY CHIRAL AUXILIARIES

Malacria has focused for some years on the use of chiral sulfur-based auxiliaries, mainly sulfoxides, because of their easy introduction, their low cost, and their versatile final functionalization. The initial approach, based on the Michael addition of a vinyl radical onto vinyl sulfoxides gave mixed results. Very high diastereoselectivities were obtained for β-alkoxy vinyl sulfoxides,[17] while the pure carbon systems have so far led to poor results, even in the case of *N*-sulfinimines.[18] Malacria preferred to modify the strategy according to the tandem reaction depicted in Scheme 21.7. The sequence would consist in a 5-*exo-trig* cyclization of a prochiral radical in an anti-Michael orientation, followed by the previously reported elimination of β-sulfinyl radicals to furnish alkylidene cyclopentane. Implying an *a priori* quite favorable α-selectivity, this radical addition should be highly diastereoselective.

To test this reaction, Malacria synthesized an *E*-precursor, bearing an isopropyl group on the vinylsulfoxyde moiety (Scheme 21.8). Under low-temperature, radical cyclization conditions, the *E*-precursor underwent an exclusive 5-*exo-trig* radical cyclization to afford a cyclopentyl derivative in 60% yield.[19] One substituent on the vinyl sulfoxide at the β-position is sufficient here to preclude the 6-*endo-trig* mode of cyclization. Moreover, no cyclopentyl derivative incorporating the sulfoxide moiety was observed, which

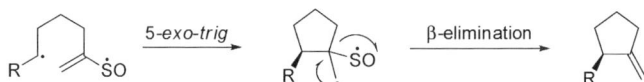

Scheme 21.7 Tandem 5-exo-trig/b-elimination reaction.

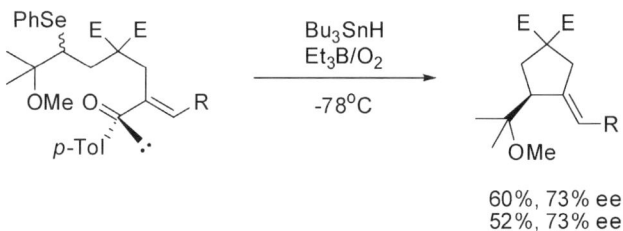

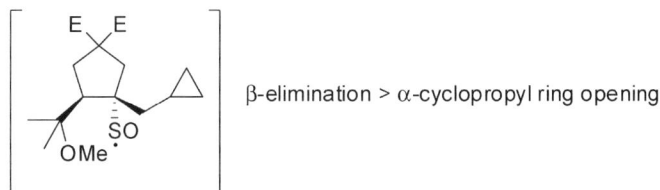

Scheme 21.8 General approaches towards 5-exo-trig radical cyclization to afford cyclopentyl derivative.

confirmed the efficiency of the β-elimination of the sulfoxide auxiliary. The promising stereoselectivity (73% ee) of this sequence was equally interesting. A similar result in terms of yield and stereoselectivity was obtained with a cyclopropyl precursor.

No sulfoxide adduct showing the opening of the cyclopropyl ring was isolated in this reaction, presumably suggesting that the β-elimination of the sulfoxide moiety is even faster than the rearrangement of the traditional radical clock, the α-cyclopropyl radical. This result is not so surprising in view of the estimated rate of about 10^9 s^{-1} for the β-elimination of the arylsulfinyl radical[20] compared to the generally admitted rate of 5×10^7 s^{-1} for the opening of related α-cyclopropyl radicals.[21]

Malacria thus decided to examine the behavior of terminally disubstituted vinylsulfoxides, anticipating that the addition of a substituent *cis* to the sulfoxide moiety should create some additional allylic strain and thus should freeze the reactive conformations and enhance the stereoselectivity. This proved correct since terminally disubstituted vinylsulfoxides afforded much higher stereoselectivities, up to 98% ee.[19] In both cases, no significant decrease of stereoselectivity was observed when running the reaction at 0 °C and the chemical yield was greatly improved (Scheme 21.9).

The compatibility of sulfoxides with aluminum-based Lewis acids[22] drove us to the utilization of methylaluminum-bis-2,6-di-tert-butyl-4-methylphenoxide (MAD) in this cyclization. The complexation of the sulfoxide moiety should modify the steric environment, and possibly reverse the group priorities. This

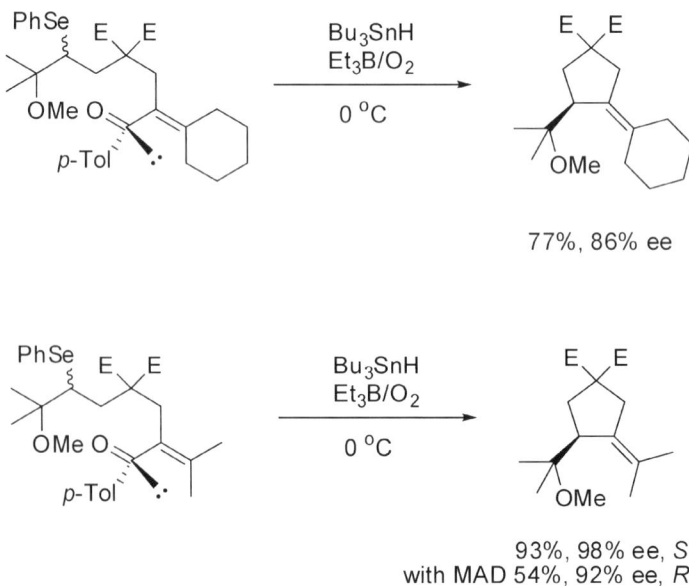

77%, 86% ee

93%, 98% ee, *S*
with MAD 54%, 92% ee, *R*

Scheme 21.9 General scheme: Cyclizations of terminally disubstituted vinyl sulfoxides.

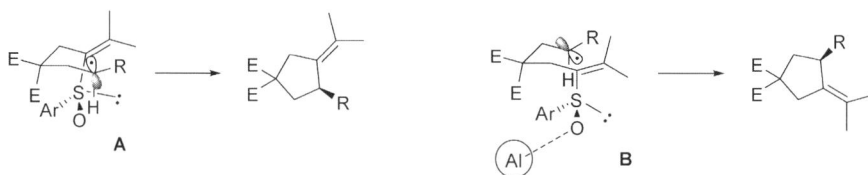

Scheme 21.10   MAD mediated free radical cyclization of sulfide moiety.

hypothesis proved correct since an almost complete inversion of induction was observed in presence of MAD (Scheme 21.10).

The above model allows a rationalization of these data. When no Lewis acid is present, the radical cyclization takes place through pseudo-chair **A**, in which the sulfoxide in the lowest energy conformer is present with the lone pair s-*cis* to the vinyl moiety. The attack takes place *syn* to the S–O bond and *anti* to the *p*-tolyl group; the bulky alkyl group is thus forced into a pseudo-equatorial position. In the presence of the very bulky MAD Lewis acid, which complexes to the oxygen atom of the sulfoxide group, the radical cyclization would occur as depicted in transition state **B**, *syn* to the bulky *p*-tolyl group. This model also applies to a large extent to monosubstituted alkenes. However, these less-rigid systems easily assume other reactive conformations which decreases the overall selectivity of the process. Malacria has studied many other precursors and have notably noticed that the enantioselectivity drops, with a less sterically demanding R group, and that MAD is the most satisfactory Lewis acid.

21.5   CONCLUSION AND SUMMARY

This part focuses on the contributions in the field of radical synthetic chemistry, focusing on cascades relying on the use of propargylic BMDMS ethers and asymmetric synthesis using enantiopure vinylsulfoxides. One very important aspect in modern drug discovery is the preparation of so-called "substance libraries" from which pharmaceutical lead structures might be selected for the treatment of different diseases. An efficient approach for the preparation of highly diversified libraries is one of the aims of the bright and green future of mature but vibrant free radical chemistry in all its power and diversity.

REFERENCES

1. L. Fensterbank, M. Malacria and S. M. Sieburth, *Synthesis,* 1997, 813–854.
2. G. Agnel and M. Malacria, *Synthesis,* 1989, 687–688.
3. (a) E. Magnol and M. Malacria, *Tetrahedron Lett.,* 1986, **27**, 2255–2256; (b) M. Journet, E. Magnol, W. Smadja and M. Malacria, *Synlett,* 1991, 58–60.

4. (a) M. Journet, E. Magnol, G. Agnel and M. Malacria, *Tetrahedron Lett.,* 1990, **31**, 4445–4448; (b) M. Journet and M. Malacria, *J. Org. Chem.,* 1992, **57**, 3085–3093; (c) L. Fensterbank, A.-L. Dhimane, S. Wu, E. Lacôte, S. Bogen and M. Malacria, *Tetrahedron,* 1996, **52**, 11405–11420.
5. M. Journet and M. Malacria, *Tetrahedron Lett.,* 1992, **33**, 1893–1896.
6. S. Bogen, M. Journet and M. Malacria, *Synlett,* 1994, 958–960.
7. S. Bogen and M. Malacria, *J. Am. Chem. Soc.,* 1996, **118**, 3992–3993.
8. S. Bogen, M. Gulea, L. Fensterbank and M. Malacria, *J. Org. Chem.,* 1999, **64**, 4920–4925.
9. G. Agnel and M. Malacria, *Tetrahedron Lett.,* 1990, **31**, 3555–3558.
10. (a) S. Bogen, L. Fensterbank and M. Malacria, *J. Am. Chem. Soc.,* 1997, **119**, 5037–5038; (b) S. Bogen, L. Fensterbank and M. Malacria, *J. Org. Chem.,* 1999, **64**, 819–825.
11. M. Journet, W. Smadja and M. Malacria, *Synlett,* 1990, 320–321.
12. S. Bogen, P. Devin, L. Fensterbank, M. Journet, E. Lacôte and M. Malacria, *Recent Res. Dev. Org. Chem.,* 1997, **1**, 385–395.
13. P. Devin, L. Fensterbank and M. Malacria, *J. Org. Chem.,* 1998, **63**, 6764–6765.
14. M. Rychlet Elliott, A.-L. Dhimane and M. Malacria, *J. Am. Chem. Soc.,* 1997, **119**, 3427–3428.
15. M. Rychlet Elliott, A.-L. Dhimane, L. Hamon and M. Malacria, *Eur. J. Org. Chem.,* 2000, 155–163.
16. *New developments in radical chemistry. Applications to total synthesis and asymmetric processes,* C. Aissa, B. Delouvrie, A.-L. Dhimane, L. Fensterbank and M. Malacria, From *Pure Appl. Chem.,* 2000, **72**(9), 1605–1613.
17. M. Zahouily, M. Journet and M. Malacria, *Synlett,* 1994, 366–368.
18. E. Lacôte and M. Malacria, *C. R. Acad. Sci. Fr., Ser. IIc: Chim.,* 1998, 191–194.
19. (a) E. Lacôte, B. Delouvrié, L. Fensterbank and M. Malacria, *Angew. Chem., Int. Ed.,* 1998, **37**, 2116–2218; (b) B. Delouvrié, L. Fensterbank, E. Lacôte and M. Malacria, *J. Am. Chem. Soc.,* 1999, **121**, 11395–11401.
20. M. Newcomb, *Tetrahedron,* 1993, **49**, 1151–1176.
21. P. J. Wagner, J. H. Sedon and M. J. Lindstrom, *J. Am. Chem. Soc.,* 1978, **100**, 2579–2580.
22. P. Renaud, N. Moufid, L. H. Kuo and D. P. Curran, *J. Org. Chem.,* 1994, **59**, 3547–3552.

CHAPTER 22

Streamlining Organic Free Radical Synthesis through Modern Molecular Technology: from Polymer-supported Synthesis to Microreactors and Beyond*

22.1 FREE RADICALS: A BRIEF INTRODUCTION AND WHY THEY ARE IMPORTANT

Free radicals are ubiquitous, reactive chemical entities. Free radical reactions are an important class of synthetic reactions that have been traditionally performed in organic solvents. In recent years, the number of reports of free radical reactions that use water and alternative media such as supercritical CO_2, ionic liquids, fluorous solvents and the solid state has increased.[1-3] The radical reaction is one of the most useful and flexible methods for organic reactions in alternative media, because most of the organic radical species are stable in water and alternative media, and they do not react with water or unusual media. In addition, by harnessing free radical reactivity within the laboratory, biological processes can be studied and controlled, leading in turn to the prevention of disease and the development of new treatments for disease states mediated by free radicals. There have been several excellent reviews on carbon–carbon (C–C) bond formation and the reactions of carbon–hydrogen (C–H) bonds in water and alternative media (supercritical CO_2, ionic liquids, fluorous solvents, microwave mediated reactions and on solid support).[1] There is a specific focus on C–H and C–C bond-forming reactions as they represent the major classes of the most useful and utilized class of free radical reactions in "streamlined free radical organic synthesis". In addition, several important electron-transfer processes as well as free radical non-chain synthetically useful reactions in a high-throughput environment and some applications will be

* Chapter written by V. Tamara Perchyonok.
Streamlining Free Radical Green Chemistry
V. Tamara Perchyonok, Ioannis Lykakis and Al Postigo
© V. Tamara Perchyonok, Ioannis Lykakis and Al Postigo 2012
Published by the Royal Society of Chemistry, www.rsc.org

discussed with a specific emphasis on the mechanistic and application aspects of these transformations (Scheme 22.1).

22.2 POLYMER-SUPPORTED REAGENTS AND FREE RADICAL SYNTHESIS: A FEW INITIAL REMARKS AND APPROACHES

The most important application of polymers in synthetic chemistry is the use of cross-linked polymers in solid-phase synthesis.[4] Two strategies are employed in the polymer-supported synthesis: (a) use of an insoluble cross-linking polymer-supported catalyst and (b) a soluble polymer-supported catalyst (Scheme 22.2).

A key difference between a strategy that uses an insoluble cross-linked polymer and a strategy that uses a soluble polymer is that the latter strategy allows a catalytic reaction to be carried out under homogenous conditions. In most cases where a cross-linked polymer is used to support a catalyst, the catalyst is necessarily separated before, during, and after the reaction because it is attached to an insoluble support. In the case of soluble polymers, the phase-separation event can occur after the reaction. The schematic diagram represents the two distinct processes very well: a process that uses soluble polymer-supported catalyst which is separated, recovered and reused after a reaction by a solid/liquid or liquid/liquid separation will be discussed (Scheme 22.2).

22.2.1 PEG-bound Reagents and Free Radical Transformations to Date: the Journey Has Begun

Poly(ethylene glycol) (PEG) is a linear polymer formed from the polymerization of ethylene oxide. Along with polystyrene, it was one of the first soluble polymers used to facilitate catalysis and synthesis.[4] Organocatalysts are playing an increasingly important role in the field of catalysis. Interest in the field of catalysis has grown due to the advantages of performing catalytic reactions in a metal-free environment, which includes working with wet

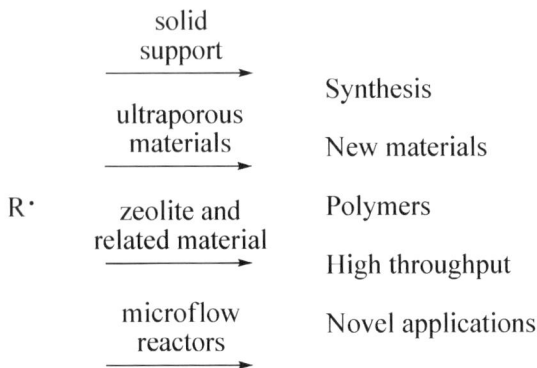

R·

        solid support → Synthesis

        ultraporous materials → New materials

        zeolite and related material → Polymers

        → High throughput

        microflow reactors → Novel applications

Scheme 22.1 General overview of radical chemistry and applications.

biphasic
reaction

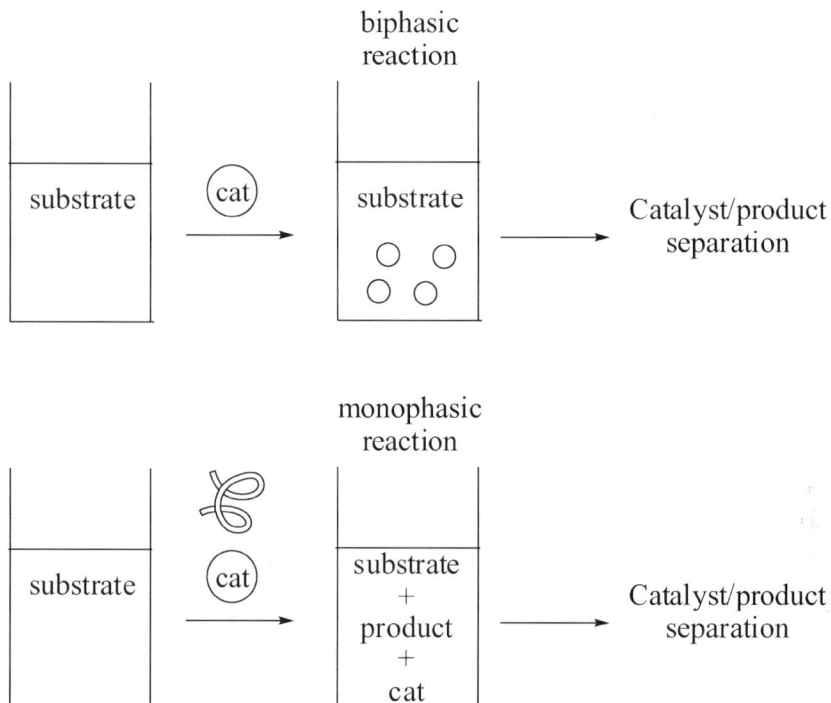

monophasic
reaction

Scheme 22.2 Strategic differences between the use of an insoluble cross-linking polymer-supported catalyst and a soluble polymer-supported catalyst.

solvents under aerobic conditions. As is true for transition metal catalysis, recycling of these homogeneous catalysts is an issue. Such recycling has been successful with a variety of polymers including PEG. The use of macroporous beads with a permanent porosity has been proposed as an alternative synthetic methodology; however, the main drawback is the poor permeability of these materials that reduces the application for organic synthesis. Solid-phase radical reactions have been developed for an important carbon–carbon bond-forming method on a solid support under mild conditions. Naito has previously reported the use of triethylborane (Et₃B/air) as a radical initiator in the intermolecular radical reactions on a solid support.[5] The work to date has been extended to tandem radical addition–cyclization of oxime ethers anchored to a polymer support. Naito and co-workers have developed the triethylborane-induced solid-phase radical reactions of oxime ethers anchored to the TentaGel OH to produce various natural and unnatural amino acid derivatives in aqueous media (Scheme 22.3 and Table 22.1).[6]

22.2.2 Solid-state Radical Reactions

Solid-state radical reactions of 1,3-cyclohexanedione with *in situ* generated imines mediated by manganese(III) acetate under mechanical milling condi-

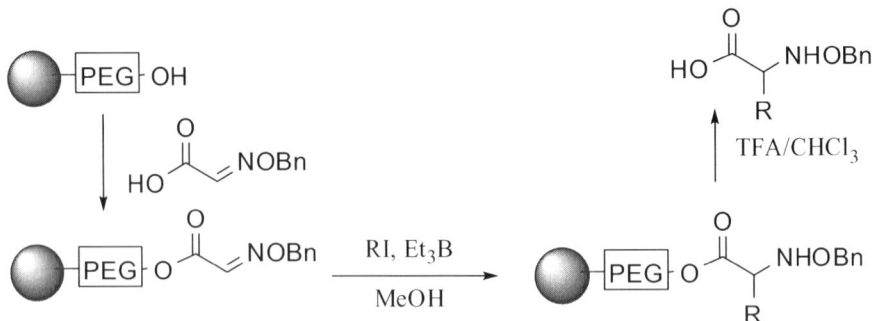

Scheme 22.3 Preparation of amino acids in aqueous media on a solid support.

tions was reported by Wang and coworkers.[7] The novel radical addition reaction to imines mediated by $Mn(OAc)_3 \cdot 2H_2O$ under solid-state conditions was reported for the first time. The high efficiency and good to excellent yield, no separation of the *in situ* generated imines, no use of any solvents in carrying out the reaction and thus the facile addition treatment make this method a potential alternative to a conventional methodology (Scheme 22.4).

Bowman *et al.* reported a solid-phase intramolecular aromatic homolytic substitution of benzoimidazole precursors.[8] The approach involves the attachment of a radical precursor to the resin *via* the radical leaving groups in the hemolytic aromatic substitution. When the radical reaction is complete, the leaving group, unaltered starting material and reduced uncyclized products remain attached to the resin, which facilitates easy separation of the cyclized products. The novel application of focussed microwave irradiation in solid-phase radical reactions is advantageous and drastically reduces reaction times. Tributylgermanium hydride (Bu_3GeH) has been used to replace the toxic and troublesome tributyl tin hydride (Bu_3SnH) in the radical reactions (Scheme 22.5).

Relatively little effort has been made to alter the properties of a polymer to influence reaction outcomes. Polymers have been mostly used as phase anchors for a catalyst, reagent or ligands. In the future the influence of polymer solubility, polyvalency or other properties should also be utilized in the optimization and further development of the synthetically useful reactions and applications.

Table 22.1 Alkyl addition to the TentaGel OH resin-bound glyoxylic oxime ether in aqueous media

Entry	Solvent	Et_3B	Yield (%)[a]
1[b]	H_2O–MeOH (2 : 1, v/v)	In hexane	No reaction
2[b]	H_2O–MeOH (2 : 1, v/v)	In THF	66
3[b]	H_2O–MeOH (2 : 1, v/v)	In MeOH	79
4[c]	CH_2Cl_2	In hexane	57

[a]Isolated yields. [b]Reactions were carried out with Et_3B (10 equiv.) at 20 °C for 15 min. [c]Reaction was carried out with Et_3B (3.6 equiv.) at 20 °C for 1 h.

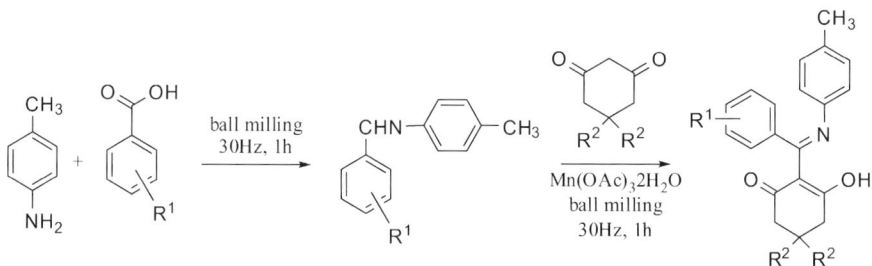

Scheme 22.4 Solid-state radical reactions of 1,3-cyclohexanedione and dimedone with *in situ* generated imine and proposed mechanistic pathway.

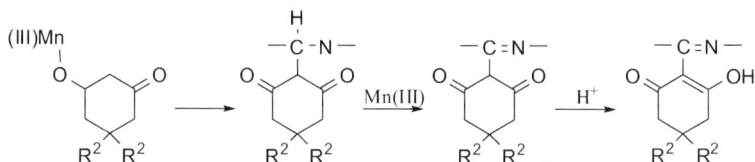

22.3 ULTRAPOROUS MATERIALS AS POSSIBLE MICROREACTORS AND FREE RADICAL SYNTHESIS

An alternative approach consists in the use of porous polymer foams having a low resistance to flow that makes them usable with low-pressure continuous-

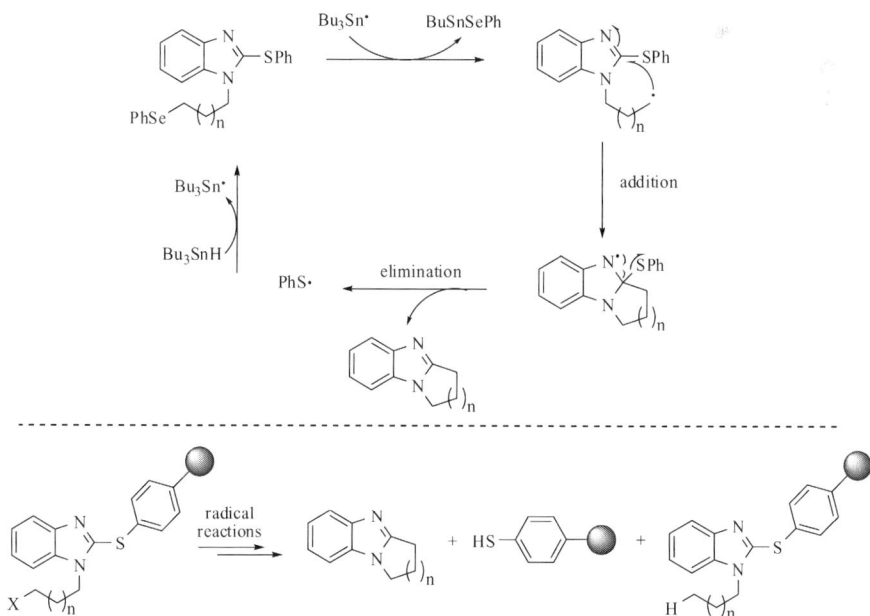

Scheme 22.5 Homolytic aromatic substitution and solid-phase radical protocol.

flow methods. These materials are well-known and have been produced by a wide variety of techniques ranging from leaching soluble fillers through gas-blowing to afford phase separation, although the structure of these materials is often irregular and difficult to control. A novel method for producing porous materials with a more regular structure has been developed based on high internal phase emulsion (HIPE, ~ 74.05 vol% internal phase, which is the maximum space occupiable by uniform spheres).[9] These emulsion-templated foams were initially developed by Unilever and called polyHIPEs.[10] By far the most widely studied polyHIPE base material is polystyrene (PS). Styrene (St) is a water-immiscible liquid, therefore water-in-oil (w/o) HIPEs are used to prepare PS polyHIPEs. Usually, varying quantities of a hydrophobic cross-linker, such as divinylbenzene (DVB), are also added to enhance structural stability. Other monomers that are used to create polyHIPEs from w/o emulsions include 2-ethylhexyl acrylate (EHA), 2-ethylhexyl methacrylate (EHMA), butyl acrylate (BA) and methyl methacrylate (MMA).[11] The characteristics and synthesis of the materials based on w/o HIPEs can be summarized as shown in Scheme 22.6. The process of preparing polyHIPEs is simple. A mixture of monomer(s) plus, usually, a cross linker and suitable surfactant serves as the organic phase, while the aqueous phase containing the initiator is added slowly. Agitation is continued during addition to break up large droplets. After all of the aqueous phase is added, the emulsion is further mixed for a period of time, then a stable HIPE is obtained. The structure of the HIPE is now analogous to soap bubbles, with thin films surrounding and separating the drops. The emulsion is cured by increasing the temperature. During the polymerization step, interconnected pores are formed in the thin films separating the droplets and an open structure is obtained. The water is removed to produce foam of the corresponding structure. The free radical polymerization of a commercial cross linker, DVB, led to a crosslinked macroporous polymer. The overall interconnected open-cellular macrostruc-

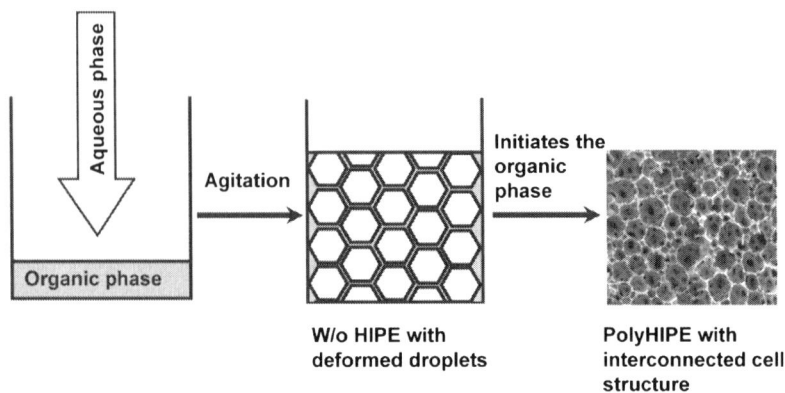

Scheme 22.6   General preparation of polyHIPE based on w/o HIPE.

ture of the materials can be clearly seen by scanning electron microscope (Scheme 22.6).

The general range of measurements/ properties of the emulsion-templated foam may be summarized as follows:[12]

- Macroscopic density as low as 0.0126 g cm^{-3}.
- Internal void volume about 98.5%, that is 98.5% pore volume/1.5% polymer.

Free radical reactions such as the dehalogenation of alkyl, vinyl or aryl halides often followed by intra- or intermolecular C–C coupling are in increasing use in organic synthesis, since a large variety of functional groups is tolerated avoiding laborious protection and deprotection sequences. These reactions are generally performed in the laboratory using tributyl tin hydride (Bu_3SnH). Recently more environmentally benign and metal free options have been explored and developments have been outstanding. However, the toxicity of this reactant is now well established and this drawback strongly limits the development of its use in the synthesis of pharmaceutically important compounds. Recently polyHIPEs either functionalized, by introduction of thiol or by introduction of tetrasubstituted ammonium salt such as cetyl trimethylammonium bromide (CTAB) have been used as functionalized media for a broad range of conventional free radical transformations in organic and aqueous media.[13]

22.3.1 A Few Words About Polarity Reversal Catalysis and its Advantages in Free Radical Transformations in PolyHIPEs

The radical-chain hydrosilylation of alkenes by triethylsilane under very mild conditions is promoted by thiols, which act as polarity-reversal catalysts for the abstraction of electron-rich hydrogen from the silane by nucleophilic α-silylalkyl radicals. This approach represents one of the alternative methodologies to overcome the toxic and unnecessary use of Bu_3SnH and its derivatives. First reported by Roberts, it consists of combination of triethylsilane, the reducing agent, with small amount of a thiol.[14] Reduction proceeds by a radical chain mechanism and the thiol acts as a polarity reversal catalyst, which mediates the hydrogen-atom transfer from the Si–H group of the silane to the alkyl radical (Scheme 22.7).

The only side effect in this methodology is the stench of thiols. The post-functionalization of a polyHIPE-supported materials leads to mercaptan functionalities after free radical addition of thioacetic acid and deprotection.

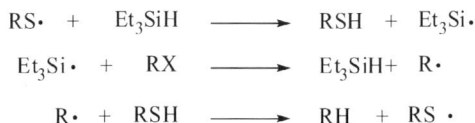

$$RS\cdot \ + \ Et_3SiH \ \longrightarrow \ RSH \ + \ Et_3Si\cdot$$

$$Et_3Si\cdot \ + \ RX \ \longrightarrow \ Et_3SiH+ \ R\cdot$$

$$R\cdot \ + \ RSH \ \longrightarrow \ RH \ + \ RS\cdot$$

Scheme 22.7 Radical chain mechanism involving the reducing agent Et_3SiH and a polarity reversal catalyst thiol.

The reduction of 1-bromoadamantane by PhSH regenerated *in situ* by triethylsilane was performed as test reaction for the proposed system (Scheme 22.8).

Once the efficiency of the supported hydrogen transfer was proven, it has appeared to be of interest to authors to study the reduction of unsaturated bromides. Then, the reduction of 6-bromohex-1-ene and 1-allyloxy-2-bromobenzene was investigated using the polyHIPE. These free radical reactions would produce two major compounds: the products of reductive cyclization and the ones arising from the direct reductions.[15] PolyHIPEs, which are highly porous, easily prepared and functionalized, have been used in conventional free radical chemistry and therefore one day could become an alternative to conventional resins in supported organic chemistry. We have recently reported on an application of polyHIPEs with a cationic surfactant (polyHIPE(CTAB)) (Figure 22.1) as a suitable medium for a broad range of fundamental free radical reactions, such as hydrogen-atom transfer, radical deoxygenations and radical cyclizations at room temperature to proceed in good to excellent yields.[13b]

The utility of the novel porous materials was tested in the presence of H_3PO_2 and TMS_3SiH as free radical hydrogen donors in order to address the so-called 'tin problem' of free radical chemistry.[16] The comparative studies of reduction 1-BrAd were conducted using H_3PO_2/polyHIPE, TMS_3SiH and TMS_3SiH/polyHIPE as environmentally friendly free radical hydrogen donors in

Radical reduction mechanism of 6-bromohex-1-ene

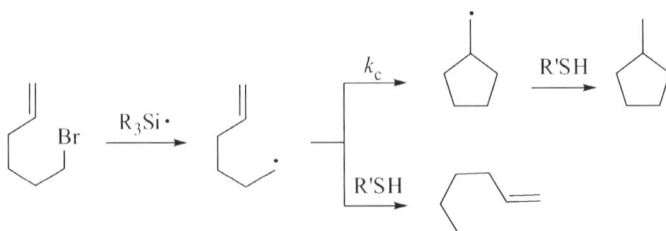

Radical reduction mechanism of 1-allyloxy-2-bromobenzene

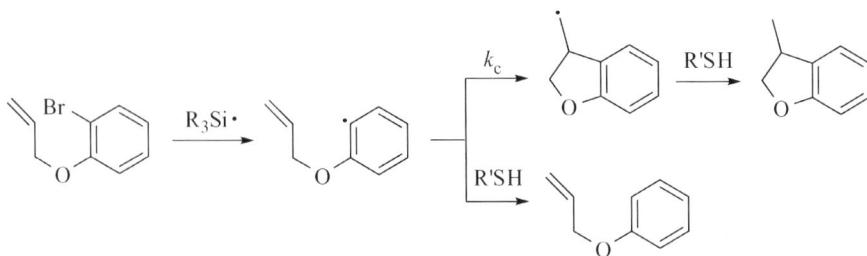

Scheme 22.8   Reduction processes with the catalytic system triethylsilane/thiol.

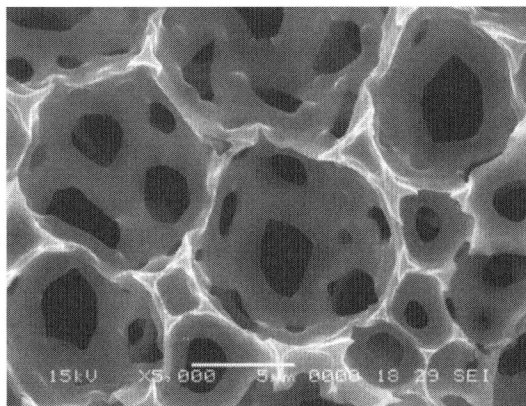

Figure 22.1 SEM image of polyHIPE (CTAB) (pore volume = 8.8 cm^{-3} g^{-1}).

benzene and water as suitable solvent systems in the presence of 10 wt%
azobisisobutyronitrile (AIBN) as the free radical initiator (Scheme 22.9).

The results obtained were encouraging since reactions proceeded with
excellent conversions in a routine free radical solvent (such as benzene or
toluene) but also equally well in water, suggesting a broad-range applicability
of these materials (polyHIPEs) in various free radical processes in a
biocompatible environment. The novel materials, polyHIPEs with cationic
surfactants such as CTAB, have been applied to a broad range of organic
halides as radical precursors in the hydrogen transfer reactions, radical
deoxygenation reactions as well as cyclizations in organic and aqueous media.

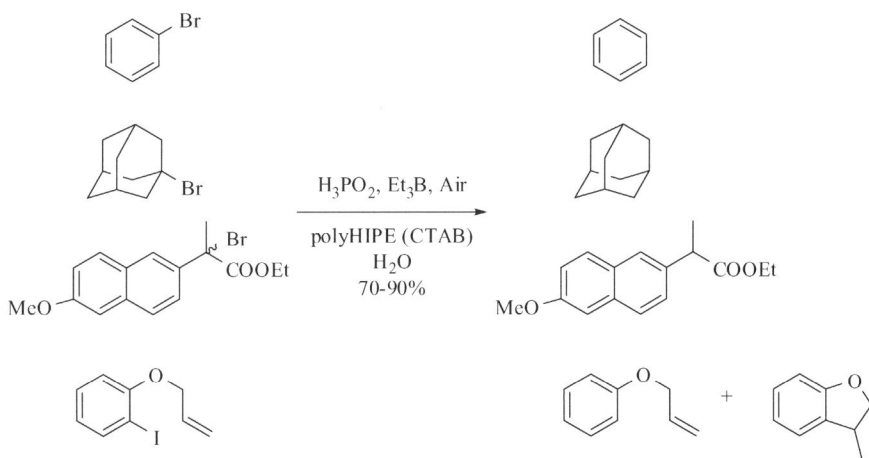

Scheme 22.9 Selected free radical transformations in the presence of polyHIPEs
(CTAB).

22.4   MICROREACTOR-CONTROLLED SELECTIVITY IN ORGANIC PHOTOCHEMICAL REACTIONS: MOLECULAR SIEVE ZEOLITES TO THE RESCUE

The term microreactor refers to organized and constrained media where the chemical reactions occur. The substrate is usually a small molecule of dimensions of several angstroms, and the microreactors often have a size of tens of angstroms or larger. These microreactors are also known as nanoreactors. Among the many classes of microreactors used in photochemical studies, molecular sieve zeolites, Nafion membranes, vesicles, and low-density polyethylene films are outstanding members. Molecular sieve zeolites represent a unique class of materials.[17] The framework of the materials contains pores, channels, and cages of different dimensions and shapes. The pores and cages can accommodate (selectively according to size/shape) a variety of organic molecules of photochemical interest, and provide restrictions on the motions of the included guest molecules and reaction intermediates.

22.4.1   Photochemistry of Phenyl Phenylacetates Included Within Zeolites and Nafion Membranes

The photochemistry of esters (Scheme 22.10) is expected to be analogous to that of phenyl acetate whose photochemistry in homogenous solutions has been well investigated and understood.[18]

Scheme 22.10 gives examples of photochemical reactions of these esters. Upon photo-irradiation, esters undergo homolytic cleavage of the C–O bond to give two paired radicals. These geminate radical pairs in the cage recombine to form the starting ester or *ortho-* and *para-*hydroxyphenones. The later reaction is known as photo-Fries rearrangement. The proposed mechanism is described in Scheme 22.10. The authors propose that the product distribution

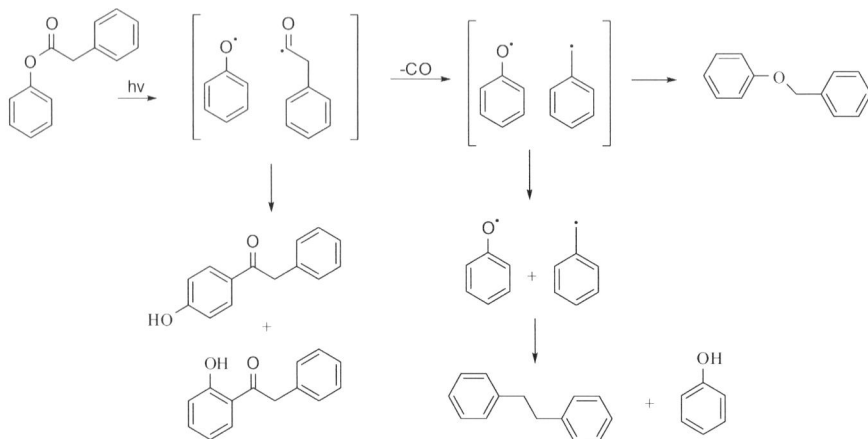

Scheme 22.10   Photochemistry of phenyl phenylacetates in zeolites.

of the reactions is directly related to the shape, size and external surface of the zeolites involved, making the transformation unique and potentially stereo and regio-selective. Using the photo-Fries rearrangements of three 1-naphthyl phenylacylates (Scheme 22.11), Ramamurthy *et al.* demonstrate that limiting the constraining space of a reaction cavity in an organized medium, such as

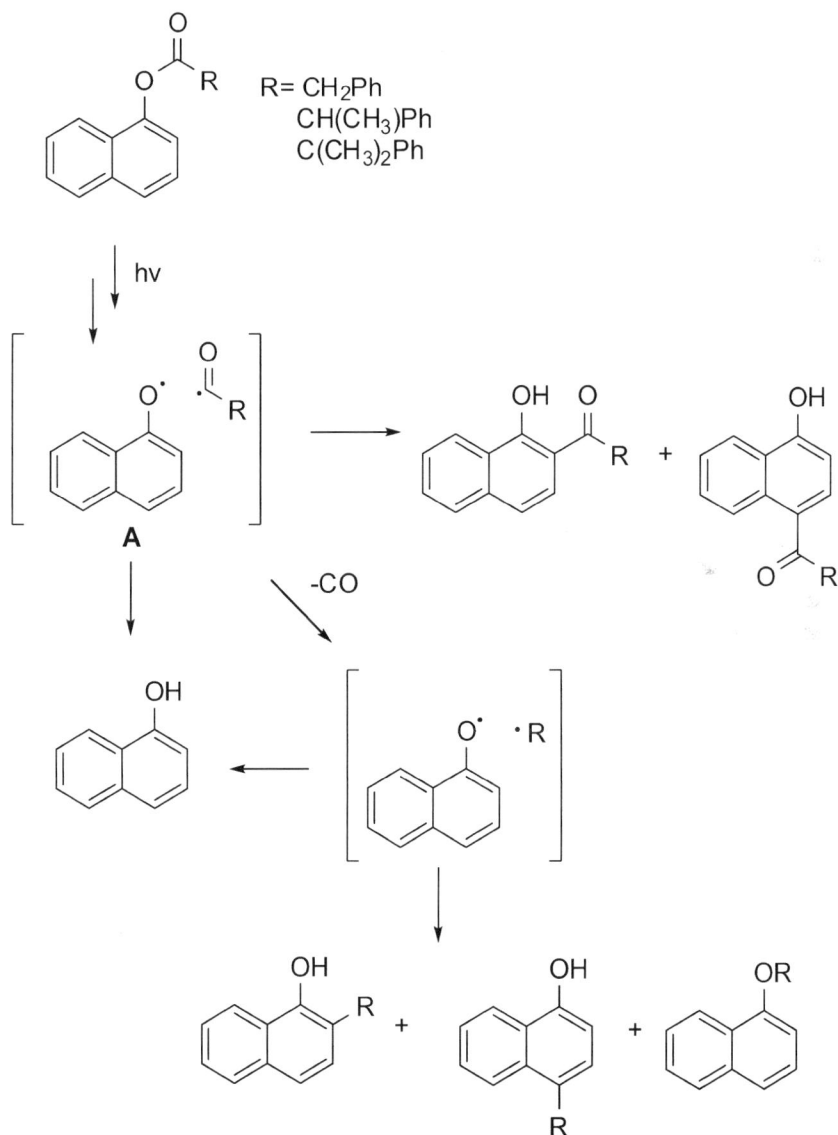

Scheme 22.11 Photo-Fries rearrangement of 1-naphthyl esters in a zeolite compartment.

zeolite, can be less important than wall–guest interactions in determining the selectivity of the reactions.[19]

Reaction cavities of the media-employed, cation-exchanged Y-zeolites, and a high-density polyethylene film of 71% crystallinity possess very different properties. Irradiation of 1-naphthyl acetate or 1-naphthyl benzoate within alkali metal cation-exchanged X and Y zeolites give a single photo-Fries photoproduct, 2-acyl-1-naphthol. In contrast, in hexane solution, the 2- and 4-isomers were formed in comparable yields. Selectivity in the zeolites was suggested to result from restrictions imposed on the naphthoxy and acyl radicals by cations along the cavity walls. Support of this observation comes from the work of Frei *et al.*, which shows that acetyl radicals (from 1-naphthyl acetate) live for $>10^{-5}$ s at room temperature within NaY zeolite.[20] The excellent yields of 2-acetyl-1-naphthol from 1-naphthyl acetate and the long radical lifetimes imply that the radical pairs **A** (R = methyl) are held tightly in place by cations before rejoining; they have the time and space, *but not the mobility,* to undergo other rearrangements (Scheme 22.11).[19]

22.4.2 Zeolites and LDPE Films as Hosts for the Preparation of Large Ring Compounds: Intramolecular Photocycloaddition of Diaryl Compounds

The construction of macrocyclic compounds continues to be an important topic of synthetic organic chemistry. A bifunctional molecule may undergo either intramolecular or intermolecular reactions. Intramolecular reactions give rise to macrocyclic ring-closure products, while intermolecular reactions result in dimers, oligomers or polymers. The rates and therefore product distribution rely heavily on the concentration and the rate of addition of the reagent. Tung *et al.* have reported a new approach towards synthesizing large ring compounds in high yields under high substrate concentration.[21] The approach involves microporous solids as templates and hosts for the cyclization reactions. The size of the micropore has been chosen to permit only one substrate molecule to fit within each. Thus intermolecular reactions are hindered, and cyclization can occur without competition under conditions of high loading, with intramolecular ring closure photocyclomers being observed as major products (Scheme 22.12).

22.4.3 Summary

Photochemical reactions of organic compounds in microreactors usually show deviation of the product distribution from their molecular photochemical reactions and, in some cases, result in the occurrence of reaction pathways that are not otherwise observed. These effects are attributed to (i) size and shape inclusion selectivities, (ii) restriction on rotational and translational motions of the included molecules and intermediates imposed by the microreactor, (iii) compartmentalization of the substrate molecules in the microreactors, and (iv) separation or close contact of the substrate with the sensitizer in a photosensitization reaction.

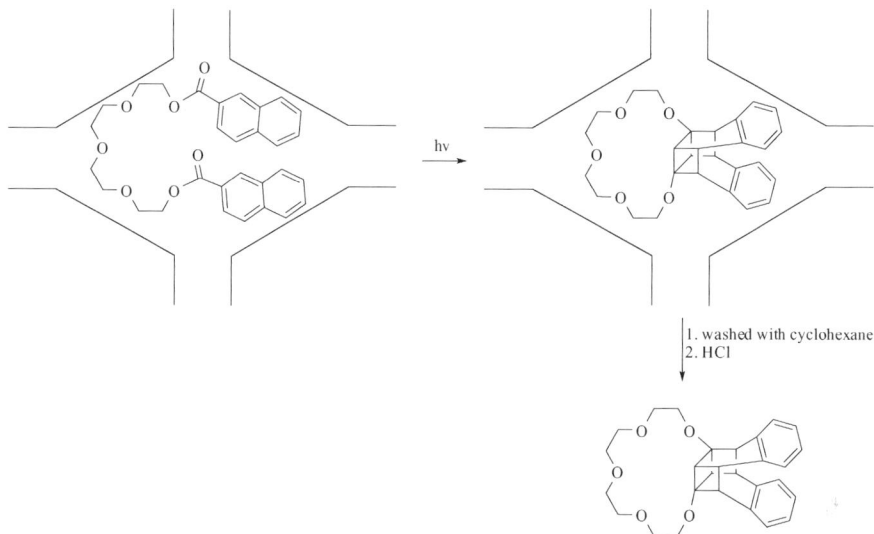

Scheme 22.12 Irradiation of N–P$_n$–N in organic solution inside the zeolite.

22.5 MICROFLOW SYSTEMS FOR PRACTICAL FREE RADICAL SYNTHESIS AND POLYMERIZATION

In the early 19th century, Liebig invented a glass-made cooler for use in glass-made batch reaction systems. The glass-made batch reactor has been a familiar tool for synthetic organic chemists ever since. Microstructured continuous-flow reactors and chip based microreactors are becoming increasingly popular to overcome hurdles and problems associated with the use of batch reactors in synthetic chemistry.[22] Traditionally, batch methods have had inherent limitations regarding the optimization of reaction conditions and scale-up. The small internal dimensions of microreactors require a minimal amount of reagent to be processed under precisely controlled conditions. Subsequently, rapid reaction screening is possible while overall process safety is improved. In addition, reactions can be conducted at elevated temperatures and under pressures which are not accessible using conventional glassware.[23] Subsequently, reactions that have been previously considered to be unattractive on a large scale for practical and/or safety reasons are now within the reach of process and industrial chemists.[24]

Radical reactions such as deoxygenations, dehalogenations, carbonylations, radical additions of azides to aldehydes, Kolbe reactions and Barton reactions, as well as a number of free radical polymerization processes, have been reported using microreactors to complete the desired transformation successfully under the hazard-free and accurate temperature control of a high-throughput system (Scheme 22.13).

Ryu and co-workers described numerous and highly established micro-reactors for use and application in free radical chemistry, the most recent of which is a highly successful and innovative application of tributyltin hydride-

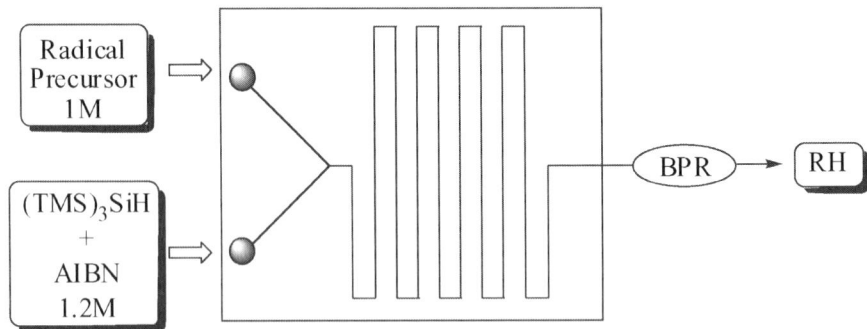

Scheme 22.13   Radical-based reductions and hydrosilylation reactions in microreactors using tris(trimethylsilyl)silane. BPR = back pressure
regulator.

mediated radical reactions of organic halides to a gram scale synthesis of
naturally occurring furofuran lignans, such as (*rac*)-paulownin and (*rac*)-
samin (Scheme 22.14).[25]

Ryu and coworkers have extended the use of continous microreaction
system for radical carbonylation reactions, in order to avoid conventionally
used stainless-steel autoclaves in the batch system (Scheme 22.15).[26]

22.6   FREE RADICAL POLYMERIZATION IN MICROREACTORS: NEW ADVANTAGES AND EXTRA CONTROL

Free radical polymerization is an important process for the synthesis of micromolecules, as free radicals are compatible with and tolerant of a wide variety of
functional groups. This represents a tremendous advantage over ionic and

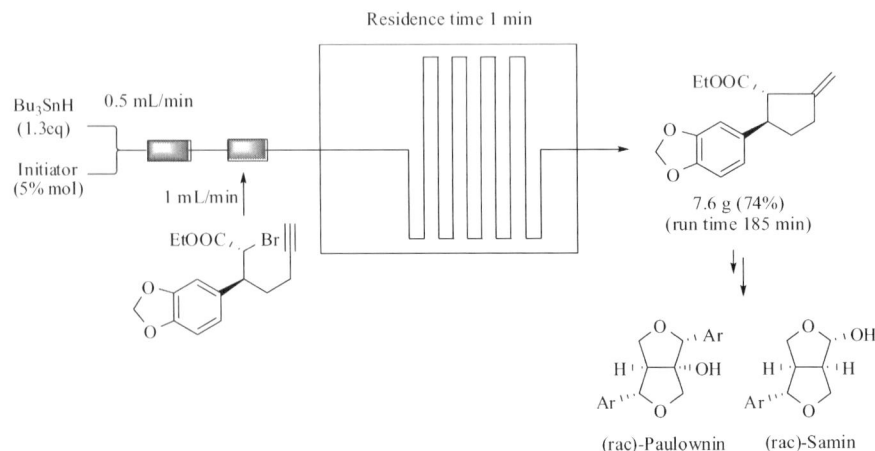

Scheme 22.14   Large scale synthesis in the microflow reactor of the key intermediate
of (*rac*)-paulownin and (*rac*)-samin.

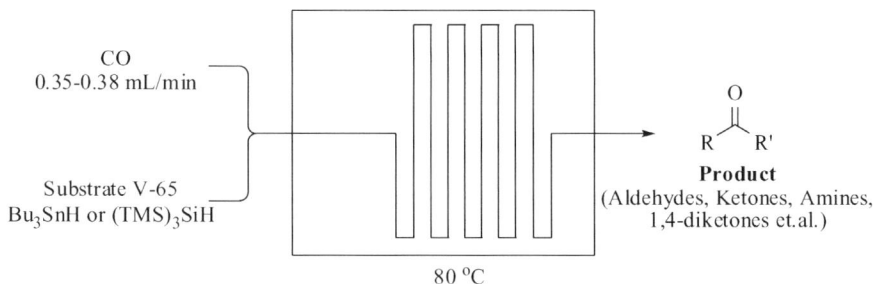

CO
0.35-0.38 mL/min

Substrate V-65
Bu_3SnH or $(TMS)_3SiH$

80 °C

Product
(Aldehydes, Ketones, Amines,
1,4-diketones et.al.)

Scheme 22.15 Radical carbonylation in a microflow system for the synthesis of aldehydes, ketones, amides, 1,4-diketones and derivatives.

metal-catalyzed polymerizations which are which are not compatible with several functional groups, hindering their application and utility. The paramount importance of the precise and efficient temperature control is essential for carrying out free radical polymerization in highly controlled manner, because free radical polymerization reactions are usually highly exothermic. Yoshida *et al.* have examined the thermal decomposition of radical initiator, AIBN in the microreactors and have found that for the polymerization of butyl acrylate (BA), the polydispersity index (PDI) of the polymers obtained using the microreactors was much smaller than that obtained with a macroscale batch reactor.[27] The explanation for the phenomenon is given by the authors in terms of much higher heat removal efficiency of the microreactor compared with the macroscale batch reactor. The results were not significantly different for the less exothermic polymerization of benzyl methacrylate (BMA) and methyl methacrylate (MMA), highlighting that microreactors are very effective for the molecular weight distribution control for highly exothermic free radical polymerizations (BA, BMA and MMA) but not as effective for less exothermic free radical polymerizations (VBz and ST) (Scheme 22.16).

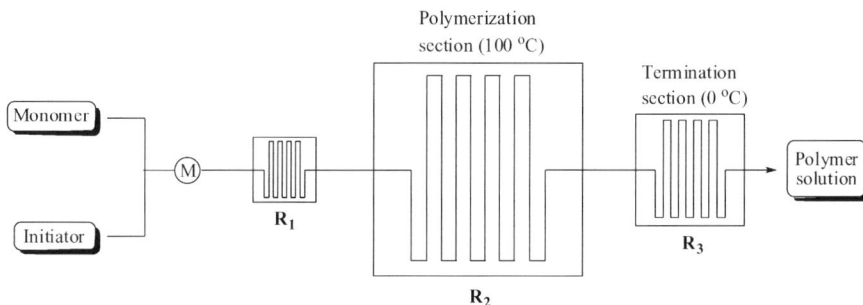

Polymerization
section (100 °C)

Termination
section (0 °C)

Monomer

Initiator

Polymer
solution

R_1

R_2

R_3

Scheme 22.16 General representation for macroreactor system for polymerization: where M is a T-shape mixer, and R_1, R_2 and R_3 are microtube reactors.

22.7 CONCLUSION

As the number of new techniques grows, organic chemists will have available a wide number of parallel synthesis options (supported synthesis, encoded chemistry objects, scavengers, polymer reagents, polymeric scavengers, solid liquid extraction (SLE), solid phase extraction (SPE), ionic liquids supported reagents as well as soluble and insoluble polymer-supported reagents) to choose from in streamlining the synthesis of pharmacologically active derivatives. Future research efforts will likely integrate a variety of these approaches and lead to the development of new generation polymeric reagents and supports for streamlined organic synthesis and a broad range of applications.

REFERENCES

1. Selected reviews: (a) C.-J. Li, Organic reactions in aqueous media with a focus on carbon–carbon bond formations: a Decade update, *Chem. Rev.*, 2005, **105**, 3095–3166; (b) R. A. Sheldon, Green solvents for sustainable organic synthesis: state of the art, *Green Chem.*, 2005, **7**, 267–278; (c) C.-J. Li and L. Chen, Organic chemistry in water, *Chem. Soc. Rev.*, 2006, **35**, 68–82; (d) V. T. Perchyonok, I. N. Lykakis and K. L. Tuck, Recent advances in C–H bond formation in aqueous media: a mechanistic perspective, *Green Chem.*, 2008, **10**, 153–163; (e) V. T. Perchyonok and I. N. Lykakis, Recent advances in free radical chemistry of C–c bond formation in aqueous media: from mechanistic origins to applications, *Mini-Rev. Org. Chem.*, 2008, **5**, 19–32.
2. Microwave chemistry and radicals review articles: A. Studer, Tin-free radical chemistry using the persistent radical effect: alkoxyamine isomerization, addition reactions and polymerizations, *Chem. Soc. Rev.*, 2004, **33**, 267 and references cited therein.
3. Ionic liquids—general reviews: R. D. Rogers and K. R. Seddon, Ionic liquids–solvents of the future? *Nature*, 2003, **302**(5646), 792 and references cited therein.
4. D. E. Bergbreiter, J. Tiam and C. Hongfa, Using soluble polymer supports to facilitate homogeneous catalysis, *Chem. Rev.*, 2009, **109**, 530–582.
5. T. Naito, Development of new synthetic reactions for nitrogencontaining compounds and their application, *Chem. Pharm. Bull*, 2008, **56**, 1367–1383, and references cited therein.
6. For reviews see: (a) R. C. D. Brown, Recent developments in solidphase organic synthesis, *J. Chem. Soc., Perkin Trans. 1*, 1998, 3293–3320; (b) a special issue devoted to combinatorial chemistry, ed. J. W. Szostak, *Chem. Rev.*, 1997, **97**, 347–510; (c) P. H. H. Hermkens, H. C. J. Ottenheijm and D. C. Rees, Solid-phase organic reactions II: a review of the literature Nov 95–Nov 96, *Tetrahedron*, 1997, **53**, 5643–5678; (d) F. Balkenhohl, C. von dem Bussche-Hunnefeld, A. Lansky and C. Zechel, Combinatorial synthesis of small organic molecules, *Angew. Chem., Int. Ed.*, 1996, **35**,

2288–2337; (e) J. S. Fruchtel and G. Jung, Organic chemistry on solid supports, *Angew. Chem., Int. Ed.*, 1996, **35**, 17–42; (f) L. A. Thompson and J. A. Ellman, Synthesis and applications of small molecule libraries, *Chem. Rev.*, 1996, **96**, 555–600; (g) P. H. H. Hermkens, H. C. J. Ottenheijm and D. Rees, Solid-phase organic reactions: a review of the recent literature, *Tetrahedron*, 1996, **52**, 4527–4554.

7. Z. Zhang, G.-W. Wang, C.-B. Miao, Y.-W. Dong and Y.-B. Shen, Solid-state radical reactions of 1,3-cyclohexanediones with *in situ* generated imines mediated by manganese(III) acetate under mechanical milling conditions, *Chem. Commun.*, 2004, 1832–1834.

8. S. M. Allin, W. R. Bowman, R. Karim and S. S. Rahman, Aromatic homolytic substitution using solid phase synthesis, *Tetrahedron*, 2006, **62**, 4306–4316.

9. (a) N. R. Cameron and D. C. Sherrington, High internal phase emulsions (HIPEs): Structure, properties and use in polymer preparation, *Adv. Polym. Sci.*, 1996, **126**, 163–214; (b) H. Zhang and A. I. Cooper, Emulsion- templated hierarchically porous silica beads using silica nanoparticles as building blocks, *Ind. Eng. Chem. Res.*, 2005, **44**, 8707–8714; (c) S. Zhang and J. Chen, Synthesis of open porous emulsion-templated monoliths using cetyltrimethylammonium bromide, *Polymer*, 2007, **48**, 3021–3025.

10. D. Barby and Z. Haq, Synthesis and ultrastructural studies of styrenedivinylbenzene polyHIPE polymers, European Patent 0060138, 1982.

11. (a) C. J. C. Edwards, D. P. Gregory and M. Sharples, Low density porous elastic cross-linked polymeric materials and their preparation, US Patent 4788225, 1988; (b) N. R. Cameron and D. C. Sherrington, Preparation and glass transition temperatures of elastomeric Poly-HIPE materials, *J. Mater. Chem.*, 1997, **7**, 2209–2212; (c) S. Zhang and J. Chen, PMMA based foams made *via* surfactant-free high internal phase emulsion templates, *Chem. Commun.*, 2009, **16**, 2217–2219.

12. (a) N. R. Cameron, High internal phase emulsion templating as a route to well-defined porous polymers, *Polymer*, 2005, **46**, 1439–1449; (b) S. Zhang, J. Chen and V. T. Perchyonok, Stability of high internal phase emulsions with sole cationic surfactant and its tailoring morphology of porous polymers based on the emulsions, *Polymer*, 2009, **50**, 1723–1731; (c) A. Richez, H. Deleuze, P. Vedrenne and R. Collier, Preparation of ultra-low-density microcellular materials, *J. Appl. Polym. Sci.*, 2005, **96**, 2053–2063; (d) A. Mercier, H. Deleuze, B. Maillard and O. Mondain-Monval, Synthesis and application of an organotin functionalised highly porous emulsion-derived foam, *Adv. Synth. Catal.*, 2002, **344**, 33–36.

13. (a) A. Mercier, H. Deleuze and O. Mondain-Monval, Preparation and functionalization of (vinyl)polystyrene polyHIPE®. Short routes to binding functional groups through a dimethylene spacer, *React. Funct. Polym.*, 2000, **46**, 67–79; (b) S. Zhang, J. Chen and V. T. Perchyonok,

PolyHIPEs as novel media for conventional free radical chemistry, *Lett. Org. Chem.,* 2008, **5**, 304–307.

14. H.-S. Dang, M. R. J. Elsegood, K.-M. Kim and B. P. Roberts, Radical-chain reductive alkylation of electron-rich alkenes mediated by silanes in the presence of thiols as polarity-reversal catalysts, *J. Chem. Soc., Perkin Trans 1*, 1999, 2061–2068.

15. H. Deluze, B. Mailard and O. Mondain-Monval, Development of a new ultraporous polymer as support in organic synthesis, *Bioorg. Med. Chem. Lett.*, 2002, **12**, 1877–1880.

16. S. Itsuno, Y. Arakawa and N. Haraguchi, Polymer-supported chiral catalysts for asymmetric reactions in water, *Nippon Gomu Kyokaishi*, 2006, **79**, 448.

17. (a) D. W. Breck, *Zeolite Molecular Sieves: Structure, Chemistry and Use*, Wiley, New York, 1974; (b) A. Dyer, *An Introduction to Zeolite Molecular Sieves*, Wiley, New York, 1988; (c) H. Van Bekkum, E. M. Flanigen and J. C. Jansen, *Introduction to Zeolite Science and Practice*, Elsevier, Amsterdam, 1991.

18. (a) C.-H. Tung and Y. M. Ying, Photochemistry of phenyl phenylacetates adsorbed on pentasil and faujasite zeolites, *J. Chem. Soc., Perkin Trans. 2*, 1997, 1319–1322; (b) C.-H. Tung and X. H. Xu, Selectivity in the photo-Fries reaction of phenyl phenylacetates included in a Nafion membrane, *Tetrahedron Lett.*, 1999, **40**, 127–130.

19. (a) K. Pitchumani, M. Warrier and V. Ramamurthy, Remarkable product selectivity during photo-fries and photo-claisen rearrangements within zeolites, *J. Am. Chem. Soc.,* 1996, **118**, 9428–9429; (b) W. Gu, M. Warrier, V. Ramamurthy and R. G. Weiss, Photo-Fries reactions of 1-naphthyl esters in cation-exchanged zeolite y and polyethylene media, *J. Am. Chem. Soc.,* 1999, **121**, 9467–9468.

20. (a) S. Vasenkov and H. Frei, Observation of acetyl radical in a zeolite by time-resolved FTIR spectroscopy, *J. Am. Chem. Soc.,* 1998, **120**, 4031–4032; (b) S. Vasenkov and H. Frei, Time-resolved study of acetyl radical in zeolite nay by step-scan FTIR spectroscopy, *J. Phys. Chem. A,* 2000, **104**, 4327–4332.

21. (a) C.-H. Tung, L. Z. Wu, Z. Y. Yuan and N. Su, Remarkable product selectivity in photosensitized oxidation of alkenes within Nafion membranes, *J. Am. Chem. Soc.,* 1998, **120**, 11874–11879; (b) C.-H. Tung, L. Z. Wu, Z. Y. Yuan and N. Su, Zeolites as templates for preparation of large-ring compounds: intramolecular photocycloaddition of diaryl compounds, *J. Am. Chem. Soc.,* 1998, **120**, 11594–11602; (c) C.-H. Tung, L. Z. Wu and H. R. Li, Reactions of singlet oxygen with olefins and sterically hindered amine in mixed surfactant vesicles, *J. Am. Chem. Soc.,* 2000, **121**, 2446–2451; (d) C.-H. Tung, L. Z. Wu and H. R. Li, Vesicle controlled selectivity in photosensitized oxidation of olefins, *Chem. Commun.,* 2000, **12**, 1085–1086.

22. For recent reviews of microreactor-based synthesis, see: (a) W. Ehrfeld, V. Hessel, H. Lowe, H., Microreactors: New Technology for Modern

Chemistry, John Wiley & Sons Inc., Weinheim, 2000; (b) K. Jahnisch, V. Hessel, H. Lowe, M. Baerns, Chemistry in Microstructured Reactors, *Angew. Chem., Int. Ed.*, 2004, **43**, 406–446; (c) P. Watts, C. Wiles, Recent advances in synthetic micro reaction technology, *Chem. Commun.*, 2007, 443–467; (c) B. P. Mason, K. E. Price, J. L. Steinbacher, A. R. Bogdan, D. T. McQuade, Greener Approaches to Organic Synthesis Using Microreactor Technology, *Chem. Rev.*, 2007, **107**, 2300–2318; (d) B. Ahmed-Omer, J. C. Brandt, T. Wirth, Advanced organic synthesis using microreactor technology, *Org. Biomol. Chem.*, 2007, **5**, 733–740; (e) K. Geyer, J. D. C. Codee, P. H. Seeberger, Microreactors as Tools for Synthetic Chemists - The Chemists' Round-Bottomed Flask of the 21st Century?, *Chem. Eur. J.*, 2006, **12**, 8434–8442; (f) T. Fukuyama, M. T. Rahman, M. Sato, I. Ryu, Adventures in Inner Space: Microflow Systems for Practical Organic Synthesis, *Synlett*, 2008, 151–163.

23. (a) E. R. Murphy, J. R. Martinelli, N. Zaborenko, S. L. Buchwald, K. F. Jensen, Accelerating Reactions with Microreactors at Elevated Temperatures and Pressures: Profiling Aminocarbonylation Reactions, *Angew. Chem., Int. Ed.*, 2007, **46**, 1734–1737; (b) V. Hessel, C. Hofmann, P. Lob, J. Lohndorf, H. Lowe, A. Ziogas, Aqueous Kolbe-Schmitt Synthesis Using Resorcinol in a Microreactor Laboratory Rig under High-p,T Conditions, *Org. Process Res. Dev.*, 2005, **9**, 479–489.

24. For recent examples of microreactor-based transformations from Seebergers group, see: (a) T. Gustafsson, F. Ponten, P. H. Seeberger, Trimethylaluminium mediated amide bond formation in a continuous flow microreactor as key to the synthesis of rimonabant and efaproxiral, *Chem. Commun.* 2008, 1100–1102; (b) F. R. Carrel, K. Geyer, J. D. C. Codee, P. H. Seeberger, Oligosaccharide Synthesis in Microreactors, *Org. Lett.*, 2007, **9**, 2285–2288; (d) K. Geyer, P. H. Seeberger, Optimization of Glycosylation Reactions in a Microreactor, *Helv. Chim. Acta*, 2007, **90**, 395–403; (e) O. Flogel, J. D. C. Codee, D. Seebach, P. H. Seeberger, Microreactor Synthesis of beta-Peptides, *Angew. Chem., Int. Ed.*, 2006, **45**, 7000–7003; (f) D. A. Snyder, C. Noti, P. H. Seeberger, F. Schael, T. Bieber, G. Rimmel, W. Ehrfeld, Modular Microreaction Systems for Homogeneously and Heterogeneously Catalyzed Chemical Synthesis, *Helv. Chim. Acta*, 2005, **88**, 1–9.

25. T. Fukuyama, M. Kobayashi, Md. Taifur Rahman, N. Kamata, I. Ruy, Spurring Radical Reactions of Organic Halides with Tin Hydride and TTMSS Using Microreactors, *Org. Lett.*, 2008, **10**, 533–536.

26. A. Sugimoto, M. Takagi, Y. Sumito, T. Fukuyama, I. Ryu, The Barton reaction using a microreactor and black light. Continuous-flow synthesis of a key steroid intermediate for an endothelin receptor antagonist, *Tetrahedron Lett.*, 2006, **47**, 6197–6200.

27. T. Iwasaki, J. Yoshida, Free Radical Polymerization in Microreactors. Significant Improvement in Molecular Weight Distribution Control *Micromolecules*, 2005, **38**, 1159–1163.

CHAPTER 23

Radical Reactions and β-cyclodextrin as a Molecular Ferrari: Is There a Hidden Advantage of Speed, Power and Class? From Fundamental Reactions to Potential Applications[*]

23.1 INTRODUCTION

Since their discovery, the β-cyclodextrins (CDs) have served as prototypes for novel host compounds and catalysts. The use of CDs as micro-vessels to perform chemical reactions has attracted the interest of chemists since the 1960s. Bender *et al.* carried out detailed and systematic studies on the CD-induced hydrolysis of phenyl acetate and discovered a significant substrate specified in the reaction rate.[1] Since then reactions catalysed by CDs have been continuously reported. Such catalytic mechanisms have been precisely clarified by the use of analytical and physicochemical information.[1,2,3]

The effects of the CDs on the organic reactions are divided mainly into two types. The first is the effect on covalent bonds where the reaction proceeds according to the Michaelis–Menten type. The CD and the reactant initially form a CD–reactant reaction intermediate involving a covalent bond, which then leads to the product. These catalytic effects have been studied and reported as the enzyme model. Various advanced "enzyme models" and "artificial enzymes" have been envisioned through the chemical modification of CDs by using detailed molecular design. The development of techniques for the precise chemical modifications of CDs has undoubtedly made an essential contribution to progress in this field. The second effect does not involve a covalent bond. The hydrophobic cavity of the CD gives the reactant access to a

* Chapter written by V. Tamara Perchyonok.
Streamlining Free Radical Green Chemistry
V. Tamara Perchyonok, Ioannis Lykakis and Al Postigo
© V. Tamara Perchyonok, Ioannis Lykakis and Al Postigo 2012
Published by the Royal Society of Chemistry, www.rsc.org

new reaction environment—an "extra reaction field" in which the reactivity such as rate or selectivity, changes. In these cases, the role of the CD is not always defined as a catalyst. More correctly, the CD mediates the reactions. There are many excellent reviews available on CDs and their catalyzed reactions. This chapter mainly concentrates on recent developments of organic reactions mediated by native or simply modified CDs and in the use of CDs as the extra reaction field rather than an enzyme model.

23.2 THE CYCLODEXTRIN REACTION MEDIA

Before describing various radical reactions mediated by CDs, some explanation of inclusion phenomena is warranted.

1. **Existence of naked reactants**. Organic reactions are dynamic processes. Therefore, the study of the organic and more specifically radical reactions mediated by the CDs cannot ignore the guest molecule outside the CD cavity. When a large excess of CD is added to the reaction system, almost all the reactant is included in CD cavities. However, the probability of a naked reactant is not quite zero since it is controlled by the concentration of both reactants and CD.

2. **Complexation mode.** Much of the interest in natural and modified CDs arises from their ability to include or encapsulate a substantial part of a guest molecule or ion inside their annuli to form complexes usually describes as either inclusion complexes, host–guest complexes or simply complexes. These complexes are unusual in that only secondary bonding occurs between the CD and the guest (G), yet their stability can be quite high depending on the nature of the CD and G. The stoichiometry of the complexes is usually encompassed by the ratios 1:1, CD.G; 1:2, CD.G_2; 2:1, CD$_2$.G; and 2:2, CD$_2$.G_2, characterized by the sequential stability constants K_{11}, K_{12}, K_{21} and K_{22} and $K_{22'}$. K_{11} = [CD.G]/{[CD][G]}, K_{12} = [CD.G_2]/{[CD][G]}, K_{21} = [CD$_2$.G]/{[CD.G][CD]}, K_{22} = [CD$_2$.G_2]/{[CD.G_2][CD]}, and $K_{22'}$ = [CD$_2$.G_2]/{[CD$_2$.G][G]}. The cooperative binding, a bimolecular process, has been observed as well as a unimolecular process as a 1:1 complex in either a "head-to-tail" or "head-to-head" state. Moreover, a study of the fluorescent decay of a number of anilinon–aphthalene sulfonates in the presence of CD supported the view that a 1:1 complex can be present in several slightly different conformations. There are, in fact various conformations and orientations of reactants as well as varying mobilities in a reactant–CD complex.

3. **Specific interactions**. Depending on the solvent and the nature of host and guest, the complexation is based on a combination of several intermolecular interactions: steric fit, van der Waals interaction, dispersion forces, dipole–dipole interactions, charge–transfer interactions, electrostatic interactions, and hydrogen bonding. The choice of reaction media used in the CD system also depends on the above

combination of interactions. The study of the organic reactions mediated by CDs means the study of weak interactions.

4. **Inclusion phenomena in solid states.** CDs can also include guest molecules in the solid state. Such host–guest complexes are clearly separated from the lattice inclusion compounds, which occur only in the solid state, and from the crown ethers, which occur in solutions. Inclusion complexes in the gas and liquid phases are not expected to be the same. In non-aqueous environments, the short-range London forces or polar interactions are responsible for host–guest complexation. Such complexes are ordered in a crystalline network. Two main structural types have been recognized: the channel- and cage-type molecular arrangements.

In the channel-type structure the molecules are piled in a "head-to-head" or "head-to tail" orientation. In the cage-type structure, the cavity of single CD molecule faces and is blocked by other CD molecules, resulting in isolated cages. The natural orientation of quest molecules in crystals and the tight molecular packing around them may be crucial to the course of the reaction. This is especially true for the photochemistry of the system as shown later.

The CD cavity is less polar than the bulk aqueous medium. It is known that the permittivity in the CD cavity is almost the same as that of dioxane. There are specific interactions between the CD and the reactant. The microenvironment around the reactant in the CD cavity is different from that in the reaction media. Three distinct microenvironmental effects are expected: (1) the micro-solvent effect, (2) the protection of unstable intermediates or products, and (3) the solubilization of the reactant. In addition, conformational effects can also be expected: (4) the control of reactant conformations, (5) control of the orientation between reactants, and (6) control of the size of the molecule. A combination of these effects is evident for organic reactions (Figure 23.1).

23.3 β-CYCLODEXTRIN-BASED MOLECULAR REACTORS FOR FREE RADICAL CHEMISTRY IN AQUEOUS MEDIA AND CHAIN REACTIONS

Radical hydrogen-atom transfer reactions have become a versatile and powerful tool for forming new bonds due to the mild and neutral nature of

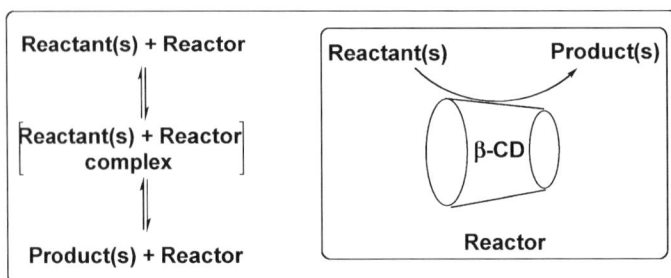

Figure 23.1   β-CD-based molecular reactor prototype.

this transformation as well as its broad functional group tolerance.[3] Considerable effort has been devoted to developing free radical reactions in aqueous media as well as introducing a degree of stereochemical control to the desired transformations.[3,4]

Phosphinic acid has been used as a radical chain carrier and radical hydrogen donor, in hydrogen atom transfers, additions to carbon–carbon and carbon–nitrogen double bonds.[3] It is proposed that quaternary ammonium hypophosphites can readily generate radicals from alkyl halides as well as playing a role as efficient surfactants thereby increasing the solubility of organic molecules with a significant degree of stereo and enantiocontrol.[5,6]

Perchyonok reported the use of naturally occurring and readily available β-cyclodextrin as a molecular reactor for a diverse range of fundamental free radical transformations in aqueous media. The unique advantage to combining "tin-free" fundamental free radical chemistry and β-cyclodextrin lies in the use of mild reaction conditions as well as the ability to recycle both reagents and reaction media. The quaternary ammonium hypophosphites are prepared by reaction of the corresponding tetrasubstituted ammonium salt or tertiary amine with an equimolar amount of phosphinic acid at room temperature in the solvent/medium of choice.[5d] Chiral and achiral novel "tin-free" hydrogen donors are prepared in quantitative yields and do not require any additional purification.[5d,6]

Perchyonok has chosen to investigate the radical hydrogen atom transfer reaction of 1-bromoadamantane (1-BrAd) in the presence of tetrabutylammonium hypophosphite as a hydrogen donor at room temperature (0.5 equivalents of Et$_3$B/air) in β-cyclodextrin (0.2 g) in water (2.5 mL) (Scheme 23.1). Reductions were carried out with 0.1 mM substrate mixed with a β-cyclodextrin/water mixture or β-cyclodextrin (neat), and initiated at ambient temperature with Et$_3$B/air as a suitable radical initiator. The reduced product was obtained in good to excellent yields and was characterized by ^{1}H NMR spectroscopy and GC-MS analysis.

Table 23.1 lists data obtained for the hydrogen atom transfer of 1-bromoadamantane as a model compound in the presence of β-cyclodextrin in water.[3] As expected, the reaction of 1-bromoadamantane in the absence of (NBu$_4$)$^+$H$_2$PO$_2$ yielded complete recovery of the unreacted starting material (1-BrAd). Reactions performed either in the absence of a free radical initiator (Et$_3$B/air) or in the presence of phenol as a free radical inhibitor and Et$_3$B/air as radical initiator failed to form the reduced product (entries 2 and 3), confirming the free radical nature of the transformation in question.

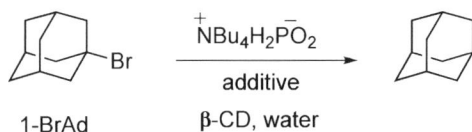

1-BrAd   NBu$_4$H$_2$PO$_2$ / additive / β-CD, water

Scheme 23.1 Reduction of 1-BrAd to Ad under various free radical conditions in water.

Table 23.1 Reduction of 1-bromoadamantane under various free radical
conditions in the presence of tetrabutylammonium hypophosphite ($(NBu_4)^+H_2PO_2^-$) in β-cyclodextrin in water.

Entry	Additive	Temperature/°C	Product	Yield (%)
1	Et₃B/air	22	Ad	83
2	None	22	Ad	0
3	Phenol Et₃B/air	22	Ad	0

Perchyonok has also attempted to recycle and reuse β-cyclodextrin (sample
was in MeOH and dried *in vacuo*) after the reaction and observed no
significant decreases in product yields after 4 runs (1st run 81%, 2nd run 80%,
3rd run 80%, 4th run 79%) using $(NBu_4)^+H_2PO_2^-$ as the hydrogen donor. In
parallel, the radical reduction of 1-BrAd was performed using Bu₃SnH as a
hydrogen donor under previously discussed experimental conditions. As
expected, the reaction proceeded smoothly, providing a crude yield of 78% for
the reduced product. All attempts to recycle β-cyclodextrin were unsuccessful
in this instance due to tin contamination.

As a test case, the aryl halide 2-iodobenzoic acid was subjected to a radical
reduction reaction at room temperature in the presence of a range of various
hypophosphinates as hydrogen donors (Scheme 23.2, Table 23.2). The reduced
product was obtained in good to excellent yield. Additionally, the molecular
reactors (β-cyclodextrins) were successfully recycled in excellent purity after
washing the β-cyclodextrin with methanol and evaporating the excess solvent.
This transformation highlights the use of β-cyclodextrin as a molecular reactor
for tin-free free radical green chemistry.

To test the scope and limitations of the tetrasubstituted hypophosphites as
hydrogen donors in β-cyclodextrin they were subjected to a broad range of
typical radical precursors: 2°, 3° halides (entries 1, 6, 7, 8, 10, and 11),
functionalized aromatic halides (entries 2, and 3) benzyl bromide (entry 5), and
the xanthate of cholesterol (entry 9). The free radical cyclization of 1-
(allyloxy)-2-bromobenzene (entry 4) proceeded smoothly with the combined
yield of cyclized and directly reduced product of 72% which are formed in the
ratio of 3:1 in favour of the cyclized product at room temperature and the
results are summarized in Table 23.3. The results obtained have exceeded
expectation as not only did they proceed with good to excellent yields but β-
cyclodextrins were recycled successfully without noticeable loss of efficiency.

Scheme 23.2 Reduction of 2-iodobenzoic acid under various free radical conditions
in water.

Table 23.2 Reduction of 2-iodobenzoic acid in the presence of a broad range of tetraalkyl hypophosphites in an aqueous β-cyclodextrin molecular reactor at 22 °C.

Entry	H-Donor	Yield (%)
1	$(Bu_4N)^+H_2PO_2^-$	91
2	$(Me_4N)^+H_2PO_2^-$	78
3	$(Et_4N)^+H_2PO_2^-$	82
4	$(isoPr_4N)^+H_2PO_2^-$	84
5	$(HDABCO)^+H_2PO_2^-$	71
6	$(PhBu_3N)^+H_2PO_2^-$	74
7	$(H(R)MePhCHNMe_2)^+H_2PO_2^-$	74
8	$(H(S)MePhCHNMe_2)^+H_2PO_2^-$	72

The β-cyclodextrins are suitable and recyclable for various free radical reactions, in particular hydrogen transfer reactions, de-oxygenation and cyclization reactions. The kinetic and mechanistic aspects of the transformations are currently being explored in our laboratories and the results of the investigations will be published in due course.

To determine if aqueous β-CD as a reaction medium is suitable for a broader range of reactions, Perchyonok investigated several other classes of free radical reactions were subjected to reaction in this medium. These reactions included radical addition to C=N bond and tandem addition/halogen transfer reactions. (Scheme 23.3, Table 23.4; Scheme 23.4, Table 23.5)

Radical addition to C=N bonds represents yet again an underutilized but extremely powerful technique for the formation of C–C bonds, while avoiding the inherent limitations of its ionic counterparts, such as low reactivity of the C=N bond towards enolates and organometallic species, as well as the stringent and basic reaction conditions that are detrimental to the nature of the C=N bond.[7] Therefore, the development of the radical addition reaction to C=N under strictly non-basic conditions makes this alternative reaction pathway more useful. The examples presented in Table 23.4 are the first examples of this transformation in β-cyclodextrin-based molecular reactors and will undoubtedly lead to the development of more elaborate and stereospecific transformations of this class.

Further exploration of the utility of β-cyclodextrin molecular reactors was *via* their application to tandem addition/bromine atom transfer addition reactions.[7,8] In this case, ethyl bromoacetate and 1-bromoadamantane underwent radical addition smoothly to yield the corresponding adducts in good yields (Table 23.5).

Fundamental free radical reactions including hydrogen transfer, carbon–carbon bond formations and deoxygenation reactions have proceeded smoothly in the above mentioned media in good to excellent yields. The advantage of this hypophosphite/β-cyclodextrin molecular reactor medium combination lies in its affordability, low toxicity, avoidance of using highly toxic "tin-based hydrogen donors", reusability and the broad range of synthetic and mechanistic applications.

23.4 ON THE USE OF β-CYCLODEXTRINS AS MOLECULAR REACTORS FOR THE RADICAL CYCLIZATIONS UNDER TIN-FREE CONDITIONS: CHAIN AND NON-CHAIN REACTIONS

Radical cyclizations of alkenes has become a valuable method for the synthesis of cyclic compounds during the past 50 years and can proceed *via* two plausible pathways: (a) reductive radical generation/cyclization/reductive termination or (b) oxidative radical generation/cyclization/oxidative termination.[9] These types of transformations have been widely used in synthesis as in general highly functionalized products are prepared from the simple precursors. Both of the methods have great advantages such as available starting materials and high yielding. The main disadvantage of these types of transformations is that the potential cyclized product may also be susceptible to further deprotonations and unwanted side reactions. This work utilizes the unique shape of the β-cyclodextrin molecules, which resemble a truncated cone, with their exterior surface being hydrophilic due to the presence of hydroxyl groups, while the annulus of the cone is hydrophobic.[10,11]

As a result, in aqueous solution, β-cyclodextrins form inclusion complexes with hydrophobic guests (Figure 23.2), which aids in the increase of the stability and tolerance of the hydrophobic substrates in water as a reaction medium. We envisage that β-cyclodextrins will be effective molecular reactors for these important types of transformations.

Perchyonok extended an application of naturally occurring and readily available β-cyclodextrin as a molecular reactor for a diverse range of fundamental free radical carbon–carbon bond formations in aqueous media. The unique advantages to combining "tin-free" fundamental free radical chemistry and β-cyclodextrin lie in the use of mild reaction conditions, the ability to recycle both reagents and reaction media, and increased reaction rates. The hydrogen donors chosen for the investigation are quaternary ammonium hypophosphites that are prepared in quantitative yields and do not require any additional purification.[12] Alternatively the novel aspects of the use of $Cu(OAc)_2$ have been uncovered in the synthesis of several useful unsaturated γ-lactones as an efficient hydrogen donor system in water in the β-cyclodextrin-based molecular reactors. Reactions proceeded smoothly and in excellent yields, with β-cyclodextrin playing a dual role as efficient molecular reactor and efficient hydrophobic cavity for the control of the reaction outcome.

23.5 RADICAL CYCLIZATIONS IN β-CYCLODEXTRINS IN AQUEOUS MEDIA UNDER PHOTOLYTIC CONDITIONS

The development of synthetic methods for the preparation of bridged systems has been stimulated by the discovery of polycyclic natural products that

Table 23.3 Various free radical reactions (hydrogen atom transfer, radical deoxygenation reactions and cyclization reactions) at ambient temperature in the presence of $(PhBu_3N)H_2PO_2$ and the β-cyclodextrin/H_2O molecular reactor.

Entry	Radical precursor	Product	Yield (%)
1			63
2			66
3			71
4		3 : 1	72
5			89
6			78
7			78
8			89

Table 23.3 Continued.

Entry	Radical precursor	Product	Yield (%)
9			72
10			83
11			76

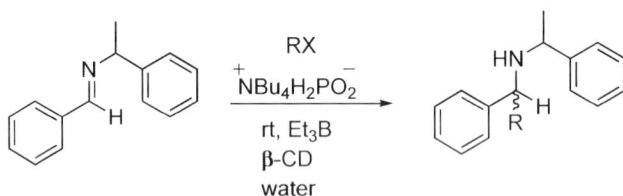

Scheme 23.3 Radical addition of RX to the C=N bond of (*E*)-*N*-benzylidene-1-phenylethanamine under various free radical conditions in water.

Table 23.4 Radical addition of RX to the C=N bond of (*E*)-*N*-benzylidene-1-phenylethanamine under various free radical conditions in the presence of the $(Bu_4N)^+H_2PO_2^-$ in β-cyclodextrin/H_2O as a molecular reactor at ambient temperature.

Entry	R–X	Yield (%)
1	CH_3Br	65
2	$CH_3CH_2CH_2Br$	57
3	$EtOC(O)CH_2Br$	54
4	$PhCH_2Br$	61

$$\text{R'-Br} + \overset{\displaystyle\equiv}{}^{R} \xrightarrow[\substack{\text{β-CD}\\\text{water}}]{\substack{\overset{+}{\text{NBu}_4\text{H}_2\overset{-}{\text{P}}\text{O}_2}\\\text{rt, Et}_3\text{B/air}}} \text{R'}\overset{R}{\underset{Br}{\diagup}}$$

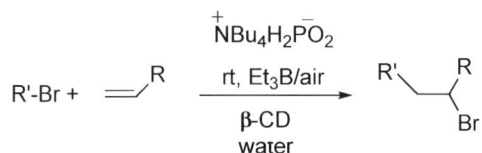

Scheme 23.4 Tandem addition/bromine atom transfer reaction under various free radical conditions in water.

Table 23.5 Tandem addition/bromine atom transfer reaction under various free radical conditions in β-cyclodextrin/H₂O as a molecular reactor at ambient temperature.

Entry	R'–Br	R	Yield (%)
1	EtOC(O)CH₂Br	CH₂OH	78
2	EtOC(O)CH₂Br	CH₂CH₂OH	67
3		CH₂OH	74
4		CH₂CH₂OH	71

Reactions of Interest: C-C bond formation through reductive radical cyclization/termination and oxidative radical cyclization/termination

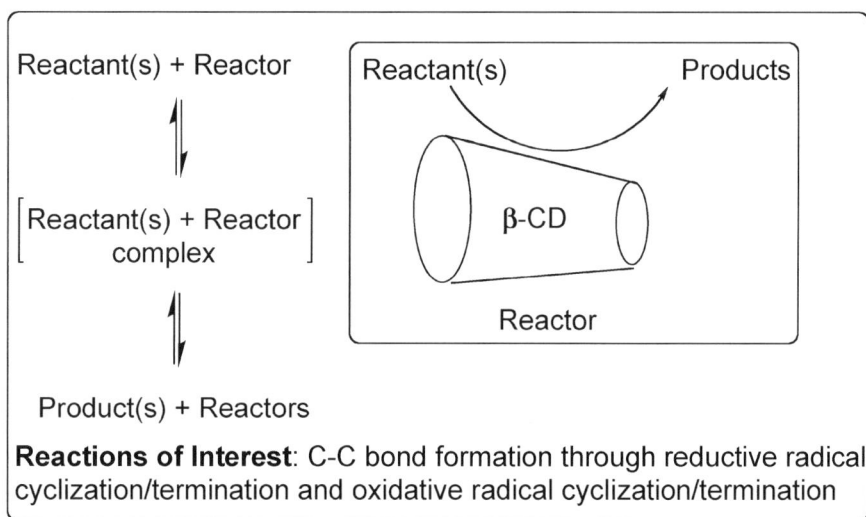

Figure 23.2 β-CD-based molecular reactor prototype and reactions of interest.

incorporate bridged systems as part structures.[13] The bicyclo[3,2,1]octane system has received a relatively large amount of attention due to its frequent presence in various sesqui- and diterpenoids.[14] In the last two decade there has been an upsurge of interest in the application of free-radical cyclization for the synthesis of a fused carbo- and heterocyclic system.[15] Even though the first report of the formation of a bridged system by a radical cyclization reaction had already appeared, until recently relatively little attention was given to the synthesis of a bridged system by transannulation radical cyclization.[16]

Perchyonok's approach towards a simple and "tin-free" methodology for the construction of chiral bicyclo[3.2.1]octanes takes its inspiration from the work of Srikrishna *et al.*[17] through the 2-step synthesis where carvone is converted into a radical precursor, by manipulation of the electron-rich olefin, and regiospecific intramolecular addition of the radical to the enone moiety. It utilizes tris(trimethyl)silyl silane (TMS$_3$SiH) and hypophosphites (as free radical hydrogen donors under a low power black light, peak wavelength: 352 nm) which allow homolytic bond cleavage/formation and therefore produce an efficient and simple route towards the synthesis of natural product cores (Scheme 23.5). This sequence has several advantages: wide commercial availability of both enantiomers of carvone, formation of the 2 new stereocentres in a regio- and stereospecific way, and the flexibility to provide more functionalize bicyclo[3.2.1] octanes. Perchyonok reported a general, efficient and preparative route to chiral bicyclo[3,2,1]octan-3-ones and described its extension to the bridgehead-substituted systems present in variety of natural products using "tin-free methodology" under photolytic conditions. The role of β-cyclodextrins as free radical molecular reactors is evaluated in the transformation (Table 23.6).

The radical precursors required for the transformations have been prepared using the methodology reported by Srikrishna *et al.* by a bromoetherification reaction involving carvone.[17] Thus, N-bromosuccinimide (NBS) bromination of the electron-rich olefin in (S)-carvone in an alcohol–methylene chloride medium yielded an unseparable mixture of 4S, 8R and 4S,8S diastereomeric alcoxy bromides in regiospecific manner. The mixture was then photolyzed in the pyrex vessel with 0.02 M acetonitrile and 1.1 equiv. of TMS$_3$SiH or H$_3$PO$_2$/ *t*Bu$_4$NCl in the presence of catalytic amount of Et$_3$B/air to produce a separable mixture of *endo* and *exo* bicyclic products.

The advantage of using black light photoinitiaton for the transformation is that a high-pressure mercury vapour lamp (300 W) radiates short wavelength

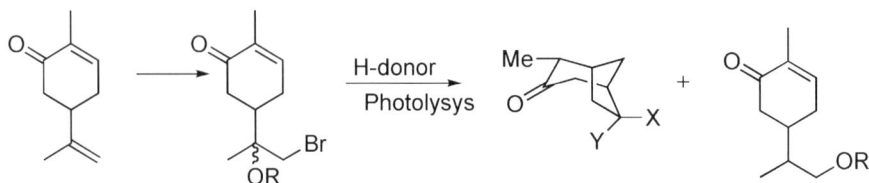

Scheme 23.5   Proposed approach and methodology.

Table 23.6 Bicyclo[3.2.1]octan-3-ones by radical cyclization reaction under photolytic conditions in the β-cyclodextrin.

R	Hydrogen donor	Products (major)	% yield isolated
Me	TMS$_3$SiH	1	65
	H$_3$PO$_2$/tBu$_4$N$^+$Br$^-$	1	62
Et	TMS$_3$SiH	1	72
	H$_3$PO$_2$/tBu$_4$N$^+$Br$^-$	1	64
Hex	TMS$_3$SiH	1	66
	H$_3$PO$_2$/tBu$_4$N$^+$Br$^-$	1	65
Octyl	TMS$_3$SiH	1	78
	H$_3$PO$_2$/tBu$_4$N$^+$Br$^-$	1	63
Isopropyl	TMS$_3$SiH	1	67
	H$_3$PO$_2$/tBu$_4$N$^+$Cl$^-$	1	62

light that can cause power loss of the light and an undesirable evolution of heat.[18] In this chapter we extend the application of the black light in the free radical carbon–carbon bond formation under "tin-free conditions". The added advantage of the use of β-cyclodextrin is the efficient recyclable reaction media.

23.6 Mn(OAc)$_3$ RADICAL CYCLIZATIONS IN β-CYCLODEXTRIN

The oxidative addition of acetic acid to alkenes reported by Heiba and Dessau provides the basis for a general approach to oxidative free radical cyclization.[19] The oxidative additions of that type have been extensively explored over the past 25 years, with recent application covering broad range of applications in organic solvents and ionic liquids.[20] In this chapter the methodology is extended to the β-cyclodextrins as molecular reactors suitable for the use in the oxidative radical cyclization of malonic acid with cinnamic acid, pentane-2,4-dione with prop-1-en-2-ylbenzene and/or (E)-N-benzylidene-1-phenylethanamine, diethyl malonate and cyclohexene in acetic acid to afford substituted γ-lactones and 2,3-dihydrooxazolederivatives in good to excellent yield in the presence of β-cyclodextrin (Scheme 23.6). The experiments were repeated in the absence of a β-cyclodextrin molecular reactor and the yields of the corresponding transformations were significantly lower, suggesting a critical role of the β-cyclodextrin in the outcome of the transformation. In-depth transformation is currently on the way in order to establish the exact of role of β-cyclodextrin as well as optimize and expand the methodology.

The results summarized in Scheme 23.6 demonstrate the successful application of Mn(III)-mediated radical cyclizations, performed for the first time in the β-cyclodextrins. The advantage of the use of the molecular reactors lies in the short reaction time (reactions were completed in 1 hour) and the recyclability of the β-cyclodextrin without losing or altering its activity.

Ph CO₂Me → [reaction]

cis:trans 100:1
68%

100%

85%

74%

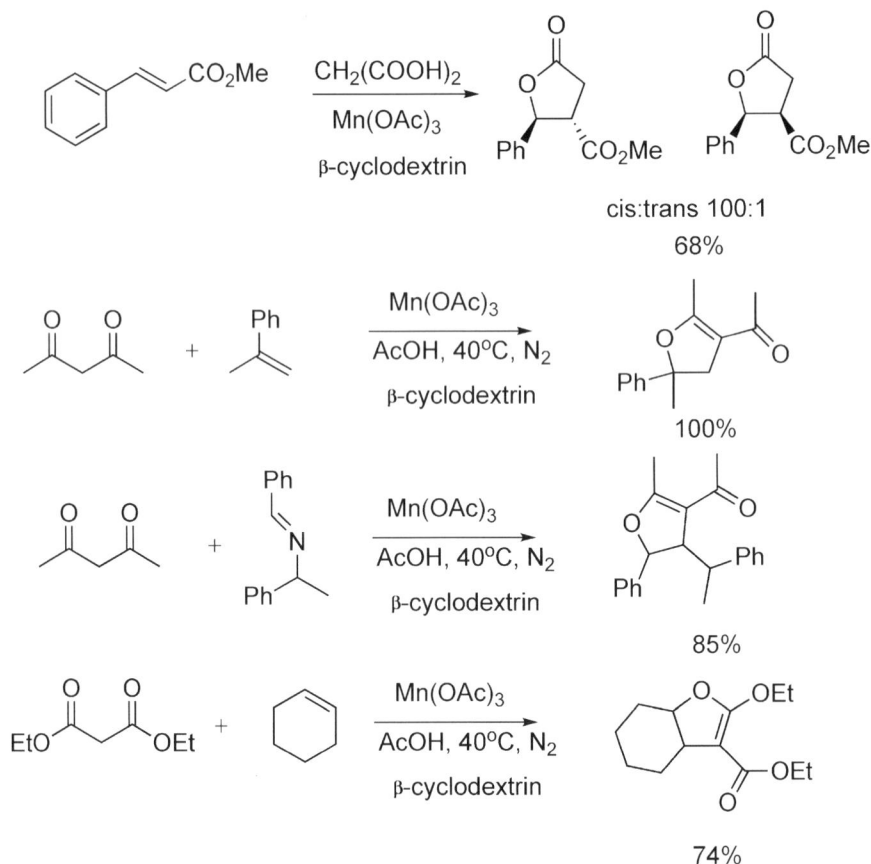

Scheme 23.6 Mn(III)-mediated radical cyclizations inside the β-cyclodextrin-based molecular reactor.

23.7 Cu(OAc)₂ RADICAL CYCLIZATIONS IN β-CYCLODEXTRINS

Several methods are reported in the literature concerning the oxidative functionalization of the carbon–carbon double bond.[21] In contrast, the oxidation of the carbon–carbon triple bond has received much lesser attention and recent reports are focused on the 2,3-dichloro-5,6-dicyano-1,4-benzoquinone (DDQ)-promoted oxidation of alkyl-acetylenes, which occurs at the propargyl carbon leading to Z-enynes in a stereoselective mode.[22] Recently Montevecchi and Navacchia have reported the Mn(III) oxidative addition to the alkyne triple bond, methodology which aims to furnish γ-lactones, which are encountered frequently in a large number of natural products, including flavour components, insect sex pheromones and antibiotics.[23]

For this work, Perchyonok used commercially available phenylacetylene (**1**) and octyne (**2**). Reactions were carried out by heating a 0.2 M solution of the appropriate alkyne in a 1:1 mixture of acetic acid–acetic anhydride in the presence of 0.3 g of β-cyclodextrin. With anhydrous Cu(OAc)$_2$ reactions were completed within 20 min. GC-MS analysis showed the disappearance of the starting alkyne and formation of 2 new peaks corresponding to the 5-acetoxy-2(5H) furanone and 2(3H)-furanone.

Phenylacetylene (**1**) after workup gave the corresponding crude 5-acetoxy-2(5H)-furanone (Scheme 23.7). Subsequent purification on silica gel column gave the products in good yield (78%). The advantage of the use of β-cyclodextrin is the major one as its controls the product formation: in the absence of β-cyclodextrin the product of the reaction is 2(3H)-furanone (minor) and 5-acetoxy-2(5H) furanone (major); however, in the presence of the β-cyclodextrin the reaction products are reversed as 2(3H)-furanone and 5-acetoxy-2(5H) furanone, suggesting that the hydrophobic nature of the β-cyclodextrin cavity is detrimental to the success of the transformation.

Octyne (**2**) after workup gave the corresponding crude 5-acetoxy-2(5H) furanone (Scheme 23.7). The results are very encouraging as in the case of the Mn(OAc)$_3$-mediated oxidative cyclization reaction of octyne: only unidentified, polymeric products were reported by Montevecchio *et al.*[22] Subsequent purification on silica gel column gave the products in good yield (75%). The advantage of the use of β-cyclodextrin is the major one as its controls the product formation: in the absence of β-cyclodextrin the product of the reaction is 2(3H)-furanone (minor) and 5-acetoxy-2(5H) furanone (major); however, in the presence of the β-cyclodextrin the reaction products are reversed as 2(3H)-furanone and 5-acetoxy-2(5H) furanone, suggesting that the hydrophobic nature of the β-cyclodextrin cavity is detrimental to the success of the transformation.

The mechanism is similar to that proposed by Montevecchi and Navacchia for the Mn(OAc)$_3$-mediated radical cyclization and is as follows (Scheme 23.8): the regioselective addition of a carboxymethyl radical to the C–C triple bond

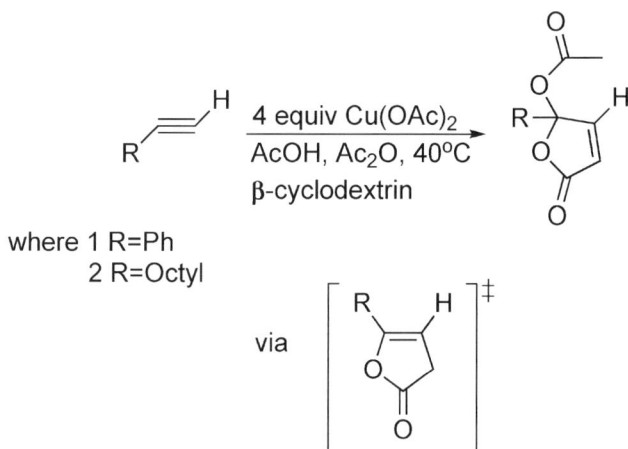

where 1 R=Ph
2 R=Octyl

via

Scheme 23.7 Cu(OAc)$_2$-mediated radical cyclization reaction both in the presence of β-cyclodextrin as a molecular reactor and without.

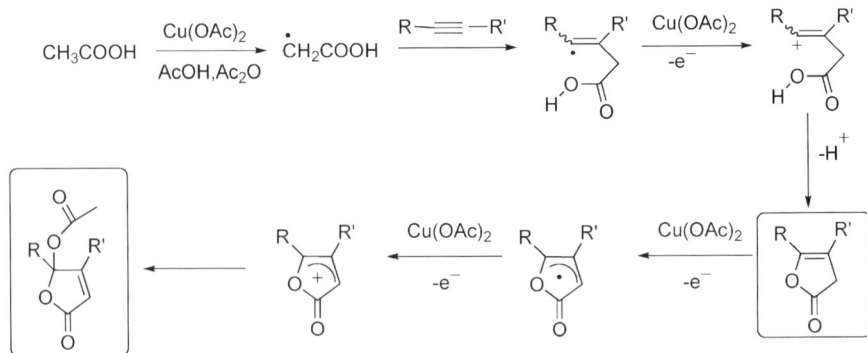

Scheme 23.8 Proposed mechanism for $Cu(OAc)_2$-promoted functionalization.

gives a vinyl radical , which affords 2(3H)-furanone (minor product in the absence of β-cyclodextrin and a major product in the presence of β-cyclodextrin) through oxidative cyclization onto the oxygen atom. 2(3H)-Furanone undergoes a further two-electron oxidation to an allyl cation, with a formation of furanone through regioselective trapping by the acetoxy counterion at the benzylic position. The role of the β-cyclodextrin is highlighted once again as the formation of the final furanone was strongly influenced by the presence and/or absence of the hydrophobic cavity inside the β-cyclodextrin. The role of molecular reactor was also highlighted as the β-cyclodextrin could be recycled after the transformation by simple washing with 10% NH_3 solution and drying *in vacuo*.

23.8 ON THE SCOPE OF β-CYCLODEXTRIN–IONIC LIQUID-BASED MOLECULAR REACTORS FOR FREE RADICAL CHEMISTRY IN BIO-COMPATIBLE AND ALTERNATIVE MEDIA

The biological damage caused by radical species is a highly interesting topic for the scientific community today since the relationship between cellular oxidative stress produced by radical activity *in vivo* and human-health-connected phenomena (aging, cancer biogenesis, cellular apoptosis) has actually been established.[24–27] Free radicals are chemical entities generated by homolytic cleavage of a bond existing between two atoms (either identical or different in nature). As the free radicals possess an incomplete electronic framework, they behave as reactive chemical intermediates, quickly interact with the chemical environment and lead to propagation of a radical chain cascade, which can produce irreversible biochemical damage.

In nature, however, biological machines perform tasks that enable life to proceed and repair the damage done. In the pursuit of nanoscale machines, chemists are inspired to mimic nature: to produce molecular vehicles that can carry drugs, efficiently cross various physiological barriers to reach disease sites and cure and/or prevent disease in a less toxic and sustained manner.[28]

Modified and unmodified cyclodextrins are suitable carriers for radical affording substances as potential pro-drugs capable to irreversibly damage the disease state at a site-specific manner in the biocompatible media.[29]

Cyclodextrins (CDs) and their chemically modified derivatives have been utilized as molecular reactors due to their wide availability and natural abundance.[7a,b] Naturally occurring CDs are homochiral cyclic oligosaccharides, the most common of which are composed of 6, 7 and 8 α-1,4-linked D-glucopyranose units.[8] The unique properties and further applications of cyclodextrins are related to the unique shape of the molecules, which resemble a truncated cone, with their exterior surface being hydrophilic due to the presence of hydroxyl groups, while the annulus of the cone is hydrophobic.[1b] As a result, in aqueous solution, cyclodextrins form inclusion complexes with hydrophobic guests. However, CDs are slightly soluble in water, which limits their sole application as a molecular carrier in aqueous media.

Room temperature ionic liquids (RTILs), which are composed entirely of ions and are fluids at room temperature, have received much interest due to their unique properties such as negligible vapour pressure, chemical stability and non-flammability, relatively high ionic conductivity and wide potential window. Among unconventional solvents, over the last decade ionic liquids have achieved the status of an efficient alternative to conventional volatile organic solvents.[30] They have already proved to be no longer an exotic medium, but instead a valuable, benign vehicle that can improve the outcome of many organic syntheses, both by enhancing separation processes (and safety) and, in many cases, by improving reactions rate and selectivity. Surprisingly, with the exception of radical polymerization, which has been quite extensively studied, the number of reported examples of radical synthetic procedures carried out in ionic liquids is extremely low:[31] the investigated reactions include the formation of carbon–carbon bonds through manganese(III) acetate[32] and ceric ammonium nitrate (CAN)-mediated oxidations of 1,3-dicarbonyl precursors,[33] triethyl borane-induced addition (or cyclization)/reduction sequences or atom-transfer cascade reactions,[34] and three-component photo-induced atom-transfer carbonylations and the addition of thiols to double (or triple) carbon–carbon bonds.[35]

Perchyonok's continuous interest in exploring the development of functional cyclodextrin-based molecular reactors in conventional free radical chemistry has prompted the exploration of the cyclodextrin-based ionic liquid (MIM-β-CDOTs) as a suitable and comparative model system for investigation of a diverse range of fundamental free radical transformations in bio-compatible and alternative reaction media[36] (Figure 23.3).

The unique advantage to combining "tin-free" fundamental free radical chemistry and β-cyclodextrin-based ionic liquids lies in the use of mild reaction conditions as well as the ability to recycle both reagents and reaction media and aims at enhancing the chiral discrimination ability of the CDs by changing their symmetric structure. The quaternary ammonium hypophosphites are prepared by reaction of the corresponding tetra-substituted ammonium salt or tertiary amine with an equimolar amount of phosphinic acid at room

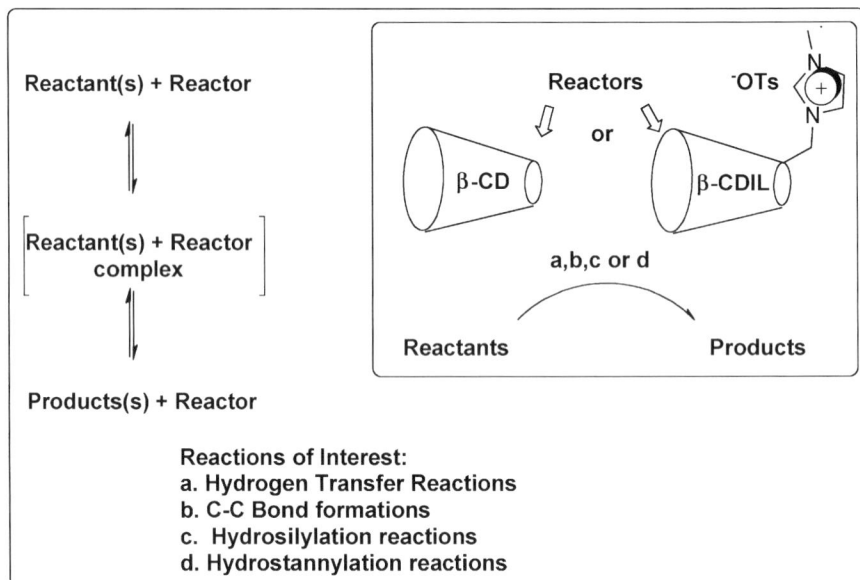

Figure 23.3   Transformations of interest.

temperature in the solvent/medium of choice.[37] Chiral and achiral novel "tin-free" hydrogen donors are prepared in quantitative yields and do not require any additional purification; they have demonstrated their efficiency in a broad range of conventional free radical transformations reported by others and us.[37,38]

23.9   β-CYCLODEXTRIN–IONIC LIQUIDS AND CONVENTIONAL FREE RADICAL REACTIONS: HYDROGEN ATOM TRANSFER REACTIONS

Perchyonok chose to investigate the radical hydrogen transfer reaction of 1-bromoadamantane (1-BrAd) in the presence of tetrabutylammonium hypophosphite as a hydrogen donor at room temperature (0.5 equiv. of Et_3B/air) in β-cyclodextrin–IL (0.2 g) in water (2.5 mL) (Scheme 23.9). Reductions were carried out with 0.1 mM substrate mixed with a β-cyclodextrin–IL/water mixture or β-cyclodextrin–IL (neat), and initiated at ambient temperature with Et_3B/air as a suitable radical initiator. The progress of reaction was monitored by thin layer chromatography. The reduced product was obtained in good to excellent yields and was characterized by ^1H NMR spectroscopy and GC/MS analysis (Scheme 23.9).

Table 23.7 lists data obtained for the hydrogen atom transfer of 1-bromoadamantane as a model compound in the presence of MIM-β-CDOTs, β-CD and [bmim][BF_4] in water (entries 1,2 and 3). As expected, reaction of 1-bromoadamantane in the absence of $(NBu_4)^+H_2PO_2$ yielded

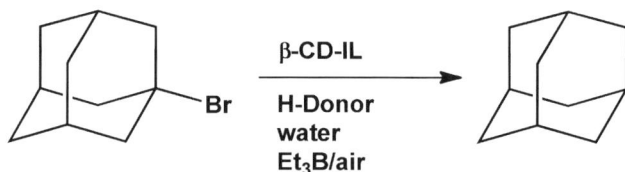

Scheme 23.9 Reduction of 1-BrAd to Ad under various free radical conditions in water.

complete recovery of the unreacted starting material (1-BrAd). Reactions performed either in the absence of a free radical initiator (Et$_3$B/air) or in the presence of phenol as a free radical inhibitor failed to form the reduced product, confirming the free radical nature of the transformation in question. We have also attempted to recycle and reuse β-cyclodextrin and MIM-β-CDOTs and [bmim][BF$_4$] after the reaction and observed no significant decreases in product yields after 4 runs using (NBu$_4$)$^+$H$_2$PO$_2^-$ as the hydrogen donor. In parallel, radical reduction of 1-BrAd was performed using Bu$_3$SnH as a hydrogen donor under previously discussed experimental conditions. As expected, the reaction proceeded smoothly providing a crude yield of 78% for the reduced product. All attempts to recycle β-cyclodextrin, MIM-β-CDOTs and [bmim][BF$_4$] were unsuccessful in this instance due to tin contamination. The role of β-cyclodextrin and MIM-β-CDOTs as an effective catalyst was demonstrated by outcomes of the reductions conducted in the presence and absence of the β-cyclodextrin. The reaction time increased slightly in the presence of β-cyclodextrin and/or MIM-β-CDOTs suggesting a potential cage effect of cyclodextrin cavity and therefore extra stabilization offered to a newly formed radical centre. In addition, the reaction yield was not greatly affected, highlighting the role of β-cyclodextrin as an effective molecular reactor and a catalyst. Interestingly to note that reactions in commercially available ionic liquid [bmim][BF$_4$] have proceeded smoothly and shorter reaction times, suggesting that polar effects involved in the stabilization of the charge separation in the transition state are responsible for a rate increase in the

Table 23.7 Reduction of 1-bromoadamantane under various free radical conditions in the presence of tetrabutylammonium hypophosphite ((NBu$_4$)$^+$H$_2$PO$_2^-$) in water.

Entry	Additive	Time	Product	Yield (%)
1	Et$_3$B/air β-CD	2.0 h	Ad	83
2	Et$_3$B/air MIM-β-CDOTs	2.0 h	Ad	78
3	Et$_3$B/air [bmim][BF$_4$]^a	45 min	Ad	76
4	Et$_3$B/air no β-cyclodextrin no MIM-β-CDOTs no [bmim][BF$_4$]	1.5 h	Ad	72

^{a}1-butyl-3-methylimidazolium tetrafluoroborate = [bmim][BF$_4$]

desired transformation in ionic liquids and therefore further investigation is required to explore the full potential of the transformation in synthesis and application.[39]

The scope and limitations of the tetrasubstituted hypophosphites as hydrogen donors in β-cyclodextrin and MIM-β-CDOTs were explored with a broad range of typical radical precursors: 2°, 3° halides, functionalized aromatic halides and benzyl bromide. Free radical cyclization of 1-(allyloxy)-2-bromobenzene (entry 4) proceeded smoothly with the combined yield of cyclized and directly reduced product of 72% which are formed in the ratio of 5:1 in favour of the cyclized product at room temperature (Table 23.8, entry 4). Some enantioselectivity has been observed in the presence of β-cyclodextrin and MIM-β-CDOTs molecular reactors, which suggests that the chirality of the environment is influencing the outcome of the free radical transformation. All reported transformations were also performed in the commercially available ionic liquid [bmim][BF$_4$] and also isolated yields are comparable to the ones observed for β-cyclodextrin and MIM-β-CDOTs, the reaction times were significantly shorter for all transformations reported (45 min *versus* 2 h in one case) suggesting that polar effects are important factors in the transformation and are needed to be evaluated in detail. The results obtained have exceeded expectations as not only did they proceed with good to excellent yields but β-cyclodextrins and MIM-β-CDOTs were recycled successfully without noticeable loss of efficiency. The β-cyclodextrins and MIM-β-CDOTs are suitable and recyclable for various free radical reactions, in particular hydrogen transfer reactions, de-oxygenation and cyclization reactions in both chiral and achiral forms. The kinetic and mechanistic aspects of the transformations are currently being explored in our laboratories and the results of the investigations will be published in due course.

23.10 β-CYCLODEXTRIN–IONIC LIQUID (MIM-β-CDOTS) AND CONVENTIONAL FREE RADICAL REACTIONS: RADICAL ADDITIONS, ATOM TRANSFER, HYDROSILYLATION AND HYDROSTANNYLATION REACTIONS IN AQUEOUS MEDIA

To determine if aqueous MIM-β-CDOTs as a reaction medium is suitable for a broader range of reactions, several other classes of free radical reactions were subjected to reaction in this medium. These reactions included radical addition to a C=N bond and tandem addition/halogen transfer reactions (Scheme 23.10, Table 23.9; Scheme 23.11, Table 23.10).

Radical addition to C=N bonds represents yet again an underutilized but extremely powerful technique for the formation of C–C bonds, while avoiding the inherent limitations of its ionic counterparts, such as the low reactivity of the C=N bond towards enolates and organometallic species, as well as the stringent and basic reaction conditions that are detrimental to the nature of the C=N bond.[7] Therefore, the development of the radical addition reaction to C=N under strictly non-basic conditions makes this alternative reaction pathway

Table 23.8 Various free radical reactions (HAT, radical deoxygenation reactions and cyclization reactions) at ambient temperature in the presence of $(Et_4N)^+H_2PO_2^-$ and the β-cyclodextrin/H_2O and MIM-β-CDOTs /H_2O molecular reactor.

Entry	Radical precursor	Additive	Product	Yield (%)
1a		β-cyclodextrin		82
1b		MIM-β-CDOTs		76
1c		[bmim][BF$_4$]^a		72
2a		β-cyclodextrin		78
2b		MIM-β-CDOTs		72
2c		[bmim][BF$_4$]^a		65
3a		β-cyclodextrin		72
3b		MIM-β-CDOTs		74
3c		[bmim][BF$_4$]^a		78
4a		β-cyclodextrin		86
4b		MIM-β-CDOTs		82
4c		[bmim][BF$_4$]^a		80
5a		β-cyclodextrin		81
5b		MIM-β-CDOTs		78
5c		[bmim][BF$_4$]^a		75
6a		β-cyclodextrin		73
6b		MIM-β-CDOTs		71
6c		[bmim][BF$_4$]^a		71
7a		β-cyclodextrin		74
7b		MIM-β-CDOTs		72
7c		[bmim][BF$_4$]^a		72
8a		β-cyclodextrin		75
8b		MIM-β-CDOTs		73
8c		[bmim][BF$_4$]^a		65

Table 23.8 Continued.

Entry	Radical precursor	Additive	Product	Yield (%)
9a		β-cyclodextrin		85
9b		MIM-β-CDOTs		83
9c		[bmim][BF₄]ᵃ		78
10a		β-cyclodextrin		72 (18% ee (S))
10b		MIM-β-CDOTs		70 (32% ee (S))
10c		[bmim][BF₄]ᵃ		60 (0% ee)
11a		β-cyclodextrin		74 (15% ee (S))
11b		MIM-β-CDOTs		76 (34% ee (S))
11c		[bmim][BF₄]ᵃ		65 (0% ee)

ᵃ1-butyl-3-methylimidazolium tetrafluoroborate = [bmim][BF₄]

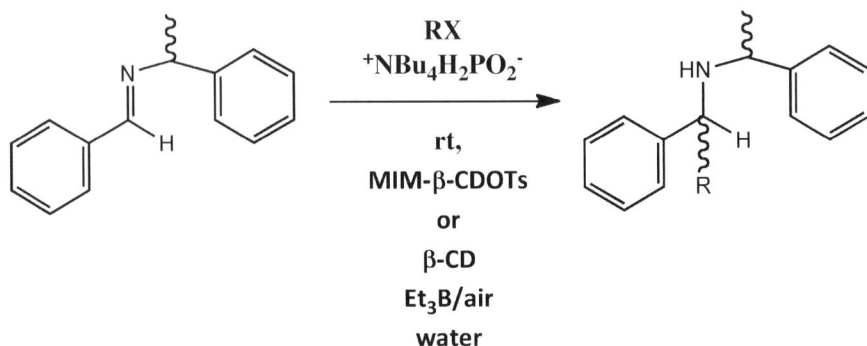

Scheme 23.10 Radical addition of RX to the C=N bond of (*E*)-*N*-benzylidene-1-phenylethanamine under various free radical conditions in water.

Table 23.9　Radical addition of RX to the C=N bond of (*E*)-*N*-benzylidene-1-phenylethanamine under various free radical conditions in the presence of the $(Et_4N)^+H_2PO_2^-$ in β-cyclodextrin/H_2O and MIM-β-CDOTs/H_2O as a molecular reactor at ambient temperature.

Entry	R–X	Additive	Yield (%)
1a	CH₃Br	β-Cyclodextrin	72
1b		MIM-β-CDOTs	80
2a	CH₃CH₂CH₂Br	β-Cyclodextrin	65
2b		MIM-β-CDOTs	73
3a	EtOC(O)CH₂Br	β-Cyclodextrin	60
3b		MIM-β-CDOTs	80
4a	PhCH₂Br	β-Cyclodextrin	57
4b		MIM-β-CDOTs	75

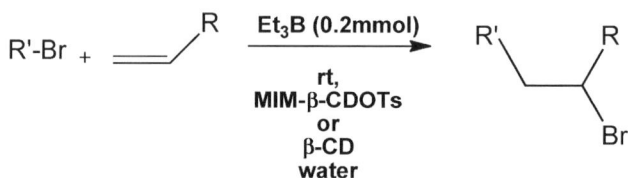

Scheme 23.11　Tandem addition/bromine atom transfer reaction under various free radical conditions in water.

Table 23.10　Tandem addition/bromine atom transfer reaction under various free radical conditions in β-cyclodextrin/H_2O or MIM-β-CDOTs/H_2O as a molecular reactor at ambient temperature.

Entry	R–Br	R'	Additive	Yield (%)
1a	EtOC(O)CH₂Br	CH₂OH	β-Cyclodextrin	78
1b			MIM-β-CDOTs	86
2a	EtOC(O)CH₂Br	CH₂CH₂OH	β-Cyclodextrin	67
2b			MIM-β-CDOTs	82
3a		CH₂OH	β-Cyclodextrin	74
3b			MIM-β-CDOTs	84
4a		CH₂CH₂OH	β-Cyclodextrin	71
4b			MIM-β-CDOTs	85

more useful The examples presented in Table 23.4 are the first examples of this transformation in MIM-β-CDOTs-based molecular reactors and will undoubtedly lead to the development of more elaborate and stereo-specific transformations of this class. The reaction times are shorter than in the presence of β-CD only (1 h *vs.* 2 h) and represent a further avenue for future in-depth investigation.

Further exploration of the utility of β-cyclodextrin and MIM-β-CDOTs ionic liquids as molecular reactors was *via* their application to tandem addition/bromine atom transfer addition reactions (Scheme 23.11).[40] In this case, ethyl bromoacetate and 1-bromoadamantane underwent radical addition smoothly to yield the corresponding adducts in good yields in both systems and therefore expanding the applicability of the easily prepared and versatile functional molecular reactors (Table 23.10).

The next free radical chain reaction which Perchyonok has investigated in some detail is the hydrosilylation reaction of an alkyne such as 1-octyne, in the presence of $(Me_3Si)_3SiH$ in β-cyclodextrin and MIM-β-CDOTs in aqueous media under Et_3B/O_2 initiation (Scheme 23.12).

Transformation yielded the respective alkenes stereoselectively in high yields of corresponding Z alkenes ($Z:E$ ratios >99:1, isolated alkene yields >95%) are formed. It is postulated that unconjugated vinyl radicals are known to be sp^2 hybridized and to invert with a very low barrier, results which are consistent with the reports of Chatgilialoglu *et al.*[42]

The next free radical chain reaction which we had investigated in some detail was the hydrostannylation reaction of an alkyne such as 1-octyne in the presence of Bu_3SnH in β-cyclodextrin/water and MIM-β-CDOTs ionic liquids under ET_3B/O_2 initiation (Scheme 23.13). Transformation yielded the respective alkenes stereoselectively in high yields of corresponding Z alkenes ($Z:E$ ratios 69:31, isolated alkene yields >95%) are formed, unable to recycle either β-cyclodextrin/water or MIM-β-CDOTs molecular reactors due to persistent organotin contamination.

In summary, a broad range of fundamental free radical transformations including hydrogen transfer, carbon–carbon bond formations, deoxygenation reactions as well as hydrosilylation and hydrostannylation reactions proceeded smoothly in good to excellent yield in the newly uncovered cyclodextrin-based molecular reactors in aqueous media as well as a commercially available ionic

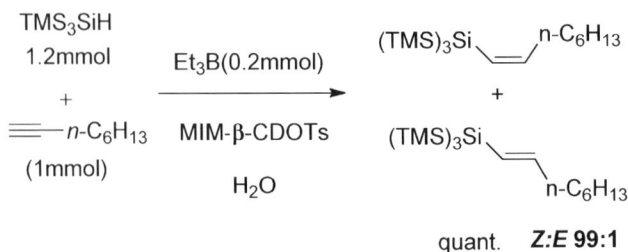

Scheme 23.12   Hydrosilylaton reaction of n-octyne in the presence of MIM-β-CDOTs.

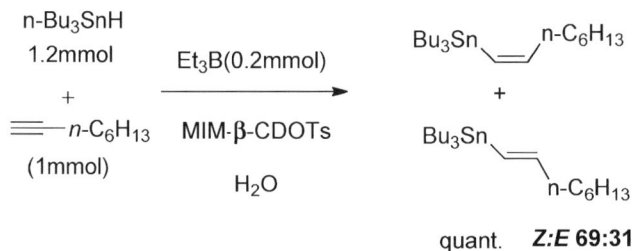

Scheme 23.13 Hydrostannylation reaction of n-octyne in the presence of MIM-β-CDOTs.

liquid for comparison. The observed enantioselectivity in free radical transformations under the reported conditions will undoubtedly lead to further exciting discoveries. The advantage of this β-cyclodextrin and MIM-β-CDOTs molecular reactor or medium combination lies in its affordability, low toxicity, avoidance of using highly toxic "tin-based hydrogen donors", reusability and broad range of synthetic, mechanistic and catalytic applications as well as scope for bio-related applications and development of novel functional materials.

23.11 POTENTIAL PRACTICAL APPLICATION: TOWARDS THE DEVELOPMENT OF NOVEL DRUG DELIVERY PROTOTYPE DEVICES FOR TARGETED-DELIVERY DRUG THERAPY AT THE MOLECULAR LEVEL IN AQUEOUS MEDIA

A novel approach in a target-specific molecular prototype drug delivery system concerns the attempt to employ radical-affording substances (RAS) or radical-quenching substances (RQS) as pro-drugs able to produce irreversible damage on the desired target and therefore to stimulate cellular apoptosis. However, the radical species once generated can react quickly within the chemical environment prior to reaching its proper site of action. Perchyonok reported investigations towards developing two alternative novel simple, flexible and effective drug delivery systems that provide the optimal dosage of respective drugs precisely where and when needed and therefore achieve and sustain a complex delivery profile. Perchyonok *et al.* have demonstrated two effective molecular prototype delivery systems able to harness free radical reactivity within the laboratory. Biological processes can be studied and controlled, leading in turn to the prevention of disease and the development of new treatments for disease states mediated by free radicals. Free radicals are chemical entities generated by homolytic cleavage of bond existing between the atom of identical or different in nature.[43] As the free radicals possess an incomplete electronic framework, they behave as reactive chemical intermediates, quickly interact with the chemical environment and lead to propagation of a radical chain cascade, which can produce irreversible biochemical damage.[44] In nature, however, biological machines perform tasks that enable life to proceed and repair the damage done.[45] In the pursuit of nano-scale machines, chemists are inspired to mimic nature, to produce molecular vehicles that can carry drugs, efficiently cross various physiological barriers to reach disease sites and cure and/or prevent disease in a less toxic and sustained manner.[46] Modified and unmodified cyclodextrins are suitable carriers for radical affording substances as potential pro-drugs capable to irreversibly damage the disease state at a site-specific manner in the biocompatible media.[47] CDs form host–guest complexes with hydrophobic guests having the appropriate size to fit inside the CD annulus and therefore the ability to regulate release and interaction with the media and other substrates and create an ideal system for site-specific pro-drug delivery for a single pharmaceutical (**Path a**) or site-specific delivery of a strong antioxidant/radical quencher to the site of a radical formation in order to prevent a free radical formation and therefore act as a molecular defence mechanism (**Path b**) (Scheme 23.14). Recently, the first β-cyclodextrin–anti-cancer pro-drug adducts model study was performed on a laboratory scale and demonstrated that β-cyclodextrin is able to perform a dual role of the solubilizing agent as well as act as a cage compound suitable to release a newly formed radical at the specific site of the delivery at the molecular level, highlighting the feasibility and significance of the newly emerging area of molecular-based drug delivery, which is site-selective and -specific.[48]

Due to Perchyonok's continuous interest in the area of free radical chemistry and cyclodextrins, in this chapter we would like to report our investigations towards developing two alternative novel simple, flexible and effective drug delivery systems that provide the optimal dosage of respective drugs precisely where and when needed and therefore achieve and sustain a complex delivery profile by bring together two important topics: (a) free radical generation in biocompatible media and (b) supramolecular host–guest recognition through miniaturizing and multi-functionalizing molecular carriers for improved drug delivery in a disease-specific manner with an aim to harnessing free radical reactivity within the laboratory with an aim of development of alternative and effective treatments for disease states mediated by free radicals.[49]

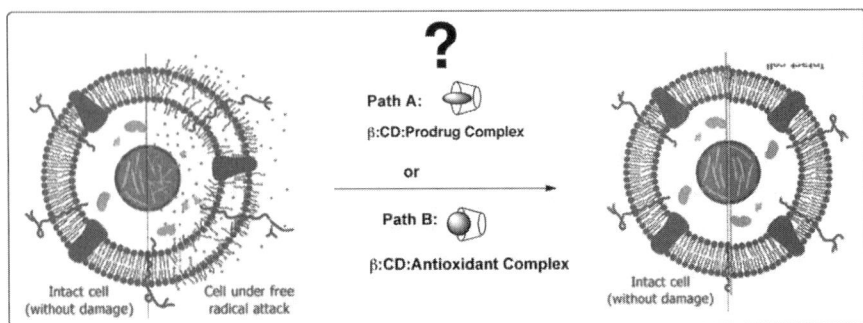

Scheme 23.14   General outline of the investigation.

23.11.1 Path A in Detail: β-Cyclodextrin–Pro-drug as an Efficient Prototype Molecular Carrier in Water Aimed at Transporting Radical-affording Species (RAS) in Aqueous Media

The β-cyclodextrin model has been shown to be a successful experimental setup as a molecular prototype carrier system suitable for in-depth evaluation of the fundamental aspects of the system under laboratory conditions. Fundamental free radical transformations such as atom transfer reaction and radical cascades were a major focus of investigation as they represent 2 major transformations commonly associated with free radical generation *in vivo* and are main culprits in cell death and damage in a disease state[50] (Scheme 23.15).

The choice and use of a broad range of radical precursors, which represent a cross-section of functional groups commonly used for pro-drug derivatization, will create depth and variety in the scope of the investigation. The selected precursors (water soluble and water insoluble) are able to tolerate several functional groups, reagents and solvent mixtures and represent the cross-section of substrates of interest for initial investigations and include (*E*)-N-benzylidene-1-phenylethanamine, 1- and 2-bromoadamantane, 1-(allyloxy)-2-bromobenzene and 1-bromo-2-(prop-2-yn-1-yloxy)benzene. Products of reactions were confirmed by ^{1}H NMR and GC-MS measurements. The reactions outcomes conducted in the presence of β-cyclodextrin being slower than corresponding reactions done in the absence of β-cyclodextrin which suggests that radicals produced in the transformation are "caged" by the host but are able to be released and react efficiently under the predetermined conditions, limiting the interaction of the radical produced with the medium. Experiments are currently under way for further investigation of the phenomenon and the scope of the transformation.

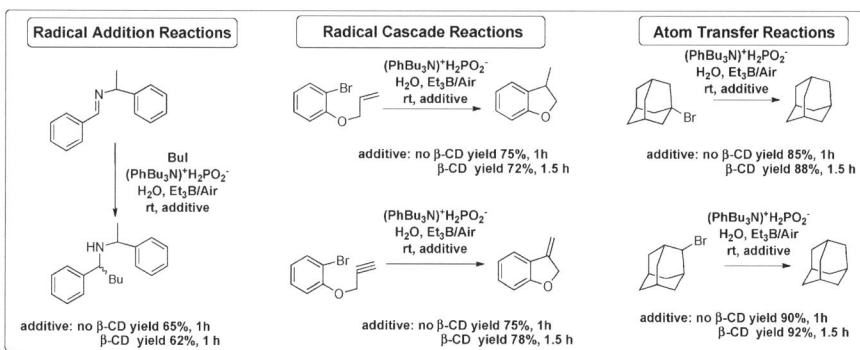

Scheme 23.15   Selected free radical transformations performed in the presence and absence of prototype system (**Path a**).

23.11.2 Path B in Detail: Investigation of Free Radical-quenching Species (RQS) from a β-Cyclodextrin–Phenol "Molecular Antioxidant Prototype" in Water as Antioxidant Delivery to the Radical Reaction Mixture

It is well established that phenolic antioxidants are useful quenchers for the excessive free radical generation and with a potential of being involved in the application in disease chemoprevention; therefore they are of particular interest for a development of the prototype model in an alternative antioxidant delivery system.[51,52–54] The extensive kinetic and mechanistic data exists on the formation and properties of phenol–β-cyclodextrin, therefore making the system ideal for the assessment of the performance of the complex in the fundamental free radical transformations such as hydrogen atom transfer reactions and radical cascade reactions, which represent some of the key free radical transformations *in vivo* responsible for DNA damage, disease state and eventual cell death.[50] The prototype of the multiple host–guest delivery model is represented in Scheme 23.16.

Three reaction types were investigated in aqueous media using derivatives of hypophosphorous acid as efficient hydrogen donors for the reduction and radical cyclization in aqueous media. For comparison purposes reactions were conducted in the aqueous media no inhibitor, phenol (1 equiv.) and phenol–β-CD (1 equiv.) under previously determined conditions for the optimal generation of free radical transformation. Reactions in the presence of phenol or β-cyclodextrin–phenol did not proceed at all which demonstrated the complete inhibition and effectiveness of the proposed methodology.

In summary: a novel approach in a target-specific molecular prototype drug delivery system concerns the attempt to employ radical-affording substances (RAS) or radical-quenching substances (RQS) as pro-drugs able to cause irreversible damage to the desired target and therefore to stimulate cellular apoptosis. However, the radical species once generated can react quickly within the chemical environment prior to reaching its proper site of action. Two alternative

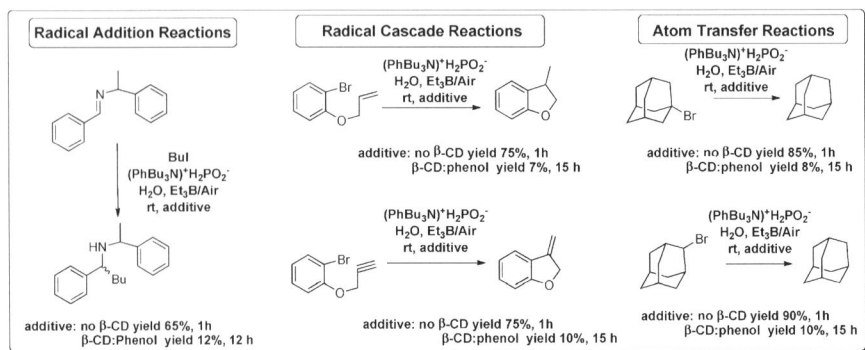

Scheme 23.16 Selected free radical transformations performed in the presence and absence of the prototype system (**Path b**).

novel simple, flexible and effective drug delivery system that provide the optimal dosage of respective drugs precisely where and when needed and therefore achieve and sustain a complex delivery profile are reported and initial results demonstrate that the two effective molecular prototype delivery systems are able to harness free radical reactivity within the laboratory. Biological processes can be studied and controlled, leading in turn to the prevention of disease and the development of new treatments for disease states mediated by free radicals.

23.12 TOWARDS STREAMLINING CONVENTIONAL RADICAL REACTIONS THROUGH THE DEVELOPMENT OF β-CYCLODEXTRIN-BASED BATCH, FLOW-THROUGH AND "TEABAG" PROTOTYPE MOLECULAR REACTORS

β-Cyclodextrin-based batch, flow-through and "teabag" prototype molecular reactors have been assembled and used in variety of conventional synthetically useful free radical transformations (hydrogen transfer reactions, radical cascade reactions and $Mn(OAc)_3$-mediated cascade/oxidation reactions). Reactions proceeded smoothly in all cases in good to excellent yields at ambient temperature in aqueous and organic media. The advantages of newly assembled reactors are ease of setup, recyclability of the β-cyclodextrin media, flexibility and scope of the transformations to be performed and ease of scale-up of the methodology.

Increasing emphasis has been placed on producing synthetic organic compounds faster and more efficiently that can be produced using a conventional methodology.[55] It is often observed that tried and true methods for performing synthesis involving workup on individual or a low number of reactions do not scale up well even to modest level of parallel synthesis. In nature, however, biological machines perform tasks that enable life to proceed.[56] For example, the contraction and expansion of the muscle fibres, caused by the simultaneous sliding of the stacked filaments of the myosin and acting upon chemical stimulation by ATP hydrolysis, enables our controlled movements.[57] In the pursuit of nanoscale machines, chemists are inspired to mimic nature, to produce synthetic structures that control the assembly of reagents to affect the outcomes of chemical transformations at the molecular level.

Free radicals are ubiquitous, reactive chemical entities. Free radical reactions are an important class of synthetic reactions that have been traditionally performed in organic solvents.[58] In recent years, the number of reports of free radical reactions that use water and alternative media such as supercritical CO_2, ionic liquids, fluorous solvents and the solid state have increased.[59] A radical reaction is one of the most useful and flexible methods for organic reactions in alternative media, because most of the organic radicals species are stable and not reactive with the alternative medium itself.

Molecular reactors are miniature reaction vessels that control the assembly of reagents to affect the outcomes of chemical reactions at the molecular level.[6]

In many ways, they are analogous to common laboratory glassware with the unique advantage that after the chemical reaction has taken place, the products are removed and the reaction vessel can be reused.[60,61] Cyclodextrins (CDs) and their chemically modified derivatives have been utilized as molecular reactors due to their wide availability and natural abundance.[62] The advantage of radical over ionic reactions lies in chemoselectivity, tolerance of a wide range of chemical functionalities without the need for protecting groups, the general absence of solvent effect, and the kinetically controlled reaction outcomes.

In line with Perchyonok's continuous interest in utilizing β-cyclodextrin as a molecular reactor in organic synthesis, the authors decided to examine the performance of the β-cyclodextrin molecular reactors under batch conditions, flow-through conditions and "teabag" conditions for the investigation of free radical hydrogen transfer reactions, radical cascade reaction and Mn(OAc)₃ cascade/oxidation conditions (Figure 23.4).[66–68]

23.12.1 β-Cyclodextrins as Molecular Batch Reactors

Perchyonok chose to investigate the 3 types of synthetically useful free radical transformations: the hydrogen transfer reaction, radical cascade reaction and Mn(OAc)₃-mediated cascade oxidation reaction. All reactions were performed in aqueous and organic media in order to expand the applicability of the

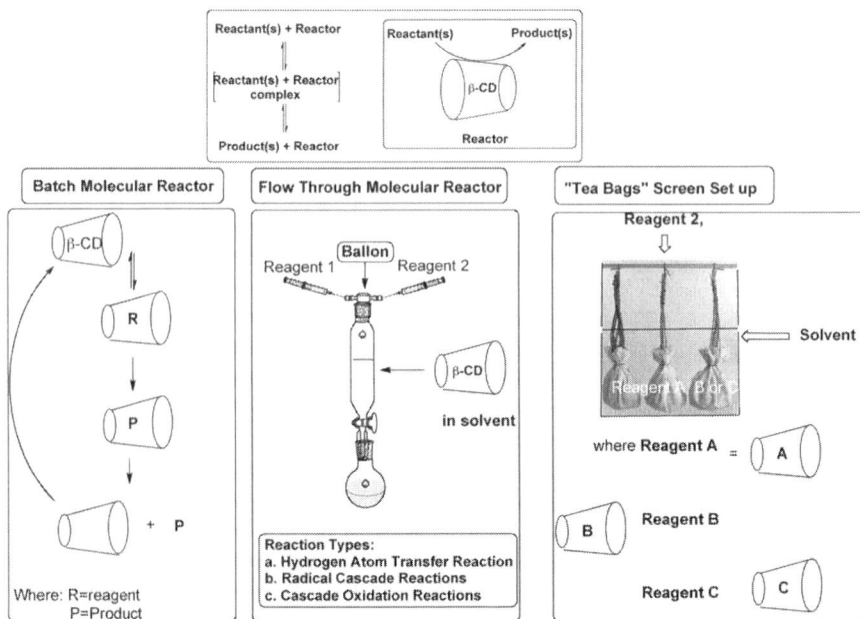

Figure 23.4 Schematic representation of the prototype molecular devices under investigation.

proposed methodology and results of the investigations are summarized in Scheme 23.17.

The β-cyclodextrins have shown to be a powerful experimental setup as a molecular batch reactor suitable for the broad range of synthetically useful transformations with the following advantages being observed in broad scope of fundamental synthetic transformations: (1) efficient substrate retention and performance in aqueous media, (2) a homogeneous distribution of substrate, reactants and products over the molecular reactor, (3) reliable operation and recyclability for 2–3 cycles at present, and (4) stable conditions under substrate-limiting conditions. These outcomes demonstrate for the first time the successful application of natural β-cyclodextrin as a functional molecular batch reactor and undoubtedly will lead to further exciting developments and applications.

23.12.2 β-Cyclodextrin Molecular Flow-through Reactor for Streamlining Organic Synthesis in a Continuous and Reusable Fashion

Flow-through reactors have come to a widespread use and application in target oriented organic synthesis due to the following advantages: there is little or no reaction workup, the supports suffered no physical damage in use, automation is relatively easy, and extension to continuous production even on a larger scale is possible.[67] In this chapter, we report a stable prototype continuous flow-through system for radical cyclization reaction of with TMS_3SiH or Bu_3SnH under $Et_3B/$ air initiation at room temperature (Scheme 23.18).

The reaction proceeded smoothly in the presence of Bu_3SnH and $(TMS)_3SiH$ as hydrogen donors yielding the cyclized product in good yields (65% and 63%

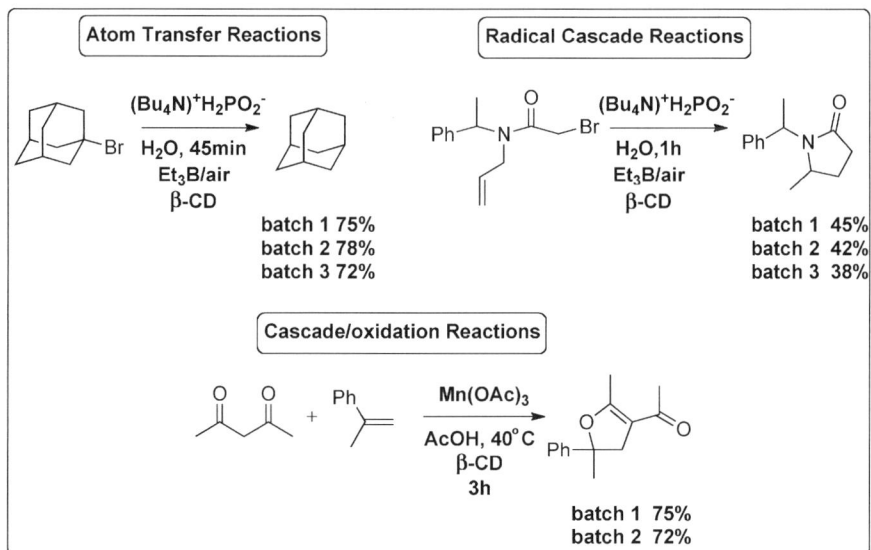

Scheme 23.17 Selected free radical transformations performed in the β-cyclodextrin batched molecular reactors.

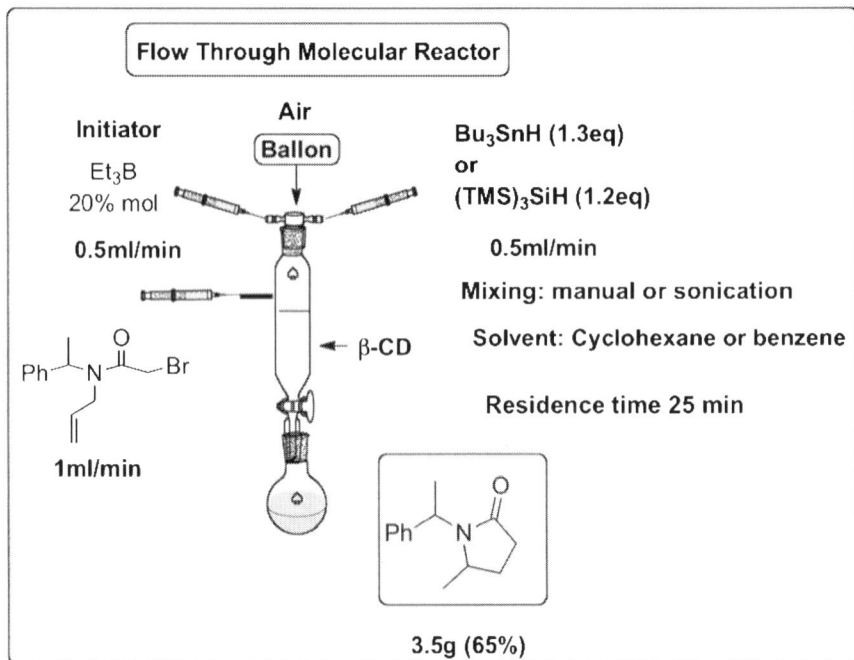

Scheme 23.18 Radical cyclization reaction under flow-through conditions.

respectively); however, in the case of the Bu₃SnH reaction we were unable to recycle the cyclodextrin as the final product was contaminated by the organotin residue. In order to establish the optimum setup for the mixing of the reaction mixture, two alternative methods were tried: (a) thorough manual mixing continued for 25 min and (b) sonication for 25 min. The reactions proceeded smoothly in both cases and yielded 64% and 66% of cyclized product. In the absence of cyclodextrin, the reaction time of the transformation was observed to be much longer and the isolated products required an additional purification step. We have successfully performed radical cyclization reaction under flow-through conditions in the presence of β-cyclodextrin as an "active molecular medium". The added advantage of the flow-through model system lies in the opportunity to complete the desired transformation successfully under the accurate temperature and mixing-controlled conditions in a high-throughput fashion with possible control of the stereospecificity of the desired transformation.

23.12.3 "Teabag" Methodology and Radical Reactions: Screening the Scope and Flexibility

Reactions using polymer-supported reagents can be carried out in the teabags so that when reaction is complete the teabag is simply removed.[67] Following the pioneering work of Ley[64] and Seebach,[65] we attempted to perform a first free radical screening of hydrogen transfer in the presence of TMS₃SiH and

$(Bu_4N)^+H_2PO_2^-$ conditions and have assessed the performance of the system. Host–guest chemistry was involved in the aqueous media. Initiation of the system was performed using Et_3B/air (Scheme 23.19).

The setup: 2 teabags filled with 0.2 g of 1 : 1 complex (cyclodextrin–1-BrAd, cyclodextrin–2-BrAd) were sealed using a nylon thread forming the "teabag" and placed into water. The reaction system was then sealed and the hydrogen donor reagent (3 equiv.) was added. The reaction mixture was flushed with nitrogen and an appropriate initiator introduced into the reaction system (Et_3B/air, 0.7 mol equiv.). The reaction was allowed to be stirred for 1 h. Teabags were removed and the excess of solvent removed *in vacuo*. Analysis of the small remaining oily residue did not yield any identifiable materials. The content of the teabags was separately washed and the products of the reaction identified as adamantine in excellent yield of 89% without need of further purification. Reactions were repeated several times with similar results observed, demonstrating the reliability and reproducibility of the transformation. Work is currently under way to expand the scope and complexity of the methodology, including stereospecific transformations.

In summary: for the first time, the excellent performance of the β-cyclodextrin-based molecular reactors under batch conditions, flow-through conditions and "teabag" methodology in organic and aqueous media has expanded the scope and application of conventional free radical chemistry in modern, potentially high-throughput conditions suitable to a broad range of applications and future developments.

23.13 CONCLUSION AND FUTURE DIRECTION

Modified and unmodified cyclodextrins are suitable carriers for radical-affording substances as potential pro-drugs capable of irreversibly damaging

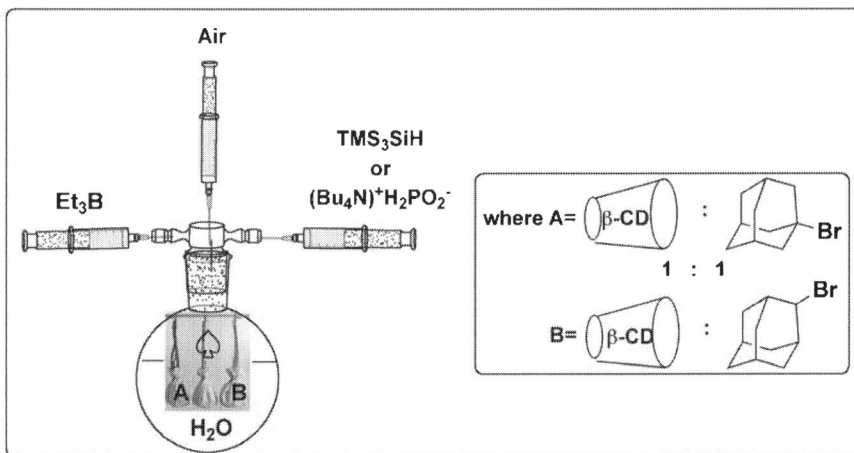

Scheme 23.19 Schematic diagram of the "teabag methodology" protocol.

the disease state in a site-specific manner in the biocompatible media. CDs form host–guest complexes with hydrophobic guests having the appropriate size to fit inside the CD annulus and therefore have the ability to regulate release and interaction with the media and other substrates, creating an ideal system for a variety of the synthetically useful transformations, including the site-specific pro-drug delivery nanosystem for a single pharmaceutical and the site-specific delivery of a strong antioxidant/radical quencher to the site of radical formation (in order to prevent a free radical formation and therefore act as a molecular defence mechanism). It is also able to perform a dual role of the solubilising agent as well as act as a cage compound suitable for releasing a newly formed radical at the specific site of the delivery at the molecular level, highlighting the feasibility and significance of the newly emerging area of molecular-based drug delivery, which is site-selective and -specific.

Since CDs are able to extend the function of pharmaceutical additives, the combination of molecular encapsulation with other carrier materials will become effective and a valuable tool in the improvement and development of novel aspects of molecular reactors, advanced pharmaceutical formulation, versatile means of constructing a new class of novel drug delivery systems like liposomes, microspheres, osmotic pumps, peptide delivery, nanoparticle and site-specific pro-drugs as well as streamlining organic synthesis through molecular batch reactors, advanced flow-through systems and "teabag" molecular reagents. Thus, there is a scope for a vibrant future in the research and development of cyclodextrin-based molecular reactors for a broad range of applications.

REFERENCES

1. (*a*) R. Breslow and S. D. Dong, Biomimetic Reactions Catalyzed by Cyclodextrins and Their Derivatives, *Chem. Rev.*, 1998, **98**, 1997; (*b*) C. J. Easton and S. F. Lincoln, *Modified Cyclodextrins: Scaffolds and Templates for Supramolecular Chemistry*, Imperial College Press, London, 1999; (*c*) L. Barr, P. G. Dumanski, C. J. Easton, J. B. Harper, K. Lee, S. F. Lincoln, A. G. Meyer and J. S. Simpson, Cyclodextrin molecular reactors, *J. Inclusion Phenom. Macrocyclic Chem.*, 2004, **50**, 19; (*d*) C. J. Easton, S. F. Lincoln, L. Barr and H. Onagi, Molecular reactors and machines: applications, potential, and limitations, *Chem.–Eur. J.*, 2004, **10**, 3120.

2. Selected references: (a) G. Wenz, B.-H. Han and A. Muller, Cyclodextrin Rotaxanes and Polyrotaxanes, *Chem. Rev.*, 2006, **106**, 782–817; (b) M. J. Frampton and H. L. Anderson, Insulated Molecular Wires, *Angew. Chem., Int. Ed.*, 2007, **46**, 1028–1064; (c) H. Onagi, B. Carrozzini, G. L. Cascarano, C. J. Easton, A. J. Edwards, S. F. Lincoln and A. D. Rae, Installation of a ratchet tooth and pawl to restrict rotation in a cyclodextrin rotaxane, *Chem.–Eur. J.*, 2003, **9**, 5971–5977; (d) M. M. Cieslinski, P. J. Steel, S. F. Lincoln and C. J. Easton, Centrosymmetric and Non-

centrosymmetric Packing of Aligned Molecular Fibres in the Solid State Self Assemblies of Cyclodextrin-based Rotaxanes, *Supramol. Chem.,* 2006, **18**, 529–536; (e) J. Terao, A. Tang, J. J. Michels, A. Krivokapic and H. L. Anderson, Synthesis of poly(*para*-phenylenevinylene) rotaxanes by aqueous Suzuki coupling, *Chem. Commun.,* 2004, 56–57.

3. Selected reviews: (a) C.-J, Li, Organic chemistry in water, *Chem. Rev.,* 2005, **105**, 3095; (b) R. A. Sheldon, Green solvents for sustainable organic synthesis: state of the art, *Green Chem.,* 2005, **7**, 267; (c) C.-J. Li and L. Chen, Organic chemistry in water, *Chem. Soc. Rev.,* 2006, **35**, 68; (d) V. T. Perchyonok, I. N. Lykakis and K. L. Tuck, *Green Chem.,* 2008, **10**, 153–163; (e) V. T. Perchyonok and I. N. Lykakis, *Mini-Rev. Org. Chem.,* 2008, **5**,19–32.

4. (a) *Radicals in Organic Synthesis*, ed. P. Renaud and M. P. Sibi, Wiley-VCH, Weinheim, 2001, vol. 1; (b) A. Postigo, C. Ferreri, M. L. Navacchia and C. Chatgilialoglu, The radical-based reduction with (TMS)$_3$SiH 'on water', *Synlett*, 2005, **18**, 2854; (c) A. Postigo, S. Kopsov, C. Ferreri and C. Chatgilialoglu, Radical Reactions in Aqueous Medium Using (Me$_3$Si)$_3$SiH, *Org. Lett.,* 2007, **9**(25), 5159–5162; (d) C. Chatgilialoglu, M. Guerra and Q. G. Mulazzani, Model Studies of DNA C5′ Radicals. Selective Generation and Reactivity of 2′-Deoxyadenosin-5′-yl Radical, *J. Am. Chem. Soc.,* 2003, **125**, 3839; (e) L. B. Jimenez, S. Encinas, M. A. Miranda, M. L. Navacchia and C. Chatgilialoglu, Solar one-way photoisomerisation of 5,8-cyclo-2-deoxyadenosine, *Photochem. Photobiol. Sci.,* 2004, **3**, 1042; (f) C. Chatgilialoglu and C. Ferreri, Geometrical trans Lipid Isomers: A New Target for Lipidomics, *Acc. Chem. Res.,* 2005, **38**, 441; (g) H. Sugiyama and Y. Xu, Photochemical Approach to Probing Different DNA Structures, *Angew. Chem., Int. Ed.,* 2006, **45**, 135.

5. Selected references on the use of organic hypophosphites as hydrogen donors in organic solvents: (a) A. F. Brigas and R. A. W. Johnstone, *J. Chem. Soc., Chem. Commun.,* 1991, 1041–1042; (b) S. R. Graham, J. A. Murphy and A. R. Kennedy, Hypophosphite mediated carbon–carbon bond formation: total synthesis of epialboatrin and structural revision of alboatrin, *J. Chem. Soc., Perkin Trans. 1*, 1999, 3071; (c) C. G. Martin, J. A. Murphy and C. R. Smith, Replacing tin in radical chemistry: *N*-ethylpiperidine hypophosphite in cyclisation reactions of aryl radicals, *Tetrahedron Lett.,* 2000, **41**, 1833–1836; (d) A. E. Johnson and V. T Perchyonok, *Curr. Org. Chem.*, 2009, **13**(9), 914–918.

6. D. H. Cho and D. O. Jang, Enantioselective radical addition reactions to the CN bond utilizing chiral quaternary ammonium salts of hypophosphorous acid in aqueous media, *Chem. Commun.,* 2006, **9**, 5045.

7. (a) U. M. Lindstrom, Stereoselective Organic Reactions in Water, *Chem. Rev.,* 2002, **102**, 2751; selected publications for enantioselectivity in free radical transformations in the organic solvents: (b) D. Dakternieks, V. T. Perchyonok and C. H. Schiesser, Single enantiomer free-radical chemistry—Lewis acid-mediated reductions of racemic halides using chiral

non-racemic stannanes, *Tetrahedron: Asymmetry,* 2003, **14**, 3057–3068; (c)
M. P. Sibi and K. Patil, Enantioselective H-atom Transfer Reactions.
Synthesis of β-Amino Acids, *Angew. Chem., Int. Ed.,* 2004, **43**, 1235 and
references cited therein.

8. Selected publications on the application of radical cascade reactions in
total synthesis: (a) A. G. Myers and K. R. Condronski, Synthesis of (±)-
7,8-Epoxy-4-basmen-6-one by a Transannular Cyclization Strategy,
J. Am. Chem. Soc., 1993, **115**, 7926; (b) A. G. Myers and K. R.
Condronski, Synthesis of (±)-7,8-Epoxy-4-basmen-6-one by a Transannular
Cyclization Strategy, *J. Am. Chem. Soc.,* 1995,**117**, 3057.

9. B. B. Snider, Manganese(III)-Based Oxidative Free-Radical Cyclizations,
Chem. Rev., 1996, **96**, 339–363.

10. (*a*) R. Breslow and S. D. Dong, Biomimetic Reactions Catalyzed by.
Cyclodextrins and Their Derivatives, *Chem. Rev.,* 1998, **98**, 1997; (*b*)
C. J. Easton and S. F. Lincoln, *Modified Cyclodextrins: Scaffolds and
Templates for Supramolecular Chemistry,* Imperial College Press, London,
1999; (*c*) L. Barr, P. G. Dumanski, C. J. Easton, J. B. Harper, K. Lee,
S. F. Lincoln, A. G. Meyer and J. S. Simpson, *J. Inclusion Phenom.
Macrocyclic Chem.,* 2004, **50**, 19; (*d*) C. J. Easton, S. F. Lincoln, L. Barr
and H. Onagi, Molecular reactors and machines: applications, potential,
and limitations, *Chem.–Eur. J.,* 2004, **10**, 3120.

11. Selected references: (a) G. Wenz, B.-H. Han and A. Muller, Cyclodextrin
Rotaxanes and Polyrotaxanes, *Chem. Rev.,* 2006, **106**, 782–817; (b)
M. J. Frampton and H. L. Anderson, Insulated molecular wires, *Angew.
Chem., Int. Ed.,* 2007, **46**, 1028–1064; (c) H. Onagi, B. Carrozzini,
G. L. Cascarano, C. J. Easton, A. J. Edwards, S. F. Lincoln and
A. D. Rae, Separated and aligned molecular fibres in solid state self-
assemblies of cyclodextrin [2] rotaxanes, *Chem.–Eur. J.,* 2003, **9**, 5971–
5977; (d) M. M. Cieslinski, P. J. Steel, S. F. Lincoln and C. J. Easton,
Centrosymmetric and non-centrosymmetric packing of alighned molecular
fibers in the solid state self assemblies of cyclodextrin-based rotaxanes,
Supramol. Chem., 2006, **18**, 529–536; (e) J. Terao, A. Tang, J. J. Michels,
A. Krivokapic and H. L. Anderson, Synthesis of poly(para-phenylenevi-
nylene) rotaxanes by aqueous Suzuki coupling, *Chem. Commun.,* 2004, 56–
57.

12. Selected references on the use of organic hypophosphites as hydrogen
donors in organic solvents: (a) A. F. Brigas and R. A. W. Johnstone,
Mechanisms in heterogeneous liquid-phase catalytic-transfer reduction:
the importance of hydrogen-donor concentration, *J. Chem. Soc., Chem.
Commun.,* 1991, 1041–1042; (b) S. R. Graham, J. A. Murphy and
A. R. Kennedy, Hypophosphite mediated carbon–carbon bond formation: total synthesis of epialboatrin and structural revision of alboatrin,
J. Chem. Soc., Perkin Trans. 1, 1999, 3071; (c) C. G. Martin, J. A. Murphy
and C. R. Smith, Replacing Tin in Radical Chemistry: N-ethylpiperidine

Hypophosphite in Cyclisation Reactions of Aryl Radicals, *Tetrahedron Lett.*, 2000, **41**, 1833–1836.

13. A. Srikrishna and P. Hemamalini, Chiral synthons from carvone, *Ind. J. Chem.*, 1990, **29B**, 201

14. J. H. Rigby and A. S. Kotnis, Synthesis of bicyclo[3.2.1]octanes by a tandem diels-alder-carbocation cyclization strategy, *Tetrahedron Lett.*, 1987, **28**, 4943 and references cited therein.

15. Review: M. Ramaiah, Radical reactions in organic synthesis, *Tetrahedron*, 1981, **43**, 3541.

16. For early reports see: G. Stork and N. H. Baine, Vinyl radical cyclization in the synthesis of natural products: seychellene, *Tetrahedron Lett.*, 1985, **26**, 5927; N. N. Marinovic and H. Ramanathan, The synthesis of fused and bridged ring systems by free radical carbocyclization. A general route to masked 1,4-diketones, *Tetrahedron Lett.*, 1983, **24**, 1871; J.-K. Choi, D.-C. Ha, D. J. Hart, C.-S. Lee, S. Ramesh and S. Wu, *J. Org. Chem.*, 1989, **54**, 279; F. MacCorguodale and J. C. Walton., *J. Chem. Soc., Perkin Trans. 1*, 1989, 347; V. Yadav and A. G. Fallis, Cyclopentane synthesis and annulation II: Radical cyclizations of oxathiolanones, *Tetrahedron Lett.*, 1989, **30**, 3283.

17. A. Srikrishna and P. Hemamalini, Chiral synthons from carvone. Part 4. Radical cyclization strategies to bridged systems. Synthesis of bicyclo[3.2.1]octan-3-ones from (S)-carvone, *J. Org. Chem.*, 1990, **55**, 4883–4887.

18. S. Tsunoi, I. Ryu, S. Iamasaki, H. Fukushima, M. Tanaka, M. Komatsu and N. Sonoda, Free Radical Mediated Double Carbonylations of Alk-4-enyl Iodides, *J. Am. Chem. Soc.*, 1996, **118**, 10670

19. E. I. Heiba and R. M. Dessau, Oxidation by metal salts. XI. Formation of dihydrofurans, *J. Org. Chem.*, 1974, **39**, 3456; E. J. Corey and M.-C. Kang, *J. Am. Chem. Soc.*, 1984, **106**, 5384; B. B. Snider, in *Transition Metals Organic Synthesis*, ed. M. Beller and C. Bolm, Wiley, Weinheim, 1998, vol. 2, pp. 439–446; E. I. Heiba, R. M. Dessau and W. J. Koehl Jr, Oxidation by metal salts. IV. A new method for the preparation of γ-lactones by the reaction of manganic acetate with olefins, *J. Am. Chem. Soc.*, 1968, **90**, 5905.

20. P. C. Montevecchi, M. L. Navacchia and P. Spagnolo, 2-Cyano-iso-propyl radical addition to alkynes, *Tetrahedron*, 1997, **53**, 7929; D. P. Curran, Radical addition reactions, in *Comprehensive Organic Synthesis*, Pergamon, Oxford, UK, 1991, vol. 4, ch. 4; G. Bar, F. Bini and A. F. Parsons, CAN-Mediated Oxidative Free Radical Reactions in an Ionic Liquid, *Synth. Commun.*, 2003, **33**(2), 213.

21. (a) M. Hudlicky, *Oxidations in Organic Chemistry*, ACS Monograph 186, Washington, 1990; (b) *Comprehensive Organic Synthesis*, Pergamon, Oxford, UK, 1991, vol. 7, ch. 3.

22. P. C. Montevecchi and M. L. Navacchia, DDQ-Promoted Functionalization of Phenylalkylacetylenes at the Propargylic Carbon, *J. Org. Chem.*, 1998, **63**, 8035.

23. P. C. Montevecchi and M. L. Navacchia, Synthesis of 5-Acetoxy-2(5H)-Furanones through Manganese(III)-Promoted Functionalization of Arylacetylenes, *Tetrahedron.*, 2000, **56**, 9339–9342.

24. P. A. Szweda, M. Camouse, K. C. Lundberg, T. D. Oberley and L. I. Szweda, Aging, lipofuscin formation, and free radical-mediated inhibition of cellular proteolytic systems, *Ageing Res. Rev.*, 2003, **2***(4)*, 383–405.

25. K. B. Beckman and B. N. Ames, The Free Radical Theory of Aging Matures, *Physiol. Rev.*, 1998, **78**, 547.

26. D. Harman, *Free radical theory of aging*, in *Free Radicals and Aging*, ed. I. Emerit and B. Chance, Birkhauser Verlag, Basel, 1992.

27. F. Okada, *Redox Rep.*, 2002, **7**, 357, and references cited therein.

28. J. Li and X. J. Loh., Cyclodextrin-based supramolecular architectures: Syntheses, structures, and applications for drug and gene delivery, *Adv. Drug Delivery Rev.*, 2008, **60**, 1000–1017 and references cited therein.

29. C. J. Easton and S. F. Lincoln, *Scaffolds and Templates for Supramolecular chemistry*, Imperial College Press, London, 1999 and references cited therein.

30. (*a*) R. Breslow and S. D. Dong, Biomimetic Reactions Catalyzed by Cyclodextrins and Their Derivatives, *Chem. Rev.*, 1998, **98**, 1997; (*b*) L. Barr, P. G. Dumanski, C. J. Easton, J. B. Harper, K. Lee, S. F. Lincoln, A. G. Meyer and J. S. Simpson, *J. Inclusion Phenom. Macrocyclic Chem.*, 2004, **50**, 19; (c) C. J. Easton, S. F. Lincoln, L. Barr and H. Onagi, Molecular Reactors and Machines: Applications, Potential, and Limitations, *Chem.–Eur. J.*, 2004, **10**, 3120.

31. Selected references: (a) G. Wenz, B.-H. Han and A. Muller, Cyclodextrin Rotaxanes and Polyrotaxanes, *Chem. Rev.*, 2006, **106**, 782–817; (b) M. J. Frampton and H. L. Anderson, Insulated molecular wires, *Angew. Chem., Int. Ed.*, 2007, **46**, 1028–1064; (c) H. Onagi, B. Carrozzini, G. L. Cascarano, C. J. Easton, A. J. Edwards, S. F. Lincoln and A. D. Rae, Separated and Aligned Molecular Fibres in Solid State Self-Assemblies of Cyclodextrin [2]Rotaxanes, *Chem.–Eur. J.*, 2003, **9**, 5971–5977; (d) M. M. Cieslinski, P. J. Steel, S. F. Lincoln and C. J. Easton, Centrosymmetric and Non-centrosymmetric Packing of Aligned Molecular Fibers in the Solid State Self Assemblies of Cyclodextrin-based Rotaxanes, *Supramol. Chem.*, 2006, **18**, 529–536; (e) J. Terao, A. Tang, J. J. Michels, A. Krivokapic and H. L. Anderson, Synthesis of poly(*para*-phenylenevinylene) rotaxanes by aqueous Suzuki coupling, *Chem. Commun.*, 2004, 56–57.

32. Not only can water be used as an alternative solvent, but it can often improve the outcome of certain radical reactions; see: (a) V. T. Perchyonok, and I. N. Lykakis, Radical Reactions in Aqueous

Media: Origins, Reason and Applications, *Curr. Org. Chem.*, 2009, **13**, 573; (b) V. T. Perchyonok, I. N. Lykakis and K. L. Tuck, Recent advances in C–H bond formation in aqueous media: a mechanistic perspective, *Green Chem.*, 2008, **10**, 153; (c) V. T. Perchyonok, K. L. Tuck, S. J. Langford and M. W. Hearn, Facile and Selective Deallylations of Esters under 'Aqueous' Free-Radical Conditions, *Tetrahedron Lett.*, 2008, **49**, 4777; (d) C.-J. Li and L. Chen, Organic chemistry in water, *Chem. Soc. Rev.*, 2006, **35**, 68 and references therein; (e) H. Yorimitsu, T. Nakamura, H. Shinokubo, K. Oshima, K. Omoto, and H. Fujimoto, Powerful Solvent Effect of Water in Radical Reaction: Triethylborane-Induced Atom-Transfer Radical Cyclization in Water, *J. Am. Chem. Soc.*, 2000, **122**, 11041. For an interesting example of use of supercritical CO_2, see: (f) S. Hadida, M. S. Super, E. J. Beckman and D. P. Curran, Radical Reactions with Alkyl and Fluoroalkyl (Fluorous) Tin Hydride Reagents in Supercritical CO_2, *J. Am. Chem. Soc.*, 1997, **119**, 7406; see also: (g) J. M. Tanko, Free-Radical Chemistry in Supercritical Carbon Dioxide, in *Green Chemistry using Liquid and Supercritical Carbon Dioxide*, ed. J. M. DeSimone and W. Tumas, Oxford University Press, New York, 2003, ch. 4, p. 64.

33. (a) *Ionic Liquids in Synthesis*, ed. P. Wasserscheid and T. Welton, Wiley-VCH, Weinheim, 2008; (b) V. I. Pârvulesku and C. Hardacre, Catalysis in Ionic Liquids, *Chem. Rev.*, 2007, **107**, 2615; (c) J. B. Harper and M. N. Kobrak, Understanding Organic Processes in Ionic Liquids: Achievements So Far and Challenges Remaining, *Mini-Rev. Org. Chem.*, 2006, **3**, 253; (d) C. Chiappe, M. Malvaldi and C. S. Pomelli, Ionic liquids: Solvation ability and polarity, *Pure Appl. Chem.*, 2009, **81**, 767.

34. (a) A. J. Carmichael, D. M. Haddleton, S. A. F. Bon and K. R. Seddon, Copper(I) mediated living radical polymerisation in an ionic liquid, *Chem. Commun.*, 2000, 1237; (b) T. Biedron and P. Kubisa, Atom-Transfer Radical Polymerization of Acrylates in an Ionic Liquid, *Macromol. Rapid Commun.*, 2001, **22**, 1237; (c) S. Harrisson, S. R. Mackenzie and D. M. Haddleton, Unprecedented solvent-induced acceleration of free-radical propagation of methyl methacrylate in ionic liquids, *Chem. Commun.*, 2002, 2850; (d) T. Biedron and P. Kubisa, Atom Transfer Radical Polymerization of Acrylates an Ionic Liquid: Synthesis of Block Copolymers, *J. Polym. Sci., Part A*, 2002, **40**, 2799; (e) H. Zhang, K. Hong, M. Jablonsky and J. W. Mays, Statistical radical copolymerization of styrene and methyl methacrylate in a room temperature ionic liquid, *Chem. Commun.*, 2003, 1356; (f) K. J. Thurecht, P. N. Gooden, S. Goel, C. Tuck, P. Licence and D. J. Irvine, Free-Radical Polymerization in Ionic Liquids: The Case for a Protected Radical, *Macromolecules*, 2008, **41**, 2814 and references therein.

35. (a) G. Bar, A. F. Parsons and C. B. Thomas, Manganese(III) acetate mediated radical reactions in the presence of an ionic liquid, *Chem. Commun.*, 2001, 1350; (b) G. Bar, F. Bini and A. F. Parsons, *Synth.*

Commun., 2003, **33**, 213; (c) T. Fukuyama, T. Inouye and I. Ryu, Atom transfer carbonylation using ionic liquids as reaction media, *J. Organomet. Chem.*, 2007, **692**, 685. For other examples concerning the behavior of radical intermediates in ionic liquids, see: (d) J. Grodkowski and P. Neta, Reaction Kinetics in the Ionic Liquid Methyltributylammonium Bis(Trifluoromethylsulfonyl)imide. Pulse Radiolysis Study of ˙CF$_3$ Radical Reactions, *J. Phys. Chem. A*, 2002, **106**, 5468; (e) D. Zhao, J. Wang and E. Zhou, Oxidative desulfurization of diesel fuel using a Brønsted acid room temperature ionic liquid in the presence of H$_2$O$_2$, *Green Chem.*, 2007, **9**, 1219.

36. H. Yorimitsu and K. Oshima, Triethyl –Borane Mediated Radical Reactions in Ionic liquids, *Bull. Chem. Soc. Jpn.*, 2002, **75**, 853–854.

37. T. Lanza, M. Minozzi, A. Monesi, D. Nanni, P. Spagnolo and C. Chiappe, Radical Additions of Thiols to Alkenes and Alkynes in Ionic Liquids, *Curr. Org. Chem.*, 2010, **17**, 234.

38. A. E. Johnson and V. T. Perchyonok, On the scope of radical reactions utilizing InCl$_3$/coreductant as an efficient hydrogen donor in water, *Curr. Org. Chem.*, 2010, **14**(17), 2007–2011.

39. A. E. Johnson, Z. Shengmiao, J. Chen and V. T. Perchyonok, On the use of β-cyclodextrins as molecular reactors for the radical cyclizations under tin free conditions, *Curr. Org. Synth.*, 2010, in press.

40. (a) A. E. Johnson and V. T. Perchyonok, β-cyclodextrin based molecular reactors for free radical chemistry in aqueous media, *Curr. Org. Chem.*, 2009, **13**(9), 914–918; (b) V. T. Perchyonok, *Radical Reactions in Aqueous Media*, ed. J. Clark and G. Kraus, RSC Publishing, Cambridge, 2009.

41. A thorough discussion on this subject can be found in: B. Giese, *Radicals in Organic Synthesis: Formation of Carbon-Carbon Bonds,* Pergamon Press, Oxford, 1986, ch. 2.

42. A. Postigo, S. Kopsov, S. S. Zlotsky, C. Ferreri and C. Chatgilialoglu, Synthetic Organometallic Radical Transformations in Water, *Organometallics,* 2009, **28**, 3282.

43. N. Zhong, H.-S. Byun and R. Bittman, An improved synthesis of 6-*O*-monotosyl-6-deoxy-ß-cyclodextrin, *Tetrahedron Lett.,* 1998, **39**, 2919–2920.

44. A. Ueno and R. Breslow, Cyclodextrin-based class I aldolase enzyme mimics to catalyze crossed aldol condensations, *Tetrahedron Lett.,* 1982, 23, 3451.

45. P. A. Szweda, M. Camouse, K. C. Lundberg, T. D. Oberley and L. I. Szweda, Ageing, liposuscin formation, and free radical-mediated inhibition of cellular proteolytic system, *Ageing Res. Rev.,* 2003, 383–405.

46. K. B. Beckman and B. N. Ames, The free radical theory of aging matures, *Physiol. Rev.,* 1998, **78**, 547.

47. D. Harman, Free radical theory of aging, in *Free Radicals and Aging*, ed. I. Emerit and B. Chance, Birkhauser Verlag, Basel, 1992.

48. F. Okada, Inflammation and free radicals in tumor development and progression, *Redox Rep.*, 2002, **7**, 357, and references cited therein.
49. J. Li and X. J. Loh, *Adv. Drug Delivery Rev.*, 2008, **60**, 1000–1017 and references cited therein.
50. (*a*) R. Breslow and S. D. Dong, Biomimetic Reactions Catalyzed by. Cyclodextrins and Their Derivatives, *Chem. Rev.*, 1998, **98**, 1997; (*b*) C. J. Easton and S. F. Lincoln, *Modified Cyclodextrins: Scaffolds and Templates for Supramolecular Chemistry*, Imperial College Press, London, 1999; (*c*) L. Barr, P. G. Dumanski, C. J. Easton, J. B. Harper, K. Lee, S. F. Lincoln, A. G. Meyer and J. S. Simpson, *J. Inclusion Phenom. Macrocyclic Chem.*, 2004, **50**, 19; (*d*) C. J. Easton, S. F. Lincoln, L. Barr and H. Onagi, Molecular reactors and machines: applications, potential, and limitations, *Chem.–Eur. J.,* 2004, **10**, 3120.
51. E. Castagnino, M. Cangiotti, S. Tongiani and M. F. Ottaviani, A study of free radical release from β-cyclodextrin-anticancer pro-drugs adducts in water, *J. Controlled Release,* 2005, **108**, 215–225.
52. I. Mawhinney, *Drug Discovery Dev.*, 2009, **12***(10)*, 32–33.
53. (a) V. T. Perchyonok, and I. N. Lykakis, Radical Reactions in Aqueous Media: Origins, Reason and Applications, *Curr. Org. Chem.*, 2009, **13**, 573; (b) V. T. Perchyonok, I. N. Lykakis and K. L. Tuck, Recent advances in C–H bond formation in aqueous media: a mechanistic perspective, *Green Chem.*, 2008, **10**, 153; (c) V. T. Perchyonok, K. L. Tuck, S. J. Langford and M. W. Hearn, Facile and Selective Deallylations of Esters under 'Aqueous' Free-Radical Conditions, *Tetrahedron Lett.,* 2008, **49**, 4777; (d) C.-J. Li and L. Chen, Organic chemistry in water, *Chem. Soc. Rev.*, 2006, **35**, 68 and references therein; (e) H. Yorimitsu, T. Nakamura, H. Shinokubo, K. Oshima, K. Omoto, and H. Fujimoto, Powerful Solvent Effect of Water in Radical Reaction: Triethylborane-Induced Atom-Transfer Radical Cyclization in Water, *J. Am. Chem. Soc.,* 2000, **122**, 11041. For an interesting example of use of supercritical CO_2, see: (f) S. Hadida, M. S. Super, E. J. Beckman and D. P. Curran, Radical Reactions with Alkyl and Fluoroalkyl (Fluorous) Tin Hydride Reagents in Supercritical CO_2, *J. Am. Chem. Soc.,* 1997, **119**, 7406; see also: (g) J. M. Tanko, Free-Radical Chemistry in Supercritical Carbon Dioxide, in *Green Chemistry using Liquid and Supercritical Carbon Dioxide*, ed. J. M. DeSimone and W. Tumas, Oxford University Press, New York, 2003, ch. 4, p. 64; (h) I. N. Lykakis and V. T. Perchyonok, Thiols as an efficient hydrogen atom donor in free radical transformations in aqueous media, *Curr. Org. Chem.,* 2010, in press.
54. Selected reference: C. Rice-Evans, N. Miller and G. Paganga, Antioxidant Properties of Phenolic Compounds, *Trends Plant Sci.*, 1997, **2***(4),* 152–159.
55. C. Easton, H. Onagi, R. Dawson, S. Maniam, R. Coulston, J. Zhang and S. Lincoln, Cyclodextrin nanoscale devices, *Chem. Aust.*, 2009, **76***(1)*, 8–12.

56. (a) R. E. Dawson, S. F. Lincoln and C. J. Easton, The foundation of a light driven molecular muscle based on stilbene and α-cyclodextrin, *Chem. Commun.*, 2008, 3980–3982; (b) R. E. Dawson, S. Maniam, S. F. Lincoln and C. J. Easton, Synthesis of α-cyclodextrin [2]-rotaxanes using chlorotriazine capping reagents, *Org. Biomol. Chem.*, 2008, **6**(10), 1814–1821; (c) S. Maniam, M. Cieslinski, S. F. Lincoln, H. Onagi, P. J. Steel, A. C. Willis and C. J. Easton, Molecular fibers and wires in solid-state and solution self-assemblies of cyclodextrin [2]rotaxanes, *Org. Lett.*, 2008, **10**(10),1885–1888.

57. (a) *Radicals in Organic Synthesis*, ed. P. Renaud and M. P. Sibi, Wiley-VCH, Weinheim, 2001, vol. 1; (b) A. Postigo, C. Ferreri, M. L. Navacchia and C. Chatgilialoglu, The radical-based reduction with (TMS)$_3$SiH on water, *Synlett*, 2005, 2854–2856; (c) A. Postigo, C. Ferreri, M. L. Navacchia and C. Chatgilialoglu, The radical-based reduction with (TMS)$_3$SiH on water, *Synlett*, 2005, 2854–2856; (d) C. Chatgilialoglu, C. Ferreri, Q. C. Mulazzani, M. Ballestri and L. Landi, Cis-trans isomerization of monounsaturated fatty acid residues in phospholipids by thiyl radicals, *J. Am. Chem. Soc.*, 2000, **122**, 4593–4601; (e) T. Yorimitsu, H. Nakamura, K. Shinokubo, K. Oshima and H. Fujimoto, Powerful solvent effect of water in radical reaction: triethylborane-induced atom-transfer radical cyclization in water, *J. Am. Chem. Soc.*, 2000, **122**, 11041–11047.

58. (a) A. E. Johnson and V. T. Perchyonok, Recent Advances in Free Radical Chemistry in unconventional medium: ionic liquids, microwaves and solid state to the rescue, review article, *Curr. Org. Chem.*, 2009, **13**(17), in press; (b) A. Postigo, S. Kopsov, S. S. Zlotsky, C. Ferreri and C. Chatgilialoglu, Hydrosilylation of C-C multiple bonds using (Me$_3$Si)$_3$SiH in water. Comparative study of the radical initiation step, *Organometallics*, 2009, **28**, 3282.

59. (a) R. Breslow and S. D. Dong, Biomimetic Reactions Catalyzed by. Cyclodextrins and Their Derivatives, *Chem. Rev.*, 1998, **98**, 1997; (b) C. J. Easton and S. F. Lincoln, *Modified Cyclodextrins: Scaffolds and Templates for Supramolecular Chemistry*, Imperial College Press, London, 1999; (c) L. Barr, P. G. Dumanski, C. J. Easton, J. B. Harper, K. Lee, S. F. Lincoln, A. G. Meyer and J. S. Simpson, *J. Inclusion Phenom. Macrocyclic Chem.*, 2004, **50**, 19; (d) C. J. Easton, S. F. Lincoln, L. Barr and H. Onagi, Molecular reactors and machines: applications, potential, and limitations, *Chem.–Eur. J.*, 2004, **10**, 3120.

60. Selected references: (a) G. Wenz, B.-H. Han and A. Muller, Cyclodextrin Rotaxanes and Polyrotaxanes, *Chem. Rev.*, 2006, **106**, 782–817; (b) M. J. Frampton and H. L. Anderson, Insulated Molecular Wires, *Angew. Chem., Int. Ed.*, 2007, **46**, 1028–1064; (c) H. Onagi, B. Carrozzini, G. L. Cascarano, C. J. Easton, A. J. Edwards, S. F. Lincoln and A. D. Rae, Installation of a ratchet tooth and pawl to restrict rotation in a cyclodextrin rotaxane, *Chem.–Eur. J.*, 2003, **9**, 5971–5977; (d)

M. M. Cieslinski, P. J. Steel, S. F. Lincoln and C. J. Easton, Centrosymmetric and Non-centrosymmetric Packing of Aligned Molecular Fibers in the Solid State Self Assemblies of Cyclodextrin-based Rotaxanes, *Supramol. Chem.,* 2006, **18**, 529–536; (e) J. Terao, A. Tang, J. J. Michels, A. Krivokapic and H. L. Anderson, Synthesis of poly(*para*-phenylenevinylene) rotaxanes by aqueous Suzuki coupling, *Chem. Commun.,* 2004, 56–57.

61. (a) V. T. Perchyonok, *Radical Reactions in Aqueous Media*, ed. J. Clark and G. Kraus, RSC Publishing, Cambridge, 2009; (b) A. E. Johnson, S. Zhang, J. Chen and V. T. Perchyonok, On the use of β-cyclodextrins as molecular reactors for the radical cyclizations under tin free conditions, *Curr. Org. Chem.,* 2010, in press; (c) A. E. Johnson and V. T. Perchyonok, β-cyclodextrin based molecular reactors for free radical chemistry in aqueous media, *Curr. Org. Chem.,* 2009, **13**(9), 914-918.

62. For recent reviews of microreactor-based synthesis, see: (a) W. Ehrfeld, V. Hessel and H. Lowe, *Microreactors: New Technology for Modern Chemistry*, John Wiley & Sons Inc., Weinheim, 2000; (b) K. Jahnisch, V. Hessel, H. Lowe and M. Baerns, Chemistry in microstructured reactors, *Angew. Chem., Int. Ed.,* 2004, **43**, 406–446; (c) P. Watts and C. Wiles, Recent advances in synthetic micro reaction technology, *Chem. Commun.,* 2007, 443–467; (d) B. P. Mason, K. E. Price, J. L. Steinbacher, A. R. Bogdan and D. T. McQuade, Greener approaches to organic synthesis using microreactor technology, *Chem. Rev.,* 2007, **107**, 2300–2318; (e) B. Ahmed-Omer, J. C. Brandt and T. Wirth, Advanced organic synthesis using microreactor technology, *Org. Biomol. Chem.,* 2007, **5**, 733–740; (f) K. Geyer, J. D. C. Codee and P. H. Seeberger, Microreactors as tools for synthetic chemists - the chemists' round-bottomed flask of the 21st century? *Chem.–Eur. J.,* 2006, **12**, 8434–8442; (g) T. Fukuyama, M. T. Rahman, M. Sato and I. Ryu, Adventures in inner space: microflow systems for practical organic synthesis, *Synlett*, 2008, 151; (h) T. Fukuyama, M. Kobayashi, Md. Taifur Rahman, N. Kamata and I. Ryu, Spurring radical reactions of organic halides with tin hydride and ttmss using microreactors, *Org. Lett.,* 2008, **10**, 533–536; (i) A. Sugimoto, M. Takagi, Y. Sumito, T. Fukuyama and I. Ryu, The Barton reaction using a microreactor and black light. Continuous flow synthesis of a key steroid intermediate for an endothelin receptor antagonist, *Tetrahedron Lett.,* 2006, **47**, 6197–6200, (j) T. Iwasaki and J. Yoshida, Free radical polymerization in microreactors significant improvement in molecular weight distribution control, *Micromolecules*, 2005, **38**, 1159–1163.

63. The first description of "teabags": R. A. Houghten, General method for the rapid solid-phase synthesis of large numbers of peptides: specificity of antigen-antibody interaction at the level of individual amino acids, *Proc. Natl. Acad. Sci. U. S. A.,* 1985, **82**, 5131.

64. (a) S. V. Ley, O. Schucht, A. W. Thomas and P. J. Murray, Synthesis of the alkaloids (+/-)-oxomaritidine and (+/-)-epimaritidine using an orchestrated multi-step sequences of polymer supported reagents, *J. Chem. Soc.,*

Perkin Trans. 1, 1999, 1251–1252; (b) J. Habermann, S. V. Ley and J. S. Scott, Synthesis of the potent analgesic compound (+/-)-epitaltidine using an orchestrated multi-step sequence of polymer supported reagents, *J. Chem. Soc., Perkin Trans. 1,* 1999, 1253–1255.

65. P. J. Comina, A. K. Beck and D. Seebach, A simple batch reactor for the Efficient Multiple Use of Polymer-Bound a,a,a',a'-Tetraaryl-1,3-dioxo-lane-4,5-dimethanol Titanates in the Nucleophilic Addition of Dialkylzinc Reagents of Aldehydes, *Org. Process Res. Dev.*, 1998, **2**, 18–26.

66. A. E. Johnson, S. Zhang, J. Chen and V. T. Perchyonok, *Curr. Org. Chem.*, 2009, **13**(17), 1746–1750.

67. V. T. Perchyonok, *Lett. Org. Chem.*, 2011, **8**(4), 292–294.

CHAPTER 24

Artificial Enzymes and Free Radicals: the Chemist's Perspective[*]

24.1 INTRODUCTION

All living things in nature maintain their internal metabolic balances quite well when they are in healthy conditions. In other words, metabolic materials are in equilibrium in each living creature primarily in a collaboration of biological catalyses by numerous enzymes. Enzymes are sophisticated proteins having catalytic groups and often require specific cofactors or coenzymes for catalytic performance. If we look at enzymatic functions from physicochemical viewpoints rather than biological ones, catalytically active amino acid residues of enzyme proteins as well as coenzyme factors are buried in hydrophobic and water-lacking reaction sites furnished by enzyme proteins and well separated from the bulk aqueous phase needed to attain thermodynamic stabilities. In consideration of such physicochemical roles of enzyme proteins, we are allowed to use man-made materials for construction of artificial enzymes that are capable of simulating catalytic functions demonstrated by enzyme proteins.[1,2]

Two types of such artificial enzymes or apoenzymes, macrocyclic compounds and molecular assemblies, are cited in this chapter as those which can provide specific microenvironments for substrate-binding and subsequent catalysis in aqueous media. Those micro-environmental properties are primarily due to hydrophobic internal cavity and the internal domain of molecular assemblies in aqueous media, and other non-covalent intermolecular interactions, such as electrostatic, charge-transfer, and hydrogen-bonding modes, between a substrate and an apoenzyme model are greatly enhanced in such microenvironments.[3] In this chapter, primarily we summarize recent studies on functional simulation of holoenzymes requiring coenzyme factors, such as vitamin B_{12}, artificial methylmalonyl-CoA mutase and glutamate mutase as representative enzymes. Most of those coenzymes are soluble in

* Chapter written by Ioannis N. Lykakis and V. Tamara Perchyonok.
Streamlining Free Radical Green Chemistry
V. Tamara Perchyonok, Ioannis Lykakis and Al Postigo
Published by the Royal Society of Chemistry, www.rsc.org

aqueous media when separated from the corresponding apoprotein and, consequently, cannot be readily incorporated into hydrophobic microenvironments provided by the artificial systems. Modified proteins which are derived by mutagenic treatments of natural enzymes are often called artificial enzymes. It must be emphasized here that artificial enzymes described in this article are not directly related to protein structures but are capable of carrying out a functional simulation of enzymatic catalysis in the overall reaction schemes.

Enzymes have had billions of years to evolve into the sophisticated three-dimensional structures of today. As chemists, we need to concentrate this period into a feasible timescale for research. In recognition of this fact, recent developments in the field of artificial enzymes have tended to move away from the rational design and multistep synthesis of complex molecules where the smallest flaw in conception can have catastrophic results. Instead, current strategies have tended to focus on selection approaches.[3a,2e]

Advances in the fields of molecular biology, biochemistry and more recently combinatorial and polymer chemistry have all furnished unique and often co-operative solutions to the synthesis of artificial enzymes, and it is the aim of this overview to discuss some of the more recent and diverse approaches taken by organic chemists towards the creation of effective enzyme mimics with particular focus on some free-radical-based mechanism enzymes.

In general these different approaches can be divided into three categories:

(a) The "design approach". A host molecule is designed with the salient functionality (often also present in the natural enzyme counterpart), which is expected to be involved in catalysis of the chosen reaction.[4] Catalytic cyclodextrins are one such example and will be discussed in some detail in this chapter.

(b) The "transition state analogue selection approach". A library of hosts is generated in the presence of a transition state analogue (TSA) and the best host is then selected from the library. This latter approach has been employed with considerable success in the field of catalytic antibodies and has more recently inspired the process of 'molecular imprinting' (*vide infra*).[5]

(c) The "catalytic activity selection approach". This takes advantage of the combinatorial chemistry revolution wherein a library of possible catalysts is generated and screened directly for enzyme-like activity. This area is justifiably deserving of review articles in their own right, and only selected examples will be discussed here. Equally, much research into enzyme mimetic systems has been carried out in the field of bio-inorganic and coordination chemistry and has been summarized elsewhere.[6] This chapter, written from the perspective of the organic chemist, will instead concentrate on less developed areas and will conclude with a discussion of some of the more recent developments in 'selection approaches' towards artificial receptors.

24.2 TRANSITION STATE THEORY: A BRIEF INTRODUCTION

The currently accepted view is that catalysis rests on the enzyme's ability to stabilize the transition state of a reaction relative to that of the ground state.

This principle is illustrated (Figure 24.1) for a uni-molecular example where the enzyme $\pm$ substrate complex is stabilized relative to the free species in solution. The activation barrier to reaction is represented by the differences $\Delta G^{\ddagger}_{cat}$ and $\Delta G^{\ddagger}_{E}$ for the enzyme-catalyzed and the uncatalyzed reactions respectively. It is clear from this picture that, for catalysis to work, the difference $\Delta G^{\ddagger}_{E}$ must be larger than $\Delta G^{\ddagger}_{ES}$. In other words, the enzyme must stabilize the transition state of the reaction more than it stabilizes the ground state of the substrate. For true catalysis, a system also needs to exhibit turnover. If the product binds to the enzyme in a significantly stronger way than the substrate does ($\Delta G_{ES} > -\Delta G_{EP}$), then product inhibition of the reaction can result. In this case substrate binding can be beneficial. The most important consequence of this picture of enzyme action is that the design of an enzyme mimic must not only consider transition state binding relative to substrate binding, but also ensure that the active site is designed such that product release is a thermodynamically favorable process. The discussion above is a simplification of the real situation since enzyme catalysis of a transformation often involves an alternative reaction pathway from that taken in the non-catalyzed process, usually taking advantage of the enzymes ability to reduce the molecularity of multi-step sequences. The situation also becomes more complicated for bimolecular processes and reactions involving covalent enzyme-bound intermediates, invoked in the initial step of the mechanism for amide bond cleavage by serine proteases. In these more complex systems, application of the above model of transition state stabilization is less straightforward.[7]

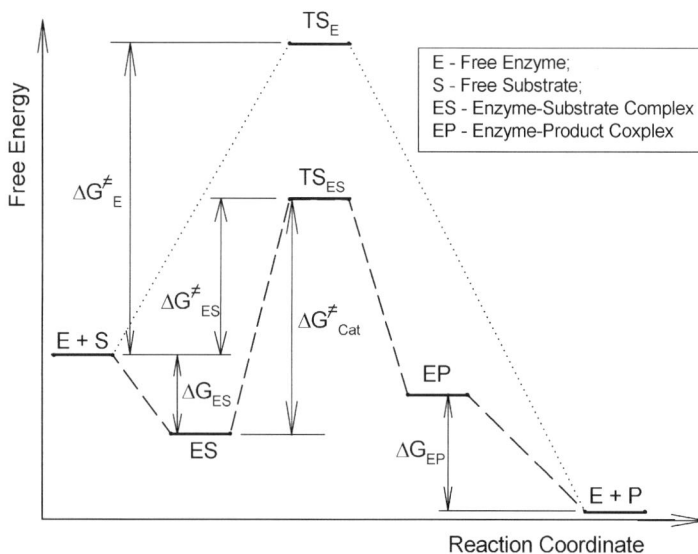

Figure 24.1 Energy diagram of an enzyme-catalyzed reaction and the corresponding uncatalyzed chemical reaction.

However, the foregoing discussion is sufficient to appreciate the basic principle behind many of the various approaches to artificial enzymes.

Perhaps the biggest obstacle to the synthesis of enzyme mimics is that in order to design an artificial enzyme it is necessary to understand how enzymes achieve this selective binding of the transition state. Many studies have been carried out in attempts to quantify the contributions of the many weak intermolecular forces involved. In general the overall binding is a product of electrostatic forces, hydrogen bonding, and cumulative hydrophobic and van der Waals influences. Since enzymes operate in water, desolvation effects and the resulting entropy changes are also very important factors.[7b] Hydrogen bonding and electrostatic interactions contribute significantly to the binding affinity between the substrate or transition state and the enzyme, although, since enzymes operate in water, the contribution of these effects is greatly moderated by solvation. In fact, in some cases, desolvation of both the polar group on the ligand and the complementary group in the enzyme may cost as much in enthalpy as is gained by bringing the two groups together. Many studies into the influence of hydrogen bonding in particular have been carried out, in an attempt to quantify the energy difference gained upon the formation of a ligand-host hydrogen bond in water. The value for a neutral $\pm$ neutral hydrogen bond has been generally found to be in the order of 1.5 kcal mol^{-1},[8] which represents a perhaps surprisingly modest energy gain.[9] As a result it has been suggested that such interactions 'may play less of a role in enhancing association of the correct ligand than they do in creating a penalty for binding the wrong ligand', *i.e.*, in determining ligand specificity. By way of contrast, the contribution of charged hydrogen bonds to binding enthalpy is more significant.

This difference between neutral and charged hydrogen bonding is elegantly illustrated by comparison of the two model receptors 2 and 3 for glutaric acid (Figure 24.2). Both receptors create the same number of hydrogen bonds with glutaric acid.[7b,7c,9-12]

However, although the neutral diamide 2 binds glutaric acid strongly in chloroform (K_{ass} = 60 000 M^{-1}), in dimethylsulfoxide (DMSO) binding is not observed,[13] whilst the receptor 3 which incorporates two charged electrostatic hydrogen bond interactions is almost as good a receptor in DMSO (K_{ass} = 50 000 M^{-1}) containing 5% THF; moreover, binding is still measurable in the presence of 25% water.[14] In quantitative terms, the presence of charged

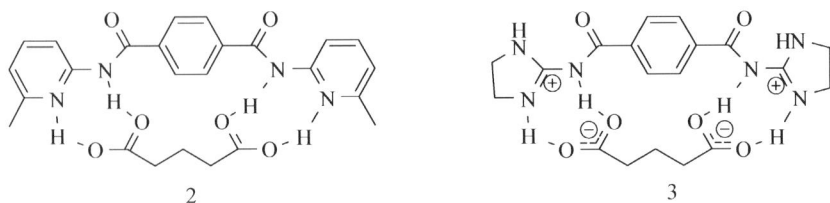

2   3

Figure 24.2

hydrogen is distinguished between an active artificial enzyme and a synthetic receptor. Although the exact nature of the transition state is unknown, the important feature is that the 'binding' of this transition state by the enzyme involves more than ordinary molecular recognition.[11] The partially formed covalent bonds at the reaction centre represent 'dynamic' binding interactions, which have no conveniently modeled ground state counterpart. It can also be expected that these interactions must make a major contribution to transition state binding and stabilization, not least because they clearly represent difference between substrate and transition state recognition. All of the factors described above contribute to the overall binding of the transition state. Despite the many reviews available on the factors which influence molecular recognition and binding in enzyme systems, the practical application of these hypotheses remains the true test of our understanding. It is thus of relevance to investigate the design and synthesis of artificial enzymes since this will hopefully also lead to a better understanding of molecular recognition itself.

24.3   THE "DESIGN APPROACH"

24.3.1   Cyclodextrins as Enzyme Mimics

Cyclodextrins (CDs) have long attracted attention in catalysis and as enzyme mimics, due to the way in which they act as hosts to complex guest molecules and induce reactions of the complex species (Scheme 24.1). The reactions exhibit kinetic characteristics, such as saturation, non-productive binding and competitive inhibition, which are typical of enzyme-catalyzed processes. In addition, the discrimination displayed by CDs in binding guests and promoting their reactions is analogous to the substrate selectivity displayed by enzymes.[15]

  With the natural CDs, hydroxy groups are the only functionality available to promote reactions of included guests. However, the introduction of a diverse range of new functional groups, through modifications of the natural CDs, results in catalysts which mimic the entire range of enzyme behaviour. The CD nucleus serves as a scaffold on which functional groups can be assembled. In some cases this has been accomplished with controlled alignment

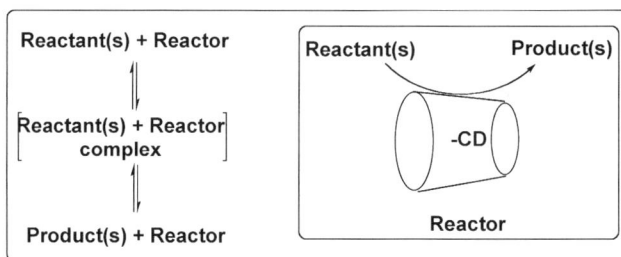

Scheme 24.1   CD-induced reaction of an included guest.

of both the functional groups and the CD annulus, to optimize the geometry for binding and reaction of a particular guest. This is an important factor in the catalytic activity of modified CDs, as it is with enzymes where the geometry at the active site is determined by the three-dimensional structure of the protein. It is the catalytic activity of CDs which is the subject of this chapter.

24.3.2 Vitamin B$_{12}$ Functions: Enzymatic Reactions

Vitamin B$_{12}$ is a cobalt complex coordinated with a tetrapyrrole ring system, namely corrin, and linked to a 5,6-dimethylbenzimidazole moiety as a heterocyclic base (Figure 24.3). There are two B$_{12}$ active forms: 5'-deoxyadenosylcobalamine and methylcobalamine. Vitamin B$_{12}$-dependent enzymes are known to catalyze two types of reactions: rearrangements as exemplified by methylmalonyl-CoA mutase and methylation by methionine synthetase. The rearrangement reactions involve the intramolecular exchange of a functional group (X) and a hydrogen atom between neighboring carbon atoms.

These reactions have attracted much attention because of their novel nature from the viewpoints of organic and organometallic chemistry. Carbon skeleton rearrangement reactions, mediated by methylmalonyl-CoA mutase, glutamate mutase, and *R*-methyleneglutarate mutase are shown by eqn (24.1)–(24.3).

Even though the real reaction mechanisms involved in the carbon-skeleton rearrangements have not been clarified up to the present time, radical

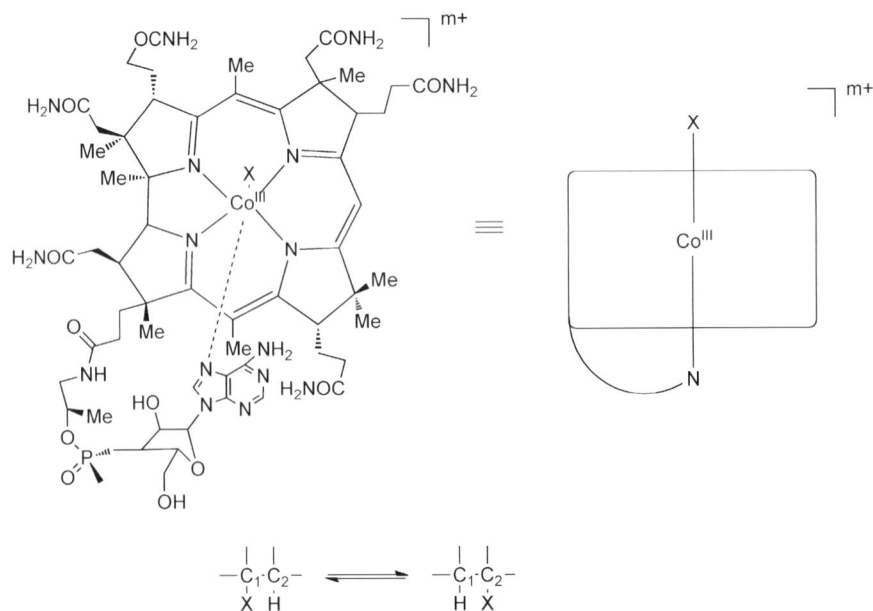

Figure 24.3 Vitamin B$_{12}$-dependent enzymes catalyze a 1,2-hydride shift in model system.

$$\underset{\underset{COS-CoA}{|}}{\overset{\overset{COOH}{|}}{H_3C-C-H}} \quad \xrightarrow[\text{mutase}]{\text{mathylmalonyl-CoA}} \quad \underset{\underset{COS\sim CoA}{|}}{\overset{\overset{COOH}{|}}{H_2C-CH_2}} \qquad (24.1)$$

$$\underset{\underset{CH_3}{|}}{\overset{\overset{NH_2}{\overset{|}{H}}}{HOOC-\overset{H}{C}-C-COOH}} \quad \xrightarrow[\text{mutase}]{\text{glutamate}} \quad \overset{\overset{H_2\,H_2\ \ NH_2}{\ \ \ \ \ \ \ \ |}}{HOOC-C\ C-\overset{}{C}-COOH} \qquad (24.2)$$

$$\underset{\underset{CH_3}{|}}{\overset{\overset{CH_2}{\overset{\|}{}}}{HOOC-\overset{H}{C}-C-COOH}} \quad \xrightarrow[\text{mutase}]{\text{a-methyleneglutamate}} \quad \overset{\overset{H_2\,H_2\ \ CH_2}{\ \ \ \ \ \ \ \ \|}}{HOOC-C\ C-\overset{}{C}-COOH} \qquad (24.3)$$

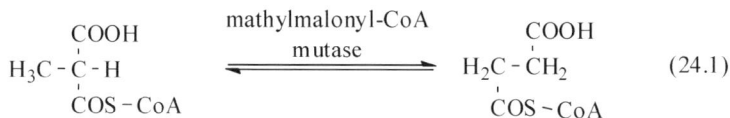

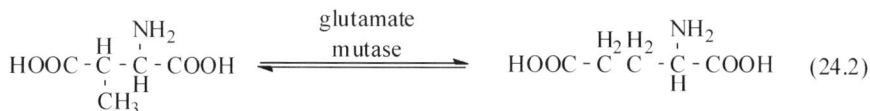

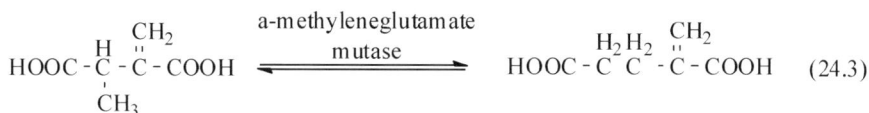

mechanisms are considered to be the most plausible ones on the basis of electron spin resonance (ESR) studies for methylmalonyl-CoA mutase, glutamate mutase, and *R*-methyleneglutarate mutase. A general feature of the radical mechanism is illustrated in Figure 24.4: the 5'-deoxyadenosyl moiety bound to cobalamin (vitamin B_{12}) undergoes homolytic cleavage to give cobalamin in the Co(II) state and the 5'-deoxyadenosyl radical upon incorporation of a substrate into a specific microenvironment provided by the corresponding apoprotein; the 5'-deoxyadenosyl radical abstracts a hydrogen atom from the incorporated substrate to afford deoxyadenosine and the substrate radical in the active site of apoprotein; the substrate radical is

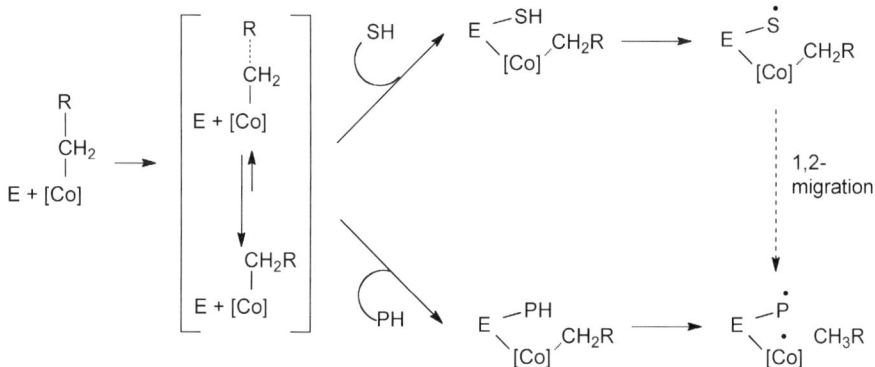

where R-CH2-, 5'-deoxyadenisyl; [Co], cobalamin

E= apoenzyme, SH= substrate, PH=product

Figure 24.4 A general feature of the radical mechanism for rearrangement reactions mediated by the 5'-deoxyadenosylcobalamin-dependent enzyme.

eventually isomerized *via* 1,2-migration of a functional group to form the corresponding product radical; the product radical abstracts a hydrogen atom from deoxyadenosine placed in its vicinity; and the deoxyadenosyl radical is then bound to cobalamin to recover the original coenzyme state.[16]

As for the role of the cobalt species, Halpern *et al.* proposed a reversible free radical carrier mechanism; coenzyme B_{12} is referred simply to a source of the 5′-deoxyadenosyl free radical that acts to generate a substrate radical by abstracting a hydrogen atom from the substrate to initiate the reaction, and behaves as a reversible free radical carrier.

24.3.3 Model Reactions with Apoenzyme Functions

It is obvious that an apoprotein plays an important role in such radical reactions as mediated by vitamin B_{12}-dependent enzymes. The model reactions, which were designed in consideration of the role of apoenzymes, are as follows. Breslow *et al.* prepared a cyclodextrin-bound B_{12}, in which cobalamin is directly linked to the primary carbon of α-cyclodextrin by a cobalt–carbon bond (Scheme 24.2).[17] They expected that a hydrophobic substrate would be incorporated into the cyclodextrin cavity in water, so that the cyclodextrinyl radical may undergo an intracomplex atom transfer to generate a substrate radical as shown in Scheme 24.2. Even though they did not mimic all steps of the B_{12}-dependent rearrangement reaction, this is an

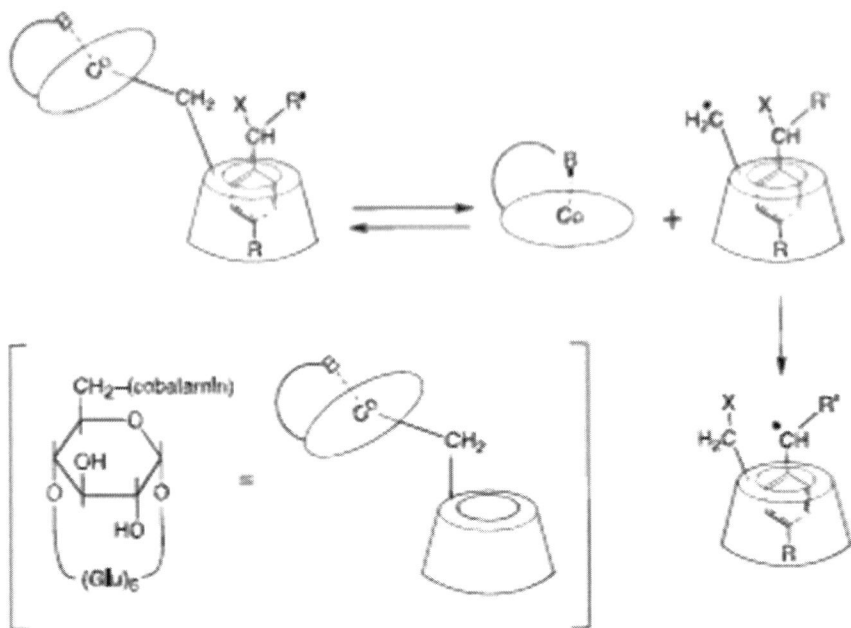

Scheme 24.2   Mode of action of a β-CD-B_{12} enzyme mimic.

interesting example showing that a substrate and B_{12} are bound together in a receptor site.

24.4   THE "TRANSITION STATE ANALOGUE SELECTION" APPROACH

24.4.1   The Transition State Analogue Selection Approach: General Introduction

The traditional approach to enzyme mimics is the design approach described above. Whilst this has furnished us with much information on the recognition processes involved in binding and the criteria required for successful catalysis, the realization of a project from original conception to experimental studies on an enzyme mimic can be a long and laborious process (Figure 24.5). A case in question is Diederich's pyruvate oxidase mimic which required an 18-step synthesis of the host.

In an attempt to move away from this linear approach, several techniques have been developed which make use of a selection strategy. This allows for the simultaneous screening of a wide range of possible candidates thus significantly reducing the time required and hopefully allowing for the detection of better hosts.

The earliest examples of a selection approach chose affinity for a transition state analogue (TSA) as the screening criteria. The logic behind this is that any macromolecule which shows strong binding to a molecule resembling the transition state of a reaction should also bind to and stabilize the real transition state. As this stabilization of the transition state is the basis behind enzyme catalysis, the hosts selected should behave as enzyme mimics for the chosen transformation.

More recently, it has been recognized that TSA binding alone may not be enough to obtain the rate accelerations needed to rival enzyme catalysis.

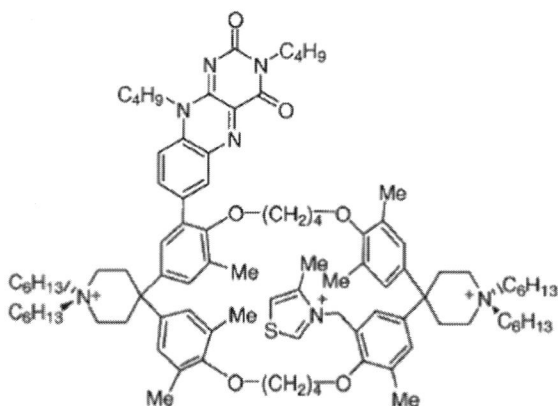

Figure 24.5

Nowadays, the incorporation of catalytic groups in the host is often a designed aspect of the selection process and it is this, in combination with the TSA host selection, which has led to some of the most impressive advances described below.[18]

24.4.2 Molecular-imprinted Polymers as a Method in the Transition State Analogue Selection Approach

Molecular imprinting is a general method for synthesizing robust, network polymers with highly specific binding sites for small molecules. Molecular imprinting is a process by which polymeric materials are synthesized with highly specific binding sites for small molecules.[19-25] Molecularly imprinted polymers (MIPs) have been developed for a variety of applications including chromatography, enzymatic catalysis, solid-phase extraction, and sensor technology.[26] Intermolecular forces that develop during polymerization between the template molecules (T), functional monomer (M) and developing polymer matrix are responsible for creating a polymer microenvironment for the template or imprint molecule.[20b,20c,21a,21b] The resulting polymer network contains synthetic receptors that are complementary in size, shape and functional group orientation to the template molecule. The polymers typically employed in imprinting are complex thermo-sets, an insoluble, highly cross-linked network polymer. Because both the morphology of the bulk polymer and the chemical microenvironment of the binding site are critical to the overall performance, the number of experimental variables which influences these factors is large. Imprinted polymers are, therefore, ideal candidates for combinatorial synthesis and its screening technologies. Recently, combinatorial methods have been used to develop highly selective MIPs. Synthetic receptors produced by this technology, in turn, have been used in the screening of libraries of small molecules. This section covers both these emerging areas of MIP technology.

Optimization of MIP formulations many variables of the imprinting process influence the selectivity and capacity of a MIP. First, complementary interactions between the template and the functional and cross-linking monomers are necessary to create short-range molecular organization at the receptor site. These interactions include hydrogen bonding, electrostatic and/or van der Waals forces. Second, the stoichiometry and concentration of the template and monomers influences both polymer morphology and MIP selectivity. Third, the solvent used in the polymerization process, also known as the 'porogen', plays a dual role. In addition to mediating the interactions between the functional groups and the template molecule, the porogen determines the timing of the phase separation during polymerization,[20-25] which is an important determinant of polymer morphology, porosity and ultimately accessibility of the binding site. Finally, the temperature of polymerization influences the timing of phase separation. Also, the temperature dependence of the equilibrium between the functional monomers and template affects MIP selectivity and capacity.

A typical imprinting protocol involves thermally or photochemically induced free-radical polymerization of a concentrated solution of monomers to produce bulk, monolithic insoluble polymers that are crushed, ground and sieved to micron size for analysis (Figure 24.6). Selectivity and binding are evaluated in the chromatographic mode by using the MIP as stationary phase for high pressure liquid chromatography (HPLC) columns, or by batch rebinding studies. These methods can be tedious and time-consuming, and obtaining an MIP with optimal binding properties can take several days to weeks, especially if the variation in the formulation is made by trial-and-error. As a result of the number of variables that affect MIP performance, there has been an overuse of certain 'standard' formulations. The process of determining an optimal MIP formulation, therefore, is an ideal candidate for a combinatorial approach for screening various formulations.

This technique offers potential for developing tailor made catalysts, perhaps with catalytic functionalities not utilized in biology. Despite the inherent heterogeneity of the molecular recognition site produced, the increased stability of MIPs against heat, chemicals and solvents when compared to natural enzymes or artificial analogues means that the attainment of MIPs remains a highly sought-after aspiration.

Mosbach *et al.* reported the first true enzyme-like catalysis of C–C bond formation using MIPs.[19] The molecular imprinting techniques was used in the development of a 4-vinylpyridine-styrene-divinylbenzene copolymers imprinted with aldol condensation intermediate analogues, dibenzoylmethane (DBM) and a Co^{2+} ion (Scheme 24.3). The imprinted polymer was able to catalyze the aldol condensation of acetophenone to benzaldehyde in manner

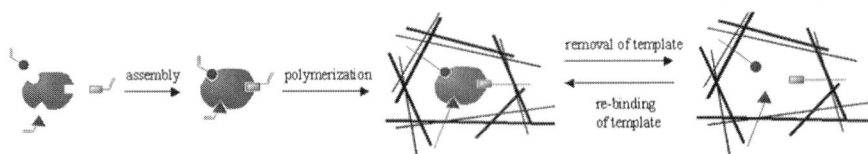

Figure 24.6 General scheme for preparation of molecular-imprinted polymers.

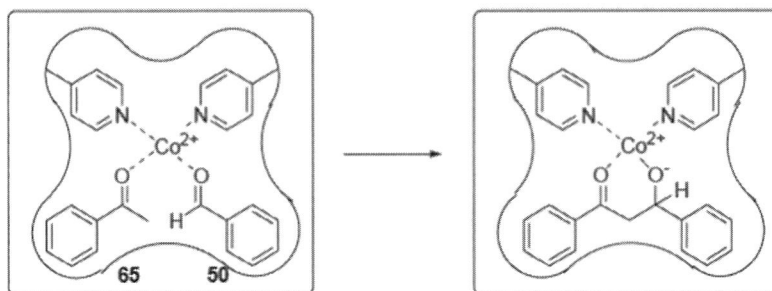

Scheme 24.3 MIP containing a Co^{2+} ion used in aldol condensation reaction.

analogous to Class II aldolases. In addition to metal coordination, the
pyridinyl residues provided the basis for the generation of the enolate of
acetophenone.

24.4.3   Imprinting an Artificial Proteinase

Another technique which utilizes the idea of 'imprinting' was reported by Suh
and Hah for the creation of an artificial aspartic proteinase. The two aspartic
carboxyl groups found within the natural enzymes are thought to act as key
catalytic groups in hydrolyzing peptide substrates.[27] In light of this fact, Suh
synthesized an organic artificial protease which contained carboxyl groups in
the active site (Scheme 24.4).

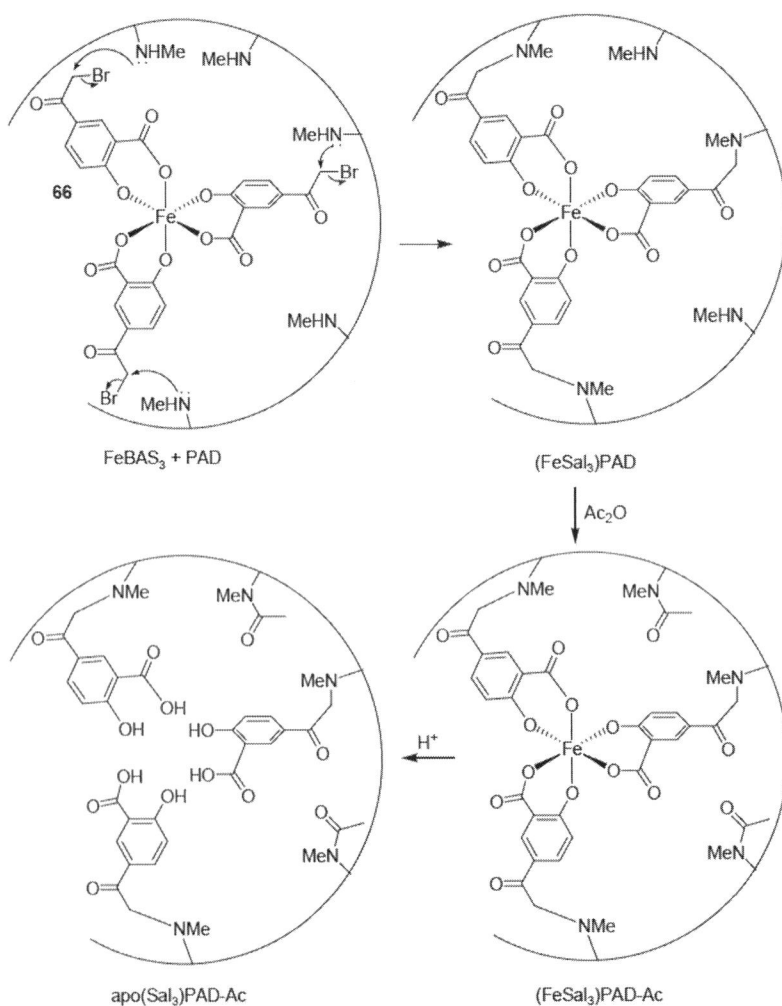

Scheme 24.4   Imprinting process for the creation of an artificial aspartic proteinase.

This involved the complexation of three molecules of 5-bromoacetylsalicylate to an Fe(III) ion to give the resultant complex (FeBAS$_3$) which was cross-linked with poly(aminomethylstyrene-co-divinylbenzene) (PAD) to obtain (FeSal$_3$)-PAD. These were subsequently capped *via* acetylation to produce (FeSal$_3$)-Ac, and the Fe(III) ions removed under acidic conditions to give the active apo(Sal$_3$)PAD-Ac protease mimics. These were obtained as insoluble catalysts which reproduced the catalytic features of aspartic proteases. The activity of apo(Sal$_3$)PAD-Ac was tested in the hydrolysis of bovine serum albumin, in which it was revealed that albumin was cleaved into fragments smaller than 2 kDa. By looking at the pH profile for this reaction, it was found that it manifested optimum activity at pH 3, which is in agreement with conditions found within natural enzymes. Since the active site of apo(Sal$_3$)PAD-Ac contained both carboxyl and phenol groups, at pH 3, phenol was thought to be acting as a general acid since its activity is likely to be lower than that of the carboxyl groups. Therefore the activity of apo(Sal$_3$)PAD-Ac at pH 3 is attributable to cooperation of two or more carboxyl groups by a mechanism analogous to that found in natural enzymes. Moreover it has a k_{cat} of over 0.17 h^{-1} at pH 3, indicating that it has a reasonably high catalytic activity. The idea of 'imprinting' has also been extended to include bio-molecules, mainly proteins, for use as efficient artificial enzymes.

24.4.4 Bioimprinting

Biomolecular imprinting or bioimprinting refers to the induction of catalytic activity in proteins by lyophilization (freeze-drying) in the presence of a transition state analogue.[28] Slade and Vulfson have demonstrated that bioimprinting proteins in the presence of a TSA leads to a conformational change which either manifests itself in the form of a new catalytic site or as improvements of the pre-existing ones, which were then able to carry out catalysis.

This process was illustrated by bioimprinting β-lactoglobulin in the presence of TSA (Scheme 24.5). β-Elimination of the substrate was studied using this novel bioimprinted protein and compared with the results of the non-imprinted control.

The imprinted β-lactoglobulin showed catalytic activity three times that of the control reaction and almost four orders of magnitude higher than spontaneous β-elimination. Although this result may seem modest when compared to rate accelerations obtained using catalytic antibodies, it was found that the rate acceleration was almost identical to that observed for molecularly imprinted polymers.

A major drawback of this method of imprinting is that the enhanced properties of these proteins can only be sustained in nearly anhydrous environments, since hydration of these proteins causes re-naturation and therefore consequent loss of the imprinted binding sites. This problem however was solved to some degree by Peissker and Fischer by combining the imprinting step with a subsequent immobilization method, resulting in the

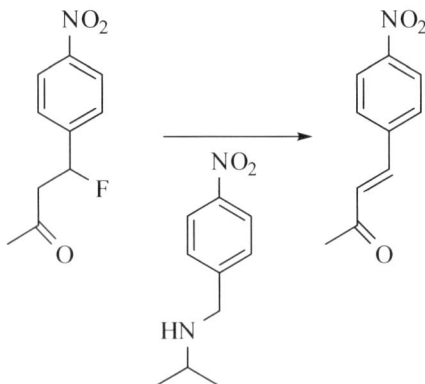

Scheme 24.5 β-Elimination of 4-fluoro-4-(*p*-nitrophenyl) butan-2-one and the structure of TSA used for protein imprinting.

retention of the imprint by the enzymes, allowing their structure to be maintained in aqueous media (Scheme 24.6).[29]

This technique was used to stabilize the ligand-induced acceptance for D-configured substrates by α-chymotrypsin or subtulisin Carlsberg. This involved the vinylation of the proteases by acylation with itaconic anhydride. Subsequent enzyme imprinting and crosslinking furnished the desired cross-linked imprinted proteins (CLIPs). Examples of the use of bioimprinting include Luo's glutathione peroxidase (GPX) mimic and those based on imprinting of myoglobin in the epoxidation of styrene.[30]

24.5 THE "CATALYTIC ACTIVITY SELECTION APPROACH": GENERAL INTRODUCTION

This approach utilizes the advances in combinatorial chemistry wherein a library of possible catalysts is generated and directly screened for enzyme-like activity. This not only provides a tool for synthesizing a large number of diverse compounds in a short amount of time but also allows for the discovery of effective catalysts, which exhibit potential activity when subjected to the relevant screening method.

24.5.1 Combinatorial Polymers as Enzyme Mimics

In the mid-1990s, Menger introduced a highly original approach towards the creation of novel artificial enzymes. This involved the use of combinatorial chemistry to attach various combinations of three or four carboxylic acids onto poly(allylamine) (PAA) or poly(ethylenimine) (PEI) (Scheme 24.7). The Scheme gives a general schematic for the synthesis of a functionalized polymeric library using poly(allylamine).[6]

Although no control was exercised over the attachment of the substituents, they were not necessarily randomly attached throughout the polymer. For example, a polymer with an octanoyl group at one site may be prone to receive

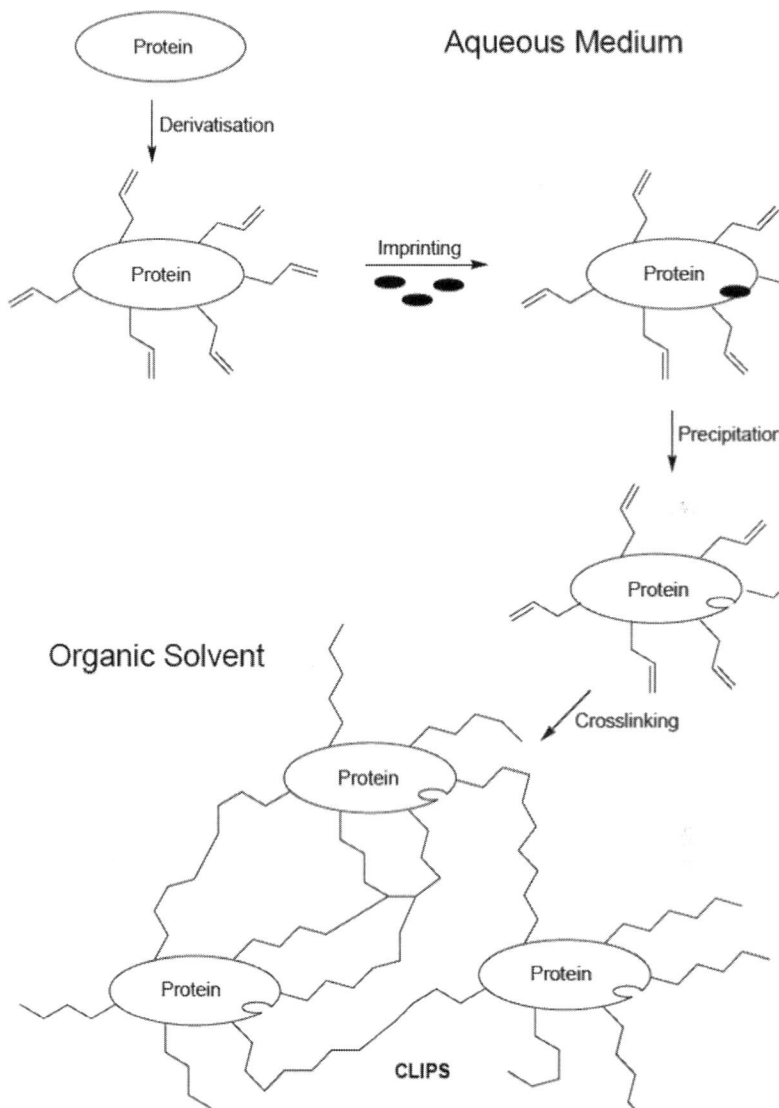

Scheme 24.6 Broadening the substrate selectivity using a combination of bio-imprinting and subsequent covalent immobilization technique.

another octanoyl group adjacent to it owing to hydrophobic interactions between the two functionalities. Although this combinatorial approach has yielded some impressive results, the major drawback is that each combinatorial polymer is a complex system, consisting of numerous polymeric variations. This not only makes the isolation of a pure component near impossible for sequencing and structural characterization but also provides very little detail to draw any significant mechanistic conclusions.

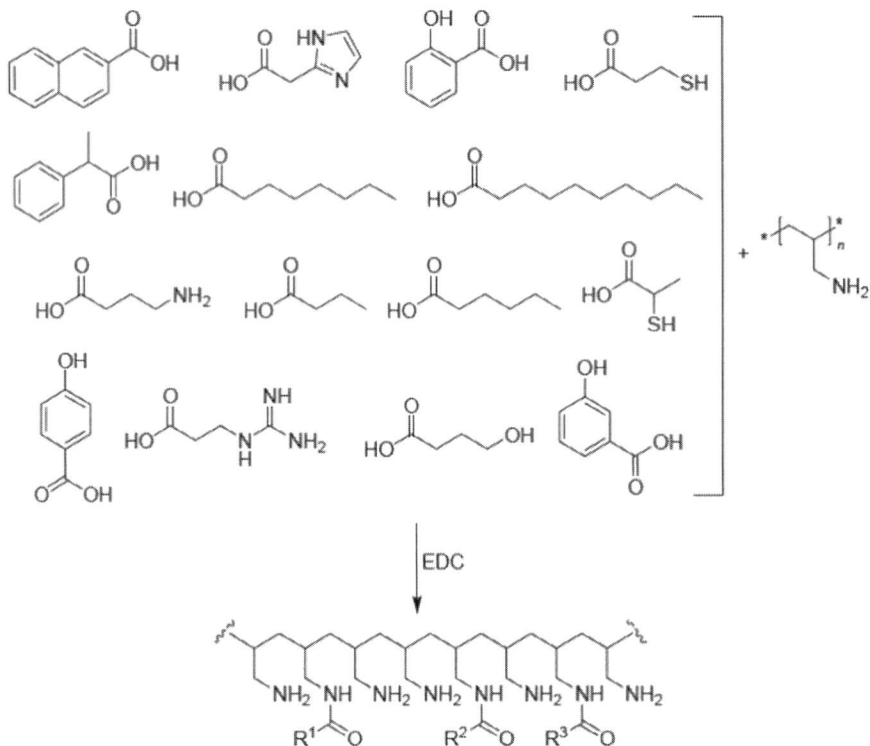

Scheme 24.7 Synthesis of functionalized polyallylamine using combinatorial chemistry.

24.5.2 Directed Evolution of Enzymes

In quite a different manner to the previous catalytic selection approaches mentioned earlier, in 1995 Reetz began developing a high-throughput screening method for assaying the enantioselectivity of thousands of biocatalysts.[31] This involved exploitation of the tools of directed evolution in the creation of enantioselective enzymes for use in organic synthesis (Scheme 24.8).

The examples in this field include artificial esterases[32a] and artificial cytochrome P450 monooxygenases.[32b,c]

24.5.3 Catalysis with Imprinted Silicas and Zeolites

24.5.3.1 *"Footprint" Catalysis*

The first attempts to obtain imprinted materials were achieved with silicas by Dickey.[33] He precipitated silica gel in the presence of dyes. After drying, the dyes were washed out, and the as-prepared silica showed an increased affinity for the template compared to similar compounds. During this approach, the

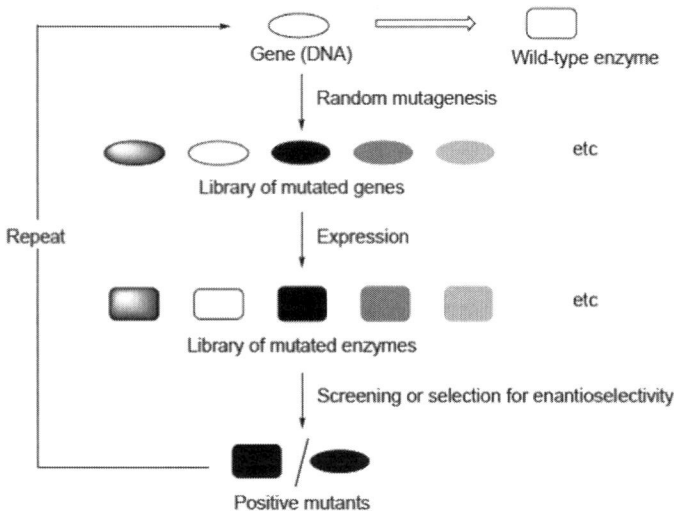

Scheme 24.8   Directed evolution of an enantioselective enzyme

formation of the rigid silica gel matrix around the template facilitates shape selectivity, and additional silanol groups might be arranged such that interactions with the template can occur. No directed interactions between template and the growing silica chains were tried in these first experiments. One should expect a high stability of these arrangements in silica, but the selectivity disappeared readily.[34] Especially, racemic resolutions have been studied for some time with imprinted silicas.[22,35]

Much later, investigations of imprinted silicas started anew. The first to use this approach for the synthesis of catalytically active silicas was Morihara.[36] He used a quite unconventional approach to obtain catalytically active silicas by surface imprinting.

This procedure was called "footprint catalysis". From 1988 onward many examples were published.[37] In Morihara's procedure, commercial silica gel is treated at pH 6.5 with Al^{3+} ions.[37d,37i,36] The incorporation of Al^{3+} in the silica matrix by isomorphous substitution of silicate with aluminate causes the formation of surface Lewis acid groups (see Scheme 24.10). The surface is then loaded at pH 4.0 with template molecules that contains Lewis base groups. The silica is then aged and dried under highly controlled conditions. It is assumed that in this way the silicate matrix rearranges around the acid–base complexes by de-polymerization of the silicate matrix and re-polymerization under thermodynamic control. The complexes are stabilized by forming a maximum number of interactions between the acid–base complex and the surrounding silicate matrix. Afterward, the template is removed with methanol. Mostly acyl transfer reactions are investigated, since substrates and transition state analogues in this case are readily available. Scheme 24.11 depicts the surface imprinting with the transition state analogue phosphonic acid diamide (*N,N'*-dibenzoylbenzenephosphone diamide).

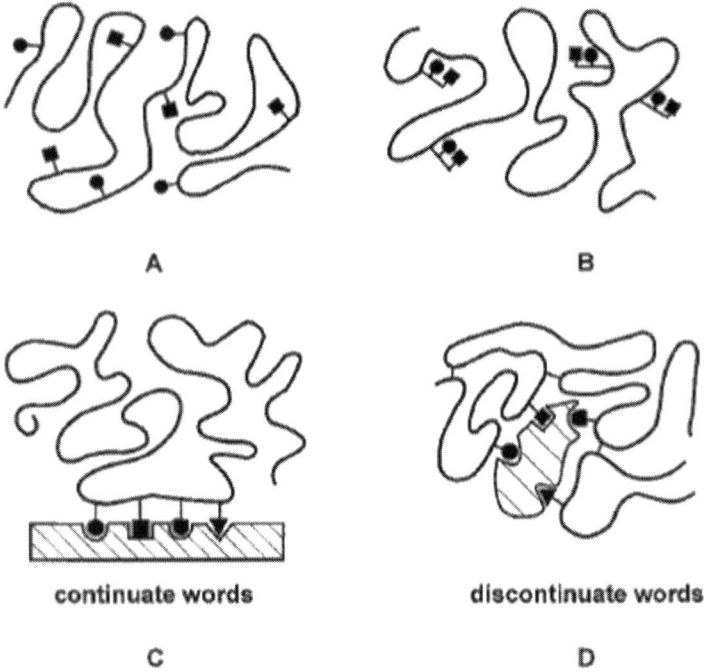

Scheme 24.9   Possible arrangements of functional groups in synthetic and natural polymers.

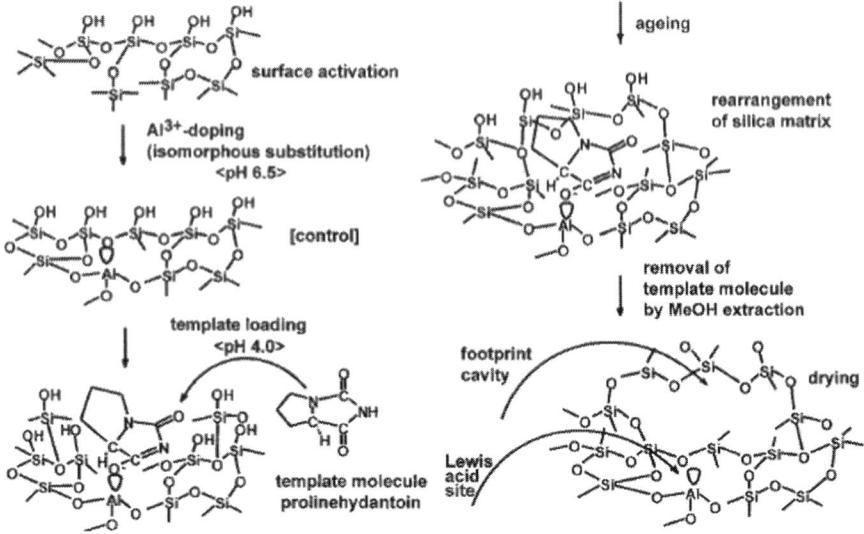

Scheme 24.10   Schematic representation of the preparation of "footprints" on the surface of silica.

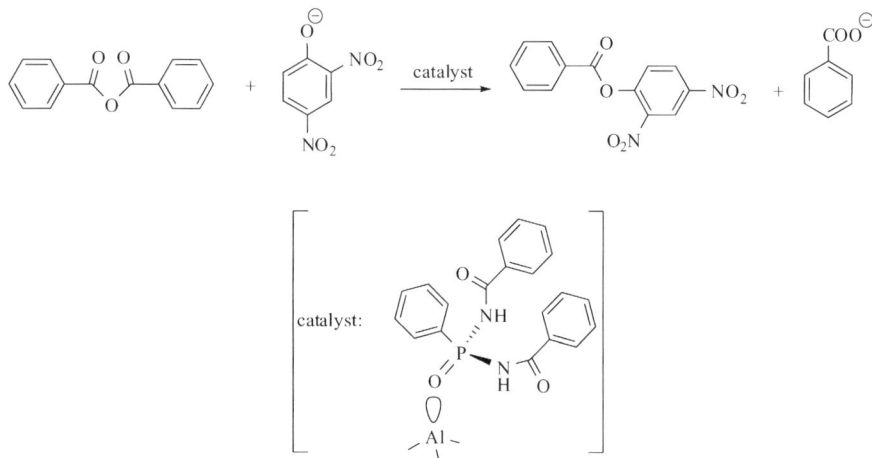

Scheme 24.11   Schematic representation of imprinting with a phosphonic diamide as a stable transition state analogue template.

The substrate was an acid anhydride, and the nucleophile was 2,4-dinitrophenolate, the consumption of which was followed photometrically. In this case (one of the best in a large series), k_{cat} increases by a factor of 10, and K_m is improved by a factor of 3 as compared to non-imprinted materials.[37d] A comparison with other imprinting methods with regard to k_{cat} and K_m is difficult; since the exact number of active sites is not incorporated in the calculations, second-order rate constants are used for k_{cat} and the reported K_m values are indeed mostly K_{ass} (M^{-1}) values. Therefore Morihara's k_{cat}/K_m values do not refer to the usual Michaelis–Menten kinetics. In comparison with the control silicas, however, the imprinted silicas show a remarkable catalytic activity.

In a number of examples, substrate specificity is shown. If an optically active template is used as a transition state analogue for imprinting, these catalysts cause enantiomers to react at different rates (kinetic racemic resolution). The substrate enantiomer corresponding to the template enantiomer reacts 2–4 times more rapidly than the other.[37e,f,h,k,m] Other reactions, such as crossed aldol condensation, enantioselective racemization, and asymmetric reductions, have also been investigated, though no details are available at present.[36]

The footprint cavity approach has reached quite promising results. At the moment it is not easy to understand the mechanism completely. Apart from the Lewis acid–Lewis base interaction, no defined interactions between silica and template are known, and no investigations in this direction have been made. Only a little shape selectivity should be possible in shallow cavities. Unfortunately, no other research groups have used this method until now. In view of the interesting results, more detailed knowledge of the method is desirable.

24.5.3.2 Other Examples for Catalytically Active Imprinted Silicas

Imprinted silicas for catalysis were also prepared by Heilmann and Meier.[38] They used 1-triethoxysiloxy(phenyl)methanephosphonic ethyl hexyl ester as a transition state analogue for a trans-esterification reaction. After a sol–gel process with this compound and tetraethoxysilane, the template was removed by calcination at 523 K, leaving behind cavities with a shape of the transition state of the trans-esterification.

An acceleration of the trans-esterification was indeed observed,[38] but more detailed investigations showed that this effect was not due to an imprinting effect but to phosphoric acid left at the polymer and in solution.[39] More similar to the usual imprinting method in polymers, silicas with defined binding sites have been prepared that were introduced by poly-condensation of functionalized silanes together with tetraethoxysilanes. In this case the polymerizable double bond has been replaced by functionalized silanes (see Figure 24.7). Selective recognition was also observed.[22a,23,40] This method has also been used to prepare catalytically active silicas. *Via* carbamate moieties to a template molecule, one, two, or three amino groups (see Figure 24.8) are introduced in cavities during a sol–gel process. Initial results (without experimental details) indicate catalytic activity for a Knoevenagel condensation.

In several recent papers Markowitz *et al.* reported on a very original molecular surface imprinting method.[41] This method involves the surface

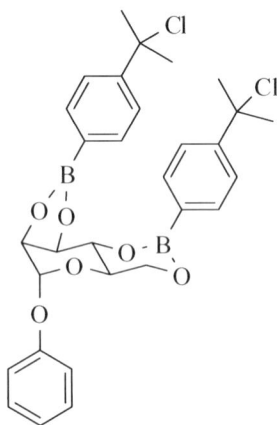

Figure 24.7

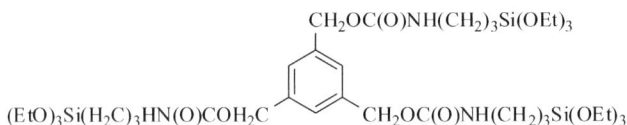

Figure 24.8

imprinting of silicas during the formation of the particles in a sol–gel process. A surfactant derivative of the imprint molecule is used as the template during the poly-condensation of tetraethoxysilane in the presence of added functionalized organosilanes. First, the surfactant imprint molecule, *e.g.*, *N*-decanoyl-L-phenylalanine-*N*-pyridin-2-ylamide (Figure 24.9), is incorporated in a water-in-oil microemulsion with added non-ionic surfactant. Tetraethoxysilane and a mixture of amine, dihydroimidazole, and carboxylate-terminated organosilanes were added to commence base-catalyzed surface imprinting and particle formation (see Scheme 24.12).

Since the silica particles are formed by a microemulsion process, the imprint molecule, which acts as the head-group of the surfactant, should be positioned at the surfactant/water interface of the reverse micelles within which the silica particles are formed. As a consequence, catalytic sites should only be formed at the surface and they should all have the same orientation with regard to the surface.

L-Phenylalanine-*N*-pyridin-2-ylamide can be regarded as a stable transition state analogue of the *R*-chymotrypsin-catalyzed amide fission of peptides. Although the catalyst was imprinted with a phenylalanine derivative, the imprinted silica catalyzed the hydrolysis of an arginine derivative best (Scheme 24.13).

An 4.8-fold enhancement in the initial rate is observed compared to a non-imprinted control having the same functional groups. The most remarkable result is the enantioselectivity as silica imprinted with an L-phenylalanine

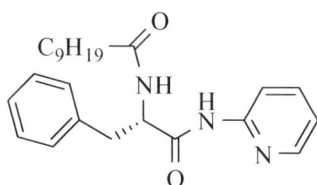

Figure 24.9

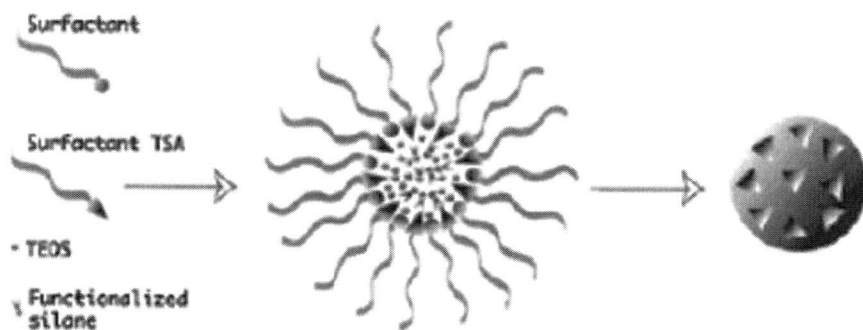

Scheme 24.12   Template-directed molecular imprinting of silica particles.

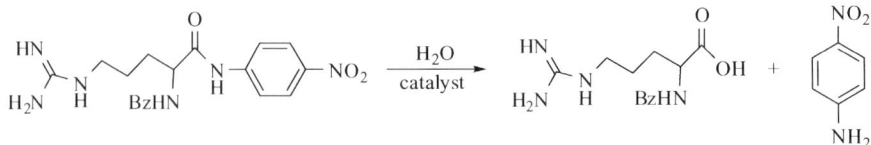

Scheme 24.13   Hydrolysis of an arginine derivative over silica.

derivative catalyzes the hydrolysis of the D-enantiomer of an arginine derivative 34 times faster than the L-enantiomer.[41a] This is by far the highest enantioselectivity for imprinted materials published to date. The reversed enantioselectivity (catalyst imprinted with L-enantiomer hydrolyses the D-substrate more actively) might be due to the surfactant moiety in the template. If the D-enantiomer is used as template instead of the L-enantiomer, the opposite enantioselectivity is observed. This result is to be expected, but it rules out possible experimental errors. Further development of this method might substantiate an important new way toward catalytic imprinted silicas and clarify some open problems.

24.5.3.3   Imprinting in Zeolites

Zeolites consist of a crystal lattice having defined pores and cavities throughout. Zeolites with very different cavity diameters are known. A very detailed review article deals with the numerous possibilities for using these compounds in catalysis.[42] Until now, no direct imprinting in zeolites has been published. There are reports on the synthesis of zeolites in the presence of certain organic molecules acting as "templates" in order to control the type of lattice being formed. This is no molecular imprinting in the original sense. Zeolites should be interesting candidates for imprinting inside the holes. This might reduce the polyclonality of the cavities. A problem can be foreseen with regard to the mass transfer inside the zeolites containing additional organic or inorganic polymers inside the pores and holes. Further possibilities of application of zeolites in the imprinting procedure are discussed by Davis et al.[21a]

24.5.3.4   Future Prospects in Catalysis

What has been reached in the preparation of catalysts by molecular imprinting in polymers or silicas? Over the last 10 years a considerable progress in the preparation of efficient catalysts has been made.

Enzymes are in every case several orders of magnitude catalytically more efficient, but in a few cases imprinted polymers have reached the activity of catalytic antibodies, *e.g.*, in the hydrolysis of carbamates.[43] This is surprising, since monoclonal antibodies are compared with "polyclonal imprinted" catalysts, and the imprinted materials are insoluble and rigid, whereas antibodies are soluble and more flexible. The binding site homogeneity in enzymes and monoclonal antibodies is high, whereas imprinted polymers, as discussed before, have a broad distribution of activity and there is no method available at the moment to really reduce this broadness.

A real advantage of imprinted catalysts is the ease of preparation and handling. They can be prepared in large quantities by suspension polymerization, and stable particles of uniform diameter can be easily obtained. Imprinted polymers can be applied directly in chemical processes. Such catalysts can also be prepared, in addition to beads or broken particles, in other very different forms, such as monoliths, microcapsules, membranes, surfaces. At the same time these materials are rather stable. Whereas enzymes and antibodies degrade under harsh conditions such as high temperature, chemically aggressive media, and high and low pH, imprinted polymers show better behavior in most cases. They have both good mechanical and thermal stability. Usually they can be used for a long time in a continuous process, or they can be reused many times. As a result of the insolubility of the materials, they can be easily filtered off after a reaction, or they can be placed in a flow reactor. All this brings a lot of advantages in the use of imprinted polymers or silicas. Though quite some progress has been made in the preparation of catalysts by molecular imprinting, for large application in industry and for broader application in research, further achievements have to be made. On one hand, the imprinting procedure has to be further improved and new approaches in the preparation of catalysts have to be used. At present, the following problems are in the forefront of investigations to improve the molecular imprinting procedure:

(a) molecular imprinting in microparticles during suspension or emulsion polymerization;
(b) imprinting procedures in aqueous solutions;
(c) imprinting with high molecular weight templates, biopolymers, or even bacteria by surface imprinting;
(d) development of new and better binding sites in molecular imprinting;
(e) improvement of the mass transfer in imprinted polymers;
(f) reduction of the "polyclonality" of cavities;
(g) increase of available active sites, especially with the usual non-covalent interaction;
(h) development of extremely sensitive detection methods for use in chemosensors; and finally
(i) development of further suitable groupings for catalysis.

24.5.4 Catalytic Antibodies and a Few Examples of Radical Transformations

The most established applications of the above TSA selection strategy lie in the field of catalytic antibodies, pioneered by Lerner and Schultz in the mid-1980s. The immune system generates a natural library of hosts, known as antibodies, in response to the introduction of a foreign molecule into the bloodstream. Advances in molecular biology techniques, notably the process of isolating monoclonal antibodies, allow the selection for a chosen antibody library members on the basis of function.

Traditionally, catalytic antibody technology focused on a purist TSA approach: the TSA was designed, synthesized and then used as a hapten in immunization (a hapten is a small molecule attached to a carrier protein which is used to stimulate the immune response). The desired monoclonal antibody was then selected from the polyclonal population on the basis of binding affinity to the TSA. Early efforts produced a range of successes in various synthetic transformations, affording artificial antibody catalysts for ester hydrolysis, the Diels–Alder reaction, cationic cyclizations, cyclopropanation, elimination reactions, the oxy-Cope rearrangement, and an allylic sulfoxide and sulfonate rearrangement, amongst others. However, rate accelerations have always fallen short of their enzyme catalyzed equivalents. Furthermore, detailed mechanistic investigations often revealed that a mechanism other than that originally assumed for the design of the TSA was involved. This has important consequences. Since the selection event is based on binding to the TSA and not on the basis of catalytic activity, the antibody selected may not be the best catalyst. The transition state is after all, not a discrete molecular entity and any TSA can only be expected to be an approximation of the true charge distribution required.

24.5.4.1 Antibody-catalyzed Enantioselective Norrish type II Cyclization

The most successful approach to limiting the broad variety of possible photoproducts has been based on solid-state reactions, in which the crystal-packing forces severely restrict the range of available conformations.[44] However, for *syn*-conformers, the Norrish type II photochemical reaction involves abstraction of a γ-hydrogen atom by an excited carbonyl oxygen atom (*e.g.* in **1**) to produce a 1,4-diradical intermediate, such as A.[45] The latter can undergo three possible reactions: (a) reverse hydrogen transfer to regenerate the ground state of **1**; (b) C–C bond cleavage to form an alkene **2** and an enol that tautomerizes to the carbonyl compound **3**; or (c) radical recombination[46] to produce the cyclobutanols **4** and **5** (Scheme 24.14). Usually, pathway (b) is

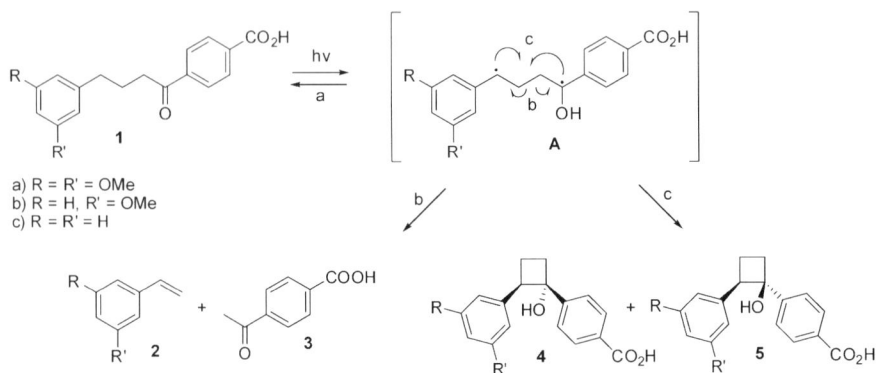

Scheme 24.14 Norrish type II reaction with possible pathways of the 1,4-diradical intermediate.

the most common route, while (c) represents a minor side reaction. Intense mechanistic studies over the past three decades have made the Norrish type II reaction one of the most well understood photochemical reactions.[45] Unfortunately, this important reaction has not yet enjoyed comparable status in synthetic chemistry, mainly as a result of a lack of control over product selectivity and, in particular, stereoselectivity.[47]

24.5.4.2 Catalysis of the Photo-Fries Reaction: Antibody-mediated Stabilization of High Energy States

A conformationally constrained hapten is presented that is capable of catalyzing the first antibody-mediated photo-Fries rearrangement.[48] In this reaction, absorption of light energy by a diphenyl ether substrate results in homolytic C–O bond cleavage followed by recombination to yield biphenyl-derived products (Scheme 24.15). The most proficient antibody studied converts 4-phenoxyaniline into 2-hydroxy-5-aminobiphenyl under high-intensity irradiation at a rate of 8.6 μM min^{-1}. These results support a recent hypothesis stating that immunization with conformationally constrained haptens provides higher titers for the acquisition of simple binding antibodies; however, in this case, the conformational constraint does not ensure the development of more efficient catalysts. Using the obtained antibodies, the presence of products resulting from escape of free radicals from the solvent cage can be suppressed, altering the excited state energy surface such that free radicals are funneled into the formation of the desired biphenyl product. However, studies also show the inactivation of the antibodies as a result of photo-decay of the biphenyl product. Using an isocyanate scavenging resin, the photo-decay product could be removed and the inactivation of the antibody drastically reduced. Furthermore, despite the observed photo-decay, turnover of the antibody was present; this represents the first case in which true

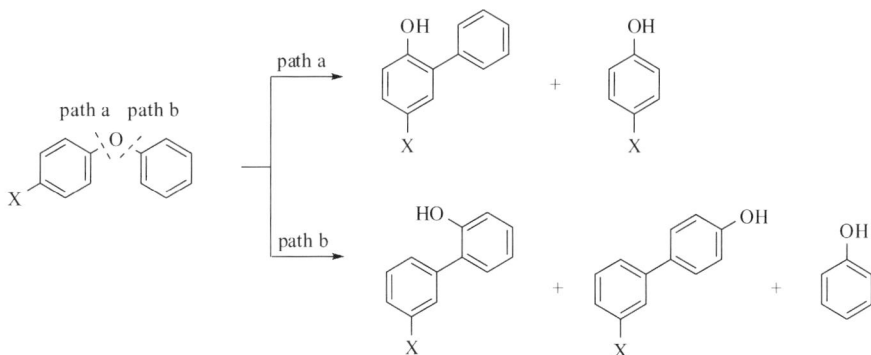

When X=EDG, (OMe, OH, NH$_2$, CH$_3$), path a >> path b
When X=EWG, (COOMe, CN), path a << path b

Scheme 24.15   Photo-Fries rearrangement.

turnover of a photochemical reaction using a catalytic antibody could be observed.

24.5.4.3   *Photochemistry of Phenyl Phenylacetates Included within Zeolites and Nafion Membranes*

The term microreactor refers to organized and constrained media where the chemical reactions occur. The substrate is usually a small molecule of dimensions of several angstroms, and the microreactors often have a size of tens of angstroms or larger. These microreactors are also known as nanoreactors. Among the many classes of microreactors used in photochemical studies, molecular-sieve zeolites, Nafion membranes, vesicles, and low-density polyethylene films are outstanding members. Molecular-sieve zeolites represent a unique class of materials.[49] The framework of the materials contains pores, channels, and cages of different dimensions and shapes. The pores and cages can accommodate, selectively according to size/shape, a variety of organic molecules of photochemical interest, and provide restrictions on the motions of the included guest molecules and reaction intermediates.

The photochemistry of the ester in Scheme 24.16 is expected to be analogous to that of phenyl acetate whose photochemistry in homogenous solutions has been well investigated and understood.[50] Scheme 24.16 gives the example of photochemical reactions of these esters. Upon photo-irradiation, esters undergo homolytic cleavage of the C–O bond to give two paired radicals. These geminate radical pairs in the cage recombine to form the starting ester or *ortho*- and *para*-hydroxyphenones. The latter reaction is known as photo-Fries rearrangement. The proposed mechanism is described in Scheme 24.16. The authors propose that the product distribution of the reactions can be directly related to the shape, size and external surface of the zeolites involved, making the transformation unique and potentially stereo and regio-selective.

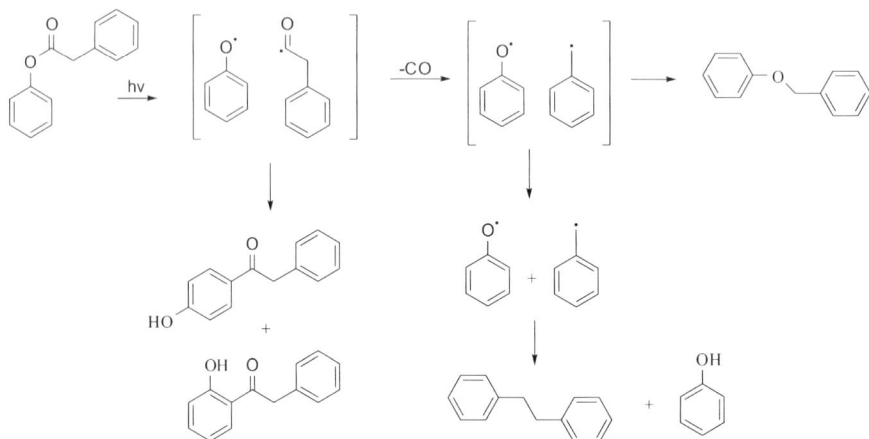

Scheme 24.16   Photochemistry of phenyl phenylacetates in zeolites.

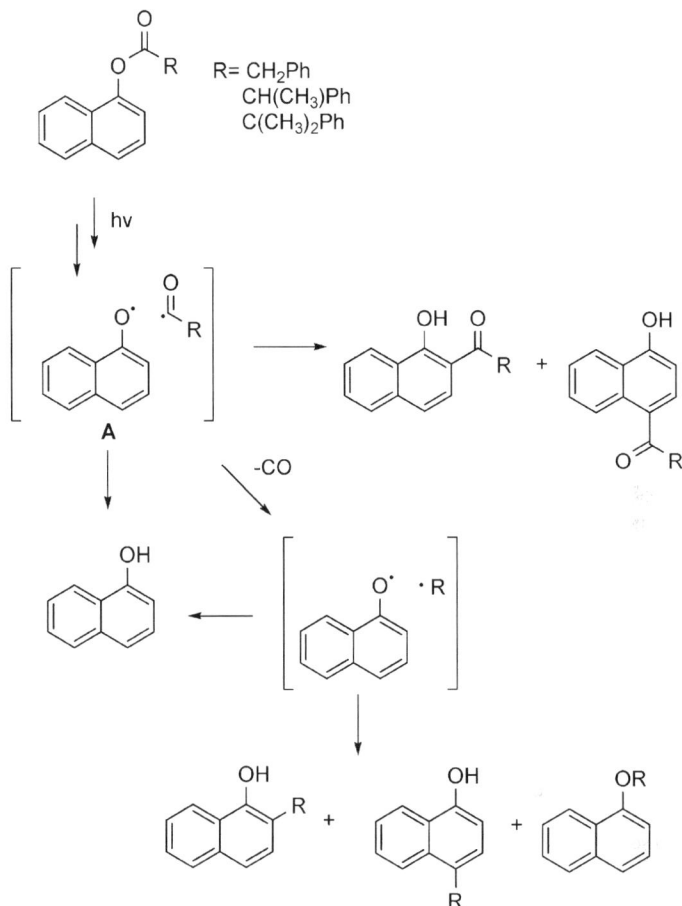

Scheme 24.17   Photo-Fries rearrangement of 1-naphtyl esters in zeolite compartment.

Using the photo-Fries rearrangements of three 1-naphthyl phenylacylates (Scheme 24.17), Ramamurthy *et al.* demonstrate that limiting the constraining space of a reaction cavity in an organized medium, such as zeolite, can be less important than wall–guest interactions in determining the selectivity of the reactions.[51] Reaction cavities of the media employed, cation-exchanged Y-zeolites, and a high density polyethylene film of 71% crystallinity possess very different properties. Irradiation of 1-naphthyl acetate or 1-naphthyl benzoate within alkali metal cation-exchanged X and Y-zeolites give a single photo-Fries photoproduct, 2-acyl-1-naphthol. In contrast, in hexane solution, the 2- and 4- isomers were formed in comparable yields. Selectivity in the zeolites was suggested to result from restrictions imposed on the naphthoxy and acyl radicals by cations along the cavity walls. Support of this observation comes from the work of Frei *et al.* that shows that acetyl radicals (from 1-naphthyl acetate) live for $>10^{-5}$ s at room temperature within NaY zeolite.[52] The excellent yields of 2-acetyl-1-naphthol from 1-naphthyl acetate and the long radical lifetimes imply that the radical pairs **A** (R = methyl) are held tightly in place by cations before rejoining; they have the time and space, *but not the mobility,* to undergo other rearrangements (Scheme 24.17).[53]

24.6 CONCLUSION

The field of artificial enzymes is a rapidly evolving subject. As the barrier between chemistry and biology becomes less distinct, a range of new methods which combine expertise from both areas are developing. In recognition of both the fact that the *de novo* design approach can be time-consuming, and that a tiny miscalculation will have a detrimental effect, a trend in all these recent techniques is the use of "selection approaches". The natural processes of selection and amplification is, after all, the way in which enzymes have evolved their sophisticated function and also the area is constantly evolving—the sky is the limit in the desired crossing and elimination of the barrier between "traditional chemistry" and "traditional biological sciences" and developments and understanding of life science at the molecular level.

REFERENCES

1. D. Voet and J. G. Voet, *Biochemistry*, J. Wiley & Sons, Hoboken, NJ, 2004; R. A. Copeland, *Enzymes, a practical introduction to structure, mechanism, and data analysis*, Wiley-VCH, New York, 2000.
2. (a) L. Pauling, The nature of forces between large molecules of biological interest, *Nature,* 1948, **161**, 707; (b) D. M. Blow, Structure and mechanism of chymotrypsin, *Acc. Chem. Res.,* 1976, **9**, 145; (c) D. M. Blow, J. J. Birktoft and B. S. Hartley, Role of a buried acid group in the mechanism of action of chymotrypsin, *Nature,* 1969, **221**, 337; (d) A. M. Davis and S. J. Teague, Hydrogen bonding, hydrophobic interactions, and failure of the rigid receptor hypothesis, *Angew. Chem., Int. Ed.,* 1999, **38**, 737; (e) W. B. Motherwell, M. J. Bingham and Y. Six, Recent progress in the design and synthesis of artificial enzymes, *Tetrahedron,* 2001, **57**, 4663.
3. For leading references see: (a) R. Breslow and S. D. Dong, Biomimetic Reactions Catalyzed by Cyclodextrins and Their Derivatives, *Chem. Rev.,* 1998, **98**, 1997–2011; (b) B. Avalle, A. Friboulet and D. J. Thomas, Enzymes and abzymes relationships, *Mol. Catal. B: Enzym.,* 2000, **10**, 39–45; (c) P. Wentworth and K. D. Janda, Catalytic antibodies, *Curr. Opin. Chem. Biol.,* 1998, **2**, 138–144; (d) D. B. Smithrud and S. J. Benkovic, The state of antibody catalysis, *Curr. Opin. Biotechnol.,* 1997, **8**, 459–466; (e) A. Kirby, The potential of catalytic antibodies, *J. Acta Chem. Scand.,* 1996, **50**, 203–210; (f) P. G. Schultz and R. A. Lerner, From molecular diversity to catalysis: lessons from the immune system, *Science,* 1995, **269**, 1835–1842.
4. R. Breslow and C. Schmuck, Goodness of Fit in Complexes between Substrates and Ribonuclease Mimics: Effects on Binding, Catalytic Rate Constants, and Regiochemistry, *J. Am. Chem. Soc.,* 1996, **118**, 6601–6605.

5. (a) A. Tramontano, K. D. Janda and R. A. Lerner, Catalytic antibodies, *Science,* 1986, **234**, 1566–1570; (b) S. J. Pollack, J. W. Jacobs and P. G Schultz, Selective Chemical Catalysis by an Antibody, *Science,* 1986, **234**, 1570–1573.

6. (a) E. M. Gordon and J. F. Kerwin, *Combinatorial Chemistry and Molecular Diversity in Drug Design,* John Wiley & Sons, New York, 1998; (b) S. R. Wilson and A. W. Czarnik, *Combinatorial Chemistry: Synthesis and Applications,* John Wiley & Sons, New York, 1997; (c) F. M. Menger, C. A. West and J. Ding, *J. Chem. Soc., Chem. Commun..* 1997, 633–634; (d) F. M. Menger, A. V. Eliseev and V. A. Mingulin, Phosphatase Catalysis Developed via. Combinatorial Organic Chemistry, *J. Org. Chem.,* 1995, **60**, 6666–6667.

7. (a) F. M. Menger, Analysis of ground-State and transition-state effects in enzyme catalysis, *Biochemistry,* 1992, **31**, 5368–5373; (b) M. M. Mader and P. A. Bartlett, Binding Energy and Catalysis: The Implications for Transition-State Analogs and Catalytic Antibodies, *Chem. Rev.,* 1997, **97**, 1281–1301.

8. A. Fersht, *Enzyme Structure and Mechanism,* W. H. Freeman and Co., New York, 2nd edn, 1985.

9. R. U. Lemieux, How water provides the impetus for molecular recognition in aqueous solution, *Acc. Chem. Res.,* 1996, **29**, 373–380.

10. J. D. Dunitz, The Entropic Cost of Bound Water in Crystals and Biomolecules, *Science,* 1994, **264**, 670.

11. D. H. Williams, M. S. Searle, J. P. Mackay, U. Gerhard and R. Maplestone, Toward an estimation of binding constants in aqueous solution: Studies of associations of vancomycin group antibiotics, *Proc. Natl. Acad. Sci. U. S. A.,* 1993, **90**, 1172–1178.

12. A. R. Fersht, J. P. Shi, J. Knill-Jones, D. A. Lowe, A. J. Wilkinson, D. M. Blow, P. Brick, P. Carter, M. M. Y. Waye and G. Winter, Hydrogen bonding and biological specificity analysed by protein engineering, *Nature,* 1985, **314**, 235–238.

13. F. Garcia-Tellado, S. Goswami, S.-K. Chang, S. J. Geib and A. D. Hamilton, Synthetic analogs of the ristocetin binding site: Neutral, multidentate receptors for carboxylate recognition, *J. Am. Chem. Soc.,* 1990, **112**, 7393–7394.

14. A. Fan, S. Van Arman, S. Kincaid and A. D. Hamilton, Molecular recognition: hydrogen-bonding receptors that function in highly competitive solvents, *J. Am. Chem. Soc.,* 1993, **115**, 369–370.

15. (a) T. Inoue, C. Weber, A. Fujishima and K. Honda, An Investigation of the Power Characteristics in Heterogeneous Electrochemical Photovoltaic Cells for Solar-energy Utilization, *Bull. Chem. Soc. Jpn.,* 1980, **53**, 334; (b) M. Komiyama and H. Hirai, General base catalyses by alpha-cyclodextrin in the hydrolyses of alkyl benzoates, *Chem. Lett.,* 1980, 1251; (c) K. Shokat, T. Uno and P. G. Schultz, Mechanistic Studies of an Antibody-Catalyzed Elimination Reaction, *J. Am. Chem. Soc.,* 1994, **116**, 2261; (d)

R. Breslow and N. Nesnas, *Tetrahedron Lett.*, 1999, **40**, 3335: (e) P. Tastan and E. U. Akkaya, A novel cyclodextrin homodimer with dual-mode substrate binding and esterase activity, *J. Mol. Catal. A: Chem.*, 2000, **157**, 261; (f) J. X. Yu, Y. Z. Zhao, M. J. Holterman and D. L. Venton, Combinatorial search of substituted β-cyclodextrins for phosphatase-like activity, *Bioorg. Med. Chem. Lett.*, 1999, **9**, 2705.

16. H.-L. Chen and B. Zha, Cyclodextrin in artificial enzyme model, rotaxane, and nano-material fabrication, *J. Inclusion Phenom. Macrocyclic Chem.*, 2006, **56**, 17–21.

17. R. Breslow, P. J. Duggen and J. P. Light, Cyclodextrin-B12, a potential enzyme-coenzyme mimic, *J. Am. Chem. Soc.*, 1992, **114**, 3982–3983.

18. W. P. Jencks, *Catalysis in chemistry and enzymology,* McGraw-Hill, New York, 1969.

19. J. Matsui, I. A. Nicholls, I. Karube and K. Mosbach, Carbon−Carbon Bond Formation Using Substrate Selective Catalytic Polymers Prepared by Molecular Imprinting: An Artificial Class II Aldolase , *J. Org. Chem.*, 1996, **61**, 5414.

20. (a) G. Wulff and A. Sarhan, On the use of enzyme-analogue-built polymers for racemic resolution, *Angew. Chem., Int. Ed. Engl.*, 1972, **11**, 341; (b) G. Wulff and A. Sarhan, German Patent, Offenlegungsschrift DE-A 2242796, 1974; *Chem. Abstr.* 1975, **83**, P 60300*w*; US Patent, continuation in part US-A 4127730, 1978; (c) G. Wulff, W. Vesper, R. Grobe-Einsler and A. Sarhan, Enzyme-analogue built polymers, 4. On the synthesis of polymers containing chiral cavities and their use for the resolution of racemates, *Makromol. Chem.*, 1977, **178**, 2799–2816.

21. (a) M. E. Davis, A. Katz and W. R. Ahmad, Rational Catalyst Design via Imprinted Nanostructured Materials, *Chem. Mater.*, 1996, **8**, 1820–1839; (b) M. E. Davis, *CATTECH,* 1997, **1**, 19–26.

22. (a) G. Wulff, Molecular Imprinting in Cross-Linked Materials with the Aid of Molecular Templates, *Angew. Chem., Int. Ed. Engl.*, 1995, **34**, 1812–1832; (b) G. Wulff, *CHEMTECH,* 1998, **28**, 19–26; (c) G. Wulff, in *Polymeric Reagents and Catalysts*, ed. W. T. Ford, ACS Symp. Ser. 308, American Chemical Society, Washington, DC, 1986, pp. 186–231; (d) G. Wulff, in *Templated Organic Synthesis*, ed. F. Diederich and P. J. Stang, Wiley-VCH, Weinheim, 1999, pp. 39–73.

23. K. Mosbach, Molecular imprinting, *Trends Biochem. Sci.*, 1994, **19**, 9–14.

24. K. Haupt and K. Mosbach, Molecularly Imprinted Polymers and Their Use in Biomimetic Sensors, *Chem. Rev.*, 2000, **100**, 2495–2504.

25. K. Mosbach and O. Ramstrom, The emerging technique of molecular imprinting and its future impact on biotechnology, *Biotechnology*, 1996, **14**, 163–170.

26. (a) S. R. Tonge and B. J. Tighe, Responsive hydrophobically associating polymers: a review of structure and properties, *Adv. Drug Delivery Rev.*, 2001, **53**, 109–122; (b) C. Alvarez-Lorenzo and A. Concheiro, Molecularly imprinted polymers for drug delivery, *J. Chromatogr., B,* 2004, **804**, 231–

245; (c) R. Langer and N. A. Peppas, Advances in biomaterials, drug delivery, and bionanotechnology, *AIChE J.,* 2003, **49**, 2990–3006; (d) D. X. Cui and H. J. Gao, Advances and prospects of bionanomaterials, *Biotechnol. Prog.,* 2003, **19**, 683–692.

27. J. Suh and S. S. Hah, S. S. Organic artificial proteinase with active site comprising three salicylate residues, *J. Am. Chem. Soc.*, 1998, **120**, 10088–10093.

28. C. J. Slade and E. N. Vulfson, Induction of catalytic activity in proteins by lyophilization in the presence of a transition state analogue, *Biotechnol. Bioeng.,* 1998, **57**, 211.

29. F. Peissker and L. Fischer, Crosslinking of imprinted proteases to maintain a tailor-made substrate selectivity in aqueous solution, *Bioorg. Med. Chem.,* 1999, **7**, 2231.

30. S. Ozawa and A. M. Klibanov, Myoglobin-catalyzed epoxidation of styrene in organic solvents accelerated by bioimprinting, *Biotechnol. Lett.,* 2000, **22**, 1269.

31. M. T. Reetz, Combinatorial and evolution-based methods in the creation of enantioselective catalysts, *Angew. Chem., Int. Ed.,* 2001, **40**, 284.

32. (a) U. T. Bornscheuer, J. Altenbuchner and H. H. Meyer, Directed evolution of an esterase: screening of enzyme libraries based on pH-indicators and a growth assay, *Bioorg. Med. Chem.,* 1999, **7**, 2169; (b) H. Joo, Z. Lin and F. H. Arnold, Laboratory evolution of peroxide-mediated cytochrome P450 hydroxylation, *Nature,* 1999, **399**, 670; (c) H. Joo, A. Arisawa, Z. L. Lin and F. H. Arnold, Directed evolution of glucose oxidase from Aspergillus niger for ferrocenemethanol-mediated electron transfer, *Chem. Biol.,* 1999, **208**, 699.

33. F. H. Dickey, The Preparation of Specific Adsorbents, *Proc. Natl. Acad. Sci. U. S. A.,* 1949, **35**, 227–229; F. H. Dickey, Molecularly imprinted polymers: a new approach to the preparation of functional materials, *J. Phys. Chem.,* 1955, **59**, 695–707.

34. S. B. Bernhard, The preparation of specific adsorbents, *J. Am. Chem. Soc.,* 1952, **74**, 4946–4947.

35. I. A. Nicholls and H. S. Andersson, in Molecularly Imprinted Polymers Man-Made Mimics of Antibodies and Their Application in Analytical Chemistry, ed. B. Sellergren, Elsevier, Amsterdam, 2001, pp. 1–19.

36. K. Morihara, in *Molecular and Ionic Recognition with Imprinted Polymers,* ed. R. A. Bartsch and M. Maeda, ACS Symp. Ser. 703, American Chemical Society, Washington, DC, 1998, pp. 300–313.

37. (a) K. Morihara, S. Kurihara and S. Suzuki, *Bull. Chem. Soc. Jpn.,* 1988, **61**, 3991–3998; (b) K. Morihara, E. Nishihata, M. Kojima and S. Miyazaki, *Bull. Chem. Soc. Jpn.,* 1988, **61**, 3999–4003; (c) K. Morihara, E. Tanaka, Y. Takeuchi, K. Miyazaki, N. Yamamoto, Y. Sagawa, E. Kawamoto and T. Shimada, *Bull. Chem. Soc. Jpn.,* 1989, **62**, 499-505; (d) T. Shimada, K. Nakanishi and K. Morihara, *Bull. Chem. Soc. Jpn.,* 1992, **65**, 954–958; (e) K. Morihara, M. Kurokawa, Y. Kamata and T. Shimada,

J. Chem. Soc., Chem. Commun., 1992, 358–360; (f) T. Matsuishi, T. Shimada and K. Morihara, *Chem. Lett.,* 1992, 1921–1924; (g) T. Shimada, R. Kurazono and K. Morihara, *Bull. Chem. Soc. Jpn.,* 1993, **66**, 836–840; (h) K. Morihara, S. Kawasaki, M. Kofuji and T. Shimada, *Bull. Chem. Soc. Jpn.,* 1993, **66**, 906–913; (i) K. Morihara, S. Doi, M. Takiguchi and T. Shimada, *Bull. Chem. Soc. Jpn.,* 1993, **66**, 2977–2982; (j) K. Morihara, T. Iijima, H. Usui and T. Shimada, *Bull. Chem. Soc. Jpn.,* 1993, **66**, 3047–3052; (k) T. Matsuishi, T. Shimada and K. Morihara, *Bull. Chem. Soc. Jpn.,* 1994, **67**, 748–756; (l) T. Shimada, R. Hirose and K. Morihara, *Bull. Chem. Soc. Jpn.,* 1994, **67**, 227–235; (m) K. Morihara, M. Takiguchi and T. Shimada, *Bull. Chem. Soc. Jpn,* 1994, **67**, 1078–1084.

38. J. Heilmann and W. F. Maier, *Angew. Chem., Int. Ed.,* 1994, **33**, 471–473.

39. (a) W. R. Ahmad and M. E. Davis, *Catal. Lett.,* 1996, **40**, 109–114; (b) W. F. Maier and W. B. Mustapha, *Catal. Lett.,* 1997, **46**, 137–140.

40. G. Wulff, B. Heide and G. Helfmeier, *J. Am. Chem. Soc.,* 1986, **108**, 1089–1091.

41. (a) M. A. Markowitz, P. R. Kust, G. Deng, P. E. Schoen, J. S. Dordick, D. S. Clark and B. P. Gaber, *Langmuir,* 2000, **16**, 1759–1765; (b) M. A. Markowitz, G. Deng and B. Gaber, *Langmuir,* 2000, **16**, 6148–6155; (c) M. A. Markowitz, P. R. Kust, J. Klaehn, G. Deng and B. P. Gaber, *Anal. Chim. Acta,* 2001, **435**, 177–185.

42. A. Dyer, An Introduction to Zeolite Molecular Sieves, Wiley, New York, 1988.

43. A. G. Strikovsky, D. Kasper, M. Grün, B. S. Green, J. Hradil and G. Wulff, *J. Am. Chem. Soc.,* 2000, **122**, 6295–6296.

44. S. Saphier, S. C. Sinha and E. Keinan, Antibody-catalyzed enantioselective Norrish type II cyclization, *Angew. Chem., Int. Ed.*, 2003, **42**, 12.

45. P. J. Wagner, in CRC Handbook of Organic Photochemistry and Photobiology, CRC, Boca Raton, FL, 1995, p. 449.

46. N. C. Yang and D. H. Yang, *J. Am. Chem. Soc.,* 1958, **80**, 2913.

47. A. G. Griesbeck and H. Heckroth, *Res. Chem. Intermed.,* 1999, **25**, 599.

48. T. J. Dickerson, M. R. Tremblay, T. Z. Hoffman, D. I. Ruiz and K. D. Janda, Catalysis of the Photo-Fries reaction: Antibody-mediated stabilization of high energy states, *J. Am. Chem. Soc.*, 2003, **125**, 15395–15401.

49. (a) D. W. Breck, *Zeolite Molecular Sieves: Structure, Chemistry and Use*, Wiley, New York, 1974; (b) A. Dyer, *An Introduction to Zeolite Molecular Sieves*, Wiley, New York, 1988; (c) H. Van Bekkum, E. M. Flanigen and J. C. Jansen, *Introduction to Zeolite Science and Practice*, Elsevier, Amsterdam, 1991.

50. (a) C.-H. Tung and Y. M. Ying, Photochemistry of phenyl phenylacetates adsorbed on pentasil and faujasite zeolites, *J. Chem. Soc., Perkin Trans. 2,* 1997, 1319–1322; (b) C.-H. Tung and X. H. Xu, Selectivity in the photo-Fries reaction of phenyl phenylacetates included in a Nafion membrane, *Tetrahedron Lett.*, 1999, **40**, 127–130.

51. (a) K. Pitchumani, M. Warrier and V. Ramamurthy, Remarkable Product selectivity during Photo-Fries and Photo-Claisen rearrangements within zeolites, *J. Am. Chem. Soc.,* 1996, **118**, 9428–9429; (b) W. Gu, M. Warrier, V. Ramamurthy and R. G. Weiss, Photo-Fries reactions of 1-naphthyl esters in cation-exchanged zeolite Y and polyethylene media, *J. Am. Chem. Soc.,* 1999, **121**, 9467–9468.
52. (a) S. Vasenkov and H. Frei, Observation of acetyl radical in a zeolite by time-resolved FT-IR spectroscopy, *J. Am. Chem. Soc.,* 1998, **120**, 4031–4032; (b) S. Vasenkov and H. Frei, Time-resolved study of acetyl radical in zeolite NaY by step-scan FT-IR spectroscopy, *J. Phys. Chem. A,* 2000, **104**, 4327–4332.
53. (a) C.-H. Tung, L. Z. Wu, Z. Y. Yuan and N. Su, Remarkable product selectivity in photosensitized oxidation of alkenes within Nafion membranes, *J. Am. Chem. Soc.,* 1998, **120**, 11874–11879; (b) C.-H. Tung, L. Z. Wu, Z. Y. Yuan and N. Su, Zeolites as templates for preparation of large-ring compounds: Intramolecular photocycloaddition of diaryl compounds, *J. Am. Chem. Soc.,* 1998, **120**, 11594–11602; (c) C.-H. Tung, L. Z. Wu and H. R. Li, Reactions of singlet oxygen with olefins and sterically hindered amine in mixed surfactant vesicles, *J. Am. Chem. Soc.,* 2000, **121**, 2446–2451; (d) C.-H. Tung, L. Z. Wu and H. R. Li, Vesicle controlled selectivity in photosensitized oxidation of olefins, *Chem. Commun.,* 2000, **12**, 1085–1086.

CHAPTER 25

Applications of Conventional Free Radicals and Advances in Total Synthesis: from the Bench to Nature through SmI₂ Radicals as an Efficient Trigger for Radical Cascades, a Journey from Orsay to the 21st Century[*]

25.1 MECHANISMS OF SmI₂-MEDIATED REACTIONS: THE BASICS

The unique place held by SmI₂ in the arsenal of the synthetic chemist is a result of its versatility in mediating numerous important organic reactions, including reductions, reductive couplings and sequential reactions, all of which are described in subsequent chapters. Although numerous applications have been discovered for SmI₂, its scope in synthesis has not been exhausted and new applications for this mild and selective single-electron transfer reagent continue to be identified at a considerable rate. The aim of this chapter is to familiarize the reader with the basic mechanistic aspects of substrate reduction by SmI₂ and apply the newly acquired knowledge to following the development of the numerous applications for SmI₂-mediated free radical transformations routinely used in natural product synthesis. This information is important for the rational design of new reactions mediated by the reagent. As most bond-forming reactions using SmI₂ generate radicals (and subsequently anions) through the reduction of organohalides, or ketyls, through reduction of carbonyl groups, the main emphasis will be on these processes as they represent the foundation for the synthetic reactions.

* Chapter written by V. Tamara Perchyonok.
Streamlining Free Radical Green Chemistry
V. Tamara Perchyonok, Ioannis Lykakis and Al Postigo
© V. Tamara Perchyonok, Ioannis Lykakis and Al Postigo 2012
Published by the Royal Society of Chemistry, www.rsc.org

25.2 RADICALS AND ANIONS FROM ORGANOHALIDES

The reduction of alkyl and aryl halides by SmI_2 provides access to radicals that can undergo a range of follow-up reactions, including dimerization, reduction to an anion (organosamarium) or hydrogen atom abstraction from solvent. The early mechanistic studies of Kagan *et al.* on the reduction of alkyl halides with SmI_2 suggested that the organosamarium species were not intermediates in the reduction.[1] These studies were carried out by heating the substrate at reflux with SmI_2 and subsequently quenching the reaction mixture with D_2O. Alkane products were obtained with no deuterium incorporation, suggesting that alkyl radical intermediates were quenched by hydrogen atom abstraction from the ketyl radical tetrahydrofuran (THF).[1] However, when the reduction of 2-bromoadamantane was carried out at room temperature in the presence of D_2O using SmI_2– hexamethylphosphoramide (HMPA), the deuterated product was obtained in 80% yield. This result led to a comprehensive mechanistic study by Curran *et al.*[2] that provided convincing evidence for the formation of organosamarium intermediates in the reduction of alkyl halides. Curran *et al.* suggested that the lack of deuterium incorporation seen in Kagan's studies was a result of organosamarium decomposition at the elevated temperatures required.

The mechanism of the reduction of primary and secondary alkyl halides with SmI_2 in the presence of THF–HMPA is now known to proceed by dissociative electron transfer from SmI_2 to the alkyl halide, generating an alkyl radical **1**, followed by rapid reduction of the radical to an organosamarium species **2** that is then protonated to provide alkane **3** upon workup (Scheme 25.1).

The reduction of other carbon–halogen bonds is dependent on the substrate, solvent milieu and the presence of additives. Curran has shown that reduction of a tertiary alkyl radical, formed from a tertiary alkyl halide, can produce a tertiary organosamarium with limited stability at room temperature.[3] Most vinyl and aryl halides are reduced to radicals by SmI_2 and subsequently abstract a hydrogen atom from THF. This led Tani *et al.* to prepare aryl organosamarium species by reduction of aryl halides using SmI_2–HMPA in benzene.[4] It is therefore clear that both aryl and alkyl radicals can be reduced by SmI_2 or SmI_2–HMPA, but hydrogen-atom abstraction from solvent and/or dimerization can be a competing process.

Key to designing reactions using SmI_2 or SmI_2–HMPA is a fundamental understanding of the rates of reaction pathways and the seminal work of Curran is fundamentally important in this regard. Curran's detailed study of the reduction of alkyl halides provided rate constants for the reduction of primary, secondary and tertiary radicals with SmI_2–HMPA (Scheme 25.2).[2,5]

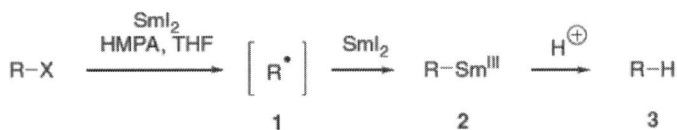

Scheme 25.1 General SmI_2 mediated transformation.

Hexenyl radicals were used as radical clocks for the indirect measurement of the rate of reduction of radicals to anions using SmI_2–HMPA. For example, the reduction of the primary iodide **4** using SmI_2–HMPA resulted in the isolation of a primary alkyl radical to a primary alkylsamarium. Reduction of the tertiary alkyl iodide **5** using SmI_2–HMPA resulted in the isolation of the cyclized–coupled product **8** as the only product. The reduction of a tertiary alkyl radical to a tertiary alkylsamarium is therefore slower ($k_E = 10^4$ M^{-1}s^{-1}) than the R–X reduction of primary radicals under the same conditions. From these data, it can be inferred that the reduction of a secondary alkyl radical to a secondary alkylsamarium is intermediate in rate ($k_E = 10^5$ M^{-1}s^{-1}).[2,5]. As the

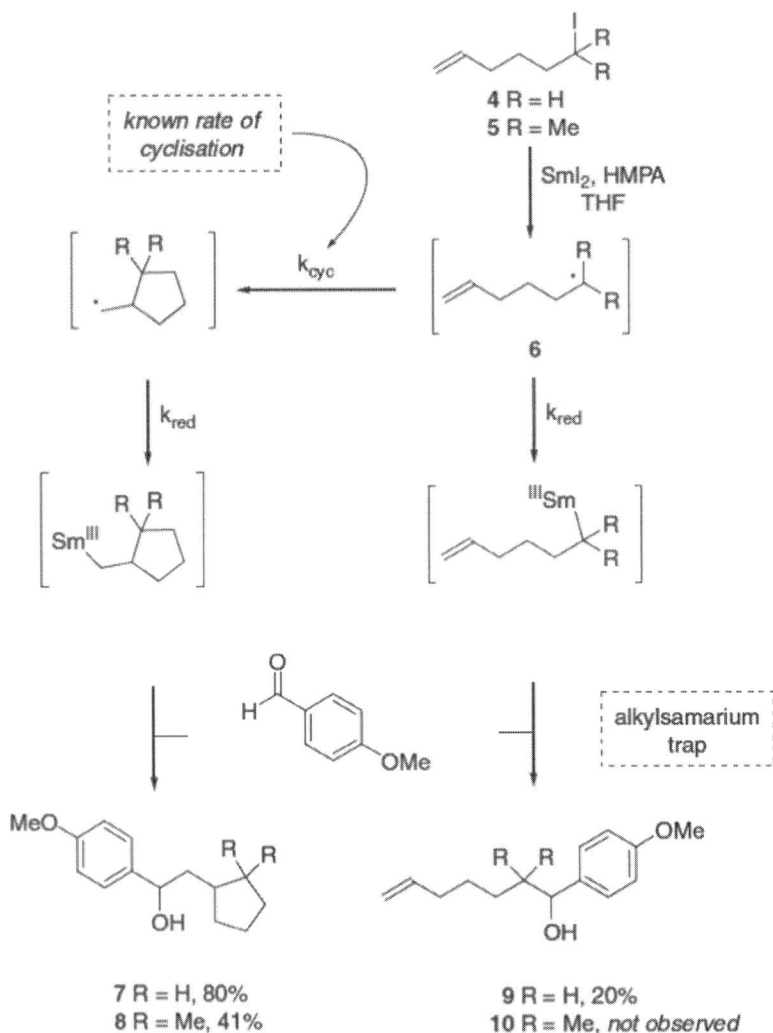

Scheme 25.2   Radical Clock for SmI2 mediated transformations.

rate of cyclization of the intermediate primary hexenyl radical **6** was known, a rate constant of $k_1 = 4 \times 10^6$ M^{-1} s^{-1} could be estimated for the reduction. It is important to note that a sufficient quantity of HMPA is required for these reductions. The efficiency of reduction of **4** (Scheme 25.2) was dependent on the amount of HMPA present. In the absence of HMPA, the reduction of **4** was too slow for the rate experiments. Appropriate reaction times were obtained after the addition of as little as 2 equiv. of HMPA (based on [SmI$_2$]) and the ratio of **9** to **7** increased upon continued addition of HMPA (up to 5 equiv.). These findings suggest that the rate of radical reduction can be fine-tuned by varying the amount of HMPA used. In practice, however, reactions that require the use of HMPA are rarely efficient if less than 4 equiv. of HMPA are used.

Curran's rate data for the reduction of alkyl radicals to the corresponding anions in the presence of SmI$_2$–HMPA is indispensable for the planning of radical and anionic transformations using the reagent: any proposed radical reaction using SmI$_2$ must be faster than the rate of reduction of the radical and, similarly, any proposed anionic process must take into account competition from fast radical reactions. Curran's approximate rate data are summarized in Scheme 25.3.

The mechanistic significance of the reduction of alkyl radicals to organosamariums is well illustrated by the continued ambiguity surrounding the mechanism of the SmI$_2$-mediated Barbier reaction. The SmI$_2$-mediated reductive couplings of an alkyl halide with a ketone or aldehyde are typically run using Barbier conditions, in which the substrates and SmI$_2$ are mixed together simultaneously in a reaction flask (Scheme 25.4).

Since both the alkyl halide and the carbonyl substrate can be reduced independently by SmI$_2$ or SmI$_2$–HMPA, mechanistic studies have been performed to determine whether the reaction occurs through the selective reduction of the alkyl halide to produce an organosamarium or whether the reaction is the result of an alternative mechanism involving the radical and

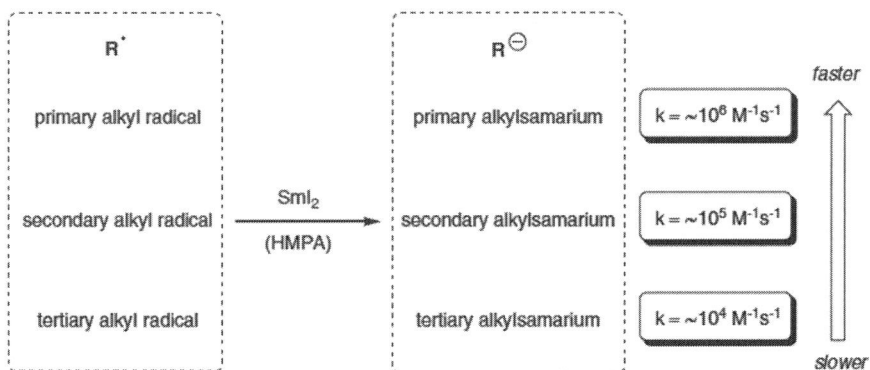

Scheme 25.3 Rate constants and trends for reduction of radicals to corresponding anions in the SmI$_2$ mediated transformations.

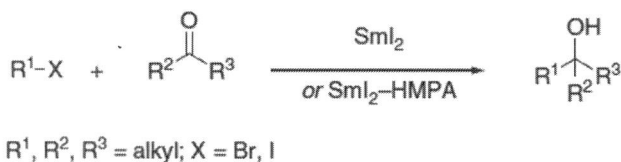

R^1, R^2, R^3 = alkyl; X = Br, I

Scheme 25.4   General reduction of alkyl halides and carbonyls under SmI$_2$ control: radical *vs.* ionic protocol.

ionic pathways. As previously mentioned, fast radical processes (such as hydrogen atom abstraction from solvent or cyclization) will compete with the reduction of a radical to an organosamarium. For example, Curran has shown that reduction of iodide with SmI$_2$–HMPA generates an aryl radical which cyclizes to provide high yields of organosamarium intermediates that can be trapped with a range of electrophiles.[6] In this case, the radical cyclization is efficient since it is several orders of magnitude faster than reduction of the intermediate aryl radical to an organosamarium intermediate (Scheme 25.5).

Although Curran's rate data for the reduction of radicals to organosamariums allow for an element of predictability,[2] problems can arise when multifunctional substrates are involved. For example, in the attempted intramolecular Barbier reaction of alkyl iodide, treatment with SmI$_2$ results in the formation of a side product in addition to the expected product cyclohexanol (Scheme 25.6).[7]

In this case, the β-keto amide motif is reduced at a rate competitive with alkyl iodide reduction, indicating that there are likely two mechanistic pathways

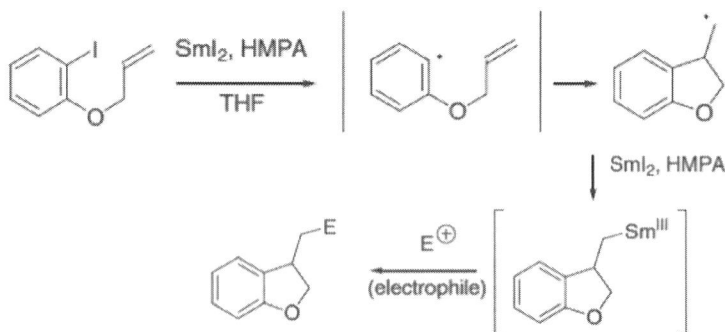

Scheme 25.5   Samarium iodide mediated and cyclization of aryl radical in detail.

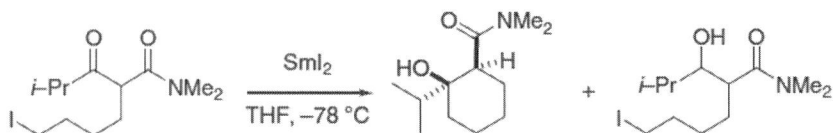

Scheme 25.6   Intramolecular Barbier reaction of functionalized alkyl iodides.

through which the reaction proceeds: a thermodynamic pathway initiated by reduction of the R–I bond providing the cyclohexanol and a kinetic pathway involving chelation of SmI$_2$ to the β-keto amide and reduction of the ketone carbonyl group to give 2-hydroxy-2-isopropyl-*N*,*N*-dimethylcyclohexanecarboxamide.[8] The recognition of structural features in substrates that can potentially alter the course of reactions is important when planning reactions employing SmI$_2$.

25.3 SmI$_2$-MEDIATED CYCLIZATIONS IN NATURAL PRODUCT SYNTHESIS

The total synthesis of natural products provides a rigorous testing ground for new reagents and new reactions. Many natural product targets have complex structures rich in functionality and sequences of highly selective reactions are required for their synthesis. The mild and selective nature of SmI$_2$ has made it an important tool for natural product synthesis and, in particular, the reagent has been used to mediate cyclizations to form a wide variety of carbocyclic and heterocyclic systems. As Procter *et al.* comprehensively reviewed the use of SmI$_2$-mediated cyclizations in natural product synthesis in 2004,[9] this section provides a selection of 'classical' applications of SmI$_2$-mediated cyclizations, alongside 'contemporary' examples from the recent literature. The examples are organized according to the size of the ring formed in the SmI$_2$-mediated cyclization.

25.4 FOUR-MEMBERED RING FORMATION USING SmI$_2$

25.4.1 A Synthesis of Paeoniflorin

In 1993, Corey reported the first synthesis of paeoniflorin.[10] The core of paeoniflorin was constructed using a SmI$_2$-mediated Reformatsky-type reaction (Scheme 25.7). Treatment of α-chloronitrile with SmI$_2$ gave

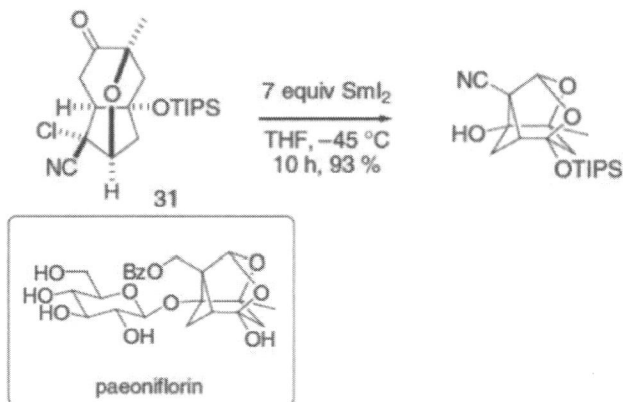

Scheme 25.7 Towards the synthesis of paeoniflorin: SmI$_2$-mediated Reformantsky type reaction.

cyclobutanol in excellent yield. The sensitivity of cyclobutanol to base precluded the use of more conventional aldol-type cyclizations.[10]

25.4.2   An Approach to the Pestalotiopsin and Taedolidol Skeletons

Pestalotiopsin A and 6-epitaedolidol are structurally related caryophyllene-type sesquiterpenes. In 2003, Procter *et al.* reported the use of a SmI_2-mediated 4-*exo-trig* carbonyl–alkene cyclization to construct the core of pestalotiopsin A.[11] Treatment of the cyclization substrate with SmI_2 in THF, MeOH and 2,2,2- trifluoroethanol gave cyclobutanol products in good yield and with moderate diastereoselectivity. The major diastereoisomer is believed to arise from a cyclization in which coordination to the silyl ether group directs addition of the ketyl radical anion to the alkene (Scheme 25.8).[11]

In the absence of 2,2,2-trifluoroethanol, low yields of cyclobutanols were obtained. It is thought that the acidic proton donor prevents elimination of the silyloxy group by rapid quenching of the samarium enolate formed in the cyclization (Figure 25.1).[12]

Procter *et al.* have recently found that the pestalotiopsin skeleton can be converted to the previously unexplored taedolidol natural products upon treatment with acid, suggesting a biosynthetic link between the two families.[13]

25.5   FIVE-MEMBERED RING FORMATION USING SmI_2: THE SYNTHESIS OF (–)-HYPNOPHILIN AND THE FORMAL SYNTHESIS OF (–)-CORIOLIN

In 1988, Curran *et al.* used a SmI_2-mediated sequential process in the synthesis of (–)-hypnophilin and the formal synthesis of (–)-coriolin.[14] Treatment of aldehyde **22** with SmI_2–HMPA gave tricyclic cyclopentanol (Scheme 25.9). Three new stereocentres were generated with complete diastereocontrol during the sequential cyclizations. Attempts to mediate the cyclization of starting precursor using Zn–TMSCl or using photolysis led to recovery of the starting material. The use of 1,3-dimethyl-3,4,5,6-tetrahydro-2(1H)-pyrimidinone (DMPU) as an alternative additive in conjunction with SmI_2 led to a decrease in the diastereoselectivity of the sequence. The reaction sequence involves 5-*exo-trig* cyclization of the ketyl radical followed by 5-*exo-dig* cyclization of the resultant radical.[14,15,16]

25.5.1   A Synthesis of Grayanotoxin III

In 1994, Matsuda reported a remarkable route to grayanotoxin III in which three of the four rings of the target are formed through the use of SmI_2.[17] In the first SmI_2-mediated step, to form the CD rings of the natural product, the hemiketal, which is in equilibrium with the hydroxy ketone, was treated with SmI_2 in the presence of HMPA (Scheme 25.10). Addition of the ketyl radical anion to the unactivated olefin proceeded well to give the final product in high yield.[17]

In the second SmI_2-mediated cyclization in Matsuda's approach, an allylic sulfide was used as a radical acceptor in a cyclization to form the A ring of the natural product (Scheme 25.11).[17] The cyclization of radical precursor resulted in

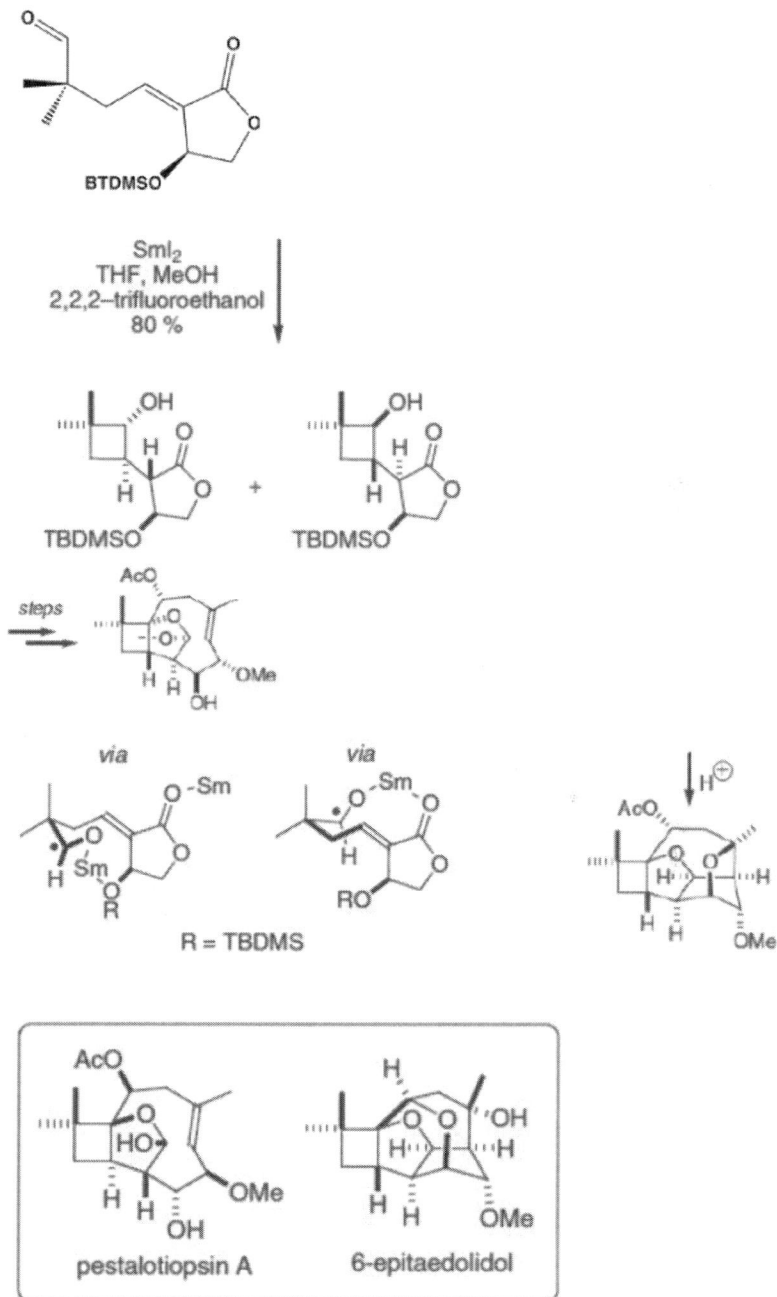

Scheme 25.8 Towards the synthesis of Pestalotiopsin A and 6-epitaedolidol through SmI₂ mediated transformation.

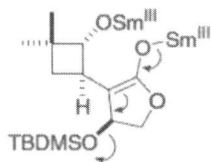

Figure 25.1   Structure of the samarium enolate **21**.

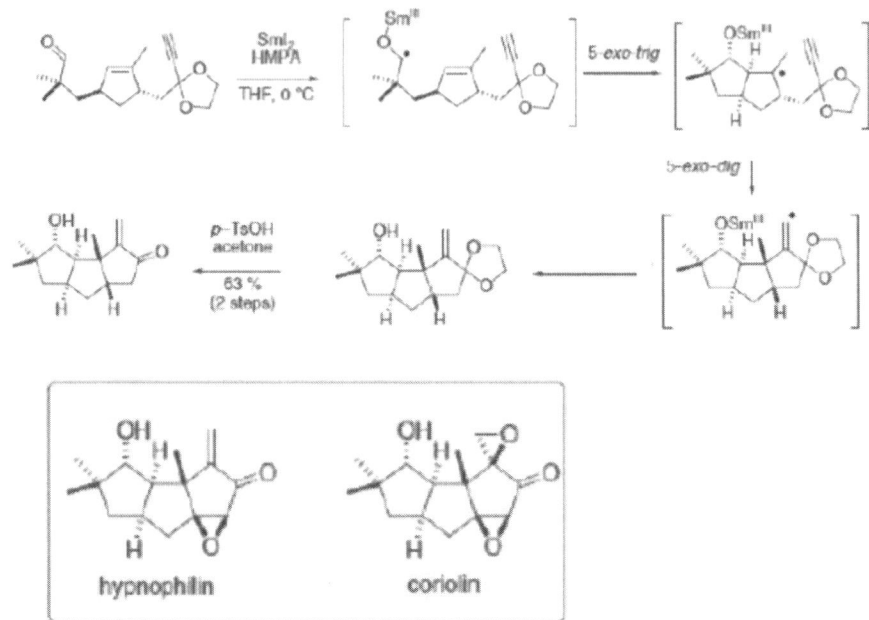

Scheme 25.9   SmI$_2$-mediated sequential process in the synthesis of (-)-coriolin.

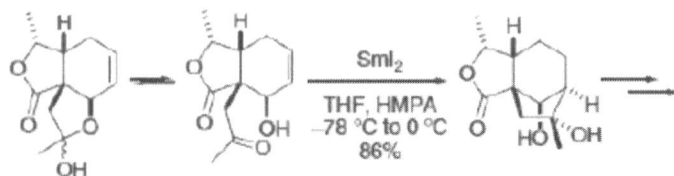

Scheme 25.10   Towards grayanotoxin III *via* SmI$_2$-mediated pathway.

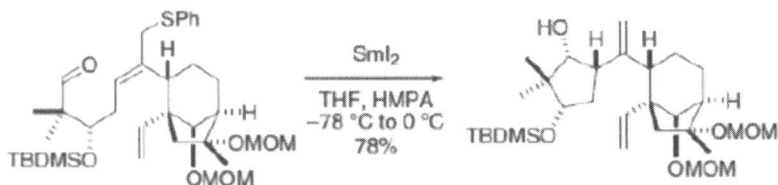

Scheme 25.11   SmI$_2$-mediated cyclization approach with allyllic sulfides as radical acceptors.

elimination of the sulfide to generate a new double bond, which was then epoxidized and reduced to form the tertiary alcohol found in the B ring of the target. The cyclization proceeded to form exclusively the *syn*-cyclopentanol in good yield. It is interesting that the stereochemistry and yield of the cyclization are independent of the initial alkene geometry of the allyl sulfide. The seven-membered ring of the target was formed using a SmI$_2$-mediated pinacol cyclization.[17]

25.5.2 A Synthesis and Structural Revision of (–)-Laurentristich-4-ol

In 2008, Wang and Li *et al.* reported the synthesis of both the proposed and the revised structure of (–)-laurentristich-4-ol using an intramolecular SmI$_2$-mediated ketyl radical addition to a benzofuran to construct the spirocyclic core of the target.[15] Treatment of the radical precursor with SmI$_2$–HMPA in THF gave a spirocycle as a single diastereoisomer in good yield (Scheme 25.12). The proposed structure of the natural product was synthesized from a spirocycle advanced intermediate in a further three steps. Modification of the approach allowed an isomer to be prepared, thus allowing the structure of the natural product to be revised.[18]

25.5.3 An Approach to (–)-Welwitindolinone A Isonitrile

The spirooxindole core of the natural product was prepared using the SmI$_2$-mediated intramolecular couplings of α,β-unsaturated ketones with isocyanates (Scheme 25.13). For example, treatment of isocyanate with SmI$_2$ and LiCl gave the final product as a single diastereoisomer. The cyclization is thought to proceed by initial reduction of the enone followed by anionic addition to the isocyanate.[19]

25.6 SIX-MEMBERED RING FORMATION USING SmI$_2$: AN APPROACH TO MARINE POLYCYCLIC ETHERS

Nakata *et al.* utilized the SmI$_2$-mediated Barbier-type cyclization of a primary iodide with an ester as part of a convergent synthesis of trans-fused 6,6,6,6-

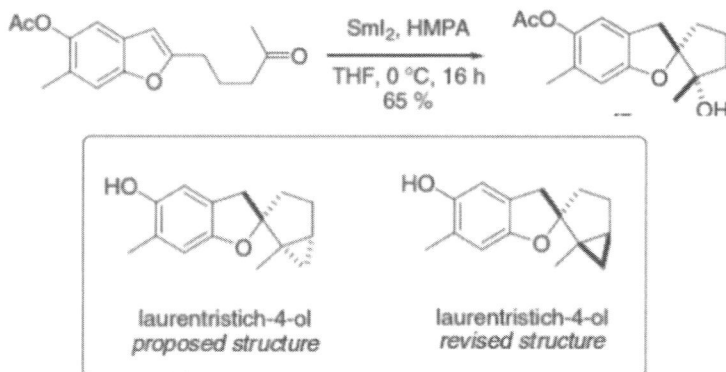

Scheme 25.12 Synthesis of (-)-laurentrich-4-ol using intramolecular SmI$_2$-mediated ketyl radical addition to a benzofuran.

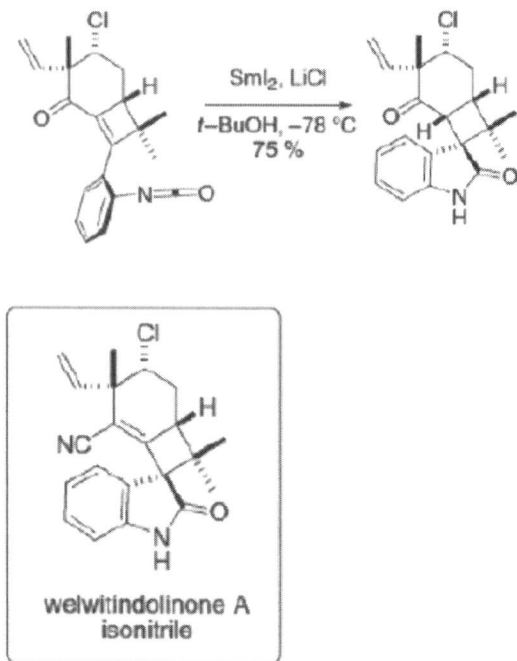

Scheme 25.13 SmI$_2$-mediated intramolecular coupling of a,b-unsaturated ketones with isocyanates.

tetracyclic ethers, which are typically found in marine polycyclic ethers.[16] Treatment of iodide with excess SmI$_2$ in the presence of 1 mol% of NiI$_2$ led to the smooth formation of an intermediate hemiacetal, which was dehydrated to give dihydropyran that was then further elaborated to give a tetracycle (Scheme 25.14).[21]

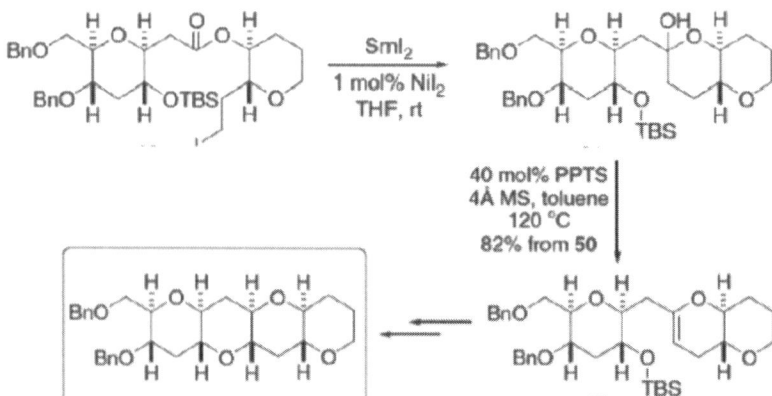

Scheme 25.14 SmI$_2$ mediated convergent synthesis of *trans*-fused 6,6,6,6-tetracyclic ethers.

25.6.1 A Synthesis of Pradimicinone

In 1999, Suzuki *et al.* completed the synthesis of pradimicinone,[22] the aromatic pentacyclic aglycone moiety common to both the pradimicin and benanomicin antibiotic classes. The key step in the approach involves a SmI_2-mediated pinacol cyclization of an axially chiral 2,20-biaryldicarbaldehyde (Scheme 25.15). This cyclization proceeds to give the trans-1,2-diol in quantitative yield and with complete transfer of the axial chirality in the starting material to the central chirality of the product, which is obtained in enantiomerically pure form. The selectivity and chiral transfer were attributed to an Re,Re-cyclization mode, giving the diequatorial product.[17]

25.6.2 A Synthesis of (+)-Microcladallene B

In 2007, Kim *et al.* reported the first asymmetric synthesis of (+)-microcladallene B.[18] In this approach, the tetrahydropyran ring of the target was prepared by a SmI_2-mediated 6-exo-trig carbonyl–alkene cyclization (Scheme 25.16). The high diastereoselectivity of the cyclization can be rationalized by invoking coordination of the aldehyde and electron-deficient alkene to samarium and formation of intermediate complex, in which the aldehyde adopts a pseudoequatorial orientation. Coordination of SmI_2 to the aldehyde facilitates electron transfer and additional coordination to the alkene

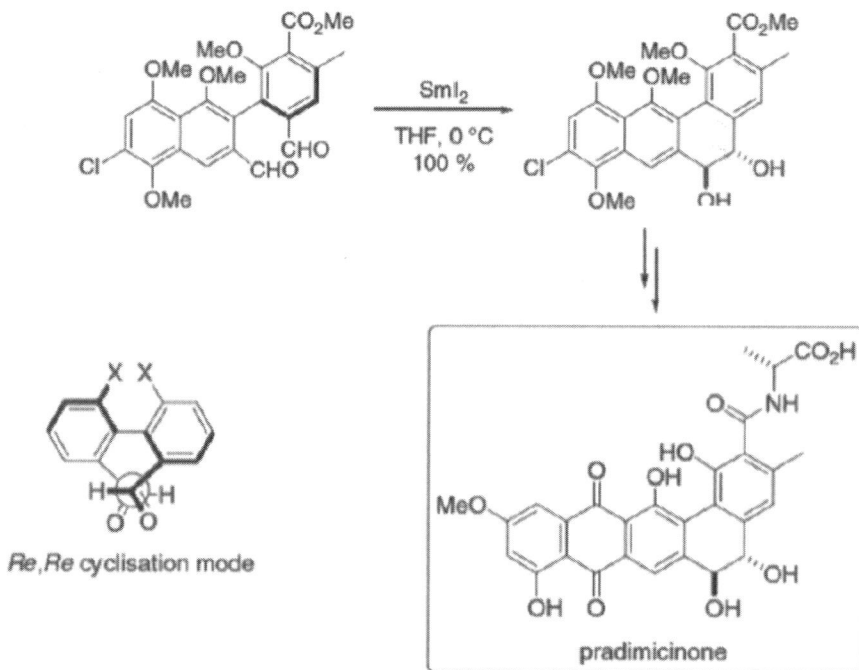

Scheme 25.15 Synthesis of pradimicinone *via* SmI_2-mediated pinacol cyclization of an axially chiral 2,20-biaryldicarbaldehyde.

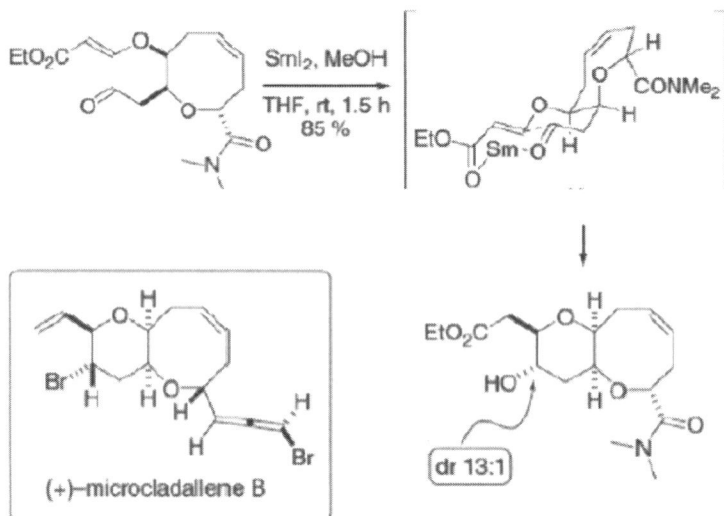

Scheme 25.16   SmI$_2$-mediated 6-exo-trig carbonyl-alkene cyclization towards synthesis of (+)-microcladallene.

radical acceptor facilitates cyclization of the resultant radical. The synthesis of (+)-microcladallene B was completed in 10 steps from the starting radical precursor.[23]

25.6.3   A Synthesis of Botcinins C, D and F

In 2008, Shiina reported the first stereoselective synthesis of botcinins C, D and F.[19,22] The botcinins contain a bicyclic heterocyclic core in which all of the carbon centers, apart from the lactone carbonyls, are stereogenic (Figure 25.2).

The bicyclic core of the targets was constructed using a SmI$_2$-mediated Reformatsky reaction (Scheme 25.17). Shiina found that the use of HMPA in the Reformatsky reaction had a dramatic effect on its outcome. When an α-bromo ester was treated with SmI$_2$ in the absence of HMPA, alcohol (containing the incorrect stereochemistry required for the natural products) was obtained as a mixture of diastereoisomers; the final product is thought to arise from a transition structure in which three oxygens in the substrate chelate to the samarium metal (Scheme 25.17).[25]

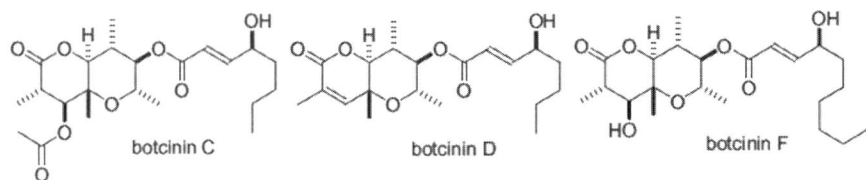

Figure 25.2   Structures of botcinins.

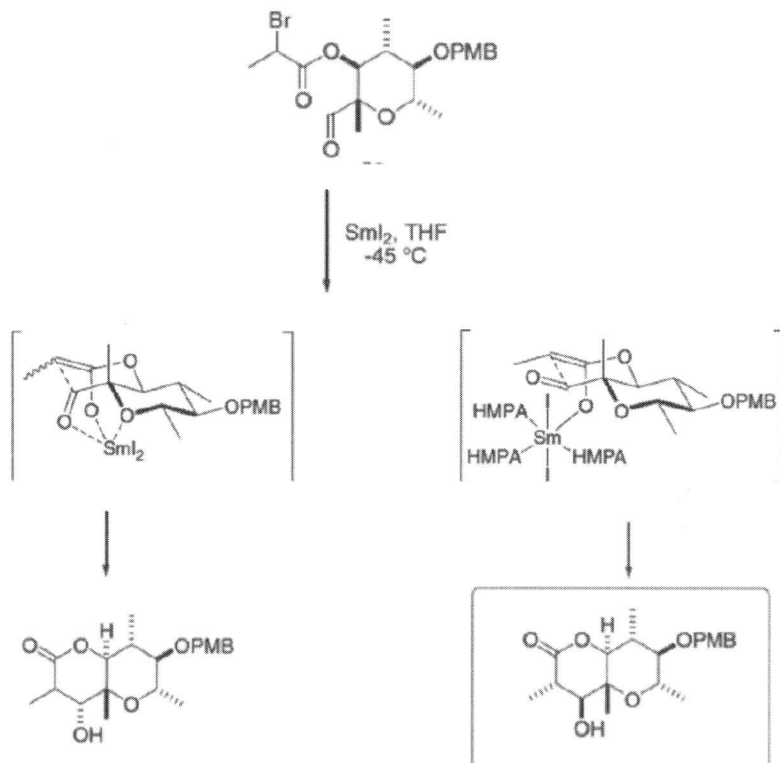

Scheme 25.17 SmI$_2$-mediated Reformatsky reaction in detail.

When HMPA was included in the reaction, final product, (3S,4S,4aR, 6S,7R,8R,8aS)-4-hydroxy-7-((4-methoxybenzyl)oxy)-3,4a,6,8-tetramethylhex-ahydropyrano[3,2-b]pyran-2(3H)-one, containing the correct stereochemistry for the target, was the major product. Pyranone is thought to arise from transition structure, in which HMPA reduces the chelation of samarium to the substrate (Scheme 25.17). Newly formed pyranone was used to prepare all three botcinins.[25]

25.7 SEVEN-MEMBERED RING FORMATION USING SmI$_2$: SYNTHESES OF (–)-BALANOL

In 1998, Naito *et al.* reported the use of SmI$_2$ to form the hexahydroazepine ring of (–)-balanol through a carbonyl–oxime coupling (Scheme 25.18).[20] Attempts to form this ring using n-Bu$_3$SnH gave only a moderate yield of final product and relatively low diastereoselectivity was observed (dr 3 : 1). The use of SmI$_2$–HMPA to mediate the cyclization was found to give an improved yield of final product, with higher diastereoselectivity (dr 7 : 1). The stereoselectivity of the cyclization was explained by invoking *anti*-transition structure. The addition of HMPA was found to be essential as no reaction was observed in its absence.[27]

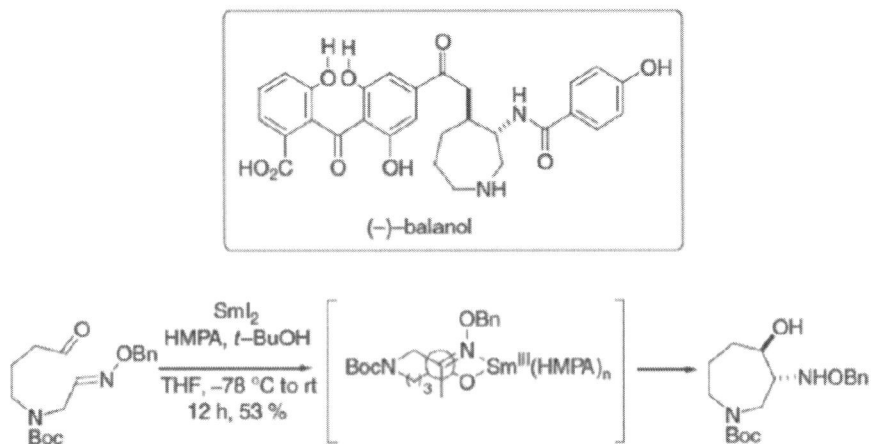

Scheme 25.18　Towards synthesis of (-)-balanol using carbonyl-oxime coupling and SmI₂ methodology.

In 2000, Skrydstrup *et al.* reported a synthesis of balanol using a similar strategy to construct the seven-membered ring: SmI₂-mediated carbonyl–hydrazone coupling of radical precursor gave the final product in good yield and with high diastereoselectivity (dr 10:1) (Scheme 25.19).[29]

The addition of HMPA was found to be crucial for efficient cyclization, as intermolecular pinacol coupling proved to be the major pathway when the aldehyde hydrazone was subjected to SmI₂ alone. Skrydstrup proposed that a chelated intermediate was not operating for the cyclizations with hydrazones and that simple steric interactions between the HMPA–SmI₂ bound ketyl

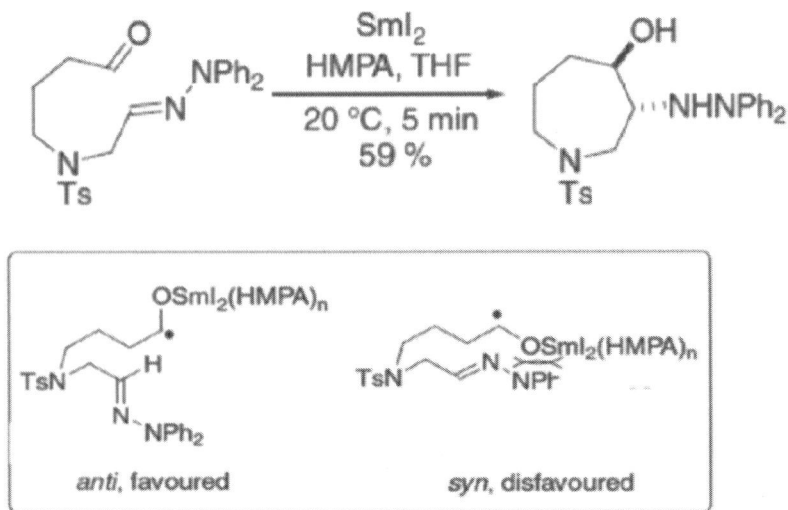

Scheme 25.19　SmI₂ mediated intermolecular pinacol coupling.

radical anion and the diphenylhydrazone unit in transition structures is the dominant factor for the *trans* selectivity observed (Scheme 25.19).[29]

25.8 EIGHT-MEMBERED RING FORMATION USING SmI$_2$: A SYNTHESIS OF PACLITAXEL (TAXOL)

Mukaiyama's group's 1999 approach to the anti-cancer natural product paclitaxel (Taxol) involved construction of the B ring using a SmI$_2$-mediated Reformatsky reaction.[22] Treatment of radical precursor with SmI$_2$ at –78 °C gave the final product in high yield and with good diastereoselectivity (Scheme 25.20).[30]

25.8.1 A Synthesis of (+)-Isoschizandrin

In 2003, Molander reported the synthesis of (+)-isoschizandrin using the SmI$_2$-mediated 8-*endo-trig* carbonyl–alkene cyclization of ketone (Scheme 25.21).[8,23] The axial chirality of the biaryl system efficiently controls the central chirality of the product. The (Z)-alkene geometry is also vital to the stereochemical outcome and the presence of HMPA in the reaction mixture helps control the conformation of the transition state by increasing the steric demands of the alkoxysamarium substituent.[31]

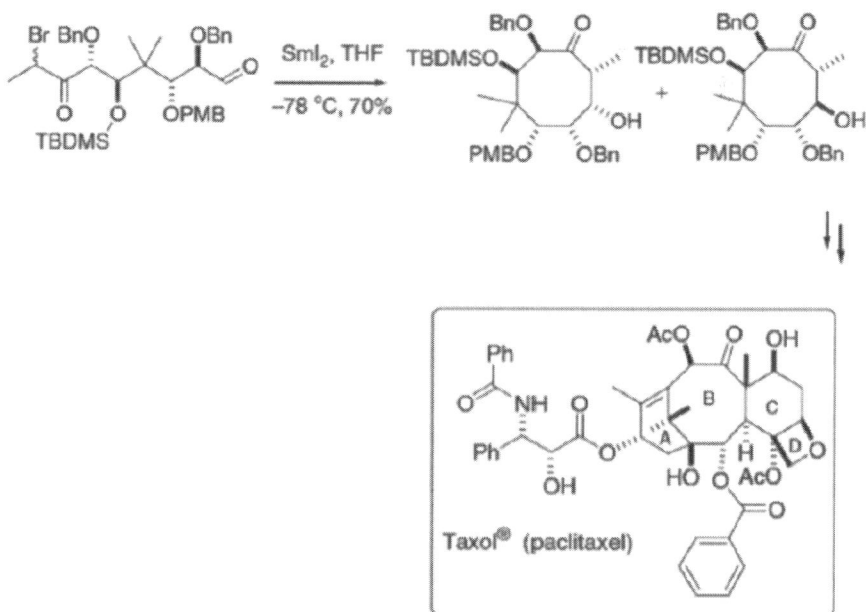

Scheme 25.20 Towards synthesis of Taxol using SmI$_2$-mediated Reformatsky reaction.

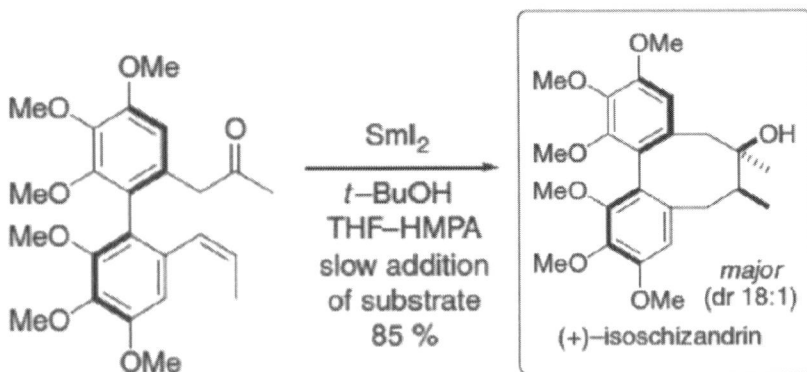

Scheme 25.21   Towards (+)-isoschizandrin using SmI$_2$ mediated 8-endo-trig carbonyl-alkene cyclization of ketone.

25.9   NINE-MEMBERED RING FORMATION USING SmI$_2$: AN APPROACH TO CIGUATOXIN

In 1998, Tachibana *et al.* used a SmI$_2$-mediated Reformatsky reaction to form
the nine-membered oxonone F ring of ciguatoxin.[24] Treatment of bromide
with SmI$_2$ gave an advanced intermediate as a single diastereoisomer after *in
situ* acetylation (Scheme 25.22). The product was converted to decacyclic
polyether, representing a model of the F–M rings of ciguatoxin.[32]

25.10   FORMING LARGER RINGS USING SmI$_2$: A SYNTHESIS OF DIAZONAMIDE A

In 2003, Nicolaou *et al.* reported a synthesis of diazonamide A that utilized a
SmI$_2$-mediated heteropinacol coupling to construct the 12-membered macrocyclic core of the target.[25] Treatment of the radical precursor with SmI$_2$ in the
presence of DMA led to sequential pinacol coupling and N–O bond cleavage.
The crude product was coupled directly with Fmoc-protected valine to give
final product in good overall yield for the three-step sequence. A di-radical
mechanism was proposed for the pivotal coupling step (Scheme 25.23).[33]

25.10.1   A Synthesis of β-Araneosene

In 2005, Corey *et al.* described a synthesis of β-araneosene that exploited a
SmI$_2$-mediated pinacol-type macrocyclization followed by a ring expansion to
build the bicyclic framework of the target.[10,26] Slow addition of the substrate
to a solution of SmI$_2$ at reflux gave 12-membered, *anti*-pinacol product as a
single diastereoisomer (Scheme 25.24). Attempts to use low-valent Ti-based
reagents to carry out the reductive coupling led to a mixture of uncyclized
reduction products.[34]

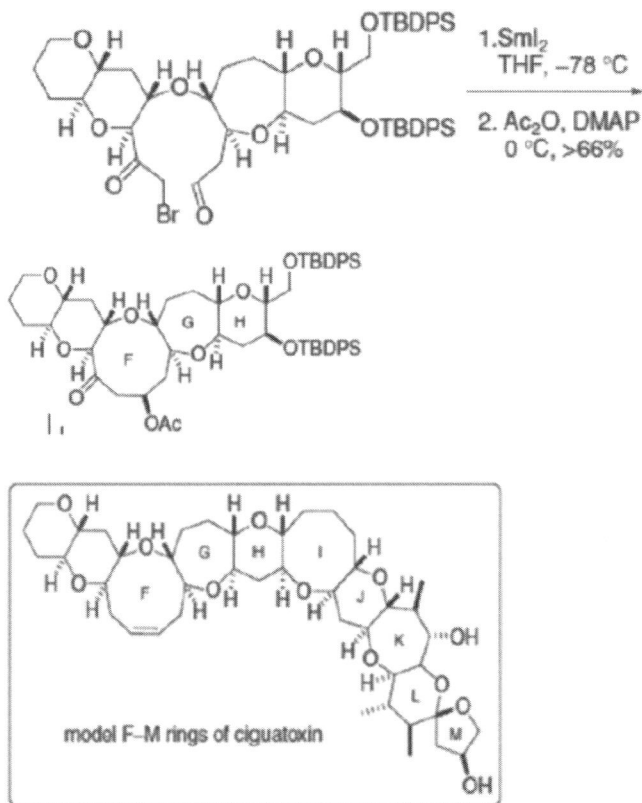

Scheme 25.22 SmI$_2$-mediated Reformatsky reaction to form the nine-membered F ring of ciguatoxin.

25.10.2 A Synthesis of Kendomycin

In 2008, Panek *et al.* reported an asymmetric synthesis of kendomycin that utilized a SmI$_2$-mediated Barbier-type cyclization to form a 16-membered ring (Scheme 25.25).[27] Treatment of benzylic bromide with SmI$_2$ gave a secondary alcohol as a single diastereoisomer. The stereochemistry of the product was not determined. Deprotection of the product and oxidation gave orthoquinone, which was converted to kendomycin in two steps.[35]

So, to draw a conclusion to date on natural product synthesis and SmI$_2$: this selective survey has showcased the use of SmI$_2$-mediated cyclization reactions in natural product synthesis and pays testament to the power of SmI$_2$ as a reagent for synthesis. The full armory of ring-forming reactions conducted by the reagent has been applied successfully in approaches to a wide range of cyclic systems in a wide range of targets. Indeed, some of the most challenging targets of recent years, such as Taxol and diazonamide A, have been prepared using pivotal cyclization steps mediated by the reagent. If natural product

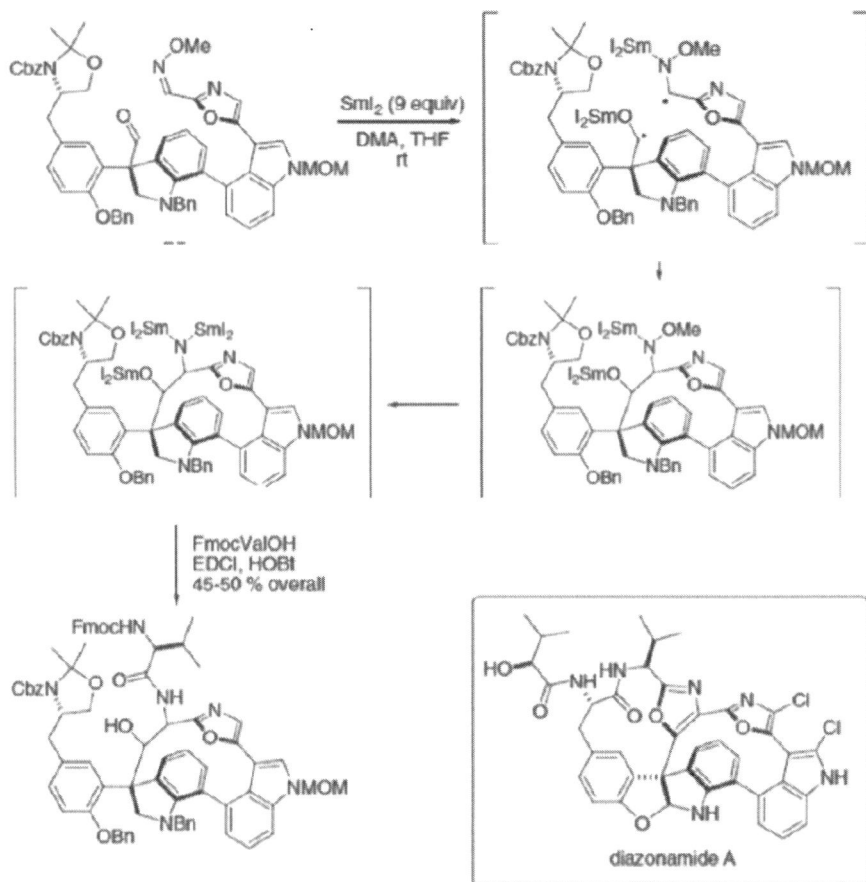

Scheme 25.23   Towards synthesis of diazonamide A using SmI$_2$-mediated hetero-
                   pinacol coupling to construct 12-membered macrocyclic core of the
                   target.

synthesis is a test for methods and reagents, then SmI$_2$ has passed and looks set
to continue meeting the challenges we put before it.

25.11   MODIFYING BIOMOLECULES USING SmI$_2$

25.11.1   Introduction

As discussed in previous chapters, SmI$_2$ promotes a wide variety of carbon–
carbon bond-forming reactions while at the same time displaying a high degree
of functional group tolerance. These properties make this low-valent
lanthanide reagent particularly suitable for carrying out structural modifications on biomolecules, such as carbohydrates and peptides, with emphasis on

β-araneosene

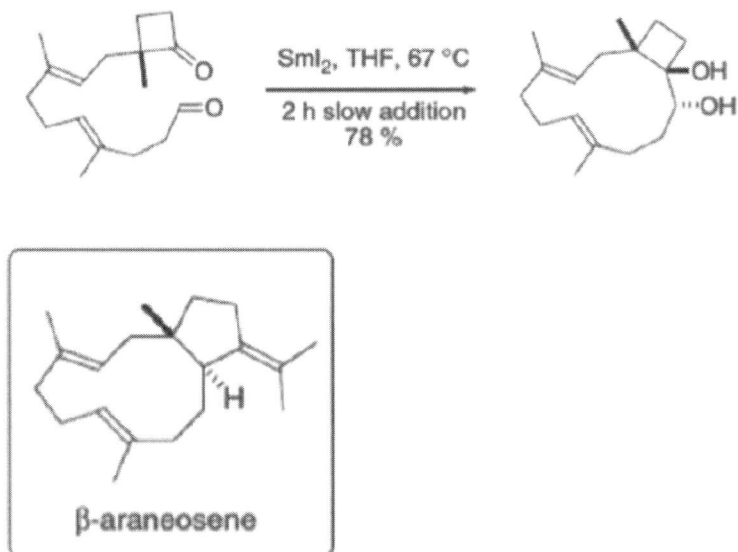

Scheme 25.24 SmI$_2$-mediated pinacol type cyclization in action.

the preparation of stable carbon-based mimics of interest for biomedical research and drug development programmes.

Performing selective chemical transformations and, in particular, carbon–carbon bond formation on such biomolecules has traditionally been challenging due to the many proximal functional groups found in such molecules (*i.e.* hydroxyl groups in sugars, amide bonds and side-chain functionalities in peptides). Nevertheless, in recent years it has been demonstrated that SmI$_2$ has a remarkable ability to promote a number of carbon–carbon bond-forming reactions on peptide and carbohydrate substrates. More specifically, SmI$_2$ has been used for the stereoselective synthesis of C-glycosides, side-chain introduction on to glycine residues of peptides, the ligation of small peptides and the direct synthesis of ketomethylene and hydroxyethylene isosteres. From a synthetic point of view, it is interesting that

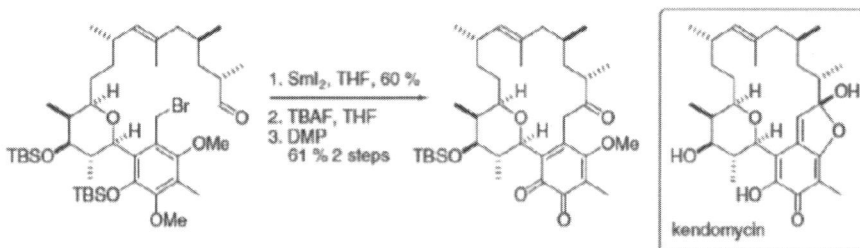

Scheme 25.25 SmI$_2$-mediated Barbier type cyclization to form 16-membered ring towards synthesis of kendomycin.

the high functional group tolerance of SmI_2 allows intact carbohydrates and peptides to be used as building blocks in the synthesis of biomolecule mimics without resort to *de novo* synthesis.

25.11.2 Modifying Carbohydrates Using SmI_2

C-Glycosides represent a popular class of carbohydrate mimics that are characterized by the replacement of the interglycosidic oxygen atom by a methylene group. These analogues are not only resistant to enzymatic and chemical hydrolysis, but also display conformational properties around the glycosidic linkage that are similar to those of the parent O-glycosides. Over the last decade, SmI_2 has been found to be an excellent reagent for the preparation of C-glycosides because, unlike most other approaches to these compounds, the intact sugar can be used in the carbon–carbon bond-forming step without problems arising from the presence of the ring substituents. Two approaches involving the generation of a reactive anomeric intermediate have been used for the SmI_2-promoted synthesis of C-glycosides, the first involving a C1-carbon centred radical and the second involving a carbanion. In 1994, Sinayan reported an extraordinary intramolecular delivery approach for the preparation of a C-disaccharide (Scheme 25.26).[36] The glycosyl sulfone was tethered to the unsaturated sugar acceptor *via* a silyl diether linkage. Upon slow addition of SmI_2 in benzene and HMPA, it underwent a 9-*endo* radical cyclization to give final product. Whereas control at the C1-position was achieved due to the anomeric effect and the large ring formed, 1,3-diaxial interactions guide the new C4-substituent of the reducing sugar into an equatorial position. Subsequent desilylation of the major C4-epimer intermediate provided the

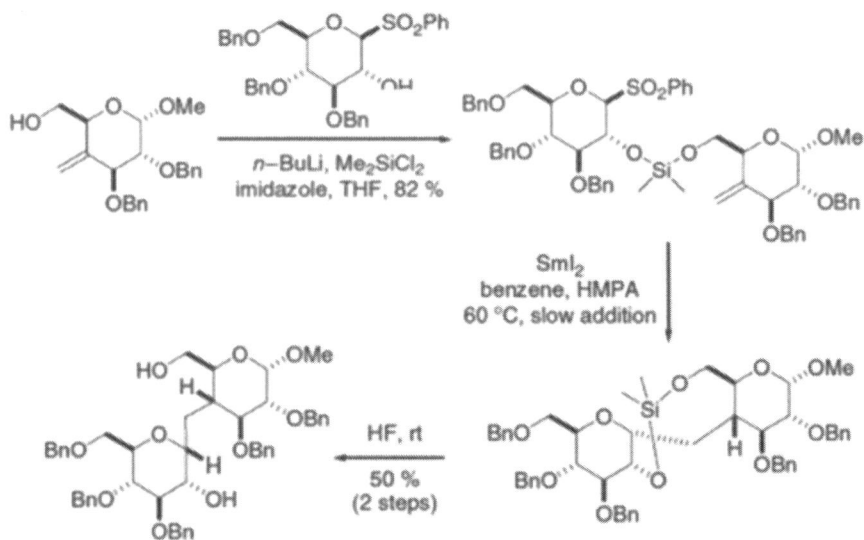

Scheme 25.26 SmI_2 promoted synthesis of glycosides.

disaccharide mimic in an overall yield of 50% for the two steps (Scheme 25.26).[37]

In an effort to avoid the use of HMPA as a cosolvent for the generation of anomeric radicals using SmI_2, Beau and Skrydstrup reported the use of the more reactive glycosyl pyridyl sulfones in conjunction with a radical acceptor tethered to the C_2-OH for the stereoselective synthesis of α-C-glucosides and β-C-mannosides.[37,38] In these examples, the installation of the C_1-alkyl group using a 5-*exo*-radical allows the stereochemistry of the C_2-substituent to dictate the stereochemical outcome in the carbon–carbon bond-forming step at the anomeric centre. As exemplified in Scheme 25.27 with the pyridyl sulfone, cyclization occurs upon addition of SmI_2 in THF alone. Subsequent desilylation, double bond hydrogenation and peracetylation provided the methyl isomaltoside mimic as a final product in an overall 48% yield (Scheme 25.28).[37,38]

Further work by Beau and Skrydstrup demonstrated the surprising stability of anomeric organosamarium species (generated from the reduction of glycosyl pyridyl sulfones with SmI_2) towards β-elimination of the protected C_2-hydroxy substitutent.[39,40] Glycosyl samariums could even be trapped with alkyl ketones and aldehydes. These Barbier-type reactions generally proceed with good stereocontrol at the anomeric centre, leading to 1,2-*trans* selectivity (in contrast to the 1,2-*cis* selectivity observed in the SmI_2-mediated radical cyclizations shown in Scheme 25.28). For example, the reductive coupling of mannosyl pyridyl sulfones with carbonyl compounds led to α-C-mannoside formation, whereas the glucosyl and galactosyl sulfones produced the corresponding β-C-glycosides (Scheme 25.29).[39,40]

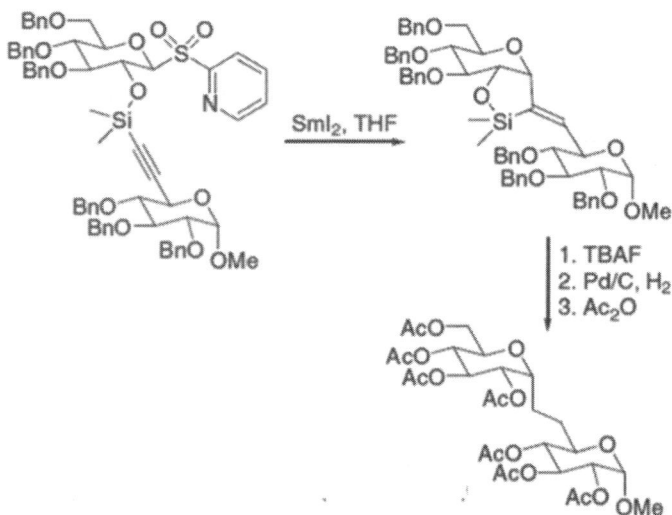

Scheme 25.27 SmI_2 mediated carbon-carbon bond formation in action.

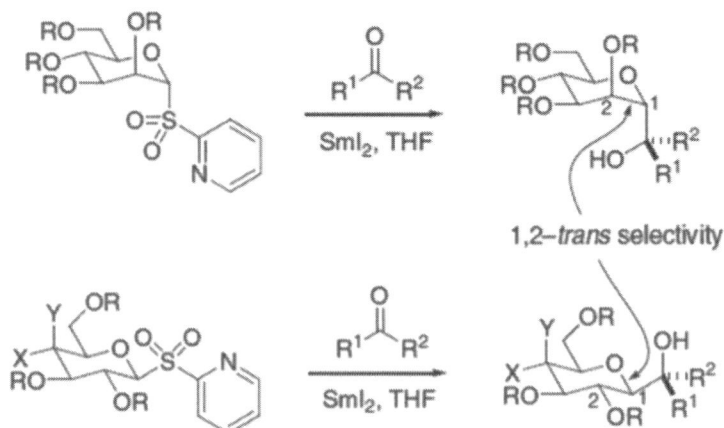

Scheme 25.28   SmI$_2$ mediated carbon-carbon bond formation of pyridyl sulfone.

The stereoselectivities observed at the C1-position of the sugar ring were explained by a consideration of the fate of the glycosyl anions formed from the reduction of the intermediate anomeric radical (Scheme 25.29). As these C1-radicals are thermodynamically more stable in the α-orientation, their reduction will lead to the kinetic axially oriented organosamarium species. Unfavourable overlap between the sC1–Sm and nO5 orbitals is relieved by either a conformational change of the ring to a skew boat, placing the C1- and C2-substituents in a diequatorial arrangement (as occurs in the manno series), or a configurational change (anomerization) to the β-organosamarium species, as observed in the gluco and galacto series (Scheme 25.29). Surprisingly, β-elimination of the C2-substituent was not generally found to be the major pathway. In subsequent work, Beau *et al.* demonstrated that the use of catalytic NiI$_2$ leads to a substantial increase in yield for these C-glycosylation reactions.[41]

In contrast, the Barbier reactions of 2-acetamido-2-deoxy sugars with ketones showed a preference for the formation of 1,2-*cis*-C-glycoside products (Scheme 25.30). This was attributed to the complexation of the C$_2$-acetamido group to the glycosyl samarium, thereby reducing the configurational lability of the organometallic species.[42–44]

Further work from the groups of Linhardt and Beau has demonstrated the viability of performing SmI$_2$-mediated C-glycosylations using derivatives of N-acetylneuraminic acid.[44–46] Most impressively, Beau has shown that simple peracetates of methyl N-acetylneuraminate perform well in couplings with carbonyl compounds by a Reformatsky-type mechanism (Scheme 25.31).[45,46] SmI$_2$-mediated C-glycosylations of this type have also been carried out on a solid phase.

Finally, two examples of the application of the above chemistry for the synthesis of a C-oligosaccharide and a C-glycosylated amino acid are illustrated in Scheme 25.32. Skrydstrup *et al.* reported a highly convergent

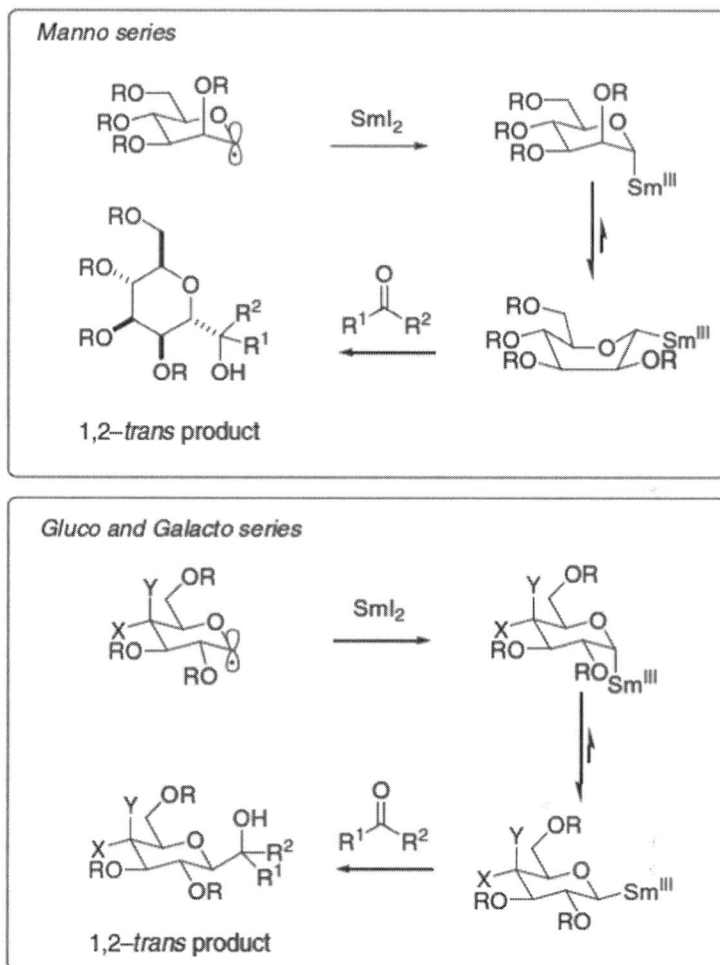

Scheme 25.29 SmI$_2$ mediated a-C-mannose formation.

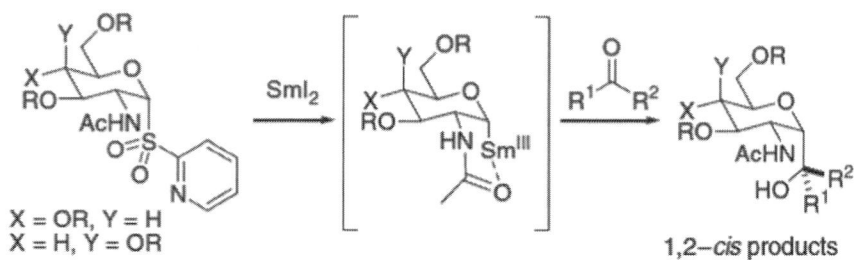

Scheme 25.30 SmI$_2$-mediated glycosilation and 1,2-cis-C-glycosides.

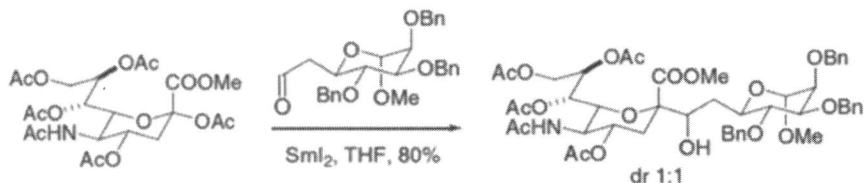

Scheme 25.31 SmI₂-mediated C-glycosilation using derivatives of N-acetylneura-
            minic acid.

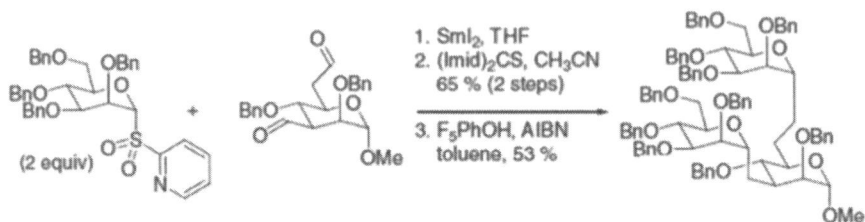

Scheme 25.32 SmI₂-mediated Barton-McCombie reaction: towards synthesis of C-
            trisaccharides.

synthesis of the C-trisaccharide that involves a double SmI₂-promoted C-
glycosylation of the dialdehyde, followed by a modified Barton–McCombie
deoxygenation step (Scheme 25.32).[46] Furthermore, the Skrydstrup–Beau team
disclosed a direct synthesis of the C-glycoside analogue of a tumour-associated
carbohydrate antigen (Tn), which exploits a SmI₂-mediated coupling of the
pyridyl sulfone with the aldehyde (Scheme 25.33).[47]

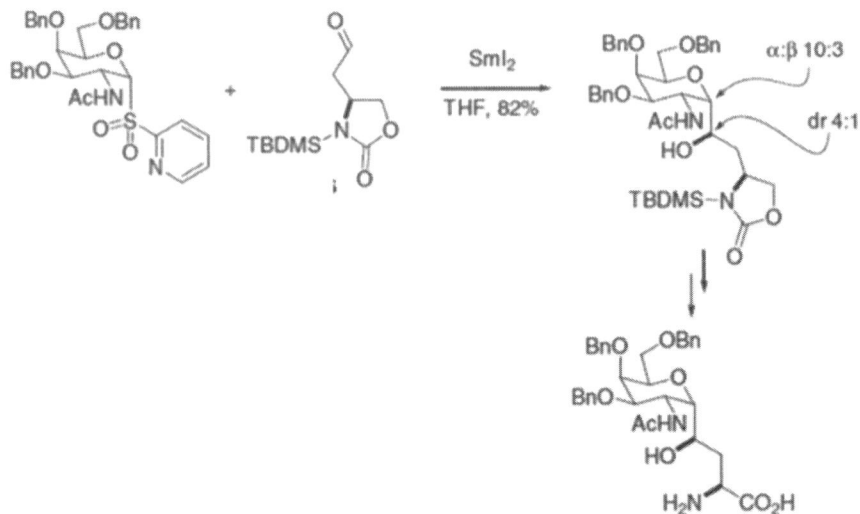

Scheme 25.33 SmI₂ mediated coupling of pyridyl sulfone and aldehyde.

25.11.3 Modifying Amino Acids and Peptides Using SmI$_2$

The ability of SmI$_2$ to promote carbon–carbon bond-forming reactions at the anomeric centre of carbohydrates, under mild coupling conditions, and the remarkable stability of the glycosyl samarium(III) species towards β-elimination inspired Skrydstrup *et al.* to examine the viability of analogous reactions on small peptide substrates for the introduction of non-natural carbinol side-chains to glycine residues.[48,49] The introduction of a reducible group on to the glycine residue was achieved in simple cases by bromination and subsequent nucleophilic displacement of the halide with 2-mercaptopyridine. However, a higher yielding and more tolerant protocol involved the oxidative fragmentation of a serine residue with Pb(OAc)$_4$ followed by substitution of the resultant acetate group with 2-mercaptopyridine. For example, the cyclic peptide and the tripeptide were converted to the corresponding sulfide and subsequent low-temperature coupling of the peptides to alkylaldehydes and ketones in the presence of SmI$_2$ and catalytic NiI$_2$ gave good yields of the modified peptide, although little stereocontrol at the α-carbon was achieved (Scheme 25.34). It is likely that the couplings involve Sm(III) enolate formation and aldol reaction

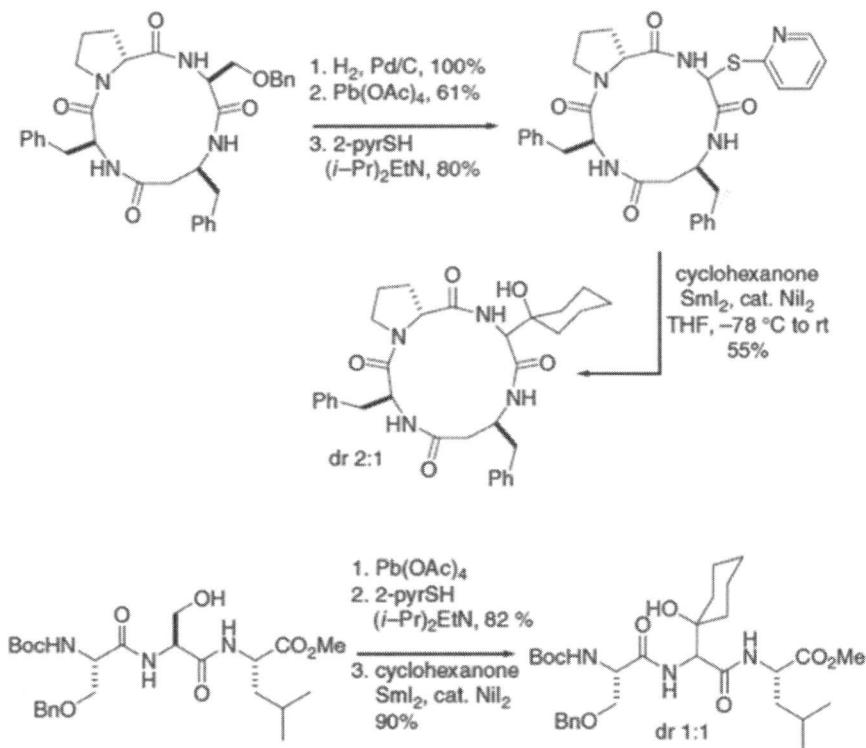

Scheme 25.34 SmI$_2$ mediated synthesis of peptides.

with the carbonyl electrophile. The low basicity of Sm(III) enolates limits proton abstraction from adjacent residues.

In 2003, a new SmI$_2$-mediated carbon–carbon bond-forming reaction was reported by Skrydstrup *et al.* for the direct synthesis of peptide mimics for evaluation as protease inhibitors.[50] For example, the low-temperature coupling of 4-thiopyridyl ester, derived from Cbz-protected phenylalanine, with the dipeptide acrylamide gave the peptide analogue in a 61% yield (Scheme 25.35). Ketone represents a ketomethylene isostere of the tetrapeptide Phe–Gly–Leu–Phe. Ketomethylene isosteres and the corresponding reduced analogues, hydroxyethylene isosteres, represent important and pharmaceutically relevant classes of protease inhibitors.[51,52]

The mechanism of these reactions was proposed to take place by initial electron transfer to the carbonyl group of the thioester, generating a ketyl-like radical anion. Subsequent radical addition to the electron-deficient alkene (acrylamide or acrylate), possibly guided by precomplexation to a Sm(III) metal ion, generates a new radical centre, which is reduced to the corresponding Sm(III) enolate by a second equivalent of SmI$_2$. The protonation of this enolate and hydrolysis of the hemithioacetal upon workup then led to the γ-ketoamide or ester.

Although the scope of the reaction using thioester-derived amino acids was found to be rather narrow with respect to the two coupling partners, further work revealed a more general protocol for accessing peptidyl ketones: SmI$_2$- mediated coupling of the readily accessible N-peptidyloxazolidinones with electron-deficient alkenes gave ketones in excellent yield.[53] For example, reductive coupling of the tripeptidyl derivative and the acrylamide alanine gave the desired peptidyl ketone in 62% yield (Scheme 25.36).[54,55] In contrast to the coupling of thioesters with electron-deficient alkenes, the coupling of oxazolidinones with electron-deficient alkenes is thought to proceed by reduction of the acrylamide

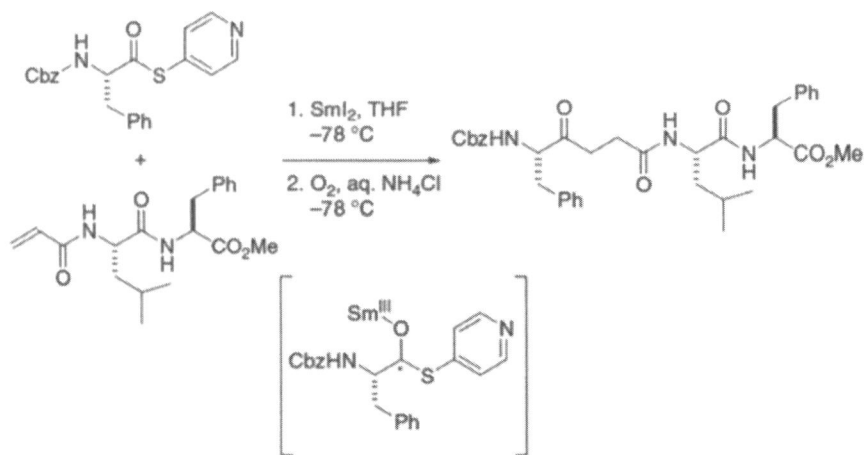

Scheme 25.35   SmI$_2$ mediated synthesis of peptide mimics for evaluation of protease inhibitors.

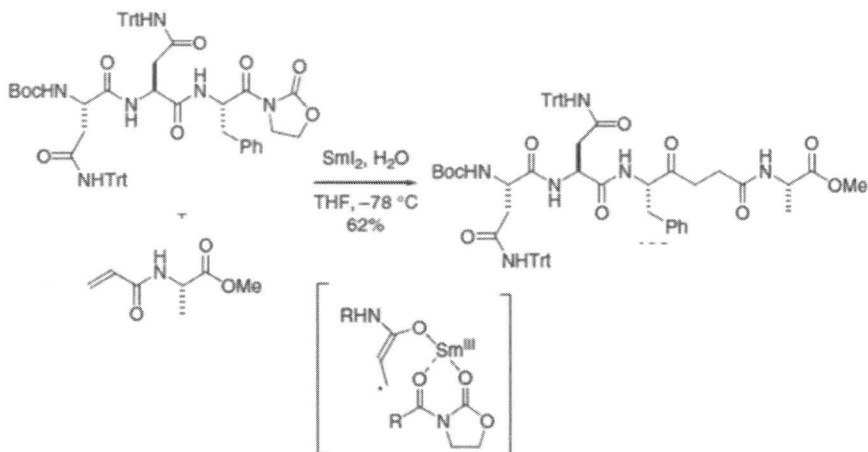

Scheme 25.36 SmI$_2$-mediated coupling of readily accessible N-pepidylozolidinones with electron-deficient alkenes.

followed by addition of the resultant radical intermediate to the exocyclic carbonyl group of the N-acyloxazolidinone (Scheme 25.36).[56]

Although the scope of the reaction using thioester-derived amino acids was found to be rather narrow with respect to the two coupling partners, further work revealed a more general protocol for accessing peptidyl ketones: SmI$_2$- mediated coupling of the readily accessible N-peptidyloxazolidinones with electron-deficient alkenes gave ketones in excellent yield.[57] For example, reductive coupling of the tripeptidyl derivative and the acrylamide alanine gave the desired peptidyl ketone in 62% yield.[58,59] In contrast to the coupling of thioesters with electron-deficient alkenes, the coupling of oxazolidinones with electron-deficient alkenes is thought to proceed by reduction of the acrylamide followed by addition of the resultant radical intermediate to the exocyclic carbonyl group of the N-acyloxazolidinone.[60]

25.12 SUMMARY

Although few groups have yet to exploit SmI$_2$ for the generation of carbon–carbon bonds within biomolecules such as carbohydrates, amino acids and peptides, the work highlighted in this section demonstrates once again the versatility of this reagent for promoting challenging anionic and radical reactions on highly functionalized substrates. There is no doubt that the reagent will be used in the future to perform selective chemistry on some of Nature's most important molecules.

REFERENCES

1. H. B. Kagan, J. -L. Namy and P. Girard, *Tetrahedron,* 1981, **37**, 175.
2. D. P. Curran, T. L. Fevig, C. P. Jasperse and M. J. Totleben, *Synlett,* 1992, 943.

3. T. Nagashima, A. Rivkin and D. P. Curran, *Can. J. Chem.,* 2000, **78**, 791.
4. M. Kunishima, K. Hioki, K. Kono, T. Sakuma and S. Tani, *Chem. Pharm. Bull.,* 1994, **42**, 2190.
5. E. Hasegawa and D. P. Curran, *Tetrahedron Lett.,* 1993, **34**, 1717.
6. D. P. Curran and M. J. Totleben, *J. Am. Chem. Soc.,* 1992, **114**, 6050.
7. E. Prasad and R. A. Flowers II, *J. Am. Chem. Soc.,* 2002, **124**, 6895.
8. G. A. Molander, J. B. Etter and P. W. Zinke, *J. Am. Chem. Soc.,* 1987, **109**, 453.
9. D. J. Edmonds, D. Johnston and D. J. Procter, *Chem. Rev.,* 2004, **104**, 3371.
10. E. J. Corey and Y. J. Wu, *J. Am. Chem. Soc.,* 1993, **115**, 8871.
11. D. J. Edmonds, K. W. Muir and D. J. Procter, *J. Org. Chem.,* 2003, **68**, 3190.
12. T.M. Baker, D. J. Edmonds, D. Hamilton, C. J. O'Brien and D. J. Procter, *Angew. Chem., Int. Ed.,* 2008, **47**, 5631.
13. T. L. Fevig, R. L. Elliot and D. P. Curran, *J. Am. Chem. Soc.,* 1988, **110**, 5064.
14. T. Kan, S. Hosokawa, S. Nara, M. Oikawa, S. Ito, F. Matsuda and H. Shirahama, *J. Org. Chem.,* 1994, **59**, 5532.
15. P. Chen, J. Wang, K. Liu and C. Li, *J. Org. Chem.,* 2008, **73**, 339.
16. K. Kawamura, H. Hinou, G. Matsuo and T. Nakata, *Tetrahedron Lett.,* 2003, **44**, 5259.
17. M. Kitamura, K. Ohmori, T. Kawase and K. Suzuki, *Angew. Chem., Int. Ed.,* 1999, **38**, 1229.
18. J. Park, B. Kim, H. Kim, S. Kim and D. Kim, *Angew. Chem., Int. Ed.,* 2007, **46**, 4726.
19. H. Fukui and I. Shiina, *Org. Lett.,* 2008, **10**, 3153.
20. H. Miyabe, M. Torieda, K. Inoue, K. Tajiri, T. Kiguchi and T. Naito, *J. Org. Chem.,* 1998, **63**, 4397.
21. D. Riber, R. Hazell and T. Skrydstrup, *J. Org. Chem.,* 2000, **65**, 5382.
22. T. Mukaiyama, I. Shiina, H. Iwadare, M. Saitoh, T. Nishimura, N. Ohkawa, H. Sakoh, K. Nishimura, Y.-i. Tani, M. Hasegawa, K. Yamada and K. Saitoh, *Chem.–Eur. J.,* 1999, **5**, 121.
23. G. A. Molander, K. M. George and L. G. Monovich, *J. Org. Chem.,* 2003, **68**, 9533.
24. M. Inoue, M. Sasaki and K. Tachibana, *Angew. Chem., Int. Ed.,* 1998, **37**, 965.
25. K. C. Nicolaou, P. B. Rao, J. Hao, M. V. Reddy, G. Rassias, X. Huang, D. Y.-K. Chen and S. A. Snyder, *Angew. Chem., Int. Ed.,* 2003, **42**, 1753.
26. J. Kingsbury and E. J. Corey, *J. Org. Chem.,* 2005, **127**, 13813.
27. J. T. Lowe and J. S. Panek, *Org. Lett.,* 2008, **10**, 3813.
28. A. Chénedé, E. Perrin, E. D. Rekai and P. Sinay, *Synlett,* 1994, 420.
29. D.Mazéas, T. Skrydstrup, O. Doumeix and J.-M. Beau, *Angew. Chem., Int. Ed. Engl.,* 1994, **33**, 1383.
30. T. Skrydstrup, J.-M. Beau, M. Elmouchir, D. Mazeas, C. Riche and Chiaroni, *Chem.–Eur. J.,* 1997, **3**, 1342.

31. D. Mazelas, T. Skrydstrup and J.-M. Beau, *Angew. Chem., Int. Ed. Engl.,*1995, **34**, 909.
32. T. Skrydstrup, O. Jarreton, D. Mazéas, D. Urban and J.-M. Beau, *Chem.–Eur. J.,* 1998, **4**, 655.
33. N. Miquel, G. Doisneau and J.-M. Beau, *Angew. Chem., Int. Ed.,* 2000, **39**, 4111.
34. D. Urban, T. Skrydstrup, C. Riche, A. Chiaroni and J.-M. Beau, *Chem. Commun.,* 1996, 1883.
35. L. Andersen, L. M. Mikkelsen, J.-M. Beau and T. Skrydstrup, *Synlett,* 1998, 1393.
36. A. Chelnedel, E. Perrin, E. D. RekaıN and P. Sinayn, *Synlett*, 1994, 420.
37. H. G. Bazin, Y. Du, T. Polat and R. J. Linhardt, *J. Org. Chem.,* 1999, **64**, 7254.
38. A. Malapelle, Z. Abdallah, G. Doisneau and J.-M. Beau, *Angew. Chem., Int. Ed.,* 2006, **43**, 6016.
39. A. Malapelle, A. Coslovi, G. Doisneau and J.-M. Beau, *Eur. J. Org. Chem.,* 2007, 3145.
40. L. M. Mikkelsen, S. L. Krintel, J. Jimenez-Barbero and T. Skrydstrup, *J. Org. Chem.,* 2002, **67**, 6287.
41. D. Urban, T. Skrydstrup and J.-M. Beau, *Chem. Commun.,* 1998, 955.
42. M. Ricci, P. Blakskjaer and T. Skrydstrup, *J. Am. Chem. Soc.,* 2000, **122**, 12414.
43. P. Blakskjaer, A. Gavrila, L. Andersen and T. Skrydstrup, *Tetrahedron Lett.,* 2004, **45**, 9091.
44. T. Skrydstrup, O. Jarreton, D. Mazel as, D. Urban and J.-M. Beau, *Chem.-Eur. J.,* 1998, **4**, 655.
45. N. Miquel, G. Doisneau and J.-M. Beau, *Angew. Chem., Int. Ed.,* 2000, **39**, 4111.
46. D. Urban, T. Skrydstrup, C. Riche, A. Chiaroni and J.-M. Beau, *Chem. Commun.,* 1996, 1883.
47. L. Andersen, L. M. Mikkelsen, J.-M. Beau and T. Skrydstrup, *Synlett,* 1998, 1393.
48. T. Mittag, K. L. Christensen, K. B. Lindsay, N. C. Nielsen and T. Skrydstrup, *J. Org. Chem.,* 2008, **73**, 1088.
49. J. Karaffa, K. B. Lindsay and T. Skrydstrup, *J. Org. Chem.,* 2006, **71**, 8219.
50. A. M. Hansen, K. B. Lindsay, P. K. Sudhadevi Antharjanam, J. Karaffa, K. Daasbjerg, R. A. Flowers II and T. Skrydstrup, *J. Am. Chem. Soc.,* 2006, **128**, 9616.
51. L. Andersen, L. M. Mikkelsen, J.-M. Beau and T. Skrydstrup, *Synlett,* 1998, 1393.
52. I. R. Vlahov, P. I. Vlahova and R. J. Linhardt, *J. Am. Chem. Soc.,* 1997, **119**, 1480.
53. H. G. Bazin, Y. Du, T. Polat and R. J. Linhardt, *J. Org. Chem.,* 1999, **64**, 7254.

54. A. Malapelle, Z. Abdallah, G. Doisneau and J.-M. Beau, *Angew. Chem., Int. Ed.,* 2006, **43**, 6016.
55. A. Malapelle, A. Coslovi, G. Doisneau and J.-M. Beau, *Eur. J. Org. Chem.,* 2007, 3145.
56. L. M. Mikkelsen, S. L. Krintel, J. Jimenez-Barbero and T. Skrydstrup, *J. Org. Chem.,* 2002, **67**, 6287.
57. D. Urban, T. Skrydstrup and J.-M. Beau, *Chem. Commun.,* 1998, 955.
58. M. Ricci, P. Blakskjaer and T. Skrydstrup, *J. Am. Chem. Soc.,* 2000, **122**, 12414.
59. P. Blakskjaer, A. Gavrila, L. Andersen and T. Skrydstrup, *Tetrahedron Lett.,* 2004, **45**, 9091.
60. P. Blakskjaer, B. H.j, D. Riber and T. Skrydstrup, *J. Am. Chem. Soc.,* 2003, **125**, 4030.

CHAPTER 26

Innovative Reactions Mediated by Zirconocene: Advantages and Scope[*]

26.1 BACKGROUND OF ZIRCONIUM IN ORGANIC SYNTHESIS

Although zirconium was discovered by Berzelius in 1824,[1] the use of inorganic Zr salts and organozirconium compounds in organic synthesis had been virtually unknown until the mid-1970s with the exception of occasional uses of Zr salts as somewhat exotic Lewis acid catalysts in the Friedel–Crafts reaction[2] and as generally inferior substitutes for Ti salts in Ziegler–Natta polymerization.[3] One significant breakthrough is Schwartz's systematic investigation[4] of hydrozirconation of alkenes and alkynes with $Cp_2Zr(H)Cl$ initiated in 1974. Especially, hydrozirconation of alkynes permits generation of synthetically attractive stereo- and regio-defined alkenylmetal intermediates (Scheme 26.1).

In addition, Negishi's discovery of the Ni- or Pd-catalyzed crosscoupling with zirconocene alkenyl chloride in 1977–1978 significantly expanded the scope of organozirconium chemistry in organic synthesis (Scheme 26.1).[5] The Zr-catalyzed carboalumination of alkyne is also noteworthy since Zr is used as a unique catalyst for the first time in organic synthesis.[6]

Together with the concurrent but seemingly independent development of the zirconium-based alkene polymerization, mainly by Kaminsky et al.,[7] the Zr-catalyzed carboalumination established the synthetic value of Zr as a catalyst component. There is one more kingfish in zirconium-mediated organic synthesis. Low-valent $Cp_2Zr(II)$ derivatives have been utilized for the preparation of five-membered zirconacycle by coupling with two molecules of unsaturated compounds (Scheme 26.2).[8]

Since the resulting zirconacycle contains two reactive C–Zr bonds (O–Zr may also be formed), various interesting transformations can be performed.

* Chapter written by V. Tamara Perchyonok.
Streamlining Free Radical Green Chemistry
V. Tamara Perchyonok, Ioannis Lykakis and Al Postigo
© V. Tamara Perchyonok, Ioannis Lykakis and Al Postigo 2012
Published by the Royal Society of Chemistry, www.rsc.org

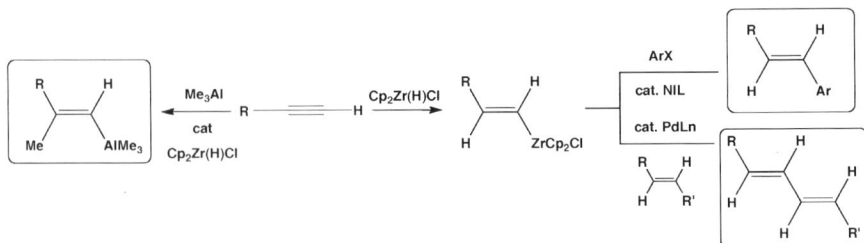

Scheme 26.1   General hydrozirconation of alkynes to yield synthetically attractive stereo- and regio-defined alkenylmetal intermediates.

Negishi's group found a quite convenient method to generate low-valent zirconocene. Treatment of Cp_2ZrCl_2 with 2 equivalent of *n*-BuLi in THF produces $Cp_2Zr(II)$ derivative. Alkene complexes such as $Cp_2Zr(II)$ derivatives are starting compounds which enjoy a variety of transformations *via* zirconacycle formation. There are many other significant reports on synthetic organic reactions concerning zirconium, which include zirconium-catalyzed ethylmagnesiation of C–C multiple bonds,[9] syntheses of vinylzirconocene from vinyl ether,[10] allylzirconocene from allyl ether, and acylzirconium *via* CO insertion,[11] and so on.[12] Zirconium-mediated reaction has been still growing.

26.2   TRIETHYLBORANE-INDUCED RADICAL REACTION WITH SCHWARTZ REAGENT

Oshima *et al.* focused on an environmentally benign radical reaction[13,14] before we started the chemistry of zirconium. The development of tin

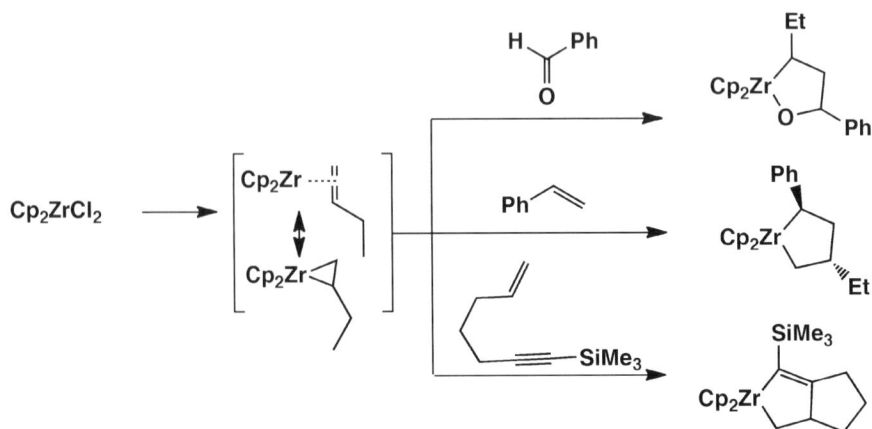

Scheme 26.2   General scheme of preparation five-membered zirconacycle by coupling with two molecules of unsaturated compounds using low-valent $Cp_2Zr(II)$.

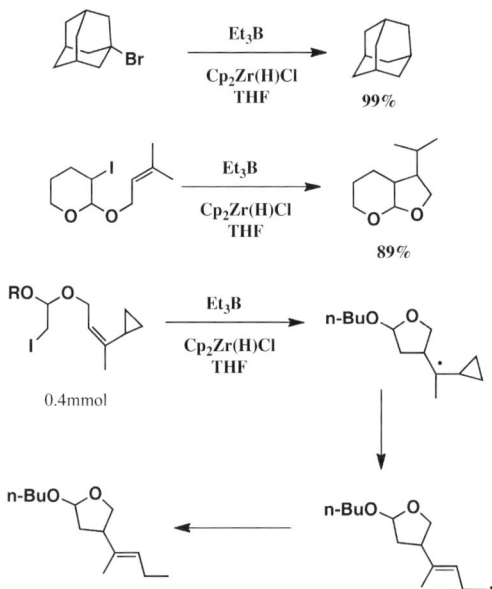

Conditions: halide (1.0 mmol), Cp₂Zr(H)Cl (1.5 mmol), Et₃B (1.0 mmol) and THF (5 ml)

Scheme 26.3   Reductive cyclization of halo acetals under Schwartz conditions.

substitutes has been one of our interests.[15] During the course of the study, it occurred that Schwartz reagent is analogous to tributyltin hydride, undoubtedly the most important reagent in radical chemistry[16] (Scheme 26.3). Tributyltin hydride has so weak a hydrogen–tin bond that it homolytically cleaves by the action of an alkyl radical. In addition, a tin-centered radical has a strong affinity toward halogen atoms. Oshima anticipated that commercially available Schwartz reagent could exhibit a hydrogen donor ability similar to tributyltin hydride and that a "zirconium-centered radical" could abstract halogen from alkyl halide.

The first attempt was indeed successful. Treatment of the prenyl ether of *o*-iodophenol with Schwartz reagent in the presence of triethylborane as a radical initiator afforded the corresponding dihydrobenzofuran derivative in 27% yield (eqn (26.1)). Further screening of reaction conditions rendered the reaction high-yielding and reliable (Scheme 26.3).

Most of the reduction reactions of alkyl iodides and bromides proceeded in satisfactory yields. It is noteworthy that the reduction of alkyl chlorides with Cp₂Zr(H)Cl, usually unreactive for the reduction reaction, completed

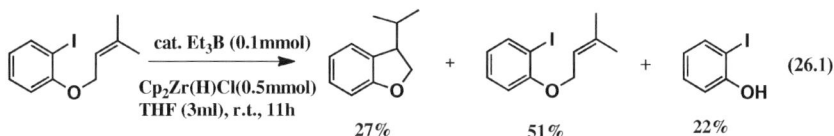

smoothly in the presence of triethylborane. The primary and secondary alkyl halides as well as tertiary ones were easily reduced in excellent yields. Reduction of aryl iodide was also very efficient. Functional groups such as ether and ester could survive under the reaction conditions. Reductive cyclization of halo acetal was also successful. The reaction of iodo acetal bearing a cyclopropyl ring provided ring-opening product in 67% yield. Ring opening of cyclopropylmethyl radical is well known[17] and is highly suggestive of the generation of alkyl radical from radical precursor. A plausible reaction mechanism is shown in Scheme 26.4 in analogy with the case of n-Bu$_3$SnH.

An ethyl radical, generated from Et$_3$B by the action of a trace amount of oxygen,[18] would abstract hydrogen homolytically from Cp$_2$Zr(H)Cl to provide a zirconium(III) radical species Cp$_2$ZrCl. Single electron transfer from Cp$_2$ZrCl to precursor furnishes the radical anion. A halide ion is immediately liberated as Cp$_2$ZrClBr and the resulting carbon-centered radical cyclizes to afford an advanced radical intermediate. The advanced radical intermediate would abstract hydrogen from Cp$_2$Zr(H)Cl to provide the final product and regenerate Cp$_2$ZrCl.

Schwartz reagent is not cheap. Oshima next investigated the preparation of Cp$_2$Zr(H)Cl from Cp$_2$ZrCl$_2$ and a reducing reagent *in situ*. As a result, Red-Al [NaAlH$_2$(OCH$_2$CH$_2$OCH$_3$)$_2$] was found to be the most effective.

Namely, treatment of Cp$_2$ZrCl$_2$ with 0.5 equivalent of Red-Al in THF at 25 °C provided Schwartz reagent. Sequential addition of 1-chloroadamantane and Et$_3$B to the solution afforded adamantane in excellent yield. As summarized in Table 26.1, not only iodoalkanes but also bromo- and chloroalkanes such as 2-bromododecane and 1-chloroadamantane were reduced to the corresponding hydrocarbons in good yields.

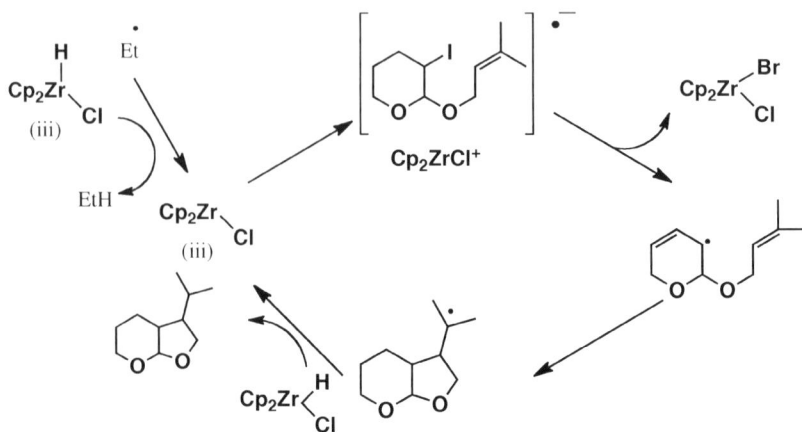

Scheme 26.4 Proposed mechanism of the free radical cascade mediated by Schwartz reagent.

Table 26.1 Reduction of organic halides with Cp$_2$Zr(H)Cl generated *in situ*.

$$\text{Cp}_2\text{ZrCl}_2 + \text{RedAl}$$

$$\downarrow$$

$$\text{RX} \xrightarrow[\substack{\text{Et}_3\text{B} \\ \text{THF}}]{\text{Cp}_2\text{ZrCl(H)}} \text{RH}$$

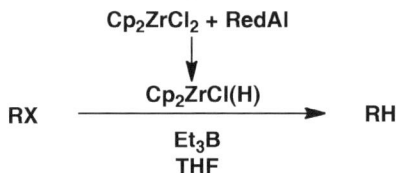

Entry	R–X	Time/h	Yield (%)
1	1-Bromoadamantane	3	89
2	2-Bromoadamantane	5	88
3	n-C$_{10}$H$_{21}$CH(Br)CH$_3$	3	94
4	n-C$_{10}$H$_{21}$CH(I)CH$_3$	3	99
5	n-C$_{12}$H$_{25}$Br	3	93
6	n-C$_{12}$H$_{25}$Cl	15	73
7	PhCOOCH$_2$CH$_2$CH$_2$Br	3	99
8	1-Iodonaphthalene	5	93

[a]R–X (1.0 mmol), Cp$_2$ZrCl$_2$ (1.5 mmol), Red-Al (0.75 mmol), Et$_3$B (1.0 mmol) and THF (5 ml) were employed.

A radical cyclization reaction also took place (eqn (26.2)). Treatment of *o*-iodoaniline derivative with Cp$_2$Zr(H)Cl afforded dihydroindole derivative in the presence of triethylborane. Unfortunately, a bromo analogue of aniline derivative did not afford the cyclized product under the same conditions (<5% yield).

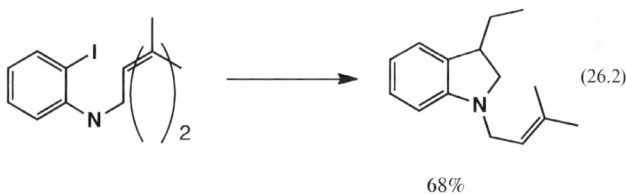

(26.2)

68%

Treatment of corresponding 2-(allyloxy)-3-bromotetrahydro-2*H*-pyran derivative that has a terminal alkene moiety with Cp$_2$Zr(H)Cl generated *in situ* in the presence of Et$_3$B in THF afforded the anticipated bicyclic acetal in 75% yield (Scheme 26.5) along with the hydrozirconation product (9%). More interestingly, we have found that the reaction path heavily depends on the reaction conditions. Treatment of corresponding 2-(allyloxy)-3-bromotetrahydro-2*H*-pyran with three equimolar amounts of purchased Cp$_2$Zr(H)Cl gave the hydrozirconation product in the absence of Et$_3$B in CH$_2$Cl$_2$ in 83% yield.

It is important to reduce the amount of Cp$_2$ZrCl$_2$ employed for the reaction. Consequently, we were delighted to discover that the reaction could function with a catalytic amount of Cp$_2$ZrCl$_2$. The reduction was performed by addition of Et3B (1.0 mmol) to a solution of 1-bromoadamantane (1.0 mmol),

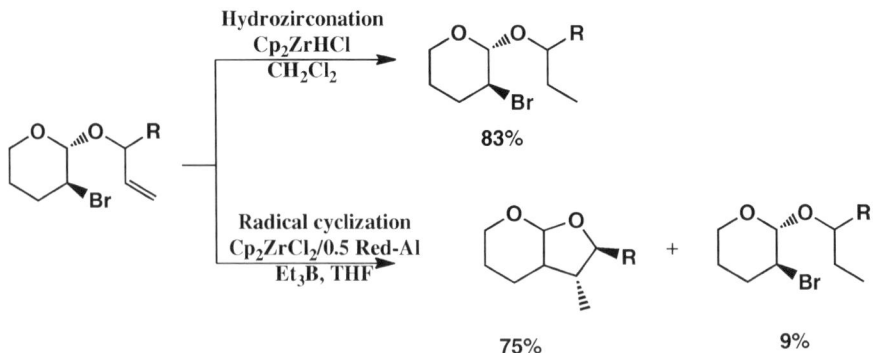

Scheme 26.5 Preparation of various 2H-pyran derivatives using Cp_2ZrCl_2 generated *in situ*.

Cp_2ZrCl_2 (0.2 mmol), and Red-Al (1.5 mmol) in THF to yield adamantane in 89% yield. Treatment of β-bromo acetal with catalytic Cp_2ZrCl_2 in THF in the presence of Red-Al and Et_3B at room temperature afforded the final product in 80% yield (eqn (26.3)).

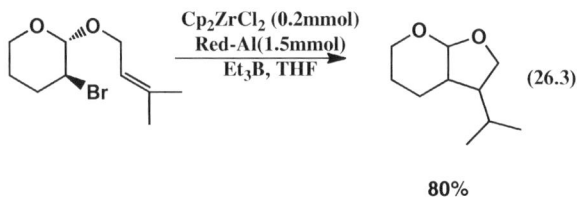

Schwartz reagent has proved to rival tributyltin hydride in efficiency and would be superior from an ecological and toxicological perspective. The key steps would be homolytic cleavage of the zirconium–hydrogen bond and halogen reduction by $Cp_2ZrCl(III)H$. Although these fundamental reactions are well established in the case of hydrosilanes,[16,19] hydrogermanes,[15] and hydrostannanes, the present results will develop a new and attractive aspect of transition metal–hydrido complexes. We have also been exploring other radical mediators such as H–Ga, H–In, H–Ge, and H–P compounds, pursuing excellent efficiency and selectivity.[15,20]

26.3 RADICAL CYCLIZATION REACTIONS WITH A ZIRCONOCENE(ALKENE) COMPLEX AS AN EFFICIENT SINGLE ELECTRON TRANSFER AGENT

Oshima *et al.* have been investigating single electron transfer from electron rich transition metal complex to alkyl halide and its application to organic synthesis.[21,22] We then expected that zirconocene(alkene) complex would effect single electron transfer. Radical cyclization of various β-halo acetals with **1a** was firstly examined. Treatment of iodo acetal **16a** with **1a** for 3 h afforded the corresponding cyclization product in 84% yield (eqn (26.4)).

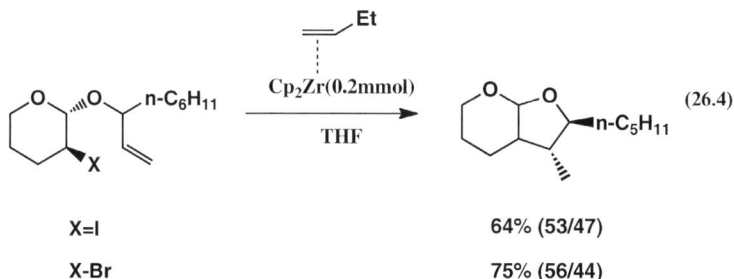

$$X=I \qquad\qquad 64\% \ (53/47)$$
$$X\text{-}Br \qquad\qquad 75\% \ (56/44)$$

Reflecting a radical mechanism, the product was obtained with high *trans* selectivity with respect to the pentyl and methyl groups.[13] Bromo acetal also underwent cyclization upon treatment with the zirconocene complex. Reduction of aryl iodides proceeded smoothly to afford the corresponding products in excellent yields.

Surprisingly, the use of 1,2-dimethoxyethane as a solvent dramatically changed the reaction pathway. In this case, tetrahydrofuranylmethylzirconium was cleanly formed. The existence of zirconium containing intermediate was unambiguously verified by deuterolysis to give final product in 70% yield with 94% deuterium incorporation (Scheme 26.6). The alkyl zirconium species could be also trapped by electrophiles such as allyl bromide and benzoyl chloride in the presence of a stoichiometric amount of CuCN.

26.4 TRIETHYLBORANE-INDUCED RADICAL ALLYLATION REACTION WITH A ZIRCONOCENE(ALKENE) COMPLEX

The radical allylation reaction is synthetically useful because the introduced allyl group serves as a versatile precursor for further functionalization.[23–27]

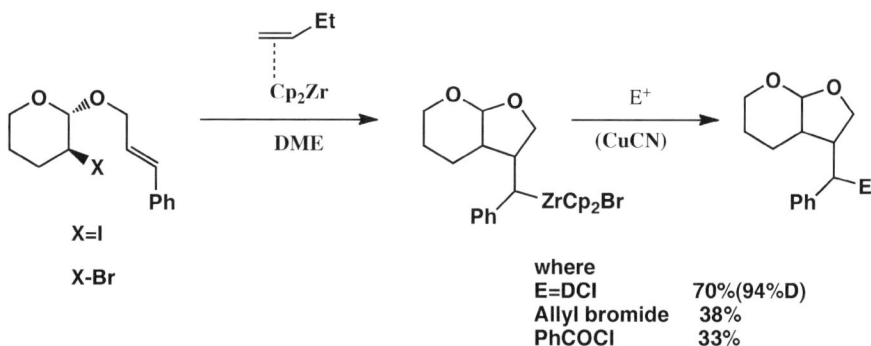

where
E=DCl   70%(94%D)
Allyl bromide   38%
PhCOCl   33%

Scheme 26.6   Schwartz-reagent mediated radical cyclization reaction.

Recent advances in the radical allylation reaction have been mainly benefited from the efficiency of allylstannanes as mediators.[28,29] The use of tin-based reagents is, however, not always convenient because of the inherent toxicity of organotin derivatives and the difficulty often encountered in removing tin residues from the product. Oshima succeeded in a Schwartz-reagent-mediated radical reduction reaction. Additionally, Oshima *et al.* have been interested in the development of surrogates for allyltin.[30] Allylic zirconium reagent seemed hopeful. Benzyl iodoacetate was subjected to radical crotylation with allylic zirconium reagent in the presence of Et₃B to provide the desired adduct quantitatively (eqn (26.5)). This protocol is also general and can be applied to a broad range of α-halo carbonyl compounds. It is worth noting that α-chloro carbonyl compounds and tertiary bromides could be allylated in good yields.

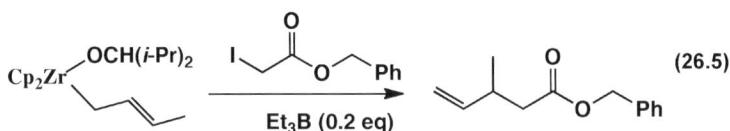

Intermolecular three-component coupling reaction of alkyl halides, alkenes, and allylzirconium, prepared from allyl Grignard reagent and Cp₂ZrCl₂, was also successful. A sequential addition of cyclohexyl iodide, *tert*-butyl acrylate, and Et₃B to a solution of allylzirconium in THF afforded the coupling product in 84% yield *via* radical intermediate intermediate (Scheme 26.7).

26.5 CONCLUSION

The entry of Schwartz reagent, zirconocene(alkene) complex, and allylated zirconocene into radical chemistry encourages chemists to reconsider radical-based synthesis. Use of these zirconium reagents provides Oshima *et al.* with a tin-free system. Oshima *et al.* showed that we can thus avoid bothersome workup and purification procedures. Moreover, the reactivity of the zirconium reagents can be easily tunable by changing the coordinating ligands. There are many reports on the ligand effect on complexes. Transition metal-mediated radical reactions will hence open a new gateway for organic synthesis to attain higher reactivity, chemoselectivity, and stereoselectivity. Also these reactions

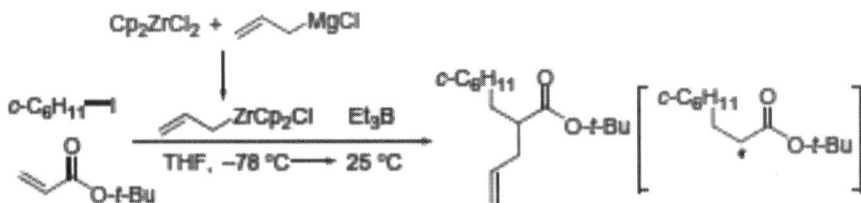

Scheme 26.7   Tin-free, transition metal-mediated synthetic transformation.

will promote researches on the reaction of alkyl halide with zirconocene in addition to conventional reaction of unsaturated compounds. Allylic C–H bond activation by zirconocene and generation of allylic anion under mild condition is quite fascinating. C–H bond activation usually requires harsh reaction conditions and/or transition metal complexes which are difficult to handle.[31] The mild conditions will allow chemists to employ functionalized alkene as an allylic zirconium precursor. In addition, the zirconium complex which realizes C–H bond activation is readily prepared *in situ*. The present method thereby offers tools applicable to organic synthesis.

REFERENCES

1. J. J. Berzelius, *Ann. Chim. Phys.*, 1824, **26**, 43.
2. G. A. Olah, *Friedel–Crafts and Related Reactions*, Interscience, New York, 1963, vol. 1, p. 1031.
3. J. Boor, *Ziegler–Natta Catalysis and Polymerization*, Academic Press, New York, 1978, p. 670.
4. D. W. Hart and J. Schwartz, *J. Am. Chem. Soc.*, 1974, **96**, 8115; P. C. Wailes and H. Weigold, *J. Organomet. Chem.*, 1970, **24**, 405; P. C. Wailes, H. Weigold and A. P. Bell, *J. Organomet. Chem.*, 1971, **27**, 373.
5. E. Negishi and D. E. Van Horn, *J. Am. Chem. Soc.*, 1977, **99**, 3168; N. Okukado, D. E. Van Horn, W. L. Klima and E. Negishi, *Tetrahedron Lett.*, 1978, **19**, 1027; E. Negishi, N. Okukado, D. E. King, D. E. Van Horn and B. I. Spigel, *J. Am. Chem. Soc.*, 1978, **100**, 2254; S. L. Wiskur, A. Korte and G. C. Fu, *J. Am. Chem. Soc.*, 2004, **126**, 82.
6. D. E. Van Horn and E. Negishi, *J. Am. Chem. Soc.*, 1978, **100**, 2252; E. Negishi, D. Y. Kondakov, D. Choueiry, K. Kasai and T. Takahashi, *J. Am. Chem. Soc.*, 1996, **118**, 9577.
7. W. Kaminsky, H. J. Vollmer, E. Heins and H. Sinn, *Makromol. Chem.*, 1974, **175**, 443; A. Andresen, H. J. Cordes, J. Herwig, W. Kaminsky, A. Merck, R. Mottweiler, J. Pein, H. Sinn and H. J. Vollmer, *Angew. Chem., Int. Ed. Engl.*, 1976, **15**, 630; H. Sinn, W. Kaminsky and H. J. Vollmer, *Angew. Chem., Int. Ed. Engl.*, 1980, **19**, 390.
8. E. Negishi, S. J. Holms, J. M. Tour, J. A. Miller, F. E. Cederbaum, D. R. Swanson and T. Takahashi, *J. Am. Chem. Soc.*, 1989, **111**, 3336; E. Negishi, *Comprehensive Organic Synthesis*, ed. L. A. Paquette, Pergamon Press, New York, 1991, vol. 5, p. 1163; T. Takahashi, N. Suzuki, M. Hasegawa, Y. Nitto, K. Aoyagi and M. Saburi, *Chem. Lett.*, 1992, 331; N. Suzuki, C. J. Rousset, K. Aoyagi, M. Kotora and T. Takahashi, *J. Organomet. Chem.*, 1994, **473**, 117; D. R. Swanson, C. J. Rousset and E. Negishi, *J. Org. Chem.*, 1989, **54**, 3521.
9. A. H. Hoveyda and J. P. Morken, *Angew. Chem., Int. Ed.*, 1996, **35**, 1262.
10. N. Chinkov, S. Mazumdar and I. Marek, *J. Am. Chem. Soc.*, 2003, **125**, 13258; N. Chinkov, S. Mazumdar and I. Marek, *J. Am. Chem. Soc.*, 2002,

124, 10282; N. Chinkov, H. Chechik, S. Mazumdar, A. Liard and I. Marek, *Synthesis,* 2002, 2473.

11. H. Ito, T. Taguchi and Y. Hanzawa, *Tetrahedron Lett.,* 1992, **33**, 1295; H. Ito, T. Nakamura, T. Taguchi and Y. Hanzawa, *Tetrahedron,* 1995, **51**, 4507; Y. Hanzawa, H. Ito and T. Taguchi, *Synlett,* 1995, 299; H. Ito, H. Kuroi, H. Ding and T. Taguchi, *J. Am. Chem. Soc.,* 1998, **120**, 6623; S. Harada, T. Taguchi, N. Tabuchi, K. Narita and Y. Hanzawa, *Angew. Chem., Int. Ed.,* 1998, **37**, 1696; Y. Hanzawa, N. Tabuchi, K. Saito, S. Noguchi and T. Taguchi, *Angew. Chem., Int. Ed.,* 1999, **38**, 2395.

12. P. Wipf and H. Jahn, *Tetrahedron,* 1996, **52**, 12853; M. Fried, D. Pérez, A. J. Peat and S. L. Buchwald, *J. Am. Chem. Soc.,* 1999, **121**, 9469; S. Yamanoi, K. Seki, T. Matsumoto and K. Suzuki, *J. Organomet. Chem.,* 2001, **624**, 143; G. J. Gorden, T. Luker, M. W. Tuckett and R. J. Whitby, *Tetrahedron,* 2000, **56**, 2113; *Titanium and Zirconium in Organic Synthesis,* ed. I. Marek, Wiley-VCH, Weinheim, 2002.

13. K. Fujita, T. Nakamura, H. Yorimitsu and K. Oshima, *J. Am. Chem. Soc.,* 2001, **123**, 3137.

14. H. Yorimitsu, H. Shinokubo and K. Oshima, *Synlett,* 2002, 674.

15. T. Nakamura, H. Yorimitsu, H. Shinokubo and K. Oshima, *Synlett,* 1999, 1415; H. Yorimitsu, H. Shinokubo and K. Oshima, *Chem. Lett.,* 2000, 104; S. Tanaka, T. Nakamura, H. Yorimitsu, H. Shinokubo and K. Oshima, *Org. Lett.,* 2000, **2**, 1911; H. Yorimitsu, H. Shinokubo and K. Oshima, 2001, **74**, 225; T. Nakamura, H. Yorimitsu, H. Shinokubo and K. Oshima, *Bull. Chem. Soc. Jpn.*, 2001, **74**, 747.

16. C. Chatgilialoglu, *Radicals in Organic Synthesis,* ed. P. Renaud and M. P. Sibi, Wiley-VCH, Weinheim, 2001, vol. 1, ch. 1.3; D. P. Curran, *Comprehensive Organic Synthesis,* ed. B. M. Trost and I. Fleming, Pergamon, Oxford, 1991, p. 715; W. P. Neumann, *Synthesis,* 1987, 665; D. P. Curran, *Synthesis,* 1988, 489.

17. *Radicals in Organic Synthesis,* ed. P. Renaud and M. P. Sibi, Wiley-VCH, Weinheim, 2001; B. Maillard, D. Forrest and K. U. Ingold, *J. Am. Chem. Soc.,* 1976, **98**, 7024; A. L. J. Beckwith and S. A. Glover, *Aust. J. Chem.,* 1987, **40**, 157.

18. H. Yorimitsu and K. Oshima, *Radicals in Organic Synthesis,* ed. P. Renaud and M. P. Sibi, Wiley-VCH, Weinheim, 2001, vol. 1, ch. 1.2; K. Nozaki, K. Oshima and K. Utimoto, *J. Am. Chem. Soc.,* 1987, **109**, 2547; K. Oshima and K. Utimoto, *J. Synth. Org. Chem. Jpn.,* 1989, **47**, 40.

19. C. Chatgilialoglu, *Radicals in Organic Synthesis,* ed. P. Renaud and M. P. Sibi, Wiley-VCH, Weinheim, 2001, vol. 1, ch. 1.3; D. P. Curran, *Comprehensive Organic Synthesis,* ed. B. M. Trost and I. Fleming, Pergamon, Oxford, 1991, p. 715; W. P. Neumann, *Synthesis,* 1987, 665; D. P. Curran, *Synthesis,* 1988, 489; C. Chatgilialoglu, *Acc. Chem. Res.,* 1992, **25**, 188.

20. S. Mikami, K. Fujita, T. Nakamura, H. Yorimitsu, H. Shinokubo, S. Matsubara and K. Oshima, *Org. Lett.,* 2001, **3**, 1853; S. Tanaka, T. Nakamura, H. Yorimitsu, H. Shinokubo and K. Oshima, *C. R. Acad.*

Sci., Ser. IIc: Chim., 2001, **4**, 461; S. Tanaka, T. Nakamura, H. Yorimitsu, H. Shinokubo, S. Matsubara and K. Oshima, *Synlett,* 2001, 1278; K. Takami, H. Yorimitsu and K. Oshima, *Org. Lett.*, 2002, **4**, 2993; K. Takami, S. Mikami, H. Yorimitsu, H. Shinokubo and K. Oshima, *Tetrahedron,* 2003, **59**, 6627; K. Takami, S. Mikami, H. Yorimitsu, H. Shinokubo and K. Oshima, *J. Org. Chem.*, 2003, **68**, 6627.

21. K. Fujita, H. Yorimitsu and K. Oshima, *Synlett,* 2002, 337.
22. H. Shinokubo and K. Oshima, *J. Synth. Org. Chem. Jpn.*, 1999, **57**, 27; K. Oshima, *J. Organomet. Chem.*, 1999, **575**, 1; D. Motoda, H. Kinoshita, H. Shinokubo and K. Oshima, *Adv. Synth. Catal.*, 2002, **344**, 261; K. Wakabayashi, H. Yorimitsu and K. Oshima, *J. Am. Chem. Soc.*, 2001, **123**, 5374; Y. Ikeda, T. Nakamura, H. Yorimitsu and K. Oshima, *J. Am. Chem. Soc.*, 2002, **124**, 6514; T. Fujioka, T. Nakamura, H. Yorimitsu and K. Oshima, *Org. Lett.*, 2002, **4**, 2257; T. Tsuji, H. Yorimitsu and K. Oshima, *Angew. Chem., Int. Ed.*, 2002, **41**, 4137.
23. K. Fujita, H. Yorimitsu, H. Shinokubo, S. Matsubara and K. Oshima, *J. Am. Chem. Soc.*, 2001, **123**, 12115.
24. V. K. Dioumaev and J. F. Harrod, *Organometallics,* 1997, **16**, 1452; E. Negishi, J. P. Maye and D. Choueiry, *Tetrahedron,* 1995, **51**, 4447.
25. K. Fujita, H. Shinokubo and K. Oshima, *Angew. Chem., Int. Ed.*, 2003, **42**, 2550.
26. K. Fujita, H. Yorimitsu, H. Shinokubo and K. Oshima, *J. Org. Chem.*, 2004, **69**, 3302.
27. P. Jones, N. Millot and P. Knochel, *Chem. Commun.*, 1998, 2405; P. Jones and P. Knochel, *Chem. Commun.*, 1998, 2407; N. Millot and P. Knochel, *Tetrahedron Lett.*, 1999, **40**, 7779; P. Jones and P. Knochel, *J. Org. Chem.*, 1999, **64**, 186.
28. K. Hirano, K. Fujita, H. Shinokubo and K. Oshima, *Org. Lett.*, 2004, **5**, 593.
29. I. J. Rosenstein *Radicals in Organic Synthesis*, ed. P. Renaud and M. P. Sibi, Wiley-VCH, Weinheim, 2001, vol. 1, ch. 1.4; G. E. Keck and J. B. Yates, *J. Org. Chem.*, 1982, **47**, 359; D. P. Curran, *Synthesis,* 1988, 489.
30. S. Usugi, H. Yorimitsu and K. Oshima, *Tetrahedron Lett.*, 2001, **42**, 4535.
31. W. D. Jones, F. Kakiuchi and S. Murai, in *Topics in Organometallic Chemistry. Activation of Unreactive Bonds and Organic Synthesis*, ed. S. Murai, Springer, New York, 1999, vol. 3, ch. 2 and 3; J. A. Labinger and J. E. Bercaw, *Nature,* 2002, **417**, 507; C. G. Jia, T. Kitamura and Y. Fujiwara, *Acc. Chem. Res.*, 2001, **34**, 633; C. H. Jun, J. B. Hong and D. Y. Lee, *Synlett,* 1999, 1; S. Murai, F. Kakiuchi, S. Sekine, Y. Tanaka, A. Kamatani, M. Sonoda and N. Chatani, *Nature,* 1993, **366**, 529; B. M. Trost, P. E. Strege, L. Weber, T. J. Fullerton and T. J. Dietsche, *J. Am. Chem. Soc.*, 1978, **100**, 3407.

CHAPTER 27

Applications of Conventional Free Radicals and Advances in Total Synthesis: Radical Cascades in Bio-inspired Terpene Synthesis[*]

27.1 INTRODUCTION

Straightforward strategies for the synthesis of natural products remain an essential goal of research in organic synthesis. At first glance, replication of the enzymatic processes in the lab using simple and usually available natural resources would be the best procedure. Within this context, terpenes have become an attractive goal for the synthetic organic community owing to their biological activities together with their structural diversity. Terpenes are a widely spread family of natural products which mainly present structures with multiples of five carbons (C5, C10, C15,..., C40). The basic C5 isoprene subunits (2-methylbutadiene) are connected in either a head-to-head or head-to-tail manner to give the corresponding natural polyprenes, the biogenetic precursors of terpenes. Subsequent enzyme-catalyzed cationic cyclizations and group migrations lead to the high structural diversity of these compounds. Noteworthily, the transformation of the acyclic polyprenic precursors to the (poly)cyclic final compounds usually takes place in only one step, despite the number of new C–C bonds and stereocenters generated in such a step. From a formal point of view, it is commonly accepted that in nature the cationic cyclization process proceeds through four stages: generation of the initial carbocation, conformational control of the acyclic polyprene, stabilization of the cationic intermediates and the transformation of the final cation, usually elimination processes or nucleophilic quenchings.[1]

* Chapter written by V. Tamara Perchyonok.
Streamlining Free Radical Green Chemistry
V. Tamara Perchyonok, Ioannis Lykakis and Al Postigo
© V. Tamara Perchyonok, Ioannis Lykakis and Al Postigo 2012
Published by the Royal Society of Chemistry, www.rsc.org

The biomimetic approach to the synthesis of polycyclic terpenes based on the cationic cyclizations of the corresponding polyprenes is nowadays regarded as the most powerful route for their preparation.[2–4] In this case, the main challenge is the efficient imitation of the cationic enzymatic-based processes by artificial cyclase-like catalysts without enzymatic assistance. Unfortunately, the lack of enzymatic assistance in the *in vitro* protocols prevents the required preorganized structure in the polyprene precursor which, together with the high reactivity of the cationic intermediates, yields 'incomplete' polycyclization processes and/or low yields of the desired products. This noticeably limits the nature of the polyprene starting materials and prevents in many cases the use of simple 'natural', usually commercially available, substrates. Moreover, the cationic nature of the intermediates normally results in the formation of endocyclic double bonds in the final steps, limiting the generality of the process Concerning the stereoselectivity of the (poly)cyclization process of simple polyprenes, cationic non-enzymatic protocols usually yield mixtures of stereoisomers.

Some forty years ago Breslow *et al.*[5,6] proposed, and subsequently dismissed, that radical cyclizations of polyprenes could be a biosynthetic alternative in nature for the synthesis of terpenes. Indeed, this seminal work showed that terpene skeletons are accessible by radical cyclizations that take place with higher yields and stereoselectivities than their cationic analogues. The main unresolved challenge at that time was the efficient generation of the initial radical from the polyprenic starting substrate. Since then, the bioinspired strategies for the straightforward synthesis of terpenes were greatly improved by a number of groups. Thus, nowadays the construction of polycyclic compounds *via* radicals without enzymatic assistance in only one step and with complete stereoselectivity from acyclic polyprenic precursors has become a reality. Therefore, these new strategies are attractive alternatives to cationic cyclizations. Interestingly in many cases they show complementary features to cationic processes. The main aim of this chapter is to discuss recent developments in this field, highlighting the substrate scope, limitations, and advantages of the contemporary radical protocols.

27.2 ANTECEDENTS

In the sixties, Breslow *et al.* presented the proof of concept for the feasibility of radical cyclizations of simple natural polyprenes to generate terpenic structures in similar yields to cationic cyclizations of the same substrates (Scheme 27.1).[5,6]

Their reaction was initiated by the regioselective addition of the benzoyloxy radical to the starting material geranyl or farnesyl acetate. Despite the limited number of examples, some key features of future radical cyclization of polyprenes already became apparent. The most useful aspect is probably the formation of an exocyclic double bond in the final product. This is complementary to the endocyclic olefins obtained in the eliminations of carbocationic intermediates. Therefore, natural terpenes with such a function-

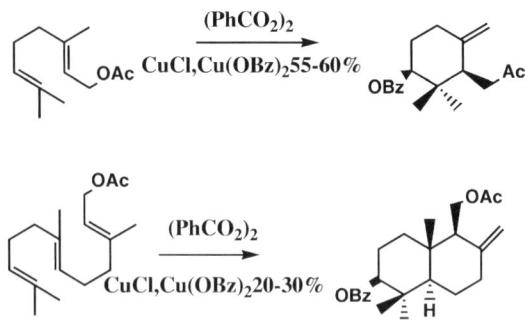

Scheme 27.1   Breslow's seminal polyene cyclizations *via* radicals.

ality, or with a functionality derived from further transformation, are obtained in a simple manner for the first time. It should be noted that cationic-derived cyclization products can be obtained from these products *via* isomerization of the double bond at the endocyclic position. The reverse transformation is not possible, however. Apparently, the presence of a transition metal is required for this *exo*-selective olefin formation. The second characteristic is that the reaction exclusively proceeds to produce trans-fused cyclohexanes, the expected products of cyclizations *via* a chair–chair conformation. Therefore, the *E* configuration of the starting polyprene is stereoselectively transferred to the final product. Substrates containing double bonds with *Z* configuration are not useful in these cyclizations. Cis-fused cyclohexanes are unavailable by this methodology.

Some years later, Julia *et al.* carried out a similar transformation of phenyl substituted polyprenes (Scheme 27.2).[7]

The yields of the polycyclization reactions, involving a final radical Friedel–Crafts-type reaction, were rather low. This fact is probably related with the reluctance of the alkyl radical to add to aromatic rings and the absence of copper salts in the reaction media. In any case, the exceptional stereochemical features of the radical reactions were again observed. Only one stereoisomer was obtained out of the sixteen possible.

Although the above-mentioned transformations seemed highly attractive from a synthetic point of view in principle, the moderate to low yields limited their applications. To this end, the development of new improved methods for the efficient generation of the initial radical was then required.

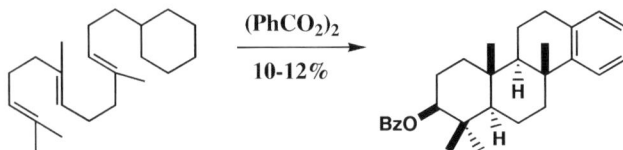

Scheme 27.2   Radical polyene cyclization featuring arenes as radical traps.

27.3 RECENT DEVELOPMENTS

27.3.1 Acyclic Terpenes

Acyclic polyprenes are not the main topic of this chapter but they are the 'natural' starting materials in bioinspired approaches. Although they are usually commercially available and/or obtained directly from natural sources, in some cases, they can also be the desired target compounds. Consequently, different routes have been developed for their synthesis. Within the context of radical chemistry, some successful methodologies have been recently reported. One of these is based on Wurtz-couplings of (poly)prenyl halides to the corresponding homodimers. Thus for example, squalene can be prepared by dimerization of farnesyl bromide in 43% yield (Scheme 27.3).[8]

Interestingly, this methodology was subsequently extended to more convenient starting materials, such as ethyl carbonates. The dimerization proceeds in 81% yield.[9]

On the other hand, selective functionalization, mainly allylic hydroxylation, of polyprenes is a rather difficult task. Hydroxylated polyprenes are valuable in their own right as natural products and as substrates for more functional terpenes. Recently, an α-regioselective Cp_2TiCl-mediated Barbier-type reaction of (poly)prenyl halides and (poly)prenylic aldehydes yielding hydroxylated polyprenes has been reported.[10] The rather rare regioselectivity has been explained by a direct addition of the corresponding (poly)prenyl radical to the aldehyde that is enabled by its coordination to Cp_2TiCl.

Rosiridol, a monoterpenoid isolated from different plants, and 12-hydroxysqualene, the major product from the biotransformation of presqualene diphosphate, catalyzed by the enzyme squalene synthase in the absence of NADPH, have been prepared using this protocol (Scheme 27.4).

It is expected that valuable starting materials of this type will be increasingly used in the preparation of highly functionalized terpenoids.

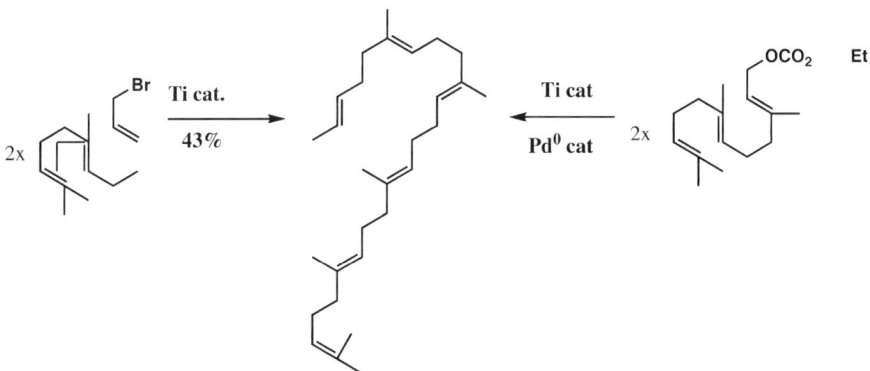

Scheme 27.3 Titanocene-catalyzed squalene synthesis.

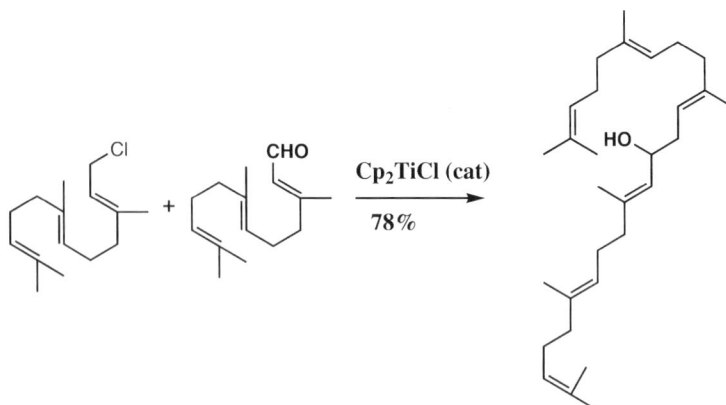

Scheme 27.4   Titanocene-catalyzed α-prenylation reactions.

27.3.2   Radical Polyprene Cyclizations

Radical chemistry has already been extensively used in the synthesis of natural products, including terpenes, mainly for the construction of the basic carbocyclic framework. The interested reader is referred to the excellent reviews devoted to this topic. Here, the focus is on radical processes that produce the terpene skeletons from bioinspired polyprenic cyclizations from simple starting materials in only one synthetic operation.

27.3.3   Photo-induced Electron Transfer (PET) Reactions as Initiation

Direct cyclization of simple polyprenic precursors could avoid lengthy and laborious construction of (poly)cyclic terpene skeletons. To this end, PET reactions[11] have been used to cyclize such starting materials to yield polycyclic structures closely related to natural terpenes. The methodology described by Demuth *et al.* is based on the regioselective generation of a cation-radical, which is subsequently trapped 'anti-Markovnikov-wise' by a nucleophile (water or methanol) as shown in Scheme 27.5.

The resulting tertiary radicals give rise to cascade polycyclizations yielding trans–anti–trans all-chair products.[12] Gratifyingly, the final products possess hydroxy groups at the C-3 position exactly as in the products of the cationic cyclization of natural epoxypolyprenes. This methodology has been successfully applied to the cyclization of geraniol, *E,E*-farnesol and *E,E,E*-geranyl

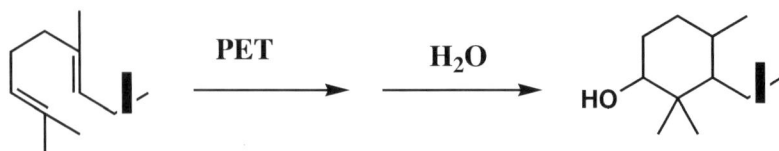

Scheme 27.5   Concept of the radical polyene cyclization *via* PET.

22%

9%

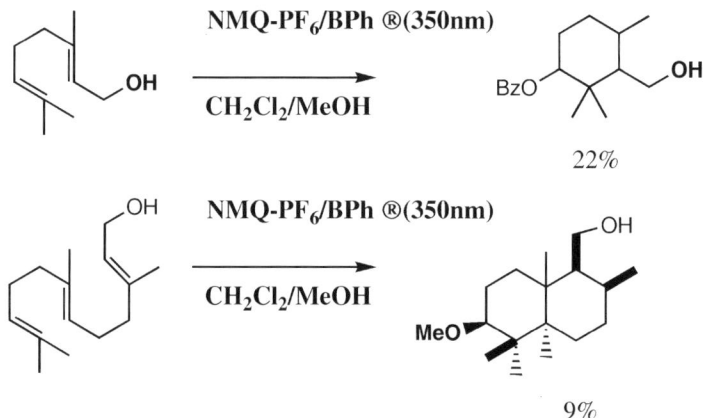

where NMQ.PF$_6$ is N-methylquinolinium hexafluorophosphate, BP is biphenyl

Scheme 27.6 Diastereoselective polyene cyclizations *via* PET.

geraniol, combining a cationic acceptor couple with a low-polarity solvent mixture. In agreement with Breslow's antecedents, only *trans*-fused cyclohexanes were obtained.[13] Unfortunately, the yields of these reactions were still low (Scheme 27.6).

Other related precursors were used for the synthesis of different steroidal products,[14] (–)-3-hydroxy-spongian-16-one and the triterpene precursor of the (–)-stypoldione (Scheme 27.7).[15,16]

In these cases the yields were very reasonable (in the range of 20–30%) in view of the creation of seven new stereocenters in a single synthetic step. This methodology was also extended to the synthesis of aromatic polycyclic terpenoids from acyclic arylpolyalkenes in a single step with moderate to good yields.[17]

Interestingly, the regioselectivity of these cascade reactions can be efficiently controlled at the final cyclization step, by the substitution pattern of the starting polyprenic precursor. The generally observed 6-*endo-trig* mode is replaced by 5-*exo-trig* if electron-withdrawing substituents at the α-terminal position are involved (Scheme 27.8).[12,18]

Although racemic products are usually obtained, the use of spyrocyclic dioxinones prepared from (–)-menthone, as chiral auxiliaries,[14,19] allows the synthesis of cyclic compounds in high diastereomeric ratios (Scheme 27.9).

Removal of the auxiliary groups yields enantiomerically pure all-*trans*-fused tricyclic terpenic skeletons possessing six new asymmetric centers. It is noteworthy that these auxiliaries are remotely located from the initiation site of the reaction.

Formally, this methodology is a non-oxidative cyclization process. However, for bioinspired terpene synthesis this constitutes a major drawback. The

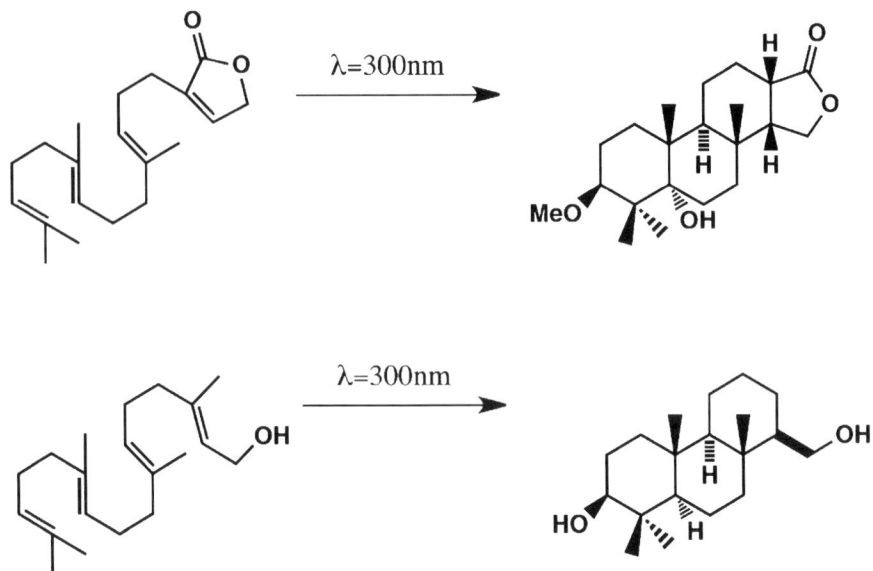

Scheme 27.7   Applications of PET reactions towards natural product synthesis.

addition of a hydrogen atom in the final reduction step prevents the formation of the pivotal olefin present in natural terpenes. This limits the use of this methodology in natural product synthesis.

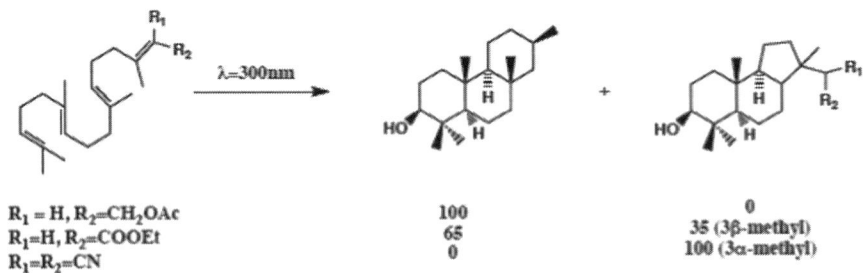

$R_1 = H, R_2 = CH_2OAc$
$R_1 = H, R_2 = COOEt$
$R_1 = R_2 = CN$

100
65
0

0
35 (3β-methyl)
100 (3α-methyl)

Scheme 27.8   Control of regioselectivity in PET cyclizations.

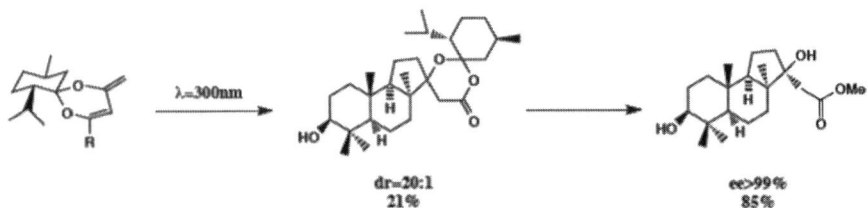

dr=20:1
21%

ee>99%
85%

Scheme 27.9   Chiral auxiliaries in the PET cyclizations.

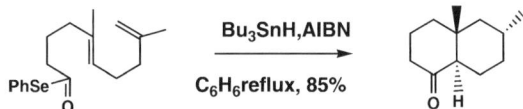

Scheme 27.10   Acyl selenides as radical precursors in polyene cyclizations.

27.3.4 Acylselenium Derivatives as Substrates

Acylselenium derivatives are well known acyl radical precursors. Pattenden and co-workers[20] have shown how acyl/alkyl radical cyclizations of polyolefin–selenoates can give rise to fused polycycles by consecutive 6-*endo-trig* cyclizations under reductive conditions (Bu₃SnH–AIBN). In an early example (Scheme 27.10), acyl radical-mediated polyene cyclization was directed toward steroid ring synthesis.

Phenylseleno ester led to decalone by successive 6-*endo-trig* radical cyclization starting from the corresponding acyl radical.[20] The alkyl group substitution and stereochemistry of the olefins also play an important role in the regio- and stereochemical outcome of these radical cyclizations of polyprenes. Thus, polyprenic subunits give essentially 6-*endo-trig* cyclizations, which mimic the natural cationic processes (Scheme 27.11).[21]

In the same way, the radical cascade polycyclization of polyprene enamide selenoesters can be used to synthesize azasteroids.[22] Therefore, treatment of the enamide selenoester as a 2:1 mixture of *E*- and *Z*-isomers at C-5 with Bu₃SnH–AIBN giving the corresponding D-homo-12-azasteroid **6** as a 4:1 mixture of diastereoisomers. The major product has a *trans,anti,trans,anti,-trans* geometry (Scheme 27.12).

The high regio- and stereoselectivity observed in these radical cascade polycyclizations leading to steroidal structures suggests that the conformation of the polyene chain must be highly ordered in the transition states of the cyclization.

This kind of reaction has also been extended to other polyprenic precursors and used in the synthesis of spongiane-type diterpenes. One of the earliest

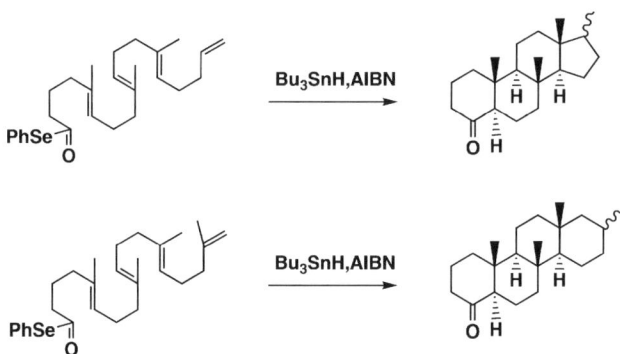

Scheme 27.11   Synthesis of polycyclic ring systems from acyl selenides.

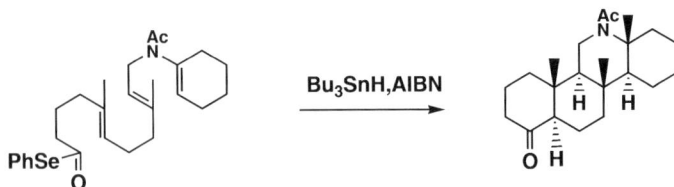

Scheme 27.12   Synthesis of azasteroids from acyl selenides.

examples was the synthesis of the marine metabolite (–)-spongian-16-one.[23] In this case, the desired *trans,anti,trans,anti,cis*-tetracyclic keto lactone **7** was constructed by three consecutive *6-endo-trig* radical cyclizations (Scheme 27.13).

Methylenation of the tetracyclic core using Lombardo's conditions, followed by a Simmons–Smith cyclopropanation to the corresponding cyclopropane and hydrogenolysis, led to the formation of synthetic (–)-spongianone.[24] It is also worth noting that, using this methodology, up to seven carbocycles can be synthesized in only one synthetic step with complete stereoselectivity to a unique all-*trans, anti*-heptacycle (Scheme 27.14).[25]

More elaborated starting materials including cyclopropyl rings in the polyene have also been used to construct terpenic structures. Thus, a sequential cascade *6-endo-trig* cyclization/ macrocyclization/transannulation reaction was developed to synthesize compounds with the unusual all *cis*-orientation of the substituents. Two carboxylate ester moieties were introduced in the designed polyene selenyl ester in order to increase the rate of *6-endo-trig* cyclization and to lower the electrophilicity of the terminal olefin, avoiding the competitive 14-*endo-trig* macrocyclization (Scheme 27.15).[26]

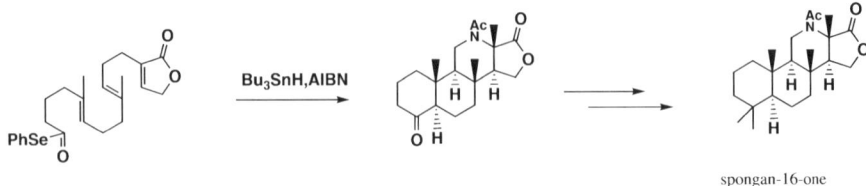

spongan-16-one

Scheme 27.13   Acyl selenide cyclization as a key step in the synthesis of spongianone.

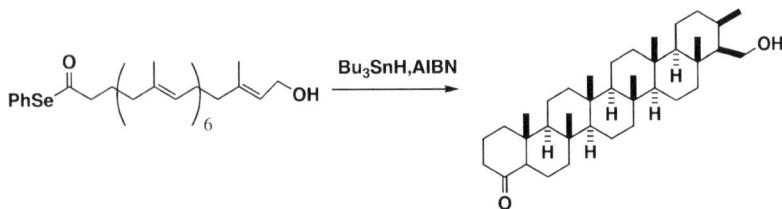

Scheme 27.14   Preparation of a heptacyclic compound from an acyl selenide.

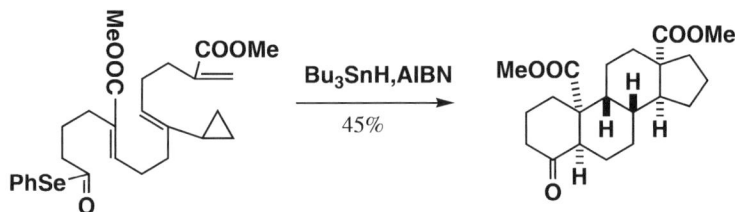

Scheme 27.15 Polyene cyclizations featuring a cyclopropylcarbinyl radical ring opening.

Despite the significant success of this approach in the synthesis of terpenic structures, two main drawbacks of this elegant methodology have become apparent. First, the preparation of the seleno esters usually involves several synthetic steps. This limits the starting materials available. Second, the reaction is terminated reductively and thus the pivotal olefins of the natural products cannot be accessed.

27.4 TRANSITION-METAL-MEDIATED TRANSFORMATIONS

As highlighted in Breslow's seminal work, the presence of transition metals is essential for an oxidative termination. Only in this manner can the olefin be generated. It is not accessible *via* standard reductive radical cyclizations.[27] At first glance it seems that two metals are required for a polycyclization, one for radical generation and a second for radical oxidation to the olefin. This chapter highlights the much more attractive chemistry of Mn(III) and Ti(III)-complexes. They can be used for radical generation and can be used for olefin synthesis from radicals. Therefore, they have been extensively used in the bioinspired synthesis of terpenes.

27.4.1 Manganese(III)-mediated Cyclizations

Manganese(III) complexes are one-electron oxidants which have found numerous synthetic applications. They have been widely used in oxidative free-radical cyclizations and intermolecular additions, and excellent reviews covering all these aspects have been published by Snider[28] and others.[29–32]

Manganese triacetate, Mn(OAc)$_3$, is able to oxidize β-keto esters and other acidic carbonyl compounds to generate the corresponding electrophilic α-carbonyl radicals, which can subsequently add to unactivated olefins to give cyclic products (Scheme 27.16).[33,34] Concerning the final step, Mn(III) is also able to transform the final alkyl radicals to alkenes. Nevertheless, Cu(II) salts, such as Cu(OAc)$_2$, are often added as co-oxidants to facilitate this termination of the radical sequence (Scheme 27.16).[35,36]

This renders the Mn(III)/Cu(II) couple a very effective tool in the synthesis of terpenic structures.[37,38] This is particularly so because an easy access to a great diversity of natural products by efficient cascade radical cyclization processes

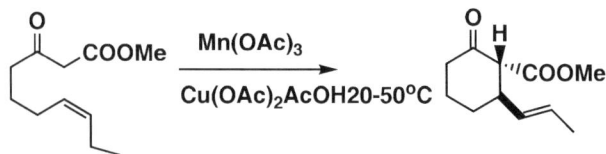

Scheme 27.16 Mn-mediated cyclization with oxidative termination.

of relatively simple polyenic compounds is opened. Thus, the cyclization reaction of polyprenes oxidized at C-1 and C-3 positions yields terpenic structures with two to four carbocycles with complete stereoselectivity and in excellent yields. One early example is the synthesis of podocarpic acid.[28] In this case, the sequence begins with the treatment of the starting β-keto ester in acetic acid with 2 equiv. of Mn(OAc)$_3$ to give the tricyclic adduct in 50% yield. The initially generated α-carbonyl radical adds to the double bond to give a tertiary radical. This intermediate can subsequently be oxidized to the corresponding carbenium ion initiating a Friedel–Crafts-type process. Alternatively, a radical addition to the aromatic ring results in a cyclohexadienyl radical which is oxidized by a second equivalent of Mn(OAc)$_3$ to the corresponding cation. In both cases, the loss of a proton results in rearomatization. Finally, this terpenic structure was converted to podocarpic acid by Clemmensen reduction of the ketone and ester saponification (Scheme 27.17).

This bioinspired approach has also been used in the synthesis of isospongiadiol, a furanoditerpene, where three carbocycles and five key stereogenic centers are stereoselectively introduced in only one step.[39] Later, the same methodology was used in the synthesis of isosteviol, possessing a beyerane skeleton,[40] D,L-norlabdane oxide (Ambrox),[41] a compound of interest for the perfume industry, and the tetracyclic bilactone, wentilactona B.[42] Interestingly, all these bioinspired cyclizations yield products possessing an exocyclic double bond exactly as described by Breslow. This strongly suggests that the alkenes are not produced by simple cationic eliminations. In fact, it has been suggested that in Mn(III)/Cu(II) promoted radical cyclization the generated radical intermediates react with Cu(OAc)$_2$ to give an alkylcopper(III) derivative which undergoes β-hydride elimination to yield the final alkene (Scheme 27.18).[36]

All these results convert the bioinspired Mn(OAc)$_3$-mediated free radical cyclization into a valuable alternative to the cationic cyclization methods. Nevertheless, asymmetric induction leading to an enantioselective synthesis of

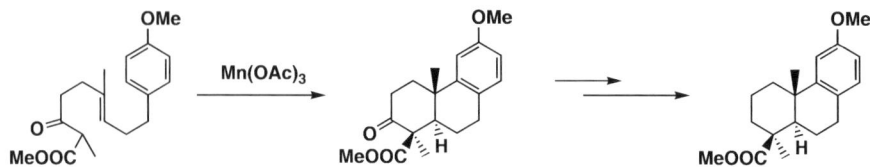

Scheme 27.17 Mn-mediated radical cyclization terminated by a Friedel–Crafts reaction.

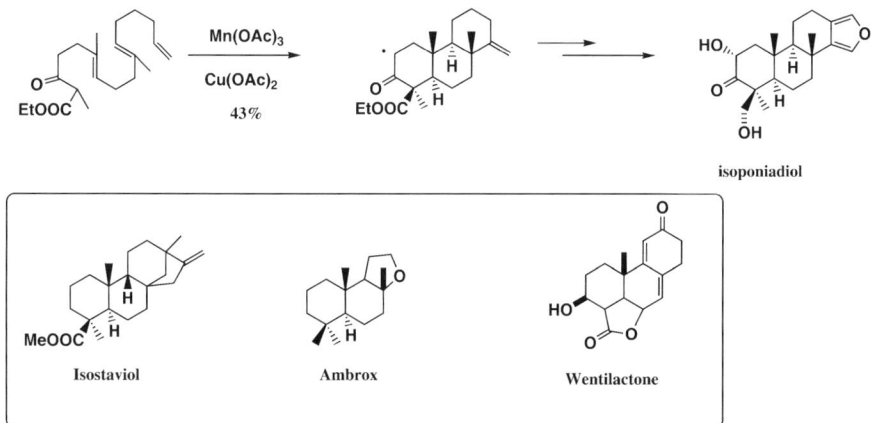

Scheme 27.18 Mn-mediated polycyclizations terminated by β-hydride elimination.

natural products constitutes a challenge. To address this issue, a number of chiral auxiliaries, such as Oppolzer's camphor sultam,[43] and several chiral β-keto sulfoxides, β-keto esters and β-keto amides have been tested.[44] Among them, the β-keto ester phenylmenthyl ester has shown the best results combining high yields and high diastereoselectivities. Snider's group has completed the synthesis of (+)-O-methylpodocarpic acid using (−)-8-phenyl-menthol and the synthesis of (−)-O-methylpodocarpic acid using (+)-8-phenylmenthol (Scheme 27.19).[44]

With the same strategy Yang's group has realized the synthesis of (−)-triptolide, (−)-triptonide, (+)-triptophenolide, and (+)-triptoquinonide in enantiomerically pure forms using (+)-8-phenylmenthol as a chiral auxiliary.[45] For the preparation of (+)-triptocallol,[46] and (+)- and (−)-wilforonide,[47] a series of epimeric 8-arylmenthyl derivatives made from the same chiral source (R)-pulegone were employed. Additionally, Yang's group has shown that lanthanide triflates catalyze the radical cyclization process, increasing the yield and the diastereomeric ratio of the final polycyclic compound. Particularly, Yb(OTf)$_3$ has shown excellent results (Scheme 27.20).

The reactions discussed above demonstrate that Mn(III)-based oxidative cyclizations are clearly advantageous compared to the corresponding cationic

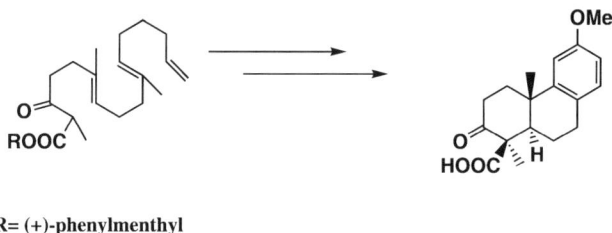

R= (+)-phenylmenthyl

Scheme 27.19 Chiral auxiliaries in Mn-mediated cyclizations.

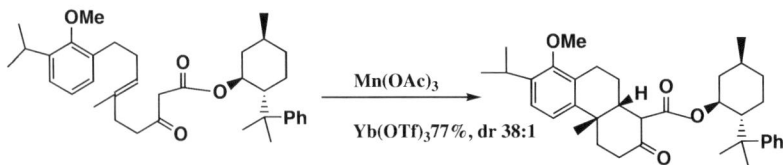

Scheme 27.20   Yb(OTf)$_3$ as cocatalyst in Mn-mediated cyclizations.

and other radical processes for the termination by olefin generation as in Breslow's proposal. The main disadvantage is again connected with the access to the starting polyprenes. A simple procedure to obtain the polyprene-based β-keto esters from natural polyprenes is not known and therefore substantial synthetic effort for the preparation of the starting materials is required. It is also worth noting that, as in cationic cyclizations, aryl substituents are incorporated into the product by a Friedel–Crafts process.

27.4.2 Ti(III)-mediated Epoxypolyprene Cyclizations

The polyprene-based α-hydroxy radicals can be efficiently used in polycyclizations. Probably, the moderate yields of the overall process were related to the efficiency of their generation. A simple solution to improve this approach would be highly desirable. As demonstrated by nature, epoxides are the ideal functional group to this end. Therefore, a regioselective homolytic opening of natural epoxypolyprenes could generate the radicals in sequences resembling cationic reactions. Within this context, Cp$_2$TiCl has emerged as a mild and highly chemoselective single electron transfer (SET) reagent which has been used as a reducing agent of different activated functional groups.[48] Thus, for example, an SET from the titanocene(III) complex to an epoxide resulting in regioselective ring opening produces a β-titanoxy radical. This intermediate is able to undergo intramolecular radical additions to alkenes and alkynes (Scheme 27.21).[49,50]

Interestingly, this characteristic reactivity of Cp$_2$TiCl converts it into a potential artificial terpene cyclase. From a practical point of view, it is possible to prepare and store such an air-sensitive complex. However, it is much more easily prepared *in situ* by stirring of commercial Cp$_2$TiCl$_2$ in the presence of Mn or Zn dust.[49] Recent developments in titanocene(III) chemistry allow the use of Cp$_2$TiCl in substoichiometric amounts with the aid of Ti(III)-regenerating agents.[50,51] Besides their use in asymmetric catalysis[52,53] and in the synthesis of small rings,[54,55] the catalytic procedures allow a control of the concentration of the active species. This is crucial for the success of many difficult cyclizations and for the Ti(III)-bioinspired protocols because premature radical trapping is prevented.

In 2001, Cp$_2$TiCl was first used to promote such bioinspired cyclization of epoxides derived from simple and commercially available geranyl and farnesyl acetate, obtaining in acceptable yield the corresponding terpenic mono (63%) and bicycles (56%) with good to excellent stereoselectivities (Scheme 27.22).[56]

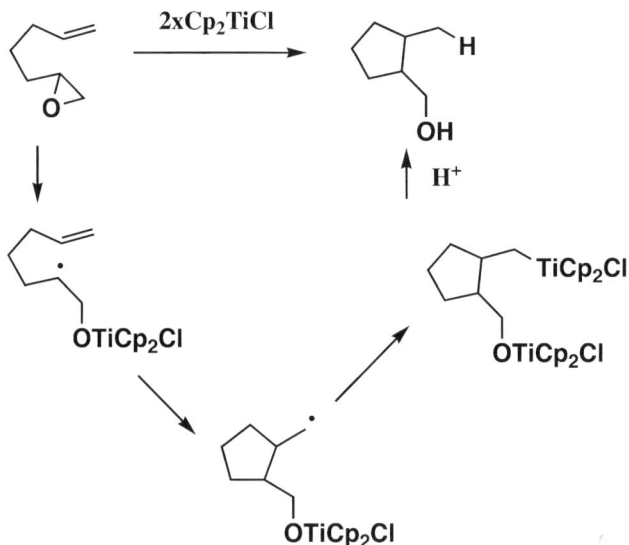

Scheme 27.21 Concept of the Cp$_2$TiCl mediated epoxy-olefin cyclization.

Although the products were expected to be similar to those described by Demuth, products with exocyclic double bonds, exactly as in Breslow's seminal papers, were obtained instead. This is most remarkable since no oxidant is present in the reaction media. Subsequent studies showed that the double bond formation arises from a mixed disproportionation reaction between the organic radicals and Cp$_2$TiCl,[57] whereas the reduced products originate from a Ti(III)-promoted unusual hydrogen atom transfer reaction from water to the corresponding carbon-centered radical.[58] In particular, the combination of Cp$_2$TiCl and a mixture of 2,4,6-collidine, TMSCl and Mn dust improved cyclizations of epoxypolyprenes substantially.[59,60]

A simple comparison between these results and those obtained under cationic conditions amply demonstrate the main differences between the two approaches: (i) endocyclic double bonds or cyclic ethers are the functionalities obtained in the termination step under acidic conditions, (ii) the cationic reactions are not stereoselective yielding mixtures of stereoisomers, (iii) in the

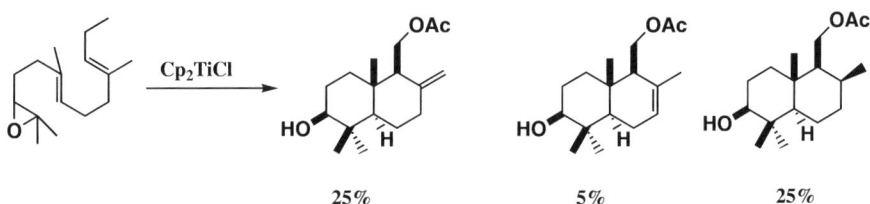

Scheme 27.22 First example of a Cp$_2$TiCl-mediated epoxypolyene cyclization.

absence of activating groups the yields of the cyclization of simple polyprenes are usually rather low.[61,62]

Since Ti(III)-mediated radical cyclizations yield almost exclusively products with exocyclic double bonds, which are present in many natural terpenes, this approach allows an easy and efficient preparation of interesting building blocks. In fact, some of the final products are natural terpenes or closely related derivatives already. Saponification of the basic drimane skeleton[13] yields natural isodrimenediol, for example. More interestingly, the final products derived from the cyclization of squalene possess the tricyclic malabaricane skeleton, which has been isolated from marine sediments. This suggests that malabaricanes could be synthesized by organisms existing under anoxic conditions *via* a pathway similar to that provided by the oxygen-free conditions of free-radical chemistry.

Radical cyclizations of the corresponding epoxides of protected geranyl and farnesyl acetone yielded mono and bicyclic compounds, which have also become relevant intermediates in terpene synthesis. Cyclohexanol has been used in the total synthesis of cyclofarnesane sesquiterpenoid isolated from *Artemisia chamaemelifolia*, and for the total synthesis of triterpenes achilleol A and B (Scheme 27.23).[60–64]

3β-Hydroxymanool, a bicyclic diterpenoid with a labdane skeleton isolated from the fern *Gleichenia japonica,* and other norditerpenes isolated from copaiba oil, could be also synthesized using a bicycle compound as a starting material (Scheme 27.24).[60]

Sclareol oxide, a labdane constituent of the essential oils of *Salvia sclarea*, is also accessible from bicyclic alcohol precursor.[65] The exocyclic double bond can be also transformed into other functional groups, which are difficult to obtain from the endocyclic olefin. An interesting example is the synthesis of 3β-hydroxydihydroconfertifolin, a drimane lactone isolated from a fungus

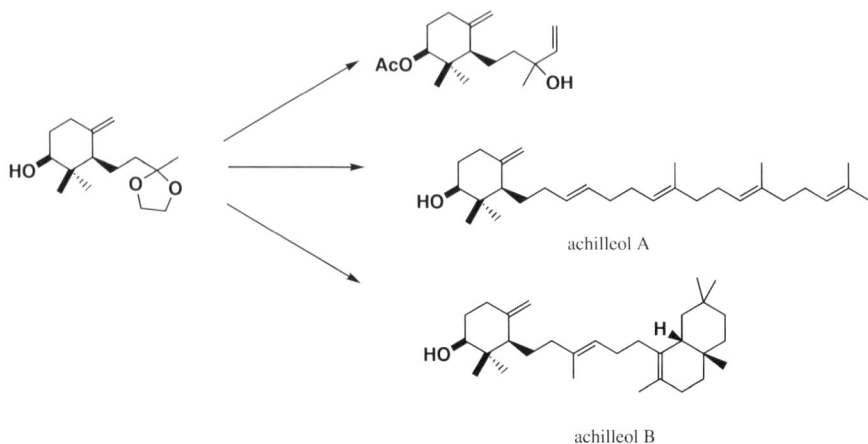

achilleol A

achilleol B

Scheme 27.23   Synthesis of achilleol A and B from products obtained by Cp$_2$TiCl-catalyzed epoxyolefin cyclization.

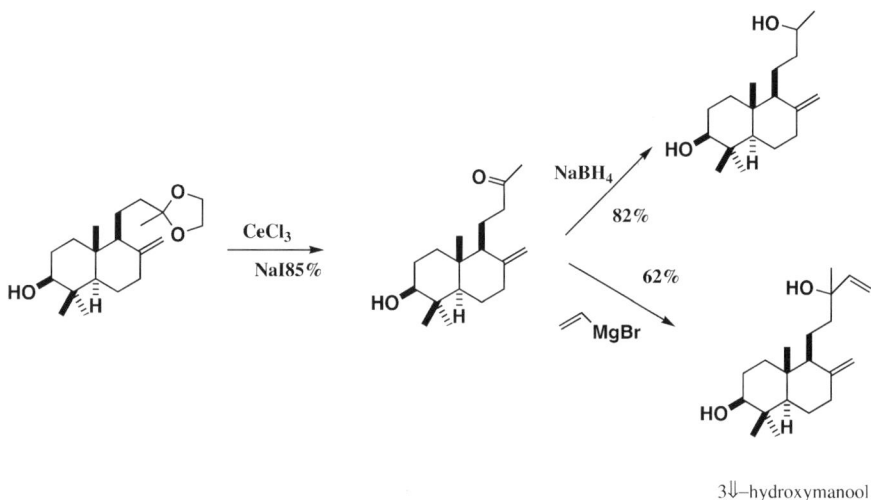

3⇂−hydroxymanool

Scheme 27.24 Synthesis of norditerpenes and hydroxymanool *via* Cp$_2$TiCl-catalyzed epoxypolyene cyclization.

associated with aspen, from isodrimenediol acetate.[66] In this case, an epoxidation reaction and a subsequent Ti(III)-mediated reduction using water as a hydrogenatom donor yielded the trihydroxylated key intermediate, which after simple manipulations could be transformed into synthetic 3β-hydroxydihydroconfertifolin (Scheme 27.25).

Beyond simple protection of the functionalities present in the starting polyprenes (alcohol or ketone groups), other transformations could lead to starting materials containing new functionalities which can be preserved in the final cyclic products. In this sense, a two-step protocol based on a regioselective allylic hydroxylation and an epoxidation reaction yields hydroxylated epoxypolyprenes. These compounds have been transformed into functionalized terpenoids with a γ-dioxygenated system on the A ring. The main limitation of this approach is the low yields obtained in the allylic hydroxylation (Scheme 27.26).

Although the radical cyclization reaction takes place with moderate stereoselectivity, the g-dioxygenated products have been used in the synthesis of smenospondiol,[67] an interesting cytotoxic sesquiterpene, and a valuable synthon for the synthesis of baccatin III C ring system (Scheme 27.27).[68]

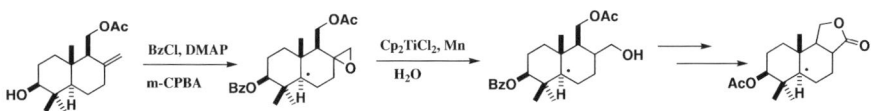

Scheme 27.25 Synthesis of 3β-hydroxydihydroconfertifolin *via* Cp$_2$TiCl-catalyzed epoxypolyene cyclization.

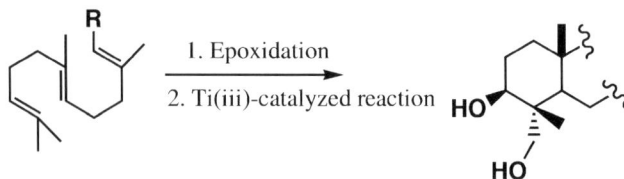

Scheme 27.26   Sharpless epoxides as radical precursors in enantioselective poly-cyclizations.

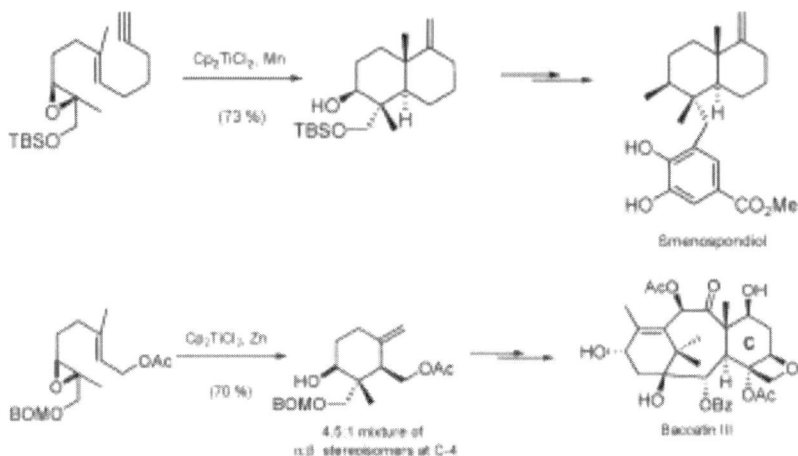

Scheme 27.27   Sharpless epoxides as starting materials in natural product synthesis.

Related to this topic, an interesting alternative to improve the stereo-selectivity has been carried out by taking advantage of the presence of the hydroxyl group at C-3. In this case, a regioselective palladium-based C–H activation of the equatorial methyl group at C-4 of different (poly)cyclic compounds could be successfully accomplished (Scheme 27.28).[69]

Taking into account that the oxidation is carried out at the end of the synthetic sequence, the use of stoichiometric amounts of palladium is not critical. This combination of titanium and palladium chemistry was used in

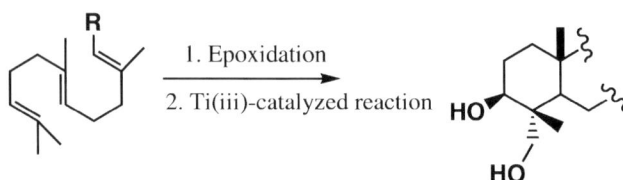

Scheme 27.28   Post-cyclization oxidation of products obtained *via* Cp₂TiCl-catalyzed epoxypolyene cyclization.

the first synthesis of rostratone, a diterpenic metabolite isolated from *Neobrachiella rostrata*.[69] Simple transformations of the nor-diterpenic ketone, synthesized previously by a bioinspired titanocene(III)-radical cyclization of a farnesyl acetone derivative, yielded the methyl ester of 3-oxoanticopalic acid, found in the needles of *Pinus strobus*. A four-step sequence (Scheme 27.29) allowed the remote functionalization affording the corresponding acetoxy ketone, which after a transesterification reaction yielded synthetic rostratone.

An additional advantage of this protocol based on the bioinspired cyclization of epoxypolyprenes is that the correct absolute configuration of final products can be introduced using the many versatile methods existing for enantioselective epoxide synthesis. Thus for example, Sharpless dihydroxylation was used in the enantioselecitve synthesis of natural terpenoids, such as achilleol B,[64] myrrhanol A,[70] and β-onocerin.[71] In this last case, it is worth noting that, besides the titanocene(III)-catalyzed cyclization of the enantioenriched epoxypolyprene, the synthesis also featured a highly efficient titanocene(III)-catalyzed Wurtz dimerization (Scheme 27.30).

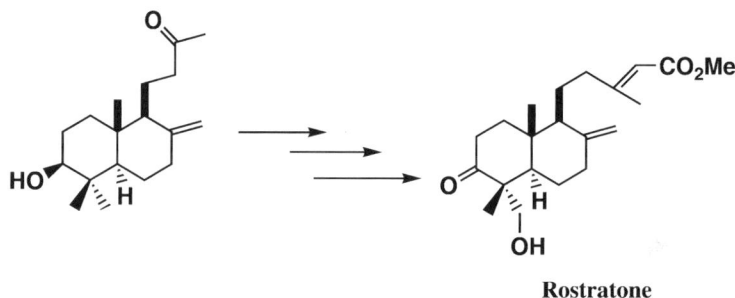

Scheme 27.29   Synthesis of rostratone from epoxypolyene cyclization products.

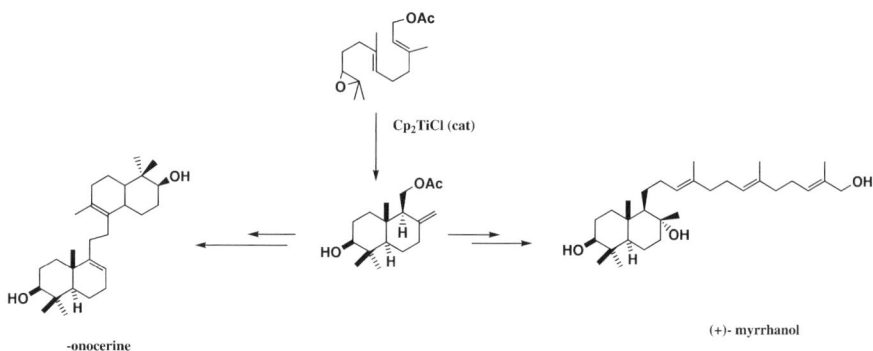

Scheme 27.30   Synthesis of onocerine and myrrhanol A *via* Cp₂TiCl-catalyzed epoxypolyene cyclization.

On the other hand, Jacobsen epoxidation of protected geranyl acetone allowed the introduction of the key stereogenic center in the synthesis of (–)-ambrinol, an odorous component of ambergris (Scheme 27.31).[72]

Sharpless epoxidation of allylic alcohols derived from polyprenes has also been used to introduce the absolute configuration in the enantioselective synthesis of labdane-type compounds.[73,74] One example, which shows the potential of this enantioselective approach, is the synthesis of the tetracyclic terpenoid fomitellic acid, a potent inhibitor of calf DNA polymerase a, rat DNA polymerase b, and human DNA topoisomerases I and II,[75] where the AB ring system was stereoselectively constructed by titanocene(III)-mediated cyclization of an enantioenriched acyclic precursor (Scheme 27.32).

A key difference between cationic and radical chemistry is related to the reluctance of alkyl radicals to add to aromatic rings. This avoids Friedel–Crafts-type processes observed with cations. Therefore, 'interrupted' cyclizations are expected when aryl-substituted polyprenes are used as starting materials. The final products with terpenic structures with pendant aryl groups are a subgroup of the meroterpene family, which seem inaccessible by cationic chemistry. They are usually prepared by a strategy based on the linkage of the terpenic and aromatic subunits. In this case, it is clear that the bioinspired radical cyclizations represent a complementary approach to classical syntheses of this kind of compounds. Different arylsubstituted epoxypolyprenes have been cyclized in the presence of Ti(III), yielding the corresponding mono to tricyclic structures with intact aromatic rings of different electronic and steric nature.[76–78] It should be noted that such aryl-substituted epoxypolyprenes can

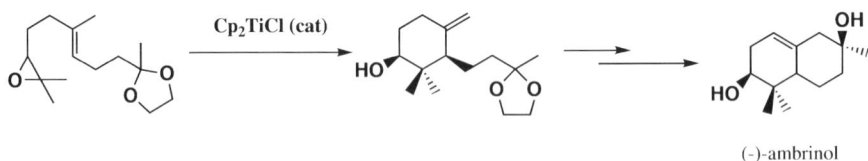

Scheme 27.31   Synthetic approach to (–)-ambrinol.

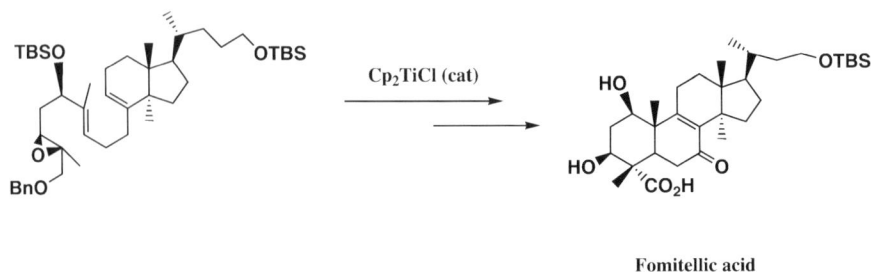

Fomitellic acid

Scheme 27.32   Approach to fomitellic acid based on a Cp$_2$TiCl-catalyzed epoxypolyene cyclization.

be easily prepared from the corresponding epoxy acetates or ethyl carbonates by Cu-catalyzed or Stille coupling reactions (Scheme 27.33).[76,77]

An excellent example of the application of this methodology is the synthesis of zonarol and zonarone. They could be prepared in a straightforward manner *via* a radical cyclization of an aryl-substituted farnesyl derivative.[77] It is also worth noting that the chemoselectivity in the radical cyclization reaction is only possible because the reaction takes place under reductive conditions. Under oxidative conditions, that is in the presence of Cu(II) or Mn(III) salts, a final oxidation step results in a termination by a cationic Friedel–Crafts-type reaction.

Although 6-*endo-trig* cyclization processes are predominant in cationic cyclizations, in nature we can find terpenes which possess seven-membered carbocycles, such as the barekane and valparane families. From a biosynthetic point of view, it is worth nothing that the biosynthesis of barekanes is proposed to take place by a sequence of 6-*endo*, 6-*endo*, 7-*endo* cationic cyclizations involving a less stable secondary cation as a key-intermediate (Scheme 27.34).

On the other hand, enzymes isolated from marine red algae genus *Laurencia* have been shown to catalyze the transformation of nerolidol to snyderol, as previously observed in the laboratory under cationic conditions.[79] The absence of the enzymatic-assisted formation of seven-membered carbocycles suggests that an alternative mechanism may be operative. In the proposed biosynthesis of valparanes,[80] other unknown cationic cyclizations have been postulated. It is also worth noting that there is no example described in the literature of carbocationic cyclizations of unfunctionalized substrates yielding seven-membered carbocycles. This precludes a simple cationic approach for the synthesis of these natural terpenes. In this context, the Ti(III)-catalyzed cyclization of epoxy derivatives of linalool (55%), nerolidol (55%) and geranyl

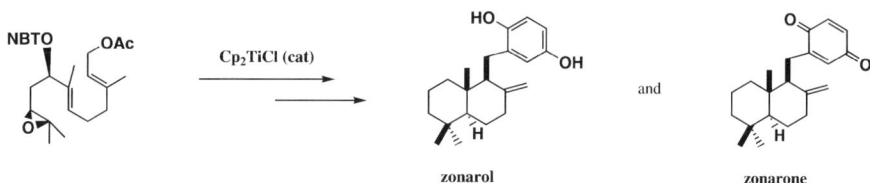

Scheme 27.33 Cu-catalyzed arylation and Cp$_2$TiCl-catalyzed epoxypolyene cyclization in the synthesis of zonarol and zonarone.

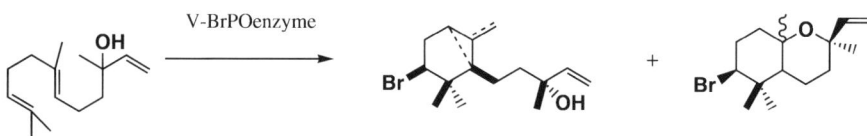

Scheme 27.34 Biosynthesis of barekane and valparane natural products.

linalool (39%) yields products with unique sevenmembered carbocycles that are derived from a rare final 7-*endo*-*trig* radical cyclization (Scheme 27.35).[81]

The similarity between the structures obtained by radical cyclization of simple epoxypolyprenes and natural barekane and valparane skeletons is intriguing. The barekane skeleton generation is biosynthetically initiated by the addition of a bromonium ion to the starting polyprene with enzymatic assistance. It is fascinating that the most efficient chemical synthesis involves addition of a bromine atom. This highlights once again the validity of Breslow's claim.[5,6] With the 7-*endo*-*trig* cyclization offering a straightforward solution for the preparation of these structures, their synthesis of barekoxide, laukarlaol and valparadiene has been described (Scheme 27.36).[81]

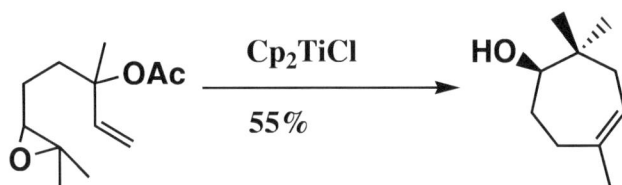

Scheme 27.35   Cp$_2$TiCl-catalyzed 7-*endo* cyclizations in epoxypolyene cyclizations.

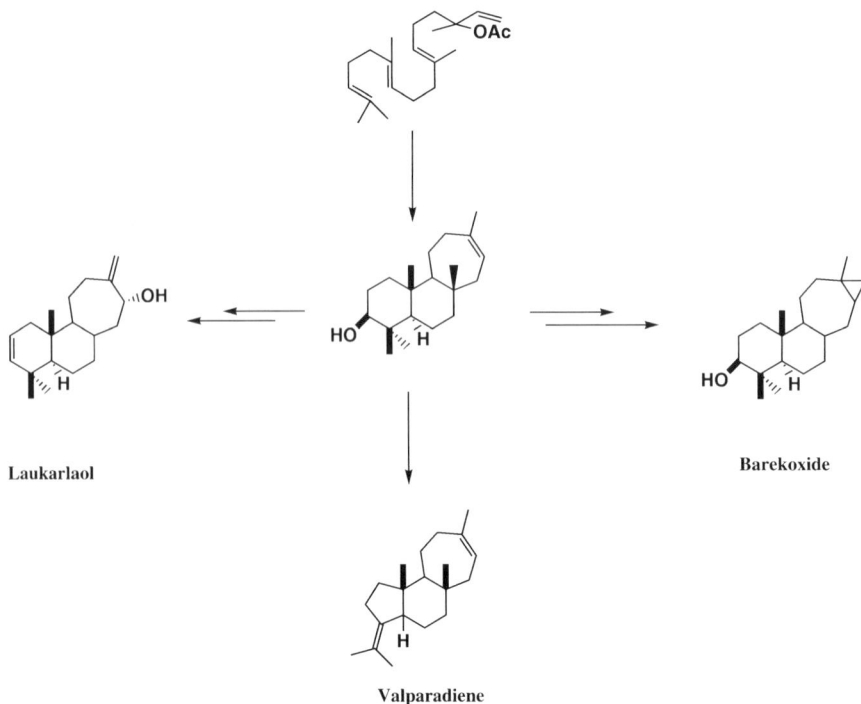

Laukarlaol

Barekoxide

Valparadiene

Scheme 27.36   Further endeavours towards the synthesis of laukarlaol, barekoxide, and valparadiene.

Thus, for example, barekoxide was obtained from the corresponding tricycle containing cycloheptene ring in only 5 steps, with an overall yield of 8% starting from geranyl linalool. A recent synthesis of barekoxide required 8 steps from commercial sclareolide, which possesses two of the three carbocycles required in the final structure.[82] Laukarlaol was also prepared from tricyclic intermediate in 5 steps, allowing the reassignment of its relative configuration at C-14. In the synthesis of valparadiene, the hydroxyl group at the C-3 position was again used to good effect for further modifications. A cationic contraction of the A-ring yielded directly the valparane skeleton, containing a five-, six- or seven-membered ring, in only four steps from commercial geranyl linalool, with an overall yield of 21% (Scheme 27.36).

A similar approach was also used for the straightforward synthesis of daucanes.[81] All these results demonstrate that bioinspired Ti(III)-mediated cyclizations have emerged into a powerful and efficient methodology to construct different terpenic structures. Further value is added to the process by the fact that in many cases a complementary selectivity compared to the more widespread cationic cyclizations was observed. A disadvantage which is shared with cationic protocols, however, is the required regioselective epoxidation of the terminal alkene functionality.

27.5 SOMO ORGANOCATALYSIS AND TERPENES

Another very attractive approach has been recently reported by MacMillan's group. In this case, an intriguing combination of an organocatalyst and a single electron oxidant, Cu(OTf)$_2$, allows an enantioselective polyene cyclization, yielding in only one synthetic step 6 new C–C bonds, 11 contiguous stereocenters, 5 all-carbon quaternary stereocenters with a 62% yield and with a level of enantiocontrol in the range of 90% ee (Scheme 27.37).[83]

The success of this approach is limited, however, by the fact that simple polyprenes are not suitable starting materials for this reaction. Instead, an alternating sequence of polarity inverted C=C double bonds in the polyprene is required. In any case, it is clear that this approach will surely find many applications in the future.

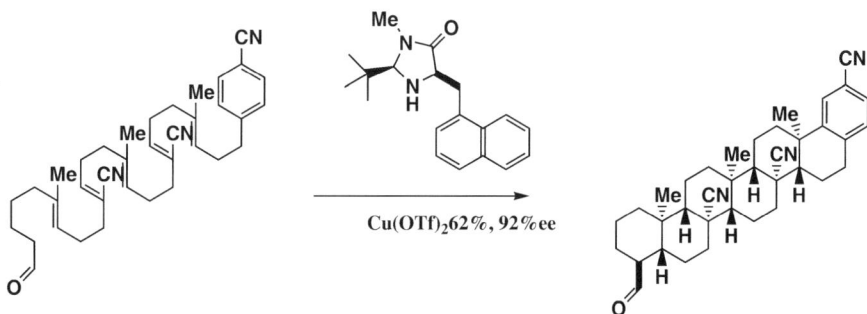

Scheme 27.37 SOMO catalysis in enantioselective polyene cyclizations.

27.6 CONCLUSIONS

Radical cyclizations have by now been applied to polyprenic starting materials in efficient reactions to yield (poly)cyclic terpenic skeletons that are valuable synthons in organic synthesis. The radical approach has in many cases resulted in better yields and stereoselectivities than the cationic equivalents. It is noteworthy that the radical bioinspired protocols are also characterized by unique chemo and regioselectivities, which are often complementary to the cationic cyclizations. These two main advantages have been exploited in the synthesis of numerous natural terpenes of large structural diversity. Nevertheless some drawbacks are intrinsic to this approach and require further research efforts. Thus, the exceptional stereochemical profile of these radical cyclizations does not allow the synthesis of *cis*-fused carbocycles, which are also present in many natural products. Moreover, the reluctance of radicals to carry out skeletal rearrangements or group migrations, which are frequent in cationic biological processes, allows a unique entrance to some structures but also renders some more common structures difficult to access.

Substantial support of Breslow's original biosynthetic proposal has been reported in recent years. This is due to the development of mild, catalytic, and highly selective methods of radical generation and radical trapping. The synthesis at room temperature of a stereo-defined terpenic skeleton without enzymatic assistance is remarkable. One is even inclined to speculate that synthetic chemists are now in a position to improve on nature's reactions because she (Nature) does not seem to be able to use these radical processes.

REFERENCES

1. I. Abe and G. D. Prestwich, in *Comprehensive Natural Products Chemistry*, ed. D. Barton and K. Nakahishi, Elsevier, Amsterdam, 1999, vol. 2.
2. G. Stork and A. W. Burgstahler, *J. Am. Chem. Soc.*, 1955, **77**, 5068–5077.
3. P. A. Sadler, A. Eschenmoser, H. Schinz and G. Stork, *Helv. Chim. Acta*, 1957, **40**, 2191–2198.
4. W. S. Johnson, M. B. Gravestock and B. E. McCarry, *J. Am. Chem. Soc.*, 1971, **93**, 4332–4334.
5. R. Breslow, E. Barrett and E. Mohacsi, *Tetrahedron Lett.*, 1962, **3**, 1207–1211.
6. R. Breslow, J. T. Groves and S. S. Olin, *Tetrahedron Lett.*, 1966, **7**, 4717–4719.
7. J. Y. Lallemand, M. Julia and D. Mansuy, *Tetrahedron Lett.*, 1973, **14**, 4461–4464.
8. A. F. Barrero, M. M. Herrador, J. F. Quilez del Moral, P. Arteaga, J. F. Arteaga, H. R. Dieguez and E. M. Sanchez, *J. Org. Chem.*, 2007, **72**, 2988–2995.
9. A. G. Campaña, B. Bazdi, N. Fuentes, R. Robles, J. M. Cuerva, J. E. Oltra, S. Porcel and A. M. Echavarren, *Angew. Chem., Int. Ed.*, 2008, **47**, 7515–7519.

10. R. E. Estealvez, J. Justicia, B. Bazdi, N. Fuentes, M. Paradas, D. Choquesillo Lazarte, J. M. Garcıaua-Ruiz, R. Robles, A. Gansauer, J. M. Cuerva and J. E. Oltra, *Chem.–Eur. J.*, 2009, **15**, 2774–2791.
11. U. Hoffmann, Y. Gao, B. Pandey, S. Klinge, K.-D. Warzecha, C. Kruger, H. D. Roth and M. Demuth, *J. Am. Chem. Soc.*, 1993, **115**, 10358–10359.
12. C. Heinemann, X. Xing, K.-D. Warzecha, P. Ritterskamp, H. Gorner and M. Demuth, *Pure Appl. Chem.*, 1998, **70**, 2167–2176.
13. M. Ozser, H. Icil, Y. Makhynya and M. Demuth, *Eur. J. Org. Chem.*, 2004, 3686–3692.
14. C. Heinemann and M. Demuth, *J. Am. Chem. Soc.*, 1999, **121**, 4894–4895.
15. F. Goeller, C. Heinemann and M. Demuth, *Synthesis*, 2001, 1114–1116.
16. X. Xing and M. Demuth, *Eur. J. Org. Chem.*, 2001, 537–544.
17. V. Rosales, J. Zambrano and M. Demuth, *Eur. J. Org. Chem.*, 2004, 1798–1802.
18. K.-D. Warzecha, X. Xing and M. Demuth, *Pure Appl. Chem.*, 1997, **69**, 109–112.
19. C. Heinemann and M. Demuth, *J. Am. Chem. Soc.*, 1997, **119**, 1129–1130.
20. L. Chen, G. B. Gill and G. Pattenden, *Tetrahedron Lett.*, 1994, **35**, 2593–2596.
21. A. Batsanov, L. Chen, G. Bryon Gill and G. Pattenden, *J. Chem. Soc., Perkin Trans. 1,* 1996, 45–55.
22. P. Double and G. Pattenden, *J. Chem. Soc., Perkin Trans. 1,* 1998, 2005–2008.
23. G. Cimino, R. Morrone and G. Sodano, *Tetrahedron Lett.*, 1982, **23**, 4139–4142.
24. G. Pattenden, L. Roberts and A. J. Blake, *J. Chem. Soc., Perkin Trans. 1,* 1998, 863–868.
25. S. Handa and G. Pattenden, *J. Chem. Soc., Perkin Trans. 1,* 1999, 843–846.
26. S. Handa, G. Pattenden and W.-S. Li, *Chem. Commun.*, 1998, 311–312.
27. A. Gansäuer and H. Bluhm, *Chem. Rev.*, 2000, **100**, 2771–2788.
28. B. B. Snider, *Chem. Rev.*, 1996, **96**, 339–364.
29. J. Iqbal, B. Bhatia and N. K. Nayyar, *Chem. Rev.*, 1994, **94**, 519–564.
30. G. G. Melikyan, *Org. React.*, 1997, **49**, 427–675.
31. G. G. Melikyan, *Aldrichim. Acta*, 1998, **31**, 50–64.
32. A. S. Demir and M. Emrullahoglu, *Curr. Org. Synth.*, 2007, **4**, 321–351.
33. E. I. Heiba and R. M. Dessau, *J. Org. Chem.*, 1974, **39**, 3456–3457.
34. J. B. Bush Jr. and H. Finkbeiner, *J. Am. Chem. Soc.*, 1968, **90**, 5903–5905.
35. J. K. Kochi, *Acc. Chem. Res.*, 1974, **7**, 351–360.
36. S. A. Kates, M. A. Dombroski and B. B. Snider, *J. Org. Chem.*, 1990, **55**, 2427–2436.
37. L. A. Paquette, A. G. Schaefer and J. P. Springer, *Tetrahedron*, 1987, **43**, 5567–5582.
38. B. B. Snider, J. E. Merritt, M. A. Dombroski and B. O. Buckman, *J. Org. Chem.*, 1991, **56**, 5544–5553.

39. P. A. Zoretic, M. Wang and Z. Shen, *J. Org. Chem.*, 1996, **61**, 1806–1813.
40. B. B. Snider, J. Y. Kiselgof and B. M. Foxman, *J. Org. Chem.*, 1998, **63**, 7945–7952.
41. P. A. Zoretic, H. Fang and A. A. Ribeiro, *J. Org. Chem.*, 1998, **63**, 4779–4785.
42. A. F. Barrero, M. M. Herrador, J. F. Quilez del Moral and M. V. Valdivia, *Org. Lett.*, 2002, **4**, 1379–1382.
43. P. A. Zoretic, X. Weng, C. K. Biggers, M. S. Biggers, M. L. Caspar and D. G. Davis, *Tetrahedron Lett.*, 1992, **33**, 2637–2640.
44. Q. Zhang, R. M. Mohan, L. Cook, S. Kazanis, D. Peisach, B. M. Foxman and B. B. Snider, *J. Org. Chem.*, 1993, **58**, 7640–7651.
45. D. Yang, X.-Y. Ye, S. Gu and M. Xu, *J. Am. Chem. Soc.*, 1999, **121**, 5579–5580.
46. D. Yang, M. Xu and M.-Y. Bian, *Org. Lett.*, 2001, **3**, 111–114.
47. D. Yang and M. Xu, *Org. Lett.*, 2001, **3**, 1785–1788.
48. A. Gansauer, T. Lauterbach and S. Narayan, *Angew. Chem., Int. Ed.*, 2003, **42**, 5556–5573.
49. T. V. RajanBabu and W. A. Nugent, *J. Am. Chem. Soc.*, 1994, **116**, 986–997.
50. A. Gansauer, H. Bluhm and M. Pierobon, *J. Am. Chem. Soc.*, 1998, **120**, 12849–12859.
51. A. F. Barrero, A. Rosales, J. M. Cuerva and J. E. Oltra, *Org. Lett.*, 2003, **5**, 1935–1938.
52. A. Gansauer, C.-A. Fan and F. Piestert, *J. Am. Chem. Soc.*, 2008, **130**, 6916–6917.
53. A. Gansauer, S. Lei and M. Otte, *J. Am. Chem. Soc.*, 2010, **132**, 11858–11859.
54. J. Friedrich, K. Walczak, M. Dolg, F. Piestert, T. Lauterbach, D. Worgull and A. Gansauer, *J. Am. Chem. Soc.*, 2008, **130**, 1788–1796.
55. A. Gansauer, D. Worgull, K. Knebel, I. Huth and G. Schankenburg, *Angew. Chem., Int. Ed.*, 2009, **48**, 8882–8885.
56. A. F. Barrero, J. M. Cuerva, M. M. Herrador and M. V. Valdivia, *J. Org. Chem.*, 2001, **66**, 4074–4077.
57. J. Justicia, T. Jimenez, S. P. Morcillo, J.M. Cuerva and J. E. Oltra, *Tetrahedron*, 2009, **65**, 10837–10841.
58. M. Paradas, A. G. Campana, T. Jimenez, R. Robles, J. E. Oltra, E. Bunuel, J. Justicia, D. J. Cardenas and J. M. Cuerva, *J. Am. Chem. Soc.*, 2010, **132**, 12748–12756.
59. S. Fuse, M. Hanochi, T. Doi and T. Takahashi, *Tetrahedron Lett.*, 2004, **45**, 1961–1964.
60. J. Justicia, A. Rosales, E. Bunuel, J. L. Oller-Lopez, M. Valdivia, A. Haidour, J. E. Oltra, A. F. Barrero, D. J. Cardenas and J. M. Cuerva, *Chem.–Eur. J.*, 2004, **10**, 1778–1788.
61. E. E. Van Tamelen, A. Storni, E. J. Hessler and M. A. Schwartz, *Bioorg. Chem.*, 1982, **11**, 133–170 and references therein.

62. R. A. Shenvi and E. J. Corey, *Org. Lett.,* 2010, **12**, 3548–3551.
63. A. F. Barrero, J. M. Cuerva, E. J. Alvarez-Manzaneda, J. E. Oltra and R. Chahboun, *Tetrahedron Lett.,* 2002, **42**, 2793–2796.
64. J. F. Arteaga, D. Victoriano, J. F. Qulez del Moral and A. F. Barrero, *Org. Lett.,* 2008, **10**, 1723–1726.
65. A. Gansäuer, D. Worgull and J. Justicia, *Synthesis*, 2006, 2151–2154.
66. J. Justicia, J. E. Oltra, A. F. Barrero, A. Guadano, A. Gonzalez- Coloma and J. M. Cuerva, *Eur. J. Org. Chem.,* 2005, 712–718.
67. H. Yamada, T. Hasegawa, H. Tanaka and T. Takahashi, *Synlett,* 2001, 1935–1937.
68. T. Doi, S. Fuse, S. Miyamoto, K. Nakai, D. Sasuga and T. Takahashi, *Chem.–Asian J.,* 2006, **1**, 370.
69. J. Justicia, J. E. Oltra and J. M. Cuerva, *J. Org. Chem.,* 2005, **70**, 8265–8270.
70. V. Domingo, L. Silva, H. R. Dieguez, J. F. Arteaga, J. F. Quilez del Moral and A. F. Barrero, *J. Org. Chem.,* 2009, **74**, 6151–6156.
71. A. F. Barrero, M. M. Herrador, J. F. Quilez del Moral, P. Arteaga, J. F. Arteaga, M. Piedra and E. M. Sanchez, *Org. Lett.,* 2005, **7**, 2301–2304.
72. J. Justicia, A. G. Campana, B. Bazdi, R. Robles, J. M. Cuerva and J. E. Oltra, *Adv. Synth. Catal.,* 2008, **350**, 571–576.
73. A. F. Barrero, J. F. Quilez del Moral, M. M. Herrador, I. Loayza, E. M. Sanchez and J. F. Arteaga, *Tetrahedron,* 2006, **62**, 5215–5222.
74. V. Domingo, H. R. Dieguez, C. P. Morales, J. F. Arteaga, J. F. Quilez del Moral and A. F. Barrero, *Synthesis*, 2010, 67–72.
75. M. Yamaoka, A. Nakazaki and S. Kobayashi, *Tetrahedron Lett.,* 2009, **50**, 6764–6768.
76. J. Justicia, J. E. Oltra and J. M. Cuerva, *J. Org. Chem.,* 2004, **69**, 5803–5806.
77. A. Gansäuer, J. Justicia, A. Rosales, D. Worgull, B. Rinker, J. M. Cuerva and J. E. Oltra, *Eur. J. Org. Chem.,* 2006, 4115–4127.
78. A. Gansäuer, A. Rosales and J. Justicia, *Synlett,* 2006, 927–929.
79. J. N. Carter-Franklin and A. Butler, *J. Am. Chem. Soc.,* 2004, **126**, 15060–15066.
80. J. G. Urones, I. S. Marcos, P. Basabe, C. A. Alonso, D. Diez, N. M. Garrido, I. M. Oliva, J. S. Rodilla, A. M. Z. Slawin and D. J. Williams, *Tetrahedron Lett.,* 1990, **31**, 4501–4504.
81. J. Justicia, J. L. Oller-Lopez, A. G. Campana, J. E. Oltra, J. M. Cuerva, E. Bunuel and D. J. Cardenas, *J. Am. Chem. Soc.,* 2005, **127**, 14911–14921.
82. Y. Lian, L. C. Miller, S. Born, R. Sarpong and H. M. L. Davies, *J. Am. Chem. Soc.,* 2010, **132**, 12422–12425.
83. S. Rendler and D. W. C. MacMillan, *J. Am. Chem. Soc.,* 2010, **132**, 5027–5029.

Subject Index

propargylamine, 277, **279**
propargylations, 267, 276
propellane, 495, 508
propenes, 262, **263,** 265, 312, 516
propionic acid, 64
propylammonium decatungstates, 473
propylenes, 8
prostaglandins, 151, 506
proteases, 627, 636–8, 684
proteinases, 636
protic
  acid, 326, 334
  additives, 319
  cleavage, 338
  conditions, 309, 321, 335
  environments, 314
  functional groups, 309
  pinacol couplings, 334
  solvents, 86
protoilludanes, 556–7
protonation titanium, 321–2, 327
protonolysis, 231
pulegone chiral sources, 498, 711
pulse radiolysis. *see* radiolysis of water
purine nucleosides, 417, 422
pyboxligands, 500
pyramidal radicals, 33–4
pyranones, 671
pyridine hydrochloride, 322, 334
pyridine-2-thione-*N*-oxycarbonyls (PTOC), 155–6, 158
pyridinyls, 636
pyrolysis, 9
pyrrolidines
  derivatives, **148,** 149
  N-funtionalized, 110
  nitroxides, 188
  radical domino reactions, 494
  synthesis, 306, **497**

pyrrolizidines, 500, 546

quanosine, 419
quinazoline, 110, 112
quinolines, 216, 371, 492, 508
quinolones, 351–3
quinoxalines, 518

R-methyleneglutarate mutases, 630–1
R-trifluoromethyl acid, 361
radical -molecular reactions
  measurements, 21
  rate constants, 19
radical and non-radical reactions, 11
radical carbonylations, 575–7, 597
radical chemistry
  general aspects, 9
  green metrics, 36
  polarity, 17
  reactivity, 11
  stereoelectronic effects, 15
  steric effects, 14
radical clocks, **16,** 256, 260, 358, **660**
radical conjugate additions (RCA)
  metal based homogeneous catalysis, 230, 244, **268–70, 273–5**
  tin-free radical reactions, 157
radical domino processes, 487–8, 490, 504, 538
radical initiations
  alternative media, 75
  black-light-initiated reactions, 426–7, 430
  enantioselective reactions, 354
  sonochemistry, 408, 411
  thiols as hydrogen donors, 175